Using Information Technology

A Practical Introduction to Computers & Communications

Second Edition

Using Information Technology

A Practical Introduction to Computers & Communications

Second Edition

Brian K. Williams

Stacey C. Sawyer

Sarah E. Hutchinson

IRWIN

Chicago • Bogotá • Boston • Buenos Aires • Caracas
London • Madrid • Mexico City • Sydney • Toronto

Publisher: Tom Casson
Sponsoring editor: Garrett Glanz
Editorial assistant: Carrie Berkshire
Marketing manager: Michelle Hudson
Developmental editor: Burrston House, Ltd., Burr Ridge, IL
Copy editor: Anita Wagner, Takoma Park, MD
Project editor, layout, & production management: Stacey C. Sawyer,
    Sawyer & Williams, Incline Village, NV
Project supervisor: Gladys True
Production supervisor: Bob Lange
Multimedia coordination: Lew Gossage, Bill Bayer, and Brian Nacik
Design coordination: Laurie Entringer, Matthew Baldwin
Designer: Ellen Pettengell
Cover Illustration: © Dick Palulian/SIS
Cover Design: Matthew Baldwin
Photo research: Monica Suder, San Francisco
Art & composition: GTS Graphics, Commerce, CA
Prepress buyer: Charlene R. Perez
Permissions: David Sweet, San Francisco
Compositor: GTS Graphics
Typeface: 10/12 Trump Mediaeval
Printer: Times Mirror Higher Education Print Group

Times Mirror
Higher Education Group

Library of Congress Cataloging-in-Publication Data

Williams, Brian K., (1938–)
    Using information technology: a practical introduction to
  computers & communications / Brian K. Williams, Stacey C. Sawyer,
  Sara E. Hutchinson. — 2nd ed.
      p.   cm.
    Includes index.
    ISBN 0-256-20981-2
    1. Computers.   2. Telecommunications systems.   3. Information
  technology.   I. Sawyer, Stacey C.   II. Hutchinson, Sarah E.
  III. Title
QA76.5.W5332 1997
004—dc20   96–22392

Photo and other credits are listed in the back of the book.

Printed in the United States of America
    3 4 5 6 7 8 9 0 WCB 3 2 1 0 9 8 7 6

Brief Contents

Preface to the Instructor

A new generation of computing is well under way. The 1980s "second wave" model of stand-alone desktop PCs, which overtook the "first wave" model of mainframes, is now giving way to a "third wave" model driven by communications technologies. For evidence we need only look to the daily headlines: The 1996 Telecommunications Act. The rise of the Internet and the World Wide Web. The browser wars. Search engines. Cable modems. Network PCs. Java. *Information technology*—the fusion of computing and communications—is creating far-reaching changes in the way we work, the way we live, and even in the way we think.

The Audience for & Promises of This Book

USING INFORMATION TECHNOLOGY: A Practical Introduction to Computers & Communications, SECOND EDITION, is intended for use as a concepts textbook to accompany a one-semester or one-quarter introductory course on computers or microcomputers. It is, we hope, a book that will make a difference in the lives of our readers.

The **key features** of *USING INFORMATION TECHNOLOGY*, SECOND EDITION, are as follows. We offer:

1. **Careful revision in response to extensive instructor feedback.**
2. **Thorough coverage of computers and information systems—in just 12 chapters.**
3. **Emphasis on unification of computer and communications systems.**
4. **Emphasis on practicality.**
5. **Emphasis throughout on ethics.**
6. **Use of techniques for reinforcing student learning.**
7. **Up-to-the-minute material—in the book and on our Web site.**

We elaborate on these features next.

Key Feature #1: Careful Revision in Response to Extensive Instructor Feedback

We were delighted to learn from our publisher that the First Edition of *USING INFORMATION TECHNOLOGY* was apparently the most successful new text in the field in 1995, with over 300 schools adopting both comprehensive and brief versions. An important reason for this success, we feel, was all the valuable contributions of the reviewers.

Both the printed version of the First Edition and the manuscript and proofs of the SECOND EDITION underwent a highly disciplined and wide-ranging reviewing process. This process of expert appraisal drew on instructors who were both users and nonusers, who were from a variety of educational institutions, and who expressed their ideas in both written form and in focus groups.

We have sometimes been overwhelmed with the amount of information, but we have tried to respond to all consensus criticisms and countless individual suggestions. It is not an exaggeration to say that every page of the SECOND EDITION has been influenced by instructor feedback. The result, we think, is **a book reflecting the wishes of most instructors.** In particular, we have addressed the following matters:

- **Old Chapters 1 and 2 combined into one:** We combined old Chapters 1 and 2 into a single chapter, so that there would be less introductory material for the student to get through. Some instructors had found some of the old introductory material too technical for a first chapter and had expressed the wish to have the "overview" material moved closer to the beginning. The new chapter reflects their wishes.
- **Storage and database material made two chapters:** We split the old "Storage & Databases" chapter into two chapters. The majority of instructors felt that our previous coverage of databases was inappropriate when combined with a discussion of storage and moreover was too scant. We have remedied these deficiencies.
- **Input and output material made two chapters:** The single chapter "Input & Output" became two chapters because of the amount of new material that has become available.
- **"Promises" chapter and "Challenges" chapter combined into one and resequenced:** Many instructors felt that our previous Chapter 11 ("The Promises of the Digital Age") and Chapter 12 ("The Challenges of the Digital Age") were too long. Moreover, a few felt that ending the book with a discussion of problems rather than promises was inappropriate. Thus, we have combined these into one chapter and reordered the material per the title "Society & the Digital Age: Challenges & Promises."

In addition to these major structural changes, we have made hundreds of line-by-line and word-by-word adjustments to conform with instructor's requests.

Key Feature #2: Thorough Coverage of Computers & Information Systems—in Just 12 Chapters

This is, of course, a book about computers and information systems, and we cover these subjects thoroughly, without abridgment. However, many instructors have told us that having the material presented in **just 12 chapters**, rather than the customary 14 or 15 or more, better suits their teaching approach.

Chapters are organized according to the topic coverage of traditional introductory computer texts. Thus, most instructors can continue to follow their present course outlines.

NOTE: The text allows for **a good deal of instructor flexibility.** After Chapter 1, the remaining 11 chapters may be taught in any sequence, or selectively omitted, at the instructor's discretion. To make this possible, the authors have occasionally **repeated the definitions of key terms throughout the text** (also a part of the book's deliberate strategy of reinforcement).

The end-of-chapter essay appearing in the Experience Box is optional material, but may be assigned if the instructor wishes. Experience Boxes, too, may be read out of sequence.

Key Feature #3: Emphasis on Unification of Computers & Communications

The text emphasizes the technological merger of the computer, communications, consumer electronics, and media industries through the exchange of information in the digital format used by computers. This is the relatively new phenomenon known as **technological convergence.**

This theme covers much of the technology currently found under such phrases as *the Information Superhighway, the Multimedia Revolution,* and *the Digital Age: mobile computing, the Internet, Web search tools, online services, workgroup computing, the virtual office, video compression, PC/TVs, information appliances,* "intelligent agents," and so on.

The theme of convergence is given in-depth treatment in five chapters—the introduction, systems software, communications, databases, and challenges and promises (Chapters 1, 3, 8, 9, 12)—and is also brought out in examples throughout other chapters.

Key Feature #4: Emphasis on Practicality

We'd like to make this book a "keeper" for students. Thus, we not only cover fundamental concepts but also offer a great deal of **practical advice.** This advice, of the sort found in computer magazines and general-interest computer books, is expressed principally in two kinds of boxes—Experience Boxes and README boxes:

- **The Experience Box:** Appearing at the end of each chapter, the Experience Box is optional material that may be assigned at the instructor's discretion. However, students will find the subjects covered are of immediate value.

  Some examples: "Becoming a Mobile Computer User"; "How to Buy Software"; "How to Buy a Multimedia System"; "Telling Computer Magazines Apart." Five of the Experience Boxes show students how to benefit from going online. They include "Finding Useful Online Databases: Directories & Search Engines" and "Online Résumés: Career Strategy for the Digital Age."

- **README boxes:** README boxes consist of optional material of two types—Practical Matters, and Case Studies:

  Practical Matters offer practical advice—such as tips for practicing online etiquette or avoiding viruses.

  Case Studies offer behind-the-scenes looks at information technology—such as how a salesperson uses a portable PC to download technical material from a mainframe, or how an expert is interviewed to construct an expert system.

Key Feature #5: Emphasis Throughout on Ethics

New to this edition! Many texts discuss ethics only once, usually in one of the final chapters. We believe this topic is too important to be treated last or lightly. Thus, **we cover ethical matters in 19 places** throughout the book, as indicated by the special sign shown here in the margin. For example, the all-important question of what kind of software can be legally copied is discussed in Chapter 2 ("Applications Software"), an appropriate place for students just starting software labs. Other ethical matters discussed are the manipulation of truth through digitizing of photographs, intellectual property rights, netiquette, censorship, privacy, and computer crime.

A list of pages of ethics coverage appears on the inside front cover. Instructors wishing to teach all ethical matters as a single unit may refer to this list.

Key Feature #6: Reinforcement for Learning

Having individually or together written over a dozen successful textbooks and scores of labs, the authors are vitally concerned with reinforcing students in acquiring knowledge and developing critical thinking. Accordingly, we offer the following *to provide learning reinforcement:*

- **Interesting writing:** Studies have found that textbooks **written in an imaginative style** significantly improve students' ability to retain information. Thus, the authors have employed a number of journalistic devices—such as the short biographical sketch, the colorful fact, the apt direct quote—to make the material as interesting as possible. We also use real anecdotes and examples rather than fictionalized ones.

- **Key terms and definitions in boldface:** Each **key term AND its definition is printed in boldface** within the text, in order to help readers avoid any confusion about which terms are important and what they actually mean.

- **Learning objectives to aid students:** *New to this edition!* Lists of learning objectives at the start of chapters are common in textbooks—and most students simply skip them. Because we believe learning objectives are excellent instruments for reinforcement, we have crafted ours to make them more helpful to students. We do this by **tying the numbered learning objectives to the end-of-chapter summary.** That is, we have numbered the objectives. Then, in the summary at the end of the chapter, we have given corresponding numbers to the terms and concepts that relate to the particular objectives.

    For example, in Chapter 2, *Learning Objective 2* is "After reading this chapter, you should be able to: 2. Discuss the ethics of copying software." Terms and concepts appearing in the end-of-chapter summary that relate to this objective—such as "copyright," "freeware," and "intellectual property"—are identified with the notation *LO 2.*

- **"Preview & Review" presents abstracts of each section for learning reinforcement:** Each main section heading throughout the book is followed by **an abstract or précis entitled Preview & Review.** This enables the student to get a *preview* of the material before reading it and then to *review* it afterward, for maximum learning reinforcement.

- **Innovative chapter summaries for learning reinforcement:** The chapter summary is especially innovative—and especially helpful to students. In fact, research through student focus groups has shown that this format was clearly first among five different choices of summary formats. Each concept is discussed under **two columns, headed "What It Is/What It Does" and "Why It's Important."**

    Each concept or term is also given a cross-reference page number that refers the reader to the main discussion within the chapter.

    In addition, as mentioned, the term or concept is also given a number (such as *LO 1, LO 2,* and so on) corresponding to the appropriate learning objective at the beginning of the chapter.

- **Cross-referencing system for key terms and concepts:** *New to this edition!* Wherever important key terms and concepts appear throughout the text that students might need to remind themselves about, we have added **"check the cross reference"** information, as in: (✓ p. 111). In student focus groups during the last two years, this device was found to rank *first* out of 20-plus study/learning aids.

- **Material in "bite-size" portions:** Major ideas are presented **bite-size form,** with generous use of advance organizers, bulleted lists, and new paragraphing when a new idea is introduced.

- **Short sentences:** Most sentences have been kept short, the majority not exceeding **22–25 words** in length.

- **Innovative use of art:** Artwork in the book is designed principally to be **didactic.** There are no unnecessary space-filling photo "galleries," for instance. To support learning concepts, photographs are often coupled with *additional* information—an elaboration of the discussion in the text, some how-to advice, an interesting quotation, or a piece of line art.

- **End-of-chapter exercises:** For practice purposes, students will benefit from *several exercises* at the end of each chapter: **short-answer questions, fill-in-the-blank questions, multiple-choice questions,** and **true-false questions.** Answers to selected exercises appear in the back of the book.

  In addition, we present several **projects/critical-thinking questions,** generally of a practical nature, to help students absorb the material.

- **Internet exercises:** *New to this edition!* In keeping with the practical and communications orientation of the book, we present **exercises on the use of the Internet** at the end of every chapter. Exercise topics include *sending and retrieving e-mail, performing research on the Web, navigating newsgroups, using FTP to download files, using Gopher and Veronica,* and more.

  Internet connection and software requirements: In general, the exercises assume an Internet setup is readily available to most college students. In Chapters 1–2, we assume students connect to the Internet using a command-line Unix interface. In Chapters 3–12, we assume students use the Navigator Netscape 2.0 or 2.01 Web browser. If these are not compatible with your setup, please check out Irwin's UIT Web site *(http://www.irwin.com/cit/uit)* for information regarding the publisher's additional Internet offerings.

Key Feature #7: Up-to-the-Minute Material—in the Text & on the Irwin Web Site

New to this edition! The number of technological developments that have occurred since we wrote the first edition has been awesome, and every day seems to bring reports of something new and important. As we write this, our August 1996 publication date is only three months away. However, because our publisher has allowed us to do several steps concurrently (writing, reviewing, editing, production), readers will find probably a hundred or more 1996 references in the notes to the book. As evidence for our being current, our text includes coverage of the following material:

Advanced TV. AT&T Internet offer. Avatars. Cable modems. Cyberdog. Cybersickness. Data mining. Data warehouses. DVD-ROMs. E-money. Genetic algorithms. Intelligent agents. Internet PC. Intranets. Java. Microsoft Exchange.

Microsoft Internet Explorer. PC/TVs. Pippin. Search engines. SIPC (Microsoft's Simply Interactive Personal Computer) standards. Spiders. Telecommunications Act of 1996. 3-D displays. 3-D sound. Vchip. VRML. VBNS. Web indexes. Windows 95 (latest information). Yahoo. Zip drives. . . . And more.

Still, we recognize that a Gutenberg-era lag exists between our last-minute scribbling and the book's publication date. And of course we also realize that fast-moving events will unquestionably overtake some of the facts in this book by the time it is the student's hands. Accordingly, after publication we are periodically offering instructors updated material and other interaction on the Irwin UIT Web Site *(http://www.irwin.com/cit/uit).*

Complete Course Solutions: Supplements That Work—Four Distinctive Offerings

It's not important how many supplements a textbook has but whether they are truly useful, accurate, and of high quality. Irwin presents **four distinctive kinds of supplement offerings** to complement the text:

1. **Application software tutorials**
2. **Interactive software**
3. **Classroom presentation software**
4. **Instructor support materials, including software support program**

We elaborate on these below.

Supplement Offering #1: Application Software Tutorials—Two Types

Our publisher, Richard D. Irwin, offers two different series of application software tutorials, or lab manuals, which present two different hands-on approaches to learning software. An Irwin sales representative can explain the specific software covered in each series.

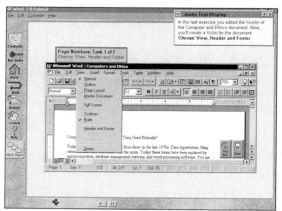

• **The Irwin Advantage Series for Computer Education:** Written by Sarah E. Hutchinson and Glen J. Coulthard, the *Irwin Advantage Series for Computer Education* covers the complete Microsoft Office Professional with your choice of either one comprehensive spiral-bound package or individual editions featuring full-color layouts and large screen captures.

Manuals are available for Microsoft *Windows 3.1* and *Windows 95*, as well as for Microsoft *Word, Excel, Access,* and *PowerPoint,* both for Windows 3.1 and for Window 95. A manual called *Integrated Microsoft Office* is also available for Windows 3.1 and Windows 95.

Each tutorial leads students through step-by-step instructions not only for the most common methods of executing commands but also for alternative methods. Each lesson begins with a case scenario and concludes with case problems, showing the real-world application of the software. Quick Reference guides summarizing important functions and shortcuts appear throughout. Annotated Toolbar screen shots provide easy and quick reference. Boxes introduce unusual functions that will enhance the user's productivity. Hands-on exercises and short-answer questions allow students to practice their skills.

- **The Irwin Effective Series:** Written specifically for the first-time computer user, by Fritz J. Erickson and John A. Vonk, this series is based on the premise that success breeds confidence and confident students learn more effectively. Exercises embedded within each lesson allow students to experience success before moving on to a more advanced topic. The "why" as well as the "how" is always carefully explained. Each lesson features several applications projects and a comprehensive problem for student solution.

   Manuals are available for Microsoft *Windows 3.1* and *Windows 95,* as well as for Microsoft *Word, Excel, Access,* and *Works* for both Windows 3.1 and Windows 95 and for *PowerPoint* for Windows 95.

Important Note—Custom Publishing: The contents of these products can be tailored to meet your course needs through *custom publishing.* Titles or specific lessons from several titles in these series can be combined. *Irwin will happily send you an examination copy of the custom-published text you want so you can see exactly what your students will get.* Ask your Irwin sales representative for details.

Supplement Offering #2: Interactive Software— Two Types

Irwin offers two types of interactive software to accompany the text—*Info Tech, Version 2.0* and *Internet: A Knowledge Odyssey.*

- **Info Tech, Version 2.0:** Developed by Irwin New Media and Tony Baxter, *Info Tech, Version 2.0* is a revised, updated, and expanded version of interactive multimedia software provided with the first edition. The CD-ROM gives students with several self-paced learning modules, on topics ranging from applications software to networks. Combining text, illustrations, and animation, the Info Tech interactive tool may also be used by instructors in a lecture setting.

   Info Tech 2.0 includes coverage of:

Data Into Information	*Data Representation*
Application Software	*Networks*
User Interfaces	*The Internet*
Processors	*Querying a Database*
Secondary Storage	*Client/Server*
Peripheral Devices	*Encryption/Decryption*
Backing Up Data	*Security* . . . and more.
Multimedia	

Each module provides three levels of learning: (1) The *introduction level* provides text and animated enhancement of computer concepts. (2) the *exploratory level* allows the user to experiment with various scenarios and see the immediate results. (3) The *practice level* poses cases and problems for which the user must provide solutions based on information learned in the first two levels.

System requirements: (a) IBM PC or compatible with at least 2 MB of RAM running Windows 3.1 or Windows 95, or (b) Macintosh with at least 2 MB of RAM running System 6.01 or later; CD-ROM drive.

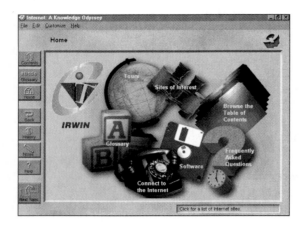

• **Internet: A Knowledge Odyssey:** A multimedia CD-ROM developed by MindQ Publishing, Inc., *Internet: A Knowledge Odyssey, Business Edition*, explains the history and workings of the Internet through 14 self-paced, interactive guided tours, 400 hyperlinked glossary terms, 90 minutes of video clips, and optional audio narration. Available to students for less than $10 (when packaged with other Irwin products), the CD-ROM also allows students to connect directly to specific sites on the Internet, automatically launching their Internet applications and inserting the correct URL address. Details and an evaluation copy are available from an Irwin representative.

System requirements: IBM PC or compatible 486SX or higher with at least 4 MB of RAM running Windows 3.1 or Windows 95, SVGA graphics card (256 colors), and CD-ROM drive.

Supplement Offering #3: Classroom Presentation Software—Two Options

To help instructors enhance their lecture presentations, Irwin makes available software in two options, in a PowerPoint version and an Astound version. Both segment the course into seven topical modules: *introductory topics, hardware, software, communications, systems development, database and information management,* and *additional topics.*

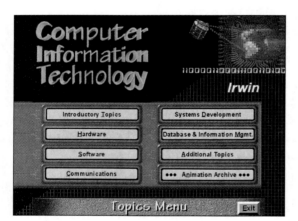

• **Irwin CIT Classroom Presentation Software—Power-Point option:** Developed by the Quality & Excellence Institute and Linda Behrens, the *PowerPoint version* of this classroom presentation software is a graphics-intensive set of lecture slides that helps instructors explain topics that may otherwise be difficult to present. The PowerPoint version, available on several 3.5-inch diskettes, offers lecture outline material with various graphics, backgrounds, and transitions.

System requirements: IBM PC or compatible with at least 2 MB of RAM running Windows 3.1. An LCD panel is needed if the images are to be shown to a large audience.

• **Irwin CIT Classroom Presentation Software—Astound option:** Also from the same developers, the *Astound version* of classroom presentation software is a CD-ROM that allows instructors to make true multimedia lecture-enhancement presentations. A flexible menu-driven tool, the Astound version offers a complete lecture outline and a navigation interface, topical menus, animations, audio, and video clips. An Irwin sales representative can provide a demonstration of this tool.

System requirements: IBM PC or compatible running Windows 3.1 or higher with a '486 processor (66 MHz or higher recommended) and at least 4 MB of RAM, SVGA graphics capability, and at least a dual-speed CD-ROM drive.

Supplement Offering #4: Instructor Support Materials

We offer the instructor the following other kinds of supplements and support to complement the text:

- **Instructor's Resource Guide:** This complete guide, prepared by Linda Behrens, supports instruction in any course environment. The *Instructor's Resource Guide* includes: *a student questionnaire, course planning and evaluation grid, suggestions for writing course objectives, suggested pace and coverage for courses of various lengths, suggestions for using the exercises in various class structures,* and *projects for small and large classes.*

   For each chapter, the IRG provides an overview, chapter outline, lecture notes, notes regarding the boxes (README boxes) from the text, solutions, and suggestions and additional information to enhance your course.

- **Color transparencies:** There are *150 full-color transparency acetates* available to the instructor. Transparencies have been specially *upsized*—enlarged and enhanced for clear projection.

- **Test bank:** The test bank, prepared by Margaret Batchelor, contains *2000 different questions,* which are directly referenced to the text. Specifically, it contains *true/false, multiple-choice,* and *fill-in questions,* categorized by difficulty and by type; *short-essay questions; sample midterm exam; sample final exam;* and *answers to all questions.*

- **Computerized testing software:** Called *Computest,* Irwin's popular computerized testing software is a *user-friendly, menu-driven, microcomputer-based test-generating system* that is free to qualified adopters. Containing all the questions from the test bank described above, Computest's Version 4 allows instructors to customize test sheets, entering their own questions and generating review sheets and answer keys.

   Available for DOS, Windows, and Macintosh formats, Computest has advanced printing features that allow instructors to print all types of graphics; Windows and Macintosh versions use easily remembered icons. All versions support over 250 dot-matrix and laser printers.

   System requirements: (a) IBM PC or compatible with at least 2 MB of RAM running Windows 3.1 or (b) Macintosh with at least 2 MB of RAM running System 6.01 or later; CD-ROM drive or 3.5-inch diskette drives.

- **Videos:** A broad selection from *21 new video segments* of the acclaimed PBS television series, *Computer Chronicles,* is available. Each video is approximately 30 minutes long. The videos cover topics ranging from computers and politics, to CD-ROM, to visual programming languages, to the Internet.

- **Instructor's data disks:** Instructor's data disks are avaliable for instructors whose students are using the tutorials for software education in the Irwin Advantage Series for Computer Education and Irwin Effective Series. These are *diskettes containing files* used in the DOS-, Windows 3.X-, and Windows 95–based software labs. Specifically, the diskettes contain the letters and memos that the student will use in the word processing labs, sample budgets and other files that the students will retrieve and modify in the spreadsheet labs, and the data and reports that the student will work with in the database labs.

- **Phone, fax, and e-mail instructor support services:** Richard D. Irwin's College New Media Department offers *telephone- or computer-linked support services to instructors* in matters related to Irwin software, such as

Computest and data disks used for the student tutorials in the Irwin Advantage Series or Effective Series. Software support analysts are available to help solve technical questions not covered in the documentation for any Irwin software product.

Three kinds of support are offered: (1) toll-free telephone numbers, available 9:00 A.M. to 5 P.M. Central, Monday through Friday (except holidays); (2) support-on-demand FAX-BACK service, available 24 hours a day, seven days a week; (3) e-mail, accessible 24 hours a day. Directions for getting this support appear in the *Instructor's Resource & Lecture Guide.*

- **Irwin Web site:** It's appropriate that a text with a strong communications focus also find a way to employ the new communications technologies available. Accordingly, a text-specific Irwin UIT Web Site has been developed as a place to go for periodic updates of text material, relevant links, downloads of supplements, an instructors' forum for sharing information with colleagues, and other value-added features.

Instructor Scenarios for Using the Text

USING INFORMATION TECHNOLOGY, SECOND EDITION, was carefully designed based on marketplace feedback. We have written the kind of book that many instructors asked for, and the materials are designed to serve a consensus kind of course.

Thus, to serve the new generation of students we are presenting a book that, we hope, reads like a magazine, offers many illustrations, and helps the reader learn through many extra pedagogical features—README boxes, Experience Boxes, section Preview & Reviews, innovative end-of-chapter summaries ("What It is/What It Does," and "Why It's Important"), and end-of-chapter exercises. *Actual material on which the student is to be tested— the general text—constitutes only slightly more than half of each chapter,* as determined from representative chapters. In Chapter 3, for instance, general text constitutes 20 of the 43 pages. (The rest consists of chapter opening, panels, boxes, suggested resources, section summaries, end-of-chapter summaries, and end-of-chapter exercises.)

Since the book consists of 12 chapters, readings may be assigned at the rate of slightly over a chapter a week in a quarter system, less than a chapter a week in a semester system. For instructors whose courses are less than 3 units or who must teach students software labs in addition to computer concepts, there are other options. Any one or combination of the following scenarios will allow instructors to teach selectively from this book without loss of continuity:

- **Scenario 1—Teach all "ethics" segments as one component:** Rather than discuss ethical matters just in one place, we have spread this topic around through the book, as indicated by the special sign shown here. All the pages of ethics coverage are indicated on the inside front cover. Instructors wishing to teach all ethical matters as a single component (as toward the end of the semester or quarter) may direct students to read the ethics material in the order shown on that list.

- **Scenario 2—Skip the Experience Boxes:** Some instructors may wish to assign all 12 chapters but not the end-of-chapter essays we call Experiences Boxes. All Experience Boxes are considered optional (not testable) material, but some instructors may wish to pick and choose which they assign, and some instructors may wish not to assign any.

- **Scenario 3—Skip chapters on systems and software development:** Some instructors may choose to forego Chapter 10, "Information Systems: Management & Development," and Chapter 11, "Software Development: Programming & Languages."

- **Scenario 4—Skip the last chapter:** Chapter 12, "Society & the Digital Age: Challenges & Promises," could be skipped. Instead, for a discussion of security and ergonomic issues, the instructor may choose to assign the Chapter 5 Experience Box: "Good Habits: Protecting Your Computer System, Your Data, & Your Health."

- **Scenario 5—skip chapters on applications and systems software:** Instructors whose courses include software labs may feel their students are already getting enough knowledge about applications and systems software that they do not need to read Chapters 2 and 3. (Chapter 2 is "Applications Software: Tools for Thinking & Working." Chapter 3 is "Systems Software: The Power Behind the Power.")

In addition, it bears repeating that, once the overview chapter (Chapter 1) has been assigned, any of the chapters that follow it may be taught out of order. Key terms have been defined anew as they appear in each chapter.

With these kinds of options, we feel sure that most instructors will be able to tailor the text to their particular course.

Acknowledgments

Three names are on the front of this book, but a great many others are powerful contributors to its development.

First among the staff of Richard D. Irwin is our sponsoring editor, Garrett Glanz, our lifeline, who did a fantastic job of supporting us and of coordinating the many talented people whose efforts on development and supplementary materials help strengthen our own. Garrett, you've been sensational. We're also grateful to Garrett's predecessor, Paul Ducham, for his initial spadework on this edition. It should go without saying that we owe a lot to our publisher, Tom Casson, who was not only the midwife on the first edition but has been one of our great fans on the second. Tom, it was good to feel your presence here.

We also appreciate the cheerfulness and efficiency of other people in Tom's and Garrett's group—the good and great Michelle Hudson, Carrie Berkshire, Linda Eiermann, and Sharon Pass. Mike Beamer did yeoman work on advertising pieces for us. Irwin's top management—the very supportive John Black, as well as David Littlehale, Jerry Saykes, and Jeff Sund—actively backed our revision, and we are extremely grateful to them. Many others at Irwin have also closely assisted us, and we would like to single out Laurie Entringer, Matt Baldwin, and Gladys True, who worked actively with us. In addition, we owe Bill Bayer, Lew Gossage, Charlie Hess, Bob Lange, Merrily Mazza, Keith McPherson, Evelyn Mosley-Harris, and Charlene Perez our special thanks. We appreciate having you all in our corner.

We would like to thank every single one of the Irwin sales reps, who did such an outstanding job on our behalf in promoting the first edition. We would particularly like to thank regional managers Jimmy Bartlett and Bob Bryan and district managers Barbara Anson, Bunny Barr, Greg Bowman, Frank Chihowski, Jr., Chad Douglas, Denise Mariani, and Jerry Swanson, plus manager for in-house sales Jean Geracie. By name we need to single out Irwin reps Rob Brown, Gary Rodgers, Steve Edmonson, Bill Firth, Liz Lindboe-Mulcahy, Julie Daniels, Stewart Mattson, Kitty Cavanaugh, Carol Preston, and Tony Noel, all of whom were honored at the national sales meeting for their spectacular efforts. Many thanks!

Outside of Irwin we were fortunate to find ourselves in a community of first-rate publishing professionals. We are ecstatic fans of the editorial development company of Burrston House, Ltd., and their active participants on this book, the highly experienced Glenn and Meg Turner and the very hard-working and always supportive Cathy Crow.

Two-thirds of the author team would like once again to sing the praises of the third, Stacey Sawyer, who not only co-wrote this book but also massaged in all the reviewers' comments, picked the photos, conceptualized and laid out much of the art, remade the pages, and directed the production of the entire enterprise under truly frantic deadlines—an exhausting business. Stacey, believe us when we say this book couldn't have happened were it not for you. Thanks for everything.

Stacey worked with freelance designer Ellen Pettengell, who bore with us through all the picky changes that we raised in order to try to get close to what we wanted. Ellen, we appreciate your efforts and patience.

Again, we can't say enough for Monica Suder, our photo researcher. Monica once again had to work with extremely tight deadlines and, miracle of miracles, came up with all the new photos that, we hope, make this book didactic and interesting. Photographer Frank Bevans fleshed out what was missing by supplying us with some outstanding photos.

Anita Wagner, an old publishing colleague and top-notch copy editor, performed her usual careful scrutiny of the manuscript. Standing behind Anita were our able proofreaders Kathy Pope and Linda Smith, who saved us from ourselves by removing inconsistencies and other potential embarrassments. Lois Oster, one of the finer talents in her line of work, did her customary unbeatable job of indexing. David Sweet did his always highly competent job of obtaining permissions.

GTS Graphics turned in their usual top-drawer performance in handling prepress production. We especially want to thank Elliott and Bennett Derman, Gloria Fontana, and their dynamite production coordination team of Donna Machado and Angie Armendarez. We also want to express our delight with the illustration program they did for us, with the creative art planning by Charlene Locke and renderings by Yvonne Welch.

Finally, the authors are grateful to a number of people for their superb work on the ancillary materials. They include Anthony Baxter, who helped develop Info Tech 2.0; MindQ Publishing, which developed Internet: A Knowledge Odyssey, Business Edition; The Quality & Excellence Institute and Linda Behrens, who developed both the PowerPoint and Astound versions of the Irwin CIT Classroom Presentation Software; Linda Behrens, again, who prepared the Instructor's Resource Guide; and Margaret Batchelor, who created the Test Bank.

Acknowledgment of Focus Group Participants, Survey Respondents, & Reviewers

We are grateful to the following people for their participation in focus groups, response to surveys, or reviews on manuscript drafts or page proofs of all or part of the book. We cannot overstate their importance and contributions in helping us to make this the most market-driven book possible.

Focus Group Participants

Patrick Callan
Concordia University

Joe Chambers
Triton College

Hiram Crawford
Olive Harvey College

Edouard Desautels
University of Wisconsin–Madison

William Dorin
Indiana University–Northwest

Bonita Ellis
Wright City College

Charles Geigner
Illinois State University

Julie Giles
DeVry Institute of Technology

Dwight Graham
Prairie State College

Stan Honacki
Moraine Valley Community College

Tom Hrubec
Waubonsee Community College

Alan Iliff
North Park College

Julie Jordahl
Rock Valley College

John Longstreet
Harold Washington College

Pattie Riden
Western Illinois University

Behrooz Saghafi
Chicago State University

Naj Shaik
Heartland Community College

Survey Respondents

Nancy Alderdice
Murray State University

Margaret Allison
University of Texas-Pan American

Angela Amin
Great Lakes Junior College

Connie Aragon
Seattle Central Community College

Gigi Beaton
Tyler Junior College

William C. Brough
University of Texas–Pan American

Jeff Butterfield
University of Idaho

Helen Corrigan-McFadyen
Massachusetts Bay Community College

James Frost
Idaho State Universtiy

Candace Gerrod
Red Rocks Community College

Julie Heine
Southern Oregon State College

Jerry Humphrey
Tulsa Junior College

Jan Karasz
Cameron University

Alan Maples
Cedar Valley College

Norman Muller
Greenfield Community College

Paul Murphy
Massachusetts Bay Community College

Sonia Nayle
Los Angeles City College

Janet Olpert
Cameron University

Pat Ormond
Utah Valley State College

Marie Planchard
Massachusetts Bay Community College

Fernando Rivera
University of Puerto Rico–Mayaguez Campus

Naj Shaik
Heartland Community College

Jack Shorter
Texas A&M University

Randy Stolze
Marist College

Ron Wallace
Blue Mountain Community College

Steve Wedwick
Heartland Community College

Reviewers

Nancy Alderdice
Murray State University

Sharon Anderson
Western Iowa Tech Community College

Bonnie Bailey
Morehead State University

David Brent Bandy
University of Wisconsin–Oshkosh

Robert Barrett
*Indiana University
Purdue University at Fort Wayne*

Anthony Baxter
University of Kentucky

Virginia Bender
William Rainey Harper College

Warren Boe
University of Iowa

Randall Bower
Iowa State University

Phyllis Broughton
Pitt Community College

J. Wesley Cain
City University, Bellevue

Judy Cameron
Spokane Community College

Kris Chandler
Pikes Peak Community College

William Chandler
University of Southern Colorado

John Chenoweth
East Tennessee State University

Ashraful Chowdhury
Dekalb College

Erline Cocke
Northwest Mississippi Community College

Robert Coleman
Pima County Community College

Glen Coulthard
Okanagan University

Robert Crandall
Denver Business School

John Durham
Fort Hays State University

John Enomoto
East Los Angeles College

Ray Fanselau
American River College

Eleanor Flanigan
Montclair State University

Ken Frizane
Oakton Community College

James Frost
Idaho State University

Jill Gebelt
Salt Lake Community College

Charles Geigner
Illinois State University

Frank Gillespie
University of Georgia

Myron Goldberg
Pace University

Sallyann Hanson
Mercer County Community College

Albert Harris
Appalachian State University

Jan Harris
Lewis & Clark Community College

Michael Hasset
Fort Hays State University

Martin Hochhauser
Dutchess Community College

James D. Holland
Okaloosa-Waltoon Community College

Wayne Horn
Pensacola Junior College

Christopher Hundhausen
University of Oregon

Jim Johnson
Valencia Community College

Jorene Kirkland
Amarillo College

Victor Lafrenz
Mohawk Valley Community College

Stephen Leach
Florida State University

Chang-Yang Lin
Eastern Kentucky University

Paul Lou
Diablo Valley College

Deborah Ludford
Glendale Community College

Peter MacGregor
Estrella Mountain Community College

Donna Madsen
Kirkwood Community College

Kenneth E. Martin
University of North Florida

Curtis Meadow
University of Maine

Marty Murray
Portland Community College

Charles Nelson
Rock Valley College

Wanda Nolden
Delgado Community College

E. Gladys Norman
Linn-Benton Community College

John Panzica
Community College of Rhode Island

Rajesh Parekh
Iowa State University

Merrill Parker
Chattanooga State Technical Community College

Leonard Presby
William Patterson State College

Delores Pusins
Hillsborough Community College

Eugene Rathswohl
Universtiy of San Diego

Jerry Reed
Valencia Community College

John Rezac
Johnson County Community College

Jane Ritter
University of Oregon

Stan Ross
Newbury College

Al Schroeder
Richland College

Earl Schweppe
University of Kansas

Tom Seymour
Minot State University

Elaine Shillito
Clark State Community College

Denis Titchenell
Los Angeles City College

Jim Vogel
Sanford Brown College

Dale Walikainen
Christopher Newport University

Reneva Walker
Valencia Community College

Patricia Lynn Wermers
North Shore Community College

Edward Winter
Salem State College

Floyd Winters
Manatee Community College

Eileen Zisk
Community College of Rhode Island

Write to Us

We welcome your response to this book, for we are truly trying to make it as useful as possible. Write to us in care of Garrett Glanz, Editor, Richard D. Irwin, 1333 Burr Ridge Parkway, Burr Ridge, IL 60521 or via e-mail: *citmail@irwin.com* (or directly to the authors at *76570.1533@ compuserve.com*).

Brian K. Williams
Stacey C. Sawyer
Sarah E. Hutchinson

Contents

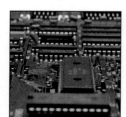

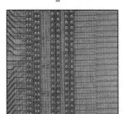

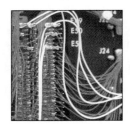

Chapter 9   Files & Databases: From Data Organizing to Data Mining   401

Chapter 11   Software Development: Programming & Languages 487

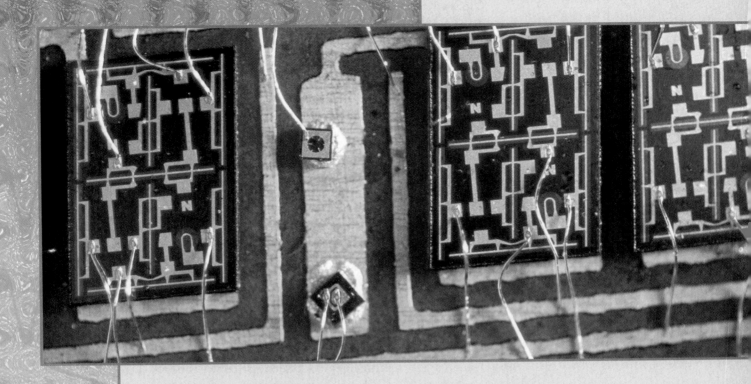

The Digital Age
Overview of the Revolution in Computers & Communications

Concepts You Should Know

After reading this chapter, you should be able to:

1. Define the terms *information technology, technological convergence, computer,* and *communications*

2. Briefly define *analog* and *digital*

3. Identify the six major elements of a computer-and-communications system

4. Describe the difference between an information technology professional and an end-user

5. Define *data* and *information*

6. Briefly explain the five operations of a computer-and-communications system: input, processing, output, storage, and communications, as well as the corresponding categories of hardware

7. Explain the difference between applications software and systems software

8. Identify the five major categories of computers

9. Discuss the important trends in computer technology

10. Define the term *Information Superhighway*

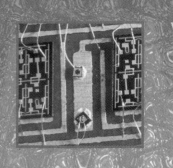

"We should all be concerned about the future," said engineer and inventor Charles Kettering, "because we will have to spend the rest of our lives there."

This book is about your future, a future rapidly becoming present. It is about a revolution that will make—indeed, is making now—profound changes in your life. The revolution has many names: The Digital Age. The Information Age. The Age of Convergence. The Interactive Revolution. The Multimedia Revolution. The Information Superhighway—or "Infobahn" or I-way or Dataway. Whatever it's called, the revolution is happening in all parts of society and in all parts of the world, and its consequences will reverberate throughout our lifetimes.

The technological systems and industries that this revolution is bringing forth may seem awesomely complex. However, the concept on which they are based is as simple as the flick of a light switch: *on* and *off*. Let us begin to see how this works.

From the Analog to the Digital Age: The "New Story" of Computers & Communications

Preview & Review: Information technology is technology that merges computers and high-speed communications links. The fusion of computer and communications technologies is producing "technological convergence"—the technological merger of several industries through various devices that exchange information in the electronic format used by computers. The industries include computers, communications, consumer electronics, entertainment, and mass media.

Computers are based on digital, binary (two-state) signals. However, most phenomena in the world are analog, representing continuously variable quantities. Some formerly analog devices are now taking digital form. To transmit a computer's digital signals over an analog telephone line requires a modem. To digitize analog sound or images, as in recording a live performance for a CD, requires sampling and averaging.

The essence of all revolution, stated philosopher Hannah Arendt, is the start of a *new story* in human experience. For us, the new story may be said to have begun in 1991. In that year, according to one report, "companies for the first time spent more on computing and communications gear . . . than on industrial, mining, farm, and construction machines." It adds: "Info tech is now as vital . . . as the air we breathe."[1]

"Info tech"—information technology—is what this book is about. **Information technology is technology that merges computing with high-speed communications links carrying data, sound, and video.**[2] The arrival of information technology is having powerful consequences, the most notable being the gradual fusion of several important industries in a phenomenon that has been called *technological convergence*.

What Is "Technological Convergence"?

Technological convergence, also known as *digital convergence*, is the technological merger of several industries through various devices that exchange information in the electronic, or digital, format used by computers. The industries are computers, communications, consumer electronics, entertain-

ment, and mass media. The direction in which digital convergence is going is illustrated by the "TV/PC," a device that combines the television and the personal computer. In the recent past, it was not possible to use your television set as a computer or to use a personal computer to watch broadcast TV programs. Now, however, the technologies of television and computing are coming together.

Technological convergence has tremendous significance. It means that, from a common electronic base, information can be communicated in all the ways we are accustomed to receiving it. These include the familiar media of newspapers, photographs, films, recordings, radio, and television. However, it can also be communicated through newer technology—satellite, fiber-optic cable, cellular phone, fax machine, or compact disk, for example. More important, as time goes on, *the same information may be exchanged among many kinds of equipment, using the language of computers.* Understanding this shift from single, isolated technologies to a unified digital technology means understanding the effects of this convergence on your life—such as:

* The increased need for continuous learning
* Adapting to less well-defined jobs as an "information worker"
* The stepped-up pace of change
* Exposure to relatively unregulated technical and social information from other cultures via global networks.

Is this consolidation of technologies an overnight phenomenon? Actually, it has been developing over several years, as we explain next. (■ *See Panel 1.1.*)

PANEL 1.1

Fusion of computer and communications technology

Today's new information environment came about gradually from two separate streams of technological development. (*Continued on pages 4–8.*)

Computer Technology

3000 BC		200 BC	
Abacus, used for arithmetic calculations, developed in Orient		Chinese artisans develop an entire mechanical orchestra	

Communications Technology

35,000 BC	4000 BC	3000 BC	1800 BC	600 BC	1453 AD
Language probably existed	Sumerian writing on clay tablets	Early Egyptian hieroglyphics	Phoenician alphabet	Book printing in China	First book printed in Europe

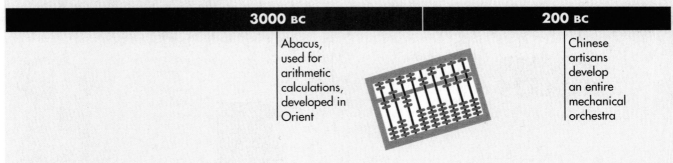

The Fusion of Computer & Communication Technologies

Technological convergence is derived from a combination of two recent technologies—*computer* and *communications*.

- **Computer technology:** Is there anyone reading this book who has not seen a computer by now? Nevertheless, let's define what it is. **A *computer* is a programmable, multiuse machine that accepts data—raw facts and figures—and processes, or manipulates, it into information we can use, such as summaries or totals.** Its purpose is to speed up problem solving and increase productivity.

  If you've actually touched a computer it's probably been a personal computer, such as the widely advertised desktop or portable models from Apple, IBM, Compaq, or Packard Bell. However, many other machines, such as microwave ovens and portable phones, use miniature electronic processing devices (microprocessors, or microcontrollers) similar to those that control personal computers.

  An example of how raw data is computer-processed into useful information is provided by the computer connected to an automated teller machine (ATM). The unseen computer processes deposit and withdrawal data to give you the total in your account, printed on the ATM receipt.

- **Communications technology:** Unquestionably you've been using communications technology for years. **Communications, or *telecommunica-***

PANEL 1.1

Continued

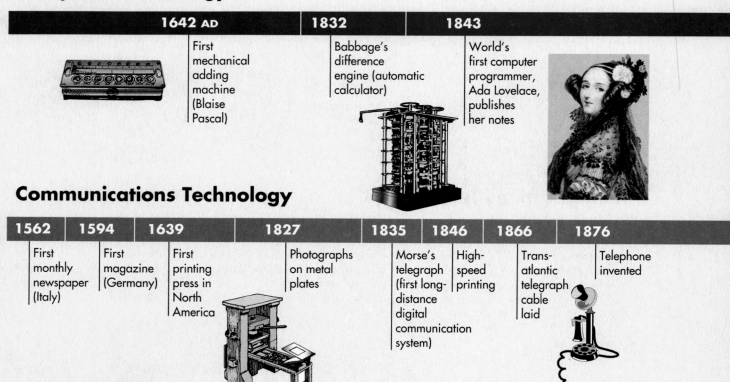

Computer Technology

	1642 AD	1832	1843
	First mechanical adding machine (Blaise Pascal)	Babbage's difference engine (automatic calculator)	World's first computer programmer, Ada Lovelace, publishes her notes

Communications Technology

1562	1594	1639	1827	1835	1846	1866	1876
First monthly newspaper (Italy)	First magazine (Germany)	First printing press in North America	Photographs on metal plates	Morse's telegraph (first long-distance digital communication system)	High-speed printing	Trans-atlantic telegraph cable laid	Telephone invented

tions, technology **consists of electromagnetic devices and systems for communicating over long distances.** The principal examples are telephone, radio, television, and cable.

Before the 1950s computer technology and communications technology developed independently, like rails in a railroad track that never merge. Since then, however, they have gradually fused together, producing a new information environment.

Why have the worlds of computers and of telecommunications been so long in coming together? The answer is this: *computers are digital, but most of the world is analog.* Let us explain what this means.

The Digital Basis of Computers

Computers may seem like incredibly complicated devices, but their underlying principle is simple. When you open up a personal computer, what you see is mainly electronic circuitry. And what is the most basic statement that can be made about electricity? It is simply this: it can be either *turned on* or *turned off*, or switched between *high voltage* and *low voltage.*

With a two-state on/off, high/low, open/closed, present/absent, positive/negative, yes/no arrangement, one state can represent a 1 digit, the other a 0 digit. People are most comfortable with the *decimal system*, which has ten digits (0, 1, 2, 3, 4, 5, 6, 7, 8, 9). Because computers are based on on/off or other two-state conditions, they use the **binary system, which consists of only two digits—0 and 1.**

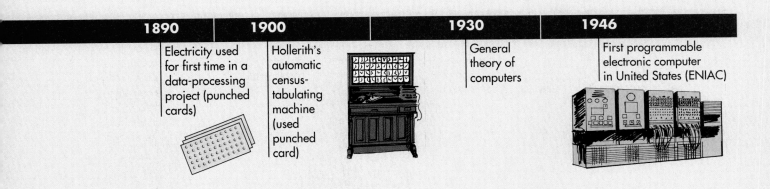

1890	1900	1930	1946
Electricity used for first time in a data-processing project (punched cards)	Hollerith's automatic census-tabulating machine (used punched card)	General theory of computers	First programmable electronic computer in United States (ENIAC)

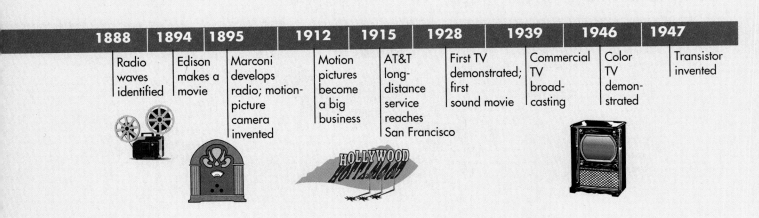

1888	1894	1895	1912	1915	1928	1939	1946	1947
Radio waves identified	Edison makes a movie	Marconi develops radio; motion-picture camera invented	Motion pictures become a big business	AT&T long-distance service reaches San Francisco	First TV demonstrated; first sound movie	Commercial TV broadcasting	Color TV demonstrated	Transistor invented

The word **digit** simply means numeral. The word *digital* is derived from "digit," referring to the fingers people used to count with. Today, however, **digital is almost synonymous with "computer-based." More specifically, it refers to communications signals or information represented in a discrete (individually distinct) form—usually in a binary or two-state way.**

In the binary system, **each 0 or 1 is called a *bit*—short for *binary digit*.** In turn, bits can be grouped in various combinations to represent characters of data—numbers, letters, punctuation marks, and so on. For example, the letter *H* could correspond to the electronic signal 01001000 (that is, off-on-off-off-on-off-off-off). In computing, **a group of eight bits is called a *byte*.**

Digital data, then, **consists of data in discrete, discontinuous form—usually 0s and 1s.** This is the method of data representation by which computers process and store data and communicate with each other.

The Analog Basis of Life

Most phenomena of the world are **analog, having continuously variable values.** Sound, light, temperature, and pressure values, for instance, can fall anywhere on a continuum or range. The highs, lows, and in-between states have historically been represented with analog devices rather than in digital form. Examples of analog devices are a humidity recorder, thermometer, and pressure sensor, which can measure continuous fluctuations. Thus, **analog data is transmitted in a continuous form that closely resembles the information it represents.** The electrical signals on a telephone line are analog-data rep-

PANEL 1.1
Continued

Computer Technology

1952	1964	1970	1971	1977
UNIVAC computer correctly predicts election of Eisenhower as U.S. President	IBM introduces 360 line of computers	Microprocessor chips come into use; floppy disk introduced for storing data	First pocket calculator	Apple II computer (first personal computer sold in assembled form)

Communications Technology

1950	1952	1957	1961	1968	1975	1976	1977
Cable TV	Direct-distance dialing (no need to go through operator); transistor radio introduced	First satellite launched (Russia)	Push-button telephones	Portable video recorders; video cassettes	Flat-screen TV	First wide-scale marketing of TV computer games (Atari)	First inter-active cable TV

resentations of the original voices. Telephone, radio, television, and cable-TV have traditionally transmitted analog data.

The differences between analog and digital transmission are apparent when you look at a drawing of a wavy analog signal, such as a voice message appearing on a standard telephone line, and an on/off digital signal. **In general, to transmit your computer's digital signals over telephone lines, you need to use a *modem* to translate them into analog signals.** (■ *See Panel 1.2.*)

The modem provides a means for computers to communicate with one another while the old-fashioned copper-wire telephone network—an analog system that was built to transmit the human voice—still exists. Our concern, however, goes far beyond telephone transmission. How can the analog realities of the world be expressed in digital form? How can light, sounds, colors, temperatures, and other dynamic values be represented so that they can be manipulated by a computer? Let us consider this.

Converting Reality to Digital Form: Sampling & Averaging

Suppose you are using an analog tape recorder to record a singer—Garth Brooks, say—during a performance. The analog process will produce a near duplicate of the sounds. This will include distortions, such as buzzings and clicks, or electronic hums if an amplified guitar is used. It will also include aberrations introduced by the recording process itself.

The digital recording process is different. The way in which music is captured for music CDs (compact disks) does not provide a duplicate of a musical performance. Rather, the digital process uses *sampling and averaging* to

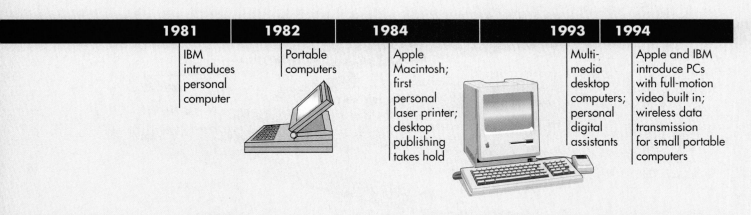

1981	1982	1984	1993	1994
IBM introduces personal computer	Portable computers	Apple Macintosh; first personal laser printer; desktop publishing takes hold	Multimedia desktop computers; personal digital assistants	Apple and IBM introduce PCs with full-motion video built in; wireless data transmission for small portable computers

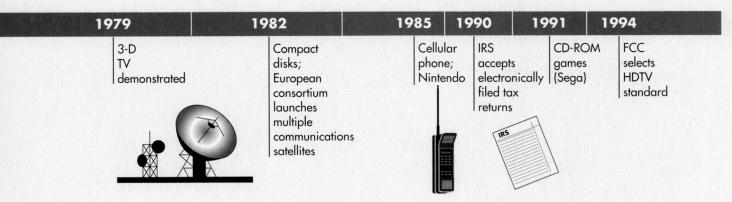

1979	1982	1985	1990	1991	1994
3-D TV demonstrated	Compact disks; European consortium launches multiple communications satellites	Cellular phone; Nintendo	IRS accepts electronically filed tax returns	CD-ROM games (Sega)	FCC selects HDTV standard

record the sounds and produce a copy that is virtually exact and free from distortion and noise. Sounds are sampled by computer-based equipment at regular intervals—nearly 44,100 times a second. They are then converted to numbers and averaged. Similarly, for visual material, values such as brightness or color can also be sampled and then averaged. The same is true of other aspects of real-life experience, such as pressure, temperature, and motion.

Are we being cheated out of experiencing "reality" by allowing computers to do sampling and averaging of sounds, images, and so on? Actually, people willingly made this compromise years ago, before computers were invented. Movies, for instance, carve up reality into 24 frames a second. Television frames are drawn at 30 lines per second. These processes happen so quickly that our eyes and brains easily jump the visual gaps. Digital processing of analog experience represents just one more degree of compromise. As one writer puts it:

> It is worth remembering that digital systems . . . couldn't intervene into media at all . . . without human cooperation. Above all else, it is our willingness to accept the limited input that our senses can deliver as "reality" that enables the digital world. The digital approach always has to ignore a certain amount of potential information in slicing up a dynamic process to produce a computer-acceptable description of it.[3]

Let us now look at how a digital computer-and-communications system works. The following sections present a brief overview that is important to an understanding of the rest of the book.

PANEL 1.1

Continued

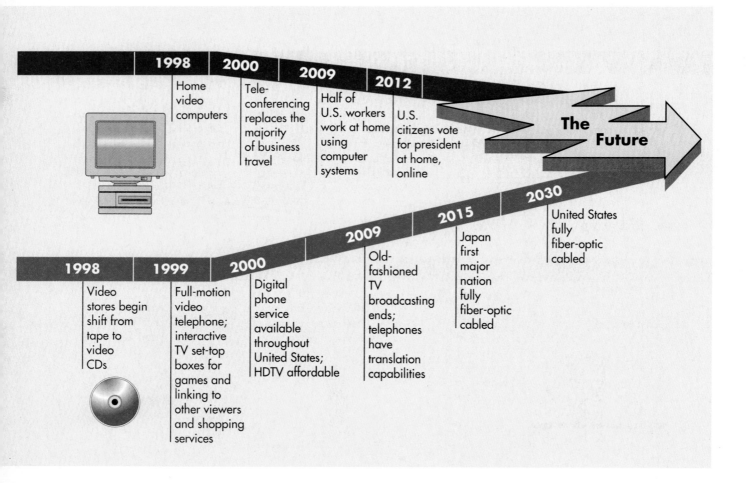

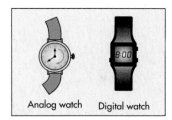

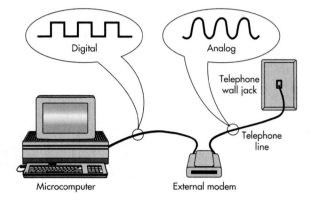

Microcomputer

Analog watch   Digital watch

Digital   Analog

Telephone
wall jack

Telephone
line

External modem   Microcomputer   External modem

PANEL 1.2

Analog versus digital signals, and the modem

(*Top*) On an analog watch, the hands move continuously around the watch face; on a digital watch, the display changes once each minute. (*Bottom*) Note the wavy line for an analog signal and the on/off (or high-voltage/low voltage) line for a digital signal. The modem shown here is outside the computer. Today most modems are inside the computer and not visible.

The Six Elements of a Computer-&-Communications System

Preview & Review: A computer-and-communications system has six elements: (1) people, (2) procedures, (3) data/information, (4) hardware, (5) software, and (6) communications.

A *system* is a group of related components and operations that interact to perform a task. A system can be many things: registration day at your college, the 52 bones in the foot, a weather storm front, the monarchy of Great Britain. Here we are concerned with a technological kind of system. **A *computer-and-communications system* is made up of six elements: (1) people, (2) procedures, (3) data/information, (4) hardware, (5) software, and (6) communications.** (■ *See Panel 1.3, next page.*) We briefly describe these elements in the next six sections and elaborate on them in subsequent chapters.

System Element 1: People

Preview & Review: People are the most important part of, and the beneficiaries of, a computer-and-communications system.

There are two types of people using information technology—professionals and end-users.

People can analyze, develop, and improve computer systems. They can also complicate the operation of a system: their assessment of information needs is sometimes faulty, their emotions may affect their performance, and their perceptions can be too slow.

Although we will not say a lot about them here, people are the most important part of a computer-and-communications system. People of all levels and

PANEL 1.3

A computer-and-communications system

The six elements of the system include people, procedures, data/information, hardware, software, and communications.

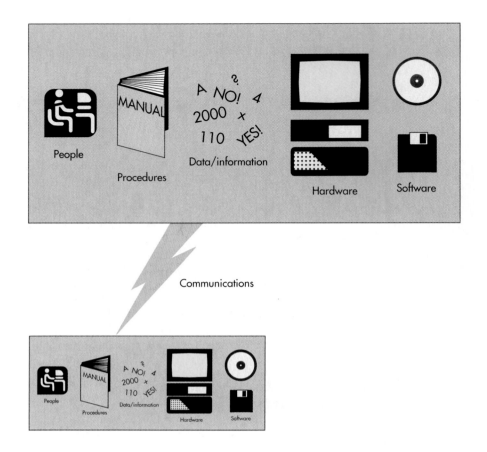

skills, from novices to programmers, are the users and operators of the system. The whole point of the system, of course, is to benefit people.

Two Types of Users: Professionals & End-Users

Two types of people use information technology—*professionals* and *"end-users."*

- **Professionals:** An *information technology professional* **is a person who has had formal education in the technical aspects of using a computer-and-communications system.** For example, a *computer programmer* creates the programs (software) that process the data in a computer system.

- **End-users:** An "end-user" is a person probably much like yourself. **An** *end-user,* **or simply a** *user,* **is someone without much technical knowledge of information technology who uses computers for entertainment, education, or work-related tasks.** The user is not a technology expert but knows enough about it to use it for his or her own purposes, such as for career advancement. Lawyers, for example, may know how to use a computer to search a huge data bank of information for legal decisions relevant to their cases.

Both computer professionals and end-users can work to improve computer systems by analyzing old systems and suggesting or developing new ones. However, people can also be a complicating factor in a computer system.

People as a Complicating Factor

When experts speak of the "unintended effects of technology," what they are usually referring to are the unexpected things people do with it. People can complicate the workings of a system in three ways:[4]

- **Faulty assessment of information needs:** Humans often are not good at assessing their own information needs. Thus, for example, many users will acquire a computer-and-communications system that either is not sophisticated enough or is far more complex than they need. If all you need is a personal computer on which to type research papers, for instance, you don't need to spend $10,000 on a state-of-the-art system. An outdated system bought used for $500 or less may do just fine. In addition, *people don't always know what information is needed to make a decision.* In other words, they don't know what they *need* to know.

- **Human emotions affect performance:** Of course, human emotions can also affect the performance of a system. For example, one frustrating experience with a computer is enough to make some people abandon the whole system. Hammering on the keyboard or bashing the display screen is certainly not going to advance the learning experience. Also, many people are afraid of computers. However, this feeling is common and diminishes with experience.

- **Human perceptions may be too slow:** Humans act on their perceptions, which in modern information environments are often too slow to keep up with the equipment. You can be so overwhelmed by information overload, for example, that decision making may be just as faulty as if you had too little information.

In summary, although people are the supposed beneficiaries of a computer-and-communications system, they can be the most complicating factor in it.

System Element 2: Procedures

Preview & Review: Procedures are steps for accomplishing a result. Some procedures may be expressed in manuals or documentation. Documentation is also available online.

Procedures **are descriptions of how things are done, steps for accomplishing a result.** Sometimes procedures are unstated, the result of tradition or common practice. You may find this out when you join a club or are a guest in someone's house for the first time. Sometimes procedures are laid out in great detail in manuals, as is true, say, of tax laws.

When you use a bank ATM—a form of computer system—the procedures for making a withdrawal or a deposit are given in on-screen messages. In other computer systems, procedures are spelled out in manuals. **Manuals, called** *documentation,* **contain instructions, rules, or guidelines to follow when using hardware or software.** When you buy a microcomputer or a software package, it comes with documentation, or procedures. Nowadays, in fact, many such procedures come not only in a book or pamphlet but also on a computer disk, which presents directions on your display screen. Many companies also offer documentation online.

System Element 3: Data/Information

Preview & Review: The distinction is made between raw data, which is unprocessed, and information, which is processed data. Units of measurement of data/information capacity include kilobytes, megabytes, gigabytes, and terabytes.

Though used loosely all the time, the word *data* has some precise and distinct meanings.

"Raw Data" Versus Information

Data can be considered the raw material—whether in paper, electronic, or other form—that is processed by the computer. In other words, **data consists of the raw facts and figures that are processed into information.**

Information **is summarized data or otherwise manipulated data that is useful for decision making.** Thus, the raw data of employees' hours worked and wage rates is processed by a computer into the information of paychecks and payrolls. Some characteristics of useful information are that it is *relevant, timely, accurate, concise,* and *complete.*

Actually, in ordinary usage the words *data* and *information* are often used synonymously. After all, one person's information may be another person's data. The "information" of paychecks and payrolls may become the "data" that goes into someone's yearly financial projections or tax returns.

Units of Measurement for Capacity: From Bytes to Terabytes

A common concern of computer users is "How much data can this gadget hold?" The gadget might be a diskette, a hard disk, or a computer's main memory. The question is a crucial one. If you have too much data, the computer may not be able to handle it. Or if a software package takes up too much storage space, it cannot be run on a particular computer.

We mentioned that computers deal with "on" and "off" (or high-voltage and low-voltage) electrical states, which are represented in the hardware in terms of 0s and 1s, called *bits.* Bits are combined in groups of eight, called *bytes,* to hold the equivalent of a character. A *character* is a single letter, number, or special symbol (such as a punctuation mark or dollar sign). Examples of characters are A, 1, and ?.

A computer system's data/information storage capacity is represented by bytes, kilobytes, megabytes, gigabytes, and terabytes:

- **Kilobyte:** A *kilobyte* **(abbreviated *K* or *KB*) is equivalent to approximately 1000 bytes** (or characters). More precisely, 1 kilobyte is 1024 (2^{10}) bytes, but the figure is commonly rounded off. Kilobytes are a common unit of measure for the data-holding (memory) capacity of personal computers. The original IBM PC, for example, could hold (in memory) 640 kilobytes, or about 640,000 bytes of data, and early home computers held only 64 K.

- **Megabyte:** A *megabyte* **(abbreviated *M* or *MB*) is about 1 million bytes.** Some personal computers can run programs requiring 16 or less megabytes, or about 16 million bytes, of memory.

- **Gigabyte:** A *gigabyte* **(*G* or *GB*) is about 1 billion bytes.** Pronounced "*gig*-a-bite" (not "*jig*-a-bite"), this unit of measure is used not only with "big iron" computers (mainframes and supercomputers) but also with newer personal computers.

- **Terabyte:** A *terabyte* (*T* or *TB*) is about 1 trillion bytes.

System Element 4: Hardware

Preview & Review: The basic operations of computing consist of (1) input, (2) processing, (3) output, and (4) storage. Communications (5) adds an extension capability to each phase.

Hardware devices are categorized according to which of these five operations they perform. (1) Input hardware includes the keyboard, mouse, and scanner. (2) Processing and memory hardware consists of the CPU (the processor) and main memory. (3) Output hardware includes the display screen, printer, and sound devices. (4) Secondary storage hardware stores data on diskette, hard disk, magnetic tape devices, and optical-disk. (5) Communications hardware includes modems.

As we said earlier, a *system* is a group of related components and operations that interact to perform a task. Once you know how the pieces of the system fit together, you can then make better judgments about any one of them. And you can make knowledgeable decisions about buying and operating a computer system.

The Basic Operations of Computing

How does a computer system process data into information? It goes through four operations: *(1) input, (2) processing, (3) output,* and *(4) storage.* (■ *See Panel 1.4, next page.*)

1. *Input operation:* In the **input operation, data is entered or otherwise captured electronically and is converted to a form that can be processed by the computer.** The means for "capturing" data (the raw, unsorted facts) is input hardware, such as a keyboard.
2. *Processing operation:* In the **processing operation, the data is manipulated to process or transform it into information** (such as summaries or totals). For example, numbers may be added or subtracted.
3. *Output operation:* In the **output operation, the information, which has been processed from the data, is produced in a form usable by people.** Examples of output are printed text, sound, and charts and graphs displayed on a computer screen.
4. *Secondary storage operation:* In the **storage operation, data, information, and programs are stored in computer-processable form.** Diskettes are examples of materials used for storage.

Often these four operations occur so quickly that they seem to be happening simultaneously.

Where does communications fit in here? In the four operations of computing, communications offers an *extension* capability. Data may be input from afar, processed in a remote area, output in several different locations, and stored in yet other places. And information can be transmitted to other computers. All this is done through a wired or wireless connection to the computer.

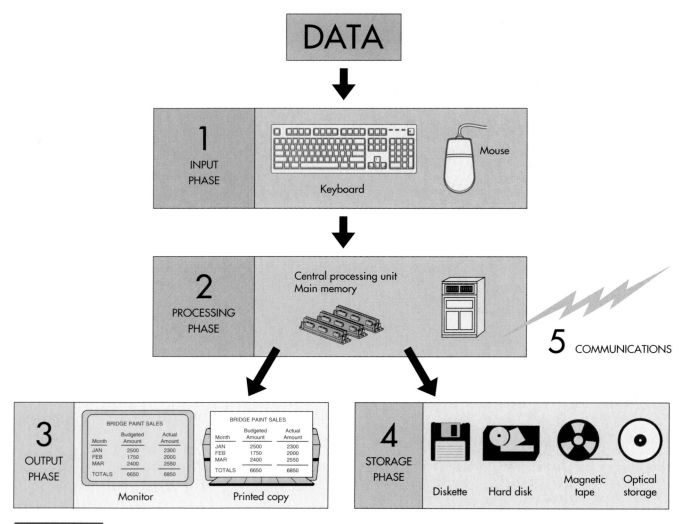

DATA

| 1 INPUT PHASE | Keyboard   Mouse |

| 2 PROCESSING PHASE | Central processing unit Main memory |

5 COMMUNICATIONS

| 3 OUTPUT PHASE | Monitor   Printed copy |

BRIDGE PAINT SALES

Month	Budgeted Amount	Actual Amount
JAN	2500	2300
FEB	1750	2000
MAR	2400	2550
TOTALS	6650	6850

BRIDGE PAINT SALES

Month	Budgeted Amount	Actual Amount
JAN	2500	2300
FEB	1750	2000
MAR	2400	2550
TOTALS	6650	6850

| 4 STORAGE PHASE | Diskette   Hard disk   Magnetic tape   Optical storage |

PANEL 1.4

The basic operations of computing

A computer goes through four operations: (1) input of data, (2) processing of data into information, (3) output of information, and (4) storage of information. Communications (5) extends the computer system's capabilities.

Hardware Categories

Hardware is what most people think of when they picture computers. *Hardware* **consists of all the machinery and equipment** in a computer system. The hardware includes, among other devices, the keyboard, the screen, the printer, and the computer or processing device itself.

As computing and telecommunications have drawn together, people have begun to refer loosely to *any* machinery or equipment having to do with either one as "hardware." This is the case whether the equipment is a "smart box," such as a cable-TV set-top controller, or (sometimes) the connecting cables, transmitters, or other communications devices.

In general, computer hardware is categorized according to which of the five computer operations it performs:

- Input
- Processing and memory
- Output
- Secondary storage
- Communications

External devices that are connected to the main computer cabinet are referred to as "peripheral devices." **A *peripheral device* is any piece of hardware that is connected to a computer.** Examples are the keyboard, mouse, monitor, and printer.

Practical Matters: Common Measurements Used in Computers & Communications

When salespeople or friends rattle on about how fast a computer's "clock speed" is or how many "dpi" the printer uses, how will you know what they're talking about? The following is a quick guide to some common measurement terms.

baud and bps Both terms are used to describe the speed at which a computer can transfer information.

- *Baud ("bawd"):* A measure of signal changes that take place during 1 second of data transfer.

- *bps (bits per second):* A measure of the actual number of bits that are transferred during that second. (A bit is a 0 or 1, the smallest unit of information used in computing.)

Baud rates are sometimes erroneously used to specify bits per second for modem speed, but only at low speeds are they the same (for example, 300 baud is equal to 300 bps—only about 6 words per second).

capacity—from bits to terabytes *Capacity* refers to how much data/information a storage device will hold. Capacity is represented by bits, bytes, kilobytes, megabytes, gigabytes, and terabytes.

- *Bit:* Short for *bin*ary dig*it;* a 0 or 1, which the computer hardware represents as an "on" or "off" (or high-voltage or low-voltage) electrical state.

- *Byte:* Usually a group of eight bits.

- *Kilobyte (K or KB):* About 1000 (1024 or 2^{10}) bytes. Bytes and their multiples are common units of measure for memory or storage capacity of personal computers.

- *Megabyte (M or MB):* About 1 million (specifically 1,048,576) bytes.

- *Gigabyte (G or GB):* pronounced *"gig-a-bite"* (not *"jig-a-bite"*); about 1 billion (1,073,741,824) bytes.

- *Terabyte (T or TB):* About 1 trillion (specifically 1,009,511, 627,776) bytes.

clock speed—Hz, KHz, and MHz *Clock speed* refers to how fast a computer processes. (The CPU, or central processing unit, is circuitry that controls the interpretation and execution of instructions.) The *CPU clock* uses a quartz crystal to generate a steady stream of pulses to the CPU to regulate the system's internal speed. The clock measures speed in hertz, kilohertz, and megahertz as frequency of electrical vibrations (cycles) per second.

- *Hertz (Hz):* A single clock cycle per second

- *Kilohertz (KHz):* 1000 cycles per second

- *Megahertz (MHz):* 1 million cycles per second

The clock speed of a microprocessor (a CPU on a single chip) in a microcomputer is measured in megahertz. For example, for '486 and later microprocessors, speeds range from 25 MHz to 166 MHz or more.

dot pitch Measurement used to describe the clarity of the image on a display screen or of a printer's output.

- *Pixels:* For display screens, dot pitch is expressed in millimeters as the distance between individual dots, or *pixels (picture elements)*. The smaller the dot pitch, the clearer the image. Typically, display screens vary from .28 to .51 millimeters.

- *dpi:* For printers, dot pitch is expressed in *dpi,* the number of dots that a printer can print in a linear inch.

dpi *Dots per inch;* dpi is a measurement used to describe the image clarity of printers and scanners. If you look closely, you will see that a printed image is made up of individual dots. The higher the number of dots per linear inch, the clearer the image. A 300 dpi printer prints 300 x 300, or 90,000, dots in 1 square inch. A 400 dpi printer produces 160,000 dots, a 500 dpi printer produces 250,000 dots. Common printer and scanner measurements range from 300 to 600 dpi and up.

fractions of a second In increasing order of rapidity:

- *Milliseconds (ms):* Thousandths of a second; measures the amount of time the computer takes to access information from a hard disk.

- *Microseconds (μs):* Millionths of a second; measures instruction execution.

- *Nanoseconds (ns):* Billionths of a second; measures the speed at which information travels through circuits, as for memory chips (70–60 ns).

- *Picoseconds (ps):* Trillionths of a second; measures transistor switching.

inch Measurement used to describe size of floppy disks and of monitors.

- *Diskette sizes:* Diskettes come in two principal sizes: 3½ (or 3.5) inches, now the more common standard; and 5¼ (or 5.25) inches, an older size, now less common.

- *Monitor sizes:* Like the screens of television sets, computer monitors, or display screens, are measured diagonally from one corner to the other. Common sizes are 14–17 inches or larger.

MIPS *million instructions per second;* MIPS are a measurement of the execution speed of a large computer.

word The standard unit of information natural to a particular system. The unit varies depending on the computer. For a computer with a microprocessor that processes 16 bits at a time, a word would be 16 bits.

We describe hardware in detail elsewhere (Chapters 4–8), but the following offers a quick overview.

Input Hardware

Input hardware consists of devices that allow people to put data into the computer in a form that the computer can use. For example, input may be by means of a *keyboard, mouse,* or *scanner.* The keyboard is the most obvious. The mouse is a pointing device attached to many microcomputers. An example of a scanner is the grocery-store bar-code scanner. (These and other input devices are discussed in detail in Chapter 5.)

- **Keyboard:** **A *keyboard* includes the standard typewriter keys plus a number of specialized keys.** The standard keys are used mostly to enter words and numbers. Examples of specialized keys are the *function keys,* labeled *F1, F2,* and so on. These special keys are used to enter commands.

- **Mouse:** **A *mouse* is a device that can be rolled about on a desktop to direct a pointer on the computer's display screen.** The pointer is a symbol, usually an arrow, on the computer screen that is used to select items from lists (menus) or to position the cursor. **The *cursor* is the symbol on the screen that shows where data may be entered next,** such as text in a word processing program.

- **Scanners:** ***Scanners* translate images of text, drawings, and photos into digital form.** The images can then be processed by a computer, displayed on a monitor, stored on a storage device, or communicated to another computer.

Processing & Memory Hardware

The brains of the computer are the *processing* and *main memory* devices, housed in the computer's system unit. **The *system unit,* or *system cabinet,* houses the electronic circuitry, called the CPU, which does the actual processing and the main memory, which supports processing.** (■ *See Panel 1.5.*) (These are discussed in detail in Chapter 4.)

- **CPU—the processor:** **The *CPU,* for *Central Processing Unit,* is the processor, or computing part of the computer. It controls and manipulates data to produce information.** In a personal computer the CPU is usually a single fingernail-size "chip" called a *microprocessor,* with electrical circuits printed on it. This microprocessor and other components necessary to make it work are mounted on a main circuit board called a *motherboard* or *system board.*

- **Memory—working storage:** ***Memory*—also known as *main memory, RAM,* or *primary storage*—is working storage. Memory is the computer's "work space," where data and programs for immediate processing are held.** Computer memory is contained on memory chips mounted on the motherboard. Memory capacity is important because it determines how much data can be processed at once and how big and complex a program may be used to process the data.

  Despite its name, memory does not remember. That is, once the power is turned off, all the data and programs within memory simply vanish. This is why data/information must also be stored in relatively permanent form on disks and tapes, which are called *secondary storage* to distinguish them from main memory's *primary storage.*

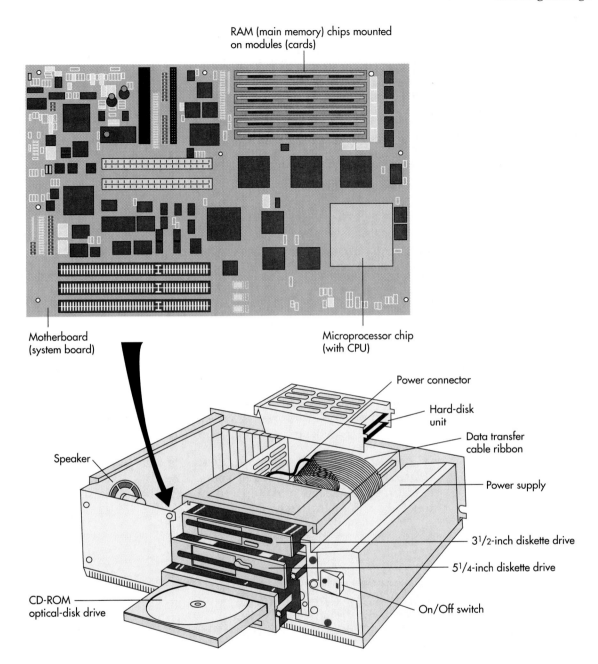

RAM (main memory) chips mounted on modules (cards)

Motherboard (system board)

Microprocessor chip (with CPU)

Power connector

Hard-disk unit

Data transfer cable ribbon

Speaker

Power supply

3¹/₂-inch diskette drive

5¹/₄-inch diskette drive

CD-ROM optical-disk drive

On/Off switch

PANEL 1.5

The system unit

Motherboard (*top*) fits inside system cabinet (*bottom*).

Output Hardware

Output hardware **consists of devices that translate information processed by the computer into a form that humans can understand.** We are now so exposed to products output by some sort of computer that we don't consider them unusual. Examples are grocery receipts, bank statements, and grade reports. More recent forms are digital recordings and even digital radio.

As a personal computer user, you will be dealing with three principal types of output hardware—*screens, printers,* and *sound output devices.* (These and other output devices are discussed in detail in Chapter 6.)

- **Screen:** **The *screen* is the display area of a computer.** A desktop computer or video terminal (such as those listing flight information in airports) will use a ***monitor*, a high-resolution screen.** The monitor is often called a ***CRT*, for *Cathode-Ray Tube*, the familiar TV-style picture tube.**

- **Printer:** **A *printer* is a device that converts computer output into printed images.** Printers are of many types, some noisy, some quiet, some able to print carbon copies, some not.

- **Sound:** Many computers emit chirps and beeps. Some go beyond those noises and contain sound processors and speakers that can play digital music or human-like speech. High-fidelity stereo sound is becoming more important as computer and communications technologies continue to merge.

Secondary Storage Hardware

Main memory (primary storage) is *internal storage.* It works with the CPU chip on the motherboard to hold data and programs for immediate processing. Secondary storage, by contrast, is *external* storage. It is not on the motherboard (although it may still be inside the system cabinet). ***Secondary storage* consists of devices that store data and programs permanently on disk or tape.**

You may hear people use the term "storage media." *Media* refers to the material that stores data, such as diskette or magnetic tape. For microcomputers the principal storage media are the *diskette (floppy disk), hard disk, magnetic tape,* and *CD-ROM.* (■ *See Panel 1.6.)* (These and other secondary storage devices are discussed in detail in Chapter 7.)

- **Diskette:** **A *diskette*, or *floppy disk*, is a removable round, flexible disk that stores data as magnetized spots.** The disk is contained in a plastic case or square paper envelope to prevent the disk surface from being touched by human hands.

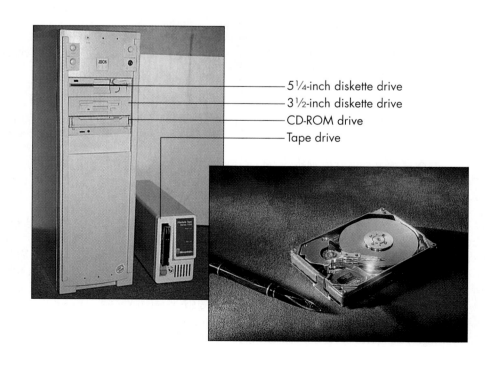

PANEL 1.6

Secondary storage for microcomputers

(*Left*) Examples of diskette, CD-ROM, and magnetic-tape drives (the hard-disk drive has no exterior opening). (*Right*) Inside of hard-disk drive. At a minimum, a personal computer will have a diskette drive.

— 5¼-inch diskette drive
— 3½-inch diskette drive
— CD-ROM drive
— Tape drive

Two sizes of diskettes are used for microcomputers. The older and larger size is *5¼ inches* in diameter. The smaller size, now by far the most common, is *3½ inches*. (■ *See Panel 1.7.*) The smaller disk, which can fit in a shirt pocket, has a compact and rigid case and actually does not feel "floppy" at all.

To use a diskette, you need a disk drive. **A *disk drive* is a device that holds and spins the diskette inside its case; it "reads" data from and "writes" data to the disk.** The words *read* and *write* are used a great deal in computing.

***Read* means that the data represented in magnetized spots on the disk (or tape) are converted to electronic signals and transmitted to the memory in the computer.**

***Write* means that the electronic information processed by the computer is recorded onto disk (or tape).**

The diskette drive may be a separate unit attached to the computer, particularly on older models. Usually, however, it is built into the system cabinet. Most newer PCs have one or two 3½-inch drives and perhaps one 5¼-inch drive.

Floppy disks

The most common size is 3½ inches (*top*); some older computers still use the 5¼-inch size (*bottom*).

- **Hard disk:** Diskettes are made out of tape-like material, which is what makes them "floppy." They are also removable. By contrast, **a *hard disk* is a disk made out of metal and covered with a magnetic recording surface. It also holds data represented by the presence (1) and absence (0) of magnetized spots.**

  Hard-disk drives *read* and *write* data in much the same way that diskette drives do. However, there are three significant differences. First, hard-disk drives can handle thousands of times more data than diskettes do. Second, hard-disk drives are usually built into the system cabinet, in which case they are not removable. Third, hard disks read and write data faster than diskettes do.

- **Magnetic tape:** Moviemakers used to love to represent computers with banks of spinning reels of magnetic tape. Indeed, with early computers, "mag tape" was the principal method of secondary storage.

  The magnetic tape used for computers is made from the same material as that used for audiotape and videotape. That is, ***magnetic tape* is made of flexible plastic coated on one side with a magnetic material; again, data is represented by the presence and absence of magnetized spots.** Because of its drawbacks (described in Chapter 7), nowadays tape is used mainly to provide low-cost duplicate storage, especially for microcomputers. A tape that is a duplicate or copy of another form of storage is referred to as a *backup*.

  Because hard disks sometimes fail ("crash"), personal computer users who don't wish to do backup using a lot of diskettes will use magnetic tape instead.

- **Optical disk—CD-ROM:** If you have been using music CDs (compact disks), you are already familiar with optical disks. **An *optical disk* is a disk that is written and read by lasers. *CD-ROM*, which stands for *Compact Disk—Read Only Memory*, is one kind of optical-disk format that is used to hold text, graphics, and sound.** CD-ROMs can hold hundreds of times more data than diskettes, and can hold more data than many hard disks.

Communications Hardware

Computers can be "stand-alone" machines, meaning that they are not connected to anything else. Indeed, many students tote around portable personal computers on which they use word processing or other programs to help them with their work. Many people are quite happy using a computer that has no communications capabilities.

However, the *communications* component of the computer system vastly extends the range of a computer. Indeed, the range is so many orders of magnitude larger that comprehending it is difficult.

In general, computer communications is of two types: *wired connections,* such as telephone wire or cable, and *wireless connections,* such as via radio waves (covered in detail in Chapter 8).

The dominant communications media that have been developed during this century use analog transmission. Thus, for many years, the principal form of direct connection was via standard copper-wire telephone lines. Hundreds of these twisted-pair copper wires are bundled together in cables and strung on telephone poles or buried underground. As mentioned, a modem is communications hardware required to translate a computer's digital signals into analog form for transmission over telephone wires. Although copper wiring still exists in most places, it is gradually being supplanted by two other kinds of direct connections: coaxial cable and fiber-optic cable. Eventually, all transmission media will accommodate digital signals.

System Element 5: Software

Preview & Review: Software comprises the step-by-step instructions that tell the computer what to do. In general, software is divided into applications software and systems software.

Applications software, which may be customized or packaged, performs useful work on general-purpose tasks.

Systems software, which includes operating systems, enables the applications software to run on the computer.

Software, **or** *programs,* **consists of the step-by-step instructions that tell the computer how to perform a task.** In most instances, the words *software* and *program* are interchangeable. Although it may be contained on disks of some sort, software is invisible, being made up of electronic blips.

There are two major types of software:

- **Applications software:** This may be thought of as the kind of software that people use to perform a general-purpose task, such as word processing software used to prepare the text for a document.
- **Systems software:** This may be thought of as the underlying software that the computer uses to manage its own internal activities and run applications software.

Although you may not need a particular applications program, you must have systems software, or you will not even be able to "boot up" your computer (make it run).

Applications Software

Applications software **is defined as software that can perform useful work on general-purpose tasks.** Examples are programs that do word processing, desktop publishing, or payroll processing.

Applications software may be either *customized* or *packaged*. *Customized software* is software designed for a particular customer. This is the kind of software that you would hire a professional computer programmer—a software creator—to develop for you. Such software would perform a task that could not be done with standard off-the-shelf packaged software available from a computer store or mail-order house.

Packaged software, or a *software package,* is the kind of "off-the-shelf" program developed for sale to the general public. This is the principal kind that will be of interest to you. Examples of packaged software that you will most likely encounter are word processing and spreadsheet programs. (We discuss these in Chapter 2.)

Systems Software

As the user, you interact mostly with the applications software. **Systems software enables the applications software to interact with the computer and manages the computer's internal resources.**

Systems software consists of several programs, the most important of which is the operating system. **The *operating system* acts as the master control program that runs the computer.** It handles such activities as running and storing programs and storing and processing data. The purpose of the operating system is to allow applications to operate by standardizing access to shared resources such as disks and memory. Examples of operating systems are MS-DOS, Windows 95, OS/2 Warp, and the Macintosh operating system (MacOS). (We discuss these operating systems in detail in Chapter 3.)

System Element 6: Communications

Preview & Review: "Communications" refers to the electronic transfer of data. The kind of data being communicated is rapidly changing from analog to digital.

Communications is defined as the electronic transfer of data from one place to another. Of all six elements in a computer-and-communications system, communications probably represents the most active frontier at this point.

We mentioned that, until now, most data being communicated has been analog data. However, as former analog methods of communication become digital, we will see a variety of suppliers, using wired or wireless connections, providing data in digital form: telephone companies, cable-TV services, news and information services, movie and television archives, interactive shopping channels, video catalogs, and more.

Developments in Computer Technology

Preview & Review: Computers have developed in three directions: smaller, more powerful, and less expensive.

Today the five types of computers are supercomputers, mainframe computers, minicomputers, microcomputers (both personal computers and workstations), and microcontrollers (embedded computers).

A human generation is not a very long time, about 30 years. During the short period of one and a half generations, computers have come from nowhere to transform society in unimaginable ways. One of the first computers, the outcome of military-related research, was delivered to the U.S. Army in 1946. ENIAC—short for *Electronic Numerical Integrator And Calculator*—weighed 30 tons, was 80 feet long and two stories high, and required 18,000 vacuum tubes. However, it could multiply a pair of numbers in the then-remarkable time of three-thousandths of a second. This was the first general-purpose, programmable electronic computer, the grandparent of today's lightweight handheld machines.

The Three Directions of Computer Development

Since the days of ENIAC, computers have developed in three directions:

- **Smaller size:** Everything has become smaller. ENIAC's old-fashioned radio-style vacuum tubes gave way to the smaller, faster, more reliable transistor. **A *transistor* is a small device used as a gateway to transfer electrical signals along predetermined paths (circuits).**

  The next step was the development of tiny integrated circuits. ***Integrated circuits (ICs)* are entire collections of electrical circuits or pathways etched on tiny squares of silicon** half the size of your thumbnail. *Silicon* is a natural element found in sand that is purified to form the base material for making computer processing devices.

- **More power:** In turn, miniaturization allowed computer makers to cram more power into their machines, providing faster processing speeds and more data storage capacity.

- **Less expense:** The miniaturized processor of a personal computer that sits on a desk performs the same sort of calculations once performed by a computer that filled an entire room. However, processor costs are only a fraction of what they were 15 years ago.

Five Kinds of Computers

Computers are categorized into five general types, based mainly on their processing speeds and their capacity to store data: *supercomputers, mainframe computers, minicomputers, microcomputers,* and *microcontrollers.* (■ *See Panel 1.8.)* They are described in detail elsewhere (Chapter 4), but let us characterize them briefly here:

- **Supercomputers:** ***Supercomputers* are high-capacity computers that cost millions of dollars, occupy special air-conditioned rooms, and are often used for research.** Among their uses are worldwide weather forecasting, oil exploration, aircraft design, and mathematical research.

- **Mainframe computers:** Less powerful than supercomputers, ***mainframe computers* are fast, large-capacity computers also occupying specially wired, air-conditioned rooms.** Mainframes are used by large organizations—banks, airlines, insurance companies, mail-order houses, universities, the Internal Revenue Service—to handle millions of transactions.

- **Minicomputers:** ***Minicomputers,* also called *midrange computers,* are generally refrigerator-size machines that are essentially scaled-down mainframes.** Because of their lesser processing speeds and data-storing capaci-

PANEL 1.8

The principal types of computers—and the microprocessor that powers them

(*Clockwise from top*) A supercomputer, a mainframe computer, a minicomputer, and two kinds of microcomputers— a personal computer (PC) and a workstation—and a microcontroller. (*Photo lower right*) A microprocessor (this one is Intel's P6) the miniaturized circuitry that does the processing in computers. A PC may have only one of these, a supercomputer thousands.

ties, they have been typically used by medium-sized companies for specific purposes, such as accounting. Minicomputers are being replaced by networks of microcomputers.

- **Microcomputers:** *Microcomputers* **are small computers that can fit on a desktop or in one's briefcase.** Microcomputers are of two types—*personal computers* and *workstations*—although the distinction is blurring rapidly.

    Personal computers (PCs) **are desktop or portable computers that can run easy-to-use programs, such as word processers or spreadsheets.** Whether desktop, laptop, notebook, or palmtop (in declining order of size), personal computers are now found in most businesses. They are also found in about one-third of American homes.

    Workstations **are expensive, powerful desktop machines used mainly by engineers and scientists for sophisticated purposes.** Providing many capabilities formerly found only in mainframes and minicomputers, workstations are used for such tasks as designing airplane fuselages or prescription drugs. Workstations are often connected to a larger computer system to facilitate the transfer of data and information.

- **Microcontrollers:** Also called **embedded computers, microcontrollers** are **installed in "smart" appliances like microwave ovens.**

Needless to say, as you progress from the microcontroller up to the supercomputer, cost, memory capacity, and speed all increase.

The Mighty Microprocessor

Computers by themselves are important. However, perhaps equally significant are the affiliated technologies made possible by the invention of the microprocessor. **A *microprocessor* is the miniaturized circuitry of the computer's processor—the part that manipulates data into information. The circuitry is etched on a sliver or "chip" of material, usually silicon.**

Microprocessors are the CPUs in personal computers. Equally important, microprocessors provide the "thinking" for most other new electronic devices, from CD players to music synthesizers to automobile fuel-injection systems. When you hear of all the things gadgetry is supposed to do for us, often you can credit the microprocessor.

Developments in Communications Technology

Preview & Review: Communications, or telecommunications, has had three important developments. They are better communications channels, better networks, and better sending and receiving devices.

Throughout the 1980s and early 1990s, telecommunications made great leaps forward. Three of the most important developments were:

* Better communications channels
* Better networks
* Better sending and receiving devices

Better Communications Channels

We mentioned that data may be sent by wired or wireless connections. The old copper-wire telephone connections have begun to yield to the more efficient coaxial cable and, more important, to fiber-optic cable, which can transmit vast quantities of information in both analog and digital form.

Even more interesting has been the expansion of wireless communication. Federal regulators have permitted existing types of wireless channels to be given over to new uses, as a result of which we now have many more kinds of two-way radio, cellular telephone, and paging devices than we had previously.

Better Networks

When you hear the word "network," you may think of a *broadcast network*, a group of radio or television broadcasting stations that cut costs by airing the same programs. Here, however, we are concerned with ***communications networks*, which connect one or more telephones or computers or associated devices.** The principal difference is that *broadcast networks transmit messages in only one direction, communications networks transmit in both directions.* Communications networks are crucial to technological convergence, for they allow information to be exchanged electronically.

A communications network may be large or small, public or private, wired or wireless or both. In addition, smaller networks may be connected to larger ones. For instance, a *local area network (LAN)* may be used to connect users located near one another, as in the same building. On some college campuses, for example, microcomputers in the rooms in residence halls are linked throughout the campus by a LAN. **A computer in a network shared by multiple users is called a** *server.*

Better Sending & Receiving Devices

Part of the excitement about telecommunications in the last decade or so has been the development of new devices for sending and receiving information. Two examples are the *cellular phone* and the *fax machine.*

- **Cellular phones:** *Cellular telephones* **use a system that divides a geographical service area into a grid of "cells." In each cell, low-powered, portable, wireless phones can be accessed and connected to the main (wire) telephone network.**

    The significance of the wireless, portable phone is not just that it allows people to make calls from their cars. Most important is its effect on worldwide communications. Countries with underdeveloped wired telephone systems, for instance, can use cellular phones as a fast way to install better communications. Such technology gives these nations—Mexico, Thailand, Pakistan, Hungary, and others— a chance to join the world economy.

    Today's cellular phones are also the forerunners of something even more revolutionary—pocket phones. Cigarette-pack-size portable phones and more fully developed satellite systems will enable people to have conversations or exchange information from anywhere on earth.

- **Fax machines:** *Fax* **stands for "facsimile," which means "a copy;" more specifically,** *fax* **stands for "facsimile transmission." A** *fax machine* **scans an image and sends a copy of it in the form of electronic signals over transmission lines to a receiving fax machine. The receiving machine re-creates the image on paper.** Fax messages may also be sent to and from microcomputers.

    Fax machines have been commonplace in offices and even many homes for some time, and new uses have been found for them. For example, some newspapers offer facsimile editions, which are transmitted daily to subscribers' fax machines. These editions look like the papers' regular editions, using the same type and headline styles, although they have no photographs. Toronto's *Globe & Mail* offers people who will be away from Canada a four-page fax that summarizes Canadian news. The *New York Times* sends a faxed edition, transmitted by satellite, to island resorts and to cruise ships in mid-ocean. (Fax machines and fax modems are covered in detail in Chapter 8.)

Computer & Communications Technology Combined: Connectivity & Interactivity

Preview & Review: Trends in information technology involve connectivity and interactivity.

Connectivity, or online information access, refers to connecting computers to one another by modem or network and communications lines. Connectivity, among other things, provides the benefits of voice mail and e-mail, telecommuting, teleshopping, databases, online services and networks, and electronic bulletin board systems.

Interactivity refers to the back-and-forth "dialog" between a user and a computer or communications device. Interactive devices include multimedia computers, personal digital assistants, and up-and-coming "smart boxes" and "Internet appliances."

Lee Taylor is what is known as a *lone eagle*. Once he was the manager of several technical writers for a California information services company. Then, taking a one-third pay cut, he moved with his wife to a tiny cabin near the ski-resort town of Telluride, Colorado. There he operates as a free-lance consultant for his old company, using phone, computer network, and fax machine to stay in touch.[5]

"Lone eagles" like Taylor constitute a growing number of professionals who, with information technology, can work almost anywhere they want, such as resort areas and backwoods towns. Although their income may be less, it is offset by such "quality of life" advantages as weekday skiing or reduced housing costs.

Taylor is one beneficiary of trends that will probably intensify as information technology continues to proliferate. These trends are:

- Connectivity
- Interactivity

Connectivity (Online Information Access)

As we discussed, small telecommunications networks may be connected to larger ones. This is called **connectivity, the ability to connect computers to one another by modem or network and communications lines to provide online information access.** It is this connectivity that is the foundation of the latest advances in the Digital Age.

The connectivity of telecommunications has made possible many kinds of activities. Although we cover these activities in more detail in Chapter 8, briefly they are as follows:

- **Voice mail and e-mail:** *Voice mail* acts like a telephone answering machine. Incoming voice messages are digitized and stored for your retrieval later. Retrieval is accomplished by dialing into your "mailbox" number from any telephone.

  The advantage of voice mail over an answering machine is that you don't have to worry about the machine running out of message tape or not functioning properly. Also, it will take messages *while* you're on the phone. You can get your own personal voice-mail setup by paying a monthly fee to a telephone company, such as AT&T.

  An alternative system is e-mail. **E-mail, or *electronic mail*, is a soft-ware-controlled system that links computers by wired or wireless connections. It allows users, through their keyboards, to post messages and to read responses on their computer screens.** Whether the network is a company's small local area network or a worldwide network, e-mail allows users to send messages anywhere on the system.

- **Telecommuting:** In standard commuting, one takes transportation (car, bus, train) from home to work and back. In *telecommuting*, one works at home and communicates with ("commutes to") the office by computer and communications technology. Already nearly 9 million people—not including business owners or independent contractors—telecommute at least part of the time. By 1998, according to Link Resources, there will be 13 million telecommuters.[6] (■ *See Panel 1.9.*)

- **Teleshopping:** Teleshopping is the computer version of cable-TV shop-at-home services. With *teleshopping*, microcomputer users dial into a telephone-linked computer-based shopping service listing prices and descriptions of products, which may be ordered through the computer. You charge the purchase to your credit card. The teleshopping service sends the merchandise to you by mail or other delivery service.

- **Databases:** A database may be a large collection of data located within your own unconnected personal computer. Here, however, we are concerned with databases located elsewhere. These are libraries of information at the other end of a communications connection that are available to you through your microcomputer. **A *database* is a collection of electronically stored data. The data is integrated, or cross-referenced, so that different people can access it for different purposes.**

    For example, suppose an unfamiliar company offered you a job. To find out about your prospective employer, you could go online to gain access to some helpful databases. Examples are Business Database Plus, Magazine Database Plus, and TRW Business Profiles. You could then study the company's products, review financial data, identify major competitors, or learn about recent sales increases or layoffs. You might even get an idea of whether or not you would be happy with the "corporate culture."[7]

- **Computer online services and networks and the Internet:** Established major commercial online services include America Online, CompuServe,

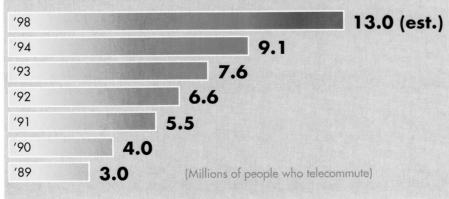

'98	13.0 (est.)
'94	9.1
'93	7.6
'92	6.6
'91	5.5
'90	4.0
'89	3.0

(Millions of people who telecommute)

PANEL 1.9

Telecommuting

The number of employees who telecommute—use computers and wired or wireless technology to work from home—for at least part of their working day has greatly increased. Telluride, Colorado, has become a testing site for telecommuting and other new communications ideas.

and Prodigy. **A *computer online service* is a commercial information service that, for a fee, makes various services available to subscribers through their telephone-linked microcomputers.**

Among other things, consumers can research information in databases, go teleshopping, make airline reservations, or send messages via e-mail to others using the service.

Through a computer online service you may also gain access to the greatest network of all, the Internet. **The *Internet* is an international network connecting approximately 36,000 smaller networks that link computers at academic, scientific, and commercial institutions.** An estimated 24 million people in the United States and Canada alone are already on the Internet—fully 11% of the North American population over age 16. The most well known part of the Internet is the World Wide Web, which stores information in multimedia form—sounds, photos, video, as well as text.

- **Electronic bulletin board systems:** **An *electronic bulletin board system (BBS)* is a centralized information source and message-switching system for a particular computer-linked interest group.** For example, there are BBSs on such varying subjects as fly-fishing, clean air, ecology, genealogy, San Diego entertainment, Cleveland city information, and adult chat. BBSs are now also generally accessible through the Internet.

Interactivity: The Examples of Multimedia Computers, Personal Digital Assistants, & Futuristic "Smart Boxes" & "Internet Appliances"

The movie rolls on your TV/PC screen. The actors appear. Instead of passively watching the plot unfold, however, you are able to determine different plot developments by pressing keys on your keyboard. This is an example of interactivity. ***Interactivity* means that the user is able to make an immediate response to what is going on and modify the processes. That is, there is a dialog between the user and the computer or communications device.** Videogames, for example, are interactive. Interactivity allows users to be active rather than passive participants in the technological process.

Among the types of interactive devices are multimedia computers, personal digital assistants, and various kinds of up-and-coming "smart boxes" that work either with a TV or a PC.

- **Multimedia computers:** The word *multimedia*, one of the buzzwords of the '90s, has been variously defined. Essentially, however, ***multimedia* refers to technology that presents information in more than one medium, including text, graphics, animation, video, music, and voice.**

Multimedia personal computers are powerful microcomputers that include sound and video capability, run CD-ROM disks, and allow users to play games or perform interactive tasks.

- **Personal digital assistants:** In 1988, handheld electronic organizers were introduced, consisting of tiny keypads and barely readable screens. They were unable to do much more than store phone numbers and daily "to do" lists.

In 1993, electronic organizers began to be supplanted by personal digital assistants, such as Apple's Newton. *Personal digital assistants (PDAs)* are small pen-controlled, handheld computers that, in their most developed form, can do two-way wireless messaging. Instead of pecking at a tiny keyboard, you can use a special pen to write out commands on the computer screen. The newer generation of PDAs can be used not only to keep an appointment calendar and write memos but also to access the Internet and send and receive faxes and e-mail. With a PDA, then, you can immediately get information from some remote location—such as your microcomputer on your desk at home—and, if necessary, change it to update it.

- **Up-and-coming "smart boxes" and "Internet appliances":** Already envisioning a world of cross-breeding among televisions, telephones, and computers, enterprising manufacturers are experimenting with developing TV/PC set-top control boxes, or *"smart boxes,"* and *"Internet appliances."* With these futuristic devices, consumers presumably could listen to music CDs, watch movies, do computing, view multiple cable channels, and go online. Set-top boxes would provide two-way interactivity not only with videogames but also with online entertainment, news, and educational programs.

  Recently a *network computer,* or "hollow personal computer," has been developed, a machine intended to cost $500 or so that would be "hollowed out." Instead of having all the complex memory and storage capabilities built in, the network PC is designed to serve as an entry point to the online world, which is supposed to contain all the resources anyone would need.[8]

  Another gadget is the *cable modem,* which will allow cable-TV subscribers to connect their personal computers to various online computer services at speeds many times faster than traditional computer modems. This represents a way for cable operators to introduce voice, data, and video services on a large scale.[9]

  The converse of this is a kind of *"Internet TV" technology* known as Intercast, produced by chip maker Intel. Intercast lets specially equipped personal computers receive data from the Internet as well as television programming. Television networks could thus broadcast not only television shows but also additional data, such as geographical information about a country that is the subject of a news story, which computer owners could then look up.[10]

All these devices seem to be leading toward a kind of *"information appliance,"* as we describe next.

The "All-Purpose Machine": The Information Appliance That Will Change Your Future

Preview & Review: In the future, we may have an "information appliance," a device that combines telephone, television, VCR, and personal computer. This device would deliver digitized entertainment, communications, and information.

The basis for the information appliance may be the personal computer, although it may come in various sizes, shapes, and degrees of portability. The device will probably become increasingly "user-friendly" and will have multimedia capability.

Computer pioneer John Von Neumann said that the computer should not be called the "computer" but rather the "all-purpose machine." After all, he pointed out, it is not just a gadget for doing calculations. The most striking thing about it is that it can be put to *any number of uses.*

More than ever, we are now seeing just how true that is.

The "Information Appliance": What Will It Be?

Recently, there has been enormous interest in what is perceived to be the coming Information Superhighway. This electronic delivery system would presumably direct a digitized stream of sound, video, text, and data to some sort of box, perhaps something called an *information appliance*. An information appliance would deliver digitized entertainment, communications, and information in a device that combines telephone, television, VCR, and personal computer. The vision inspired by this futuristic gadget has caused furious activity in the communications world. Telephone, cable, computer, consumer electronics, and entertainment companies have rushed to position themselves to take advantage of these developments.

So what, exactly, will the information appliance (a term coined by Apple Macintosh researcher Jef Raskin in 1978) turn out to be? Perhaps it could be the under-$500 network computer proposed by Oracle and others. Or it might be a variation on the TV set, like Gateway's Destination, a home-entertainment setup built around a personal computer and a 31-inch TV. Or it could be a new wrinkle on the videogame player, like the Pippin Atmark combination game player–Internet browser. Or it might be the grand fusion of the Internet and household appliances envisioned by Microsoft in its 1996 SIPC (Simply Interactive Personal Computer) standards. Whatever its final form, the information appliance will no doubt be adapted from a machine now present everywhere—the microcomputer.[11] (■ *See Panel 1.10, pp. 32–33*) Clearly, then, anyone who learns to use a microcomputer now is getting a head start on the revolution.

The Ethics of Information Technology

 Ethical questions pervade all aspects of the use of information technology, as will be pointed out with a special symbol—**E**—throughout this book.

Every reader of this book at some point will have to wrestle with ethical issues related to computers-and-communications technology. *Ethics* is defined as a set of moral values or principles that govern the conduct of an individual or group. Indeed, ethical questions arise so often in connection with information technology that we have decided to earmark them wherever they appear in this book—with the special "E-for-ethics" symbol you see in the margin.

Here, for instance, are some important ethical concerns pointed out by Tom Forester and Perry Morrison in their book *Computer Ethics:*[12]

- **Speed and scale:** Great amounts of information can be stored, retrieved, and transmitted at a speed and on a scale not possible before. Despite the benefits, this has serious implications "for data security and personal privacy (as well as employment)," they say, because information technology can never be considered totally secure against unauthorized access.

- **Unpredictability:** Computers and communications are pervasive, touching nearly every aspect of our lives. However, compared to other pervasive technologies—such as electricity, television, and automobiles—information technology is a lot less predictable and reliable.

- **Complexity:** The on/off principle underlying computer systems may be simple, but the systems themselves are often incredibly complex. Indeed, some are so complex that they are not always understood even by their creators. "This," say Forester and Morrison, "often makes them completely unmanageable," producing massive foul-ups or spectacularly out-of-control costs.

Onward: The Gateway to the Information Superhighway

The term *Information Superhighway* has roared into the nation's consciousness in recent times. Some say it promises what might be called a communications cornucopia—a *"communicopia"* of electronic interactive services. Others say it is surrounded "by more hype and inflated expectations than any technological proposal of recent memory."[13] What, in fact, is this electronic highway? Does it or will it really exist?

The *Information Superhighway* is a vision or a metaphor for a fusion of the two-way wired and wireless capabilities of telephones and networked computers with cable-TV's capacity to transmit hundreds of programs. The resulting interactive digitized traffic would include movies, TV shows, phone calls, databases, shopping services, and online services. This superhighway, it is hoped, would link all homes, schools, businesses, and governments.

Parts of this idea have been raised before. Indeed, in many ways the Information Superhighway is a 1990s dusting off of earlier concepts of "the wired nation." In 1978, for example, James Martin wrote *The Wired Society*, which considered the social impacts of various telecommunications technologies.

At present, this electronic highway remains a vision, much as today's interstate highway system was a vision in the 1950s. It is as though we still had old-fashioned Highway 40s and Route 66s, along with networks of one-lane secondary and gravel backroads. These, of course, have largely been replaced by high-speed blacktop and eight-lane freeways. In 40 years, will the world be as changed by the electronic highway as North America has been by the interstate highways of the last four decades? It is the thesis of this book that it will be—and that we should prepare for it.

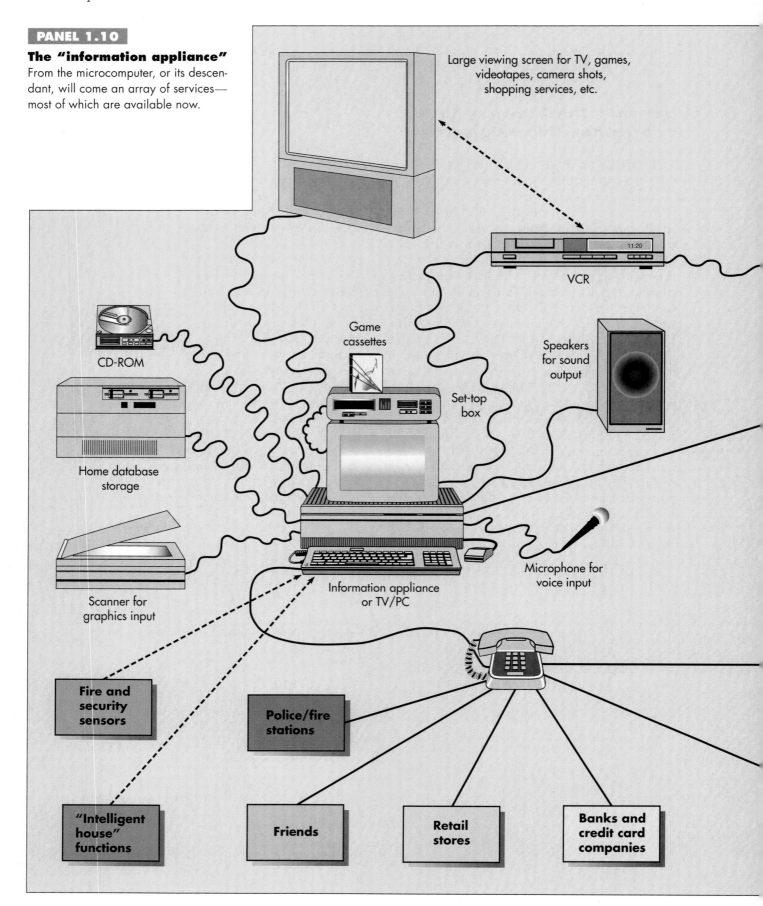

The "information appliance"
From the microcomputer, or its descendant, will come an array of services—most of which are available now.

Large viewing screen for TV, games, videotapes, camera shots, shopping services, etc.

VCR

CD-ROM

Game cassettes

Speakers for sound output

Home database storage

Set-top box

Scanner for graphics input

Information appliance or TV/PC

Microphone for voice input

Fire and security sensors

Police/fire stations

"Intelligent house" functions

Friends

Retail stores

Banks and credit card companies

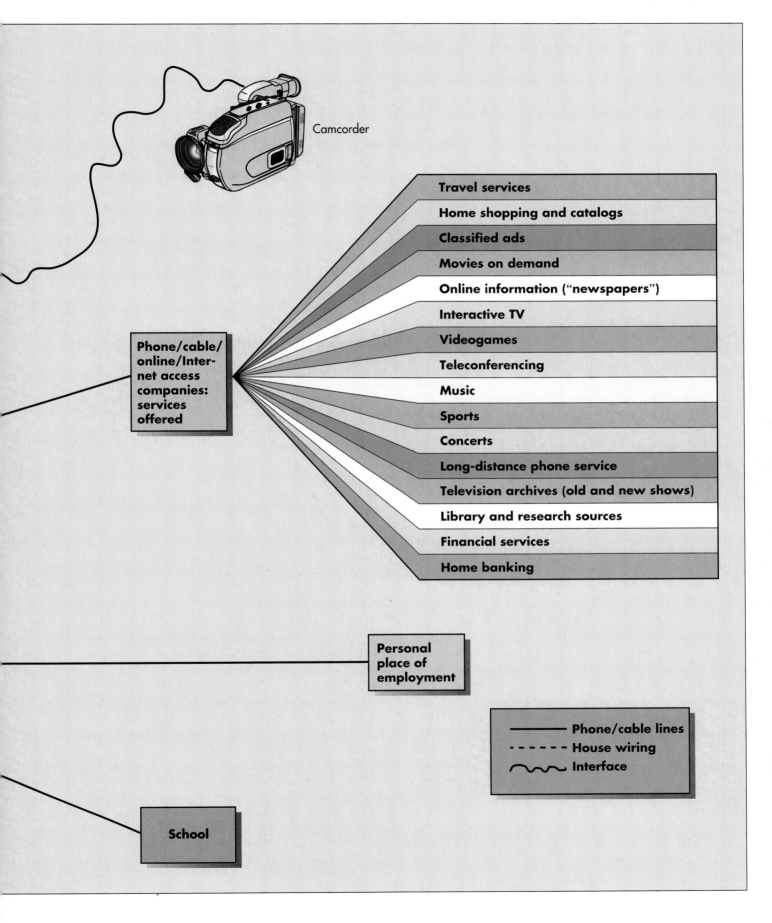

Camcorder

Phone/cable/ online/Internet access companies: services offered

- Travel services
- Home shopping and catalogs
- Classified ads
- Movies on demand
- Online information ("newspapers")
- Interactive TV
- Videogames
- Teleconferencing
- Music
- Sports
- Concerts
- Long-distance phone service
- Television archives (old and new shows)
- Library and research sources
- Financial services
- Home banking

Personal place of employment

School

———— Phone/cable lines
- - - - - House wiring
~~~ Interface

Better Organization & Time Management in College: Dealing with the Information Deluge

One great problem most of us face now—and that will probably only increase in the future—is how to handle the information glut. A solution is to develop better organization and time-management skills. This first Experience Box illustrates skills that will benefit you in college, in this course and others.

At one time in human history it was possible to keep up with the fund of knowledge. In the year 1300, for instance, there were a mere 1338 volumes in the Sorbonne library in Paris. Thus, the great Italian poet Dante Alighieri, for example, could have absorbed the entire body of knowledge of the then-known world. Three hundred and seventy years later, the library at Oxford University alone held over 25,000 books—too much knowledge for any one person to absorb.

Today there is so much information that even people with narrow specialties are hard pressed to keep up. In 1994, the Library of Congress in Washington, D.C., for instance, held 104 million books, manuscripts, and documents. More new information was produced in the last 30 years than in all the previous 5000. Information today is doubling every five years. By the end of the century, it will double every 20 months. This explosion in information is produced or made available by the onrushing juggernaut of information technology.[14]

For you, the question is simple: "How on earth am I going to be able to keep up with what's required of me?" The answer is: by *learning how to learn.* By building your skills as a learner, you certainly help yourself do better in college. More than that, however, you also train yourself to be an information manager in the future.

The Art of Learning

How does one become a good learner in college? By deciding to seriously participate in the game of higher education. "Winning the game of higher education is like winning any other game," say learning experts Debbie Longman and Rhonda Atkinson. "It consists of the same basic process. First, you decide if you really want to play. If you do, then you gear your attitudes and habits to learning. Next you learn the rules. To do this, you need a playbook, a college catalog. Third, you learn about the other players—administration, faculty, and other students. Finally, you learn specific plans to improve your playing skills."[15]

Unfortunately, many students come to college with faulty study skills. They are not entirely to blame. In high school and earlier, much of the emphasis is on *what* is to be studied rather than *how* to study it.

"The secret to controlling time is to remember that there is always enough time to do what is really important," say Mervill Douglass and Donna Douglass. "The difficulty is knowing what is really important."[16]

What *is* important in college? Studying, going to classes, writing papers, and taking tests compete with social life, extracurricular activities, and perhaps part-time work. All must somehow fit into the same 24 hours available each day. Yet—unlike high school or many paying jobs—time in college is often very unstructured. For students, the clash of college demands can lead to several personal problems: sleep disturbances, alcohol and drug abuse, eating disorders, money difficulties, procrastination. Let us discuss ways to improve academic performance, which should be your top priority.

Developing Study Habits: Finding Your "Prime Study Time"

How can you use knowledge about your own body and mind to improve your academic performance? Here's one fact: in the hours you *feel best,* you'll *study best.* What is called "prime study time" is the time of day when you are at your best for learning and remembering.

Each of us has a different energy cycle. For example, two roommates may have different patterns. One (the "day person") may be an early riser who prefers to work on difficult tasks in the morning. The other (the "night person") may start slowly but be at the peak of his or her form during the evening hours.

The trick, then, is to effectively *use* your daily energy cycle. Here are some important actions to take:

Make a Study Schedule First make a master schedule that shows all your regular obligations. You'll include classes and work, of course, but you may also wish to list meals and exercise times. This schedule should be indicated for the *entire school term.*

Now insert the times during which you plan to study. As mentioned, it's best if these study periods correspond to times when you are most alert and can best concentrate. However, don't forget to schedule in hourly breaks, since your concentration will flag periodically.

Next write in major academic events. Examples are when term papers and other assignments are due, when quizzes and exams take place, any holidays and vacations.

At the beginning of every week, schedule your study sessions. Write in the specific tasks you plan to accomplish during each session. It's best to try to study something connected with every class every day. If the subject is difficult for you, try to spend an hour a day on it. This is more effective than 5 hours in one day.

In addition, don't put off major projects, such as term papers, thinking you'll do them in one concentrated period of effort. It's more efficient to break the task into smaller steps that you can handle individually. This prevents you from delaying so long that you finally have to pull an all-nighter to complete the project.

Find Some Good Places to Study Studying means first of all avoiding distractions. No doubt you know several people who study while listening to the radio or watching television. Indeed, maybe you've done this yourself. The fact is, however, that most people *are* distracted by these activities.

Avoid studying in places that are associated with other activities, particularly comfortable ones. That is, don't do your academic reading lying in bed or sitting at a kitchen table. Studying should be an intense, concentrated activity.

You may wish to designate two or three sites as regular areas for studying. Assuming they are free of distractions, two good places are at a desk in your room or at a table in the library. As these places become associated with studying, they will reinforce better studying behavior.[17]

Make sure the place you study is free of clutter, which can affect your concentration and make you feel disorganized. Your desktop should contain the material you are studying and nothing else.

Avoid Time Wasters, but Reward Your Studying
Certainly it's much more fun to hang out with your friends or to watch television than to study. Moreover, these pleasures are real and immediate. Getting an A in a course, let alone getting a degree, seems to be in the distant future.

Thus, clearly you need to learn to avoid distractions so that you can study. However, you must also give yourself frequent rewards so that you will indeed be *motivated* to study. You should study with the notion that, after you finish, you will "pleasure yourself." The reward need not be elaborate. It could be a walk, a snack, a television show, a videogame, a conversation with a friend, or some similar treat.

Improving Your Memory

Memorizing is, of course, one of the principal requirements of staying in college. And distractions are a major impediment to remembering (as they are to other forms of learning). *External distractions* are those you have no control over—noises in the hallway, an instructor's accent, people whispering in the library. If you can't banish the distraction by moving, you might try to increase your interest in the subject you are studying. *Internal distractions* are daydreams, personal worries, hunger, illness, and other physical discomforts. Small worries can be shunted aside by listing them on a page for future handling. Large worries may require talking with a friend or counselor.

Beyond getting rid of distractions, there are certain techniques you can adopt to enhance your memory.

Space Your Studying, Rather Than Cramming
Cramming—making a frantic, last-minute attempt to memorize massive amounts of material—is probably the least effective means of absorbing information. Indeed, it may actually tire you out and make you even more anxious prior to the test. Research shows that it is best to space out your studying of a subject on successive days. This is preferable to trying to do it all during the same number of hours on one day.[18] It is *repetition* that helps move information into your long-term memory bank.

Review Information Repeatedly—Even "Overlearn" It By repeatedly reviewing information—"rehearsing" it—you can improve both your retention and your understanding of it.[19] Overlearning can improve your recall substantially. Overlearning is continuing to repeatedly review material even after you appear to have absorbed it. We said that "cramming" is not an effective way to learn. However, reviewing the material right before an examination can help counteract any forgetting that occurred since the last time you studied it.

Use Memorizing Tricks There are several ways to organize information so that you can retain it better. Longman and Atkinson mention the following methods of establishing associations between items you want to remember:[20]

• *Mental and physical imagery:* Use your visual and other senses to construct a personal image of what you want to remember. Indeed, it helps to make the image humorous, action-filled, sexual, bizarre, or outrageous in order to establish a personal connection. How, for instance, would you go about trying to remember the name of the 21st president of the United States, Chester Arthur? Perhaps you could visualize an author writing the number "21" on a wooden chest. This mental image helps you associate *chest, author* (Arthur), and *21* to recall that Chester Arthur was the 21st president.

You can also make your mental image a physical one by, for example, drawing or diagramming. Thus, to assist your recall of the parts of a computer system, you could draw a picture and label the parts.

- *Acronyms and acrostics:* An acronym is a word created from the first letters of items in a list. For instance, *Roy G. Biv* helps you remember the colors of the rainbow in order: *red, orange, yellow, green, blue, indigo, violet.* An acrostic is a phrase or sentence created from the first letters of items on a list. For example, *Every Good Boy Does Fine* helps you remember that the order of musical notes is *E-G-B-D-F.*

- *Location:* Location memory occurs when you associate a concept with a place or imaginary place. For example, you could learn the parts of a computer system by imagining a walk across campus. Each building you pass could be associated with a part of the computer system.

- *Word games:* Jingles and rhymes are devices frequently used by advertisers to get people to remember their products. You may recall the spelling rule "*I before E except after C or when sounded like A as in neighbor or weigh.*" A stalactite hangs from the top of a cave, whereas a stalagmite forms on the cave's floor. To recall the difference, you might remember that the *t* in *stalactite* signifies "top." Or you might recall that a stalactite has to hold on "tight." You can also use narrative methods, such as making up a story.

How to Benefit From Lectures

Are lectures really a good way of transmitting knowledge? Perhaps not always, but the fact remains that most colleges rely heavily on this method. Attending them makes a difference. Students with the highest grades tend to have the fewest absences.[21] (■ *See Panel 1.11.*)

Most lectures are reasonably well organized, but you probably will attend some that are not. Even so, they will indicate what the instructor thinks is important, which will be useful to you on the exams.

Regardless of the strengths of the lecturer, here are some tips for getting more out of lectures.

Take Effective Notes by Listening Actively Research shows that good test performance is related to good note taking.[22] And good note taking requires that you *listen actively*—that is, participate in the lecture process. Here are some ways to take good lecture notes:

- *Read ahead and anticipate the lecturer:* Try to anticipate what the instructor is going to say, based on your previous reading (text or study guide). Having background knowledge makes learning more efficient.

- *Listen for signal words:* Instructors use key phrases such as "The most important point is . . . ," "There are four

Successful students

Always or almost always in class	84%
Sometimes absent	8%
Often absent	8%

Unsuccessful students

Always or almost always in class	47%
Sometimes absent	8%
Often absent	45%

PANEL 1.11

Class attendance and grade success

Students with grades of B or above were more apt to have better class attendance than students with grades of C− or below.

reasons for . . . ," "The chief reason . . . ," "Of special importance . . . ," "Consequently . . . " When you hear such signal phrases, mark your notes with an asterisk (*) or write *Imp* (for "Important").

- *Take notes in your own words:* Instead of just being a stenographer, try to restate the lecturer's thoughts in your own words. This makes you pay attention to the lecture and organize it in a way that is meaningful to you. In addition, don't try to write everything down. Just get the key points.

- *Ask questions:* By asking questions during the lecture, you necessarily participate in it and increase your understanding. Although many students are shy about asking questions, most professors welcome them.

Review Your Notes Regularly The good news is that most students, according to one study, do take good notes. The bad news is that they don't use them effectively. That is, they wait to review their notes until just before final exams, when the notes have lost much of their meaning.[23] Make it a point to review your notes regularly, such as the afternoon after the lecture or once or twice a week. We cannot emphasize enough how important this kind of reviewing is.

How to Improve Your Reading Ability: The SQ3R Method

We cannot teach you how to speed-read. However, perhaps we can help you make the time you do spend reading more efficient. The method we will describe here is known as the *SQ3R method,* in which "SQ3R" stands for *survey, question, read, recite,* and *review.*[24] The strategy for this method is to break down a reading assignment into small segments and master each before moving on.

The five steps of the SQ3R method are as follows:

1. *Survey* the Chapter Before You Read It. Get an overview of the chapter or other reading assignment before you begin reading it. If you have a sense of what the material is about before you begin reading it, you can predict where it is going. You can also bring your own experience to it and otherwise become involved in ways that will help you retain it.

2. *Question* the Segment in the Chapter Before You Read It. This step is easy to do, and the point, again, is to get yourself involved in the material. After surveying the entire chapter, go to the first segment—section, subsection, or even paragraph, depending on the level of difficulty and density of information. Look at the topic heading of that segment. In your mind, restate the heading as a question. After you have formulated the question, go to steps 3 and 4 (read and recite). Then proceed to the next segment and restate the heading there as a question.

For instance, consider the section heading in this chapter that reads "Developments in Computer Technology" You could ask yourself, "What developments are occurring in computer technology?" For the heading of the subsection "Converting Reality to Digital Form: Sampling & Averaging," ask "What do the terms *sampling* and *averaging* mean?"

3. *Read* the Segment About Which You Asked the Question. Now read the segment you asked the question about. Read with purpose, to answer the question you formulated. Underline or highlight sentences you think are important, if they help you answer the question. Read this portion of the text more than once, if necessary, until you can answer the question. In addition, determine whether the segment covers any other significant questions, and formulate answers to these, too. After you have read the segment, proceed to step 4.

Perhaps you can see where this is all leading. If you read in terms of questions and answers, you will be better prepared when you see exam questions about the material later.

4. *Recite* the Main Points of the Segment. Recite means "say aloud." Thus, you should speak out loud (or softly) the answer to the principal question about the segment and any other main points. State these points in your own words, to better enhance your understanding. If you

wish, make notes on the principal ideas so you can look them over later.

Now that you have actively studied the first segment, move on to the second segment and do steps 2–4 for it. Continue this procedure through the rest of the segments until you have finished the chapter.

5. *Review* the Entire Chapter by Repeating Questions. After you have read the chapter, go back through it and review the main points. Then, without looking at the book, test your memory by repeating the questions.

Clearly the SQ3R method takes longer than simply reading with a rapidly moving color marker or underlining pencil. However, the technique is far more effective because it requires your *involvement and understanding.* This is the key to all effective learning.

How to Become an Effective Test Taker

The first requirement of test taking is, of course, knowledge of the subject matter. That is what our foregoing discussion has been intended to help you obtain. You should also make it a point to *ask* your instructor what kinds of questions will be asked on tests. Beyond this, however, there are certain skills one can acquire that will help during the test-taking process. Here are some suggestions offered by the authors of *Doing Well in College:*[25]

Reviewing: Study Information That Is Emphasized and Enumerated Because you won't always know whether an exam will be an objective or essay test, you need to prepare for both. Here are some general tips.

- *Review material that is emphasized:* In the lectures, this consists of any topics your instructor pointed out as being significant or important. It also includes anything he or she spent a good deal of time discussing or specifically advised you to study.

  In the textbook, pay attention to key terms (often emphasized in *italic* or **boldface** type), their definitions, and their examples. Also, of course, material that has a good many pages given over to it should be considered important.

- *Review material that is enumerated:* Pay attention to any numbered lists, both in your lectures and in your notes. Such lists often provide the basis for essay and multiple-choice questions.

- *Review other tests:* Look over past quizzes, as well as the discussion questions or review questions provided at the end of chapters in many textbooks.

Prepare by Doing Final Reviews & Budgeting Your Test Time Learn how to make your energy and time work for you. Whether you have studied methodically or must cram for an exam, here are some tips:

- *Review your notes:* Spend the night before the test reviewing your notes. Then go to bed without interfering with the material you have absorbed (as by watching television). Get up early the next morning and review your notes again.

- *Find a good test-taking spot:* Make sure you arrive at the exam with pencils or other materials you need. Get to the classroom early, or at least on time, and find a quiet spot. If you don't have a watch, sit where you can see a clock. Again review your notes. Avoid talking with others, so as not to interfere with the information you have learned or to increase your anxiety.

- *Read the test directions:* Many students don't do this and end up losing points because they didn't understand precisely what was required of them. Also, listen to any verbal directions or hints your instructor gives you before the test.

- *Budget your time:* Here is an important point of test strategy: Before you start, read through the entire test and figure out how much time you can spend on each section. There is a reason for budgeting your time, of course. You would hate to find you have only a few minutes left and a long essay still to be written.

  Write the number of minutes allowed for each section on the test booklet or scratch sheet and stick to the schedule. The way you budget your time should correspond to how confident you feel about answering the questions.

Objective Tests: Answer Easy Questions & Eliminate Options

Some suggestions for taking objective tests, such as multiple-choice, true/false, or fill-in, are as follows:

- *Answer the easy questions first:* Don't waste time stewing over difficult questions. Do the easy ones first, and come back to the hard ones later. (Put a check mark opposite those you're not sure about.) Your unconscious mind may have solved them in the meantime, or later items may provide you with the extra information you need.

- *Answer all questions:* Unless the instructor says you will be penalized for wrong answers, try to answer all questions. If you have time, review all the questions and make sure you have written the responses correctly.

- *Eliminate the options:* Cross out answers you know are incorrect. Be sure to read all the possible answers, especially when the first answer is correct. (After all, other answers could also be correct, so that "All of the above" may be the right choice.) Be alert that subsequent questions may provide information pertinent to earlier ques-

tions. Pay attention to options that are long and detailed, since answers that are more detailed and specific are likely to be correct. If two answers have the opposite meaning, one of the two is probably correct.

Essay Tests: First Anticipate Answers & Prepare an Outline

Because time is limited, your instructor is likely to ask only a few essay questions during the exam. The key to success is to try to anticipate beforehand what the questions might be and memorize an outline for an answer. Here are the specific suggestions:

- *Anticipate ten probable essay questions:* Use the principles of reviewing lecture and textbook material that is *emphasized* and *enumerated.* You will then be in a position to identify ten essay questions your instructor may ask. Write out these questions.

- *Prepare and memorize informal essay answers:* For each question, list the main points that need to be discussed. Put supporting information in parentheses. Circle the key words in each main point and put the first letter of the key word below the question. Make up catch phrases, using acronyms, acrostics, or word games, so that you can memorize these key words. Test yourself until you can recall the key words the letters stand for and the main points the key words represent.

When you receive the questions for the essay examination, read all the directions carefully. Then start with the *least demanding question.* Putting down a good answer at the start will give you confidence and make it easier to proceed with the rest. Make a brief outline, similar to the one you did for your anticipated question, before you begin writing.

The Peak-Performing Student

Good students are made, not born. We have listed some of the studying, reading, and test-taking skills that will help you be a peak-performing student. The practice of these skills is up to you.

Suggestions for Further Reading

Lakein, Alan. *How to Get Control of Your Time and Your Life.* New York: Signet, 1978. One of the classic books on time management.

Longman, Debbie G., and Atkinson, Rhonda H. *College Learning and Study Skills* (2nd ed.). St. Paul, MN: West, 1992. One of the best of several books on study skills.

Wahlstrom, Carl, and Williams, Brian K. *Learning Success: Being Your Best at College & Life.* Belmont, CA: Wadsworth, 1996. A new book on college study skills and personal and career strategies, co-authored by one of the authors of this book.

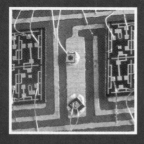

SUMMARY

What It Is / What It Does	**Why It's Important**

analog *(p. 6, LO 2*)* Refers to nondigital (noncomputer-based), continuously variable forms of data transmission, including voice and video. Most current telephone lines and radio, television, and cable-TV hookups are analog transmissions media. Analog is the opposite of digital.

You need to know about analog and digital forms of communication to understand what is required for you to connect your computer to other computer systems and information services. Computers cannot communicate over analog lines. A modem and communications software are usually required to connect a microcomputer user to other computer systems and information services.

analog data *(p. 6, LO 2)* See analog.

applications software *(p. 20, LO 7)* Software that can perform useful work on general-purpose tasks.

Applications software such as word processing, spreadsheet, database manager, graphics, and communications packages have become commonly used tools for increasing people's productivity.

binary digit *(p. 6, LO 2)* See bit.

binary system *(p. 5, LO 2)* Two-state system.

Computer systems use a binary system for data representation—two digits, 0 and 1, to refer to the presence or absence of electrical current or a pulse of light.

bit *(p. 6, LO 2)* Short for *binary digit,* which is either a 1 or a 0 within the binary system of data representation in computer systems.

The bit is the fundamental element of all data and information stored and manipulated in a computer system; 8 bits are combined according to a coding scheme to form a character, such as A. The bit is also the basic unit of measure for describing the capacity of computer hardware storage units and the processing speeds of some other hardware components.

byte *(p. 6, LO 2)* A group of 8 bits.

Bytes—such as 01101110 and 11101100—represent characters according to the particular coding scheme used in a computer system. Bytes are also used to describe computer storage hardware capacities.

CD-ROM (compact disk—read only memory) *(p. 19, LO 3)* Compact optical disk that holds text, graphics, and sound.

CD-ROM disks are used in computer systems to create fancy presentations, store multimedia presentations, and provide reference materials for research.

cellular phone *(p. 25, LO 9)* Mobile, wireless telephone.

Cellular phones further the availability of instant communication, no matter where you are.

central processing unit (CPU) *(p. 16, LO 3)* The processor; it controls and manipulates data to produce information. In a microcomputer the CPU is usually contained on a single integrated circuit or chip called a microprocessor. This chip and other components that make it work are mounted on a circuit board called a motherboard. In larger computers the CPU is contained on one or several circuit boards.

The CPU is the "brain" of the computer; without it, there would be no computers.

*Note to the reader: "LO" refers to Learning Objective; see the first page of the chapter. The number ties the summary term to the appropriate Learning Objective.

What It Is / What It Does	**Why It's Important**

communications *(pp. 4, 21, LO 3)* The sixth element of a computer-and-communications system; the electronic transfer of data from one place to another.

Communications systems have helped to expand human communication beyond face-to-face meetings to electronic connections.

communications network *(p. 24, LO 9)* System of inter-connected computers, telephones, or other communications devices that can communicate with one another.

Communications networks allow users to share applications and data; without networks, information could not be electronically exchanged.

communications technology *(p. 4, LO 1)* Consists of electromagnetic devices and systems for communicating over long distances; also called *telecommunications*.

Communications technology enables computers and people to be connected in order to share information resources.

computer *(p. 4, LO 2)* Programmable, multiuse machine that accepts raw data—facts and figures—and processes (manipulates) it into useful information, such as summaries and totals.

Computers greatly speed up the process whereby people solve problems and accomplish many tasks and thus increases their productivity.

computer-and-communications system *(p. 9, LO 3)* System made up of six elements: people, procedures, data/information, hardware, software, and communications.

Users need to understand how the six elements of a computer-and-communications system relate to one another in order to make knowledgeable decisions about buying and using a computer system.

computer online service *(p. 27, LO 9)* Commercial information service that, for a fee, makes available to subscribers various services through their telephone-linked microcomputers.

Online services allow users to, among many other things, make airline reservations, research databases, check on the weather, send e-mail, shop, and bank—all through their keyboards.

connectivity *(p. 26, LO 9)* Ability to connect devices by telecommunications lines to other devices and sources of information.

Connectivity is the foundation of the latest advances in the Digital Age. It provides online access to countless types of information and services.

CRT (cathode-ray tube) *(p. 18, LO 3)* Familiar TV-style picture tube; vacuum tube used as a computer display screen on desktop computers.

CRT display screens provide one of the principal types of output.

cursor *(p. 16, LO 3)* Also called a *pointer;* the movable symbol on the screen that shows where data may be entered next.

All applications software packages use cursors to show users where their current "work location" is on the screen.

data *(p. 12, LO 3, 5)* Consists of the raw facts and figures that are processed into information; third element in a computer-and-communications system.

Users need data to create useful information.

database *(p. 27, LO 9)* Collection of integrated, or cross-referenced, electronically stored data that different people may access to use for different purposes.

Users with online connections to database services have enormous research resources at their disposal. In addition, businesses and organizations build databases to help them keep track of and manage their affairs.

digit *(p. 6, LO 2)* "Number."

Computer users need to understand what a binary digit is in order to understand how a computer works.

digital *(p. 6, LO 2)* Term used synonymously with *computer;* refers to communications signals or information represented in a binary, or two-state, way—1s and 0s, on and off.

The whole concept of an information superhighway is based on the existence of communications in digital form, which allows computers to transmit voice, text, sound, graphics, color, and animation.

digital data *(p. 6, LO 2)* Data represented by discrete (individually distinct), discontinuous transmission bursts of power or light—0s and 1s.

Computers transmit data in digital form, as opposed to the analog form of data transmitted by regular telephone lines.

disk drive *(p. 19, LO 3)* Computer hardware device that holds, spins, reads from, and writes to magnetic or optical disks.

Users need disk drives in order to use their disks. Disks drives can be internal (built into the computer system cabinet) or external (connected to the computer by a cable).

What It Is / What It Does	**Why It's Important**

diskette (floppy disk) *(p. 19, LO 3)* Secondary storage medium; removable round, flexible disk that stores data as magnetized spots. The most common sizes are 5¼ inches, the older size, and 3½ inches, now the most popular size. The disk is contained in a square paper envelope or plastic case to prevent the disk from being touched by human hands.

Diskettes are used on all microcomputers.

documentation *(p. 11, LO 3)* Also called *manuals*; set of instructions, rules, or guidelines (procedures) to follow when using hardware or software.

When users buy a microcomputer or software package, it comes with documentation, in booklet form and often also on disk. This documentation serves as reference material to help users learn how to use a product.

electronic bulletin board system (BBS) *(p. 28, LO 9)* Centralized information source and message-switching system for a particular computer-linked interest group.

BBSs enable users from wide geographic areas to share similar interests, get advice, and exchange information without having to go anywhere.

electronic mail *(p. 26, LO 9)* Also called *e-mail;* software-controlled system linking computers by wired or wireless connections that allow users, through their keyboards, to post messages and to read responses on their computer screens.

E-mail allows businesses and organizations to quickly and easily send messages to employees and outside people without having to use and distribute paper messages.

end-user *(p. 10, LO 3)* Also called user; a person without much technical knowledge of information technology who uses computers for entertainment, education, and/or work-related tasks.

End-users are the people for whom most computer-and-communications systems are created (by information technology professionals).

fax *(p. 25, LO 3)* Stands for *facsimile transmission;* the communication of text or graphic images between remote locations.

See fax machine.

fax machine *(p. 25, LO 9)* Device that scans an image and sends a copy of it as electronic signals over transmission lines to a receiving fax machine, which recreates the image on paper.

Fax availability has increased the pace at which business can be conducted and thus has improved productivity. Also, fax machines enable people in isolated areas to obtain copies of newspapers and printed information they may otherwise not be able to get.

gigabyte (G or GB) *(p. 12, LO 3)* Unit for measuring storage capacity; equals 1 billion bytes.

This unit of measure is used with supercomputer, mainframe, and even with microcomputer secondary storage devices.

hard disk *(p. 19, LO 3)* Secondary storage medium; generally nonremovable disk made out of metal and covered with a magnetic recording surface. It holds data represented by the presence (1) and absence (0) of magnetized spots. Hard disks, which hold much more data than floppy disks do, are usually built into the computer's system cabinet.

Nearly all microcomputers now use hard disks as their principal secondary storage medium.

hardware *(p. 14, LO 3)* Fourth element in a computer-and-communications system; refers to all machinery and equipment in a computer system. Hardware is classified into five categories: input, processing and memory, output, secondary storage, and communications.

Hardware design determines the type of commands the computer system can follow. However, hardware runs under the control of software and is useless without it.

information *(p. 12, LO 3)* In general, refers to summarized data or otherwise manipulated data. Technically, data comprises raw facts and figures that are processed into information. However, information can also be raw data for the next person or job. Thus sometimes the terms are used interchangeably. Information/data is the third element in a computer-and-communications system.

The whole purpose of a computer (and communications) system is to produce (and transmit) usable information.

What It Is / What It Does	Why It's Important
information superhighway *(p. 31, LO 10)* Vision or metaphor for a fusion of the two-way wired and wireless capabilities of telephones and networked computers with cable-TV's capacity to transmit hundreds of programs; the resulting interactive digitized traffic would include movies, TV shows, phone calls, databases, shopping services, and online services.	The information superhighway would fundamentally change the nature of communications and hence society, business, government, and personal life.
information technology *(p. 2, LO 1)* Technology that merges computing with high-speed communications links carrying data, sound, and video.	Information technology is bringing about the gradual fusion of several important industries in a phenomenon called *digital convergence* or *technological convergence*.
information technology professional *(p. 10, LO 3)* Person who has had formal education in the technical aspects of using computer-and-communications systems.	Information technology professionals create and manage the software and systems that enable users (end-users) to accomplish many types of business, professional, and educational tasks and increase their productivity.
input hardware *(p. 16, LO 3)* Devices that allow people to put data into the computer in a form that the computer can use; that is, they perform *input operations*. Input may be by means of a keyboard, pointer, scanner, or voice-recognition device.	Useful information cannot be produced without input data.
input operation *(p. 13, LO 6)* The phase of information processing in which data is captured electronically and converted to a form that can be processed by the computer. The means for entering data is an input device such as a keyboard or scanner.	During this phase, the raw data for producing useful information is put into the computer system for processing.
integrated circuit (IC) *(p. 22, LO 8)* Collection of electrical circuits, or pathways, etched on tiny squares, or chips, of silicon half the size of a person's thumbnail.	The development of the IC enabled the manufacture of the small, powerful, and relatively inexpensive computers used today.
interactivity *(p. 28, LO 9)* Situation in which the user is able to make an immediate response to what is going on and modify processes; that is, there is a dialog between the user and the computer or communications device.	Interactive devices allow the user to actively participate in the ongoing processes instead of just reacting to them.
Internet *(p. 27, LO 9)* International network connecting approximately 36,000 smaller networks that link computers at academic, scientific, and commercial institutions.	The Internet makes possible the sharing of all types of information and services for millions of people all around the world.
keyboard *(p. 16, LO 2)* Input hardware device that uses standard typewriter keys plus a number of specialized keys to input data and issue commands.	Microcomputer users will probably use the keyboard more than any other input device.
kilobyte (K or KB) *(p. 12, LO 3)* Unit for measuring storage capacity; equals 1024 bytes (often rounded off to 1000 bytes).	The sizes of stored electronic files are often measured in kilobytes.
magnetic tape *(p. 19, LO 3)* Secondary storage medium made of flexible plastic coated on one side with magnetic material; data is represented by the presence and absence of magnetized spots.	Nowadays tape is used mainly to provide duplicate (backup) storage, especially for microcomputers.
mainframe computer *(p. 22, LO 8)* Second-largest type of computer available, after the supercomputer; occupies a specially wired, air-conditioned room and is capable of great processing speeds and data storage.	Mainframes are used by large organizations—such as banks, airlines, insurance companies, and colleges—for processing millions of transactions.

What It Is / What It Does	**Why It's Important**
megabyte (M or MB) *(p. 12, LO 3)* Unit for measuring storage capacity; equals 1 million bytes.	The storage capacities of most microcomputer hard disks are measured in megabytes. Users need to know how mach data their hard disks can hold and how much space new software programs will take so that they do not run out of disk space.
memory *(p. 16, LO 3)* Also called *main memory, primary storage, RAM;* the computer's "work space," where data and programs for immediate processing are held. Memory is contained on chips mounted on the system board.	Memory size determines how much data can be processed at once and how big and complex a program may be used to process it. Memory is usually measured in megabytes.
microcomputer *(p. 23, LO 8)* Small computer that fits on a desktop; used either as a personal computer or a workstation.	Microcomputers are used in virtually every area of modern life. People going into business or professional life today are often required to have basic knowledge of the microcomputer.
microcontroller *(p. 23, LO 8)* Also called an *embedded computer;* the smallest category of computer.	Microcontrollers are built into "smart" electronic devices as controlling agents.
microprocessor *(p. 24, LO 8)* Short for "microscopic processor," the miniaturized circuitry of the computer's processor—the part that manipulates data into information—which is etched on a sliver, or "chip," of material such as silicon.	Without the microprocessor, we would not have the microcomputer.
minicomputer *(p. 22, LO 8)* Third-largest type of computer; refrigerator-sized computer that is essentially a scaled-down mainframe; overlaps high-end microcomputers and low-end mainframe computers in price and performance.	Typically used by medium-sized companies for specific purposes, such as accounting.
modem *(p. 7, LO 2)* Communications hardware device for converting a computer's digital signals to analog signals—for transmission over copper telephone wires—and then back to digital signals.	Without modems, computers would not be able to transmit data over most existing telephone wiring.
mouse *(p. 16, LO 3)* Input hardware device that can be rolled about on a desktop to direct a pointer (cursor) on the computer's display screen.	With microcomputers, a mouse is needed to use most graphical user interface programs and to draw illustrations.
multimedia *(p. 28, LO 9)* Refers to technology that presents information in more that one medium, including text, graphics, animation, video, music, and voice.	Use of multimedia is becoming more common in business, the professions, and education as a means of improving the way information is communicated.
operating system *(p. 21, LO 7)* A component of systems software; it acts as the master control program that runs the computer.	The operating system sets the standards for the applications software programs. All programs must "talk to" the operating system.
optical disk *(p. 19, LO 3)* Disk that is written to and read by lasers.	Optical disks hold much more data than many types of magnetic disks.
output hardware *(p. 17, LO 3)* Consists of devices that translate information processed by the computer into a form that humans can understand; that is, they perform *output operations.* Common output devices are monitors (softcopy output) and printers (hardcopy output). Sound is also a form of computer output.	Without output devices, computer users would not be able to view or use their work.
output operation *(p. 13, LO 6)* The phase of data processing in which information, which has been processed from data, is produced in a form usable by people.	The output phase represents the productive aspect of computer-based information processing.

What It Is / What It Does	Why It's Important
peripheral device *(p. 14, LO 3)* Any hardware device that is connected to a computer. Examples are keyboard, mouse, monitor, printer, and disk drives.	Most of a computer system's input and out-put functions are performed by peripheral devices.
personal computer (PC) *(p. 23, LO 8)* Desktop or portable (laptop, notebook, or palmtop) microcomputer that can run easy-to-use, personal-assistance programs such as word processing or spreadsheets.	The PC has enabled people to speed up many of their work and learning tasks, thus improving their productivity.
printer *(p. 18, LO 3)* Hardware device that converts computer output into printed images.	Printers provide one of the principal forms of computer output.
procedures *(p. 11, LO 3)* Descriptions of how things are done; steps for accomplishing a result.	Procedures are the second element in a computer-and-communications system. In the form of documentation, procedures help users learn to use hardware and software.
processing operation *(p. 13, LO 6)* The phase of data processing in which data is manipulated to process or transform it into information.	The processing phase represents the critical core of computer-based data processing. It enables people to solve problems quickly and improve their productivity. It also allows them to view information in new ways, such as through special effects.
read *(p. 19, LO 3)* Refers to the computer obtaining data from a secondary storage medium, such as disk or tape. It means that the data represented in the magnetized or laser-created spots on the disk (or tape) are converted to electronic signals and transmitted to the computer's memory (RAM).	Reading is an essential computer operation.
scanner *(p. 16, LO 3)* Input device that translates images of text, drawings, and photos into digital form.	Scanners simplify the input of complex data. The images can be processed by the computer, manipulated, displayed on a monitor, stored on a storage device, and/or communicated to another computer.
screen *(p. 18, LO 3)* Display area of a computer—also called a *monitor;* either CRT or flat-panel.	*See CRT.*
secondary storage *(p. 18, LO 3)* Refers to devices and media that store data and programs permanently—such as disks and disk drives, tape and tape drives. These devices perform *storage operations.* Storage capacity is measured in kilobytes, megabytes, gigabytes, and terabytes.	Without secondary storage media, users would not be able to save their work.
secondary storage operation *(p. 13, LO 6)* The phase of data processing in which data, information, or programs are stored in computer-processable form.	The storage phase enables people to save their work for later retrieval, manipulation, and output.
server *(p. 24, LO 9)* Computer shared by several users in a network.	Servers enable users to share data and applications.
software *(p. 20, LO 3)* Also called *programs;* step-by-step instructions that tell the computer hardware how to perform a task. Software represents the fifth element of a computer-and-communications system.	Without software, hardware would be useless.
supercomputer *(p. 22, LO 8)* Largest, fastest, and most expensive type of computer available, costing millions of dollars.	Supercomputers are used for research, weather forecasting, oil exploration, airplane building, complex mathematical operations, and movie special effects, for example.
system *(p. 9, LO 3)* Group of related components and operations that interact to perform a task.	Computer hardware and software cannot function unless they work together as a system.

What It Is / What It Does	Why It's Important
system unit *(p. 16, LO 3)* Also called the *system cabinet;* housing that includes the electronic circuitry (CPU), which does the actual processing, and main memory, which supports processing.	The microcomputer was born when processing, memory, and power supply were made small enough to fit into a cabinet that would fit on a desktop.
systems software *(p. 21, LO 7)* Software that controls the computer and enables it to run applications software. Systems software, which includes the operating system, allows the computer to manage its internal resources.	Applications software cannot run without systems software.
technological convergence *(p. 2, LO 1)* Also called *digital convergence;* refers to the technological merger of several industries through various devices that exchange information in the electronic, or digital, format used by computers. The industries are computers, communications, consumer electronics, entertainment, and mass media.	From a common electronic base, the same information may be exchanged among many organizations and people.
terabyte (T or TB) *(p. 13, LO 3)* Unit for measuring storage capacity; equals 1 trillion bytes.	The storage capacities of supercomputers are measured in terabytes.
transistor *(p. 22, LO 8)* Small electronic device (gateway) used to transfer electrical signals along predetermined paths.	Transistors enabled the manufacture of computers that were much smaller than the original ones, which used vacuum tubes.
workstation *(p. 23, LO 8)* Expensive, powerful desktop microcomputer used mainly by engineers and scientists for sophisticated purposes.	The development of high-performance workstations has speeded up and improved the previously laborious and time-consuming processes of design, testing, and manufacturing in many industries.
write *(p. 19, LO 3)* Refers to the electronic information processed by the computer being recorded magnetically or by laser onto disk (or tape).	Being able to electronically "write" to disk or tape enables users to save their work.

EXERCISES

(Selected answers appear at the back of the book.)

Short-Answer Questions

1. What does the term *digital* mean?
2. What is a modem used for?
3. What are a kilobyte, a megabyte, a gigabyte, and a terabyte?
4. Briefly describe the five categories of computer hardware.
5. What is the difference between systems and applications software?
6. What is the purpose of the microprocessor in a computer system?
7. To what does the term *connectivity* refer?
8. What do the terms *access* and *online* mean?
9. What is multimedia?
10. What is the Internet?

Fill-in-the-Blank Questions

1. A(n) _____ is a programmable, multiuse machine that accepts data and processes it into information.
2. A(n) _____ allows two computers to communicate with each other over phone lines.
3. A person who doesn't have much technical knowledge of computers but who uses computers for entertainment, education, or work-related purposes is called a(n) _____.
4. A(n) _____ is equal to approximately 1 million bytes.
5. Telephones send _____ signals, whereas computers send _____ signals.
6. The largest, most powerful type of computer is called a _____.
7. _____ _____ consists of hardware devices that store data and programs permanently on disk or tape.
8. A(n) _____ _____ stores data on a metal disk and typically stores more than a removable _____ _____.

9. In _____, one works at home and communicates with the office via a computer and other technology.
10. _____ refers to the presentation of information in more than one medium, including text, graphics, animation, video, music, and voice.

Multiple-Choice Questions

1. In a(n) _____, information is produced in a form usable by people.
   a. input operation
   b. processing operation
   c. output operation
   d. storage operation
   e. All of the above
2. _____ refers to the raw facts that are processed into _____.
   a. Data, information
   b. Information, data
   c. Input, output
   d. Primary storage, secondary storage
   e. None of the above
3. All the machinery and equipment in a computer system is referred to as _____.
   a. software
   b. hardware
   c. the system cabinet
   d. the central processing unit
   e. All of the above
4. Which of the following houses the central processing unit of a microcomputer?
   a. keyboard
   b. monitor
   c. system unit
   d. main memory
   e. None of the above
5. Which of the following stores data and programs for immediate processing?
   a. keyboard
   b. monitor
   c. system unit
   d. main memory
   e. None of the above
6. Which of the following isn't used to connect computers?
   a. copper wire
   b. coaxial cable
   c. fiber-optic cable
   d. radio waves
   e. All of the above

7. Which of the following types of software does the computer use to manage its internal resources?
   a. custom-written software
   b. packaged software
   c. applications software
   d. systems software
   e. All of the above

8. Which of the following best describes the direction of computer development over the past 50 years?
   a. smaller size
   b. more power
   c. less expense
   d. All of the above

9. The _____ controls and manipulates data to produce information.
   a. system unit
   b. monitor
   c. central processing unit
   d. CRT
   e. None of the above

10. A _____ is a small pen-controlled, handheld computer that can perform two-way messaging.
    a. smart box
    b. information appliance
    c. personal digital assistant
    d. minicomputer
    e. None of the above

True/False Questions

T F 1. A byte is made up of 8 bits.

T F 2. Most of today's phone lines can carry digital signals.

T F 3. In an input operation, data is transformed into information.

T F 4. A terabyte is bigger than a gigabyte.

T F 5. A piece of hardware that is connected to your computer is commonly referred to as a *peripheral device*.

T F 6. A mouse can translate images of drawings and photos into digital form.

T F 7. Diskettes are removable flexible disks that store data as magnetized spots.

T F 8. CD-ROMs typically store more data than diskettes.

T F 9. Computers are typically categorized into four general types: supercomputers, mainframe computers, minicomputers, and "smart boxes."

T F 10. Three categories of hardware devices that provide interactivity are multimedia computers, personal digital assistants, and futuristic "smart boxes."

Projects/Critical-Thinking Questions

1. Determine what types of computers are being used where you work or go to school. Are microcomputers being used? Minicomputers? Mainframes? All types? What are they being used for? How are they connected, if at all?

2. Based on what you have learned so far, how do you think the five operations of a computer-and-communications system would be represented in your chosen profession? For example, how would data be input? Output? Under what circumstances? What kind of processing activities would take place? What kinds of communications activities?

3. Describe the computer you use at school, work, or home. Provide details about any input, processing and memory, output, secondary storage, and communications components that are part of your computer system. If you were to spend your own money to improve this computer system, what would you spend it on and why? Do you think you should buy a new computer system instead?

4. In an article for the *Harvard Business Review* (September/October 1992, p. 97), Peter Drucker predicted that, in the next 50 years, schools and universities will change more drastically than they have since they assumed their present form more than 300 years ago, when they reorganized themselves around the printed book. What will force these changes is, in part, new computer-and-communications technology and, in part, the demands of a knowledge-based society in which organized learning must become a life-long process for knowledge workers.

   How do you feel about the prospect of bookless reading and learning? What advantages and disadvantages can you see in using computers instead of books? And how do you feel about perhaps having to renew your fund of knowledge about your job or profession every 4 or 5 years?

5. Although more new information has been produced in the last 30 years than in the previous 5000, information is not knowledge. In our quest for knowledge in the Information Age, we are often overloaded with information that doesn't tell us what we want to know. Richard Wurman identified this problem in his book *Information Anxiety*. John Naisbett, in his books *Megatrends* and *Megatrends 2000*, said that

uncontrolled and unorganized information is no longer a resource in an information society. Instead, it becomes the enemy of the information worker.

Identify some of the problems of information overload in one or two departments in your school or place of employment—or in a local business, such as a real estate firm, health clinic, pharmacy, or accounting firm. What types of problems are people having? How are they trying to solve them? Are they rethinking their use of computer-related technologies?

Using the Internet

Objective: *In this exercise we describe how to log onto the Internet and use some basic commands.*

Before you continue: *We assume that you have at least partial access to the Internet through your college, business, or commercial service provider and have a user ID and password. For this exercise, we assume your Internet connection uses: (1) a command-line interface, (2) VT100 terminal emulation, and (3) the Unix operating system. If any of these assumptions are incorrect, you may still be able to perform this Internet exercise; however some of the commands and screen output may be different. If necessary, ask your instructor or system administrator for assistance.*

1. Access your Internet account using your user name and password. The procedure you use depends on your particular system. If necessary, ask your instructor or lab assistant.

   Sample procedure:

   TYPE: *your user name*
   PRESS: [Enter]
   TYPE: *your user password*
   PRESS: [Enter]

   Next, depending on your system, you may be prompted to choose a terminal emulation mode. If possible, choose VT100 terminal emulation. Otherwise, simply press [Enter] to bypass this option.

2. Guess what? You're on the Internet! At this point you are either presented with a command prompt such as $, %, or > and a blinking cursor, or a menu. If a menu appears, choose the option that exits you to command-line, or shell, mode.

3. The command prompt you see is referred to as the Unix prompt. As you work with the Internet, you will type commands after this prompt and then press [Enter] to execute the commands. Be forewarned that Unix is case sensitive. As a general rule, most Unix commands must be typed using lowercase letters, whereas filenames can use a combination of both.

For example, to find out about yourself:

TYPE: `who am i`
PRESS: [Enter]

The output of this command shows your login name, the name of the terminal or line you are using, and the date and time you logged onto the Internet. Your screen may display information that is similar in format to the following:

```
$ who am i
sarahc      ttyp6      Jan 19 12:58      (router-1)
```

4. If you make an error when typing a command, Unix displays an error message followed by the Unix prompt. You can then try the command again.

   For example:

   TYPE: `Who am I`
   PRESS: [Enter]

   Because you didn't use all lowercase letters, a message similar to the following will appear:

   ```
   $ Who am I
   Who: not found
   ```

   (*Note:* Depending on the version of Unix that you're using on your computer, the error message you see may be slightly different.)

5. Your home directory, or working directory, is active when you log onto the Internet. To see the name of the working directory, you use the *pwd command* (print working directory):

   TYPE: `pwd`
   PRESS: [Enter]

   Your screen may display information that is similar in format to the following:

   ```
   $ pwd
   /users/home/sarahc
   ```

6. To list the files, if any, in the working directory, you use the *ls command* (list). The -l option provides additional information about any files stored in the working directory.

   TYPE: `ls -l`
   PRESS: [Enter]

7. Now that you know how to log onto the Internet and use some basic commands, let's disconnect from the Internet. You can exit by typing "exit" or "logout."

   For example:

   TYPE: `exit`
   PRESS: [Enter]

   Note: If a menu appears, choose the option to disconnect, or quit.

CHAPTER 2

Applications Software
Tools for Thinking & Working

Concepts You Should Know

After reading this chapter, you should be able to:

1. Distinguish between the two principal kinds of software: applications and systems.

2. Discuss the ethics of copying software.

3. Describe the four types of applications software: entertainment, education and reference, productivity, and business and specialized.

4. Explain key features shared by many types of applications software packages.

5. Identify the key functions of word processors, spreadsheets, database managers, graphics programs, communications programs, desktop accessories and personal information managers, integrated programs and suites, groupware, and Internet Web browsers.

6. Briefly describe the key functions of programs for desktop publishing, personal finance, project management, computer-aided design, drawing and painting, and of hypertext.

"**T**hink of it as a map to the buried treasures of the Information Age."

That's how one writer described a particular kind of software named Mosaic when it first came out.[1] Mosaic is designed to help computer users find their way around the Internet. The global "network of networks," the Internet is rich in information but can be baffling to navigate. The developers of Mosaic had tried to remove that difficulty. Indeed, they had hoped their program might be the first "killer app"—killer application—of network computing. That is, it would be a breakthrough development that would help millions of people become comfortable using computer networks, a technology formerly used by only a relative few.

The computer industry puts great stock in history-making "killer apps." The last big one was the early 1980s development of the electronic spreadsheet program, software for manipulating numbers in financial documents. Spreadsheet software transformed the personal computer, until then used mainly by technicians and computer buffs, into an essential business tool. The application led to the widespread acquisition of desktop computers in offices all over the country.

Mosaic was not to become the software that would make the Internet available to everyone, being overtaken in a matter of months by another program called Netscape Navigator. Indeed, as this is written, developers are engaged in a titanic struggle to come up with the defining tool that will simplify users' abilities to summon text, as well as sound and images, from among the Internet's many information sources.

Nevertheless, the search for highly useful applications shows how truly important software is. Without software, a computer on a desk is about as useful as half of a pair of bookends. Furthermore, the easier the software is to use, the greater the number of people who will use the hardware.

How to Think About Software

Preview & Review: Software consists of the step-by-step instructions that tell the computer how to perform a task.

Software is of two types, systems software and applications software. Systems software enables the applications software to interact with the computer. Applications software, which may be custom-written or packaged, enables users with computers to perform work on general-purpose tasks.

Improvements to a software package may come out as a new version, representing a major upgrade, or as a new release, a minor upgrade. New versions and releases are usually upward compatible, able to work with documents from earlier software, but rarely is older software downward compatible with new software.

Mosaic and Netscape Navigator illustrate a trend: software is continually getting easier to use. At one time—and even now, with some software—every user had to learn cryptic commands such as "format a: /n:9 /t:40." Now software is available that lets computer users perform operations by simply pointing to words and images on the display screen and clicking a mouse button. Needless to say, ease of use would probably greatly influence your choice of software.

The Most Popular Uses of Software

Let's get right to the point: What do most people use software for? The answer hasn't changed in a decade. If you don't count games, by far the most popular applications are (1) *word processing* and (2) *spreadsheets*, according to the Software Publishers Association. Moreover, studies show, most people use only a few basic features of these programs, and they use them for rather simple tasks. For example, 70% of all documents produced with word processing software are one-page letters, memos, or simple reports. And 70% of the time people use spreadsheets simply to add up numbers.[2]

This is important information. If you are this type of user, you may have no more need for fancy software and hardware than an ordinary commuter has for an expensive Italian race car. On the other hand, you may be in a profession in which you need to become a "power user," learning almost all the software features available in order to keep ahead in your career. Moreover, in the Multimedia Age, you may wish to do far more than current software and hardware allow, in which case you need to be continually learning what computing and communications can do for you.

Let us now look at the various types and uses of software.

The Two Kinds of Software: Applications & Systems

As we've said, *software,* or *programs,* **consists of the step-by-step instructions that tell the computer how to perform a task.** Software is of two types. (■ *See Panel 2.1.*)

- **Applications software:** *Applications software* **is software that can perform useful work on general-purpose tasks,** such as word processing and creating spreadsheets.

- **Systems software:** As the user, you interact with the applications software. In turn, *systems software* **enables the applications software to interact with the computer and helps the computer manage its internal resources,** and you can instruct it in some of those tasks.

Software may be either *custom-written* or *packaged:*

- **Custom-written software:** *Custom-written software* **is software designed for a particular customer.** It is the kind of software written by a computer programmer to fulfill a highly specialized task. Unless you go on to become a programmer, you will probably not be required to know the intricacies of this kind of software.

- **Packaged software:** *Packaged software,* **or a** *software package,* **is an "off-the-shelf" program available on disk for use by the general public,**

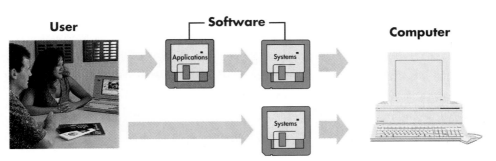

User **Software** **Computer**

PANEL 2.1

Applications software and systems software

You interact with both the applications and systems software. The systems software interacts with the computer.

such as word processing or spreadsheets. This is the kind of software discussed in this chapter and the next.

If you buy a new microcomputer in a store, you will find that some software has already been installed on it—that is, is "bundled" with it. This includes systems software and various types of applications software that are compatible with the systems software. We discuss systems software in Chapter 3. At this point, however, you are no doubt more concerned about what you can use a computer *for*. In this chapter, therefore, we describe applications software.

Versions, Releases, & Upward & Downward Compatibility

Every year or so, software developers find ways to enhance their products and put forth a new *version* or new *release:*

- **Version:** **A *version* is a major upgrade in a software product.** Versions are usually indicated by numbers such as 1.0, 2.0, 3.0, and so forth. The higher the number preceding the decimal point, the more recent the version.

- **Release:** **A *release* is a minor upgrade.** Releases are usually indicated by a change in number after the decimal point—3.0, then 3.1, then perhaps 3.11, then 3.2, and so on.

Some software developers have departed from this system. Microsoft, for instance, decided to call its new operating system, launched in 1995, "Windows 95" instead of "Windows 4.0."

Most software products are upward compatible (or "forward compatible"). *Upward compatible* **means that documents created with earlier versions of the software can be processed successfully on later versions.** Thus, you can use the new version of a word processing program, for instance, to get into and revise the file of a term paper you wrote on an earlier version of that program. However, downward-compatible ("backward-compatible") software is less common. *Downward compatible* **means that applications developed for a new version of a software product can be run on older versions.** For example, if you can run your new word processing program on your old operating system, then the new package is downward compatible.

Before we continue with the details about various types of applications software, we need to raise the ethical issue of copying intellectual property, including software.

Ethics & Intellectual Property Rights: When Can You Copy?

Preview & Review: Intellectual property consists of the products of the human mind. Such property can be protected by copyright, the exclusive legal right that prohibits copying it without the permission of the copyright holder.

Software piracy, network piracy, and plagiarism violate copyright laws.

Public domain software, freeware, and shareware can be legally copied, which is not the case with proprietary software.

Information technology has presented legislators and lawyers—and you—with some new ethical questions regarding rights to intellectual property. *Intellectual property* consists of the products, tangible or intangible, of the human mind. There are three methods of protecting intellectual property. They are *patents* (as for an invention), *trade secrets* (as for a formula or method of doing business), and *copyrights* (as for a song or a book).

What Is a Copyright?

Of principal interest to us is copyright protection. **A *copyright* is the exclusive legal right that prohibits copying of intellectual property without the permission of the copyright holder.** Copyright law protects books, articles, pamphlets, music, art, drawings, movies—and, yes, computer software. Copyright protects the *expression* of an idea but not the idea itself. Thus, others may copy your idea for, say, a new shoot-'em-up videogame but not your particular variant of it. Copyright protection is automatic and lasts a minimum of 50 years; you do not have to register your idea with the government (as you do with a patent) in order to receive protection.

These matters are important because the Digital Age has made the act of copying far easier and more convenient than in the past. Copying a book on a photocopier might take hours, so people felt they might as well buy the book. Copying a software program onto another floppy disk, however, might take just seconds.

Digitization threatens to compound the problem. For example, current copyright law doesn't specifically protect copyrighted material online. Says one article:

> Copyright experts say laws haven't kept pace with technology, especially digitization, the process of converting any data—sound, video, text—into a series of ones and zeros that are then transmitted over computer networks. Using this technology, it's possible to create an infinite number of copies of a book, a record, or a movie and distribute them to millions of people around the world at very little cost. Unlike photocopies of books or pirated audiotapes, the digital copies are virtually identical to the original.[3]

Piracy, Plagiarism, & Ownership of Images & Sounds

Three copyright-related matters deserve our attention: software and network piracy, plagiarism, and ownership of images and sounds.

- **Software and network piracy:** It may be hard to think of yourself as a pirate (no sword or eyepatch) when all you've done is make a copy of some commercial software for a friend. However, from an ethical standpoint, an act of piracy is like shoplifting the product off a store shelf—even if it's for a friend.

  Piracy is theft or unauthorized distribution or use. A type of piracy is to appropriate a computer design or program. This is the kind that Apple Computer claimed in a suit (since rejected) against Microsoft and Hewlett-Packard alleging that items in Apple's interface, such as icons and windows, had been copied.

  Software piracy is the unauthorized copying of copyrighted software. One way is to copy a program from one diskette to another. Another is to download (transfer) a program from a network and make a copy of it. **Network piracy is using electronic networks for the unauthorized distribution of copyrighted materials in digitized form.** Record companies, for example, have protested the practice of computer users' sending unauthorized copies of digital recordings over the Internet.[4] Both types of piracy are illegal.

The easy rationalization is to say that "I'm just a poor student, and making this one copy or downloading only one digital recording isn't going to cause any harm." But it is the single act of software piracy multiplied millions of times that is causing the software publishers a billion-dollar problem. They point out that the loss of revenue cuts into their budget for offering customer support, upgrading products, and compensating their creative people. Piracy also means that software prices are less likely to come down; if anything, they are more likely to go up.

In time, anti-copying technology may be developed that, when coupled with laws making the disabling of such technology a crime, will reduce the piracy problem. Regardless, publishers, broadcasters, movie studios, and authors must be persuaded to take chances on developing online and multimedia (✓ p. 28) versions of their intellectual products. Such information providers need to be able to cover their costs and make a reasonable return. If not, says one writer, the Information Superhighway will remain "empty of traffic because no one wants to put anything on the road."[5]

- **Plagiarism:** *Plagiarism* **is the expropriation of another writer's text, findings, or interpretations and presenting it as one's own.** Information technology puts a new face on plagiarism in two ways. On the one hand, it offers plagiarists new opportunities to go far afield for unauthorized copying. On the other hand, the technology offers new ways to catch people who steal other people's material.

  Electronic online journals are not limited by the number of pages, and so they can publish papers that attract a small number of readers. In recent years, there has been an explosion in the number of such journals and of their academic and scientific papers. This proliferation may make it harder to detect when a work has been plagiarized, since few readers will know if a similar paper has been published elsewhere.[6]

  Yet information technology may also be used to identify plagiarism. Scientists have used computers to search different documents for identical passages of text. In 1990, two "fraud busters" at the National Institutes of Health alleged after a computer-based analysis that a prominent historian and biographer had committed plagiarism in his books. The historian, who said the technique turned up only the repetition of stock phrases, was later exonerated in a scholarly investigation.[7]

- **Ownership of images and sounds:** Computers, scanners, digital cameras, and the like make it possible to alter images and sounds to be almost anything you want. What does this mean for the original copyright holders? An unauthorized sound snippet of James Brown's famous howl can be electronically transformed by digital sampling into the background music for dozens of rap recordings.[8] Images can be appropriated by scanning them into a computer system, then altered or placed in a new context.

  The line between artistic license and infringement of copyright is not always clearcut. In 1993, a federal appeals court in New York upheld a ruling against artist Jeff Koons for producing ceramic art of some puppies. It turned out that the puppies were identical to those that had appeared in a postcard photograph copyrighted by a California photographer.[9] But what would have been the judgment if Koons had scanned in the postcard, changed the colors, and rearranged the order of the puppies to produce a new postcard?

In any event, to avoid lawsuits for violating copyright, a growing number of artists who have recycled material have taken steps to protect themselves. This usually involves paying flat fees or a percentage of their royalties to the original copyright holders.

These are the general issues you need to consider when you're thinking about how to use someone else's intellectual property in the Digital Age. Now let's see how software fits in.

Public Domain Software, Freeware, & Shareware

No doubt most of the applications programs you will study in conjunction with this book will be commercial software packages, with brand names such as Microsoft Word or Lotus 1-2-3. However, there are a number of software products—many available over communications lines from the Internet—that are available to you as *public domain software, freeware,* or *shareware.*[10]

- **Public domain software:** ***Public domain software* is software that is not protected by copyright and thus may be duplicated by anyone at will.** Public domain programs—usually developed at taxpayer expense by government agencies—have been donated to the public by their creators. They are often available through sites on the Internet (or electronic bulletin boards) or through computer users groups. A users group is a club, or group, of computer users who share interests and trade information about computer systems.

  You can duplicate public domain software without fear of legal prosecution. (Beware: Downloading software through the Internet may introduce some problems—bad code called *viruses*—into your system. We discuss this problem, and how to prevent it, in Chapter 8.)

- **Freeware:** ***Freeware* is software that is available free of charge.** Freeware is distributed without charge, also usually through the Internet or computer users groups.

  Why would any software creator let the product go for free? Sometimes developers want to see how users respond, so they can make improvements in a later version. Sometimes it is to further some scholarly purpose, such as to create a standard for software on which people are apt to agree because there is no need to pay for it. An example of freeware is Mosaic, mentioned earlier.

  Freeware developers often retain all rights to their programs, so that technically you are not supposed to duplicate and distribute it further. Still, there is no problem about your making several copies for your own use.

- **Shareware:** ***Shareware* is copyrighted software that is distributed free of charge but requires users to make a contribution in order to receive technical help, documentation, or upgrades.** Shareware, too, is distributed primarily through communications connections such as the Internet.

  Is there any problem about making copies of shareware for your friends? Actually, the developer is hoping you will do just that. That's the way the program gets distributed to a lot of people—some of whom, the software creator hopes, will make a "contribution" or pay a "registration fee" for advice or upgrades.

  Though copying shareware is permissible, because it is copyrighted you cannot use it as the basis for developing your own program in order to compete with the developer.

Proprietary Software & Types of Licenses

Proprietary software **is software whose rights are owned by an individual or business,** usually a software developer. The ownership is protected by the copyright, and the owner expects you to buy a copy in order to use it. The software cannot legally be used or copied without permission.

Software manufacturers don't sell you the software so much as sell you a license to become an authorized user of it. What's the difference? In paying for a *software license,* **you sign a contract in which you agree not to make copies of the software to give away or for resale.** That is, you have bought only the company's permission to use the software and not the software itself. This legal nicety allows the company to retain its rights to the program and limits the way its customers can use it.[11] The small print in the licensing agreement allows you to make one copy (working copy or archival copy) for your own use.

Two types of licenses are *shrink-wrap licenses* and *site licenses:*

- **Shrink-wrap licenses:** *Shrink-wrap licenses* **are printed licenses inserted into software packages and visible through the clear plastic wrap.** The use of shrink-wrap licenses eliminates the need for a written signature, since buyers know they are entering into a binding contract by merely opening the package.[12]
- **Site licenses:** **A** *site license* **permits a customer to make multiple copies of a software product for use just within a given facility,** such as a college computer lab or a particular business. Usually the license stipulates a maximum number of copies.

The Software Police

Industry organizations such as the Software Publishers Association (hotline for reporting illegal copying: 800-388-7478) are going after software pirates large and small. Commercial software piracy is now a felony, punishable by up to five years in prison and fines of up to $250,000 for anyone convicted of stealing at least ten copies of a program or more than $2500 worth of software. Campus administrators are getting tougher with offenders and are turning them over to police.

The Four Types of Applications Software

Preview & Review: The four types of applications software may be considered to be (1) entertainment software, (2) education and reference software, (3) productivity software, and (4) business and specialized software.

The first contact many people have with applications software is with entertainment software, especially videogames. Students may find use for educational and reference software—for example, encyclopedias and library searches. Basic productivity tools include word processing, spreadsheet, database manager, presentation graphics, communications, desktop accessories and personal information managers, integrated programs and suites, groupware, and Internet Web browsers.

Software can change the way we act, even the way we think. Some readers may intuitively understand this because they grew up playing videogames.

Indeed, some observers hold that videogames are not quite the time wasters we have been led to believe. However, these forms of entertainment are only a way station to something else. Videogames are training wheels for using more sophisticated software that can help us learn better and be more productive.

There are four types of applications software. (■ *See Panel 2.2.*) They are:

- Entertainment software
- Education and reference software
- Basic productivity software
- Business and specialized software

Entertainment Software: The Serious Matter of Videogames

Whatever else may come about during the convergence of computers and communications, you can bet one kind of software will be available: entertainment—videogames in particular. A $6.5-billion-a-year industry in the United States, *videogames* are interactive electronic games that may be played at home through a television set or personal computer or in entertainment arcades of the sort found in shopping malls.[13]

Electronic spreadsheets may have put microcomputers on office desks; however, it was Pong—an electronic version of table tennis introduced by Atari in 1972—that popularized computers in the home. "Pong was the first time people saw computers as friendly and approachable," states one technology writer. "It launched a videogame boom that made thousands of kids want to become computer programmers, and prepared an entire generation for interaction with a blinking and buzzing computer screen."[14]

Pong was followed by Space Invaders and Pac-Man, and then by Super Mario, which begot Sonic the Hedgehog, which led to Mortal Kombat I and II. In 1986 Nintendo began to reshape the market when it introduced 8-bit entertainment systems. *Bit numbers* measure how much data a computer

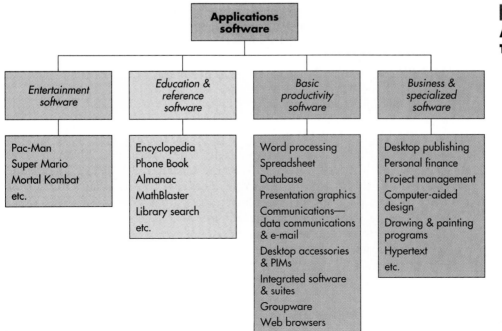

PANEL 2.2

Applications software: The four types

Case Study: How Videogames Are Training the New Work Force

A generation that played the early Nintendo [videogame] machines in high school is just now entering the work force. David Shpilberg, Ernst & Young's director of financial services consulting, finds that new hires who grew up playing videogames are more at ease with the company's sophisticated computer programs, including business simulations, than staffers with years of experience. And discussions with the new hires reveal that they acquired a better feel for the technology from playing videogames than from working on PCs.

"Think of reengineering a company," says Shpilberg. "The first thing you have to do is represent the obstacle course that the company is now. For instance, how a docu-

ment gets from an application to an actual loan—through credit reviews, checking of income statements and balance sheets, approval from superior officers, and so on. If you think of it, it's just like a Mario Brothers game. Our new hires understand this—that you can get through those obstacles only so fast before you have to redesign the game so that the obstacles go away."

—Rick Tetzeli,
"Videogames:
Serious Fun,"
Fortune

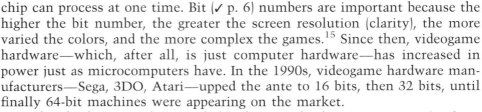

chip can process at one time. Bit (✓ p. 6) numbers are important because the higher the bit number, the greater the screen resolution (clarity), the more varied the colors, and the more complex the games.[15] Since then, videogame hardware—which, after all, is just computer hardware—has increased in power just as microcomputers have. In the 1990s, videogame hardware manufacturers—Sega, 3DO, Atari—upped the ante to 16 bits, then 32 bits, until finally 64-bit machines were appearing on the market.

 So far the biggest sales punch has been in hack-'em-up games (such as Mortal Kombat) and sports games.[16] Apart from ethical questions about the effects of violence on immature personalities, such games have perhaps ignored the interests of half the population—namely, females. More recently, however, videogame makers have introduced nonviolent games such as McKenzie & Company, essentially an interactive movie aimed at preadolescent girls.[17] Focusing on emotions rather than action, this game has the viewer, as the main character, try to solve problems of the heart.

Educational & Reference Software

Because of the popularity of videogames, many educational software companies have been blending educational content with action and adventure—as in MathBlaster or the problem-solving game Commander Keen. They hope this marriage will help students be more receptive to learning. After all, as one writer points out, players of Nintendo's Super Mario Brothers must "become intimately acquainted with an alien landscape, with characters, artifacts, and rules completely foreign to ordinary existence. . . . Children assimilate this essentially useless information with astonishing speed."[18] Why not, then, design software that would educate as well as entertain?

Computers alone won't boost academic performance, but they can have a positive effect on student achievement in all major subject areas, preschool through college, according to an independent consulting firm, New York's

Interactive Educational Systems Design. Skills improve when students use programs that are self-paced or contain interactive video. This is particularly true for low-achieving students. The reason, says a representative of the firm, which analyzed 176 studies done over the past 5 years, is that this kind of educational approach is "a different arena from the one in which they failed, and they have a sense of control."[19]

In addition to educational software, library search and reference software have become popular. For instance, there are CD-ROMs with encyclopedias, phone books, voter lists, mailing lists, maps, home-remodeling how-to information, and reproductions of famous art. With the CD-ROM encyclopedia Microsoft Encarta, for example, you can search for, say, music in 19th-century Russia, then listen to an orchestral fragment from Tchaikovsky's *1812 Overture*.[20]

Basic Productivity Software

Basic productivity software consists of programs found in most offices and probably on all campuses, on personal computers and on larger computer systems. Their purpose is simply to make users more productive at performing general tasks.

The most popular kinds of productivity tools are:

- Word processing software
- Spreadsheet software
- Database software
- Presentation graphics software
- Communications software—both data communications and e-mail
- Desktop accessories and personal information managers
- Integrated software and suites
- Groupware
- Internet Web browsers

It may be possible to work in an office somewhere in North America today without knowing any of these programs. However, that won't be the case in the 21st century. We describe common productivity software shortly.

Business & Specialized Software

Whatever your occupation, you will probably find it has specialized software available to it. This is so whether your career is as an architect, building contractor, chef, dairy farmer, dance choreographer, horse breeder, lawyer, nurse, physician, police officer, tax consultant, or teacher.

Some business software is of a general sort used in all kinds of enterprises, such as accounting software, which automates bookkeeping tasks, or payroll software, which keeps records of employee hours and produces reports for tax purposes. Other software is more specialized. Some programs help lawyers or advertising people, for instance, keep track of hours spent on particular projects for billing purposes. Other programs help construction estimators pull together the costs of materials and labor needed to estimate the costs of doing a job.

In this chapter we describe the following kinds of specialized software: *desktop publishing, personal finance, project management, computer-aided design, drawing and painting programs,* and *hypertext.*

The User Interface & Other Basic Features

Preview & Review: Applications software packages share some basic features and functions. They use special-purpose keys, function keys, and a mouse to issue commands and choose options. Their graphical user interfaces use menus, Help screens, windows, icons, and dialog boxes to make it easy for people to use the program.

Software packages are also accompanied by tutorials and documentation.

Offering just a handheld controller, a videogame machine allows you to make only limited moves. A microcomputer, by contrast, has a full-fledged keyboard and often a mouse or trackball, which allow software (including videogames) to be far more versatile.

Features of the Keyboard

We will describe the keyboard as an input device in Chapter 5. Here, however, we need to explain some aspects of the keyboard because it and the mouse are the means for manipulating software.

Besides a typewriter-like layout of letter, number, and punctuation keys and often a calculator-style numeric keypad, computer keyboards have special-purpose and function keys. Sometimes keystrokes are used in combinations called *macros.*

Let us explain:

- **Special-purpose keys:** *Special-purpose keys* **are used to enter, delete, and edit data and to execute commands.** An example is the Esc (for "Escape") key. The most important is the Enter key, which you will use often to tell the computer to execute commands entered with other keys. *Commands* are instructions that cause the software to perform specific actions. For example, pressing the Esc key commands the computer, via software instructions, to cancel an operation or leave ("escape from") the current mode of operation.

  Special-purpose keys are generally used the same way regardless of the applications software package being used. Most IBM-style keyboards include the following special-purpose keys: Esc, Ctrl, Alt, Del, Ins, Home, End, PgUp, PgDn, Num Lock, and a few others. (*Ctrl* means Control, *Del* means Delete, *Ins* means Insert, for example.)

- **Function keys:** *Function keys,* **labeled F1, F2, and so on, are positioned along the top or left side of the keyboard. They are used for commands specific to the software being used.** For example, one applications software package may use F6 to exit a file, whereas another may use F6 to underline a word.

  Many software packages come with templates that you can attach to the keyboard. Like the explanation of symbols on a roadmap, the template explains the purpose of each function key and certain combinations of keys. For example, in one word processing program, pressing Alt and F6 at the same time means "position these lines flush right on the page."

- **Macros:** Sometimes you may wish to reduce the number of keystrokes required to execute a command. To do this, you use a macro. **A *macro* is a single keystroke or command—or a series of keystrokes or commands— used to automatically issue a longer, predetermined series of keystrokes or commands.** Thus, you can consolidate several keystrokes for a command into only one or two keystrokes. The user names the macro and stores the corresponding command sequence; once this is done, the macro can be used repeatedly.

Although many people have no need for macros, others who find themselves continually repeating complicated patterns of keystrokes say they are quite useful.

The User Interface: GUIs, Menus, Help Screens, Windows, Icons, & Dialog Boxes

The first thing you look at when you call up any applications software on the screen is the user interface. **The *user interface* is the part of the software that displays information and presents on the screen the various commands by which you communicate with it.** The type of user interface is usually determined by the systems software (discussed in the next chapter). However, because this is what you see on the screen before you can begin using the applications software, we will briefly describe it here.

Some user interfaces require that you indicate your commands by typing in characters and text. However, the kind of interface now used by most people is the graphical user interface. **With a *graphical user interface*, or *GUI* (pronounced "gooey"), you may use graphics (images) and menus as well as keystrokes to choose commands, start programs, and see lists of files and other options**.

Common features of GUIs are *menus, Help screens, windows, icons, buttons*, and *dialog boxes*. (■ *See Panel 2.3, pp. 62–63.*)

- **Menus:** **A *menu* is a list of available commands presented on the screen.** Menus may appear as menu bars or pull-down menus.

  A *menu bar* is a line or two of command options across the top or bottom of the screen. Examples of commands, which you activate with a mouse or with key combinations, are File, Edit, and Print.

  A *pull-down* menu is a list of command options that "drops down" from a selected menu bar item at the top of the screen. For example, you might use the mouse to "click on" (activate) a command (for example, File) on the menu bar, which in turn would yield a pull-down menu offering further commands. These other commands might be Save, Open, Print, Close, Copy, and Delete. Choosing one of these options may produce further menus called *pop-up menus*, which seem to appear out of nowhere on the screen.

  A particularly useful type of menu is the **Help menu, or Help screen, which offers assistance on how to perform various tasks,** such as printing out a document. Having a set of Help screens is like having a built-in electronic instruction manual.

- **Windows:** A particularly interesting feature of GUIs is the use of windows. **A *window* is a rectangle that appears on the screen and displays information from a particular part of a program.** A display screen may show more than one window—for instance, one showing information from a word processing program, another information from a spreadsheet.

  A window (small w) should not be confused with *Microsoft Windows* (capital W), which is the most popular form of systems software. However, as you might expect, Windows features extensive use of windows.

- **Icons:** **An *icon* is a picture used in a GUI to represent a command, a program, or a task.** For example, a picture of a floppy disk might represent the command "Save (store) this document." Icons are activated by a mouse or other pointing device.

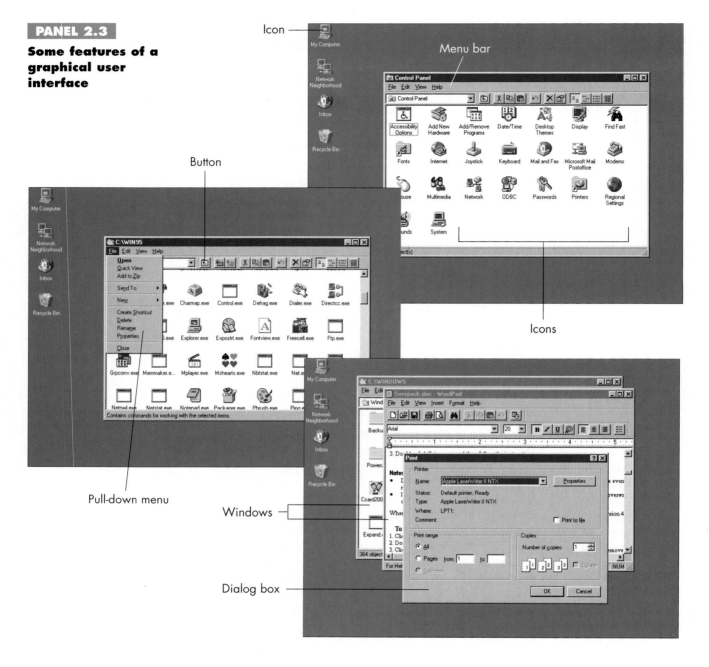

PANEL 2.3

Some features of a graphical user interface

Icon

Menu bar

Button

Pull-down menu

Windows

Dialog box

Icons

- **Buttons:** A *button* is a simulated on-screen button (kind of icon) that is activated ("pushed") by a mouse or other pointing device to issue a command, such as "Print document."

- **Dialog box:** A *dialog box* is a box that appears on the screen and displays a message requiring a response from you, such as pressing Y for "Yes" or N for "No" or typing in the name of a file. For example, when you're saving changes you've written in a document, the program might display a dialog box asking if you want to replace the previous version of the document.

Tutorials & Documentation

How are you going to learn a given software program? Most commercial packages come with tutorials and documentation.

PANEL 2.3

Continued

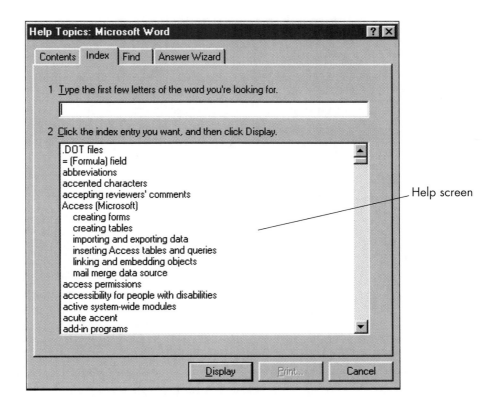

Help screen

- **Tutorials:** **A** *tutorial* **is an instruction book or program that takes you through a prescribed series of steps to help you learn how to use the product.** For instance, our publisher offers several how-to books, known as the Irwin Advantage Series, that enable you to learn different kinds of software. Tutorials can also be on-screen, provided as part of the software package.

- **Documentation:** *Documentation* **is a user manual (book) or reference manual that is a narrative and graphical description of a program.** Documentation may be instructional, but features and functions are usually grouped by category for reference purposes. For example, in word processing documentation, all cut-and-paste features are grouped together so you can easily look them up if you have forgotten how to perform them.

Often you can ask your software for directions on how to use the software. That is, some software makers (using a technique known as *natural language processing,* described in Chapter 11) equip their programs with features that allow you to ask questions in plain English. Thus, if you type "How do I add up the numbers in this column?" the software will respond by directing you to an interactive tutor or "coach" that can help you through the procedure. Lotus WordPro, for instance, has a feature called Ask the Expert, which asks you to complete a "How Do I?" question box. Microsoft Office 95 offers the same thing in a feature called Answer Wizard. In a few years, some predict, you'll be able to ask for help by speaking out loud, with voice-recognition software.[21]

Now let us consider the various forms of applications software used as productivity tools, plus a few specialized tools.

Word Processing

Preview & Review: Word processing software allows you to use computers to format, create, edit, print, and store text material.

One of the first typewriter users was Mark Twain. However, the typewriter, that long-lived machine, has gone to its reward. Indeed, if you have a manual typewriter, it is becoming as difficult to get it repaired as it is to find a blacksmith. What, then, are the alternatives?

The Different Kinds of Word Processors

Today your choice is generally to buy (1) a word processing typewriter, (2) a personal word processor, or (3) a microcomputer-plus-printer that runs word processing software.[22]

- **Word processing typewriter:** The *word processing typewriter* is like the old typewriter in that it can be made to type directly on paper. Yet it can also let you see and edit your words on a small display screen before they are printed on paper. It can automatically check your spelling. It can also store a few pages of text that you can retrieve and print later. This machine prints with a daisy wheel, the petals of which stamp characters directly onto the paper. Word processing typewriters cannot handle graphics.

  Priced at around $200, the word processing typewriter is probably fine, experts say, if you do only short reports and routine correspondence. Models are available from Brother, Sears, and Sharp.

- **Personal word processor:** The *personal word processor* is really a personal computer with a built-in word processing program, but it usually cannot run other types of programs. The machine is dedicated to creating, editing, and printing documents, and it can also store your written materials on diskettes. Display screens are usually easier to read than those on word processing typewriters.

  Prices are higher than for word processing typewriters, starting at about $300 (models are available from Brother, Panasonic, and AEG Olympia). Personal word processors are preferable to word processing typewriters if you handle lots of correspondence or do long reports.

- **Microcomputer, printer, and word processing software:** Microcomputers run not only word processing programs but also many other kinds of software. This is their principal advantage over personal word processors. **Word processing software allows you to use computers to format, create, edit, print, and store text material.** (■ *See Panel 2.4.*) Three common word processing programs for IBM-style Windows computers are WordPerfect, Microsoft Word, and Ami Pro. For Macintoshes they are Word and MacWrite.

Word processing software allows microcomputers to do what the other two types of machines do—namely, maneuver through a document and *delete, insert,* and *replace* text, the principal correction activities. However, word processing software offers additional features that the other methods often lack: *formatting, creating, editing, printing,* and *saving.* Let us consider these.

Formatting Documents

Formatting means determining the appearance of a document. There are many choices here.

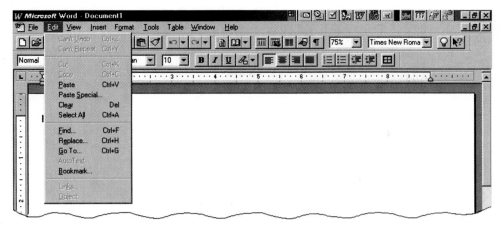

PANEL 2.4

Word processing

This Microsoft Word 7.0 for Windows program has a pull-down Edit menu that offers, among other things, several options for moving text around in a document and finding and replacing words.

- **Type:** You can decide what *typeface* and *type size* you wish to use. You can specify what parts of it should be <u>underlined</u>, *italic,* or **boldface.**

- **Spacing and columns:** You can choose whether you want the lines to be *single-spaced* or *double-spaced* (or something else). You can specify whether you want text to be *one column* (like this page), *two columns* (like many magazines and books), or *several columns* (like newspapers).

- **Margins and justification:** You can indicate the dimensions of the *margins*—left, right, top, and bottom—around the text.

  You can specify whether the text should be *justified* or not. *Justify* means to align text evenly between left and right margins, as, for example, is done with most newspaper columns and this text. *Left-justify* means to not align the text evenly on the right side, as in many business letters ("ragged right").

- **Pages, headers, footers:** You can indicate *page numbers* and *headers* or *footers.* A *header* is common text (such as a date or document name) that is printed at the top of every page. A *footer* is the same thing printed at the bottom of every page.

- **Other formatting:** You can specify *borders* or other decorative lines, *shading, tables,* and *footnotes.* You can even pull in ("import") *graphics* or drawings from files in other software programs.

It's worth noting that word processing programs (and indeed most forms of applications software) come from the manufacturer with *default settings.* **Default settings are the settings automatically used by a program unless the user specifies otherwise, thereby overriding them.** Thus, for example, most word processing programs will automatically prepare a document single-spaced, left-justified, with 1-inch right and left margins unless you alter these default settings.

Creating Documents

Creating a document means entering text, using the keyboard. Word processing software has three features that you will not encounter with a typewriter—the *cursor, scrolling,* and *word wrap.* (■ *See Panel 2.5.)*

- **Cursor:** **The *cursor* is the movable symbol on the display screen that shows you where you may enter data or commands next.** The symbol is often a blinking rectangle or I-beam. You can move the cursor on the screen using the keyboard's directional arrow keys or an electronic mouse.

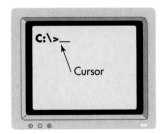

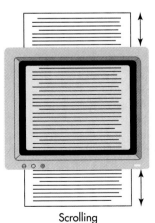

Scrolling

PANEL 2.5

Cursor and scrolling

- **Scrolling:** *Scrolling* **is the activity of moving quickly upward or downward through the text or other screen display.** A standard computer screen displays only 20–22 lines of standard-size text. Of course, most documents are longer than that. Using the directional arrow keys or a mouse, you can move ("scroll") through the display screen and into the text above and below it.

- **Word wrap:** *Word wrap* automatically continues text on the next line when you reach the right margin. That is, the text "wraps around" to the next line. You do not need to press a carriage-return key when you reach the right margin, as you would on a typewriter.

Editing Documents

Editing is the act of making alterations in the content of your document. Some features of editing that can't be found on a typewriter are *insert and delete, undelete, search and replace, block and move, spelling checker, grammar checker,* and *thesaurus.*

- **Insert and delete:** *Inserting* is the act of adding to the document. You simply place the cursor wherever you want to add text and start typing; the existing characters will move aside.

  Deleting is the act of removing text, usually using the Delete or Backspace keys.

  The *Undelete command* allows you to change your mind and restore text that you have deleted. Some word processing programs offer as much as 100 layers of "undo," allowing users who delete several blocks of text, but then change their minds, to reinstate one or more of the blocks.

- **Search and replace:** **The** *Search command* **allows you to find any word, phrase, or number that exists in your document. The** *Replace command* **allows you to automatically replace it with something else.**

- **Block and move:** Typewriter users were accustomed to using scissors and glue to "cut and paste" to move a paragraph or block of text from one place to another in a manuscript. With word processing, you can exercise the **Block command** **to indicate the beginning and end of the portion of text you want to move.** Then you can use the **Move command** **to move it to another location in the document.**

  You can also use the *Copy command* to copy the block of text to a new location while also leaving the original block where it is.

- **Spelling checker, grammar checker, thesaurus:** Many writers automatically run their completed documents through a **spelling checker, which tests for incorrectly spelled words.** (Some programs, such as Microsoft Word 6.0, have an "Auto Correct" function that automatically fixes such common mistakes as transposed letters—"teh" instead of "the.") Another feature is a **grammar checker, which flags poor grammar, wordiness, incomplete sentences, and awkward phrases.**

  If you find yourself stuck for the right word while you're writing, you can call up an on-screen **thesaurus, which will present you with the appropriate word or alternative words.**

Printing Documents

Most word processing software gives you several options for printing. For example, you can print *several copies* of a document. You can print *indi-*

vidual pages or *a range of pages.* You can even preview a document before printing it out. *Previewing (print previewing)* means viewing a document on screen to see what it will look like in printed form before it's printed. Whole pages are displayed in reduced size.

Some word processors even come close to desktop-publishing programs in enabling you to prepare professional-looking documents, with different type-faces and sizes. However, as we shall see later, desktop-publishing programs do far more.

Saving Documents

Saving means to store, or preserve, the electronic files of a document permanently on diskette, hard disk, or magnetic tape. Saving is a feature of nearly all applications software, but anyone accustomed to writing with a typewriter will find this activity especially valuable. Whether you want to make small changes or drastically revise your word processing document, having it stored in electronic form spares you the arduous chore of having to retype it from scratch. You need only call it up from disk or tape and make just those changes you want, then print it out again.

The key features of word processors are summarized in the box on the next page. (■ *See Panel 2.6.)*

Spreadsheets

Preview & Review: Spreadsheet software allows users to create tables and financial schedules by entering data into rows and columns arranged as a grid on a display screen. If one (or more) numerical value or formula is changed, the software automatically calculates the effect of the change on the rest of the spreadsheet.

Spreadsheet software also allows users to create analytical graphics charts to present data.

What is a spreadsheet? Traditionally, it was simply a grid of rows and columns, printed on special green paper, that was used by accountants and others to produce financial projections and reports. A person making up a spreadsheet often spent long days and weekends at the office penciling tiny numbers into countless tiny rectangles. When one figure changed, all the rest of the numbers on the spreadsheet had to be recomputed—and ultimately there might be wastebaskets full of jettisoned worksheets.

In the late 1970s, Daniel Bricklin was a student at the Harvard Business School. One day he was staring at columns of numbers on a blackboard when he got the idea for computerizing the spreadsheet. The result, VisiCalc, was the first of the electronic spreadsheets. **An *electronic spreadsheet,* also called simply a *spreadsheet,* allows users to create tables and financial schedules by entering data into rows and columns arranged as a grid on a display screen.**

The electronic spreadsheet quickly became the most popular small-business program. As we mentioned, it has been held directly responsible for making the microcomputer a widely used business tool. Unfortunately for Bricklin, VisiCalc was shortly surpassed by Lotus 1-2-3, a sophisticated program that combines the spreadsheet with database and graphics programs. Today the principal spreadsheets are Microsoft Excel, Lotus 1-2-3, and Quattro Pro.

PANEL 2.6 **Summary of common word processing features**

Formatting Features

Boldface / italic / underline: Word processing software makes it easy to emphasize text by using **bold,** *italic,* or underlining.

Font choice: Many packages allow you to change the font, or the typeface and the size of the characters, to improve the document's appearance.

Justify / unjustify: This feature allows you to print text aligned on both right and left margins or unaligned.

Word Processing Functions

Footnote placement: This feature allows the user to build a footnote file at the same time he or she is writing a document; the program then places the footnotes at appropriate page bottoms when the document is printed.

Mail merge: Most word processing programs allow the user to combine parts of different documents (files) to make the production of form letters much easier, faster, and less tedious than doing the same thing using a typewriter. For example, you can combine address files with a letter file that contains special codes where the address information is supposed to be. The program will insert the different addresses in copies of the letter and print them out.

Outlining: Some packages automatically outline the document for you; you can use the outline as a table of contents.

Split screen: This feature allows you to work on two documents at once—one at the top of the screen and one at the bottom. You can scroll each document independently and move and copy text between the documents.

Thesaurus: Thesaurus programs allow the user to pick word substitutions. For example, if you are writing a letter and want to use a more exciting word than *impressive,* you can activate your thesaurus program and ask for alternatives to that word—such as *awe-inspiring* or *thrilling.*

Word wrap: As you type, the text insertion point automatically moves to the beginning of the next line when the end of the current line is reached. In other words, you may type a paragraph continuously without pressing Return or Enter.

Editing Features

Correcting: Deleting and inserting; simply place the cursor where you want to correct a mistake and press either the Delete key or the Backspace key to delete characters. You can then type in new characters.

Check grammar: Word processing packages often include programs that check and highlight, for example, incomplete sentences, awkward phrases, wordiness, over-long sentences, and poor grammar.

Check spelling: Many packages come with a spelling checker program that, when executed, will alert you to misspelled words and offer correct versions.

Cut, Copy, and Paste (block move, block copy): Selecting and changing the position of a block (one or more characters) of text; this can be done within the same document or between different documents. To move a paragraph, for example, select it and then "cut" it from the document. The "cut" text is removed from the document and temporarily stored in memory. Then move the cursor to the new location and "paste" or insert the text from memory into the document. (The procedure to "copy" text is the same except that the original text isn't removed from the document.)

Scrolling: This feature allows you to move up or down through a document until you can see the text you're interested in. Most packages allow you to "jump" over many pages at a time—for example, from the beginning of a document straight to the end.

Search and replace: You can easily search through a document for a particular word—for example, a misspelled name—and replace it with another word.

Printing and Saving

Previewing: You can view a document on screen to see what it will look like when it is printed out.

Printing: You can print individual pages or a range of pages, as well as several copies of a document.

Saving: All programs allow you to store, or preserve, electronic documents on a diskette, hard disk, or magnetic tape.

Principal Features

The arrangement of a spreadsheet is as follows. (■ *See Panel 2.7.*)

- **Columns, rows, and labels:** *Column headings* appear across the top ("A" is the name of the first column, "B" the second, and so on). *Row headings* appear down the left side ("1" is the name of the first row, "2" the second, and so forth). Labels are any descriptive text, such as APRIL, PHONE, or GROSS SALES.

- **Cells, cell addresses, values, and spreadsheet cursor:** **The place where a row and a column intersect is called a *cell*, and its position is called a *cell address*.** For example, "A1" is the cell address for the top left cell, where column A and row 1 intersect. A number entered in a cell is called a *value*. The values are the actual numbers used in the spreadsheet—dollars, percentages, grade points, temperatures, or whatever. A *cell pointer*, or *spreadsheet cursor*, indicates where data is to be entered. The cell pointer can be moved around like a cursor in a word processing program.

PANEL 2.7

Electronic spreadsheet

The Lotus 1-2-3 for Windows electronic spreadsheet (*top*) is a computerized version of the traditional paper spreadsheet (*bottom*). The beauty of the electronic version, however, is its *recalculation* feature: When a number is changed, all related numbers on the spreadsheet are automatically recomputed.

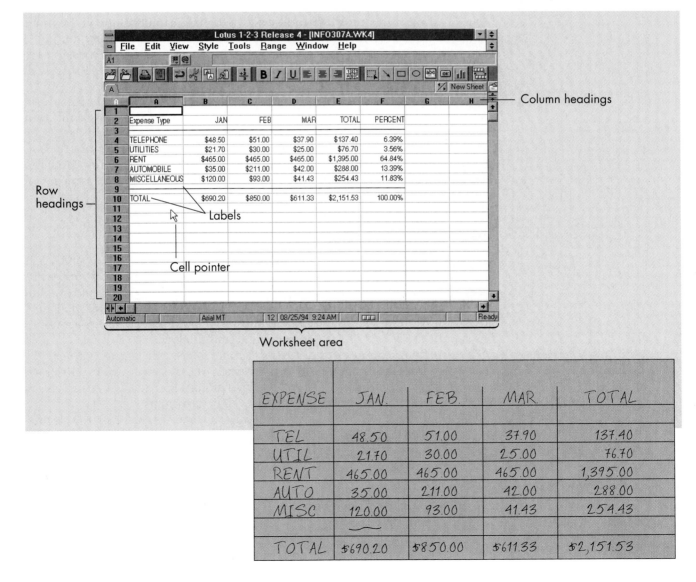

- **Formulas, functions, and recalculation:** Now we come to the reason the electronic spreadsheet has taken offices by storm. *Formulas* **are instructions for calculations.** For example, a formula might be @SUM(A5..A15), meaning "Sum (add) all the numbers in the cells with cell addresses A5 through A15."

  Functions are stored formulas that perform common calculations. For instance, a function might average a range of numbers or round off a number to two decimal places.

  After the values have been plugged into the spreadsheet, the formulas and functions can be used to calculate outcomes. What is revolutionary, however, is the way the spreadsheet can easily do recalculation. *Recalculation* is the process of recomputing values *automatically*, either as an ongoing process as data is being entered or afterward, with the press of a key. With this simple feature, the hours of mind-numbing work required in manually reworking paper spreadsheets became a thing of the past.

- **The "what if?" world:** The recalculation feature has opened up whole new possibilities for decision making. As a user, you can create a plan, put in formulas and numbers, and then ask yourself, "What would happen if we change that detail?"—and immediately see the effect on the bottom line. You could use this if you're considering buying a new car. Any number of things can be varied: total price ($10,000? $15,000?), down payment ($2,000? $3,000?), interest rate on the car loan (7%? 8%?), or number of months to pay (36? 48?). You can keep changing the "what if" possibilities until you arrive at a monthly payment figure that you're comfortable with.

Spreadsheets can be linked with other spreadsheets. The feature of *dynamic linking* **allows data in one spreadsheet to be linked to and automatically update data in another spreadsheet.** Thus, the amount of data being manipulated can be enormous. For instance, Frank Austin, a computer consultant who developed a program for rehabilitating the sewers of Houston, says his average worksheets are a demanding 700 kilobytes to 1 megabyte (✓ p. 12), representing thousands of cells.[23]

Analytical Graphics: Creating Charts

A nice feature of spreadsheet packages is the ability to create analytical graphics. *Analytical graphics,* **or** *business graphics,* **are graphical forms that make numeric data easier to analyze** than when it is in the form of rows and columns of numbers, as in electronic spreadsheets. Whether viewed on a monitor or printed out, analytical graphics help make sales figures, economic trends, and the like easier to comprehend and analyze.

The principal examples of analytical graphics are *bar charts, line graphs,* and *pie charts.* (■ *See Panel 2.8.*) Quite often these charts can be displayed or printed out so that they look three-dimensional. Spreadsheets can even be linked to more exciting graphics, such as digitized maps.

The key features of spreadsheet software are summarized in the accompanying box. (■ *See Panel 2.9.*)

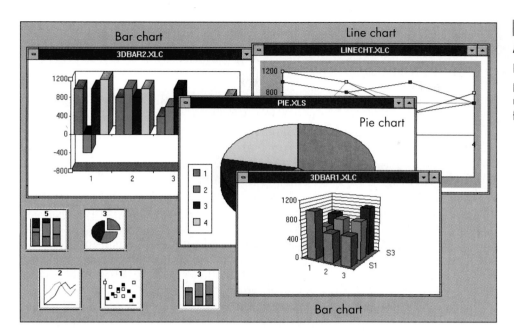

Bar chart

Line chart

Pie chart

Bar chart

PANEL 2.8

Analytical graphics

Bar charts, line graphs, and pie charts are used to display numerical data in graphical form.

PANEL 2.9   **Summary of common spreadsheet features**

Cell: The intersection of a column and a row; a cell holds a single unit of information.

Cell address: The location of a cell. For example, B3 is the address of the cell at the intersection of column B and row 3.

Cell pointer: Indicates the position where data is to be entered or changed; the user moves the cell pointer around the spreadsheet, using the particular software package's commands.

Column labels: The column headings across the top of the worksheet area.

Formula: Instructions for calculations; these calculations are executed by the software based on commands issued by the user. For example, a formula might be SUM(A5..A15), meaning "Sum (add) all the numbers in the cells with cell addresses A5 through A15."

Graphics: Most spreadsheets allow users to display data in graphic form, such as bar, line, and pie charts.

Recalculation: Automatic reworking of all the formulas and data according to changes the user makes in the spreadsheet.

Row labels: The row headings that go down the left side of the worksheet area.

Scrolling: Moving the cell pointer up and down, and right and left, to see different parts of the spreadsheet.

Value: The number within a cell.

Database Software

Preview & Review: A database is a computer-based collection of interrelated files. Database software is a program that controls the structure of a database and access to the data.

In its most general sense, a database is any electronically stored collection of data in a computer system. In its more specific sense, **a *database* is a collection of interrelated files** in a computer system. These computer-based files are organized according to their common elements, so that they can be retrieved easily. (Databases are covered in detail in Chapter 9.) Sometimes called a *database manager* or *database management system (DBMS)*, **data-**

base software **is a program that controls the structure of a database and access to the data.**

The Benefits of Database Software

Because it can access several files at one time, database software is much better than the old file managers (also known as flat-file management systems) that used to dominate computing. A *file manager* is a software package that can access only one file at a time. With a file manager, you could call up a list of, say, all students at your college majoring in English. You could also call up a separate list of all students from Wisconsin. But you could not call up a list of English majors from Wisconsin, because the relevant data is kept in separate files. Database software allows you to do that.

Databases are a lot more interesting than they used to be. Once they included only text. The Digital Age has added new kinds of information—not only documents but also pictures, sound, and animation. (■ *See Panel 2.10.*) It's likely, for instance, that your personnel record in a future company database will include a picture of you and perhaps even a clip of your voice. If you go looking for a house to buy, you will be able to view a real estate agent's database of video clips of homes and properties without leaving the realtor's office.[24] Today the principal database software packages are dBASE, Access, Paradox, Filemaker Pro for Windows, FoxPro for Windows, Q&A for Windows, and Approach for Windows.

Databases have gotten easier to use, but they still can be difficult to set up. Even so, the trend is toward making such programs easier for both database creators and database users.

Principal Features of Database Software

Some features of databases are as follows:

- **Organization of a database:** A database is organized—from smallest to largest items—into *fields, records,* and *files.*

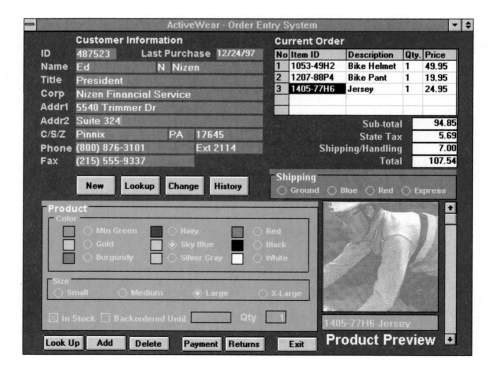

A *field* is a unit of data consisting of one or more characters. An example of a field is your name, your address, or your driver's license number.

A *record* is a collection of related fields. An example of a record would be your name *and* address *and* driver's license number.

A *file* is a collection of related records. An example of a file could be one in your state's Department of Motor Vehicles. The file would include everyone who received a driver's license on the same day, including their names, addresses, and driver's license numbers.

- **Select and display:** The beauty of database software is that you can locate records in the file quickly. For example, your college may maintain several records about you—one at the registrar's, one in financial aid, one in the housing department, and so on. Any of these records can be called up on a computer display screen for viewing and updating. Thus, if you move, your address field will need to be changed in all records. The database is quickly corrected by finding your name field. Once the record is displayed, the address field can be changed.

- **Sort:** With database software you can easily change the order of records in a file. Normally, records are entered into a database in the order they occur, such as by the date a person registered to attend college. However, all these records can be sorted in different ways. For example, they can be rearranged by state, by age, or by Social Security number.

- **Calculate and format:** Many database programs contain built-in mathematical formulas. This feature can be used, for example, to find the grade-point averages for students in different majors or in different classes. Such information can then be organized into different formats and printed out.

Some of the principal vocabulary associated with databases and database software is summarized in the accompanying box. (■ *See Panel 2.11.*)

PANEL 2.11   **Summary of common features of database software**

Field: A *field* is a unit of data consisting of one or more characters. An example of a field is your name, your address, or your driver's license number.

Record: A *record* is a collection of related fields. An example of a record would be your name and address and driver's license number.

File: A *file* is a collection of related records. An example of a file could be one in your state's Department of Motor Vehicles. The file would include all the people who received driver's licenses on the same day, including their names, addresses, and driver's license numbers.

Calculate and format: Many database management programs contain built-in mathematical formulas. This feature can be used, for example, to find the grade-point averages for students in different majors or in different classes. Such information can then be organized into different formats and printed out.

Select and display: The beauty of database management programs is that you can select specific records in a file quickly and display them. For example, suppose you use a database manager to keep track of names and addresses that were originally stored in your address book. To select and then display a list of those friends who live in Texas, for example, you could type "Texas" into the State field. After completing the command, a list of those friends who live in Texas would be displayed on the screen.

Sort: Data is entered into the database in a random fashion; however, the user can easily change the order of records in a file—for example, alphabetically by employee last name, chronologically according to date hired, or by zip code. The field according to which the records are ordered is the *key field*, or *key*.

Presentation Graphics Software

Preview & Review: Presentation graphics software allows people to create graphical representations of data to present to other people. This type of graphics is more sophisticated than the analytical graphics produced by spreadsheet packages.

Computer graphics can be highly complicated, such as those used in special effects for movies (such as *Toy Story* or *Jurassic Park)*. Here we are concerned with just one kind of graphics called presentation graphics.

***Presentation graphics* are graphics used to communicate or make a presentation of data to others,** such as clients or supervisors. Presentations may make use of bar, line, and pie charts, but they usually look much more sophisticated, using, for instance, different texturing patterns (speckled, solid, cross-hatched), color, and three-dimensionality. (■ *See Panel 2.12.)* Examples of well-known presentation graphics packages are Microsoft PowerPoint, Aldus Persuasion, Lotus Freelance Graphics, and SPC Harvard Graphics.

In general, these graphics are presented as *slides,* which can be projected on a screen or displayed on a large monitor. Presentation graphics packages often come with *slide sorters,* which group together a dozen or so slides in miniature. The person making the presentation can use a mouse or keyboard to bring the slides up for viewing.

Some presentation graphics packages provide artwork ("clip art") that can be electronically cut and pasted into the graphics. These programs also allow you to use electronic painting and drawing tools for creating lines, rectangles, and just about any other shape. Depending on the system's capabilities, you can add text, animated sequences, and sound. With special equipment you can do graphic presentations on slides, transparencies, and videotape. With all these options the main problem may be simply restraining yourself.

Communications Software

Preview & Review: Communications software manages the transmission of data between computers. It also enables users to send and receive electronic mail.

In the past, many microcomputer users felt they had all the productivity they needed without ever having to hook up their machines to a telephone. One

PANEL 2.12

Presentation graphics

of the major themes of this book, however, is that having communications capabilities vastly extends your range. This great leap forward is made possible with communications software. Two types of communications software are *data communications software* and *electronic mail software.*

Data Communications Software

***Data communications software* manages the transmission of data between computers.** For most microcomputer users this sending and receiving of data is by way of a modem and a telephone line. As described in Chapter 1 (✓ p. 7), a *modem* is an electronic device that allows computers to communicate with each other over telephone lines. The modem translates the digital signals of the computer into analog signals that can travel over telephone lines to another modem, which translates the analog signals back to digital. When you buy a modem, you often get communications software with it. Popular microcomputer communications programs are Crosstalk and Procomm Plus.

Data communications software gives you these capabilities:

- **Online connections:** You can connect to electronic bulletin board systems (BBSs) organized around special interests, to online services, and to the Internet.

- **Use of financial services:** With communications software you can order discount merchandise, look up airline schedules and make reservations, follow the stock market, and even do some home banking and bill paying.

- **Automatic dialing services:** You can set your software to answer for you if someone tries to call your computer, to dial certain telephone numbers automatically, and to automatically redial after a certain time if a line is busy.

- **Remote access connections:** While traveling you can use your portable computer to exchange files via modem with your computer at home.

Electronic Mail Software

***Electronic mail (e-mail) software* enables users to send letters and documents from one computer to another.** Many organizations have "electronic mailboxes." If you were a sales representative, for example, such a mailbox would allow you to transmit a report you created on your word processor to a sales manager in another area. Or you could route the same message to a number of users on a distribution list.

Desktop Accessories & Personal Information Managers

Preview & Review: Desktop accessory software provides an electronic version of tools or objects commonly found on a desktop: calendar, clock, card file, calculator, and notepad.

Personal information manager (PIM) software combines some features of word processing, database manager, and desktop accessory programs to organize specific types of information, such as address books.

Pretend you are sitting at a desk in an old-fashioned office. You have a calendar, clock, calculator, Rolodex-type address file, and notepad. Most of these items could also be found on a student's desk. How would a computer and software improve on this arrangement?

Many people find ready uses for types of software known as *desktop accessories* and *personal information managers (PIMs).*

Desktop Accessories

A *desktop accessory,* or *desktop organizer,* is a software package that provides an electronic version of tools or objects commonly found on a desktop: calendar, clock, card file, calculator, and notepad.

Some desktop-accessory programs come as standard equipment with some systems software (such as with Microsoft Windows). Others, such as Borland's SideKick or Lotus Agenda, are available as separate programs to run in your computer's main memory at the same time you are running other software. Some are principally *scheduling and calendaring programs;* their main purpose is to enable you to do time and event scheduling.

Personal Information Managers

A more sophisticated program is the ***personal information manager (PIM),* a combination word processor, database, and desktop accessory program that organizes a variety of information.**[25] Examples of PIMs are Commence, Dynodex, Ecco, Lotus Organizer, and Franklin Planner.

Lotus Organizer, for example, looks much like a paper datebook on the screen—down to simulated metal rings holding simulated paper pages. The program has screen images of section tabs labeled Calendar, To Do, Address, Notepad, Planner, and Anniversary. The Notepad section lets users enter long documents, including text and graphics, that can be called up at any time.[26] Whereas Lotus Organizer resembles a datebook, the PIM called Dynodex resembles an address book, with spaces for names, addresses, phone numbers, and notes.

◤ Integrated Software & Suites

Preview & Review: Integrated software packages combine the features of several applications programs—for example, word processing, spreadsheet, database manager, graphics, and communications—into one software package.

What if you want to take data from one program and use it in another—say, call up data from a database and use it in a spreadsheet? You can try using separate software packages, but one may not be designed to accept data from the other. Two alternatives are the collections of software known as *integrated software* and *software suites.*

Integrated Software: "Works" Programs

Integrated software packages **combine the features of several applications programs—such as word processing, spreadsheet, database, graphics, and communications—into one software package.** These so-called "works" collections—the principal representatives are AppleWorks, ClarisWorks, Lotus Works, Microsoft Works, and PerfectWorks—give good value because the entire bundle often sells for $100 or less.

Some of these "works" programs have "assistants" that help you accomplish various tasks. Thus, Microsoft's Works for Windows 95 helps you create new documents with the help of 39 "task wizards." The wizards lead you through the process of creating a letter, for example, that permits you to customize as many features as you want.[27]

Integrated software packages are less powerful than separate programs used alone, such as a word processing or spreadsheet program used by itself. But that may be fine, because single-purpose programs may be more complicated and demand more computer resources than necessary. You may have no need, for instance, for a word processor that will create an index. Moreover, Microsoft Word takes up about 20 megabytes on your hard disk, whereas Microsoft Works takes only 7 megabytes, which leaves a lot more room for other software.[28]

Software Suites: "Office" Programs

Software suites, **or simply** *suites,* **are applications—like spreadsheets, word processing, graphics, communications, and groupware—that are bundled together and sold for a fraction of what the programs would cost if bought individually.** (■ *See Panel 2.13.*)

"Bundled" and "unbundled" are jargon words frequently encountered in software and hardware merchandising. *Bundled* **means that components of a system are sold together for a single price.** *Unbundled* **means that a system has separate prices for each component.**

The principal suites, sometimes called "office" programs, are Microsoft Office from Microsoft, SmartSuite from Lotus, and PerfectOffice from Corel.[29] Microsoft's Office 95 consists of programs that separately would cost perhaps $1500 but as a suite cost roughly $500 to $600. Special pricing makes some suites available to students for less than $200.

PANEL 2.13 **Suite deals**

Applications packages bundled in suites from Microsoft, Lotus, and Corel.

	Microsoft Office	**Lotus SmartSuite**	**Corel PerfectOffice Select**
Word processor	Word	Ami Pro	WordPerfect
Spreadsheet	Excel	1-2-3	Quattro Pro
Database	Access	Approach	Paradox
Graphics	PowerPoint	Freelance	Presentations
Other	Microsoft Mail (e-mail)	Organizer (Personal planner)	Infocentral (Personal planner), Groupwise (e-mail), two other programs

Although cost is what makes suites attractive to many corporate customers, they have other benefits as well. Software makers have tried to integrate the "look and feel" of the separate programs within the suites to make them easier to use. "The applications mesh more smoothly in the package form," says one writer, "and the level of integration is increasing. More and more, they use the same commands and similar icons in the spreadsheet, word processor, graphics, and other applications, making them easier to use and reducing the training time."[30]

A tradeoff, however, is that such packages require a lot of hard-disk storage capacity. Microsoft Office 95, for instance, comes on 24 or more floppy disks (there is also a CD-ROM version) and occupies *at least 89 megabytes* of hard disk space—quite a lot if your hard disk holds only 200 megabytes.

Software for the "Digital Office"

Software suites are probably only a way station to something else. The push is on to link *everything*—to achieve the complete "digital office." For instance, Microsoft is spearheading a standard, using infrared technology, by which microcomputers are integrated with other office technologies. This union will include printers, fax machines, photocopiers, and telephones. Microsoft is working in partnership with more than 60 companies in the computer, office machinery, and telecommunications industries to establish connectivity and common linkages. As Karen Ann Hargrove, general manager of digital office systems for Microsoft, explains, "There's no good reason why what you fax and what you print and what you copy and what you view on the computer shouldn't be the same. The only difference is today a copier can't accept a digital original. We're going to change that."[31]

Groupware

Preview & Review: Groupware is software used on a network that serves a group of users working together on the same project.

Most microcomputer software is written for people working alone. **Groupware is software that is used on a network and serves a group of users working together on the same project.** Groupware improves productivity by keeping you continually notified about what your colleagues are thinking and doing, and they about you. "Like e-mail," one writer points out, "groupware became possible when companies started linking PCs into networks. But while e-mail works fine for sending a message to a specific person or group—communicating one-to-one or one-to-many—groupware allows a new kind of communication: many-to-many."[32]

Groupware is essentially of four types:[33]

* **Basic groupware:** Exemplified by Lotus Notes and Microsoft Exchange, this kind of groupware uses an enormous database containing work records, memos, and notations and combines it with a messaging system. Thus, a company like accounting giant Coopers & Lybrand uses Lotus Notes to let co-workers organize and share financial and tax information. It can also be used to relay advice from outside specialists, speeding up audits and answers to complex questions from clients.[34]

* **Workflow software:** Workflow software, exemplified by ActionWorkflow System and ProcessIt, helps workers understand and redesign the steps that make up a particular process. It also routes work automatically

among employees and helps organizations reduce paper-jammed bureau-cracies.

- **Meeting software:** An example of meeting software is Ventana's Group-Systems V, which allows people to have computer-linked meetings. With this software, people "talk," or communicate, with one another at the same time by typing on microcomputer keyboards. As one writer describes it, "Because people read faster than they speak, and don't have to wait for others to finish talking, the software can dramatically speed progress toward consensus."[35]
- **Scheduling software:** Scheduling software uses a microcomputer net-work to coordinate co-workers' electronic datebooks or appointment cal-endars so they can figure out a time when they can all get together. An example is Network Scheduler 3 from Powercore.

Groupware has changed the kind of behavior required for success in an organization. For one thing, it requires workers to take more responsibility. Ethically, of course, when you are contributing to a group project of any kind, you should try to do your best. However, when your contribution to the pro-ject is clearly visible to all, as happens with groupware, you *have* to do your best. In addition, using e-mail or groupware means you need to use good man-ners and be sensitive to others while you're online.

Internet Web Browsers

Preview & Review: Web browsers are software programs that allow people to view information at Web sites in the form of colorful, on-screen magazine-style "pages" with text, graphics, and sound.

The Internet, that network of thousands of interconnected networks, "is just a morass of data, dribbling out of [computers] around the world," says one writer. "It is unfathomably chaotic, mixing items of great value with cyber-trash." This is why so-called *browsers* have caught people's imaginations, he states. "A browser cuts a path through the tangled growth and even creates a form of memory, so each path can be retraced."[36]

We cover the Internet in detail elsewhere (especially Chapter 8). Here let us consider just a part of a part of it, one that you may find particularly useful.

The World Wide Web

The most exciting part of the Internet is probably that fast-growing region or subset of it known as the World Wide Web. The *World Wide Web*, or sim-ply *"the Web,"* consists of hundreds of thousands of intricately interlinked sites called "home pages" set up for on-screen viewing in the form of color-ful magazine-style "pages" with text, images, and sound.

To be connected to the World Wide Web, you need an automatic setup with an online service or Internet access provider (described in the Chapter 3 Experience Box and also in Chapter 8), who will then give you a "browser" for actually exploring the Web. (The reverse is also true: If you buy some Web browsers, they will help you find an access provider.) **A *Web browser*, or sim-ply *browser*, is software that enables you to "browse through" and view Web sites.** You can move from page to page by "clicking on" or selecting an icon or by typing in the address of the page. The accompanying drawing explains what the parts of a Web electronic address mean. (■ *See Panel 2.14.*)

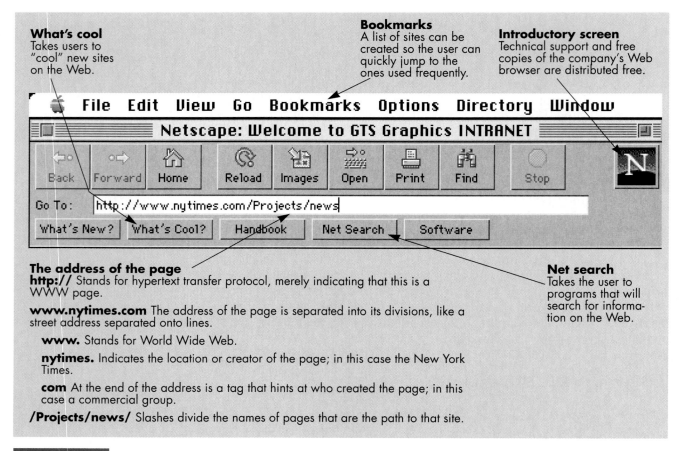

What's cool
Takes users to "cool" new sites on the Web.

Bookmarks
A list of sites can be created so the user can quickly jump to the ones used frequently.

Introductory screen
Technical support and free copies of the company's Web browser are distributed free.

The address of the page
http:// Stands for hypertext transfer protocol, merely indicating that this is a WWW page.

www.nytimes.com The address of the page is separated into its divisions, like a street address separated onto lines.

www. Stands for World Wide Web.

nytimes. Indicates the location or creator of the page; in this case the New York Times.

com At the end of the address is a tag that hints at who created the page; in this case a commercial group.

/Projects/news/ Slashes divide the names of pages that are the path to that site.

Net search
Takes the user to programs that will search for information on the Web.

PANEL 2.14

What's a Web browser?

This "tool bar" from Netscape's Navigator shows what you see at the top of each home page (Web site). From here you can move from one page to the next.

There are a great many browsers, including some unsophisticated ones offered by Internet access providers and some by the large commercial online services such as America Online, CompuServe, and Prodigy. However, the recent battle royal to find the "killer app" browser has been between Netscape, which produces Navigator, and Microsoft, which developed a browser called Internet Explorer. As of mid-1996, Netscape was winning with 81% of the browser market; Microsoft had 71%.[37] However, America Online has made Internet Explorer its "preferred" browser on its main online service (in return for which Microsoft includes AOL software in its new releases of its Windows 95 operating system.

Search Tools on the Web: Directories & Indexes

Once you're on your browser, you need to know how to find what you're looking for. Search tools are of two basic types—*directories* and *indexes*. (■ *See Panel 2.15.*)

- **Directories:** **Web directories are search tools classified by topic.** One of the foremost examples is Yahoo (*http://www.yahoo.com*), which provides you with an opening screen offering 14 general categories.

- **Indexes:** **Web indexes allow you to find specific documents through keyword searches.** An example of one useful index tool is Lycos (*http://www.lycos.com*).

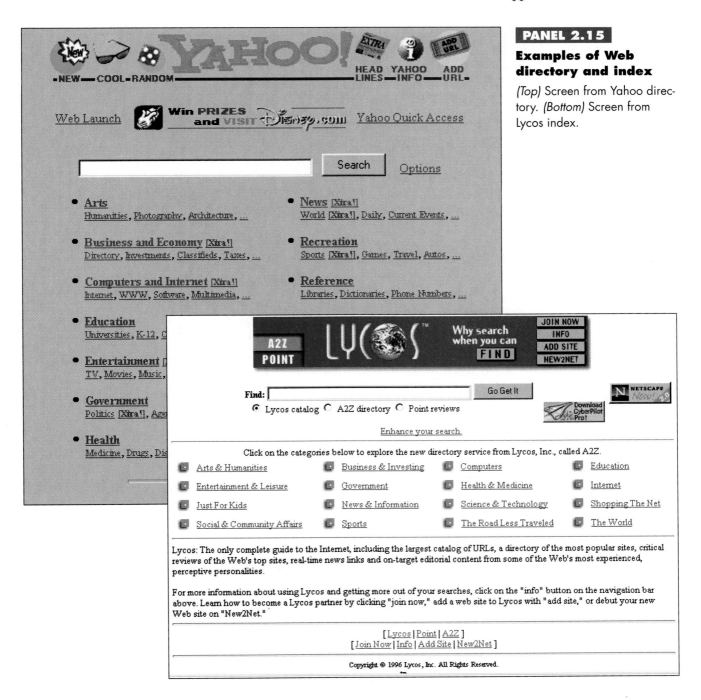

PANEL 2.15

Examples of Web directory and index

(Top) Screen from Yahoo directory. *(Bottom)* Screen from Lycos index.

Although the promise of spectacular services delivered via the Web is still mostly just that, a promise, the fact remains that only 6% to 8% of American households are now connected to the Internet. By 2000, that is expected to rise to 22%.[38] Thus, Web browsers may come to be one of our most important software productivity tools.

Specialized Software

Preview & Review: Specialized software tools include programs for desktop publishing, personal finance, project management, computer-aided design, drawing and painting, and hypertext.

After learning some of the productivity software just described, you may wish to extend your range by becoming familiar with more specialized programs. For example, you might first learn word processing and then move on to desktop publishing, the technology used to prepare much of today's printed information. Or you might learn spreadsheet programs and then go on to master personal-finance, tax, and investment software. Let us consider some of these specialized tools. We describe the following, although these are but a handful of the thousands of programs available:

- Desktop-publishing programs
- Personal finance programs
- Project management programs
- Computer-aided design
- Painting and drawing programs
- Hypertext

Desktop Publishing

Once you've become comfortable with a word processor, could you then go on and learn to do what Margaret Trejo did? When Trejo, then 36, was laid off from her job in 1987 because her boss couldn't meet the payroll, she was stunned. "Nothing like that had ever happened to me before," she said later. "But I knew it wasn't a reflection on my work. And I saw it as an opportunity."[39]

Today Trejo Production is a successful desktop-publishing company in Princeton, New Jersey, using Macintosh equipment to produce scores of books, brochures, and newsletters. "I'm making twice what I ever made in management positions," says Trejo, "and my business has increased by 25% every year."

Not everyone can set up a successful desktop-publishing business, because many complex layouts require experience, skill, and a knowledge of graphic design. Indeed, use of these programs by nonprofessional users can lead to rather unprofessional-looking results. Nevertheless, the availability of microcomputers and reasonably inexpensive software has opened up a career area formerly reserved for professional typographers and printers. ***Desktop publishing*, abbreviated *DTP*, involves using a microcomputer and mouse, scanner, laser printer, and DTP software for mixing text and graphics to produce high-quality printed output.** Often the laser printer is used primarily to get an advance look before the completed job is sent to a typesetter for even higher-quality output. Principal desktop-publishing programs are Aldus Page-Maker, Ventura Publisher, Quark XPress, and First Publisher. Microsoft Publisher is a low-end DTP package. Some word processing programs, such as Word and WordPerfect, also have many DTP features.

Desktop publishing has the following characteristics:

- **Mix of text with graphics:** Unlike traditional word processing programs, desktop-publishing software allows you to manage and merge text with graphics. Indeed, while laying out a page on screen, you can make the text "flow," liquid-like, around graphics such as photographs.

 Software used by many professional typesetters shows display screens full of formatting codes rather than what you will see when the job is printed out. By contrast, DTP programs can display your work in WYSI-WYG form. ***WYSIWYG* (pronounced "wizzy-wig") stands for "What You See Is What You Get." It means that the text and graphics appear on the display screen exactly as they will print out.** (■ *See Panel 2.16.*)

- **Varied type and layout styles:** DTP programs provide a variety of *fonts,* **or typestyles,** from readable Times Roman to staid Tribune to wild Jester and Scribble. You can also create all kinds of rules, borders, columns, and

PANEL 2.16

WYSIWYG

(Top) How a desktop-published document—a restaurant menu—looks on a display screen. *(Bottom)* How the page looks when printed out. Because the display screen shows that "What You See Is What You Get" (WYSIWYG), the user can make changes in the appearance before printing the document out.

---

page numbering styles. A *style sheet* in the DTP program enables you to choose and record the settings that determine the appearance of the pages. This may include defining size and typestyle of text and headings, numbers of columns of type on a page, and width of lines and boxes.

- **Use of files from other programs:** Most DTP programs don't have all the features of full-fledged word processing or computerized drawing and painting programs. Thus, text is usually composed on a word processor, artwork is created with drawing and painting software, and photographs are scanned in using a scanner. Prefabricated art may also be obtained from disks containing **clip art, or "canned" images that can be used to illustrate DTP documents.** The DTP program is used to integrate all these files. You can look at your work on the display screen as one page or as two facing pages in reduced size. Then you can see it again after it is printed out on a printer. (■ *See Panel 2.17, next page.*)

- **Page description language:** Once you have finished your composition and layout, you can send the document to the printer. Much of the shaping of text characters and graphics is done within the printer rather than in the computer. For instance, instead of sending the complete image of a circle from the computer to the printer, you send a command to the printer to draw a circle. **A *page description language* is software used to describe**

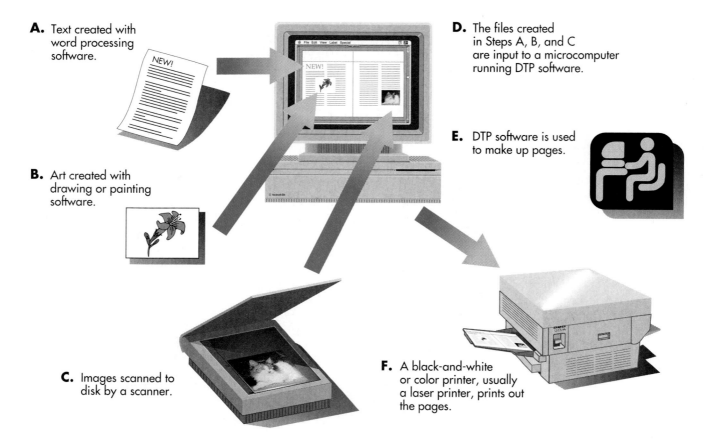

A. Text created with word processing software.

B. Art created with drawing or painting software.

C. Images scanned to disk by a scanner.

D. The files created in Steps A, B, and C are input to a microcomputer running DTP software.

E. DTP software is used to make up pages.

F. A black-and-white or color printer, usually a laser printer, prints out the pages.

PANEL 2.17

How DTP uses other files

Text is composed on a word processor, graphics are drawn with drawing and painting programs, and photographs and other artwork are scanned in with a scanner. Data from these files is integrated using desktop-publishing software, then printed out on a laser printer.

to the printer the shape and position of letters and graphics. An example of a page description language is Adobe's PostScript, which is used with Aldus PageMaker.

As time goes on, the distinction between word processing and desktop-publishing programs seems to be blurring. For instance, the word processing programs Word and WordPerfect can do many DTP functions, such as inserting graphics and wrapping text around them.

Personal Finance Programs

Personal finance software **lets you keep track of income and expenses, write checks, and plan financial goals.** Whether or not you learn how to use electronic spreadsheet programs, you'll probably find it useful to use personal finance software. Such programs don't promise to make you rich, but they can help you manage your money, maybe even get you out of trouble.

Certainly that was the case for Nick Ryder. An airline pilot from Marietta, Georgia, Ryder credited the best-selling financial software Quicken with saving his marriage by keeping his finances afloat. When Ryder and his wife, Penny, were married, after many years of being single, they found themselves "deep into the credit-card hole." Despite two healthy paychecks, they never seemed to have enough money.

In 1991 the Ryders acquired Quicken and began entering everything into the program's various account categories: checking, credit cards, utility bills, all incidentals over a dollar. After a few months of tracking expenses, some patterns began to emerge. "We were spending way too much on eating out," Ryder said. "Day to day, it doesn't look like much, but it adds up." The incidentals category also turned up a shocking number of impulse buys—magazines, snacks—that were out of line. With the knowledge acquired from Quicken, the Ryders began to cut back on expenses and even saved enough to set up investment accounts—managed by Quicken.[40]

Many personal finance programs include a calendar and a calculator, but the principal features are the following:

- **Tracking of income and expenses:** The programs allow you to set up various account categories for recording income and expenses, including credit card expenses.

- **Checkbook management:** All programs feature checkbook management, with an on-screen check writing form and check register that look like the ones in your checkbook. Checks can be purchased to use with your computer printer. Some programs even offer a nationwide electronic payment service that lets you pay your regular bills automatically, even depositing funds electronically into the accounts of the people owed.

- **Reporting:** All programs compare your actual expenses with your budgeted expenses. Some will compare this year's expenses to last year's.

- **Income tax:** All programs offer tax categories, for indicating types of income and expenses that are important when you're filing your tax return. Most personal finance programs also are able to interface with a tax-preparation program.

- **Other:** Some of the more versatile personal finance programs also offer financial-planning and portfolio-management features.

Quicken (there are versions for DOS, Windows, and Macintosh) seems to have generated a large following, but other personal finance programs exist as well. They include Kiplinger's CA-Simply Money, Managing Your Money, Microsoft Money, and WinCheck. Some offer enough features that you could use them to manage a small business.

In addition, there are tax software programs, which provide virtually all the forms you need for filing income taxes. Tax programs make complex calculations, check for mistakes, and even unearth deductions you didn't know existed. (Principal tax programs are Andrew Tobias' TaxCut, Kiplinger TaxCut, TurboTax/MacInTax, Personal Tax Edge, and CA-Simply Tax.) Finally, there are investment software packages, such as StreetSmart from Charles Schwab and Online Xpress from Fidelity, as well as various retirement planning programs.

Project Management Software

A desktop accessory or PIM can help you schedule your appointments and do some planning. That is, it can help you manage your own life. But what if you need to manage the lives of others to accomplish a full-blown project, such as steering a political campaign or handling a nationwide road tour for a band? Strictly defined, a *project* is a one-time operation consisting of several tasks that must be completed during a stated period of time. The project can be small, such as an advertising campaign for an in-house advertising department, or large, such as construction of an office tower or a jetliner.

Project management software **is a program used to plan, schedule, and control the people, costs, and resources required to complete a project on time.** For instance, the associate producer on a feature film might use such software to keep track of the locations, cast and crew, materials, dollars, and schedules needed to complete the picture on time and within budget. The software would show the scheduled beginning and ending dates for a particular task—such as shooting all scenes on a certain set—and then the date that task was actually completed. Examples of project management software are Harvard Project Manager, Microsoft Project for Windows, Project Scheduler 4, SuperProject, and Time Line.

Project management software has evolved into new forms. For example, a program called ManagePro for Windows is designed to manage not only goals and tasks but also the people charged with achieving them. "I use it to track projects, due dates, and the people who are responsible," says the manager of management information systems at a Lake Tahoe, Nevada, time-share condominium resort. "And then you can get your reports out either on project information, showing progress on all the steps, or a completely different view, showing all the steps that have to be taken by a given individual."[41] The software also offers built-in expert tips and strategies on human management for dealing with employees involved in the project.[42]

Computer-Aided Design

Computers have long been used in engineering design. *Computer-aided design (CAD) programs* **are software programs for the design of products and structures.** CAD programs, which are now available for microcomputers, help architects design buildings and work spaces and engineers design cars, planes, and electronic devices. One advantage of CAD software is that the product can be drawn in three dimensions and then rotated on the screen so the designer can see all sides. (■ *See Panel 2.18.*)

Examples of CAD programs for beginners are Autosketch, EasyCAD2 (Learn CAD Now), and TurboCAD. One CAD program, Parametric, allows engineers to do "what if" overhauls of designs, much as users of electronic spreadsheets can easily change financial data. This feature can dramatically cut design time. For instance, using Parametric, Motorola was able to design its Micro Tac personal cellular telephone in 9 months instead of the usual

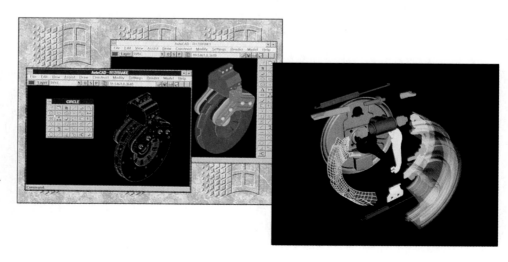

PANEL 2.18

CAD: example of computer-aided design

(Left) Screens from Autodesk CAD system. *(Right)* CAD screen of automobile brake assembly. A designer can draw in three dimensions and rotate the figure on the screen.

18.[43] Yet not all CAD programs are used by technical types; a version is available now, for example, that a relatively unskilled person can use to design an office. Other programs are available for designing homes. These programs include "libraries" of options such as cabinetry, furniture, fixtures, and, in the landscaping programs, trees, shrubs, and vegetables.[44]

A variant on CAD is **CADD, for *computer-aided design and drafting*, software that helps people do drafting.** CADD programs include symbols (points, circles, straight lines, and arcs) that help the user put together graphic elements, such as the floor plan of a house. Examples are Autodesk's AutoCAD and Intergraph's Microstation.

***CAD/CAM*—for *computer-aided design/computer-aided manufacturing*—software allows products designed with CAD to be input into an automated manufacturing system that makes the products.** For example, CAD, and its companion, CAM, brought a whirlwind of enhanced creativity and efficiency to the fashion industry. Some CAD systems, says one writer, "allow designers to electronically drape digital-generated mannequins in flowing gowns or tailored suits that don't exist, or twist imaginary threads into yarns, yarns into weaves, weaves into sweaters without once touching needle to garment."[45] The designs and specifications are then input into CAM systems that enable robot pattern-cutters to automatically cut thousands of patterns from fabric, with only minimal waste. Whereas previously the fashion industry worked about a year in advance of delivery, CAD/CAM has cut that time to 8 months—a competitive edge for a field that feeds on fads.

Drawing & Painting Programs

It may be no surprise to learn that commercial artists and fine artists have begun to abandon the paintbox and pen-and-ink for software versions of palettes, brushes, and pens. The surprise, however, is that an artist can use mouse and pen-like stylus to create computer-generated art as good as that achievable with conventional artist's tools. More surprising, even *nonartists* can be made to look good with these programs.

There are two types of computer art programs: drawing and painting.

- **Drawing programs:** **A *drawing program* is graphics software that allows users to design and illustrate objects and products.** CAD and drawing programs are similar. However, CAD programs provide precise dimensioning and positioning of the elements being drawn, so that they can be transferred later to CAM programs. Also, CAD programs lack the special effects for illustrations that come with drawing programs.[46] Some drawing programs are CorelDraw, Illustrator, Freehand, and Sketcher.

- **Painting programs:** Whereas drawing programs are generally gray-scale programs, painting programs add color. ***Painting programs* are graphics programs that allow users to simulate painting on screen.** A mouse or a tablet stylus is used to simulate a paintbrush. The program allows you to select "brush" sizes, as well as colors from a color palette.

  The difficulty with using painting programs is that a powerful computer system is needed because color images take up so much main memory and disk storage space. In addition, these programs require sophisticated color printers, of the sort found in specialized print shops called *service bureaus*.

R E A D M E

Case Study: A CAD Program for Designing Your Office

What time is it when you need to move people around or create a new floor plan in a small or medium-size business? Amateur hour. Office Layout, by Autodesk of Sausalito, California, is a great alternative to cardboard cutouts and scissors. This software program has many of the features of CAD (computer-aided design) products favored by architects and engineers, yet just about anyone can use it successfully. Start by measuring your office space, and enter the dimensions into the program to create a diagram of the exterior walls. Next, add panels to divide the space into cubicles or work areas. Now you're ready to furnish your new spaces using nearly 200 symbols representing desks, tables, chairs, computers, even plants. Each symbol can be labeled for a particular occupant, department, manufacturer, price, and size, as well as up to five other attributes you choose. When finished, you can give a diskette or print to your architect or builder to modify into working drawings, or create lists of furniture and equipment to be purchased. Office Layout runs on an IBM-compatible PC with a mouse. . . .

—Alison L. Sprout, "Here's How to Design Your Own Office," *Fortune*

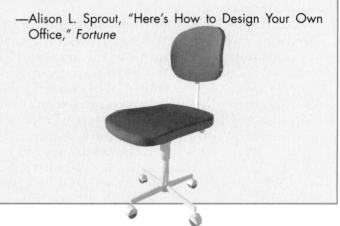

Hypertext

Hypertext **is software that allows users to have fast and flexible access to information in large documents, constructing associations among data items as needed.** Database managers have made the retrieval and organization of facts infinitely easier than was possible in the old index-card and file-cabinet days. Hypertext goes beyond the restrictive search-and-retrieval methods of traditional database systems and encourages people to follow their natural train of thought as they discover information. That is, hypertext works the way people think, allowing them to link facts into sequences of information in ways that resemble those that people use to obtain new knowledge. The term *hypertext* was coined by computer visionary Ted Nelson, who is still working on an advanced version that he calls Xanadu, named after Kublai Khan's legendary pleasure dome.

Hypertext is often used in Help systems, a staple of many programs. Help systems allow users to query the software for help or additional information on how a program or system works. Another well-known example of the use of hypertext is that found in HyperCard, introduced for the Apple Macintosh in 1987. Since then other companies have introduced similar products, such as SuperCard from Silicon Beach Software. HyperCard is based on the concept of *cards* and *stacks* of cards—just like notecards, only they are electronic. A card is a screenful of data that makes up a single record; cards are organized into related files called stacks. On each card there may be one or more *buttons*, which, when clicked on by a mouse, can pull up another card or stack. By clicking buttons, you can make your way through the cards and stacks to find information or discover connections between ideas. (■ *See Panel 2.19.)*

Recently, hypertext has come into its own as a means of accessing Web sites. The hidden codes of hypertext allow users to use a mouse to click on a highlighted word or phrase to automatically access a related site. As we

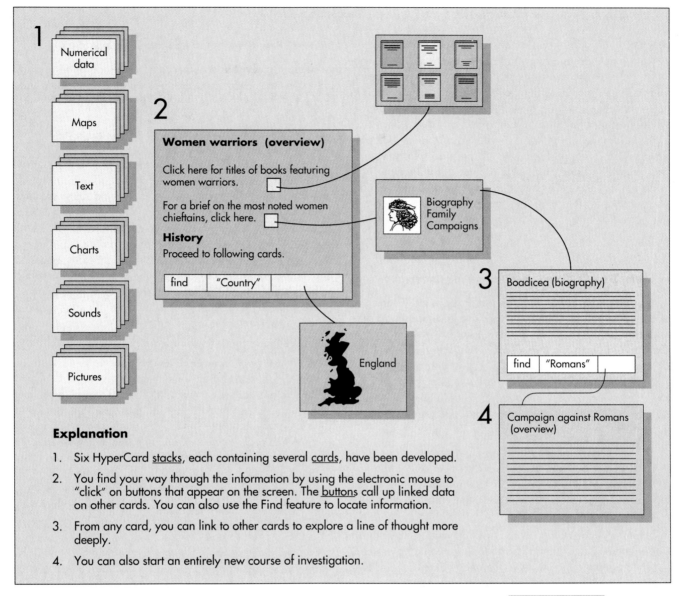

Explanation

1. Six HyperCard <u>stacks</u>, each containing several <u>cards</u>, have been developed.

2. You find your way through the information by using the electronic mouse to "click" on buttons that appear on the screen. The <u>buttons</u> call up linked data on other cards. You can also use the Find feature to locate information.

3. From any card, you can link to other cards to explore a line of thought more deeply.

4. You can also start an entirely new course of investigation.

Hypertext: how Hyper-Card works

The subject here is women warriors.

will discuss in detail elsewhere (Chapter 3 Experience Box and Chapter 8), *hypertext markup language (HTML)* is the language in which Web pages are written and linked; *hypertext transfer protocol (HTTP)* is how information is sent over the Web.

Onward

In this chapter we have described what the more elementary forms of software are. Still to be discussed, however, are some of the truly exciting software developments of the Digital Age. Later in the book, we describe software technology such as virtual reality, multimedia, simulation, expert systems, and information-seeking "agents," as well as software affecting the Internet.

Getting Started With Computers in College & Going Online

You may still hear the sounds of late-night typing in a college residence hall. However, it's certainly not the smart way to work anymore. Indeed, coping with a typewriter actually detracts from learning. You're worrying about making mistakes and avoiding retyping a whole paper rather than concentrating on educational principles. Using a computer to write your papers not only makes life easier, it also opens up new areas of freedom and knowledge and helps prepare you for the future.

Thus, students who come to college with a personal computer as part of their luggage are certainly ahead of the game. If you don't have one, however, there are other options.

If You Don't Own a Personal Computer

If you don't have a PC, you can probably borrow someone else's sometimes. However, if you have a paper due the next day, you may have to defer to the owner, who may also have a deadline. When borrowing, then, you need to plan ahead and allow yourself plenty of time.

Virtually every campus now makes computers available to students, either at minimal cost or essentially for free as part of the regular student fees. This availability may take two forms:

- *Library or computer labs:* Even students who have their own PCs may sometimes want to use the computers available at the library or campus computer lab. These may have special software or better printers that they don't have themselves.

- *Dormitory computer centers or dorm-room terminals:* Some campuses provide dormitory-based computer centers (for example, in the basement). Even if you have your own PC, it's nice to know about these for backup purposes.

  More and more campuses are also providing computers or terminals within students' dormitory rooms. These are usually connected by a campuswide local area network (LAN) to lab computers and administrative systems. Often, however, they also allow students to communicate over phone lines to people in other states.

Of course, if the system cannot accommodate a large number of students, all the computers may be in high demand come term-paper time. Clearly, owning a computer offers you convenience and a competitive advantage. Not having one, however, does not mean you are destined to fail in school.

If You Do Own a Personal Computer

Perhaps someone gave you a personal computer, or you acquired one, before you came to college. It will probably be one of three types: (1) an IBM or IBM-compatible, such as a Compaq, Packard Bell, AST, Radio Shack, Zenith, or Dell; (2) an Apple Macintosh or compatible; (3) other, such as an Apple II or Commodore.

If all you need to do is write term papers, nearly any microcomputer, new or used, will do. Indeed, you may not even need to have a printer, if you can find other ways to print out things. The University of Michigan, for instance, offers "express stations" or "drive-up windows." These allow students to use a diskette (floppy disk) or connect a computer to a student-use printer to print out their papers.[47] Or, if a friend has a compatible computer, you can ask to borrow it and the printer for a short time to print your work.

You should, however, take a look around you to see if your present system is appropriate for your campus and your major.

- *The fit with your campus:* Some campuses are known as "IBM" (or IBM-compatible) schools, others as "Mac" (Macintosh) schools. Apple IIs and Commodores, still found in elementary and high schools, are not used much at the college level.

  Why should choice of machine matter? The answer is that diskettes generally can't be read interchangeably among the two main types of microcomputers. Thus, if you own the system that is out of step for your campus, you may find it difficult to swap files or programs with others. Nor will you be able to borrow their equipment to finish a paper if yours breaks down. (There are some conversion programs, but these take time and may not be readily available.)

  Most campuses favor either Macintoshes or IBMs (and IBM-compatibles). You should call the dean of students or otherwise ask around to find which system is most popular.

- *The fit with your major:* Speech communications, foreign language, physical education, political science, biology, and English majors probably don't need a fancy computer system (or even any system at all). Business, engineering, architecture, and journalism majors may have special requirements. For instance, an architecture major doing computer-aided design (CAD) projects or a journalism major doing desktop publishing will need reasonably powerful systems. A history or nursing major, who will mainly be writing papers, will not.

Of course you may be presently undeclared or undecided about your major. Even so, it's a good idea to find out what kinds of equipment and programs are being used in the majors you are contemplating.

How to Buy a Personal Computer

Buying a personal computer, like buying a car, often requires making a tradeoff between power and expense.

Power Many computer experts try to look for a personal computer system with as much power as possible. The word *power* has different meanings when describing software and hardware:

- *Powerful software:* Applied to software, "powerful" means that the program is *flexible.* That is, it can do many different things. For example, a word processing program that can print in different typestyles (fonts) is more powerful than one that prints in only one style.

- *Powerful hardware:* Applied to hardware, "powerful" means that the equipment (1) is *fast* and (2) has *great capacity.*

  A fast computer will process data more quickly than a slow one. With an older computer, for example, it may take several seconds to save, or store on a disk, a 50-page term paper. On a newer machine, it might take less than a second.

  A computer with great capacity can run complex software and process voluminous files. *This is an especially important matter if you want to be able to run the latest releases of software.*

Will computer use make up an essential part of your major, as it might if you are going into engineering, business, or graphic arts? If so, you may want to try to acquire powerful hardware and software. People who really want (and can afford) their own desktop publishing system might buy a new Macintosh Quadra with LaserWriter printer, scanner, and PageMaker software. This might well cost $7000. Most students, of course, cannot afford anything close to this.

Expense If your major does not require a special computer system, a microcomputer can be acquired for relatively little. You can probably buy a used computer, with software thrown in, for under $500 and a printer for under $200.

What's the *minimum* you should get? Probably an IBM-compatible or Macintosh system, with 4 megabytes of memory (✓ p. 12) and two diskette drives or one diskette and one hard-disk drive. However, up to 16 megabytes of memory is preferable if you're going to run many of today's programs. Dot-matrix printers are still in use on many campuses (24-pin printers are preferable to 9-pin). To be sure, the more expensive laser and inkjet printers produce a better image. However, you can always use the dot-matrix for drafts and print out the final version on a campus student-use printer.

Where to Buy New Fierce price wars among microcomputer manufacturers and retailers have made hardware more affordable. One reason IBM-compatibles have become so widespread is that non-IBM manufacturers early on were able to copy, or "clone," IBM machines and offer them at cut-rate prices. For a long time, Apple Macintoshes were considerably more expensive. In part this was because other manufacturers were unable to offer inexpensive clones. In recent times, however, Apple has felt the pinch of competition and has dropped its prices. It also has licensed parts of its technology to others so that we are now seeing Macintosh "clones."

When buying hardware, look to see if software, such as word processing or spreadsheet programs, comes "bundled" with it. In this case, *bundled* means that software is included in the selling price of the hardware. This arrangement can be a real advantage, saving you several hundred dollars.

Because computers are somewhat fragile, it's not unusual for them to break down, some even when newly purchased. Indeed, nearly 25% of 45,000 PC users surveyed by one computer magazine reported some kind of problem with new computers.[48] (The failure rates were: hard drive—21%, motherboard—20%, monitor—12%, diskette drive—11%, and power supply—10%.) The PCs (Apple was not included) that had the fewest problems included those from AST Research, Compaq, Epson, Hewlett-Packard, IBM, and NCR (AT&T). Most troublesome were low-cost PCs.

There are several sources for inexpensive new computers, such as student-discount sources, computer superstores, and mail-order houses. (■ *See Panel 2.20.*)

Where to Buy Used Buying a used computer can save you a minimum of 50%, depending on its age. If you don't need the latest software, this can often be the way to go. The most important thing is to buy *recognizable* brand names, examples being Apple and IBM or well-known IBM-compatibles: Compaq, Hewlett-Packard, NCR, Packard Bell, Tandy, Toshiba, Zenith. Obscure or discontinued brands may not be repairable.

Among the sources for used computers are the following:

- *Student discount sources:* With a college ID card, you're probably entitled to a student discount (usually 10 to 20%) through the campus bookstore or college computer resellers. In addition, during the first few weeks of the term, many campuses offer special sales on computer equipment. Campus resellers also provide on-campus service and support and can help students meet the prevailing campus standards while satisfying their personal needs.
- *Computer superstores:* These are big chains such as Computer City, CompUSA, and Microage. Computers are also sold at department stores, warehouse stores (such as Costco and Price Club), Wal-Mart, Circuit City, Radio Shack, and similar outlets.
- *Mail-order houses:* Companies like Dell Computer and Gateway 2000 found they could sell computers inexpensively while offering customer support over the phone. Their success inspired IBM, Compaq, and others to plunge into the mail-order business.

    The price advantage of mail-order companies has eroded with the rise of computer superstores. Moreover, the lack of local repair and service support can be a major disadvantage. Still, if you're interested in this route, look for a copy of the phone-book-size magazine *Computer Shopper,* which carries ads from most mail-order vendors.

- *Retail sources:* A look in the telephone-book Yellow Pages under "Computers, Used" will produce several leads. Authorized dealers (of IBM, Apple, Compaq, and so on) may shave prices on demonstration (demo) or training equipment.
- *Used-computer brokers:* There are a number of used-computer brokers, such as American Computer Exchange, Boston Computer Exchange, Damark, and National Computer Exchange.
- *Individuals:* Classified ads in local newspapers, shopper throwaways, and (in some localities) free computer newspapers/magazines provide listings of used computer equipment. Similar listings may also appear on electronic bulletin board systems (BBSs).

    One problem with buying from individuals is that they may not feel obligated to take the equipment back if something goes wrong. Thus, you should inspect the equipment carefully. (■ *See Panel 2.21.)* For a small fee, a computer-repair shop can check out the hardware for damage before you buy it.

How much should you pay for a used computer? This can be tricky. Some sellers may not be aware of the rapid depreciation of their equipment and price it too high. The best bet is to look through back issues of the classified ads for a couple of newspapers in your area until you have a sense of what equipment may be worth.

Checklist Here are some decisions you should make before buying a computer:

- *What software will I need?* Although it may sound backward, you should select the software before the hardware. This is because you want to choose software that will perform the kind of work you want to do. First find

the kind of programs you want—word processing, spreadsheets, communications, graphics, or whatever. Check out the memory and other hardware requirements for those programs. Then make sure you get a system to fit them.

The advice to start with software before hardware has always been standard for computer buyers. However, it is becoming increasingly important as programs with extensive graphics come on the market. Graphics tend to require a lot of memory, hard-disk storage, and screen display area.

- *Do I want a desktop or a portable?* Look for a computer that fits your work style. For instance, you may want a portable if you spend a lot of time at the library. Some students even use portables to take notes in class. If you do most of your work in your room, you may find it more comfortable to have a desktop PC. Though not portable, the monitors of desktop computers are usually easier to read.

    It's possible to have both portability and a readable display screen. Buy a laptop, but also buy a monitor that you can plug the portable into. Computers are also available with "docking" systems that permit a portable to fit inside a desktop computer or monitor.

    Also keep in mind that portable computers are more expensive to maintain than desktop computers, and portable keyboards are smaller. We consider portables in more detail in the Experience Box at the end of Chapter 4.

- *Is upgradability important?* The newest software being released is so powerful (meaning flexible) that it requires increasingly more powerful hardware. That is, the software requires hardware that is faster and has greater main memory and storage capacity. If you buy

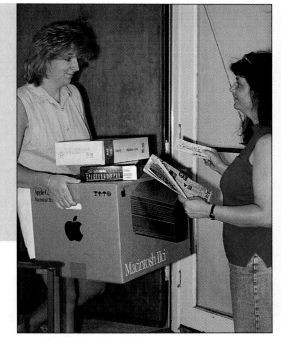

"Buying from an individual means you have little recourse if something goes wrong. The following tips should help you to buy carefully:

- If possible, take someone who knows computers with you
- Turn the computer on and off a few times to make sure there are no problems on startup.
- Use the computer and, if possible, try the software you want to use. Listen for strange sounds in the hard drive or the floppies.
- Turn the computer off and look for screen burn-in, a ghost image on the screen after the machine has been turned off. It can be a sign of misuse.
- Ask about the warranty. Some companies, including Apple and IBM, permit warranties to be transferred to new owners (effective from the date of the original purchase). A new owner can usually have the warranty extended by paying a fee."

—Richard Williams, "On the Hunt for a Used Computer," *Globe & Mail* (Toronto)

PANEL 2.21

Tips for buying used computers

an outdated used computer, you probably will not be able to *upgrade* it. That is, you will be unable to buy internal parts, such as additional memory, that can run newer software. This limitation may be fine if you expect to be able to afford an all-new system in a couple of years. If, however, you are buying new equipment right now, be sure to ask the salesperson how the hardware can be upgraded.

- *Do I want an IBM-style or a Macintosh?*   Although the situation is changing, until recently the division between IBM and IBM-compatibles on the one hand and Apple Macintoshes on the other was fundamental. Neither could run the other's software or exchange files of data without special equipment and software. We mentioned that some campuses and some academic majors tend to favor one type of microcomputer over the other. Outside of college, however, the business world tends to be dominated by IBM and IBM-compatible machines. In a handful of areas—graphic arts and desktop publishing, for example—Macintoshes are preferred.

  If you think there's a chance you will need to become familiar with both IBMs and Macs, you might consider buying a high-end Macintosh PowerMac. This microcomputer can handle disks and programs for both systems.

Getting Started Online

Computer networks have transformed life on campuses around the country, becoming a cultural and social force af-

fecting everybody, not just nerds and wonks. How do you join this vast world of online information and interaction?

Hardware & Software Needed   Besides a microcomputer with a hard disk (any made in the last 5 years will do), you need a modem, to send messages from one computer to another via a phone line. The speed at which a modem can transmit data is generally measured in bps (bits per second). A slow modem, 2400 bps, can be bought for less than $50. For less than $100 you can get a faster modem—9600, 14,400, or 28,800 bps—which allows data to flow faster and reduces your phone charges. About 40% of personal computers these days have a modem. If yours doesn't, you can have a store install an internal modem as an electronic circuit board on the inside of the computer, or buy an external modem, which appears as a box outside the computer.

To go online, you'll need communications software, which may come bundled with any computer you buy or is sold on floppy disks in computer stores. (Popular brands are Microphone and White Knight for Macintosh computers. IBM-style computers running Windows and DOS may use ProComm Plus, HyperAccess, QModem Pro, or Crosstalk.) However, many modems come with communications software when you buy them. Or, if you sign up for an online service, it will supply the communications program you need to use its network.

Your modem will connect to a standard telephone wall jack. (When your computer is in use, it prevents you from using the phone line, and callers trying to reach you will hear a busy signal.)

Getting Connected: Starting with an Online Service Unless you already have access to a campus network, probably the easiest first step for using this equipment is to sign up with a commercial online service, such as one of the Big Three: America Online, CompuServe, or Prodigy. (Another is the Microsoft Network.)

You'll need a credit card in order to join, since online services charge a monthly fee, typically $10 to $20. Most also charge for the time spent online—from the time you dial and connect to the time you disconnect. Charges are billed to your credit card. Note: It's easy to get carried away and run up charges by staying online too long. However, you can keep your costs down by going online only during off-hours (evenings and weekends). You can also not do your reading (of articles, say) online but download (save to your hard disk) material unread, then read it later.

All online services have introductory offers that allow you a free trial period. You can get instructions and free start-up communications disk by phoning their toll-free 800 numbers. (■ See Panel 2.22.) You'll also find promotional offers at computer stores or promotional diskettes shrink-wrapped inside computer magazines on newsstands.

PANEL 2.22

The Big Three online services

Listed here are the leading online services, number of users in 1996, costs, phone numbers, and other details. Rates are subject to change. All have special introductory offers.

AMERICA Online	Number of users:	5 million
	Cost:	$9.95/month for 5 hours for all services, then $2.95/hour.
	How to connect:	800-827-6364; startup software, with 10 free hours of online time, available in computer stores or from America Online. 14,400 or 28,800 bps modem.
	Features:	Friendly, mouse-driven graphical interface; low-cost access makes this service ideal for families and hobbyists. America Online encourages interaction among members, leading one writer to call it "an online service with a personality." Other features include SeniorNet, Small Business Center, and locally oriented versions (such as Chicago Online). Offers Internet access. Unique features: animation.
CompuServe	Number of users:	4.2 million
	Cost:	$9.95 for 5 hours, then $2.95/hour. 20–hour plan: $19.95, then $2.00/hour. Premium services cost extra.
	How to connect:	800-848-8199; membership kits with user's guide and software available through computer stores, mail-order outlets, or CompuServe. 28,800 bps modem.
	Features:	CompuServe offers huge libraries of publicly available software and 350 forums. Its graphics-based software makes navigating easy for novices. However, array of services and databases makes it also the choice of more advanced computer users and professionals, who can gain access with any general communications software. Offers Internet access. Unique features: White pages, WorldsAway.
PRODIGY Service	Number of users:	1.4 million
	Cost:	$9.95 for 5 hours, then $2.95/hour. 30–hour plan: $29.95, then $2.95/hour. Plus Service costs extra.
	How to connect:	Phone: 800-776-3449. Startup software is also available in computer stores. 14,400 or 28,800 bps modem.
	Features:	The online service claiming the most users. Also the simplest, suitable for people who want to get the most out of an online service with the least effort. Offers friendly graphics for beginners, although you'll have to put up with scrolling advertisements. Strong features: bulletin boards on special topics; coverage of popular culture (movie reviews, TV schedules, entertainment news). Offers Internet access. Unique features: Web-based content.

SUMMARY

What It Is / What It Does

analytical graphics *(p. 70, LO 5)* Also called *business graphics*; graphical forms representing numeric data. The principal examples are bar charts, line graphs, and pie charts. Analytical graphics programs are a type of applications software.

applications software *(p. 51, LO 1)* Software that can perform useful work on general-purpose tasks—for example, word processing or spreadsheet software.

block command *(p. 66, LO 5)* Software command that allows the user to indicate beginning and end portions of text to be moved or otherwise manipulated.

bundled *(p. 77, LO 5)* Describes a system whose components are sold together for a single price. (*Unbundled* means that a system has separate prices for each component.)

button *(p. 62, LO 4)* Simulated on-screen button (kind of icon) that is activated ("pushed") by a mouse or other pointing device to issue a command.

cell *(p. 69, LO 5)* In an electronic spreadsheet, the rectangle where rows and columns intersect.

cell address *(p. 69, LO 5)* In an electronic spreadsheet, the position of a cell—for example, "A1," where column A and row 1 intersect.

clip art *(p. 84, LO 5)* "Canned" images that come on disks, which users can copy and place as desired to illustrate documents.

computer-aided design (CAD) *(p. 86, LO 6)* Applications software programs for designing products and structures.

computer-aided design and drafting (CADD) *(p. 86, LO 6)* Applications software that helps people do drafting.

computer-aided design/computer-aided manufacturing (CAD/CAM) *(p. 87, LO 6)* Applications software that allows products designed with CAD to be input into a computer-based manufacturing system (CAM) that makes the products.

copyright *(p. 53, LO 2)* Body of law that prohibits copying of intellectual property without the permission of the copyright holder.

Why It's Important

Numeric data is easier to analyze in graphical form than in the form of rows and columns of numbers, as in electronic spreadsheets.

Applications software such as word processing, spreadsheet, database manager, graphics, and communications packages are used to increase people's productivity.

The block command makes it easy to rearrange and reformat documents.

Bundled software and hardware cost less than the same items sold separately.

Buttons make it easier for users to enter commands.

The cell is the smallest working unit in a spreadsheet. Data and formulas are entered into the cells.

Cell addresses provide location references for spreadsheet users.

Clip art enables users to illustrate their word-processed or desktop-published documents without having to hire an artist.

CAD programs help architects design buildings and work spaces and engineers design cars, planes, and electronic devices. With CAD software, a product can be drawn in three dimensions and then rotated on the screen so the designer can see all sides.

CADD programs include symbols (points, circles, straight lines, and arcs) that help the user put together graphic elements, such as the floor plan of a house.

CAD/CAM systems have greatly enhanced creativity and efficiency in many industries.

Copyright law aims to prevent people from taking credit for and profiting unfairly from other people's work.

What It Is / What It Does

cursor *(p. 66, LO 4)* Also called a *pointer;* the movable symbol on the display screen that shows the user where data may be entered next. The cursor is moved around with the keyboard's directional arrow keys or an electronic mouse.

custom-written software *(p. 51, LO 1)* Applications software designed for a particular customer and written for a highly specialized task by a computer programmer.

database *(p. 71, LO 5)* Collection of interrelated files in a computer system that is created and managed by database manager software. These files are organized so that those parts with a common element can be retrieved easily.

database software *(p. 71, LO 5)* Applications software for maintaining a database. It controls the structure of a database and access to the data.

data communications software *(p. 75, LO 5)* Applications software that manages the transmission of data between computers.

default settings *(p. 65, LO 5)* Settings automatically used by a program unless the user specifies otherwise, thereby overriding them.

desktop accessory *(p. 76, LO 5)* Also called *desktop organizer;* software package that provides electronic counterparts of tools or objects commonly found on a desktop: calendar, clock, card file, calculator, and notepad.

desktop publishing (DTP) *(p. 82, LO 6)* Applications software that, along with a microcomputer, mouse, scanner, and laser printer (usually), is used to mix text and graphics, including photos, to produce high-quality printed output. Some word processing programs also have many DTP features. Text is usually composed first on a word processor, artwork is created with drawing and painting software, and photographs are scanned in using a scanner. Prefabricated art and photos may also be obtained from disks (CD-ROM and/or floppy) containing clip art.

dialog box *(p. 62, LO 4)* With graphical user interface (GUI) software, a box that appears on the screen and displays a message requiring a response from you—for example, Y for "Yes" or N for "No."

documentation *(p. 63, LO 4)* User's manual or reference manual that is a narrative and graphical description of a program. Documentation may be instructional, but usually features and functions are grouped by category.

downward compatible *(p. 52, LO 1)* Means that applications developed for a new version of a software product can be run on older versions; also called *backward compatible.*

drawing programs *(p. 87, LO 6)* Applications software that allows users to design and illustrate objects and products.

Why It's Important

All applications software packages use cursors to show users where their current work location is on the screen.

When packaged applications software can't do the tasks a user requires, then software must be custom-written by a professional computer programmer.

Online database services provide users with enormous research resources. Businesses and organizations use databases to keep track of transactions and increase people's efficiency.

Database manager software allows users to organize and manage huge amounts of data.

Communications software is required to transmit data via modems in a communications system.

Users need to know how to change default settings in order to customize their documents.

Desktop accessories help users to streamline their daily activities.

Desktop publishing has reduced the number of steps, the time, and the money required to produce professional-looking printed projects.

Dialog boxes are only one aspect of GUIs that make software easier for people to use.

Documentation helps users learn software commands and use of function keys, solve problems, and find information about system specifications.

Downward compatible software is less common than upward compatible software. It is useful because it allows users, for example, to run new applications software on an older operating system.

Drawing programs and CAD are similar. However, drawing programs provide special effects that CAD programs do not.

What It Is / What It Does	**Why It's Important**

dynamic linking *(p. 70, LO 5)* Feature of electronic spreadsheet software that allows data in one spreadsheet to be linked to and automatically update data in another spreadsheet.

The linking feature increases the amount of spreadsheet data that can be manipulated at the same time.

electronic mail (e-mail) software *(p. 75, LO 5)* Software that enables computer users to send letters and documents from one computer to another.

E-mail allows businesses and organizations to quickly and easily send messages to employees and outside people without resorting to paper messages.

electronic spreadsheet *(p. 67, LO 5)* Also called *spreadsheet;* applications software that simulates a paper worksheet and allows users to create tables and financial schedules by entering data and/or formulas into rows and columns displayed as a grid on a screen. If data is changed in one cell, values in other cells specified in the spreadsheet will automatically recalculate.

The electronic spreadsheet became such a popular small-business applications program that it has been held directly responsible for making the microcomputer a widely used business tool.

font *(p. 83, LO 6)* Set of type characters in a particular typestyle and size.

Desktop publishing programs, along with laser printers, have enabled users to dress up their printed projects with many different fonts.

formula *(p. 70, LO 5)* In an electronic spreadsheet, instructions for calculations that are entered into designated cells. For example, a formula might be SUM CELLS(A5:A15), meaning "Sum (add) all the numbers in the cells with cell addresses A5 through A15."

The use of formulas enables spreadsheet users to change data in one cell and have all the cells linked to it by formulas automatically recalculate their values.

freeware *(p. 55, LO 2)* Software that is available free of charge.

Freeware is usually distributed through the Internet. Users can make copies for their own use but are not free to make unlimited copies.

function keys *(p. 60, LO 4)* Computer keyboard keys that are labeled F1, F2, and so on; usually positioned along the top or left side of the keyboard.

Function keys are used to issue commands. These keys are used differently, depending on the software.

grammar checker *(p. 66, LO 5)* Software feature that flags poor grammar, wordiness, incomplete sentences, and awkward phrases.

A grammar checker allows users to improve their prose for both style and accuracy.

graphical-user interface (GUI) *(p. 61, LO 4)* User interface that uses images to represent options. Some of these images take the form of icons, small pictorial figures that represent tasks, functions, or programs.

GUIs are easier to use than command-driven interfaces and menu-driven interfaces; they permit liberal use of the electronic mouse as a pointing device to move the cursor to a particular icon or place on the display screen. The function represented by the icon can be activated by pressing ("clicking") buttons on the mouse.

groupware *(p. 78, LO 2)* Applications software that is used on a network and serves a group of users working together on the same project.

Groupware improves productivity by keeping users continually notified about what colleagues are thinking and doing, and vice versa.

Help menu *(p. 61, LO 4)* Also called *Help screen;* offers on-screen instructions for using software. Help screens are accessed via a function key or by using the mouse to select Help from a menu.

Help screens provide a built-in electronic instruction manual.

hypertext *(p. 88, LO 6)* Applications software that allows users to link information in large documents, constructing associations among data items as needed.

Hypertext goes beyond the restrictive search-and-retrieval methods of traditional database systems and encourages people to follow their natural train of thought as they discover information. Hypertext is used to link Web pages.

icon *(p. 61, LO 4)* In a GUI, small pictorial figure that represents a task, function, or program.

The function represented by the icon can be activated by pointing at it with the mouse pointer and pressing ("clicking") on the mouse. The use of icons has simplified the use of computers.

integrated software *(p. 77, LO 5)* Applications software that combines several applications programs into one package—usually electronic spreadsheets, word processing, database management, graphics, and communications.

Integrated software packages offer greater flexibility than separate single-purpose programs.

macro *(p. 60, LO 4)* Software feature that allows a single keystroke or command to be used to automatically issue a predetermined series of keystrokes or commands.

Macros increase productivity by consolidating several command keystrokes into one or two.

menu *(p. 61, LO 4)* List of available commands displayed on the screen.

Menus are used in graphical-user interface programs to make software easier for people to use.

move command *(p. 66, LO 5)* Software command that allows users to move any highlighted, or blocked, items in a document to any other designated location.

See block command.

network piracy *(p. 53, LO 2)* The use of electronic networks for unauthorized distribution of copyrighted materials in digitized form.

If piracy is not controlled, people may not want to let their intellectual property and copyrighted material be dealt with in digital form.

packaged software *(p. 51, LO 1)* Also called *software package;* an "off-the-shelf" program available on disk for sale to the general public.

Most software used by general microcomputer users comes in software packages available at local stores.

page description language *(p. 84, LO 6)* Software used in desktop publishing that describes the shape and position of characters and graphics to the printer.

Page description languages, used along with laser printers, gave birth to desktop publishing. They allow users to combine different types of graphics with text in different fonts, all on the same page.

painting programs *(p. 87, LO 6)* Applications programs that simulate painting using a mouse or tablet stylus like a paintbrush and that use colors. A powerful computer system is required to use these programs.

Painting programs can render sophisticated illustrations.

personal finance software *(p. 85, LO 6)* Applications software that helps users track income and expenses, write checks, and plan financial goals.

Personal finance software can help people manage their money more effectively.

personal information manager (PIM) *(p. 76, LO 5)* Applications software that combines a word processor, database, and desktop accessory program to organize a variety of information.

PIMs offer an electronic version of an appointment calendar, to-do list, address book, notepad, and similar daily office tools, all in one place.

plagiarism *(p. 54, LO 2)* Expropriation of another writer's text, findings, or interpretations and presenting them as one's own.

Information technology offers plagiarists new opportunities to go far afield for unauthorized copying, yet it also offers new ways to catch these people.

presentation graphics *(p. 74, LO 5)* Graphical forms used to communicate or make a presentation of data to others, such as clients or supervisors. Presentation graphics programs are a type of applications software.

Presentation graphics programs may make use of analytical graphics but look much more sophisticated, using texturing patterns, complex color, and dimensionality.

project management software *(p. 86, LO 6)* Applications software used to plan, schedule, and control the people, costs, and resources required to complete a project on time.

Project management software increases the ease and speed of planning and managing complex projects.

proprietary software *(p. 56, LO 2)* Software whose rights are owned by an individual or business.

Ownership of proprietary software is protected by copyright. This type of software must be purchased to be used. Copying is restricted.

public domain software *(p. 55, LO 2)* Software that is not protected by copyright and thus may be duplicated by anyone at will.

Public domain software offers lots of software options to users who may not be able to afford a lot of commercial software. Users may make as many copies as they wish.

What It Is / What It Does	**Why It's Important**

release *(p. 52, LO 1)* Refers to a minor upgrade in a software product.

Releases are usually indicated by a change in the number after a version's decimal point—such as 3.1, 3.2. The higher the number, the more recent the release.

replace command *(p. 66, LO 5)* Software command that allows users to automatically replace with a new item any existing item identified using the search command.

All occurrences of an item in a document can be replaced automatically, using just a single command.

scrolling *(p. 66, LO 4)* The activity of moving quickly upward or downward through text or other screen display, using directional arrow keys or mouse.

Normally a computer screen displays only 20–22 lines of text. Scrolling enables users to view an entire document, no matter how long.

search command *(p. 66, LO 5)* Software command that allows users to find any item known to exist in the document.

The search command saves users from having to read an entire document to find a particular item they want to check or change.

shareware *(p. 55, LO 2)* Copyrighted software that is distributed free of charge, usually over the Internet, but that requires users to make a contribution in order to receive technical help, documentation, or upgrades.

Along with public domain software and freeware, shareware offers yet another inexpensive way to obtain new software.

shrink-wrap license *(p. 56, LO 2)* Printed licenses inserted into software packages and visible through the clear, plastic wrap.

The use of shrink-wrap licenses eliminates the need for a written signature, since buyers know they are entering a binding contract by opening the package.

site license *(p. 56, LO 2)* License that permits a customer to make multiple copies of a software product for use only within a given facility.

Site licenses eliminate the need to buy many copies of one software package for use in, for example, an office or a computer lab.

software *(p. 51, LO 1)* Also called *programs;* step-by-step instructions that tell the computer how to perform a task. Software instructions are written by programmers. In most instances, the words *software* and *program* are interchangeable. Software is of two types: applications software and systems software.

Without software, hardware would be useless.

software license *(p. 56, LO 2)* Contract by which users agree not to make copies of proprietary software to give away or to sell.

Software manufacturers don't sell people software so much as sell them licenses to become authorized users of the software.

software piracy *(p. 53, LO 2)* Unauthorized copying of copyrighted software—for example, copying a program from one floppy disk to another or downloading a program from a network and making a copy of it.

Software piracy represents a serious loss of income to software manufacturers and is a contributor to high prices in new programs.

software suite *(p. 77, LO 5)* Several applications software packages—like spreadsheets, word processing, graphics, communications, and groupware— bundled together and sold for a fraction of what the programs would cost if bought individually.

Software suites can save users a lot of money.

special-purpose keys *(p. 60, LO 4)* Computer keyboard keys used to enter and edit data and execute commands— for example, Esc, Alt, and Ctrl.

All computer keyboards have special-purpose keys. The user's software program determines how these keys are used.

spelling checker *(p. 66, LO 5)* Word processing software feature that tests for incorrectly spelled words.

Although spelling checkers cannot flag words that are correctly spelled but incorrectly used (like "for" instead of "four"), they are helpful in assisting users to proofread documents before printing them out.

systems software *(p. 51, LO 1)* Software that controls the computer and enables it to run applications software. Systems software, which includes the operating system, allows the computer to manage its internal resources.

Applications software cannot run without systems software.

What It Is / What It Does

thesaurus *(p. 66, LO 5)* Word processing software feature that provides a list of similar words and alternative words for any word specified in a document.

tutorial *(p. 63, LO 4)* Instruction book or program that takes users through a prescribed series of steps to help them learn the product.

unbundled *(p. 77, LO 5)* Describes a system whose components are sold separately.

upward compatible *(p. 52, LO 1)* Means that applications and documents created with earlier versions of particular software will run on later versions; also called *forward compatible.*

user interface *(p. 61, LO 4)* Part of a software program that presents on the screen the alternative commands by which you communicate with the system and that displays information.

version *(p. 52, LO 1)* Refers to a major upgrade in a software product.

Web browser *(p. 79, LO 5)* Software that enables people to view Web sites on their computers.

Web directories *(p. 81, LO 5)* Software search tools that allow Web users to search for items classified by topic.

Web index *(p. 81, LO 5)* Software search tools that allow Web users to search for items through keyword searches.

window *(p. 61, LO 4)* Feature of graphical user interfaces; rectangle that appears on the screen and displays information from a particular part of a program.

word processing software *(p. 64, LO 5)* Applications software that enables users to create, edit, revise, store, and print text material.

WYSIWYG *(p. 83, LO 6)* Abbreviation for "What You See Is What You Get"; this desktop publishing term means that the text and graphics appear on the display screen exactly as they will print out. (Software used by many professional typesetters shows display screens full of formatting codes, which means that what you see on the screen is not what you will see when the job is printed out.)

Why It's Important

The thesaurus feature helps users find the right word and avoid repetitiveness (using the same word again and again).

Tutorials, which accompany applications software packages, enable users to practice new software in a graduated fashion, thereby saving them the time they would have used trying to teach themselves.

See bundled.

When software versions are upward compatible, users don't have to discard material created with early versions.

Some user interfaces are easier to use than others. Most users prefer a graphical user interface.

Versions are usually indicated by numbers, such as 1.0, 2.0, 3.0, and so on. The higher the number before the decimal point, the more recent the version.

Without browser software, users cannot use the part of the Internet called the World Wide Web.

Web directories make it easy for users to find Web sites they may be interested in.

See Web directories.

Using the windows feature, an operating system (or operating environment) can display several windows on a computer screen, each showing a different application program such as word processing, spreadsheets, and graphics.

Word processing software allows a person to use a computer to easily create, edit, copy, save, and print documents such as letters, memos, reports, and manuscripts.

Most popular desktop publishing programs are WYSIWYG, thereby making them relatively easy for people to use.

(Selected answers appear at the end of the book.)

Short-Answer Questions

1. What is the difference between freeware and share-ware?

2. What does it mean when a software application is *downward compatible*?

3. What is software piracy? Network piracy?

4. What are the four categories of applications software?

5. What would a good use be for database software?

6. Why are software suites useful?

7. Why is the World Wide Web one of the fastest-growing subsets of the Internet?

8. What do the abbreviations CAD, CADD, and CAM mean? What do these programs do?

9. What is the purpose of data communications software?

10. What is the difference between analytical graphics and presentation graphics?

Fill-in-the-Blank Questions

1. _____ is a collection of related programs designed to enable users to perform work on general-purpose tasks.

2. _____ offers capabilities that enable the user to create and edit documents easily.

3. If you need to develop a report that involves the use of extensive mathematical, financial, or statistical analysis, you would use _____ software.

4. _____ software enables you to combine near-typeset-quality text and graphics on the same page in a professional-looking document.

5. A _____ is a box that appears on a GUI screen that requires a response from you.

6. In a computer system, a _____ is organized into fields, records, and files.

7. Using _____ software, you can send letters and documents from one computer to another.

8. _____ software packages combine the features of several applications into one software package.

9. _____ enables applications software to interact with the computer.

10. A _____ is software that lets you move through sites on the World Wide Web.

Multiple-Choice Questions

1. Which of the following software types may be copied by anyone for any purpose without penalty?
   a. applications software
   b. systems software
   c. shareware
   d. public domain software
   e. groupware

2. The most popular uses of software are:
   a. word processing and personal information managers
   b. word processing and spreadsheet
   c. spreadsheet and database
   d. word processing and graphics
   e. word processing and desktop publishing

3. Which of the following types of software is free to use but requires that you pay a fee before you can receive technical help or documentation?
   a. freeware
   b. shareware
   c. proprietary software
   d. entertainment software
   e. None of the above

4. Which of the following is a pictorial representation of a command or task that is used in graphical user interfaces?
   a. window
   b. icon
   c. button
   d. dialog box
   e. None of the above

5. Which of the following isn't a feature of word processing software?
   a. what-if analysis
   b. thesaurus
   c. spell-checker
   d. justification
   e. headers and footers

6. Which of the following types of applications software does dynamic linking relate to?
   a. word processing
   b. electronic spreadsheet
   c. database
   d. personal information manager
   e. None of the above

7. Which of the following is a capability of data communications software?
   a. online connections
   b. use of financial services
   c. automatic dialing services
   d. remote access communications
   e. All of the above

8. Which of the following combines the capabilities of several applications into a single application?
   a. software suite
   b. shareware
   c. integrated software
   d. groupware
   e. None of the above

9. Which of the following is used on a network and serves multiple users working on the same project?
   a. software suite
   b. shareware
   c. integrated software
   d. groupware
   e. None of the above

10. Which of the following isn't typically part of a desktop-publishing system?
    a. scanner
    b. laser printer
    c. mouse
    d. software suite
    e. desktop-publishing software

True/False Questions

T F   1. Function keys are used the same way with all applications software packages.

T F   2. Tutorials are the same as documentation.

T F   3. Most users of communications software also use a modem and the phone line.

T F   4. A personal information manager combines some of the capabilities of a word processor, database, and spreadsheet.

T F   5. To price an unbundled computer system, you must price each system component individually.

T F   6. Hyperlinks of the World Wide Web allow users to browse through information by jumping from topic to topic.

T F   7. Spreadsheet software offers presentation graphics capabilities.

T F   8. CADD stands for computer-aided design and drafting.

T F   9. Hypertext is commonly used in Help systems to allow users to go to related helpful information.

T F   10. A document appearing on the screen in WYSIWYG form must be printed before you see how it will look in final form.

Projects/Critical-Thinking Questions

1. Locate a department at your school or place of work that is using some custom-written software. What does this software do? Who uses it? Why couldn't it have been purchased off the shelf? How much did it cost? Do you think there is an off-the-shelf program that can be used instead? Why/why not?

2. Prepare a short report about how you would use an electronic spreadsheet to organize and manage your personal finances and to project outcomes of changes. What column headings (labels) would you use? Row headings? What formula relationships would you want to establish among the cells? (For example, if your tuition increased by $2000, how would that affect the monthly amount you set aside to buy a car or take a trip?)

3. Attend a meeting of a computer users group in your area. What is the overall purpose of the group? Software support? Hardware support? In what ways is support available? Does it cost money to be a member? How many members are there? How does the group get new members? If you were planning to join a users group, would you be interested in joining this group? Why/why not?

4. Industry observers believe that although only 1% of today's computer users have actually purchased something over the Internet, this number will increase in the future. Do you think that more people will teleshop in the future? What do retailers think about the future for teleshopping? Do you think retailers have a reason to be concerned? Why/why not? How do you think the Web figures into the future of shopping over the Internet?

5. Research the state of the art in copy protection for software. What are software companies doing now to prevent the unauthorized copying of copyrighted software? Do you think these companies should do more? Why/why not? If you were the owner of a software company, what would you do to help ensure that users obtain your software through legal channels?

net Using the Internet

Objective: *In this exercise we describe how to send and receive mail using the Pine mail program.*

Before you continue: *We assume that your Internet connection provides access to the Pine mail program. We also assume your Internet connection uses: (1) a command-line interface, (2) VT100 terminal emulation, and (3) the Unix operating system. If any of these assumptions are incorrect, you may still be able to perform this Internet exercise; however, some of the commands and screen output may be different. If necessary, ask your instructor or system administrator for assistance.*

1. Access your Internet account using your user name and password. The procedure you use depends on your particular system. Next, depending on your system, you may be prompted to choose a terminal emulation mode. If possible, choose VT100 terminal emulation. Otherwise, simply press **Enter** to by-pass this option.

2. At this point you are either presented with a command prompt such as $, %, or > and a blinking cursor, or a menu. If a menu appears, choose the option that exits you to command-line, or shell, mode.

3. To start the Pine mail program:

   TYPE: pine
   PRESS: **Enter**

   The following menu should appear:

4. In this step you will use the Compose menu option to send a message to yourself. To choose the Compose option:

   TYPE: C

The Compose Message screen should appear:

5. The cursor is currently positioned in the message header. Although you must type an e-mail address, or user name, into the "To:" information, the rest of the information (Cc, Attchmnt, and Subject) is optional. However it is considered good "netiquette" to include an appropriate subject line.

   The following is a sample e-mail address: kjones@is.college.edu (*Note:* If the person you're sending a message to uses the same computer system that you're logged into, you don't have to type "@" or any of the information to the right in the address.)

   To send a message to yourself:
   TYPE: *your e-mail address (user name)*

6. To move the cursor to the Subject area:
   PRESS: **Enter** *three times*

7. In the Subject area:
   TYPE: About Shareware
   PRESS: **Enter**

8. In the Message Text area, type the following message to yourself:
   TYPE: Shareware is copyrighted software that is widely available on the Internet. You can download the software for free and distribute it legally to other users. However, you must pay a fee to receive technical help, documentation, or upgrades.

Your Compose Message screen may appear similar to the following:

9. As indicated on the bottom of the screen, to send the message you hold down Ctrl and type X.

   PRESS: Ctrl +X

10. At the "Send message?" prompt, you can simply press Enter to send the message.

    PRESS: Enter

    The Pine main menu should appear and the words "Message Sent" should appear above the commands at the bottom of the screen. At this point, wait about one minute before proceeding with the next step.

11. You should be viewing the main menu and the Folder List option should be highlighted in bold letters. To locate the message you sent to yourself:

    PRESS: Enter to select the Folder List option

    PRESS: Enter to display the contents of your Inbox
    The following screen shows the Inbox with three messages:

12. The message you sent to yourself should be highlighted. If it isn't, press ↓ until the message is highlighted. To view the message you sent to your-

self, you can use the ViewMsg option that is listed on the bottom of the screen, or simply press Enter.

TYPE: V

The screen should appear similar to the following:

13. To redisplay the main menu, choose the Main Menu option:

    TYPE: M

14. To quit the Pine mail program, you must use the Quit option:

    TYPE: Q

    PRESS: Enter

15. Now that you know how to send and receive e-mail, let's disconnect from the Internet. You can exit by typing "exit" or "logout."

    For example:
    TYPE: exit
    PRESS: Enter

    Note: If a menu appears, choose the option to disconnect, or quit.

CHAPTER 3

Systems Software
The Power Behind the Power

Concepts You Should Know

After reading this chapter, you should be able to:

1. Explain the difference between applications software and systems software.
2. Explain the three types of systems software.
3. Describe six functions of the operating system.
4. Define and describe the three types of user interfaces, including the GUI.
5. Explain the key features of the principal microcomputer operating systems and operating environments.
6. List and describe the principal external utility programs.
7. Explain how new developments in communications could eliminate the need for users to concern themselves with systems software.

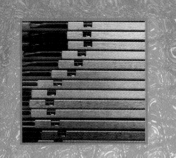

ou're gonna know in the first 10 minutes whether you are going to like that person."

That's how we react to meeting someone for the first time, says Tim Eckles, a consumer-product manager for a microcomputer maker. And, he says, many novices react to their first encounter with a computer in the same way.[1]

Not too many years ago, manufacturers could get away with sending out personal computers that required arcane commands to operate. That's because most customers were businesses, usually with some technical expertise on staff. Today, however, sales of microcomputers are exploding among non-business users, as people become attracted to multimedia capabilities (✓ p. 28) that add video and sound. These consumers may not have the motivation to figure out typical computer commands. "PC makers are afraid home users will pack up their new computers and send them back if they get too frustrated," says one journalist.[2]

Accordingly, computer makers are working hard to make sure that the first meeting consumers have with a computer is a friendly one. Indeed, a great deal of effort has gone into trying to make interfaces (✓ p. 61) as intuitive as possible for newcomers. For instance, Microsoft put 3 years into a project that went on sale in March 1995 with the name of *Bob.* Instead of having pull-down menus, dialog boxes, and arcane error messages, Bob is what Microsoft called a "social interface," with animated characters (a dog named Rover, a rat named Scuzz) that act as guides, asking users what they want to do and giving them tips on how to do it. A similar program is Computer Associates' Simply Village, which has the look and feel of a three-dimensional village and responds to spoken commands. Another friendly interface is Packard Bell's Navigator. (■ *See Panel 3.1.*)

In time, as interfaces are refined, computers will probably become no more difficult to use than a car. Until then, however, for smooth encounters you need to know something about how systems software works. Today people communicate one way, computers another. People speak words and phrases; computers process bits and bytes (✓ p. 5). For us to communicate with these machines, we need an intermediary, an interpreter. This is the function of systems software. We interact mainly with the applications software, which interacts with the systems software, which controls the hardware.

Three Types of Systems Software

Preview & Review: Systems software is of three basic types: operating systems, utility programs, and language translators.

As we've said, **software, or programs, consists of the step-by-step instructions that tell the computer how to perform a task.** Software is of two types—*applications software* and *systems software.* **Applications software is software that can perform useful work on general-purpose tasks,** such as word processing or spreadsheets. (Applications software was covered in Chapter 2.) **Systems software enables the applications software to interact with the computer and helps the computer manage its internal and external resources.** Systems software is required to run applications software; however, the reverse is not true. Buyers of new computers will find the systems software has already been installed by the manufacturer.

PANEL 3.1

Friendly interfaces

Some microcomputer makers provide interface software that uses familiar objects or cartoon characters in order to make systems software easy for novices to use. *(Top)* A screen from Microsoft's Bob shows the dog Rover, who acts as a user's guide. *(Bottom)* Packard Bell's Navigator.

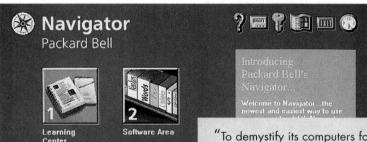

"To demystify its computers for . . . novices, Packard Bell has designed a program called Navigator 2.0. . . . The software displays a home hallway on the screen and fills each 'room' with a different kind of software application, such as children's entertainment or home-office programs. Click with a mouse on a particular room and a light goes on, letting the user pick from a selection of 'books' or 'file-folders' containing the programs that are on shelves or in drawers. Fun stuff is included, too, like a spider and a lizard that appear out of nowhere when the user is rummaging through the children's room. . . .

Compaq has a system called TabWorks that displays each program inside a three-ring binder. The tabs indicate categories of software applications, which can be activated by clicking with a mouse. As new programs are loaded into the machine, tabs can be added."

—Jim Carlton, "Computer Firms Try to Make PCs Less Scary," *Wall Street Journal*

There are three basic types of systems software—*operating systems, utility programs*, and *language translators.*

- **Operating systems:** An operating system is the principal piece of systems software in any computing system. We describe it at length in the next section.

- **Utility programs:** **Utility programs are generally used to support, enhance, or expand existing programs in a computer system.** Many operating systems have utility programs built in for common purposes such as merging two files into one file. Other external, or nonresident, utility programs (such as The Norton Utilities) are available separately to, for example, recover damaged files. We describe external utility programs later in this chapter.

- **Language translators:** *A language translator* is software that translates a program written by a programmer in a language such as C, for example a word processing applications program, into machine language (0s and 1s), which the computer can understand. (See Chapter 11, "Software Development: Programming & Languages.")

The types of systems software are diagrammed below. (■ *See Panel 3.2.*)

The Operating System

Preview & Review: The operating system manages the basic operations of the computer. These operations include booting and housekeeping tasks. Another feature is the user interface, which may be a command-driven, menu-driven, or graphical user interface. Other operations are managing computer resources and managing files. The operating system also manages tasks, through multi-tasking, multiprogramming, time-sharing, or multiprocessing.

The *operating system (OS)* consists of the master system of programs that manage the basic operations of the computer. These programs provide resource management services of many kinds, handling such matters as the control and use of hardware resources, including disk space, memory, CPU time allocation, and peripheral devices. The operating system allows you to concentrate on your own tasks or applications rather than on the complexities of managing the computer.

PANEL 3.2

The three types of systems software

An operating system is required for applications software to run on your computer. The user usually works with the applications software but can bypass it to work directly with the systems software for certain tasks.

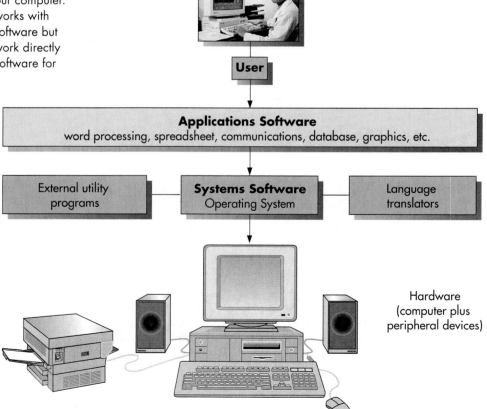

User

Applications Software
word processing, spreadsheet, communications, database, graphics, etc.

External utility programs

Systems Software
Operating System

Language translators

Hardware (computer plus peripheral devices)

Different sizes and makes of computers have their own operating systems. For example, Cray supercomputers use UNICOS and COS, IBM mainframes use MVS and VM, Data General minicomputers use AOS and DG, and DEC minicomputers use VAX/VMS. Pen-based computers have their own operating systems—PenRight, PenPoint, Pen DOS, and Windows for Pen Computing—that enable users to write scribbles and notes on the screen. *These operating systems are not compatible with one another.* That is, in general, an operating system written for one kind of hardware will not be able to run on another kind of machine.

Microcomputer users may readily experience the aggravation of such incompatibility when they buy a new microcomputer. Should they get an Apple Macintosh with Macintosh Systems Software, which won't run IBM-compatible programs? Or should they get an IBM or IBM-compatible (such as Compaq, Dell, or Zenith), which won't run Macintosh programs? And, if the latter, should they buy one with DOS with Windows, Windows 95, OS/2, Windows NT, or Unix? Should they also be concerned with an operating system such as NetWare that will link several computers on a local area network? Should they wait for a new operating system to be introduced that may resolve some of these differences?

Before we try to sort out these perplexities, we should see what operating systems do that deserves our attention. We consider:

- Booting
- Housekeeping tasks
- User interface
- Managing computer resources
- Managing files
- Managing tasks

Booting

The operating system begins to operate as soon as you turn on, or "boot," the computer. The term ***booting* refers to the process of loading an operating system into a computer's main memory from diskette or hard disk.** This loading is accomplished by a program (called the *bootstrap loader* or *boot routine*) that is stored permanently in the computer's electronic circuitry. When you turn on the machine, the program obtains the operating system from your diskette or hard disk and loads it into memory. Other programs called *diagnostic routines* also start up and test the main memory, the central processing unit (✓ p. 16), and other parts of the system to make sure they are running properly. As these programs are running, the display screen may show the message "Testing RAM" (main memory). Finally, other programs (indicated on your screen as "BIOS," for basic input-output system) will be stored in main memory to help the computer interpret keyboard characters or transmit characters to the display screen or to a diskette.

All these activities may create a jumble of words and numbers on your screen for a few seconds before they finally stop. Then a guide may appear, such as "A:\>" or "C:\>." This is the system prompt. **The *system prompt* indicates the operating system has been loaded into main memory and asks ("prompts") you to enter a command.** You may now enter a command. The operating system remains in main memory until you turn the computer off. With newer operating systems, the booting process puts you into a graphically designed starting screen, from which you choose the applications programs you want to run.

Housekeeping Tasks

If you have not entered a command to start an applications program, what else can you do with the operating system? One important function is to perform common repetitious "housekeeping tasks."

One example of such a housekeeping task is formatting blank diskettes. Before you can use a new diskette that you've bought at a store, you may have to format it. **Formatting, or initializing, electronically prepares a diskette so it can store data or programs.** (On IBM-style computers, for example, you might insert your blank disk into drive A and type the command *Format a:*. You can also buy preformatted disks.)

User Interface

Many operating-system functions are never apparent on the computer's display screen. What you do see is the user interface. **The *user interface* is the part of the operating system that allows you to communicate, or interact, with it.**

There are three types of user interfaces, for both operating systems and applications software—*command-driven, menu-driven,* and *graphical.* (■ *See Panel 3.3.*) The latter two types of user interface are often called a *shell.*

- **Command-driven:** **A *command-driven interface* requires you to enter a command by typing in codes or words.** An example of such a command

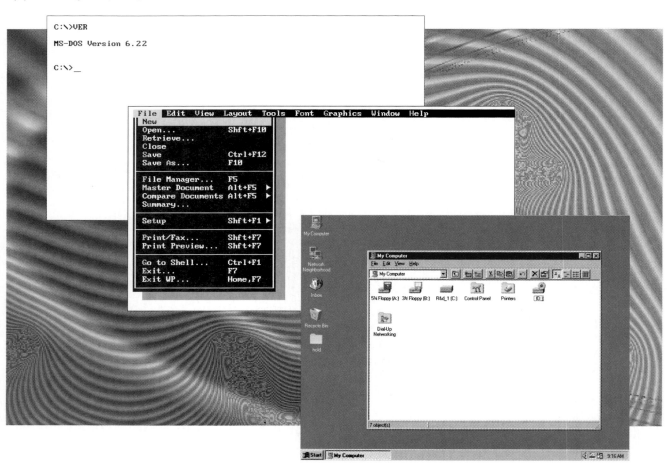

might be DIR (for "directory"). This command instructs the computer to display a directory list of all file names on a disk.

You type a command at the point on the display screen where the cursor follows the prompt (such as following "C:\>"). Then you press the Enter key to execute the command. You'll recall that a *cursor* is a symbol (such as a blinking rectangle of light) that cues where you may type data or enter a command.

The command-driven interface is seen on IBM and IBM-compatible computers with the MS-DOS operating system (discussed shortly).

- **Menu-driven:** **A *menu-driven interface* allows you to choose a command from a menu.** Like a restaurant menu, **a software *menu* offers you options to choose from—in this case, commands available for manipulating data,** such as Print or Edit.

  Menus are easier to use than command-driven interfaces, especially for beginners. Their disadvantage, however, is that they are slower to use. Thus, some software programs offer both features—menus for novice users and keyboard codes for experienced users.

- **Graphical:** The easiest interface to use, the ***graphical user interface (GUI), uses images to represent options.*** Some of these images take the form of icons. ***Icons* are small pictorial figures that represent tasks, functions, or programs**—for example, a trash can for a delete-file function.

  Another feature of the GUI (pronounced "gooey") is the use of windows. ***Windows* divide the display screen into sections.** Each window may show a different display, such as a word processing document in one and a spreadsheet in another.

  Finally, the GUI permits liberal use of the mouse. The mouse is used as a pointing device to move the cursor to a particular place on the display screen or to point to an icon or button. The function represented by the icon can be activated by pressing ("clicking") buttons on the mouse. Or, using the mouse, you can move ("drag") an image from one side of the screen to the other or change its size.

  Microcomputer users first became aware of the graphical user interface in Apple Macintosh computers (although Apple got the idea from Xerox). Later Microsoft made a graphical user interface available for IBM and IBM-compatible computers through its Windows program. Now most operating systems on microcomputers feature a GUI.

Managing Computer Resources

Suppose you are writing a report using a word processing program and want to print out a portion of it while continuing to write. How does the computer manage both tasks?

Behind the user interface, the operating system acts like a police officer directing traffic. This activity is performed by the ***supervisor*, or *kernel*, the central component of the operating system. The supervisor, which manages the CPU, resides in main memory while the computer is on and directs other programs to perform tasks to support applications programs.** Thus, if you enter a command to print your document, the operating system will select a printer (if there is more than one). It will then notify the computer to begin executing instructions from the appropriate program (known as a *printer driver*, because it controls, or "drives," the printer). Meanwhile, many operating systems allow you to continue writing. Were it not for this supervisor program, you would have to stop writing and wait for your document to print out before you could resume.

The operating system also manages memory—it keeps track of the locations within main memory where the programs and data are stored. It can swap portions of data and programs between main memory and secondary storage, such as your computer's hard disk. This capability allows a computer to hold only the most immediately needed data and programs within main memory. Yet it has ready access to programs and data on the hard disk, thereby greatly expanding memory capacity.

There are several ways operating systems can manage memory. Some use *partitioning*—that is, they divide memory into separate areas called *partitions*, each of which can hold a program or data. Large computer systems often divide memory into *foreground* and *background* areas. High-priority programs are executed in foreground memory, and low-priority programs are executed in background memory. For example, if a user is interacting with a program, that program will be in foreground memory. While the user is entering data, the CPU will be unused. Thus, during that time, the CPU can be made available to process something in background memory, such as printing a spreadsheet. Programs wait on disk in *queues* for their turn to be executed.

Managing Files

Files of data and programs are located in many places on your hard disk and other secondary-storage devices. The operating system allows you to find them. If you move, rename, or delete a file, the operating system manages such changes and helps you locate and gain access to it. For example, you can **copy, or duplicate, files and programs from one disk to another.** You can **back up, or make a duplicate copy of, the contents of a disk.** You can **erase, or remove, from a disk any files or programs** that are no longer useful. You can **rename, or give new filenames, to the files on a disk.**

Managing Tasks

A computer is required to perform many different tasks at once. In word processing, for example, it accepts input data, stores the data on a disk, and prints out a document—seemingly simultaneously. Some computers' operating systems can also handle more than one program at the same time—word processing, spreadsheet, database searcher—displaying them in separate windows on the screen. Others can accommodate the needs of several different users at the same time. All these examples illustrate *task management*—a "task" being an operation such as storing, printing, or calculating.

Among the ways operating systems manage tasks in order to run more efficiently are *multitasking, multiprogramming, time-sharing,* and *multiprocessing.* Not all operating systems can do all these things.

- **Multitasking—executing more than one program concurrently:** *Multitasking* **is the execution of two or more programs by one user concurrently—not simultaneously—on the same computer with one central processor.** You may be writing a report on your computer with one program while another program searches an online database for research material. How does the computer handle both programs at once?

    The answer is that the operating system directs the processor (CPU) to spend a predetermined amount of time executing the instructions for each program, one at a time. In essence, a small amount of each program is processed, and then the processor moves to the remaining programs, one at a time, processing small parts of each. This cycle is repeated until processing is complete. The processor speed is usually so fast that it may seem as if all the programs are being executed at the same time. However, the processor is still executing only one instruction at a time, no matter how it may appear to the user.

- **Multiprogramming—concurrent execution of different users' programs:** *Multiprogramming* **is the execution of two or more programs on a *multi-user* operating system.** As with multitasking, the CPU spends a certain amount of time executing each user's program, but it works so quickly, it seems as though all the programs are being run at the same time.

- **Time-sharing—round-robin processing of programs for several users:** *Time-sharing* **is a single computer's processing of the tasks of several users at different stations in round-robin fashion.** Time-sharing is used when several users are linked by a communications network to a single computer. The computer will first work on one user's task for a fraction of a second, then go on to the next user's task, and so on.

  How is this done? The answer is through *time slicing.* Computers operate so quickly that it is possible for them to alternately apportion slices of time (fractions of a second) to various tasks. Thus, the computer's operating system may rapidly switch back and forth among different tasks, just as a hairdresser or dentist works with several clients or patients concurrently. The users are generally unaware of the switching process.

  Multitasking and time-sharing differ slightly. With multitasking, the processor directs the programs to take turns accomplishing small tasks or events within the programs. These events may be making a calculation, searching for a record, printing out part of a document, and so on. Each event may take a different amount of time to accomplish. With time-sharing, the computer spends a *fixed amount* of time with each program before going on to the next one.

- **Multiprocessing—simultaneous processing of two or more programs by multiple computers:** *Multiprocessing* **is processing done by two or more computers or processors linked together to perform work simultaneously**—that is, at precisely the same time. This can entail processing instructions from different programs or different instructions from the same program.

  Multiprocessing goes beyond multitasking, which works with only one microprocessor. In both cases, the processing should be so fast that, by spending a little bit of time working on each of several programs in turn, a number of programs can be run at the same time. With both multitasking and multiprocessing, the operating system keeps track of the status of each program so that it knows where it left off and where to continue processing. But the multiprocessing operating system is much more sophisticated than multitasking.

  Multiprocessing can be done in several ways. One way is *coprocessing,* whereby the controlling CPU works together with specialized microprocessors called *coprocessors,* each of which handles a particular task, such as display-screen graphics or high-speed mathematical calculations. Many sophisticated microcomputer systems have coprocessing capabilities.

  Another way to perform multiprocessing is by *parallel processing,* whereby several full-fledged CPUs work together on the same tasks, sharing memory. Parallel processing is often used in large computer systems designed to keep running if one of the CPUs fails. These systems are called *fault-tolerant* systems; they have several CPUs and redundant components, such as memory and input, output, and storage devices. Fault-tolerant systems are used, for example, in airline reservation systems.

Operating system functions are summarized below. (■ *See Panel 3.4.*)

Microcomputer Operating Systems & Operating Environments

Preview & Review: The principal microcomputer operating systems and operating environments are DOS, Macintosh Operating System, Windows 3.X (for DOS), OS/2 Warp, Windows 95, Windows NT, Unix, and NetWare.

Operating systems are not just a topic of academic interest. As a microcomputer user, you'll have to learn not only whatever applications software you want to use but also, to some degree, the operating system with which they work. Moreover, when you buy a PC, it comes with an operating system. You have to know which one you want.

In this section, we describe the following:

- DOS
- Macintosh Operating System
- Windows 3.X for DOS
- OS/2 Warp
- Windows 95
- Windows NT
- Unix
- NetWare
- Coming attractions

PANEL 3.4

Some operating system functions

Booting	House-keeping Tasks	User Interface	Managing Computer Resources	Managing Files	Managing Tasks
Loads operating system into computer's main memory. Uses diagnostic routines to test system for equipment failure. Stores BIOS programs in main memory	Formats diskettes. Displays information about operating system version. Displays disk space available	Provides a way for user to interact with the operating system—can be command-driven, menu-driven, or graphical	Via the supervisor, manages the CPU and directs other programs to perform tasks to support applications programs. Keeps track of locations in main memory where programs and data are stored (memory management). Moves data and programs back and forth between main memory and secondary storage (swapping)	Copies files/programs from one disk to another. Backs up files/programs. Erases (deletes) files/programs. Renames files	May be able to perform multitasking, multiprogramming, time-sharing, or multiprocessing

Operating Environments Add a Graphical User Interface

Before we proceed, we need to define what an *operating environment* is, since some people have trouble distinguishing it from an *operating system.* **An** *operating environment*—**also known as a** *windowing environment* **or** *shell*—**adds a graphical user interface or a menu-driven interface as an outer layer to an operating system.** The most well-known operating environment is the Windows 3.X program sold by Microsoft, which adds a graphical user interface to DOS. Another is IBM's Workplace Shell, which provides a GUI for OS/2. Similar operating environments are available for Unix.

Common features of these operating environments are use of an electronic mouse, pull-down menus, and icons and other graphic displays. They also have the ability to run more than one application (such as word processing and spreadsheets) at the same time and the ability to exchange data between these applications.

Let's now examine the principal operating systems (and operating environments) you will probably encounter.

DOS

There are reportedly over 100 million users of DOS. This makes it the most popular software of any sort ever adopted, and certainly the most popular systems software.[3] **DOS—for Disk Operating System—runs primarily on IBM and IBM-compatible microcomputers,** such as Compaq, Zenith, AST, Dell, Tandy, and Gateway.

There are now two main operating systems calling themselves DOS.

- **Microsoft's MS-DOS:** DOS is sold under the name MS-DOS by software maker Microsoft. The "MS" stands for Microsoft. Microsoft launched its original version, MS-DOS 1.0, in 1981, and there have been several upgrades since then.

- **IBM's PC-DOS:** Microsoft licenses a version to IBM called PC-DOS. The "PC" stands for "Personal Computer." The most recent version is PC-DOS 7, released March 1995.

What do the numbers in the names mean? The number before the period refers to a *version.* The number after the period refers to a *release,* which has fewer refinements than a version. The most recent versions are all backward compatible. For operating systems, **backward compatible means that users can run the same applications on the later versions of the operating system that they could run on earlier versions.** *(See Panel 3.5, next page.)*

Recent versions of DOS have expanded the range of the operating system. For example, Version 4.0 of MS-DOS offered the options of a command-driven interface and a menu-driven interface. Version 5.0 added a graphics-based interface. Version 6.0 added features that took advantage of a computer's main memory. Version 6.22 added a data-compression feature to double the amount of information that could be stored on a hard-disk drive.

No doubt DOS will be around for years. After all, there are a great many old but still useful microcomputers running it and a great many application programs written for it. And many IBM mainframe systems use DOS-VSE. Nevertheless, as a command-driven, single-user program, DOS is probably a fading product. Although satisfactory for many uses, it will unquestionably be succeeded by other, more versatile operating systems. As Ken Wasch, president of the Software Publishers Association, says, "Nobody was forced to

Backward compatability

If a new version of an operating system is backward compatible, users can run the same applications program on it that they could run on older versions.

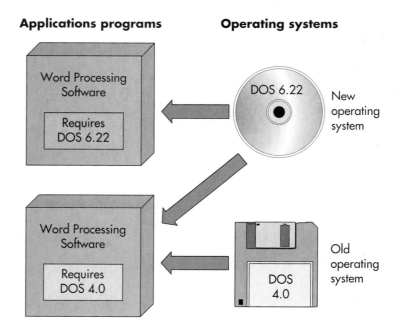

upgrade to electric lights when they still could read by kerosene lamps, but still most people found it advantageous to upgrade."[4] (■ *See Panel 3.6.*)

Macintosh Operating System (Mac OS)

The Apple Macintosh has always had one outstanding feature: it is easy to use. To be sure, it can't do as much as some other operating systems. Still, the easy-to-use interface has generated a strong legion of fans. (■ *See Panel 3.7.*)

In the past, however, Apple kept Macintosh prices high, a deciding factor for many people in picking a personal computer. As computer journalist David Kirkpatrick described it:

> If you wanted a friendly machine and could afford a premium price, you bought an Apple Macintosh. . . If you needed lots of cheap computing power for complicated tasks, or didn't care so much about user-friendliness, you bought an IBM-style [DOS-based] PC. There were dozens of brands to choose from. . .[5]

Unfortunately, IBM-style and Macintosh microcomputers were designed around different microprocessors, so it was impossible to combine the best of both. IBM and IBM-compatible computers used microprocessors built by Intel. These were the Intel 80286 (called the *'286 chip*), 80386 (*'386 chip*), 80486 (*'486 chip*), the Pentium (the successor to the '586 chip), and most recently the Pentium Pro. Macintoshes were built around microprocessors made by Motorola—the 68000, 68020, 68030, 68040, and PowerPC chips. Intel chips could not run Macintosh programs and Motorola chips could not run DOS programs.

Because of price, and because in pre-PC times businesses were already comfortable with IBM equipment, DOS-equipped microcomputers have ruled the day. Compared to DOS's 100 million users, Macintosh has only about 8.5 million users. The situation was complicated by the appearance of Windows, Microsoft's IBM-compatible operating environment, and even more so by Windows 95, both of which have many Mac-like qualities, as we shall describe.

The aging system of DOS

With DOS, you work with precise, typed commands.

"Like a turn-of-the-century building constructed according to venerable principles of design, DOS perpetuates principles designed for prior generations of computers. That's because DOS was developed before computers had their present speed and the ability to generate elaborate on-screen images.

The instructions you use to run a program, retrieve previous work, and the like are adaptations of the typewritten instructions used to communicate with early computers. To use DOS, you must type in an idiosyncratic string of words or letters or (in later versions) choose a written command from an on-screen menu."

—"The Basic Choice in Computers," *Consumer Reports*

For the love of Mac

The icons, pull-down menus, and windows of the Apple Macintosh System 7 operating system give the Mac a look and feel that makes many users say they love their machines.

"What makes the Mac such a magnet for affection? . . . Consider one of the most mundane actions a user performs: inserting a floppy disk. When you put one into a Mac's drive, the computer notes that the disk is there: a little icon appears on the screen, showing the name you've given the disk. Eject the disk and insert a different one, and the Mac registers that as well. If you absent-mindedly try to call up a file stored on the disk you've just removed, the Mac will name the disk and ask you to reinsert it. By contrast, you have to tell your [Version 3.1] Windows PC when you add or change a floppy. Make the same mistake in looking for a file, and you'll get the cryptic response BAD COMMAND OR FILE NAME. It's up to you to figure out your error."

—David Kirkpatrick, "Mac vs. Windows," *Fortune*

Still, the Mac, introduced in 1984, set the standard for icon-oriented graphical user interfaces. Indeed, the Macintosh Operating System (Mac OS) is easy to use because Apple designed its hardware and software together, from the start. Kirkpatrick, for one, believes that the Macintosh operating system is the best in terms of elegance and ease of use. So do evaluators from *Consumer Reports,* which studied a panel of users performing typical home-use tasks on computers equipped with the latest versions of Mac OS, DOS, and DOS with Windows software.[6]

The Macintosh System 7.5 operating system has an important program, a file manager called the *Finder,* which manages the desktop screen and its icons. System 7.5 also enables users to read MS-DOS and Windows files, even if they don't have the software to create such files. In addition, System 7.5 has a feature called Apple Guide, which offers "active assistance." Active assistance helps users accomplish different tasks on the computer—for example, explaining how to share files with other users.

After almost a decade of Macintoshes costing more than comparable IBM-style personal computers, Apple cut its prices. The result was to bring Mac prices more into line with equipment from IBM, Compaq, Dell, and others. In addition, in September 1994 Apple finally agreed to license its operating system to other computer makers. With this development, we are now seeing the appearance of Macintosh clones and further price drops.

Although the Macintosh is easy to use, not as many programs have been written for it as for DOS/Windows-based systems. Only about 6900 commercial applications packages have been written for Macs, according to BIS, a Norwell, Massachusetts, market research firm.[7] By contrast, some *29,400* applications packages are available for DOS computers. However, its graphics capabilities make the Macintosh a popular choice for people working in commercial art, desktop publishing, multimedia, and engineering design.

Windows for DOS (Windows 3.X)

As we introduce the Windows environment, it's worth expanding on what we said in Chapter 2 about how windows differ from *Windows* (✓ p. 61). A *window* (lowercase "w") is a portion of the video display area dedicated to some specified purpose. An operating system (or operating environment) can display several windows on a computer screen, each showing a different applications program, such as word processing and spreadsheets. However, *Windows* (capital "W") is something else. **Windows is an operating environment made by Microsoft that lays a graphical user interface shell around the MS-DOS or PC-DOS operating system.** Like Mac OS, Windows contains windows, which can display multiple applications. (■ *See Panel 3.8.*)

Note: It's important to realize that Windows 3.X ("3.X" represents versions 3.0, 3.1, and 3.11) is different from *Windows 95* (discussed shortly), which is not just an operating environment but a true operating system.

Microsoft's Windows 3.X is designed to run on IBM-style microcomputers with Intel microprocessors—the '386 and '486 chips. Earlier versions of Windows could not make full use of '386 and '486 chips, but later versions can. To effectively use Windows 3.X, one should have a reasonably powerful microcomputer system. This would include a '386 microprocessor or better, much more main memory than is required for DOS (a minimum of 4 megabytes), and a hard-disk drive.

Microsoft released Windows 3.0 (the first really useful version) in May 1990 and promoted it as a way for frustrated DOS users not to have to switch to more user-friendly operating systems, such as Macintosh. Indeed, Windows has about 80% of the Macintosh features. Although Windows is far eas-

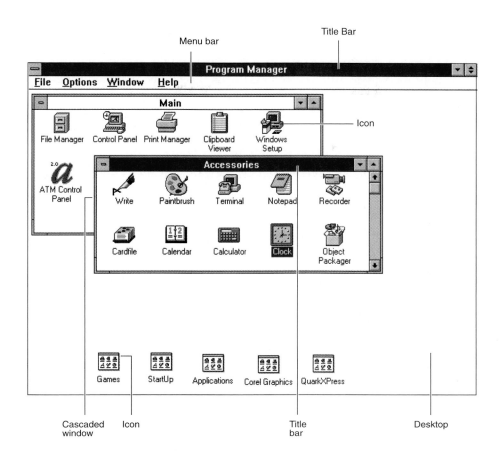

PANEL 3.8

Windows 3.X's graphical user interface

ier to use than DOS, its earlier versions have not been as easy to use as the Mac operating system. This is because Windows sat atop the 11-year-old command-driven DOS operating system, which required certain compromises on ease of use. Indeed, the system had something of a split personality. In handling files, for example, after passing through the Macintosh-style display of icons the user had to deal with the DOS file structures beneath. In addition, many users complained that installing peripherals, such as a hard-disk drive, was somewhat difficult with DOS and Windows.

Yet, even if the various Windows 3.X versions were a bit creaky, they were certainly usable by most people. And when Windows 95 was rolled out, most of the objections vanished.

Before describing Windows 95, we will introduce another of its predecessors and competitors, OS/2.

OS/2 & OS/2 Warp

OS/2 (there is no OS/1) was initially released in April 1987 as IBM's contender for the next mainstream operating system. *OS/2—for Operating System/2—is designed to run on many recent IBM and compatible microcomputers.* Unlike Windows 3.X, OS/2 does not require DOS to run underneath it, so it generally processes more efficiently. Like Windows, it has a graphical user interface, called the Workplace Shell (WPS), which uses icons resembling documents, folders, printers, and the like. OS/2 can also run most DOS, Windows, and OS/2 applications simultaneously. This means that users don't have to throw out their old applications software to take advantage of new features. In addition, OS/2 is the first microcomputer operating system to take full advantage of the power of the newer Intel microprocessors, such as

486 and Pentium chips. Lastly, this operating system is designed to connect everything from small handheld personal computers to large mainframes.[8]

Unfortunately, because of an array of management and marketing disasters, IBM slipped far behind Microsoft. (In fact, IBM and Microsoft were once partners in developing OS/2. Then Microsoft abandoned it to put all its efforts into backing Windows 3.X and later Windows NT.) By mid-1994, an estimated 50 million copies of Microsoft's Windows had been sold versus 5 million of OS/2.[9]

Nevertheless, OS/2 can perform some advanced feats. It can, for example, receive a fax and run a video while at the same time recalculating a spreadsheet. This is the kind of multitasking (and even multimedia) activity that is increasingly important for networked computers. It is also the first operating system created just for today's new "workgroup" environments. In workgroups, individuals work in groups sharing electronic files and databases over communications lines. Yet, despite its seeming complexity, OS/2 has been considered reasonably easy for beginners to use. (■ *See Panel 3.9.*) Indeed, its file management resembles Macintosh's System 7 (allowing long file names, for example), making it easier to use than the older versions of Windows. Finally, IBM also offers a version called OS/2 for Windows.

In late 1994 IBM unveiled a souped-up version of OS/2 called *Warp.* Despite spending $2 billion on OS/2 in its long struggle against Windows for DOS—IBM even claimed that OS/2 could run Windows better than Windows itself—the company failed to increase its market share. OS/2 remained hopelessly mired with only 4% of the market for desktop operating systems, versus roughly 80% for Windows 3.X—and Microsoft had not even released its much ballyhooed Windows 95 yet. It is expected that IBM will abandon OS/2.

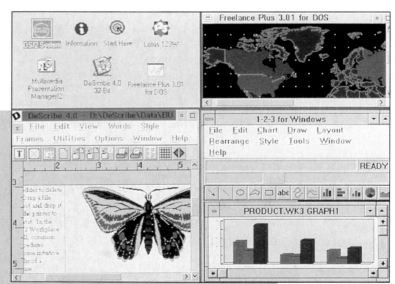

PANEL 3.9

Friendly OS/2

Computer columnist John Dvorak tells a reader that even an early version OS/2 is big but friendly to beginners.

"**Q:** *I have been in the market for a new computer and I thought I knew what I wanted until I started reading ads for IBM's newest operating system—OS/2. The ad says that it's better than DOS or Windows, but does that mean it's compatible? Is it something I should consider?*
A: I use OS/2 exclusively except on an old laptop which won't run OS/2. OS/2 is a compatible operating system that can run both Windows and DOS programs and also any programs designed specifically for OS/2. I find the program to be faster and more interesting than either DOS or Windows. By "more interesting" I mean there are a lot of interesting ways to do simple functions such as the ability to use either the command line or a graphical user interface interchangeably.

The problem with OS/2 is that it's a large complex system that requires the user have a modern computer with a lot of memory. Luckily, memory prices have dropped in the past few years. . . .

I should also mention that while OS/2 is complex and powerful that doesn't mean that it is not suited for the beginner. In fact, it provides an outstanding intuitive interface for newcomers which is very appealing and quite easy to use. Many beginners prefer it to other operating systems."

—John C. Dvorak, "Ask Dvorak," *San Francisco Examiner*

Case Study: Making Icons

If Susan Kare ever gets a retrospective of her art, it will include a trash can, a watch and a portrait of a computer with a sly Mona Lisa smile.

Kare is the Matisse of computer icons, the screen symbols that users click on twice to tell the machine what to do. Despite her ponderous job title, "a user interface graphic designer," she spends her days striving for simplicity, keeping in mind the computer neophyte who cares less about how things work and more about getting things done.

During the past 10 years, she has drawn more than 2000 icons for computers, coming up with dozens of symbols representing the commands "print," "merge" and "quit." Her clients have included the leaders of the computer age—Apple Computer, Autodesk, Electronic Arts, IBM, Intuit, Sony Pictures, Motorola, and Microsoft . . . .

Some of her creations—such as a phone to represent instructions for dialing up a modem—may seem like no-brainers. But Kare said making icons for computers is more complicated than it appears.

"Some icons are easy because they're nouns—a calendar, for example," Kare said. "But verbs are hard to do. Undo is especially hard. I struggle year in and year out about [the command] undo."

Execute, she said, is another difficult one. "Some people have guns for execute, which doesn't seem good. I had dominoes falling over and have tried running shoes."

It was a fluke Kare ended up in the computer industry. After receiving a doctorate in fine arts from New York University in 1978 . . . , she was heading for a museum or academic career. . . .

[W]hen Andy Hertzfeld, a high school friend from Philadelphia, recruited Kare to work at Apple Computer on the first Macintosh, she jumped at the chance. She learned how to draw on the computer using bit map design, a process that involves turning on and off a series of dots or pixels on the screen to make an image.

At the time, graphical user interfaces were considered no more than window dressing on computers. Xerox Corp. had released some graphics, but the first computer with pleasing aesthetics—the Macintosh—didn't hit the market until 1984.

The result was that cheerful characters have become part of the computing daily life of Mac users—the trash can (for discarding files), the Mona Lisa face (which means the machine is working) and the clock (which means the computer is busy and can't be used).

—Michelle Quinn, "Art That Clicks," *San Francisco Chronicle.*

Susan Kare has drawn more than 2000 icons for computers, including those pictured below.

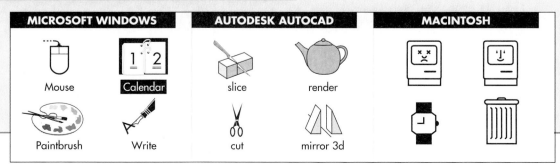

MICROSOFT WINDOWS		AUTODESK AUTOCAD		MACINTOSH	
Mouse	Calendar	slice	render		
Paintbrush	Write	cut	mirror 3d		

Windows 95 & Later

"Thank God it's finally over," wrote a reporter. "Or thank Bill Gates."[10]

He was referring to Gates's and Microsoft's months of promotional buildup that ended in a spectacular display of excess when Windows 95 finally made its debut on August 24, 1995. **Windows 95, the successor to Windows 3.1 for DOS, is a true operating system for IBM-style personal computers rather than just an operating environment.** As part of the $200 million spent on advertising and marketing, Microsoft acquired the Rolling Stones's "Start Me Up" as a kind of theme song, spotlighted New York's Empire State Building in the company colors, and paid to give away 1.5 million copies of the London *Times* with Windows 95 supplements. "It's unbelievable," said one industry observer about the global extravaganza. "You'd think we had world peace."[11] Said another: "Thus did a software 'upgrade'—usually the most routine of events—transcend the realm of high-tech to become the biggest consumer-product launch of any kind. Ever."[12]

Is Windows 95—and its modest upgrade, variously called *Nashville* or *Windows 96*—worth all the hoopla? Could you not get the essential elements using something else? Macintosh enthusiasts, for instance, sneeringly refer to Windows 95 as "Mac 88" because so many of its features were available on Macintoshes 7 years earlier.[13] OS/2 users also might well say "Been there, done that."

The first thing to realize is that the best product does not always prevail, despite the conventional wisdom that competition and free markets produce optimal solutions. "It's naive to believe that efficient engineering solutions win in the marketplace," declares Stanford and Oxford economist Paul A. David. "That view is not supported by the study of history."[14] (As an example he cites the prevalence of the standard QWERTY arrangement of keys on most typewriter and computer keyboards, even though other layouts have proven to be more efficient.)

The personal computer technology finally culminating in Windows 95—the keyboard and mouse, the box housing a computer, the screen dividing into cascading windows, the little on-screen pictures called icons, the point-and-click software—was actually invented more than 20 years ago by researchers at Xerox's Palo Alto Research Center (PARC) in northern California. Xerox executives asked about the profitability of marketing the PARC personal computer, which in those days, says one former researcher, "had the same ring to it as 'personal nuclear reactor' would have today."[15] When PARC couldn't answer that, the PC was relegated to in-house use. At about that time, PARC gave a tour to Steve Jobs, who saw the PC. Later, as a founder of Apple Computer, Jobs used the PARC technology to create the Lisa PC, which was succeeded by the Macintosh, launched in 1984. Microsoft's Windows for DOS—in many ways inferior to the Mac technology—then followed in 1990. Now we have Windows 95, which after two decades finally catches up to the Xerox PARC PC.

In sum: Microsoft has not been an innovative organization in technology so much as a hugely successful *marketing* organization. In particular, it has been able to persuade most personal computer makers to install its systems software—first DOS, then Windows 3.X, now Windows 95—on their hardware, so that it is a Microsoft product that the purchaser of a new computer encounters when he or she turns it on. (Windows is installed on 8 out of 10 new PCs.) Microsoft also benefited from the fact that Apple chose not to license its software to possible manufacturers of cheap Macintosh clones, whereas there was a rapidly expanding market of cloners of IBM-style computers, all using Microsoft's systems software. Finally, Microsoft also actively cultivated developers of applications software, sharing information about its forthcoming systems software, so that when its operating system came to market there would soon be plenty of applications programs to run under it.

Following are just some of the features of Windows 95:

- **Clean "Start":**   Instead of encountering a confusing array of similar program groups (as with Windows 3.X), you'll first see a clean "desktop" with a "Taskbar" of important icons at the bottom of the screen and one button labeled START. (■ *See Panel 3.10 on the next page.*)

- **Better menus:**   Windows 3.1's quirky Program Manager and File Managers have been replaced by more accessible features called THE EXPLORER and MY COMPUTER, which let you quickly see what's stored on your disk drives and make tracking and moving files easier.

- **Long file names:**   File names can now be up to 256 characters instead of the 8 characters (plus 3-character extension) of DOS and Windows 3.X. This means you can now have a file name for your resume, for example, of "Resume—January 15, 1997, version" instead of "RES11597." (Macintosh OS and OS/2 have always permitted long file names.)

- **The "Recycle Bin":**   This feature allows you to delete complete files and then get them back if you change your mind.

- **32-bit instead of 16-bit:**   The new software is a 32-bit program, whereas most Windows 3.1 software is 16-bit. ***Bit numbers* refer to how many bits of data a computer chip, and software written for it, can process at one time.** Such numbers are important because they refer to the amount of information the hardware and software can use at any one time. This doesn't mean that 32-bit software will necessarily be twice as fast as 16-bit software, but it does promise that new 32-bit applications software will offer better speed and features once software developers take advantage of the design.[16]

- **Plug and play:**   It has always been easy to add new hardware components to Macintoshes. It used to be extremely difficult with IBM-compatible PCs. ***Plug and play* refers to the ability to add a new hardware component to a computer system and have it work without needing to perform complicated technical procedures.**

   More particularly, *Plug and Play* (abbreviated *PnP*) is a standard developed for IBM-style PCs by Microsoft and chip maker Intel and incorporated into Windows 95 to eliminate user frustration when one is adding new components. Now when you add a new printer or modem, your PC will recognize the model and set it up.

Windows 95 was expected to be superseded in 1996 by an upgrade called *Nashville,* or *Windows 96.* Most of the improvements represented fine-tuning. However, the most significant improvement is intended to make the Web browser disappear as a separate piece of software. Instead, the new version of Internet Explorer, recently a standard feature of Windows, allows users to use the same commands not only to access their own computer but also to browse the Internet and other networks.[17] This could ultimately threaten Netscape's dominance.

So far we have described operating systems (and operating environments) pretty much in the chronological order in which they appeared.

PANEL 3.10

Windows 95 screen

Start button: Click for an easy way to start using the computer.

Microsoft Network: Click here to connect to the Microsoft Network, the company's online service.

My Briefcase: Allows you to synchronize files in two computers—say, an office PC and a laptop.

Recycle Bin: Allows you to dispose of files—or retrieve them later.

Network Neighborhood: If your PC is linked to a network of PCs, click here to get a glimpse of everything available on the network.

My Computer: Gives you a quick overview of all the files and programs installed in your PC.

Document: New multitasking capabilities allow people to smoothly run more than one program at once.

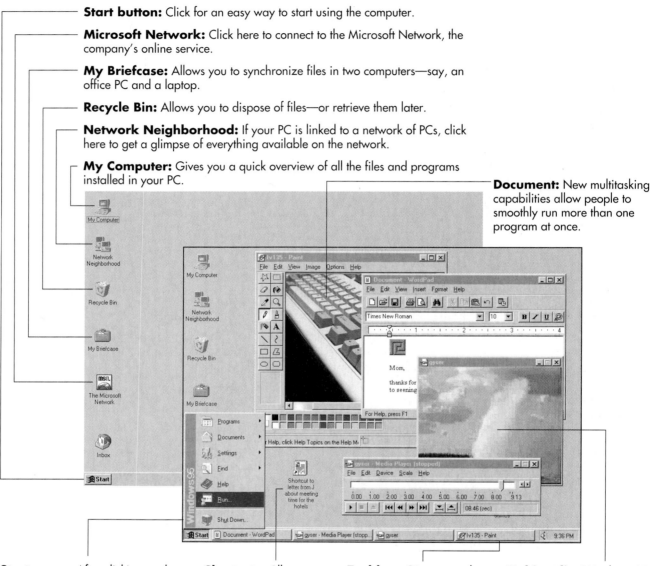

Start menu: After clicking on the start button, a menu appears, giving you a quick way to handle common tasks. You can launch programs, call up documents, change system settings, get help, and shut down your PC.

Shortcuts: Allows you to immediately launch often-used files and programs.

Taskbar: Gives you a log of all programs you have opened. To switch programs, click on the buttons that appear in the taskbar.

Multimedia: Windows 95 features sharper graphics and improved video capabilities.

"Windows has long since stretched the definition of *operating* system past the breaking point. The original DOS was little more than a thin (and clumsy) layer of hooks that applications could use for reading and writing data to memory, screen and disks. Windows 95 not only provides a rich environment for controlling many programs at once; it also offers, built in, a word processor, communications software, a fax program, an assortment of games, screen savers, a telephone dialer, a paint program, back-up software and a host of other housekeeping utilities and, of course, Internet software. By historical standards, you get a remarkable bargain."

—James Gleick, "Making Microsoft Safe for Capitalism," *New York Times Magazine*

Except for OS/2, these operating systems were principally designed to be used with stand-alone desktop machines, not large systems of networked computers. However, Windows NT, described next, actually preceded Windows 95. We consider it separately, along with Unix and NetWare, because it was mainly designed to work with networks.

Windows NT

Unveiled by Microsoft in May 1993, **Windows NT, for *New Technology*, is an operating system intended to support large networks of computers.** Examples of such networks are those used in airline reservations systems. Unlike the early Windows operating *environment* (which ran with DOS), Windows NT is a true operating *system*, interacting directly with the hardware. Although on a screen it looks identical to Windows 3.1 and can run on the most modern PCs, it is primarily designed to run on workstations or other more powerful computers. Indeed, NT will run on the most powerful of microprocessors, both Intel's Pentium and Motorola's PowerPC.

Most PC owners will find Windows NT far more than they need. Its power benefits engineers and others who use workstations and who do massive amounts of computing at their desks. A second category of users consists of those tied together in "client/server" networks with "file server" computers. A *client/server network* is a type of local area network (LAN). The "client" is the requesting PC or workstation and the "server" is the supplying file-server or mainframe computer, which maintains databases and processes requests from the client. A *file server* is a high-speed computer in a LAN that stores the programs and files shared by the users.

Customers often are more concerned about protecting their past investments in technology than about flashy new computer developments. Microsoft has tried to appeal to such customers by promoting NT's compatibility with earlier versions of DOS and Windows. Even so, users have been cautious about putting NT in charge of the large computer networks that are the lifeblood of their businesses. Some companies, for instance, favor IBM's OS/2 because, despite delays in its introduction, it is considered a thoroughly tested product. Others remain with Unix, a prime competitor of Windows NT.

In the long run, it is suggested, maintaining Windows 95 and Windows NT as separate operating systems will become a strain even for a company with the resources of Microsoft. Eventually, the company will probably merge the two. Indeed, in 1996 the company proposed standards called *SIPC*, for *Simply Interactive Personal Computer*, which would migrate the Windows 95 operating system so that it could link up stereos, videodisk players, and other entertainment appliances while communicating with the Internet.[18]

Unix

Unix was invented more than two decades ago by American Telephone & Telegraph, making it one of the oldest operating systems. **Unix is an operating system for multiple users and has built-in networking capability, the ability to run multiple tasks at one time, and versions that can run on all kinds of computers.** Because it can run with relatively simple modifications on different types of computers—from micros to minis to mainframes—Unix is called a "portable" operating system. The primary users of Unix are government agencies, large corporations, and banks, which use the software for everything from airplane-parts design to currency trading.

For a long time, AT&T licensed Unix to scores of companies that make minicomputers and workstations. As a result, the operating system was modified and resold by several companies, producing several versions of Unix. Recent versions include Solaris, by Sun Microsystems, and Motif, by a coalition of companies (including IBM, Digital Equipment Corporation, and Hewlett-Packard) called the Open Software Foundation. Repeated attempts to unify the various Unix versions into one standard have come to naught.

Finally galvanized by the threat that Microsoft's Windows NT might become the standard for large microcomputer networks, Unix suppliers struggled toward agreement. Over 50 companies have agreed to adhere to a single standard for a common interface linking Unix to applications like word processing and spreadsheet programs.[19] More significant, Novell purchased the Unix trademark from AT&T. To promote standardized approaches to Unix, Novell gave away the Unix brand name to an independent foundation, which declared that it would be responsible for maintaining the Unix specifications.[20] In September 1995, Novell's Unix business was bought by Santa Cruz Operation, a Unix software maker. In April 1996, seven computer makers agreed to use this version of Unix in server computers.

Will Unix endure? It is a popular operating system in Europe, where users have discovered that its applications can survive changes in hardware, so that business is not unduly disrupted when new hardware is introduced. Perhaps, with agreement on Unix standards, the same thing will be discovered in North America. (■ *See Panel 3.11 for an example of a Unix screen.*)

NetWare

Novell, of Orem, Utah, is the maker of NetWare, the software Microsoft is trying to beat with its Windows NT. Developed during the 1980s, **NetWare has become the most popular operating system for coordinating microcomputer-based local area networks (LANs) throughout a company or campus.** (LANs allow PCs to share data files, printers, and other devices.) NetWare controlled 45% of the market for corporate network operating system software in 1994.[21] By another yardstick, it presently represents 60 to 70% of the installed base and sales of networking software, roughly 10 times that of Windows NT.[22]

Can you continue to use, say, MS-DOS on your office personal computer while it is hooked up to a LAN running NetWare? Indeed you can. NetWare provides a shell around your own operating system. If you want to work "off network," you respond to the usual prompt (for example, the DOS-based A:\>, B:\>, or C:\>) and run the PC's regular operating system. If you want to work "on network," you respond to another prompt (for instance, F:\>) and type in whatever password will admit you to the network.

Novell also offers NetWare Directory Services, which allows employees to access files and services on a corporate network no matter where they are, without searching mysterious lists of electronic addresses and without having to go through multiple requests for passwords and authorization. Novell plans to offer its directory services over the Internet, as well as through NetWare Connect Services, the private alternative to the Internet that it is developing with AT&T.[23]

The long-term vision Novell has, according to its chairman, Robert Frankenberg, is to make possible a network for "connecting people with other people and the information they need, enabling them to act on it anytime, anyplace."[24] In this highly ambitious view, the network will extend beyond office networks of PCs, even beyond the global Internet. It envisions wireless networks linking automobiles, appliances, vending machines, electronic

PANEL 3.11
Unix screen

cash registers, factory automation, security systems, and other nontraditional computing devices, as well as telephones, fax machines, and copiers. Novell has also developed ways to exchange information over ordinary electric power lines. With this technology, even standard electrical outlets could become NetWare connections.[25]

The principal characteristics of the operating systems described thus far are summarized in the accompanying table. (■ *See Panel 3.12.*)

Operating System	Types of Microprocessors	Single User	Multitask	Multiuser
MS-DOS PC-DOS	Intel 8088,8086, 80286, 80386 80486, Pentium	Yes		
Macintosh System Software	Motorola 8030, 8040, 68030, 68040, PowerPC	Yes	Yes	
Windows 3.1	Intel 80286, 80386, 80486, Pentium	Yes	Yes	Yes
OS/2	Intel 80286, 80386, 80486, Pentium, PowerPC	Yes	Yes	Yes
Windows 95	Intel 80386, 80486, Pentium	Yes	Yes	Yes
Windows NT	Intel 80386, 80486, Pentium, PowerPC, DEC Alpha RISC		Yes	Yes
Unix	Almost all processors		Yes	Yes
NetWare	Intel 80386, 80486, Pentium		Yes	Yes

PANEL 3.12

Microcomputer operating systems compared

External Utility Programs

Preview & Review: External utility programs provide services not performed by other systems software. They often include screen savers, data recovery, backup, virus protection, file defragmentation, data compression, and memory management. Multiple-utility packages are available.

"You wouldn't take a cruise on a ship without life preservers, would you?" asks one writer. "Even though you probably wouldn't need them, the terrible *what if* is always there. Working on a computer without the help and assurance of utility software is almost as risky."[26]

The "what if" being referred to is an unlucky event, such as your hard-disk drive "crashing" (failing), risking loss of all your programs and data; or your computer system being invaded by someone or something (a virus) that disables it.

External utility programs **are special programs that provide specific useful services not performed or performed less well by other systems software programs.** Examples of such services are backup of your files for storage, recovery of damaged files, virus protection, data compression, and memory management. Some of these features are essential to preventing or rescuing you from disaster. Others merely offer convenience.

Some Specific Utility Tasks

Some of the principal services offered by utilities are the following:

- **Screen saver:** A *screen saver* **is a utility that supposedly prevents a monitor's display screen from being etched by an unchanging image ("burn-in").** Some people believe that if a computer is left turned on without keyboard or mouse activity, whatever static image is displayed may burn into the screen. Screen savers automatically put some moving patterns on the screen, supposedly to prevent burn-in. Actually, burn-in doesn't happen on today's monitors. Nevertheless, people continue to buy screen savers, often just to have a kind of "visual wallpaper." Some of these can be quite entertaining, such as flying toasters.

- **Data recovery:** One day in the 1970s, so the story goes, Peter Norton was doing a programming job when he accidentally deleted an important file. This was, and is, a common enough error. However, instead of re-entering all the information, Norton decided to write a computer program to recover the lost data. He called the program *The Norton Utilities.* Ultimately it and other utilities made him very rich.[27]

   A *data recovery utility* is used to *undelete* a file or information that has been accidentally deleted. **Undelete** **means to undo the last delete operation that has taken place.** The data or program you are trying to recover may be on a hard disk or a diskette.

- **Backup:** Suddenly your hard-disk drive fails, and you have no more programs or files. Fortunately, you have (we hope) used a utility to make a backup, or duplicate copy, of the information on your hard disk. DOS has commands to help you make backups on diskettes, but they are not easy to use. Other utilities are more convenient. Moreover, they also condense (compress) the data, so that fewer diskettes are required.

   Examples of backup utilities are Norton Backup from Symantec, Backup Exec from Arcada Software, Colorado Backup, and Fastback Plus from Fifth Generation Systems.

- **Virus protection:** Few things can make your heart sink faster than the sudden failure of your hard disk. The exception may be the realization that your computer system has been invaded by a virus. **A *virus* consists of hidden programming instructions that are buried within an applications or systems program. They copy themselves to other programs, causing havoc.** Sometimes the virus is merely a simple prank that pops up a message. Sometimes, however, it can destroy programs and data. Viruses are spread when people exchange diskettes or download (make copies of) information from computer networks or the Internet.

  Fortunately, antivirus software is available. ***Antivirus software* is a utility program that scans hard disks, diskettes, and the microcomputer's memory to detect viruses.** Some utilities destroy the virus on the spot. Others notify you of possible viral behavior, in case the virus originated after the antivirus software was released.

  Examples of antivirus software are Anti-Virus from Central Point Software, Norton AntiVirus from Symantec, McAfee virus protection software, and ViruCide from Parsons Technology.

- **File defragmentation:** Over time, as you delete old files from your hard disk and add new ones, something happens: the files become *fragmented.* ***Fragmentation* is the scattering of portions of files about the disk in non-adjacent areas, thus greatly slowing access to the files.**

  When a hard disk is new, the operating system puts files on the disk contiguously (next to one another). However, as you update a file over time, new data for that file is distributed to unused spaces. These spaces may not be contiguous to the older data in that file. It takes the operating system longer to read these fragmented files. By using a utility program, you can "defragment" the file and speed up the drive's operation.

  An example of a program for unscrambling fragmented files is Norton SpeedDisk utility.

- **Data compression:** As you continue to store files on your hard disk, it will eventually fill up. You then have three choices: You can delete old files to make room for the new. You can buy a new hard disk with more capacity and transfer the old files and programs to it. Or you can buy a data compression utility.

  ***Data compression* removes redundant elements, gaps, and unnecessary data from a computer's storage space so less space is required to store or transmit data.** With a data compression utility, files can be made more compact for storage on your hard-disk drive. The files are then "stretched out" again when you need them.

  Examples of data compression programs are Stacker from Stac Electronics, Double Disk from Verisoft Systems, and SuperStor Pro from AddStor.

- **Memory management:** Different microcomputers have different types of memory, and different applications programs have different memory requirements. *Memory-management* utilities are programs that determine how to efficiently control and allocate memory resources.

  Memory-management programs may be activated by software *drivers.* **A *driver* is a series of program instructions that standardizes the format of data transmitted between a computer and a peripheral device,** such as a mouse or printer. Electrical and mechanical requirements differ among peripheral devices. Thus, software drivers are needed so that the computer's operating system will know how to handle them. Many basic drivers come with the operating system. If you buy a new peripheral device, however, you need to install the appropriate software driver so the computer can operate it.

Other examples of utilities are file conversion, file transfer, and security. A *file conversion utility* converts files between any two applications or systems formats—such as between WordPerfect and Word for Windows or between Windows and Mac OS. A *file transfer utility* allows files from a portable computer to be transferred to a desktop computer or a mainframe computer and vice versa. A *security utility* protects unauthorized people from gaining access to your computer without using a password, or correct code. Other utilities also exist.

External utility programs are often originally offered by companies other than those making operating systems. Later the operating system developers may incorporate these features as part of the upgrades of their products.

Multiple-Utility Packages

Some utilities are available singly, but others are available as "multipacks." These multiple-utility packages provide several utility disks bundled in one box, affording considerable savings. Examples are Symantec's Norton Desktop (for DOS, Windows, or Macintosh), which provides data-recovery, defragmenting, memory-management, screen-saving, and other tools. Similar combination-utility packages are 911 Utilities from Microcom, and PC Tools from Central Point Software.

The Future: What's Coming?

Preview & Review: Upgrades in operating systems are planned by Microsoft, IBM, and Apple, as well as some joint ventures. Some future computers might be "network PCs," without their own operating systems, and dominated by Web browsers.

Nothing stands still. The major systems software developers toil on the versions to come, those works in progress to which they have given fanciful code names such as Memphis and Gershwin. However, almost without warning, the Internet and the World Wide Web have dramatically changed the picture. Since this book focuses on computers and *communications* technology, we will briefly discuss this issue here.

Coming Attractions

Here are some developments to keep your eye on:

- **From Microsoft:** Code-named *Cairo* (pronounced "*Kay*-roh" and named for the city in southern Illinois), this successor to Windows NT is due out from Microsoft in 1997. Beyond that is a successor named *Daytona*, which will feature speed improvements (and is therefore named for the Daytona 500 stock-car race). A major upgrade for Windows 95 code-named *Memphis* is planned. We have already mentioned the *SIPC* standards proposed for the next wave of PC/TV-type devices.
- **From IBM:** One of IBM's long-range strategies is to reduce the importance of operating systems altogether. In late 1995 it announced a new kind of universal-software code called *Microkernel*, a form of software even more basic than an operating system, which would enable software to work on different computers, whether mainframes or microcomputers.

Microkernel offers the opportunity for the long-desired "open computing," in which software developers and businesses could use one kind of programming code to run applications on many different machines. However, in a world dominated by "*Wintel*"—the slang name given to *Win*dows software allied with In*tel*, maker of the most popular line of PC microprocessors—IBM clearly will have an uphill marketing battle. One analyst for market-research firm International Data compared the technology of Microkernel to the development of a great automobile transmission that has yet to be turned into a truck or car.[28]

- **From Apple:** The secret name for the original Macintosh was *Mozart.* Now Apple's engineers are working on something code-named after another composer, Aaron Copland. *Copland* (another program called *Gershwin* is next) will be the first rewrite of the Macintosh operating system since 1984. It promises to be a complete retooling that will, like Windows NT and OS/2, be better suited for corporate networking. Far behind schedule, Copland was supposed to come out in time to compete with Windows 96 or Nashville, but some predict it won't arrive until 1997.[29]

  Copland will unquestionably be an improvement over the original Mac OS, although *San Jose Mercury News* computing editor Dan Gillmor, for one, believes "it won't offer some things Windows 95, Windows NT, and OS/2 offer today."[30] Among other features, Copland will have a screen that can be designed to fit any taste, an "intelligent" interface that makes computers easier to use, better ways to find and organize files, and software "agents" that users can order to do routine tasks (such as print out your e-mail every morning).[31–33]

  In addition, Copland uses an IBM-Apple software standard called *Open-Doc,* which allows software developers to create applications in mix-and-match components. (Windows 95, OS/2, and Unix systems can also employ OpenDoc.) Thus, software developers will have greater flexibility and customers will have more choices in putting together their programs.

  Finally, in January 1996 Apple launched a programming system called *Cyberdog* that allows users to set up their own custom-made Internet interfaces and to easily add Internet links to their favorite applications. Thus, you could interrupt your word processing program, for example, to execute a Cyberdog software module that will call up a weather report from the World Wide Web.

- **Joint efforts:** To stop the Microsoft juggernaut, companies have tried several kinds of joint ventures. For instance, in 1991—in an event described as "the equivalent of General Motors and Ford agreeing to offer cars sharing the same engines and designs"—IBM and Apple (joined later by Hewlett-Packard) created a company named *Taligent* to develop a universal operating system.[34] Another Apple-IBM joint venture called *Kaleida Labs* was designed to produce software that would let consumers play any kind of multimedia program on PCs, game players, and TV set-top boxes.[35] In late 1995, because of lack of progress, conflicting cultures, and divergent timetables, Kaleida was disbanded, and Taligent became just an engineering subsidiary of IBM.

  Still, new joint efforts go forward. For instance, in July 1995, Oracle teamed up with Apple and IBM to jointly market a software tool, Oracle Power Objects, that would help software makers develop programs for computer networks that could run interchangeably on different operating systems, including those for Windows, Mac OS, and OS/2.[36] And in November 1995, Apple, IBM, and chip-maker Motorola developed a blueprint for a microprocessor, called the PowerPC, around which they hoped to build computers that could run Mac OS, OS/2, and Windows NT.[37]

The Problem with Personal Computing Today

Will any of the foregoing efforts achieve the broad-based ease of use that customers are looking for? We are in an era that resembles the early days of videocassette recorders, before VHS finally triumphed over Betamax as the standard for VCRs. What is needed, as Bill Gates observes, is computer software "designed to take everyday tasks and make them automatic, and to take complex tasks and make them easier."[38]

Today personal computing is complicated because of conflicting standards. Could it be different tomorrow as more and more people join the trend toward networked computers and access to the World Wide Web?

As we've seen, there are different hardware and software standards, or "platforms." ***Platform* means the particular hardware or software standard on which a computer system is based.** Examples are the Macintosh platform versus the IBM-compatible platform, or Unix versus Windows NT. Developers of applications software, such as word processors or database managers, need to make different versions to run on all the platforms.

Networking complicates things even further. "Text, photos, sound files, video, and other kinds of data come in so many different formats that it's nearly impossible to maintain the software needed to use them," points out one writer. "Users must steer their own way through the complex, upgrade-crazy world of computing."[39]

Today microcomputer users who wish to access online data sources must provide not only their own computer, modem, and communications software but also their own operating system software and applications software. (■ *See Panel 3.13, top.*)

Could this change in the future?

Personal Computing Tomorrow

Today you must take responsibility for making sure your computer system will be compatible with others you have to deal with. (For instance, if a Macintosh user sends you a file to run on your IBM PC, it's up to you to take the trouble to use special software that will translate the file so it will work on your system.) What if the responsibility for ensuring compatibility between different systems were left to online service providers?

In this future model, you would use your Web browser (✓ p. 79) to access the World Wide Web and take advantage of applications software anywhere on the network. (■ *See Panel 3.13, bottom.*) It would not matter what operating system you used. Applications software would become nearly disposable. You would download applications software and pay a few cents or a few dollars for each use. You would store frequently used software on your own computer. You would not need to worry about buying the right software, since it could be provided online whenever you needed to accomplish a task.[40]

Bloatware or the Network Computer?

A new concept has entered the language, that of "bloatware." *Bloatware* is a colloquial name for software that is so crowded ("bloated") with features that it requires a powerful microprocessor and enormous amounts of main memory and hard-disk storage capacity to run efficiently. Bloatware, of course, fuels the movement toward upgrading, in which users must buy more and more powerful hardware to support the software. Windows 95 and the various kinds of software suites, or "officeware," are examples of this kind of software.

Personal computing today

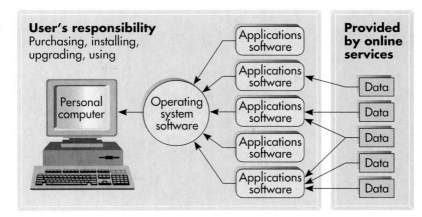

Personal computing tomorrow

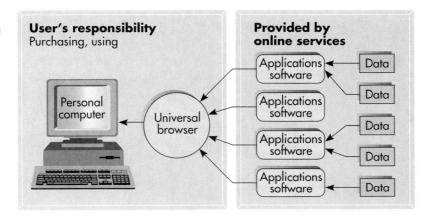

PANEL 3.13

Online personal computing—today and tomorrow

(Top) Today users provide their own operating system software and their own applications software and are usually responsible for installing them on their personal computers. They are also responsible for any upgrades of hardware and software. Data can be input or downloaded from online sources. *(Bottom)* Tomorrow, according to this model, users would not have to worry about operating systems or even about having to acquire and install (and upgrade) their own applications software. Using a universal Web browser, they could download not only data but also different kinds of applications software from an online source.

Against this, engineers have proposed the idea of the "network computer" or "hollow PC." This view—which not everyone accepts—is that the expensive PCs with muscular microprocessors and operating systems would be replaced by network computers costing perhaps $500 or so.[41] Also known as the *Internet PC*, the *network computer (NC)* would theoretically be a "hollowed out" computer, perhaps without even a hard disk, serving as a mere terminal or entry point to the online universe. The computer thus becomes a peripheral to the Internet.

A number of companies are touting the $500 network computer: Sun Microsystems, Netscape Communications, Oracle, IBM, Apple Computer, Hewlett-Packard, AT&T, Silicon Graphics, Toshiba, and several others.[42,43] Why would companies like these, some of them hardware manufacturers, support the notion of selling an inexpensive box? The answer is: to keep Microsoft from further dominating the world of computing.

The concept of the "hollow PC" raises some questions:[44–46]

- **Would the browser really become the OS?** Would a Web browser become the operating system? Or will existing operating systems expand, as in the past, taking over browser functions?

- **Would communications functions really take over?** Would communications functions become the entire computer, as proponents of the network PC contend? Or would they simply become part of the personal computer's existing repertoire of skills?

- **Would an NC really be easy to use?** Would a network computer really be user friendly? At present, features such as graphical user interfaces require lots of hardware and software.

R E A D M E

Practical Matters: Computer Books for "Dummies"

FOSTER CITY [California]—John Kilcullen is clever. He thinks like a dummy.

His understanding of the word, which he calls a term of endearment, has helped take his "For Dummies" instructional book series to more than 18 million books in print and revenue of $50 million [in 1994].

"Everybody is a dummy at something," said Kilcullen, co-founder, president and chief executive officer of the "Dummies" publishing house, IDG Books Worldwide Inc. "Eight-year-old kids send in letters saying I learned OS/2. Prisoners write who want free books. Elderly people say, 'Finally, I understand what my kids are doing.'" . . .

On subjects ranging from surfing the Internet to cutting taxes, 150 different "Dummies" books follow a proven formula: humor, attitude and back-to-basics instructions.

Chapters are arranged to anticipate readers' frustrations. To prevent embarrassment among those sensitive to the title, "Macs for Dummies" author David Pogue even designed a fake cover to go under the original. If you rip off the classic black and yellow "Dummies" front, the book becomes "Macintosh Methodologies in Theory and Practice."

The earliest books, which focused on personal computers, sold well because Kilcullen had done his homework, said *Publishers Weekly* technology editor Paul Hilts.

Kilcullen had observed that most computer books were boring and written for users familiar with the technology. These books were displayed in the techie "ghetto" of bookstores, not where the average book buyers hung out.

Most "Dummies" books sell for under $20, feature an eye-catching yellow cover with a distinctive chalkboard design and are displayed outside computer sections of most bookstores. . . .

Waldenbooks, the largest mall bookseller in the country, initially declined to carry the books, saying readers would be insulted by the title. Even after "DOS for Dummies" became a runaway hit—2 million copies now sold—critics said "Dummies" was a one-book wonder. Not any more.

"They are always among the best-selling books at our 1100 stores nationwide," said Susan Arnold, spokesperson for Waldenbooks.

—Lisa Alcalay Klug, "Dumb Luck," *San Francisco Examiner*

- **Aren't high-speed connections required?** Even users equipped with the fastest modems would find downloading even small programs ("applets") time-consuming. Doesn't the network computer ultimately depend on faster connections than are possible with the standard telephone lines and modems now in place?

- **Doesn't the NC run counter to computing trends?** Most trends in computing have moved toward personal control and access, as from the mainframe to the microcomputer. Wouldn't a network computer that leaves most functions with online sources run counter to that trend?

- **Would users go for it?** Would computer users really prefer scaled-down generic software that must be retrieved from the Internet each time it is used? Would a pay-per-use system tied to the Internet really be cheaper in the long run?

Onward: Toward Compatibility

The push is on to make computing and communications products compatible. Customers are demanding that computer companies work together to create products that will make it easy to access and use great amounts of information. As technological capabilities increase, so will the demand for simplicity.

Whether compatibility and simplicity will be provided by a proprietary system like Windows 95 or Windows NT or by "open standards" of some sort of Web software, perhaps, finally, the best products *will* triumph.

Using Software to Access the World Wide Web

What's the easiest way to use the *Internet* ("the Net"), that international conglomeration of thousands of smaller networks? Getting on that part known as the *World Wide Web* ("the Web") is no doubt the best choice. Increasingly, systems software is coming out with features for accessing and exploring the Net and the Web (as OS/2 already has). This Experience Box, however, describes ways to tour both the Net and the Web independent of whatever systems software you have.

The Web resembles a huge encyclopedia filled with thousands of general topics or so-called *Web sites* that have been created by computer users and businesses around the world. The entry point to each Web site is a *home page,* which may offer cross-reference links to other pages. Pages may theoretically be in *multimedia* form—meaning they can appear in text, graphical, sound, animation, or video form. At present, however, the Web is dominated by lots of pictures and text but little live, moving content, although that is changing.

To get on the Internet and its World Wide Web, you need a microcomputer, a modem, a telephone line, and communications software. (For details about the initial setup, see the Experience Box at the end of Chapter 2.) You then need to gain access to the Web and, finally, to get a browser. Some browsers—software programs that help you navigate the Web—come in kits that handle the setup for you, as we will explain.[47–51]

Gaining Access to the Web

There are three principal ways of getting connected to the Internet: (1) through school or work, (2) through commercial online services, or (3) through an Internet service provider.

Connecting Through School or Work The easiest access to the Internet is available to students and employees of universities and government agencies, most colleges, and certain large businesses. If you're involved with one of these, you can simply ask another student or co-worker with an Internet account how you can get one also. In the past, college students have often been able to get a free account through their institutions. However, students and faculty living off-campus may not be able to use the connections of campus computers.

Connections through universities and business sites are called *dedicated connections* and consist of phone lines (called *T1* or *T3 carrier lines*) that typically cost thousands of dollars to install and maintain every month. Their main advantage is their high speed, so that the graphic images and other content of the Web unfold more quickly.

Connecting Through Online Services The large commercial online services—such as America Online (AOL), CompuServe, Microsoft Network, or Prodigy—also offer access to the Internet. (See the Experience Box at the end of Chapter 2 for information.) Some offer their own Web browsers, but some (such as AOL) offer Netscape Navigator and Internet Explorer. Commercial online services may also charge more than independent Internet service providers, although they are probably better organized and easier for beginners.

Web access through online services is usually called a *dial-up connection.* As long as you don't live in a rural area, there's no need to worry about long-distance telephone charges; you can generally sign on ("log on") by making a local call. When you receive membership information from the online service, it will tell you what to do.

Connecting Through Internet Service Providers Internet service providers (ISPs) are local or national companies that will provide public access to the Internet for a fee. Examples of national companies include PSI, UUNet, Netcom, and Internet MCI. Telephone companies such as AT&T and Pacific Bell have also jumped into the fray by offering Internet connections. Most ISPs offer a flat-rate monthly fee for a set number of hours of service, typically $20 a month for about 10 hours of use. Additional hours are billed at about $2.00 an hour. The connections offered by ISPs (called *SLIP/PPP connections*—discussed in Chapter 8) may offer faster access to the Internet than those of commercial online services. However, setting up a basic system on a microcomputer can require considerable fussing. Generally, ISPs are better for Internet experts than for beginners.

The whole industry of Internet connections is so new that many ISP users have had problems with uneven service (such as busy signals or severing of online "conversations"). Often ISPs signing up new subscribers aren't prepared to handle traffic jams caused by a great influx of newcomers. Some suggestions for choosing an ISP are given in the accompanying box. (■ *See Panel 3.14.*)

You can also ask someone who is already on the Web to access for you the worldwide list of ISPs at *http://www.thelist.com.* Besides giving information about each provider in your area, "thelist" provides a rating (on a scale of 1 to 10) by users of different ISPs.

Accessing the Web: Browser Software

Once you're connected to the Internet, you then need a Web browser. This software program will help you to get whatever information you want on the Web by clicking

Tips for choosing an Internet service provider (ISP)

ISPs may be less expensive and faster than online services, but they can also cause problems. Here are some questions to which you'll want answers before you sign up.

- **Is the ISP connection a local call?** Some ISPs are local, some are national. Be sure the ISP is in your local calling area, or the telephone company will charge you by the minute for your ISP connection. To find out if your ISP is local, call the phone company's directory assistance operator, and provide the prefix of your (modem's) phone number and the prefix of the ISP. The operator can tell you if the call to the ISP is free.
- **How much will it cost?** Ask about setup charges. Ask what the fee is per month for how many hours. Ask if any software (such as browsers) is included when you join.
- **How good is the service?** Ask how long the ISP has been in business, how fast it has been growing, what the peak use periods are, and how frequently busy signals occur. Ask if customer service (a help line) is available evenings and weekends as well as during business hours.

  Ask about the ratio of subscribers to ISP modems. If the service's modems are all in use, you will get a busy signal when you dial in. A ratio of 15 or 20 subscribers to every one modem probably means frequent busy signals (a 10-to-1 ratio is better).

your mouse pointer on words or pictures the browser displays on your screen.

There are all sorts of Web browsers, the best known being Netscape Navigator (the most popular), Mosaic, Microsoft's Internet Explorer, and Netcom's NetCruiser. (See Panel 3.15.)

Features of Browsers What kinds of things should you consider when selecting a browser? Here are some features:

- *Price:* Some browsers are free (freeware, ✓ p. 55), such as the original Mosaic. Some come free with membership in an online service or ISP—America Online, for instance, offers browsers for both IBM-style computers and Macintoshes. Some may be acquired for a price separately from any online connection: the original Mosaic is free, but Spyglass offers Enhanced Mosaic for $29.95.

  You can get a kit that offers other features besides a browser. For instance, Macintosh users can buy the Apple Internet Connection Kit ($59), which contains the browser Netscape Navigator and several other programs. (They include Claris E-mailer Lite, News Watcher, Fetch, Alladin Stuffit Expander, NCSA Telnet, Adobe Acrobat Reader, Sparkle, Real Audio, MacTCP, MacPPP, and Apple Quicktime VR Player).[52] The kit comes with an Apple Internet Dialer application that

helps you find an Internet service provider. Another kit, for Windows, is Internet In a Box ($149.95), which includes the browser Enhanced Mosaic and coupons for various ISPs and for CompuServe.

- *Ease of setup:* Especially for a beginner, the browser should be easy to set up. Ease of setup favors the university/business dedicated lines or commercial online services, of course, which already have browsers. If not provided by your online service or ISP, the browser should be compatible with it. Most online services allow you to use other browsers besides their own.
- *Ease of use:* If you have a multimedia PC, the browser should allow you to view and hear all of the Web's multimedia—not only text and images but also sounds and video. It should be easy to use for saving "hot lists" of frequently visited Web sites and for saving text and images to your hard disk. Finally, the browser should allow you to do "incremental" viewing of images, so that you can go on reading or browsing while a picture is slowly coming together on your screen, rather than having to wait with browser frozen until the image snaps into view.[53]

Surfing the Web

Once you are connected to the Internet and have used your browser to access the Web, you begin by clicking on the *Home* button found on most browsers. This will take you to

Product	Price	Company & contact
Mosaic	Free	NCSA, Champaign, IL; 217-244-0072, *ftp.ncsa.uiuc.edu*
Netscape Navigator 2.0 (Windows and Mac versions)	Free for education and nonprofit uses if downloaded; otherwise $49	Netscape Communications Corp., Mountain View, CA; 415-528-2555, *http://www.mcom.com*
NetCruiser (Windows only)	Free with subscription	Netcom, Inc., San Jose, CA; 800-501-8649
Internet Explorer	Free with Windows 95	Microsoft Corp., Bellevue, WA; 800-386-5550
Enhanced Mosaic	$29.95	Spyglass, Inc., Naperville, IL; 800-505-1010, *info@spyglass.com*, *http://www.spyglass.com*

a predefined home page, established by the software maker that developed the browser. (■ *See Panel 3.16.)*

Web Untanglers Where do you go from here? You'll find that, unlike a book, there is no page 1 where everyone is supposed to start reading and, unlike an encyclopedia, the entries are not in alphabetical order. Moreover, there is no definitive listing of everything available.

There are, however, a few search tools for helping you find your way around, which can be classified as directories and indexes. (■ *See Panel 3.17.)*

Home button

Web search tools

You may wish to use more than one search tool, since some data will be found on some sources but not others.

Directories

These are lists of Web sites classified by topic.

- **Yahoo** *(http://www.yahoo.com)* is supposedly the largest index of Web places, with more than 36,000 entries, and is used by more than 400,000 people a day. It is supported by on-screen advertising. The opening screen offers 14 general categories. You can "drill down" through layers of increasingly specialized lists to find what you're looking for or you can enter a keyword, which Yahoo will try to match.
- **World Wide Web Virtual Library** *(http://info.cern.ch/hypertext/ DataSources/by-Subject/Overview.html)* is a subject index of Web pages run by volunteers all over the world.
- **Yanoff's Special Internet Connections** *(http://www.uwm.edu/Mirror/ inet.services.html)* is an interesting though subjective list of selected Web locations, organized by topic.

Indexes

These search tools allow you to use keywords or menu choices to look for documents.

- **WebCrawler** *(http://webcrawler.com* or *http://webcrawler.cs.washington. edu/WebCrawler/Home.html)* is one of the best search tools and easier to use than Lycos or InfoSeek, though more limited.
- **Lycos** *(http://www.lycos.com* or *http://lycos2.cs.cmu.edu/)* is the Carnegie-Mellon search page, now a commercial product. Its searching mechanism takes some effort to learn.
- **InfoSeek** *(http://www.infoseek.com)* includes access to proprietary data, such as news from the Associated Press. Unlike other index-type search tools, it charges for its services (20 cents per transaction).

- *Directories:* Directories are lists of Web sites classified by topic. Perhaps the best known example is the Yahoo directory, but there are others.
- *Indexes:* Indexes allow you to find specific documents through keyword searches or menu choices. Examples are WebCrawler, Lycos, and InfoSeek.

We describe directories and indexes in more detail in the Experience Box at the end of Chapter 9.

Web Addresses: URLs Getting to a directory, index, or any other Web location is easy if you know the address. Just click your mouse pointer on the *Open* button. This will open the address, or *URL* (for *Uniform Resource Locator*). Web addresses usually start with *http* (for Hypertext Transfer Protocol) and are followed by a colon and double slash (://). For example, to reach the home page of Yahoo, you would type the address *(http://www.yahoo.com.* Your browser uses the address

to connect you to the computer of the Web site; then it downloads (transfers) the Web page information to display it on your screen.

If you get lost on the Web, you can return to your home page by clicking on the *Home* button.

Visiting Web Sites

Nowadays you see Web site addresses appearing everywhere, in all the mass media, and some of it's terrific and some of it's awful. For a sample of the best, try Yahoo's What's Cool list *(http://www.yahoo.com/Entertainment/COOL_Links)*. For less-than-useful information, try Worst of the Web *(http://turnpike.net/metro/mirsky/ Worst.html)*.

Other interesting Web sites you may want to try are presented in the accompanying box (although some may no longer exist when you read this). (■ See Panel 3.18.)

Some interesting Web sites

Web sites are volatile. Some of the entertainment URL addresses in particular may no longer exist by the time you read this.

Weather

The Interactive Weather Browser *(http://rs560.cl.msu.edu/weather/interactive/html)* offers the latest weather information.

U.S. Government

Federal agencies ranging from the White House to the Supreme Court, from the CIA to the IRS, have their own Web sites. Try accessing some of the following:

- **The White House** *(http://whitehouse.gov)* or *(http://sunsite.unc.edu/white-house/white-house.html)* offers information and details about the White House, including e-mail system, press conferences, and the federal budget.
- **Supreme Court Decisions** *(http://www.law.cornell.edu/supct)* allows you to search by topics or key words for summaries of High Court decisions since 1990.
- **The Library of Congress** *(http://lcweb.loc.gov/homepage/lchp.html)* provides links to tons of interesting material, including the library's vast card catalog.
- **The Smithsonian** *(http://www.si.edu)* has several collections. The Smithsonian gem and mineral collection *(http://galaxy.einet.net/images/gems/gems-icons.html)*, for example, offers nearly 50 images and descriptions of different types of gems and minerals.
- **U.S. Bureau of the Census** *(http://www.census.gov)* provides up-to-the-minute population projections and some access to census data. Try the POPclock (population clock) Projection at *http://www.census/gov/cgi-bin/pop-clock* to see the latest U.S. population estimates.
- **Internal Revenue Service** *(http://www.ustreas.gov/treasury/bureau/irs/irs.html)*—Get advice about filing taxes.
- **Central Intelligence Agency** *(http://www.ic.gov/)* offers a look at declassified spy photos from the 1960s, CIA maps, and other facts.

Canadian Government

This Web site *(http://debra.dgbt.doc.ca:80/opengov/)* enables you to explore (in both English and French) segments of the Canadian federal government: House of Commons, the Senate, the Supreme Court, and federal departments and agencies.

News

The news offerings of the commercial online services—America Online, CompuServe, Prodigy—are generally superior to those on the Web, but you might want to try the following:

- **Online Newspapers** *(http://marketplace.com/e-papers.list.www/e-papers.links.html)* provides a list of links to the major newspapers that are available online.
- **Daily Sources of Business and Economic News** *(http://www.helsinki.fi/~lsaarine/news/html)* offers links to free sources of daily news on the Internet.
- **Sports Information Server** *(http://www.netgen.com/sis/sports.html)* covers football, basketball, and hockey.
- **Canadian Broadcasting Corporation (CBC) Radio Trial** *(http://debra.dgbt.doc.ca/cbc.html)* offers CBC radio transcripts of daily news broadcasts, program listings, and sample radio programs.

Some interesting Web sites *continued*

Science

Because the Internet and World Wide Web originated in science and research, there is an abundance of Web sites having to do with science.

- **National Academy of Sciences**   *(http://www.nas.edu)* includes information and news from the National Research Council, Institute of Medicine, and National Academy of Engineering.
- **National Science Foundation**   *(http://www.nsf.gov)* presents NSF news and information on science trends, research, statistics, education, and grants.
- **NASA Information Services**   *(http://www.gsfc.nasa.gov/NASA_homepage.html)* offers links to the space centers and research labs of the National Aeronautics and Space Administration.
- **National Centers for Environmental Prediction** *(http://grads.iges.org/pix/head.html)* offers weather maps, satellite images, and movies of the Earth.
- **Physics News**   *(http://www.het.brown.edu/news/index.html)* presents news from physical sciences.
- **National Center for Atmospheric Research**   *(http://ucar.edu/metapage.html)* presents research on climate, weather, and the atmosphere.
- **Project Bluebook**   *(http://www.cis.ksu.edu/psiber/substand/bluebook/html)* presents UFO reports.

Culture and the Arts

Although a lot of "culture" on the Web is just silly entertainment, here are some alternatives to try:

- **Internet Underground Music Archive**   *(http://www.iuma.com/)* music archives concentrating on up-and-coming bands.
- **Cardiff's Movie Database Browser**   *(http://www.msstate.edu/Movies/)* is a large database (that Internet users can add to themselves) in which you can find out almost any movie-related fact.
- **World Arts Resources**   *(http://www.cgrg.ohio-state.edu/Newark/artsres.html)* is a guide to online art exhibitions, galleries, and museums.
- **Shakespeare Homepage**   *(http://the-tech.mit.edu/Shakespeare/works.html)* offers the complete works of Shakespeare, for searching or downloading.
- **Philosophy on the Web**   *(http://www.phil.ruu.nl/philosophy-sites.html)* is a listing of sites having to do with philosophical big questions.

Suggested Resources About the Internet

Magazines

Internet World. P.O. Box 713, Mt. Morris, IL 61054-9965. Telephone: 203-226-6967. A magazine for Internet users. Subscription: $29.00 a year (10 issues).

Books

Filo, David, and Yang, Jerry. *Yahoo! Unplugged: Your Discovery Guide to the Web.* Foster City, CA: IDG Books Online, 1995. (Contains CD-ROM with Mosaic Web browser.)

Hahn, Harley, and Stout, Rick. *The Internet Yellow Pages* (3rd ed.) Berkeley, CA: Osborne McGraw-Hill, 1996.

Levine, John. *Internet for Dummies* (2nd ed.). Foster City, CA: IDG, 1995.

Seiter, Charles. *Internet for Macs for Dummies.* Foster City, CA: IDG, 1995.

SUMMARY

What It Is / What It Does	**Why It's Important**

driver *(p. 120, LO 5)* Series of program instructions that standardizes the format of data between a computer and a peripheral device.

Drivers are needed so that the computer's operating system will know how to handle the data and run the peripheral device. If you buy some new peripheral hardware and hook it up to your system, you will also probably have to install that hardware's driver software on your hard disk.

erase *(p. 112, LO 3)* Operating system housekeeping feature that removes from a disk files or programs that are no longer useful.

Erasing is essential to keep hard disks from becoming overcrowded.

external utility programs *(p. 128, LO 6)* Special programs that provide specific useful services not provided or performed less well by other system software programs.

Some of these programs are essential for helping rescue users from disaster—for example, recovery of damaged files, virus protection, data compression, and memory management.

formatting *(p. 110, LO 3)* Also called *initializing;* a computer process that electronically prepares a diskette so it can store data or programs.

Before you can use a new diskette, you usually have to format it.

fragmentation *(p. 129, LO 6)* Uneven distribution of data on a hard disk.

Fragmentation causes operating systems to run slower; to solve this problem, users can buy a file defragmentation software utility.

graphical user interface (GUI) *(p. 111, LO 4)* User interface that uses images to represent options. Some of these images take the form of icons, small pictorial figures that represent tasks, functions, or programs.

GUIs are easier to use than command-driven interfaces and menu-driven interfaces; they permit liberal use of the electronic mouse as a pointing device to move the cursor to a particular icon or place on the display screen. The function represented by the icon can be activated by pressing ("clicking") buttons on the mouse.

icon *(p. 111, LO 4)* Small pictorial figure that represents a task, function, or program.

The function represented by the icon can be activated by pointing at it with the mouse pointer and pressing ("clicking") on the mouse. The use of icons has simplified the use of computers.

language translator *(p. 108, LO 2)* Systems software that translates a program written in a computer language written by a computer programmer (such as BASIC) into the language (machine language) that the computer can understand.

Without language translators, software programmers would have to write all programs in machine language, which is difficult to work with.

Macintosh Operating System (Mac OS) *(p. 116, LO 5)* Operating system used on Apple Macintosh computers.

Although not used in as many offices as DOS and Windows, the Macintosh operating system is easier to use.

menu *(p. 111, LO 4)* List of available commands displayed on the screen.

Menus are used in graphical user interface programs to make software easier to use.

menu-driven interface *(p. 111, LO 4)* User interface that allows users to choose a command from a menu.

Like a restaurant menu, a software menu offers you options to choose from—in this case commands available for manipulating data. Two types of menus are available, menu bars and pull-down menus. Menu-driven interfaces are easier to use than command-driven interfaces.

multiprocessing *(p. 113, LO 3)* Operating system software feature that allows two or more computers or processors linked together to perform work simultaneously. Whereas "concurrently" means at almost the same time, "simultaneously" means at precisely the same time.

Multiprocessing is faster than multitasking and time-sharing. Microcomputer users may encounter an example of multiprocessing in specialized microprocessors called *coprocessors.* Working simultaneously with a computer's CPU microprocessor, a coprocessor will handle such specialized tasks as display screen graphics and high-speed mathematical calculations.

What It Is / What It Does	Why It's Important

multiprogramming *(p. 113, LO 3)* Operating system software feature that allows the execution of two or more programs on a multiuser system. Program execution occurs concurrently, not simultaneously.

See multitasking.

multitasking *(p. 112, LO 3)* Operating system software feature that allows the execution of two or more programs by one user concurrently on the same computer with one central processor.

Allows the computer to rapidly switch back and forth among different tasks. The user is generally unaware of the switching process and is able to use more than one applications program at the same time.

NetWare *(p. 126, LO 5)* Most popular operating system, from Novell, for orchestrating microcomputer-based local area networks (LANs) throughout a company or campus.

NetWare allows PCs to share data files, printers, and file servers.

operating environment *(p. 115, LO 5)* Also known as a *windowing environment* or *shell;* adds a graphical user interface as an outer layer to an operating system. Common features of these operating environments are use of an electronic mouse, pull-down menus, icons and other graphic displays, the ability to run more than one application (such as word processing and spreadsheets) at the same time, and the ability to exchange data between these applications.

Operating environments make systems software easier to use.

operating system (OS) *(p. 109, LO 2)* Principal piece of systems software in any computer system; consists of the master set of programs that manage the basic operations of the computer. The operating system remains in main memory until the computer is turned off.

These programs act as an interface between the user and the computer, handling such matters as running and storing programs and storing and processing data. The operating system allows users to concentrate on their own tasks or applications rather than on the complexities of managing the computer.

OS/2 (Operating System/2) & OS/2 Warp *(p. 119, LO 5)* Microcomputer operating system designed to run on many recent IBM and compatible microcomputers.

Unlike traditional Windows, OS/2 and its most recent version, Warp, do not require DOS to run underneath and so generally process more efficiently. Like Windows, they have a graphical user interface, called the *Workplace Shell (WPS).* OS/2 and Warp can also run most DOS, Windows, and OS/2 applications programs simultaneously, which means users don't have to throw out their old applications to take advantage of new features.

platform *(p. 132, LO 7)* Refers to the particular hardware or software standard on which a computer system is based—for example IBM platform or Macintosh platform.

Users need to be aware that, without special arrangements or software, different platforms are not compatible.

plug and play *(p. 123, LO 5)* Refers to the ability to add a new hardware component to a computer system and have it work without needing to perform complicated technical procedures.

Plug and play greatly simplifies the process of expanding and modifying systems.

rename *(p. 112, LO 3)* Operating system housekeeping feature that allows users to give new filenames to the files on a disk.

Renaming is an essential systems software function.

screen saver *(p. 128, LO 6)* Software utility that supposedly prevents a monitor's display screen from being etched by an unchanging image.

Although new monitors are not usually damaged by unchanging images, many users buy screen savers because their displays are entertaining.

What It Is / What It Does	**Why It's Important**

software *(p. 106, LO 1)* Also called *programs;* the step-by-step coded instructions that tell the computer how to perform a task. Software is of two types: applications software and systems software.

Without software, hardware would be useless.

supervisor *(p. 111, LO 3)* Also called *kernel;* central component of the operating system. It resides in main memory while the computer is on and directs other programs to perform tasks to support applications programs.

Were it not for the supervisor program, users would have to stop one task—for example, writing—and wait for another task to be completed—for example, printing out of a document.

system prompt *(p. 109, LO 3)* A screen display—such as "A:\>" or "C:\>"—that indicates the operating system has been loaded into main memory and asks ("prompts") you to enter a command.

Users need to be familiar with their system's prompts in order to be able to use their software effectively.

systems software *(p. 106, LO 1)* Software that enables applications software to interact with the computer and helps the computer manage its internal resources.

Applications software cannot run without systems software.

time-sharing *(p. 113, LO 3)* Operating system software feature whereby a single computer processes the tasks of several users at different stations in round-robin fashion. Time-sharing and multitasking differ slightly. With time-sharing, the computer spends a fixed amount of time with each program before going on to the next one. With multitasking the computer works on each program until it encounters a logical stopping point, as in waiting for more data to be input.

Time sharing is used when several users are linked by a communications network to a single computer. The computer will work first on one user's task for a fraction of a second, then go on to the next user's task, and so on.

undelete *(p. 128, LO 6)* Data recovery software utility feature that allows the user to undo the last delete operation that has taken place.

This feature allows users who accidentally deleted some data or a file to get it back without having to type it in all over again.

Unix *(p. 125, LO 5)* Operating system for multiple users, with built-in networking capability, the ability to run multiple tasks at one time, and versions that can run on all kinds of computers.

Because it can run with relatively simple modifications on many different kinds of computers, from micros to minis to mainframes, Unix is said to be a "portable" operating system. The main users of Unix are large corporations and banks that use the software for everything from designing airplane parts to currency trading.

user interface *(p. 110, LO 4)* Also called *shell;* part of the operating system that allows users to communicate, or interact, with it. There are three types of user interfaces, for both operating systems and applications software: command-driven, menu-driven, graphical-user.

User interfaces are necessary for users to be able to use a computer system.

utility programs *(p. 107, LO 2)* Systems software generally used to support, enhance, or expand existing programs in a computer system.

Many operating systems have utility programs built in for common purposes such as copying the contents of one disk to another. Other external utility programs are available on separate diskettes to, for example, recover damaged or erased files.

virus *(p. 129, LO 6)* Hidden programming instructions that are buried within an applications or systems program and that copy themselves to other programs, often causing damage.

Viruses can cause users to lose data or files or even shut down entire computer systems.

windows *(p. 118, LO 4)* Feature of graphical user interfaces; causes the display screen to divide into sections. Each window is dedicated to a specific purpose.

Windows *(p. 118, LO 5)* Operating environment made by Microsoft that places a graphical user interface shell around the MS-DOS and PC-DOS operating systems.

Windows 95 *(p. 122, LO 5)* Successor to Windows 3.X for DOS; this is a true operating system for IBM-style computers, rather than just an operating environment.

Windows NT (New Technology) *(p. 125, LO 5)* Operating system intended to support large networks of computers, such as those involved in airline reservations systems.

Using the windows feature, an operating system (or operating environment) can display several windows on a computer screen, each showing a different application program, such as word processing, spreadsheets, and graphics.

The Windows operating environment made DOS easier to use; far more applications have been written for Windows than for DOS alone.

Windows 95 and later versions may become the most common systems software used on microcomputers.

Unlike the traditional Windows operating environment, Windows NT is a true operating system, eliminating the need for DOS and interacting directly with the hardware. It is primarily designed to run on workstations or other more powerful computers.

(Selected answers appear at the back of the book.)

Short-Answer Questions

1. Why does a computer need systems software?
2. What are utility programs?
3. What does the term *booting* mean?
4. What does a language translator do? Why is such a program included in systems software?
5. Can an operating system designed for a mainframe run on a microcomputer?
6. Which is faster: multiprocessing, multitasking, or time-sharing?
7. Why have data compression utilities become necessary for some users?
8. Why do microcomputer users have to format their diskettes before using them?
9. What is NetWare? What does it do?
10. What is a computer virus?

Fill-In-the-Blank Questions

1. The three types of systems software are

   _____, _____,

   and _____.
2. The _____ consists of the master programs that manage the basic operations of the computer.
3. Before you can use a new diskette on a microcomputer, you must _____ it.
4. A graphical user interface uses pictures, or

   _____, to represent processing functions.
5. Programs that are used to expand the existing capabilities of programs in a computer system are called

   _____.
6. In a command-driven user interface, the user types commands after the _____.
7. The main component of the operating system is the

   _____; it directs other programs to perform tasks.

8. Name three microcomputer operating systems besides DOS and the Macintosh Operating System:

   a. _____

   b. _____

   c. _____
9. _____ refers to the ability to add new hardware to a computer system and use it immediately without performing complicated installation procedures.
10. To scan a computer's hard disk, diskette, and memory for viruses, you should use

    _____ software.

Multiple-Choice Questions

1. Which of the following must you use to enable applications software to interact with the computer?
   a. software utilities
   b. systems software
   c. command-driven interface
   d. operating environment
   e. none of the above
2. Which of the following best describes the process of loading an operating system into a computer's memory?
   a. system prompt
   b. formatting
   c. backing up
   d. booting
   e. none of the above
3. Which of the following are you probably using if you're viewing windows and icons?
   a. command-driven interface
   b. menu-driven interface
   c. graphical user interface
   d. none of the above
4. Which of the following is the central component of an operating system?
   a. supervisor
   b. system prompt
   c. operating environment
   d. icons
   e. all of the above

5. _____ is the processing of tasks from several users at different locations in a round-robin fashion.
   a. multitasking
   b. multiprogramming
   c. time-sharing
   d. multiprocessing
   e. none of the above

6. _____ is the simultaneous processing of more than one program by multiple processors.
   a. multitasking
   b. multiprogramming
   c. time-sharing
   d. multiprocessing
   e. none of the above

7. Which of the following will you likely find in a company that uses networks of microcomputers?
   a. DOS
   b. Windows
   c. Unix
   d. NetWare
   e. all of the above

8. A popular service offered by an external utility program is:
   a. screen-saver assistance
   b. data recovery
   c. backup assistance
   d. virus protection
   e. all of the above

9. Which of the following should you use to optimize the amount of space that is used on a disk drive?
   a. antivirus software
   b. defragmentation software
   c. data compression software
   d. memory-management utility
   e. none of the above

10. Which of the following do you need to transmit data between a computer and a peripheral device?
    a. file conversion utility
    b. driver
    c. memory-management software
    d. peripheral utility
    e. all of the above

True/False Questions

T F 1. Applications software starts up the computer and functions as the principal coordinator of all hardware components.

T F 2. A menu-driven user interface is the easiest type of operating system interface to use.

T F 3. Fragmentation is the uneven distribution of a file on the disk.

T F 4. Viruses can be found in both applications and systems programs.

T F 5. Screen burn-in is a bad problem with today's monitors.

T F 6. DOS is a 32-bit operating system.

T F 7. Multitasking is processing done by two or more processors at the same time.

T F 8. The central portion of the operating system is called the *supervisor*.

T F 9. All microcomputer diskettes must be formatted before you can use them.

T F 10. Applications software helps the computer manage its internal resources.

Projects/Critical-Thinking Questions

1. If you have been using a particular microcomputer for 2 years and are planning to upgrade the version of systems software you are using, what issues must you consider before you go ahead and buy the new version?

2. By the time this textbook goes to print, Microsoft will have released Windows 96. What is new in this latest release? What utilities are included in Windows 96? Which of these utilities were also included in Windows 95? Would you advise a Windows 95 user to upgrade to Windows 96? Why/why not?

3. PC-DOS 7 was released by IBM in March of 1995. Why would IBM spend money developing a new version of DOS if most experts believed that DOS would soon be obsolete? Who is using PC-DOS 7 now? Do you think IBM's decision to release a new version of DOS was a good one? Why/why not?

4. If you were in the market for a new microcomputer today, what software would you want to use on it? What systems software would you choose? Applications software? Why? How would you go about making your choices?

5. Pen-based computing uses its own particular type of systems software. Check some articles in computer magazines to find out what makes this type of systems software different from regular microcomputer systems software. Is pen-based systems software compatible with DOS? Windows? How would pen-based systems software limit a traditional microcomputer user?

net **Using the Internet**

Objective: *In this exercise we describe how to navigate the Web using Netscape Navigator 2.0 or 2.01.*

Before you continue: *We assume you have access to the Internet through your university, business, or commercial service provider and to the Web browser tool named Netscape Navigator. Additionally, we assume you know how to connect to the Internet and then load Netscape Navigator. If necessary, ask your instructor or system administrator for assistance.*

1. The home page for Netscape Navigator should appear on your screen (see below).

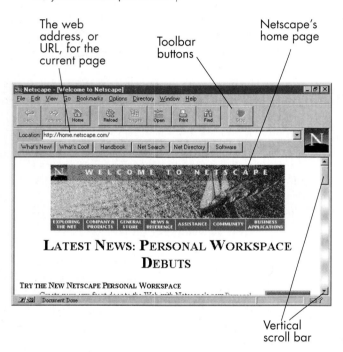

If at any time during the current work session you want to return to this home page, simply click the *Home icon* (🏠) on the toolbar. The labeled toolbar appears below:

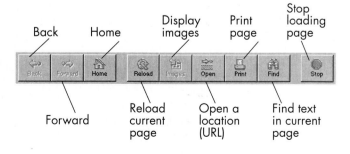

2. All underlined phrases are linked to other pages of information. When you click an underlined phrase, Netscape takes you to another page. For example:
   DRAG: the vertical scroll bar downward until you see the "Welcome to Netscape!" message and underlined phrases below
   CLICK: *an underlined phrase of your choice*
   A new page should appear.

3. Locate another underlined phrase of interest and then click. A new page should appear.

4. What if you want to go back to the previous page?
   CLICK: Back icon (🔙) on the toolbar
   The page you displayed last should appear.

5. To go to the next page:
   CLICK: Forward icon (🔜) on the toolbar

6. To return to the home page:
   CLICK: Home button (🏠)

7. To look at where you've been on the Web so far, you can use the Go menu. A list of the Web sites you've visited will appear at the bottom of the menu.
   CHOOSE: Go from the Menu bar
   Note the list of Web sites at the bottom of the pull-down menu. To go to one of the listed Web sites, simply click it with the mouse. To exit the pull-down menu:
   CHOOSE: Go from the Menu bar

8. To go directly to a specific page by typing in a web address, or URL (Uniform Resource Locator), you use the Open button (📂) on the toolbar.
   CLICK: Open button (📂)
   TYPE: *a web address listed in the Experience Box of this chapter*
   For example, http://www.yahoo.com will take you to the home page for Yahoo, a popular Web library. Remember that web addresses often change, so some of the addresses listed in this chapter might no longer be in use.
   PRESS: [Enter]
   A new page should appear.

9. Now that you're familiar with some navigating basics, practice navigating the Web on your own. When you're finished, be sure to exit Netscape by choosing File, Exit from the Menu bar.

CHAPTER 4

Processors
Hardware for Power & Portability

Concepts You Should Know

After reading this chapter, you should be able to:

1. Explain the miniaturization of processors and its link to micromachines and mobility.
2. Identify and describe the major types of computer systems.
3. Describe the function and operations of the CPU, including the machine cycle, and of main memory.
4. Define the different ways of measuring processing speeds.
5. Discuss how data and programs are represented in the computer.
6. Identify the components of the system unit and explain their uses.
7. Describe some adverse effects information technology has had on the environment.

Anticipointment" is a common experience for buyers of information technology.

Anticipointment, as Berkeley, California, editor Hank Roberts explained in an online computer conference, is a word coined "to describe always finding that, just as the techie-toy I've been dreaming about getting for six months has become affordable, there's something so much better on the horizon that I guess I have to wait just a bit longer."[1]

This feeling of anticipation-plus-disappointment could have been experienced by anyone observing trends in *portability* or *mobility* in electronic devices. For example, in 1955, Zenith ran ads showing a young woman holding a television set. The caption read: IT DOESN'T TAKE A MUSCLE MAN TO MOVE THIS LIGHTWEIGHT TV. That "lightweight" TV weighed a hefty 45 pounds. Today, by contrast, there is a handheld Casio color TV weighing a mere 6.2 ounces.

Similarly, tape recorders went from RCA's 35-pound machine in 1953 to today's Sony microcassette recorder of 3.5 ounces. Video cameras for consumers went from two components weighing 18.8 pounds in RCA's 1979 model to JVC's 18-ounce digital video camcorder today. Portable computers began in 1982 with Osborne's advertised "24 pounds of sophisticated computing power," a "luggable" size that most people would consider too unwieldy today. Since then portable computers have rapidly come down in weight and size. Now there are laptops (8–20 pounds), notebooks (4–7.5 pounds), subnotebooks (2.5–4 pounds), and pocket PCs (1 pound or less).

All this goes to show how relative the term *portability* is—and how much existing sizes are subject to obsolescence.[2]

Microchips, Miniaturization, & Mobility

Preview & Review: Computers used to be made from vacuum tubes. Then came the tiny switches called *transistors,* followed by integrated circuits made from the common mineral silicon. Integrated circuits called *microchips,* or *chips,* are printed and cut out of "wafers" of silicon. The microcomputer microprocessors, which process data, are made from microchips. They are also used in other instruments, such as phones and TVs.

Had the transistor not arrived, as it did in 1947, the Age of Portability and consequent mobility would never have happened. To us a "portable" telephone might have meant the 40-pound backpack radio-phones carried by some American GIs through World War II, rather than the 6-ounce shirt-pocket cellular models available today.

From Vacuum Tubes to Transistors to Microchips

Old-time radios used vacuum tubes—lightbulb-size electronic tubes with glowing filaments. The last computer to use these tubes, the ENIAC, which was turned on in 1946, employed 18,000 of them. Unfortunately, a tube failure occurred every 7 minutes, and it took more than 15 minutes to find and replace the faulty tube. Thus, it was difficult to get any useful computing work done. Moreover, the ENIAC was enormous, occupying 1500 square feet and weighing 30 tons.

The transistor changed all that. **A *transistor* is essentially a tiny electrically operated switch that can alternate between "on" and "off" many millions of times per second.** The first transistors were one-hundredth the size of a vacuum tube, needed no warmup time, consumed less energy, and were faster and more reliable. (■ *See Panel 4.1.*) Moreover, they marked the beginning of a process of miniaturization that has not ended yet. In 1960 one transistor fit into an area about a half-centimeter square. This was sufficient to permit Zenith, for instance, to market a transistor radio weighing about 1 pound (convenient, they advertised, for "pocket or purse"). Today more than 3 million transistors can be squeezed into a half centimeter, and a Sony headset radio, for example, weighs only 6.2 ounces.

In the old days, transistors were made individually and then formed into an electronic circuit with the use of wires and solder. Today transistors are part of an ***integrated circuit;* that is, an entire electronic circuit, including wires, is all formed together on a single chip of special material, silicon,** as part of a single manufacturing process. An integrated circuit embodies what is called *solid-state technology.* **Solid state means that the electrons are traveling through solid material**—in this case silicon. They do not travel through a vacuum, as was the case with the old radio vacuum tubes.

What is silicon, and why use it? ***Silicon* is an element that is widely found in clay and sand.** It is used not only because its abundance makes it cheap but also because it is a *semiconductor.* **A *semiconductor* is material whose electrical properties are intermediate between a good conductor of electricity and a nonconductor of electricity.** (An example of a good conductor of electricity is copper in household wiring; an example of a nonconductor is

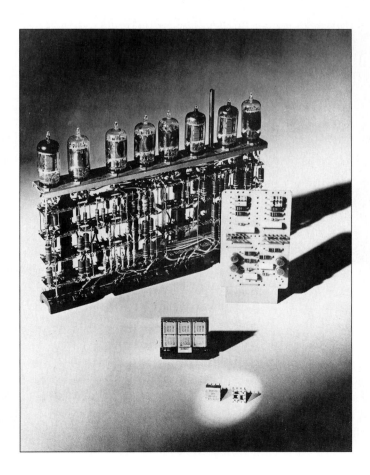

Shrinking components

The lightbulb-size 1940s vacuum tube was replaced in the 1950s by a transistor one-hundredth its size. Today's transistors are much smaller, being microscopic in size.

the plastic sheath around that wiring.) Because it is only a semi-conductor, silicon has partial resistance to electricity. As a result, when good-conducting metals are overlaid on the silicon, the electronic circuitry of the integrated circuit can be created.

How is such microscopic circuitry put onto the silicon? In brief, like this:[3]

1. A large drawing of electrical circuitry is made that looks something like the map of a train yard. The drawing is photographically reduced hundreds of times so that it is of microscopic size.
2. That reduced photograph is then duplicated many times so that, like a sheet of postage stamps, there are multiple copies of the same image or circuit.
3. That sheet of multiple copies of the circuit is then printed (in a printing process called *photolithography*) and etched onto an 8-inch-diameter piece of silicon called a *wafer.* (■ *See Panel 4.2.*)
4. Subsequent printings of layer after layer of additional circuits produce multilayered and interconnected electronic circuitry built above and below the original silicon surface.
5. Later an automated die-cutting machine cuts the wafer into separate *chips,* which are usually less than 1 centimeter square and about half a millimeter thick. **A *chip,* or *microchip,* is a tiny piece of silicon that contains millions of microminiature electronic circuit components,** mainly

PANEL 4.2

Making of a chip

(Top) Wafer imprinted with many microprocessors. *(Bottom)* Microprocessor chip mounted in protective frame with pins that can be connected to an electronic device such as a microcomputer.

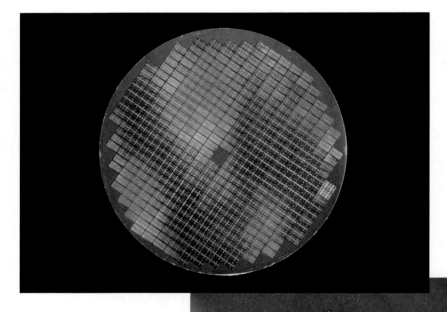

transistors. An 8-inch silicon wafer will have a grid of nearly 300 chips, each with as many as 5.5 million transistors.

6. After being tested, each chip is then mounted in a protective frame with extruding metallic pins that provide electrical connections through wires to a computer or other electronic device. (■ *Refer back to Panel 4.2.*)

Chip manufacture requires very clean environments, which is why chip manufacturing workers appear to be dressed for a surgical operation. Such workers must also be highly skilled, which is why chip makers are not found everywhere in the world.

Miniaturization Miracles: Microchips, Microprocessors, & Micromachines

Microchips are "industrial rice," as the Japanese call them, the force behind the miniaturization that has revolutionized consumer electronics, computers, and communications.[4] They are the devices that store and process data in all the electronic gadgetry we've become accustomed to. This covers a range of things from microwave ovens to videogame controllers to music synthesizers to cameras to automobile fuel-injection systems to pagers to satellites.

There are different kinds of microchips—for example, microprocessor, memory, logic, communications, graphics, and math coprocessor chips. We discuss some of these later in this chapter. Perhaps the most important is the microprocessor chip. **A *microprocessor* ("microscopic processor" or "processor on a chip") is the miniaturized circuitry of a computer processor—the part that processes, or manipulates, data into information.** When modified for use in machines other than computers, microprocessors are called *microcontrollers*, or *embedded computers* (✓ p. 23).

The microprocessor, says Michael Malone, author of *The Microprocessor: A Biography*, "is the most important invention of the 20th century."[5] Quite a bold claim, considering the incredible products that have issued forth during the past nearly 100 years. Part of the reason, Malone argues, is the pervasiveness of the microprocessor in the important machines in our lives, from computers to transportation. (■ *See Panel 4.3.*) However, pervasiveness isn't the whole story. "The microprocessor is, intrinsically, something special," he says. "Just as [the human being] is an animal, yet transcends that state, so too the microprocessor is a silicon chip, but more." Why? Because it can be programmed to recognize and respond to patterns in their environment, as humans do. Malone writes:

> Implant [a microprocessor] into a traditional machine—say an automobile engine or refrigerator—and suddenly that machine for the first time can learn, it can adapt to its environment, respond to changing conditions, become more efficient, more responsive to the unique needs of its user. That machine now evolves, not from generation to generation but within itself.[6]

Mobility

Smallness—in TVs, phones, radios, camcorders, CD players, and computers—is now largely taken for granted. In the 1980s portability, or mobility, meant trading off computing power and convenience in return for smaller size and weight. Today, however, we are getting close to the point where we don't have to give up anything. As a result, experts predict that small, powerful, wireless personal electronic devices will soon transform our lives far more than the personal computer has done so far.[7] (■ *See Panel 4.4.*)

The pervasiveness of the microprocessor

"Everybody knows that the microprocessor is the brains of the personal computer, the video game, and the automated teller machine. But it also made possible the revolution in graphic workstations that gives us everything from 3D product design to *Jurassic Park*. That's just computation. The microprocessor also has led to a renaissance in test and measurement, bringing intelligence to a vast array of products from fetal monitors to gas chromatographs.

The microprocessor has transformed control and automation as well: all of those new industrial robots are run by microprocessors, so is the air conditioning, the gas pump and maybe even all the lights in your house. The modern airplane would have a hard time flying without hundreds of microprocessors. Nor would your automobile run safely and efficiently—many of the new models contain dozens running everything from the fuel injection to the windshield wipers."

—Michael S. Malone, *The Microprocessor: A Biography*

"By the end of the decade, experts say, the new breed of portable electronics will transform our lives more than the personal computer did during the 1980s. For the truth is that the personal computer didn't turn out to be personal: It stayed put while its users moved around.

By contrast, the new generation of machines will be truly personal computers, designed for our mobile lives. We will view office video presentations on the homeward train ride, and reply to electronic mail during boring movies on television. We will read office memos between strokes on the golf course, and answer messages from our children in the middle of business meetings.

Portability, clearly, will profoundly change the way we live and work, says Richard Shaffer, who writes a newsletter about high-tech developments. 'People will soon be able to structure their work around their lives,' he predicts, 'rather than the reverse.'"

—Laurence Hooper, "No Compromises," *Wall Street Journal*

The small-is-beautiful revolution

Portability will change the way we work and live.

Five Types of Computer Systems

Preview & Review: Computers are classified into microcontrollers, microcomputers, minicomputers, mainframe computers, and supercomputers.

Microcontrollers are embedded in machines such as cars and kitchen appliances.

Microcomputers may be personal computers (PCs) or workstations. PCs include desktop and floor-standing units, laptops, notebooks, subnotebooks, pocket PCs, and pen computers. Workstations are sophisticated desktop microcomputers used for technical purposes.

Minicomputers are intermediate-size machines.

Mainframes are the traditional size of computer and are used in large companies to handle millions of transactions.

The high-capacity machines called supercomputers are the fastest calculating devices and are used for large-scale projects. Supercomputers have two designs: vector processing and massively parallel processing.

Any of these types of computers may be used as a server, a central computer in a network.

Generally speaking, the larger the computer, the greater its processing power. As we mentioned in Chapter 1 (✓ p. 22), computers are often classified into five sizes—tiny, small, medium, large, and superlarge. (■ *See Panel 4.5.*)

- Microcontrollers—embedded in "smart" appliances
- Microcomputers—both personal computers and workstations
- Minicomputers
- Mainframe computers
- Supercomputers

In this chapter, we give you a bit more information about these categories.

Microcontrollers

Microcontrollers, also called *embedded computers*, are the tiny, specialized microprocessors installed in "smart" appliances and automobiles. These microcontrollers enable, for example, microwave ovens to store data about how long to cook your potatoes and at what temperature.

Microcomputers: Personal Computers

Microcomputers are small computers that can fit on or beside a desk or are portable. Microcomputers are considered to be of two types: personal computers and workstations.

Personal computers (PCs) are desktop, tower, or portable computers that can run easy-to-use programs such as word processing or spreadsheets. PCs come in several sizes, as follows.

- **Desktop and tower units:** Even though many personal computers today are portable, buyers of new PCs often opt for nonportable systems, for reasons of price, power, or flexibility. For example, the television-tube-like

Different sizes of computers

Shown are several types of microcomputers—pen computer, pocket PC, subnotebook, notebook, laptop, desktop, workstation, and tower—plus a minicomputer, a mainframe, and a supercomputer.

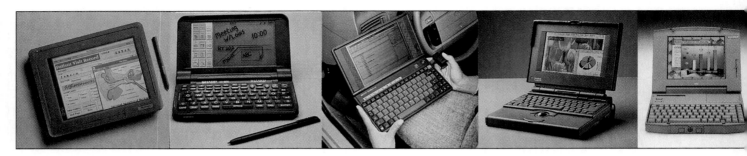

(CRT, or cathode-ray tube) monitors that come with desktops have display screens that are easier to read than those of many portables. Moreover, you can stuff a desktop's roomy system cabinet with add-on circuit boards and other extras, which is not possible with portables.

Desktop PCs are those in which the system cabinet sits on a desk, with keyboard in front and monitor often on top. A difficulty with this arrangement is that the system cabinet's "footprint" can deprive you of a fair amount of desk space. *Tower PCs* are those in which the system cabinet sits as a "tower" on the desk or on the floor next to the desk, giving you more usable desk space.

- **Laptops:** A *laptop computer* **is a portable computer equipped with a flat display screen and weighing 8–20 pounds.** The top of the computer opens up like a clamshell to reveal the screen.

   We describe the differences between display screens elsewhere. Here we will simply say that flat screens don't provide the quality of the monitors found with desktop computers (although that is changing). However, most laptops can be hooked up to standard desktop-type monitors so that you don't lose display quality.

- **Notebooks:** A *notebook computer* **is a portable computer that weighs 4–7.5 pounds and is roughly the size of a thick notebook,** perhaps 8½ by 11 inches. Notebook PCs can easily be tucked into a briefcase or backpack or simply under your arm.

   Notebook computers can be just as powerful as some desktop machines. However, because they are smaller, the keys on the keyboards are closer together and harder to use. Also, as with laptops, the display screens are more difficult to read.

- **Subnotebooks:** A *subnotebook computer* **weighs 2.5–4 pounds.** Clearly, subnotebooks have more of both the advantages and the disadvantages of notebooks.

- **Pocket PCs:** *Pocket personal computers,* **or** *handhelds,* **weigh about 1 pound or less.** These PCs are useful in specific situations, as when a driver of a package-delivery truck must feed hourly status reports to company headquarters. Another use allows police officers to check out suspicious car license numbers against a database in a central computer. Other pocket PCs have more general applications as electronic diaries and pocket organizers.

Pocket PCs may be classified into three types:

(1) *Electronic organizers* **are specialized pocket computers that mainly store appointments, addresses, and "to do" lists. Recent versions feature wireless links to other computers for data transfer.**

(2) *Palmtop computers* **are PCs that are small enough to hold in one hand and operate with the other.**

(3) *Pen computers* **lack a keyboard or a mouse but allow you to input data by writing directly on the screen with a stylus, or pen.** Pen computers are useful for inventory control, as when a store clerk has to count merchandise; for package-delivery drivers who must get electronic signatures as proof of delivery; and for more general purposes, like those of electronic organizers and PDAs.

Personal digital assistants (PDAs), **or** *personal communicators,* **are small, pen-controlled, handheld computers that, in their most developed form, can do two-way wireless messaging.**

We explain more about notebooks, subnotebooks, and pocket PCs, and their usefulness, in the Experience Box at the end of this chapter.

What is the one thing besides their light weight that makes portable computers truly portable? The answer: batteries. A typical notebook's batteries will keep it running about 3–5 hours, a subnotebook's about 3 hours. The record holder seems to be 8 hours and 45 minutes on a standard nickel metal hydride battery.[8] Then the PC must be plugged into an AC outlet and charged up again. Some travelers carry spare battery packs.

In the works is a zinc-air battery that can run a laptop for up to 12 hours without a recharge. However, refinements are still being made.[9]

Microcomputers: Workstations

Workstations look like desktop PCs but are far more powerful. Traditionally, *workstations* **were sophisticated machines that fit on a desk, cost $10,000–$150,000, and were used mainly by engineers and scientists for technical purposes.** However, workstations have long been used for computer-aided design and manufacturing (CAD/CAM, ✓ p. 87), software development, and scientific modeling. Workstations have caught the eye of the public

Workstation

Workstations were used to create the special 3D effects for the movie *Toy Story*.

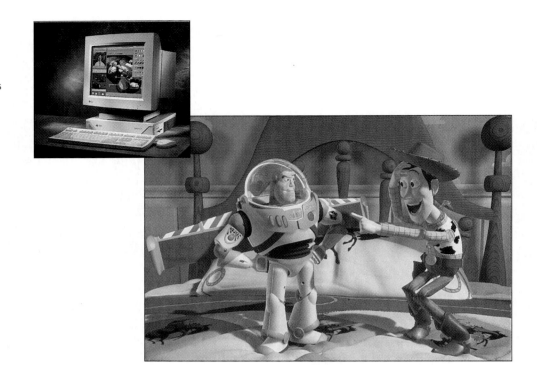

mainly for their graphics capabilities, such as those used to breathe three-dimensional life into toys for the movie *Toy Story*. (■ *See Panel 4.6.*)

Two recent developments have altered the differences between workstations and PCs:[10,11]

- **Decline in workstation prices:** A workstation that not long ago cost $15,000 or more is now available for under $5000, which puts it within range of many PC buyers.

- **Increase in PC power:** In 1993 Intel introduced the Pentium chip; in 1994 Motorola (with IBM and Apple) introduced its PowerPC chip. Both of these very powerful microprocessors are now found in PCs. In addition, Microsoft introduced Windows NT (✓ p. 125), the first operating system designed to take advantage of more powerful microprocessors.

You might deduce from this that, if PCs are becoming more powerful, then workstations are becoming more powerful still—and indeed they are. Over the past 15 years the fastest workstations have increased in speed a thousandfold.[12] They have been cutting into the sales not only of minicomputers and mainframes but even of supercomputers. These large machines have become vulnerable particularly since workstations can now be harnessed in "clusters" to attack a problem simultaneously.

Minicomputers

Minicomputers **are machines midway in cost and capability between microcomputers and mainframes. They can be used either as single workstations or as a system tied by network to several hundred terminals for many users.** (■ *See Panel 4.7.*) The minicomputer overlaps with other categories of computers. A low-end minicomputer may be about as powerful as a high-end

Minicomputer

The VAX is made by Digital Equipment Corporation.

microcomputer and cost about the same. A high-end minicomputer may equal a low-end mainframe.

Traditionally, minicomputers have been used to serve the needs of medium-size companies or of departments within larger companies, often for accounting or design and manufacturing (CAD/CAM). Now many are being replaced by systems of networked microcomputers.

Mainframes

The large computers called *mainframes* are the oldest category of computer system. The word "mainframe" probably comes from the metal frames, housed in cabinets, on which manufacturers mounted the computer's electronic circuits.

Occupying specially wired, air-conditioned rooms and capable of great processing speeds and data storage, *mainframes* are water- or air-cooled computers that are about the size of a Jeep and that range in price from $50,000 to $5 million. (■ *See Panel 4.8.*) Such machines are typically operated by professional programmers and technicians in a centrally managed department within a large company. Examples of such companies are banks, airlines, and insurance companies, which handle millions of transactions.

Today, one hears, "mainframes are dead," being supplanted everywhere by small computers connected together in networks, a trend known as "downsizing." Is this true? The world has an estimated $1 trillion invested in this kind of "big iron"—perhaps 50,000 mainframes, 60% of them made and sold by IBM.[13] But what are the future prospects for people working with mainframes? Although mainframe manufacturers will probably promote new uses for their equipment, there appear to be three trends:[14,15]

Mainframe
The IBM ES/9000.

- **Old mainframes will be kept for some purposes:** Massive and repetitive computing chores, such as maintaining a company's payroll, may best be left on a mainframe rather than moved to a new system.

- **Networks of smaller computers will grow:** Mainframes usually cannot be reprogrammed quickly to develop new products and services, such as pulling together information about single customers from different divisions of a bank. Networks offer the flexibility that mainframes lack because networks are not burdened with an accumulation of out-of-date programming.

- **Mainframes will be reinvented:** IBM has worked to redesign mainframes, which formerly were essentially custom-built. Now they are being manufactured on an assembly-line basis, making them less expensive. In addition, the automobile-size machines will be reduced to the size of a desk. Encompassing more recent technology (called parallel processing, described shortly), new mainframes will not require water cooling. As a result, a $1 million machine will come down in price to only $100,000.

Despite the trend toward downsizing and using networks of smaller computers, mainframe makers such as IBM and Amdahl have continued to ship "big iron" in record amounts. One reason may be that the costs of maintaining a mainframe are actually cheaper (averaging $2300 per user per year, according to one study) than networks of PCs ($6400).[16] Moreover, networks of PCs can't match mainframes for reliability or security of data.[17]

R E A D M E

Case Study: Using the Micro-Mainframe Connection

Al Cohn, 38, is an engineer who is also a marketing manager for a division of Akron-based Goodyear Tire & Rubber. Part of his job is to call on big customers to sell tires used for fleets of vehicles. One of his tools is a 7.6-pound portable IBM ThinkPad with software that helps connect him via phone line to the company's mainframe. He told an interviewer how this arrangement helps him with sales calls.

How can a mainframe help land a sale?

We've got six years' worth of performance data, such as statistics on tread wear, durability, and fuel economy, as well as information on pricing, tire sizes, and competitors' products. When you have that much information, you need a mainframe: no portable system can crunch the numbers and generate charts and graphs that fast.

During a sales call we hook into the mainframe by phone and can answer specific questions from the customer—for example, how a particular tire performs on Mack trucks that haul coal in Kentucky with an average load of 6000 pounds per tire. In 8 to 10 seconds we can generate a graph with the answer and print it out. In the old days people would ask questions like these, and we'd say, "Sure, let us look in our files, and we'll get back to you in two weeks."

What hardware and software lets you do this?

The equipment I carry depends on what we're trying out. Right now I have an IBM ThinkPad 702C, which is a note-book computer with a color screen. We chose the ThinkPad in part because of its removable hard-disk-drive, which makes it possible to carry around several different drives loaded with presentation material. The ThinkPad's optional credit-card-size modem lets me call the mainframe back in Akron.

We use software called Attachmate EXTRA! for Windows to make the ThinkPad function like a mainframe terminal. Attachmate lets you switch from PC to mainframe programs with ease, and also sends files back and forth. . . .

To put everything in nice presentation form I use Microsoft Powerpoint for Windows or IBM Storyboard LIVE!. Storyboard LIVE! also lets me add animation. Sometimes I bring an nVIEW MediaPro LCD display panel that fits on an overhead projector so I can show everything I'm doing on a big screen. If the customer has a VCR and TV, I'll use that to show clips of a truck rolling down the highway. It's my own kind of multimedia show.

—Alison Sprout, "Getting Mileage from a Mainframe," *Fortune*

Supercomputers

Gregory Chudnovsky, 39, with the help of his older brother, David, built a supercomputer in the living room of his New York City apartment. Seven feet tall and 8 feet across, the supercomputer was put together from mail-order parts and cost only $70,000 (compared to $30 million for some commercial supercomputers).

The brothers, both mathematicians and former citizens of the Soviet Union, have found some drawbacks to their homemade machine, which they named "m zero." They must keep the computer, along with 25 fans, running day and night. They must make sure the apartment's lights are turned off as much as possible, to prevent blowing the wiring in the living unit. "The building superintendent doesn't know that the Chudnovsky brothers have been using a supercomputer in Gregory's apartment," reports journalist Richard Preston, "and the brothers haven't expressed an eagerness to tell him."[18] Still, the machine makes their lives more convenient. The "m zero" performs computations that make up the basis of many of the scholarly papers and books they write on number theory and mathematical physics.

Most supercomputer users aren't as resourceful as the Chudnovskys and must buy their equipment from manufacturers. **Typically priced from $225,000 to over $30 million, *supercomputers* are high-capacity machines that require special air-conditioned rooms and are the fastest calculating devices ever invented.** (■ *See Panel 4.9.*)

Supercomputer users are those who need to model complex phenomena. Examples are automotive engineers who simulate cars crashing into walls and airplane designers who simulate air flowing over an airplane wing. "Supers," as they are called, are also used for oil exploration and weather forecasting. They can also help managers in department-store chains decide what to buy and where to stock it. Finally, they have been used to help redesign parachutes, which are surprisingly complex from the standpoint of aerodynamics. The supercomputer simulates the flow of air in and around the parachute during its descent. In 1995 Intel announced plans to build a supercomputer that would enable scientists to simulate the explosion of a nuclear bomb.[19,20]

Supercomputers are designed in two ways:

PANEL 4.9

Supercomputer

- **Vector processors:** The traditional design, now 20 years old, is vector processing. In *vector processing*, a relatively few (1–16) large, highly specialized processors run calculations at high speeds. The drawback is that tasks are accomplished by a single large processor (or handful of processors) one by one, creating potential bottlenecks. In addition, the processors are costly to build, and they run so hot that they need elaborate cooling systems.

- **Massively parallel processors:** The newer design is called *massively parallel processing (MPP)*, **which spreads calculations over hundreds or even thousands of standard, inexpensive microprocessors** of the type used in PCs. Tasks are parceled out to a great many processors, which work simultaneously. The reason the Chudnovsky brothers were able to build their super so cheaply was that they used standard microprocessors available for PCs in an MPP design.

   A difficulty is that MPP machines are notoriously difficult to program, which has slowed their adoption.[21] Still, with the right software, 100 small processors can often run a large program in far less time than the largest supercomputer running it in serial fashion, one instruction at a time.[22]

Massively parallel processing might seem as powerful as one could expect. However, fiber-optic (✓ p. 20) communications lines have made possible supercomputing power that is truly awesome. In 1995 the National Science Foundation and MCI Communications, the nation's No. 2 long-distance provider, established a Very-high-speed Backbone Network Service (VBNS) that links the five most important concentrations of supercomputers into what they call a nationwide "metacenter." Each of these locations has more than one supercomputer (Cornell and Champaign-Urbana have six each). With this arrangement a scientist sitting at a terminal or workstation anywhere in the country could have access to all the power of these fast machines simultaneously.[23–25] (■ *See Panel 4.10.*)

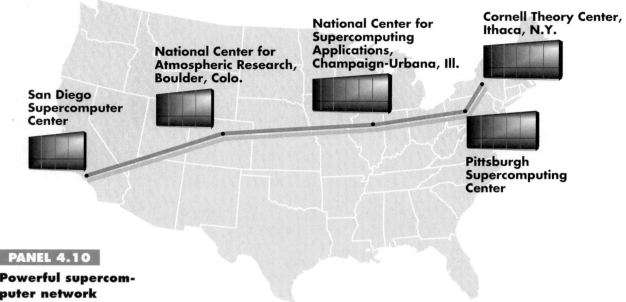

San Diego Supercomputer Center

National Center for Atmospheric Research, Boulder, Colo.

National Center for Supercomputing Applications, Champaign-Urbana, Ill.

Cornell Theory Center, Ithaca, N.Y.

Pittsburgh Supercomputing Center

PANEL 4.10

Powerful supercomputer network

In the National Science Foundation–MCI arrangement, supercomputers are linked at five locations into a "metacenter." This is a national network that allows scientists anywhere in the United States to have access to the computers' combined processing power.

Servers

The word "server" does not describe a size of computer but rather a particular way in which a computer is used. Nevertheless, because of the principal concerns of this book—the union of computers and communications—servers deserve separate discussion here. (This topic is also included in Chapter 8.)

A *server*, or *network server*, is a central computer that holds databases and programs for many PCs, workstations, or terminals, which are called *clients*. These clients are linked by a wired or wireless network. The entire network is called a *client/server network*. In small organizations, servers can store files and transmit electronic mail between departments. In large organizations, servers can house enormous libraries of financial, sales, and product information. The surge in popularity of the World Wide Web has also led to an increased demand for servers at tens of thousands of Web sites.[26]

Network servers appear in a variety of sizes. As one writer points out, they may consist of "everything from souped-up PCs selling for $10,000 to mainframes and supercomputer-class systems costing millions."[27] For more than a decade, he points out, the computer industry was driven by a rush to put stand-alone microcomputers in offices and homes. Now that we are far along in putting a PC on every desktop, the spotlight is shifting to computers that can do work for many different people at once.

On the one hand, then, this puts "big iron"—minis, mainframes, supers—back in the picture. Recognizing this, IBM has combined formerly separate personal, midrange, and mainframe computer units into an umbrella organization called the Server Group.[28] On the other hand, the demand for servers based on microcomputers has made souped-up PCs and Macintoshes a growth industry—and is bringing these machines "close to the power of more expensive minicomputers and mainframes," according to some PC makers.[29]

Now let's move on to look *inside* computers to see how they work.

The CPU & Main Memory

Preview & Review: The central processing unit (CPU) consists of the control unit and the arithmetic/logic unit (ALU). Main memory holds data in storage temporarily; its capacity varies in different computers. Registers are staging areas in the CPU that store data during processing.

The operations for executing a single program instruction are called the *machine cycle*, which has an instruction cycle and an execution cycle.

Processing speeds are expressed in three ways: fractions of a second, MIPS, and flops.

How is the information in "information processing" in fact processed? As we indicated, this is the job of the circuitry known as the microprocessor. This device, the "processor-on-a-chip" found in a microcomputer, is also called the CPU. The CPU works hand in hand with other circuits known as main memory to carry out processing.

CPU

The *CPU*, for *central processing unit*, follows the instructions of the software to manipulate data into information. The CPU consists of two parts: (1) the control unit and (2) the arithmetic/logic unit. The two components are connected by a kind of electronic "roadway" called a *bus*. (■ *See Panel 4.11.*)

- **The control unit:** The *control unit* tells the rest of the computer system how to carry out a program's instructions. It directs the movement of electronic signals between main memory and the arithmetic/logic unit. It also directs these electronic signals between main memory and the input and output devices.

- **The arithmetic/logic unit:** The *arithmetic/logic unit*, or *ALU*, performs arithmetic operations and logical operations and controls the speed of those operations.

  As you might guess, *arithmetic* operations are the fundamental math operations: addition, subtraction, multiplication, and division.

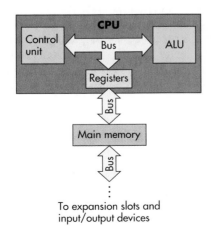

PANEL 4.11

The CPU and main memory

The two main components (control unit and CPU), the registers, and main memory are connected by a kind of electronic "roadway" called a *bus*.

Logical operations are comparisons. That is, the ALU compares two pieces of data to see whether one is equal to (=), greater than (>), or less than (<) the other. (The comparisons can also be combined, as in "greater than or equal to" and "less than or equal to.")

Main Memory

Main memory—variously known as memory, primary storage, internal memory, or RAM (for random access memory)—is working storage. It has three tasks. (1) It holds data for processing. (2) It holds instructions (the programs) for processing the data. (3) It holds processed data (that is, information) waiting to be sent to an output or secondary-storage device. Main memory is contained on special microchips called *RAM chips,* as we describe in a few pages. This memory is in effect the computer's short-term capacity. It determines the total size of the programs and data files it can work on at any given moment.

There are two important facts to know about main memory:

• **Its contents are temporary:** Once the power to the computer is turned off, all the data and programs within main memory simply vanish. This is why data must also be stored on disks and tapes—called "secondary storage" to distinguish them from main memory's "primary storage."

Thus, main memory is said to be *volatile.* **Volatile storage is temporary storage; the contents are lost when the power is turned off.** Consequently, if you kick out the connecting power cord to your computer, whatever you are currently working on will immediately disappear. This impermanence is the reason you should *frequently save* your work in progress to a secondary-storage medium such as a diskette. By "frequently," we mean every 3–5 minutes.

• **Its capacity varies in different computers:** The size of main memory is important. It determines how much data can be processed at once and how big and complex a program may be used to process it. This capacity varies with different computers, with older machines holding less.

For example, the original IBM PC, introduced in 1979, held only about 64,000 bytes (characters) of data or instructions. By contrast, new microcomputers can have 24 million bytes or more of memory.

Registers: Staging Areas for Processing

The control unit and the ALU also use registers, or special areas that enhance the computer's performance. (■ *Refer back to Panel 4.11.*) **Registers are high-speed storage areas that temporarily store data during processing.** It could be said that main memory holds material that will be used "a little bit later." Registers hold material that is to be processed "immediately." The computer loads the program instructions and data from main memory into the registers just prior to processing, which helps the computer process faster. (There are several types of registers, including *instruction register, address register, storage register,* and *accumulator register.*)

Machine Cycle: How an Instruction Is Processed

How does the computer keep track of the data and instructions in main memory? Like a system of post-office mailboxes, it uses addresses. **An *address* is the location, designated by a unique number, in main memory in** which a character of data or of an instruction is stored during processing. To process each character, the control unit of the CPU retrieves that character from its address in main memory and places it into a register. This is the first step in what is called the *machine cycle.*

The *machine cycle* is a series of operations performed to execute a single program instruction. The machine cycle consists of two parts: an instruction cycle, which fetches and decodes, and an execution cycle, which executes and stores. (■ *See Panel 4.12.*)

- **The instruction cycle:** In the *instruction cycle,* or *I-cycle,* the control unit (1) fetches (gets) an instruction from main memory and (2) decodes that instruction (determines what it means).

- **The execution cycle:** During the execution cycle, or *E-cycle,* the arithmetic/logic unit (3) executes the instruction (performs the operation on the data) and (4) stores the processed results in main memory or a register.

The details of the machine cycle are actually a bit more involved than this, but our description shows the general sequence. What's important for you to know is that the entire operation is synchronized by a *system clock,* as we will describe. The microprocessor clock speed, measured in megahertz, is an important factor to consider when you are buying a computer.

Processing Speeds

With transistors switching off and on perhaps millions of times per second, the tedious repetition of the machine cycle occurs at blinding speeds.

There are several ways in which processing speeds are measured:

- **Time to complete one machine cycle, in fractions of a second:** The speeds for completing one machine cycle are measured in milliseconds for older and slower computers. They are measured in microseconds for most microcomputers and in nanoseconds for mainframes. Picosecond measurements occur only in some experimental machines.

  A *millisecond* is one-thousandth of a second. A *microsecond* is one-millionth of a second. A *nanosecond* is one-billionth of a second. A *picosecond* is one-trillionth of a second.

PANEL 4.12

The machine cycle

(Left) The machine cycle executes instructions one at a time during the instruction cycle and execution cycle. *(Right)* Example of how the addition of two numbers, 50 and 75, is processed and stored in a single cycle.

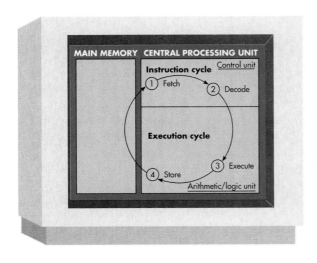

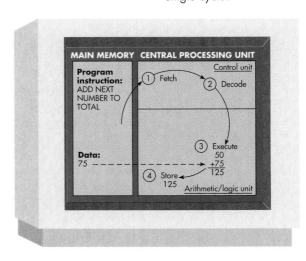

- **Time in millions of machine cycles per second (megahertz):** Microprocessor speeds are usually expressed in **megahertz (MHz), millions of machine cycles per second,** which, as we mentioned, is also the measure of a microcomputer's clock speed. For example, a 166-MHz Pentium-based microcomputer processes 166 million machine cycles per second.

- **Time to complete instructions, in millions of instructions per second (MIPS):** Another measurement is the number of instructions per second that a computer can process, which today is in the millions. **MIPS is a measure of a computer's processing speed; it stands for millions of instructions per second that the processor can perform.** A microcomputer (with an 80486 chip) might perform 54 MIPS, a mainframe 240 MIPS.

- **Time in floating-point operations per second (flops):** The abbreviation **flops** stands for **floating-point operations per second,** a floating-point operation being a special kind of mathematical calculation. This measure, usually expressed in *megaflops*—millions of floating-point operations per second—is used mainly with supercomputers.

Now that you know *where* data and instructions are processed, we need to review how those data and instructions are represented in the CPU, registers, buses, and RAM.

How Data & Programs Are Represented in the Computer

Preview & Review: Computers use the two-state 0/1 binary system to represent data.

A computer's capacity is expressed in bits, bytes, kilobytes, megabytes, gigabytes, or terabytes.

One common binary coding scheme is ASCII-8.

Parity-bit schemes are used to check for accuracy.

Human-language-like programming languages are processed as 0s and 1s by the computer in machine language.

As we explained in Chapter 1 (✓ p. 6), electricity is the basis for computers and communications because electricity can be either *on* or *off* (or low-voltage or high-voltage). This two-state situation allows computers to use the *binary system* to represent data and programs.

Binary System: Using Two States

The decimal system that we are accustomed to has 10 digits (0, 1, 2, 3, 4, 5, 6, 7, 8, 9). By contrast, **the binary system has only two digits: 0 and 1.** Thus, in the computer the 0 can be represented by the electrical current being off (or low voltage) and the 1 by the current being on (or high voltage). All data and programs that go into the computer are represented in terms of these binary numbers. (■ *See Panel 4.13.*) For example, the letter *H* is a translation of the electronic signal 01001000, or off-on-off-off-on-off-off-off. When you press the key for *H* on the computer keyboard, the character is automatically converted into the series of electronic impulses that the computer can recognize.

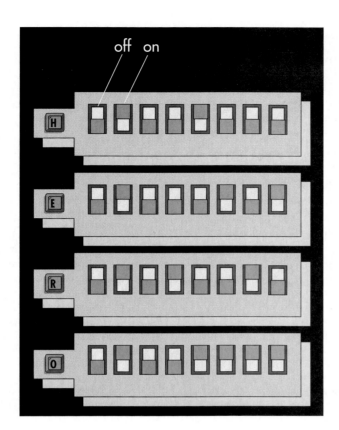

Binary data representation

How the letters *H-E-R-O* are represented in one type of off/on, 0/1 binary code (ASCII-8).

How Capacity Is Expressed

How many 0s and 1s will a computer or a storage device such as a hard disk hold? To review what we covered in Chapter 1, the following terms are used to denote capacity.

- **Bit:** In the binary system, **each 0 or 1 is called a *bit*, which is short for "binary dig*it*."**

- **Byte:** To represent letters, numbers, or special characters (such as ! or *), bits are combined into groups. **A group of 8 bits is called a *byte*, and a byte represents one character, digit, or other value.** (As we mentioned, in one scheme, 01001000 represents the letter *H*.) The capacity of a computer's memory or a diskette is expressed in numbers of bytes or multiples such as kilobytes and megabytes.

- **Kilobyte:** **A *kilobyte (K, KB)* is about 1000 bytes.** (Actually, it's precisely 1024 bytes, but the figure is commonly rounded.) The kilobyte was a common unit of measure for memory or secondary-storage capacity on older computers.

- **Megabyte:** **A *megabyte (M, MB)* is about 1 million bytes** (1,048,576 bytes). Many measures of microcomputer capacity today are expressed in megabytes.

- **Gigabyte:** **A *gigabyte (G, GB)* is about 1 billion bytes** (1,073,741,824 bytes). This measure is often used with "big iron" types of computers and now also with smaller systems.

- **Terabyte:** **A *terabyte (T, TB)* represents about 1 trillion bytes** (1,009,511,627,776 bytes).

Binary Coding Schemes

Letters, numbers, and special characters are represented within a computer system by means of *binary coding schemes.* That is, the off/on 0s and 1s are arranged in such a way that they can be made to represent characters, digits, or other values. One popular binary coding scheme ASCII-8 uses 8 bits to form each byte. (■ *See Panel 4.14.*)

Pronounced "*as*-key," **ASCII stands for American Standard Code for Information Interchange and is the binary code most widely used with microcomputers.**

ASCII originally used seven bits, but a zero was added in the left position to provide an 8-bit code, which offers more possible combinations with which to form characters, such as math symbols and Greek letters.

When you type a word on the keyboard (for example, *HERO*), the letters are converted into bytes—eight 0s and 1s for each letter. The bytes are represented in the computer by a combination of eight transistors, some of which are closed (representing the 0s) and some of which are open (representing the 1s).

PANEL 4.14

ASCII-8 binary code

There are many more characters than those shown here, such as punctuation marks, math symbols, Greek letters, and other foreign-language symbols.

Character	ASCII-8	Character	ASCII-8
A	0100 0001	N	0100 1110
B	0100 0010	O	0100 1111
C	0100 0011	P	0101 0000
D	0100 0100	Q	0101 0001
E	0100 0101	R	0101 0010
F	0100 0110	S	0101 0011
G	0100 0111	T	0101 0100
H	0100 1000	U	0101 0101
I	0100 1001	V	0101 0110
J	0100 1010	W	0101 0111
K	0100 1011	X	0101 1000
L	0100 1100	Y	0101 1001
M	0100 1101	Z	0101 1010
0	0011 0000	5	0011 0101
1	0011 0001	6	0011 0110
2	0011 0010	7	0011 0111
3	0011 0011	8	0011 1000
4	0011 0100	9	0011 1001
!	0010 0001	;	0011 1011

The Parity Bit

Dust, electrical disturbance, weather conditions, and other factors can cause interference in a circuit or communications line that is transmitting a byte. How does the computer know if an error has occurred? Detection is accomplished by use of a parity bit. **A *parity bit*, also called a *check bit*, is an extra bit attached to the end of a byte for purposes of checking for accuracy.**

Parity schemes may be *even parity* or *odd parity*. In an even-parity scheme, for example, the ASCII letter *H* (01001000) contains two 1s. Thus, the ninth bit, the parity bit, would be 0 in order to make the sum of the bits come out even. With the letter *O* (01001111), which has five 1s, the ninth bit would be 1 to make the byte come out even. (■ *See Panel 4.15.*) The systems software in the computer automatically and continually checks the parity scheme for accuracy. (If the message "Parity Error" appears on your screen, you need a technician to look at the computer to see what is causing the problem.)

Machine Language

Why won't word processing software that runs on an Apple Macintosh run (without special arrangements) on an IBM microcomputer? It's because each computer has its own machine language. **Machine language is a binary-type programming language that the computer can run directly.** To most people an instruction written in machine language is incomprehensible, consisting only of 0s and 1s. However, it is what the computer itself can understand, and the 0s and 1s represent precise storage locations and operations.

How do people-comprehensible program instructions become computer-comprehensible machine language? Special systems programs called *language translators* rapidly convert the instructions into machine language—language that computers can understand (discussed in Chapter 11). This translating occurs virtually instantaneously, so that you are not aware it is happening.

Because the type of computer you will most likely be working with is the microcomputer, we'll now take a look at what's inside the microcomputer's system unit.

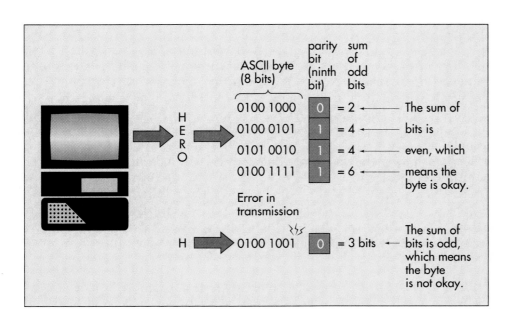

PANEL 4.15

Parity bit

This example uses an even-parity scheme.

The Microcomputer System Unit

Preview & Review: The system unit, or cabinet, contains the following electrical components: the power supply, the motherboard, the CPU chip, specialized processor chips, the system clock, RAM chips, ROM chips, other forms of memory (cache, VRAM, flash), expansion slots and boards, bus lines, ports, and PC (PCMCIA) slots and cards.

What is inside the gray or beige box that we call "the computer"? **The box or cabinet is the *system unit*; it contains the electrical and hardware components that make the computer work.** These components actually do the processing in information processing.

The system unit of a desktop microcomputer does not include the keyboard or printer. Quite often it also does not include the monitor or display screen (although it does in early Apple Macintoshes and some Compaq Presarios). It usually does include a hard-disk drive and one or two diskette drives, and sometimes a tape drive. We describe all these and other *peripheral devices—hardware that is outside the central processing unit—*in other chapters on input, output, and secondary-storage devices. Here we are concerned with 12 parts of the system unit, as follows:

- Power supply
- Motherboard
- CPU
- Specialized processor chips
- System clock
- RAM chips
- ROM chips
- Other forms of memory—cache, VRAM, flash
- Expansion slots and boards
- Bus lines
- Ports
- PC (PCMCIA) slots and cards

These are terms that appear frequently in advertisements for microcomputers. After reading this section, you should be able to understand what these ads are talking about.

Power Supply

The electricity available from a standard wall outlet is alternating current (AC), but a microcomputer runs on direct current (DC). **The *power supply* is a device that converts AC to DC to run the computer.** (■ *See Panel 4.16.*) The on/off switch in your computer turns on or shuts off the electricity to the power supply. Because electricity can generate a lot of heat, a fan inside the computer keeps the power supply and other components from becoming too hot.

Electrical power drawn from a standard AC outlet can be quite uneven. For example, a sudden surge, or "spike," in AC voltage can burn out the low-voltage DC circuitry in your computer ("fry the motherboard"). Instead of plugging your computer directly into the wall electrical outlet, it's a good idea to plug it into a power protection device. The two principal types are *surge protectors* and *UPS units.* One protects against surges, the other against power loss.

- **Surge protector:** **A *surge protector,* or *surge suppressor,* is a device that protects a computer from being damaged by surges (spikes) of high voltage.** The computer is plugged into the surge protector, which in turn is plugged into a standard electrical outlet.

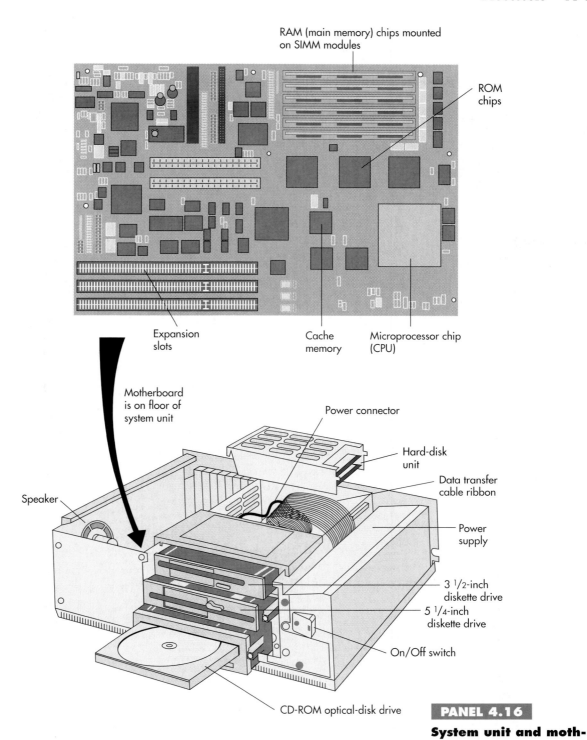

RAM (main memory) chips mounted on SIMM modules

ROM chips

Expansion slots

Cache memory

Microprocessor chip (CPU)

Motherboard is on floor of system unit

Power connector

Hard-disk unit

Data transfer cable ribbon

Power supply

Speaker

3 1/2-inch diskette drive

5 1/4-inch diskette drive

On/Off switch

CD-ROM optical-disk drive

System unit and motherboard components

(Top) Motherboard.
(Bottom) System unit.

- **UPS:** **A** *UPS,* **for** **u**ninterruptible **p**ower **s**upply, **is a battery-operated device that provides a computer with electricity if there is a power failure.** The UPS will keep a computer going from 5–30 minutes or more. It goes into operation as soon as the power to your computer fails.

Motherboard

The *motherboard,* or *system board,* is the main circuit board in the system unit. (■ *Refer to Panel 4.16.*)

READ ME

Practical Matters: Guarding Against Electrical Damage

Like all electrical appliances, a [microcomputer] can be damaged by changes to the power source. Here are a few things you can do to keep it safe:

• Unplug all components, including phone lines, during thunder and lightning storms. If lightning strikes the house or the power lines, the equipment can be ruined.

• When you turn on your computer, turn on the devices that use the most power first to avoid a power drain on smaller devices. This is the most common order:
1. Turn on external peripherals.
2. Turn on the system unit.
3. Turn on the monitor.
4. Turn on the printer.

• When you turn off the computer, follow the reverse order to avoid a power surge to smaller devices.

• Always use a power protection device [such as the surge protector shown here].

—Suzanne Weixel, *Easy PCs*, 2nd ed.

The motherboard consists of a flat board that fills the bottom of the system unit. (It is accompanied by the power-supply unit and fan and probably one or more disk drives.) This board contains the "brain" of the computer, the CPU or microprocessor; electronic memory that assists the CPU, known as RAM; and some sockets, called *expansion slots*, where additional circuit boards, called *expansion boards*, may be plugged in. The processing is handled by the CPU and memory (RAM), as we explain next.

CPU Chip

Most personal computers today use CPU chips (microprocessors) of two kinds—those made by Intel and those by Motorola—although that situation may be changing. (■ *See Panel 4.17.*) Workstations generally use RISC chips.

• **Intel-type "86"-series chips:** Intel makes chips for IBM and IBM-compatible computers such as Compaq, Dell, Gateway, Tandy, Toshiba, and Zenith. Variations of Intel chips are made by other companies—for example, Advanced Micro Devices (AMD), Cyrix, and Chips and Technologies.

Intel has identified its chips by numbers—8086, 8088, 80286, 80386, 80486—and is now marketing its newest chips under the names Pentium and Pentium Pro. The higher the number, the newer and more powerful the chip and the faster the processing speed, which means that software runs more efficiently. The chips are commonly referred to by their last three digits, such as '386 and '486.

Some chips have different versions—for example, "386SX" or "486DX." SX chips are usually less expensive than DX chips and run more slowly. Thus, they are more appropriate for home use, whereas DX chips are more appropriate for business use. SL chips are designed to reduce power consumption and so are used in portable computers. DX2 and DX4 chips are usually used for heavy-duty information processing.

• **Motorola-type "68000"-series chips:** Motorola makes chips for Apple Macintosh computers. These chip numbers include the 68000, 68020,

Manufacturer and Chip	Date Introduced	Systems Chip	Clock Speed (MHz)	Bus Width
Intel 8088	1979	IBM PC, XT	4–8	8
Motorola 68000	1979	Macintosh Plus, SE; Commodore Amiga	8–16	16
Intel 80286	1981	IBM PC/AT, PS/2 Model 50/60; Compaq Deskpro 286	8–28	16
Motorola 68020	1984	Macintosh II	16–33	32
Sun Microsystems RISC	1985	Sun Sparcstation 1, 300	20–25	32
Intel 80386DX	1985	IBM PS/2; IBM-compatibles	16–33	32
Motorola 68030	1987	Macintosh IIx series, SE/30	16–50	32
Intel 80486DX	1989	IBM PS/2; IBM-compatibles	25-66	32
Motorola 68040	1989	Macintosh Quadras	25–40	32
IBM RISC 6000	1990	IBM RISC/6000 workstation	20–50	32
Sun Microsystems MicroSpar	1992	Sun Sparcstation LX	50	32
Intel Pentium	1993	Compaq Deskpro; IBM-compatibles	60–166	64
IBM/Motorola/Apple PowerPC RISC	1994	Power Macintoshes; Power Computing PowerWave	60–150	64
Intel Pentium Pro	1995	Compaq Proliant; Data General server	150–200	64

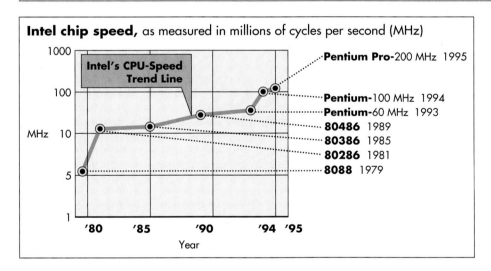

Intel chip speed, as measured in millions of cycles per second (MHz)

Intel's CPU-Speed Trend Line

Pentium Pro-200 MHz 1995
Pentium-100 MHz 1994
Pentium-60 MHz 1993
80486 1989
80386 1985
80286 1981
8088 1979

MHz

Year

PANEL 4.17

Microcomputers and microprocessors

(Top) Some widely used microcomputer systems and their chips. *(Bottom)* Intel chip speed has exploded. The Pentium Pro, which packs 5.5 million transistors onto a single chip, can process more than 200 million cycles per second.

68030, and 68040. In 1994, Motorola joined forces with IBM and Apple and produced the PowerPC chip.

- **RISC chips:** Sun Microsystems, Hewlett-Packard, and Digital Equipment use RISC chips in their desktop workstations, although the technology is also showing up in some portables.

  RISC **stands for** *reduced instruction set computing.* With RISC chips a great many needless instructions are eliminated. Thus, a RISC computer system operates with fewer instructions than those required in conventional computer systems. RISC-equipped workstations have been found to work 10 times faster than conventional computers. A problem, however, is that software has to be modified to work with them.

Most new chips are "downward compatible" with older chips. *Downward compatible,* or *backward compatible,* **means that you can run the software written for computers with older chips on a computer with a newer chip.** For example, the word processing program and all the data files that you used for your '386 machine will continue to run on a '486 machine. However, the reverse is not usually true. *Upward compatible* **means that software written for a machine with a newer chip will run on a machine with an older chip.** Thus, if you are using software written for your '486-powered desktop PC and buy an old '286 portable, you probably won't be able to run your software on both.

The capacities of CPUs are expressed in terms of *word size.* **A *word,* also called *bit number,* is the number of bits that may be manipulated or stored at one time by the CPU.** Often the more bits in a word, the faster the computer. An 8-bit-word computer will transfer data within each CPU chip itself in 8-bit chunks. A 32-bit-word computer is faster, transferring data in 32-bit chunks.

Specialized Processor Chips

A motherboard usually has slots for plugging in specialized processor chips. (■ *Refer back to Panel 4.16.*) Two in particular that you may encounter are math and graphics coprocessor chips. **A *math coprocessor chip* helps programs using lots of mathematical equations to run faster. A *graphics coprocessor chip* enhances the performance of programs with lots of graphics and helps create complex screen displays.** Specialized chips significantly increase the speed of a computer system. (To do co-processing, your operating system must be able to handle it—✓ p. 113.)

System Clock

When people talk about a computer's "speed," they mean how fast it can do processing—turn data into information. Every microprocessor contains a system clock. **The *system clock* controls how fast all the operations within a computer take place.** The system clock uses fixed vibrations from a quartz crystal to deliver a steady stream of digital pulses to the CPU. The faster the clock, the faster the processing, assuming the computer's internal circuits can handle the increased speed.

As we mentioned earlier in the chapter, processing speeds are expressed in megahertz (MHz), with 1 MHz equal to 1 million cycles per second. An old IBM PC had a clock speed of 4.77 MHz, whereas computers with '486 chips may run at 66 MHz. The high-end Macintosh-compatible PowerWave computer, from Power Computing, uses a PowerPC microprocessor running at 150 MHz. The most recent Intel Pentium Pro chip, used in workstations, runs at speeds up to 200 MHz.

RAM Chips

***RAM,* for *random access memory,* is memory that temporarily holds data and instructions that will be needed shortly by the CPU.** RAM is what we have been calling *main memory, internal memory,* or *primary storage;* it operates like a chalkboard that is constantly being written on, then erased, then written on again. (The term *random access* comes from the fact that data can be stored and retrieved at random—from anywhere in the electronic RAM chips—in approximately equal amounts of time, no matter what the specific data locations are.)

Like the microprocessor, RAM consists of circuit-inscribed silicon chips attached to the motherboard. **RAM chips are often mounted on a small circuit**

board, such as a *SIMM* (for *single inline memory module*),
which is plugged into the motherboard. (■ *Refer back to
Panel 4.16.)* The two principal types of RAM chips are *DRAM*
(for *dynamic random access memory*) chips, used for most
main memory, and *SRAM* (for *static random access memory*)
chips, used for some specialized purposes within main
memory.

Microcomputers come with different amounts of RAM.
In many cases, additional RAM chips can be added by plugging a memory-expansion card into the motherboard, as we
will explain. The more RAM you have, the faster the computer operates, and the better your software performs. If, for
instance, you type such a long document in a word processing program that it will not all fit into your computer's RAM, the computer will put part of the document onto your disk (either hard or floppy).
This means you have to wait while the computer swaps data back and forth
between RAM and disk.

Having enough RAM has become a critical matter! Before you buy a software package, look at the outside of the box to see how much RAM is
required. Windows 95 supposedly will run with 4 megabytes of RAM, but a
realistic minimum is 8–12 megabytes, and 16 is preferable.

ROM Chips

Unlike RAM, which is constantly being written on and erased, **ROM, which
stands for read-only memory and is also known as *firmware*, cannot be written on or erased by the computer user.** (■ *Refer back to Panel 4.16.)* ROM
chips contain programs that are built in at the factory; these are special
instructions for basic computer operations, such as those that start the computer or put characters on the screen.

There are variations of the ROM chip that allow programmers to vary
information stored on the chip and also to erase it.

Other Forms of Memory

The performance of microcomputers can be enhanced further by adding other
forms of memory, as follows.

- **Cache memory:** Pronounced "cash," **cache memory is a special high-
  speed memory area that the CPU can access quickly.** Cache memory can
  be located on the microprocessor chip or elsewhere on the motherboard.
  (■ *Refer back to Panel 4.16.)* Cache memory is used in computers with
  very fast CPUs. The most frequently used instructions are kept in cache
  memory so the CPU can look there first. This allows the CPU to run faster
  because it doesn't have to take time to swap instructions in and out of
  main memory. Large, complex programs benefit the most from having a
  cache memory available.

  At least 8 kilobytes of cache memory generally come with '486 processors. If you plan to run large spreadsheets or database management programs, you may want to have greater amounts of cache, such as 16 or 32
  kilobytes.

- **Video memory:** ***Video memory* or *video RAM (VRAM)* chips are used to
  store display images for the monitor.** The amount of video memory determines how fast images appear and how many colors are available. Video
  memory chips are particularly desirable if you are running programs that
  display a lot of graphics.

- **Flash memory:** Used primarily in notebook and subnotebook computers, *flash memory,* or *flash RAM,* cards consist of circuitry on credit-card-size cards that can be inserted into slots connected to the motherboard. Unlike standard RAM chips, flash memory is *nonvolatile.* That is, it retains data even when the power is turned off. Flash memory can be used not only to simulate main memory but also to supplement or replace hard-disk drives for permanent storage.

Expansion Slots & Boards

Today all new microcomputer systems can be expanded. **Expandability refers to a computer's capacity for adding more memory or peripheral devices.** Having expandability means that when you buy a PC you can later add devices to enhance its computing power. This spares you from having to buy a completely new computer.

Expandability is made possible with expansion slots and expansion boards. **Expansion slots are sockets on the motherboard into which you can plug expansion cards. *Expansion cards,* or *add-on boards,* are circuit boards that provide more memory or control peripheral devices.** (■ *Refer back to Panel 4.16.)* The words *card* and *board* are used interchangeably. Some slots may be needed right away for ordinary functions, but if your system unit leaves enough slots open, you can use them for expansion later.

Among the types of expansion cards are the following. (■ *See Panel 4.18.)*

- **Expanded memory:** Memory expansion cards (or SIMMs) allow you to add RAM chips, giving you more main memory.
- **Display adapter or graphics adapter cards:** These cards allow you to adapt different kinds of color video display monitors for your computer.
- **Controller cards:** *Controller cards* **are circuit boards that allow your CPU to work with the computer's various peripheral devices.** For example, a disk controller card allows the computer to work with different kinds of hard-disk and diskette drives.
- **Other add-ons:** You can also add special circuit boards for modems, fax, sound, and networking, as well as math or graphics coprocessor chips. Of particular interest for Macintosh users who also want to run DOS or Windows 3.1 programs is the Power Macintosh DOS Compatible System, popularly known as "Houdini." The setup consists of a DOS/Windows card with a '486 chip that will run on certain Macintoshes, permitting users to run programs for both Macs and IBM-style computers.

Bus Lines

A *bus line,* or simply *bus,* is an electrical pathway through which bits are transmitted within the CPU and between the CPU and other devices in the system unit. There are different types of buses (address bus, control bus, data bus), but for our purposes the most important is the *expansion bus,* **which carries data between RAM and the expansion slots.** To obtain faster performance, some users will use a bus that avoids RAM altogether. **A bus that connects expansion slots directly to the CPU is called a *local bus.* (**■ *See Panel 4.19.)*

A bus resembles a multilane highway: The more lanes it has, the faster the bits can be transferred. The old-fashioned 8-bit bus of early microprocessors had only eight pathways. It was therefore four times slower than the 32-bit bus of later microprocessors, which had 32 pathways. Intel's

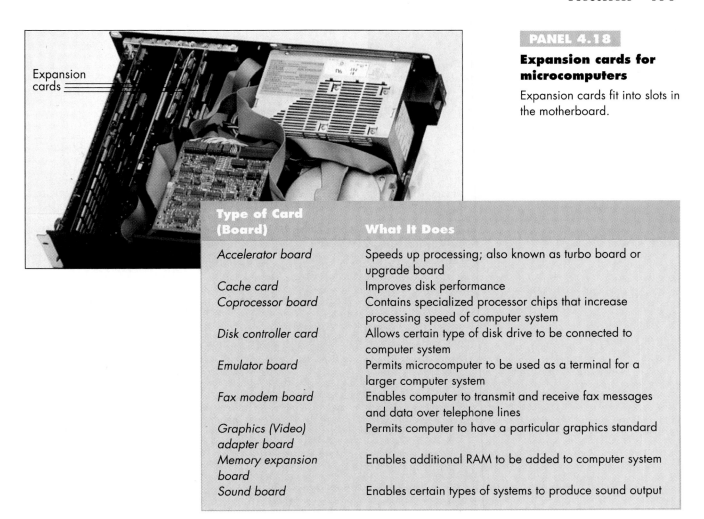

Expansion cards for microcomputers

Expansion cards fit into slots in the motherboard.

Type of Card (Board)	What It Does
Accelerator board	Speeds up processing; also known as turbo board or upgrade board
Cache card	Improves disk performance
Coprocessor board	Contains specialized processor chips that increase processing speed of computer system
Disk controller card	Allows certain type of disk drive to be connected to computer system
Emulator board	Permits microcomputer to be used as a terminal for a larger computer system
Fax modem board	Enables computer to transmit and receive fax messages and data over telephone lines
Graphics (Video) adapter board	Permits computer to have a particular graphics standard
Memory expansion board	Enables additional RAM to be added to computer system
Sound board	Enables certain types of systems to produce sound output

Pentium chip is a 64-bit processor. Some supercomputers contain buses that are 128 bits. Today there are several principal expansion bus standards, or "architectures," for microcomputers.

Ports

A *port* is a socket on the outside of the system unit that is connected to an expansion board on the inside of the system unit. A port allows you to plug in a cable to connect a peripheral device, such as a monitor, printer, or modem, so that it can communicate with the computer system.

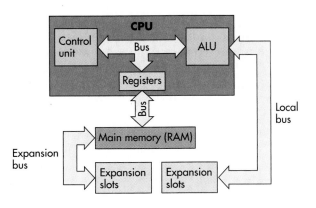

Buses

Buses are the electrical pathways that carry bits within the CPU and from the CPU to peripheral devices. Expansion buses connect RAM with expansion slots. Local buses avoid RAM and connect expansion slots directly with the CPU.

Ports are of five types. (■ *See Panel 4.20.*)

- **Parallel ports:** **A *parallel port* allows lines to be connected that will enable 8 bits to be transmitted simultaneously,** like cars on an eight-lane highway. Parallel lines move information faster than serial lines do, but they can transmit information efficiently only up to 15 feet. Thus, parallel ports are used principally for connecting printers.

- **Serial ports:** **A *serial port,* or *RS-232 port,* enables a line to be connected that will send bits one after the other on a single line,** like cars on a one-lane highway. Serial ports are used principally for communications lines, modems, and mice. (They are frequently labeled "COM" for communications.)

- **Video adapter ports:** ***Video adapter ports* are used to connect the video display monitor outside the computer to the video adapter card inside the system unit.** Monitors may have either a 9-pin plug or a 15-pin plug. The plug must be compatible with the number of holes in the video adapter card.

- **SCSI ports:** Pronounced "scuzzy" (and **short for *Small Computer System Interface*), a *SCSI port* provides an interface for transferring data at high speeds for up to eight SCSI-compatible devices.** These devices include external hard-disk drives, CD-ROM drives, and magnetic-tape backup units.

- **Game ports:** ***Game ports* allow you to attach a joystick or similar game-playing device to the system unit.**

Why are so many ports needed? Why can't plugging peripherals into a computer be as easy as plugging in a lamp in your living room? The reason, says an Intel engineer, is that "The PC industry has evolved ad hoc. We were always adding one more piece of equipment."[30] As a result, connecting a new device, such as a scanner or a second printer, "is about as straightforward as triple-bypass surgery," says one writer. In many cases, it involves opening the computer and inserting a circuit board, installing or modifying pertinent software, and fiddling with little switches.[31]

Fortunately, help is on the way. Several companies (Intel, Microsoft, IBM, Compaq, others) have agreed on something called the *Universal Serial Bus,* which would be applicable to future IBM-style computers. Today different types of connections, or buses, support different devices. Tomorrow, however, many peripheral devices (though not all) will be attached—one device after the other—through one connection port. The Universal Serial Bus connecting the port inside the computer will interpret the signals from each peripheral device and tell the computer to recognize it.

Plug-In Cards: PC (PCMCIA) Slots & Cards

Although its name doesn't exactly roll off the tongue, PCMCIA is changing mobile computing more dramatically than any technology today.

Short for *Personal Computer Memory Card International Association,* *PCMCIA* is a completely open, relatively new bus standard for portable computers. When talking about this standard, one pronounces every letter: "P-C-M-C-I-A." ***PC cards*—renamed because it's easier to say—are 2-by-3-inch cards that may be used to plug peripherals into slots in portable computers.** Slots are now being designed into desktop-size machines as well. The PC cards may be used to hold credit-card-size modems, sound boards, hard disks, extra memory, and even pagers and cellular communicators.

IBM-compatible

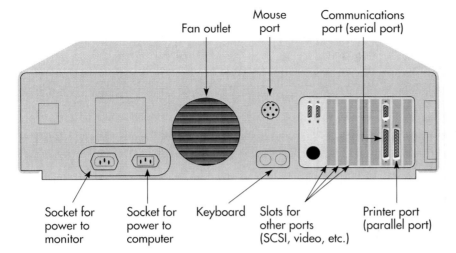

Fan outlet — Mouse port — Communications port (serial port)

Socket for power to monitor — Socket for power to computer — Keyboard — Slots for other ports (SCSI, video, etc.) — Printer port (parallel port)

Apple Macintosh

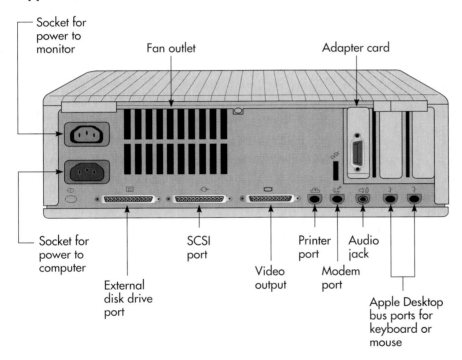

Socket for power to monitor — Fan outlet — Adapter card

Socket for power to computer — External disk drive port — SCSI port — Video output — Printer port — Modem port — Audio jack — Apple Desktop bus ports for keyboard or mouse

At present there are three official PCMCIA slot sizes—I (thin), II (thick), and III (thicker)—and an unofficial one (IV, thickest). Type I is used primarily for flash memory cards. Type II, the kind you'll find most often, is used for fax modems, adapters for local area networks (LANs), and other slim media. Type III is for rotating disk devices, such as hard-disk drives. Type IV, not official but found on Toshiba portables, is for large-capacity hard-disk drives.

Unfortunately, implementation of PCMCIA standards is still shaking down. Still, these cards are helpful to portable computer users and are worth learning to use. You should, however, find out whether a given PC card *really* is compatible with your existing PC.

Computers & Environmental Questions

 Preview & Review: Information technology has had some adverse effects on the environment, including energy consumption and environmental pollution.

As the upgrade merry-go-round continues, as it has since the birth of the computer industry, more and more people are joining the Digital Revolution—32% of American households now have home PCs and 18% have modems, showing that huge numbers remain to be plugged in.[32] The worldwide personal computer market is predicted to double in 5 years, with 100 million PCs expected to be shipped in 1999, twice the 1994 figure of 48 million units.[33]

Everyone hopes, of course, that the principal effects of this growth will be beneficial. But you need not be anti-technology to wonder just what negative impact computerization will have. How, for instance, will it affect the environment—energy consumption and environmental pollution, for example?

Energy Consumption & "Green PCs"

All the computers and communications devices discussed in this book run on electricity. Much of this is simply wasted. Computers themselves have in the past been built in ways that used power unnecessarily. An office full of computers also generates a lot of heat, so that additional power is required to run air-conditioning systems to keep people comfortable. Finally, people leave their computer systems on even when they're not sitting in front of them—not just during the day but overnight and weekends as well. (■ *See Panel 4.21.*)

In recent years, the U.S. Environmental Protection Agency launched Energy Star, a voluntary program to encourage the use of computers that consume a minimum amount of power. The goal of Energy Star is to reduce the amount of electricity microcomputers and monitors use from the typical 150 watts of power to 60 watts or less. This goal is about the power requirement for a moderately bright lightbulb.[34] (Half the wattage would be for the system unit, the other half for the monitor.) As a result, manufacturers are now coming forth with Energy Star–compliant "green PCs." For example, the system unit for one thrifty microcomputer, the PC&C Iguana EG-2000, uses just under 22 watts at full power.

"If you use your PC for 8 hours a day but always leave it on," says one writer, "a green PC could save about $70 a year. If everyone used only green PCs, $2 billion could be saved annually."[35]

Environmental Pollution

Communities like to see computer manufacturers move to their areas because they are viewed as being nonpolluting. Is this true? Actually, in the past, chemicals used to manufacture semiconductors polluted air, soil, and groundwater. Today, however, computer makers are literally cleaning up their act.

However, another problem is that the rush to obsolescence has produced numbers of computers, printers, monitors, fax machines, and so on, that have wound up as junk in landfills, although some are stripped by recyclers for valuable metals. More problematic is the disposal of batteries, as from portable computers. Nickel-cadmium batteries contain the toxic element cadmium, which, when buried in a landfill garbage dump, can leach into groundwater supplies. Disposal of such batteries should be through local toxic-waste disposal programs. Newer battery technology, such as nickel-hydride and lithium cells, may eventually replace nickel-cadmium.

If you have an old-fashioned computer system, consider donating it to an organization that can make use of it. Don't abandon it in a closet. Don't dump it in the trash. "Even if you have no further use for a machine that

"[R]educing power consumption could open up new avenues of computing, because it will allow us to run our computers night and day without fear of PC burnout or killer utility bills.

　　The obvious question is: Why leave the PC on at all? An ever-ready PC begins to fulfill the promise of personal computer as information appliance. This starts with 24-hour telephony operations such as taking phone messages, connecting for incoming faxes, and transmitting faxes when the phone rates are lowest. But this concept extends to having a PC always ready to receive commands from you remotely via a dial-in program for transferring files or reading e-mail. That way, when you're on the road, you can always be a phone call away from your desktop PC and all that it contains.

　　The all-night PC can periodically scan the overnight news and financial market wire services and, at daybreak, prepare a summary based on your areas of interest. This computer summary would be more up-to-date than the morning newspaper, would have more depth than the radio or television news, and would be customizable to your specific interests. And what your PC finds on its late-night sojourn may have a great impact on your mood the next morning."

—"Power to the PC," *PC Magazine*

PANEL 4.21

Why leave computers turned on all the time?

Energy Star PCs are being promoted by the Environmental Protection Agency.

seems horribly antiquated," writes *San Jose Mercury News* computing editor Dan Gillmor, "someone else will be grateful for all it will do."[36]

Several nonprofit groups exist that accept or pass along used computer equipment to deserving organizations. However, there are some practical steps you should take to ensure that the system ends up doing some good. (■ *See Panel 4.22, next page.*)

Near & Far Horizons: Processing Power in the Future

Preview & Review: On the near horizon are ultra-tiny multimedia superchips, billion-bit memory chips, teraflop supercomputers, stripped-down Internet PCs, and Intercast TV/Internet PCs. On the far horizon are technologies using gallium arsenide, superconducting materials, optical processing, nanotechnology, and DNA.

How far we have come. The onboard guidance computer used in 1969 by the Apollo 11 astronauts—who made the first moon landing—had 2 kilobytes of RAM and 36 kilobytes of ROM, ran at a speed of 1 megahertz, weighed 70 pounds, and required 70 watts of power. Even the Mission Control computer on the ground had only 1 megabyte of memory. "It cost $4 million and took up most of a room," says a space physicist who was there.[37] Today you can easily buy a personal computer with 90 times the processing power and 10 times the memory for just a couple of thousand dollars.

Future Developments: Near Horizons

The old theological question of how many angels could fit on the head of a pin has a modern counterpart: the technological question of how many *circuits* could fit there. Computer developers are obsessed with speed and power, constantly seeking ways to promote faster processing and more main memory in a smaller area. Some of the most promising directions, already discussed, are *RISC chips* and *parallel processing.* Some other research-and-development paths being explored in the near term include the following:

● **Ultra-tiny multimedia superchips:** The general-purpose microprocessor we've described in this chapter, such as Intel's Pentium, is about to be

PANEL 4.22 **Groups accepting used computers**

"If you're going to donate a used machine, don't just leave it on the doorstep somewhere. A little planning will ensure that your well-meaning gift actually goes to a good cause.

And before you give it away, make sure you've moved your personal data—letters, financial information, etc.—onto your new computer, using machine-to-machine software such as LapLink or copying the data to floppies and then to the new computer.

If you plan to keep the software you were using before, you should delete it from the machine you give away. Otherwise you're committing piracy; even in a good cause that's not a good idea.

When you're ready to give the computer away, call the school, church or organization first. Some will be unable to use the model you're offering even if it works well.

Some groups, however, welcome computers of any age and in almost any condition (though you should still call them before you donate)."

—Dan Gillmore, "Old Computer Will Mean a Lot to Those in Need," *San Jose Mercury News*

California
Plugged In
1923 University Ave.
East Palo Alto, CA 94303
800-225-PLUG

Detwiler Foundation Computer for Schools Program
470 Nautilus St., Suite 300
La Jolla, CA 92037
800-939-6000

The Computer Recycling Center
1245 Terra Bella Ave.
Mountain View, CA 94043
415-428-3700

CompuMentor
89 Stillman St.
San Francisco, CA 94107
415-512-7784

The Shareware Project
410 Townsend St., Suite 408
San Francisco, CA 94107
415-543-0500

Connecticut
The National Christina Foundation
591 West Putnam Ave.
Greenwich, CT 06830
800-274-7846

Illinois
Information Technology Resource Center
59 East Van Buren, Suite 2020
Chicago, IL 60605-1219
312-939-8050

Massachusetts
CONNECT
Technical Development Corporation
30 Federal St., 5th floor
Boston, MA 02110
617-728-9151

East-West Education Development Foundation
55 Temple Place
Boston, MA 02111
617-542-1234

New York
Nonprofit Computer Exchange
Fund for the City of New York
121 Sixth Ave., 6th floor
New York, NY 10013
212-925-5101

Virginia
Gifts in Kind America
700 North Fairfax St., Suite 300
Alexandria, VA 22314
703-836-2121

replaced. Several companies (Intel, IBM, MicroUnity, Chromatic Research, Philips) have announced they are working on versions of a new breed of chip called a *media processor.* [38–40]

As we stated in Chapter 1, *multimedia* refers to technology that presents information in more than one medium, including text, graphics, animation, video, music, and voice. A *media processor*, or so-called "multimedia accelerator," is a chip with a fast processing speed that can do specialized multimedia calculations and handle several multimedia functions at once, such as audio, video, and three-dimensional animation.

MicroUnity, for example, is using tricks that will pack perhaps three times as many transistors on a chip as there are on a standard Pentium, which is about the same size. With this process, the company expects to obtain multimedia chips that will operate at *1 billion cycles per second*— five times the speed of a 200-megahertz Pentium Pro.

Superior processing speeds are necessary if the media and communications industries are to realize their visions for such advances as realistic videogame animation and high-quality video phones. An all-digital TV, for

Media processors make it easier and cheaper to deliver:

- *Audio*   Better than CD-quality sound in software games and communications
- *Video*   Smooth, movie-like images in game software and video conferences
- *Graphics*   Ultra-realistic computer-generated images
- *Online services*   Send and receive complex video and data files over cable-TV lines
- *Wireless data*   Send and receive computerized video on handheld devices
- *Telephony*   Smaller, cheaper cellular phones; PCs with voice mail and caller ID

PANEL 4.23

The world of media processors

A new generation of specialized chips handles many multimedia tasks—audio, video, three-dimensional animation, and the like. Installed in microcomputers and other devices, the processors use software that can be upgraded as technologies change.

instance, needs media processors to perform the calculations for the million or more dots that make up one frame of video—and 30 such video frames race by each second.[41] (■ *See Panel 4.23.*)

- **Billion-bit memory chips:** In 1995 two sets of companies—Hitachi and NEC on the one hand, and Motorola, Toshiba, IBM, and Siemens on the other—announced plans to build plants to make memory chips capable of storing 1 billion bits (a gigabit) of data. This is 60 times as much information as is stored on the DRAM (dynamic random access memory) chips used in today's latest personal computers. One thumbnail-size piece of silicon could then store 10 copies of the complete works of Shakespeare, 4 hours of compact-disk quality sound, or 15 minutes of video images. Engineering samples of such chips are expected in 1998.[42,43]

- **Teraflop supercomputers:** Intel announced in 1995 that it was building a new supercomputer that would be the first to achieve the goal of calculating more than a trillion floating-point operations a second, known as a *teraflop*. Using 9000 Pentium Pro microprocessors in the configuration known as massively parallel processing, the machine would be applied to the study of nuclear weapons safety, among other things.[44]

- **Stripped-down Internet PCs:** The reverse of supercomputers is the stripped-down *Internet PC*, or "hollow PC" (✓ p. 29). This appliance—built by Oracle and England's Acorn Computer Group—is designed as an inexpensive device for cruising the Internet and World Wide Web and for doing basic computing.[45–47]

The Internet PC doesn't have CD-ROM drives and will not be able to use store-bought software (but software applications can presumably be extracted from the Web). It includes 4 megabytes of main memory, a microprocessor similar to that used in Apple Computer's handheld Newton devices, a keyboard, mouse, and network connections.

A variation being licensed by Apple is Pippin, a game-player Internet connector that plugs into a TV. Expected to cost about $500, Pippin could boost demand for the Macintosh operating system.[48]

- **Intercast TV/Internet PC:** Another new technology, developed by Intel, is *Intercast,* which links the Internet and televisions to microcomputers.[49] Intercast allows PCs equipped with modems to receive broadcast data from the Internet as well as television programming. Thus, you could watch a television news show about Bosnia on your computer screen and then, if you wished, look up related historical and geographical information broadcast by the television network.

Future Developments: Far Horizons

Silicon is still king of semiconductor materials, but researchers are pushing on with other approaches. Most of the following, however, will probably take some time to realize:

- **Gallium arsenide:** A leading contender in chip technology is *gallium arsenide,* which allows electrical impulses to be transmitted several times faster than silicon can. Gallium arsenide also requires less power than silicon chips and can operate at higher temperatures. However, chip designers at present are unable to squeeze as many circuits onto a chip as they can with silicon.

- **Superconductors:** Silicon, as we stated, is a semiconductor: Electricity flows through the material with some resistance. This leads to heat buildup and the risk of circuits melting down. A *superconductor,* by contrast, is material that allows electricity to flow through it without resistance.

  Until recently superconductors were considered impractical because they have to be kept at subzero temperatures in order to carry enough current for many uses. In 1995, however, scientists at Los Alamos National Laboratory in New Mexico succeeded in fabricating a high-temperature, flexible, ribbon-like superconducting tape that could carry current at a density of more than 1 million amperes per square centimeter, considered a sort of threshold for wide practical uses.[50-52]

  While the material is still very cold, it is hot compared to earlier extremely chilly superconductors. Now, perhaps, superconducting wire will find widespread applications. In computers it could produce circuitry 100 times faster than today's silicon chips.

- **Opto-electronic processing:** Today's computers are electronic; tomorrow's might be *opto-electronic*—using light, not electricity. With optical-electronic technology, a machine using lasers, lenses, and mirrors would represent the on-and-off codes of data with pulses of light.

  Except in a vacuum, light is faster than electricity. Indeed, fiber-optic networks, which consist of hair-thin glass fibers, can move information at speeds up to 3000 times faster than conventional networks. However, the signals get bogged down when they have to be processed by silicon chips. Opto-electronic chips would remove that bottleneck.

- **Nanotechnology:** Nanotechnology, nanoelectronics, nanostructures, nanofabrication—all start with a measurement known as a nanometer. A *nanometer* is a billionth of a meter, which means we are operating at the level of atoms and molecules. A human hair is approximately 100,000 nanometers in diameter.

  Nanotechnology is a science based on using molecules to create tiny machines to hold data or perform tasks. Experts attempt to do "nanofabrication" by building tiny "nanostructures" one atom or molecule at a time. When applied to chips and other electronic devices, the field is called "nanoelectronics."

- **Biotechnology—using DNA molecules:** Now we get to the *real* science fiction—or is it?

  Not long ago, University of Southern California computer science professor Leonard Adleman watched associates in a research lab do experiments with DNA, the chain of molecules that make up the genetic code of living things. "Adleman was amazed at the intricacy of the DNA strands," reports science writer Jane Allen. "And he was struck by how similar the laboratory cutting, splicing, and copying of these strands were to the manipulations of numbers he performed with computers."

  All of a sudden the classic lightbulb went on: Perhaps, Adleman thought, DNA could be used to perform calculations just like a computer.[53] Says another writer, "It's like Stallone and Schwarzenegger teaming up in the ultimate buddy picture—biology and electronic computers, together at last. The future may never be the same."[54]

  The code used in silicon chip–based calculations is binary, having two states. DNA, however, carries information in four molecules designated A, T, C, G. To perform calculations, these four molecules can be combined together to form numbers or words, which then combine to make larger words. Biological calculations, which take place by letting the molecules react in a test tube, are not very fast and a single operation may well take 30 minutes. However, because there are trillions of molecules, they can—in a display of massively parallel processing—do billions of calculations at once. (Similarly, a supercomputer may do millions of calculations simultaneously, but it may take hours or even days to solve a problem.)[55, 56]

  Adleman successfully tested his theory using a form of the Traveling Salesman problem, which requires finding the shortest flight route to connect seven cities with only one stop at each. Although, because of the time it takes, DNA computing will not replace laptops, it could be applied to the kinds of problems that even the fastest supercomputers have difficulty with, such as making or breaking highly complex encryption codes.

Perhaps sometime in the future these various avenues will come together. Imagine millions of nanomachines grown from microorganisms processing information at the speed of light and sending it over far-reaching pathways. What would the world be like with such technology?

Onward

New work habits have led to changes in how computers are used, and new computer uses have also changed work habits. For instance, employers have been seeking to trim costs and to respond to employees' demands for more flexibility about when and where they work—at home, on weekends, or out of the office. This situation has led to greater use of portable computers that can be taken anywhere. Conversely, distributing portables to employees has altered ways of doing business. For example, Wilsons The Leather Experts, a Minneapolis leather retailer, distributed notebook computers to its district sales managers, who formerly had used desktop computers in an office. Now the managers, keeping in touch with headquarters through modems on their notebooks, spend more time on the road. And their offices have been eliminated.[57] More and more, it is not so much computing as communications-linked *mobile* computing that is transforming our lives.

Becoming a Mobile Computer User: What to Look for in Notebooks, Subnotebooks, & Pocket PCs

Question: What do the following colleges and universities have in common: Bentley, Case Western, Columbia, Drew, Hartwick, Mississippi State, Nichols, UCLA, and the University of Minnesota's Crookston campus?

Answer: They *require* some or all of their students to have portable computers.

For instance, all students in the freshman class at Hartwick College, a small liberal-arts institution in Oneonta, New York, are given notebook-size PCs. (■ *See Panel 4.24.*) Through a special arrangement the college made with computer-maker Zenith, each student pays only $650 more a year in tuition to cover the computer, printer, and software.

Although currently the vast majority of American students who buy computers get the larger desktop models, many are finding that mobile computing makes a difference. Because they can take the computers to their dormitory rooms, to the cafeteria, to a campus bench, or on a bus, Hartwick students praise the portables for their convenience. Freshman Amy Grenier, for example, says she worked on a paper during a trip of several hours to see friends. "I did some work in the car, which I couldn't do with a large computer," she said. "You get to utilize more time to get your work done."[58]

Your First Decision: Should You Go Mobile?

Having a personal computer that you can carry around offers tremendous benefits. However, compared to the nonmobile desktop PCs, portables have some limitations: They have smaller screens and keyboards. They are more expensive—about $1000 more than a desktop with similar performance. They are more vulnerable to theft or loss (so be sure you get replacement insurance).

You have four choices, then:

1. Forget about mobility. Just get a desktop.
2. Get a portable computer, but make sure you're comfortable with the keyboard and screen when you buy it.
3. Get *both* a desktop and a portable, if you can afford them. They should be compatible with each other, of course, so that you can easily swap files and programs between them.
4. As a compromise, get a portable but also get a full-size (101-key) keyboard and a desktop-size color monitor. You can then plug the more comfortable keyboard and monitor into the portable when you're at your regular desk. A variation on this is to get a *docking station,* as we describe in a few pages.

PANEL 4.24

Portable computing in college

First-year students at Hartwick College in Oneonta, New York, are required to buy notebook computers.

Going Mobile: What to Look For

Mobile computers come in four sizes: *laptops, notebooks, subnotebooks,* and *pocket PCs.* Portables, especially the 7-pound notebooks and under-4-pound subnotebooks, represent the best intersection of power and convenience. The 8-pound-plus laptops probably have no advantages over the notebook size, except perhaps in price. The under-1-pound pocket PCs, although practical for many applications, may be too limited for regular student use. This is because they have no diskette drive, no hard-disk drive, and no easy way to swap files with other computers.

Among the factors to consider in buying a new notebook or subnotebook are: price, display screen, keyboard and mouse, portability, battery life and weight, disk drive, software, and expandability.

Price

Affordability is always important, of course. New desktops generally cost far less than new portables. Desktops can now be had for approximately $1000, whereas a new portable may set you back at least $2000. For example, the Apple Macintosh Performa models, which are desktops, start at about $1200 for machines with 8 megabytes of RAM (main memory). Portables start at about $2200 for a PowerBook 520 with monochrome (noncolor) screen or $2900 for a color screen. IBM-style desktops start at about $1000 for any '486 machine running Windows; a model equipped with a Pentium microprocessor will push the price into the $2000 range. Portables start at around

$1500; however, you have to watch that you don't get an underpowered unit with too little memory and a tiny hard disk. Deluxe portables top out at around $7500.

If you buy a subnotebook, a high-end version (with 24 megabytes of main memory and a 720-megabyte hard drive) might set you back around $5000. A cheaper version costs about $3000 (with 8 megabytes of memory and a 340-megabyte hard drive).

Note that Windows 95 requires at least 8 megabytes of main memory—16 megabytes is better—compared with 4 megabytes for Windows 3.1. This is important to know, points out one expert, "because memory is roughly a third of the cost of the PC and, as other costs of the computer drop, memory accounts for a bigger portion of the total."[59]

Display Screen

Price of a portable is affected by the kind of display screen: color or no-color (called *monochrome*). Color is nice but not necessary unless you plan on doing desktop publishing or lots of graphics work. Still, color is becoming an increasingly popular feature in notebooks and subnotebooks, and deals may be had: by the time you read this, you may be able to get a notebook with a Pentium chip and a 10.4-inch color screen for $2000.

Lightweight portables have two principal types of color screens:[60]

- *Active-matrix color:* In this version each little dot on the screen is controlled by its own transistor. The advantage of active-matrix screens is that colors are much brighter than those in the other version, passive matrix.

- *Passive-matrix color:* In this version a transistor controls a whole row or column of dots. The advantage of these screens is that they are less expensive—you can save $500–$1000 over the active-matrix screen. They are also less power-hungry than active-matrix screens. (Technical detail: Dual-scan screens rate better than single-scan screens. In dual-scan the tops and bottoms of the screens are "refreshed" independently at twice the rate of single-scan; thus, they deliver richer colors, though not usually as rich as active-matrix screens.)

The display screens of most notebooks and subnotebooks measure 9.5 inches diagonally, and some are as small as 7.4 inches. This compares with the much more readable 14- or 15-inch (or even 17- or 20-inch) monitors that are standard with most desktop computers. Thus, if you do a lot of graphics work, you will probably want to get a portable with an external video connector that will attach it to a big color monitor on your desk.

Several portables come with optional docking stations, as we discuss below.

Keyboard & Mouse

You should try out the keyboard on the notebook or subnotebook to see if you can realistically touch-type on it. (This is very much a problem with the pocket PCs.)

Notebooks and subnotebooks offer many variants on the mouse, the pointing device. Some devices must be unlimbered and clipped onto the side of the keyboard, which means you have yet another piece of paraphernalia to tote along. The IBM ThinkPad's pointing controller resembles a pencil eraser stuck among the G, H, and B keys. Other portables use a trackball built into the right side of the screen or centered below the space bar.

Portability

How portable is that advertised "portable" computer? That depends on how deep a groove you're willing to let the strap put in your shoulder. It also depends on what kind of hardware you're toting.

- *Notebook versus subnotebook:* If you opt for desktop capability and bright screens, you will probably want the notebook, which will weigh about 7.5 pounds. Bright screens, however, require heavy batteries. Subnotebooks are lighter and smaller, although they may not provide some other features you want.

- *Accessories and AC adapter:* Notebooks usually have built-in diskette drives. Subnotebooks do not, which means you may want to carry an external drive as an accessory. This could add about a pound.

  For both notebooks and subnotebooks, you will also always want to carry an AC adapter. Weighing perhaps a pound or less, the adapter enables you to plug your machine into a wall plug for recharging after the battery runs down; it also allows you to keep working. You can probably count on 6–8.5 pounds with the adapter.

- *Battery life and weight:* Portables vary in how long their batteries will hold a charge. Thus, battery life may be a short 1 hour and 20 minutes or a long 8 hours and 45 minutes. Testing batteries is not an exact science, but the average among batteries evaluated by *PC Magazine* was 2 hours.[61]

  It's possible to take a backup battery—provided you've charged it first, of course—on, say, a long plane trip. So far, planes, buses, and cars offer no way of plugging in an AC adapter so that you can continue working after your battery runs down. (People used to try to recharge their batteries in airplane lavatories, but the airlines do not allow this.) Amtrak trains offer AC outlets, but they're best used for recharging your unit rather than for computing.

An essential question to ask: How long do you need to plug in to charge the battery up? Some computers take only a couple of hours, but others need overnight.

Another essential question: Are the batteries nickel-cadmium (once the standard, now obsolete), nickel/metal hydride, lithium-ion, or zinc-air (still being developed)? Most portables today rely on nickel/metal hydride, which last about 2½–3 hours between charges; a spare battery pack costs about $100. Lithium-ion batteries typically provide 3 hours or more of cordless operation each and cost $125–$150 for a spare. The old nickel-cadmium battery lasted only about 2 hours. The new zinc-air battery, still undergoing refinement, lasts 12 hours between recharges but is heavy (4.5 pounds) and pricey (about $400).[62]

Diskette Drive, Docking Station, or PC Cards?

One way manufacturers were able to make notebooks into the even lighter subnotebooks was, as we mentioned, by leaving out the internal diskette drive.

What if you need a diskette to transfer a file between computers or to install a new program? Or what if you want to make a backup copy of some material in case your hard disk fails?

There seem to be three choices:

• *Separate drive:* You buy a separate, external diskette drive that plugs directly into your subnotebook. Make sure this is at least available for the machine you buy, in case you decide to get it later.

• *Docking station or port replicator:* You buy a *docking station* that includes a diskette drive. Docks have been described as "home bases the subnotebooks can attach to."[63] Docking stations cost around $500 to $800.

Some docks are far more complicated than the subnotebooks. They have not only diskette drives but also their own hard disks, CD-ROM drives, network interfaces, add-on slots for expanding the computer's features, and so on.

Other docking stations, called *mini-docks, ramps,* or *port replicators,* available for $200–$300, add an inch or so to the back of the subnotebook and provide additional connectors. These may be used to connect external diskette drives, networks, color monitors, and the like. (They also enable you to connect into a local area network without rearranging cables.)

• *PC slots and cards:* You buy a subnotebook that has one or two slots to hold *PC (PCMCIA) cards.* These credit-card-size circuit boards can hold extra memory, programs, hard drives, modems, sound cards, and even pagers. PC hard-drive cards may provide you with some additional backup storage that you can use in lieu of diskettes.

There are different sizes of cards, and issues of compatibility are still being worked out. Thus, you may not be able to exchange files with your friends. However, most experts agree that "PCMCIA owns the foreseeable future of both portable and desktop computing."[64]

Software, Performance, & Expandability

If you're buying an IBM-compatible portable, at minimum you'll want to run Windows 3.1 applications, and you'll probably want to run Windows 95 programs.

How fast a processor you want for your portable is up to you. Presentation graphics and financial modeling may require more advanced chips. Notebooks and subnotebooks generally now come with '486 chips, and Pentium portables are also in evidence.

You should also check out how expandable the portable is. That is, find out how easily it will take add-ons such as extra storage or devices such as a modem or larger hard-disk drive. Some subnotebooks in particular allow add-ons only in the form of PC cards.

Some other issues to consider in buying a notebook or subnotebook are addressed in the accompanying checklist. (■ *See Panel 4.25.*)

Pocket PCs: Electronic Organizers & Personal Digital Assistants (PDAs)

It's the perfect gift for a student—a pocket-size electronic gadget that allows you to play videogames while killing time at a bus stop or to transmit (via infrared light beams) messages to a similar device across the classroom. Yet it also contains an appointment calendar with scheduled dates for all your papers and exams, a "to do" list, an address book, and a notebook of key facts for studying.

These so-called *toy organizers* combine game capabilities with personal organizing tools. Made by Casio, Sharp, Sega Enterprises, and some toy makers and costing $30–$200, such devices have been popular for some time among teens and preteens in Japan. They have been available in North America for the past couple of years.[65,66]

Toy organizers are only one type of device in a category we call *pocket personal computers.* These gadgets are variously called "electronic organizers," "palmtop PCs," "personal digital assistants," "handheld data communicators," and "personal communicators." Offered under different brand names, they differ in a number of ways. Some have keyboards, some a pen-like stylus, and some both. Some have fax capability, some have pagers and cellular phones.

Let's look closely at some of these devices.

	Yes	No
1. Is the device lightweight enough that you won't be tempted to leave it behind when you travel?	___	___
2. Does it work with lightweight nickel-hydride batteries instead of heavier nickel-cadmium batteries?	___	___
3. Is the battery life sufficient for you to finish the jobs you need to do, and is a hibernation mode available to conserve power?	___	___
4. Can you type comfortably on the keyboard for a long stretch?	___	___
5. Is the screen crisp, sharp, and readable in different levels of light?	___	___
6. Does the system have enough storage for all your software and data?	___	___
7. Can the system's hard disk and memory be upgraded to meet your needs?	___	___
8. Does the system provide solid communications options, including a fast modem, so you can send files, retrieve data, and plug into a local area network?	___	___

PANEL 4.25

Going mobile: a buyer's checklist

Intelligent Electronic Books

An *intelligent electronic book* reads material on small inter-changeable cartridges, slip-in cards, or disks (CD-ROMs). Three recent examples are the following:

- *Digital Book System:* Franklin Electronic Publishers' shirt-pocket-size Digital Book weighs less than 5 ounces (including batteries and two data cartridges), has a QWERTY-style keyboard, and can store 10 megabytes of information (the equivalent of 10 Bibles) on match-book-size, interchangeable cartridges. Cartridges include a dictionary and thesaurus, a word game, and a listing of information on 7000 movies available on videotape.

    The Digital Book can operate two cartridges simulta-neously. An audio feature is also available, so that users can listen to speech and sound from cartridges. In addi-tion, the unit lets you download (transfer) information to a microcomputer.

- *The Bookman:* Also from Franklin, the Bookman is a pocket-size device that comes with a choice of built-in databases and has a slot for a second cartridge. The built-in databases include dictionaries, Bibles, encyclo-pedias, cookbooks, and other reference titles.

- *Data Discman:* The Data Discman from Sony is a portable CD-ROM-based reference product. It weighs just 1 pound, has a backlit LCD screen displaying 30 characters across and 10 lines down, and features a QWERTY-type keyboard. Each 3-inch disk can hold 200 megabytes of information.

Among the compact disks available are the World Traveler Translator, which stores a dictionary of words and phrases and translates them into different lan-guages. Also available are *Compton's Concise Ency-clopedia* and the *Wellness Encyclopedia,* based on the *Wellness Letter* from the University of California at Berkeley.

Electronic Organizers or Low-End PDAs

Electronic organizers are small key-boards that are programmable for tasks such as data collection and ac-counting. Some of these devices are also called *palmtop computers.* Here we'll call them *low-end per-sonal digital assistants.* Their main characteristic is that they make use of just tiny keyboards; there is no pen input and no communications capability. Three exam-ples are as follows.

- *Casio Boss:* The Casio Boss SF-R20 contains a built-in spreadsheet and accepts memory cards that offer extra functions, such as spell-checker and expense tracking.

- *Hewlett-Packard 200LX Palmtop:* This pocket PC weighs 11 ounces, measures 6.3 by 3.4 inches, has the spreadsheet software Lotus 1-2-3

built in, and runs other microcomputer programs. The main difficulty for users is coping with the small keyboard and screen.

- *Psion Series 3a:* This pocket PC weighs 9.7 ounces and comes with word processor, spreadsheet, calculator, and world clock built in. It, too, has a tiny keyboard that makes you feel your hands are oversized.

Some students may find electronic organizers useful, and certainly their lower prices relative to high-end PDAs are attractive. (Still, at $500–$700, they aren't cheap.) Those who, for whatever reason, want to investigate more sophisticated pocket PCs should read on.

High-End Personal Digital Assistants (PDAs)

Apple Computer's Newton was the first personal digital assistant (PDA) to make headlines, but since then a slew of others have come along. PDAs are principally useful for businesspeople who travel frequently and who need to do mobile communicating and other on-the-road tasks. Prices new range from around $400 up to more than $1600, depending on the amount of memory installed and the kind of wireless services ordered. (Older versions may be found at discount for as little as $200.) Many have pen-based input only.

One experienced computer journalist, Phillip Robinson of the *San Jose Mercury News,* divides PDA tasks into three main categories: (1) delivering information (as a reference source), (2) communicating information (faxes and e-mail in and out), and (3) handling information (accepting your input and organizing it). "Any PC or Mac can do all three, if you add the right software and training," he says. "No PDA can."[67] This is because—being the size of a videotape cassette or smaller—they don't have the processing power or the software options.

Most PDAs depend on pen input, although a few have miniature keyboards. All run on AA or AAA batteries or on their own special battery packs. Beyond that, general features of high-end PDAs are as follows:[68]

1. All PDAs contain an address book.
2. PDAs also contain a calendar or appointment manager for scheduling appointments. Most have alarms that will signal you before a scheduled event.
3. PDAs usually contain spreadsheet programs or make them available.
4. Probably the most important feature: All high-end PDAs have a communications link. Some are wireless; others

provide a cable modem, so that you can exchange electronic mail and faxes, connect with information services such as America Online, get news and weather reports, and so on. Be sure there's a connectivity kit (usually a cable and a software program) that will work with your Macintosh or PC. It's a lot easier, for example, to type your address lists on a full-size keyboard, then transfer them to your PDA.

What about that most famous feature, handwriting recognition? This activity takes time. First you have to "train" the machine to your handwriting. Then you have to write (or hand-print) slowly enough to input your words. Instead, suggests one journalist, it's best just to "capture your notes as digital ink, which is just a fancy way of saying that whatever you write on the screen will be stored in your own handwriting."[69]

Should You Buy a PDA?

Would a student benefit from having a high-end personal digital assistant? The question is: Do you really need the communications features enough to pay the price? The low-end PDAs may provide the kind of electronic organizer, scheduler, and address book that you might find truly useful.

Some factors to be aware of are these:

- *Software features:* Does the device contain the programs you want, such as spreadsheet, word processing, games?
- *Communications features:* How easily can you swap information and programs between your PDA and your PC? Do you need to be able to send and receive computer data and faxes? Is wireless communication important to you?
- *Handwriting features:* It might be nice to take lecture notes in handwriting and have your PDA translate them into typewritten words and numbers, but this is not easily done. Some PDAs cannot translate handwriting—they can only store your scrawls as you scribble them. Others have not yet been proven to adequately and accurately translate handwriting to text—at least not without days of "training" the devices to recognize your writing.

A final point to keep in mind is that, whether equipped with a pen or a keyboard, PDAs may not be as easy to use as they look. Perhaps the best thing to do is to wait for new developments in portable computers. These may well shrink almost to the size of PDAs but have keyboards and other features that are easier to use.

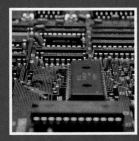

SUMMARY

What It Is / What It Does	Why It's Important
address *(p. 164, LO 3)* The location in main memory, designated by a unique number, in which a character of data or of an instruction is stored during processing.	To process each character, the control unit of the CPU retrieves it from its address in main memory and places it into a register. This is the first step in what is called the *machine cycle*.
American Standard Code for Information Interchange (ASCII) *(p. 168, LO 3)* Binary code used in microcomputers; ASCII originally used seven bits to form a character, but a zero was added in the left position to provide an eight-bit code, providing more possible combinations with which to form other characters and marks.	ASCII is the binary code most widely used with microcomputers.
arithmetic/logic unit (ALU) *(p. 163, LO 3)* The part of the CPU that performs arithmetic operations and logical operations and that controls the speed of those operations.	Arithmetic operations are the fundamental math operations: addition, subtraction, multiplication, and division. Logical operations are comparisons, such as is equal to (=), greater than (>), or less than (<).
binary system *(p. 166, LO 5)* A two-state system.	Computer systems use a binary system for data representation; two digits, 0 and 1, to refer to the presence or absence of electrical current or a pulse of light.
bit *(p. 167, LO 5)* Short for *binary digit,* which is either a 1 or a 0 in the binary system of data representation in computer systems.	The bit is the fundamental element of all data and information stored in a computer system.
bus *(p. 176, LO 3)* Electrical pathway through which bits are transmitted within the CPU and between the CPU and other devices in the system unit. There are different types of buses (address bus, control bus, data bus, input/output bus).	The larger a computer's buses, the faster it operates.
byte *(p. 167, LO 5)* A group of 8 bits.	A byte holds the equivalent of a character—such as a letter or a number—in computer data-representation coding schemes. It is also the basic unit used to measure the storage capacity of main memory and secondary storage devices.
cache memory *(p. 175, LO 3)* Special high-speed memory area on a chip that the CPU can access quickly. A copy of the most frequently used instructions is kept in the cache memory so the CPU can look there first.	Cache memory allows the CPU to run faster because it doesn't have to take time to swap instructions in and out of main memory. Large, complex programs benefit the most from having a cache memory available.
central processing unit (CPU) *(p. 163, LO 3)* The processor; it controls and manipulates data to produce information. In a microcomputer the CPU is usually contained on a single integrated circuit or chip called a *microprocessor*. This chip and other components that make it work are mounted on a circuit board called a *system board*. In larger computers the CPU is contained on one or several circuit boards. The CPU consists of two parts: (1) the control unit and (2) the arithmetic/logic unit. The two components are connected by a bus.	The CPU is the "brain" of the computer.

What It Is / What It Does	**Why It's Important**

chip (microchip) *(p. 152, LO 1)* Microscopic piece of silicon that contains thousands of microminiature electronic circuit components, mainly transistors.

Chips have made possible the development of small computers.

client/server network *(p. 162, LO 2)* Network consisting of central computer (server) that holds databases and programs for PCs, workstations, or terminals (clients).

Client/server networks are the most common type of local area network.

control unit *(p. 163, LO 3)* The part of the CPU that tells the rest of the computer system how to carry out a program's instructions.

The control unit directs the movement of electronic signals between main memory and the arithmetic/logic unit. It also directs these electronic signals between the main memory and input and output devices.

controller card *(p. 176, LO 6)* Circuit board that allows the CPU to work with the computer's different peripheral devices.

For example, a disk controller card allows the computer to work with different kinds of hard-disk and diskette drives.

downward compatible *(p. 174, LO 3)* Also called *backward compatible;* means that users can run the software written for computers with older chips on a computer with a new chip.

When upgrading their computer systems, users must check compatibility to be sure they can continue to run their old software and files.

electronic organizer *(p. 157, LO 2)* Specialized pocket computer that mainly stores appointments, addresses, and "to do" lists; recent versions feature wireless links to other computers for data transfer.

Puts in electronic form the kind of day-to-day personal information formerly in paper form.

execution cycle (E-cycle) *(p. 165, LO 3)* Part of the CPU machine cycle during which the ALU executes the instruction and stores the processed results in main memory or a register.

The completion time of the execution cycle determines how fast data is processed. The execution cycle is preceded by the instruction cycle.

expandability *(p. 176, LO 6)* Refers to the amount of room available in a computer for adding more memory or peripheral devices. Expandability is made possible with expansion slots and expansion boards.

If a microcomputer has expandability, it means that users can later add devices to enhance its computing power, instead of having to buy a new computer.

expansion bus *(p. 176, LO 3)* Bus that carries data between RAM and the expansion slots.

Without buses, computing would not be possible.

expansion card *(p. 176, LO 6)* Add-on circuit board that provides more memory or a new peripheral-device capability. (The words *card* and *board* are used interchangeably.) Expansion cards are inserted into expansion slots inside the system unit.

Users can use expansion cards to upgrade their computers instead of having to buy entire new systems.

expansion slot *(p. 176, LO 6)* Socket on the motherboard into which users may plug an expansion card.

See expansion card.

flash memory *(p. 176, LO 3)* Used primarily in notebook and subnotebook computers; flash memory, or flash RAM cards, consist of circuitry on credit-card-size cards that can be inserted into slots connecting to the motherboard.

Unlike standard RAM chips, flash memory is nonvolatile—it retains data even when the power is turned off. Flash memory not only can be used to simulate main memory but also to supplement or replace hard-disk drives for permanent storage.

floating-point operations per second (flops) *(p. 166, LO 3,4)* A floating-point operation is a kind of mathematical calculation. This measure, usually expressed in megaflops—millions of floating-point operations per second—is mainly used with supercomputers.

Floating-point methods are used for calculating a large range of numbers quickly.

game port *(p. 178, LO 6)* A port is a pathway in and out of a computer where peripheral devices can be connected; a game port allows users to attach a joystick or similar game-playing device to the system unit.

A game port allows a microcomputer to be made into a game machine.

gigabyte (G, GB) *(p. 167, LO 5)* Approximately 1 billion bytes (1,073,741,824 bytes); a measure of storage capacity.

Gigabyte is used to express the storage capacity of large computers, such as mainframes, although it is also applied to some microcomputer secondary-storage devices.

What It Is / What It Does	Why It's Important

graphics coprocessor chip *(p. 174, LO 3)* Secondary, "assistant" processor that enhances the performance of programs with lots of graphics and helps create complex screen displays.

Specialized chips such as these can significantly increase the speed of a computer system.

instruction cycle(I-cycle) *(p. 165, LO 3)* Part of the CPU machine cycle in which a single computer instruction is retrieved from memory, put into a register, and decoded.

"Decoding" means that the control unit alerts the circuits in the microprocessor to perform the specified operation. The instruction cycle is followed by the execution cycle.

integrated circuit (IC) *(p. 151, LO 1)* Collection of electrical circuits, or pathways, etched on tiny squares, or chips, of silicon half the size of a person's thumbnail. In a computer, different types of ICs perform different types of operations. An integrated circuit embodies what is called *solid-state technology.*

The development of the IC enabled the manufacture of the small, powerful, and relatively inexpensive computers used today.

kilobyte (K, KB) *(p. 167, LO 5)* Unit for measuring storage capacity; equals 1024 bytes (usually rounded off to 1000 bytes).

The sizes of stored electronic files are often measured in kilobytes.

laptop computer *(p. 156, LO 2)* Portable computer equipped with a flat display screen and weighing 8–20 pounds. The top of the computer opens up like a clamshell to reveal the screen.

Laptop and other small computers have provided users with computing capabilities in the field and on the road.

local bus *(p. 176, LO 3)* Bus that connects expansion slots to the CPU, bypassing RAM.

A local bus is faster than an expansion bus.

machine cycle *(p. 165, LO 3)* Series of operations performed by the CPU to execute a single program instruction; it consists of two parts: an instruction cycle and an execution cycle.

The machine cycle is the essence of computer-based processing.

machine language *(p. 169, LO 5)* Binary code (language) that the computer uses directly. The 0s and 1s represent precise storage locations and operations.

For a program to run, it must be in the machine language of the computer that is executing it.

mainframe *(p. 159, LO 2)* Second-largest computer available, after the supercomputer; occupies a specially wired, air-conditioned room, is capable of great processing speeds and data storage, and costs $50,000–$5 million.

Mainframes are used by large organizations (banks, airlines) that need to process millions of transactions.

main memory *(p. 164, LO 3)* Also known as *memory, primary storage, internal memory,* or *RAM* (for *random access memory*). Main memory is working storage that holds (1) data for processing, (2) the programs for processing the data, and (3) data after it is processed and is waiting to be sent to an output or secondary-storage device.

Main memory determines the total size of the programs and data files a computer can work on at any given moment.

massively parallel processing (MPP) *(p. 161, LO 2)* Type of supercomputer design that spreads calculations over hundreds or thousands of standard microprocessors of the type used in PCs. Tasks are parceled out to a great many processors, which work simultaneously.

Several small processors can run large programs in far less time than even large traditional supercomputers can run in serial fashion, one instruction at a time.

math coprocessor chip *(p. 174, LO 3)* Specialized microprocessor that helps programs that use lots of mathematical equations to run faster.

Math coprocessors can significantly increase the speed of processing of a computer system.

megabyte (M, MB) *(p. 167, LO 5)* About 1 million bytes (1,048,576 bytes).

Most microcomputer main memory capacity is expressed in megabytes.

megahertz (MHz) *(p. 166, LO 3,4)* Measurement of transmission frequency; 1 MHz equals 1 million beats (cycles) per second.

Generally, the higher the megahertz rate, the faster a computer can process data.

microcomputer *(p. 155, LO 2)* Small computer that can fit on or beside a desktop or is portable; uses a single microprocessor for its CPU. A microcomputer may be a workstation, which is more powerful and is used for specialized purposes, or a personal computer (PC), which is used for general purposes.	The microcomputer has lessened the reliance on mainframes and has enabled more ordinary users to use computers.
microcontroller *(p. 155, LO 2)* Also called an *embedded computer;* the smallest category of computer.	Microcontrollers are built into "smart" electronic devices, as controlling devices.
microprocessor *(p. 153, LO 1)* A CPU (processor) consisting of miniaturized circuitry on a single chip; it controls all the processing in a computer.	Microprocessors enabled the development of microcomputers.
microsecond *(p. 165, LO 3,4)* One-millionth of a second; used as a measure of computer processing speed.	The speed of a computer's machine cycle is often measured in microseconds.
millions of instructions per second (MIPS) *(p. 166, LO 3,4)* Another measure of a computer's execution speed; for example, .5 MIPS is 500,000 instructions per second.	This measure is often used for large, relatively powerful computers and new sophisticated microcomputers.
millisecond *(p. 165, LO 3,4)* One-thousandth of a second; a measure of storage access and transmission speed.	The speed of disk access time is measured in milliseconds.
minicomputer *(p. 158, LO 2)* Computer midway in cost and capability between a microcomputer and a mainframe and costing $20,000–$250,000.	Minicomputers can be used as single workstations or as a system tied by network to several hundred terminals for many users. Minicomputers are being replaced by networked microcomputers.
motherboard *(p. 171, LO 3)* Also called *system board;* the main circuit board in the system unit of a microcomputer.	It is the interconnecting assembly of important components, including CPU, main memory, other chips, and expansion slots.
nanosecond *(p. 165, LO 3,4)* One-billionth of a second; a measure of storage access and transmission speed.	A computer's memory access time is often measured in nanoseconds.
notebook computer *(p. 156, LO 2)* Type of portable computer weighing 4–7.5 pounds and measuring about 8½ × 11 inches.	Notebooks have more features than many subnotebooks yet are lighter and more portable than laptops.
palmtop computer *(p. 157, LO 2)* Type of pocket personal computer, weighing less than 1 pound, that is small enough to hold in one hand and operate with the other.	Unlike other pocket PCs, palmtops use the same software as IBM microcomputers and so are compatible with larger computers.
parallel port *(p. 177, LO 6)* Part of the computer through which a parallel device, which transmits 8 bits simultaneously, can be connected.	Enables microcomputer users to connect to a printer using a cable.
parity bit *(p. 169, LO 3)* Also called a *check bit;* an extra bit attached to the end of a byte.	Enables a computer system to check for errors during transmission (the check bits are organized according to a particular coding scheme designed into the computer).
pen computer *(p. 157, LO 2)* Type of portable computer; it lacks a keyboard or mouse but allows users to input data by writing directly on the display screen with a pen (stylus).	Pen computers are useful for specific tasks, such as for signatures to show proof of package delivery, and some general purposes, such as those fulfilled by electronic organizers and personal digital assistants.
peripheral devices *(p. 170, LO 6)* Hardware that is outside the central processing unit, such as input/output and secondary storage devices.	These devices are used to get data into and out of the CPU and to store large amounts of data that cannot be held in the CPU at one time.
personal computer (PC) *(p. 155, LO 2)* Type of microcomputer; desktop, floor-standing, or portable computer that can run easy-to-use programs, such as word processing or spreadsheets.	The PC is designed for one user at a time and so has boosted the popularity of computers.

What It Is / What It Does	**Why It's Important**

Personal Computer Memory Card International Association (PCMCIA) *(p. 178, LO 6)* Completely open, nonproprietary bus standard for portable computers.

This standard enables users of notebooks and subnotebooks to insert credit-card-size peripheral devices, called *PC cards*, such as modems and memory cards, into their computers.

personal digital assistant (PDA) *(p. 157, LO 2)* Also known as a *pocket communicator*; type of handheld pocket personal computer, weighing 1 pound or less, that is pen-controlled and in its most developed form can do two-way wireless messaging.

PDAs may supplant book-style personal organizers and calendars, as well as allow transmission of personal messages.

picosecond *(p. 165, LO 3,4)* One-trillionth of a second; a measure of storage access and transmission speed.

The speed of transistor switching may be measured in picoseconds.

pocket personal computer *(p. 156, LO 2)* Also known as a *handheld computer*; a portable computer weighing 1 pound or less. Three types of pocket PCs are electronic organizers, palmtop computers, and personal digital assistants.

Pocket PCs are useful to help workers with specific jobs, such as delivery people and parking control officers.

port *(p. 177, LO 6)* Connecting socket on the outside of the computer system unit that is connected to an expansion board on the inside of the system unit. Ports are of five types: parallel, serial, video adapter, SCSI, and game ports.

A port enables users to connect by cable a peripheral device such as a monitor, printer, or modem so that it can communicate with the computer system.

power supply *(p. 170, LO 3)* Device in the computer that converts AC current from the wall outlet to the DC current the computer uses.

The power supply enables the computer (and peripheral devices) to operate.

random-access memory (RAM) *(p. 174, LO 3)* Also known as *main memory* or *primary storage*; type of memory that temporarily holds data and instructions needed shortly by the CPU. RAM is a volatile type of storage.

RAM is the working memory of the computer; it is the workspace into which applications programs and data are loaded and then retrieved for processing.

read-only memory (ROM) *(p. 175, LO 6)* Also known as *firmware*; a memory chip that permanently stores instructions and data that are programmed during the chip's manufacture. Three variations on the ROM chip are PROM, EPROM, and EEPROM. ROM is a nonvolatile form of storage.

ROM chips are used to store special basic instructions for computer operations such as those that start the computer or put characters on the screen.

reduced instruction set computing (RISC) *(p. 173, LO 3)* Type of design in which the complexity of a microprocessor is reduced by reducing the amount of superfluous or redundant instructions.

With RISC chips, a computer system gets along with fewer instructions than those required in conventional computer systems. RISC-equipped workstations work 10 times faster than conventional workstations.

register *(p. 164, LO 3)* High-speed circuit that is a staging area for temporarily storing data during processing.

The computer loads the program instructions and data from the main memory into the staging areas of the registers just prior to processing.

semiconductor *(p. 151, LO 1)* Material, such as silicon (in combination with other elements), whose electrical properties are intermediate between a good conductor and a nonconductor. When good-conducting materials are laid on the semiconducting material, an electronic circuit can be created.

Semiconductors are the materials from which integrated circuits (chips) are made.

serial port *(p. 178, LO 6)* Also known as *RS-232 port*; a port to which a cable is connected that transmits 1 bit at a time.

Serial ports are used principally for connecting communications lines, modems, and mice to microcomputers.

server *(p. 162, LO 2)* Computer in a network that holds databases and programs for multiple users.

The server enables many users to share equipment, programs, and data.

silicon *(p. 151, LO 1)* Element widely found in sand and clay; it is a semiconductor.

Silicon is used to make integrated circuits (chips).

single inline memory module (SIMM) *(p. 174, LO 3)* Small circuit board plugged into the motherboard.

A SIMM holds RAM chips and can be used to increase a computer's main memory capacity.

Small Computer System Interface (SCSI) port *(p. 178, LO 6)* Pronounced "scuzzy"; an interface for transferring data at high speeds for up to eight SCSI-compatible devices.

SCSI ports are used to connect external hard-disk drives, magnetic-tape backup units, and CD-ROM drives to the computer system.

What It Is / What It Does	**Why It's Important**

solid state device *(p. 151, LO 1)* Electronic component made of solid materials with no moving parts, such as an integrated circuit.

Solid-state integrated circuits are far more reliable, smaller, and less expensive than electronic circuits made from several components.

subnotebook computer *(p. 156, LO 2)* Type of portable computer, weighing 2.5–4 pounds.

Subnotebooks are lightweight and thus extremely portable; however, they may lack features found on notebooks and other larger portable computers.

supercomputer *(p. 161, LO 2)* High-capacity computer that is the fastest calculating device ever invented; costs $225,000–$30 million. It may have a vector processing design or massively parallel processing design.

Used principally for research purposes, airplane design, oil exploration, weather forecasting, and other activities that cannot be handled by mainframes and other less powerful machines.

surge protector *(p. 170, LO 6)* Also called *surge suppressor;* device that protects a computer from being damaged by surges of high voltage.

A surge protector is an inexpensive investment compared to a new motherboard.

system unit *(p. 170, LO 6)* The box or cabinet that contains the electrical components that do the computer's processing; usually includes processing components, RAM chips (main memory), ROM chips (read-only memory), power supply, expansion slots, and disk drives but not keyboard, printer, or often even the display screen.

The system unit protects many important processing and storage components.

system clock *(p. 174, LO 3)* Internal timing device that uses a quartz crystal to generate a uniform electrical frequency from which digital pulses are created.

The system clock controls the speed of all operations within a computer. The faster the clock, the faster the processing.

terabyte (T, TB) *(p. 167, LO 5)* Approximately 1 trillion bytes (1,009,511,627,776 bytes); a measure of capacity.

Some forms of mass storage, or secondary storage for mainframes and supercomputers, are expressed in terabytes.

transistor *(p. 151, LO 1)* Semiconducting device that acts as a tiny electrically operated switch, switching between "on" and "off" many millions of times per second.

Transistors act as electronic switches in computers. They are more reliable and consume less energy than their predecessors, electronic vacuum tubes.

uninterruptible power supply (UPS) *(p. 171, LO 6)* Battery-operated device that provides a microcomputer with electricity if there is a power failure.

A UPS can keep a microcomputer going long enough to allow the user to save data files before shutting down.

upward compatible *(p. 174, LO 3)* Refers to the situation wherein software written for a microcomputer with a new chip will run on a machine with an older chip.

When upgrading their computer systems, users must check compatibility to be sure they can continue to run their software and files. They must also remember that new software may require more main memory to run, in addition to other requirements.

video adapter port *(p. 178, LO 6)* Part of the computer used to connect the video display monitor outside the computer to the video adapter card inside the system unit.

The video adapter port enables users to have different kinds of monitors, some having higher resolution and more colors than others.

video memory *(p. 175, LO 3)* Video RAM (VRAM) chips are used to store display images for the monitor.

The amount of video memory determines how fast images appear and how many colors are available on the display screen. Video memory chips are useful for programs displaying lots of graphics.

volatile storage *(p. 164, LO 3)* Temporary storage, as in main memory (RAM).

The contents of volatile storage are lost when power to the computer is turned off.

word *(p. 174, LO 3)* Also called *bit number;* group of bits that may be manipulated or stored at one time by the CPU.

Often the more bits in a word, the faster the computer. An 8-bit word computer will transfer data within each CPU chip in 8-bit chunks. A 32-bit word computer is faster, transferring data in 32-bit chunks.

workstation *(p. 157, LO 2)* Type of microcomputer; desktop or floor-standing machine that costs $10,000–$150,000 and is used mainly for technical purposes.

Workstations are used for scientific and engineering purposes and also for their graphics capabilities.

(Selected answers appear at the back of the book.)

Short-Answer Questions

1. What is the purpose of the parity bit in a binary coding scheme?
2. What is the main difference between a 66 MHz computer and a 133 MHz computer?
3. What are the two main parts of the CPU? What is each part responsible for?
4. Describe why having more main memory, or RAM, in your computer (as opposed to less) is useful.
5. What is the function of the ALU in a microcomputer system?
6. What is a stripped-down Internet PC?
7. What are PC cards used for?
8. Name three of the five kinds of ports discussed in this chapter. What are they used for?
9. What are surge protectors and UPSs used for?
10. Supercomputers use either vector processing or massively parallel processing. Which is faster? Why?

Fill-in-the-Blank Questions

1. The _____ links the CPU to every hardware device in the computer system.
2. List five of the components located in the system unit.
   a. _____
   b. _____
   c. _____
   d. _____
   e. _____
3. The _____ is located inside the system unit and controls how fast all the operations within the computer take place.
4. The _____ is located inside the system unit and converts AC current to DC current to run the computer.
5. The machine cycle is a series of operations performed to execute a single program instruction; it consists of the _____, which fetches and decodes, and a(n) _____, which executes and stores.
6. The binary coding scheme most commonly used in microcomputers is _____.

7. Name four types of portable computers:
   a. _____
   b. _____
   c. _____
   d. _____
8. The main circuit board in a system unit is called the _____.
9. To attach a joystick or similar game-playing device to a computer, the system unit must include a _____.
10. _____ is the semiconductor material commonly used to make integrated circuits (chips).

Multiple-Choice Questions

1. A(n) _____, or character, is composed of 8 _____.
   a. byte, bits
   b. bit, bytes
   c. megabyte, kilobytes
   d. kilobyte, megabytes
   e. none of the above
2. Microcontrollers are also called:
   a. chips
   b. embedded computers
   c. personal computers
   d. microcomputers
   e. none of the above
3. Which of the following memory types can be used in a computer system?
   a. main memory
   b. cache memory
   c. video memory
   d. flash memory
   e. all of the above
4. Which of the following is attached to the end of a byte for purposes of error checking?
   a. parity check
   b. bit number
   c. character
   d. parity bit
   e. none of the above
5. Main memory is also known as:
   a. primary storage
   b. RAM
   c. memory
   d. all of the above

6. Which of the following would you want inside your computer's system unit if you plan to perform lots of mathematical calculations?
   a. graphics coprocessor
   b. vector processing
   c. math coprocessor
   d. extended memory
   e. none of the above

7. Which of the following would your computer need to use a modem?
   a. parallel port
   b. serial port
   c. SCSI port
   d. video adapter port
   e. none of the above

8. A sophisticated microcomputer that fits on a desktop and is used mainly by engineers and scientists for technical purposes is called a(n)

   _____.

   a. workstation
   b. minicomputer
   c. mainframe computer
   d. supercomputer
   e. none of the above

9. Which of the following is used as a central computer that holds databases and programs for networks of PCs, workstations, or terminals?
   a. massively parallel computer
   b. supercomputer
   c. mainframe computer
   d. server
   e. none of the above

10. Which of the following technologies is based on using molecules to create very tiny machines that hold data or perform tasks?
    a. gallium arsenide computing
    b. opto-electronic processing
    c. nanotechnology
    d. biotechnology
    e. none of the above

True/False Questions

T F   1. The amount of main memory in your computer system is an important factor in determining what software the system can run.

T F   2. A printer is sometimes contained inside the system unit of a microcomputer.

T F   3. To connect a peripheral device, users plug a cable into an expansion slot on the outside of the system unit.

T F   4. RAM is volatile.

T F   5. Registers are used to transmit bits between the CPU and other devices in the system unit.

T F   6. Data and instructions in ROM are lost when the computer is turned off.

T F   7. A gigabyte is smaller than a terabyte.

T F   8. It is rare that a microprocessor chip houses more than ten transistors.

T F   9. The CPU consists of the control unit and the ALU.

T F   10. In machine language, data and instructions are represented with 0s and 1s.

Projects/Critical-Thinking Questions

1. Can you envision yourself using a supercomputer in your (planned) profession or job? How?

2. Computer magazines often sponsor tests to compare PCs based on their speed. For example, in March of 1995, PC World determined that the Micron Millennia was the fastest PC. By reviewing current computer magazines, identify the fastest PC today. What processing components make this the fastest PC? How much does the PC cost? What retail products would make this PC even faster?

3. Research is underway at Carnegie Mellon University to develop a PC that you can wear. The wearable PC might consist of a head band and visor, and a necklace that functions as a video input device. What is the rationale behind the wearable PC? What processing components are in the wearable PC? How long have researchers been working on this project? When do researchers predict that the wearable PC will be available for purchase? What do you think the future holds for the wearable PC?

4. Identify a manual or computerized procedure, such as buying postage or servicing your car, that you think could be improved with a new or improved computerized procedure. How is the procedure currently performed? By whom? Do computers currently perform or help with the procedure? What processing hardware do you think is needed to improve upon the existing procedure? Why do you think your idea hasn't been implemented already?

5. Most industry observers agree that in the years ahead computer chip makers will continue to develop faster and faster microprocessors. What impact do you think this will have on the software industry? Will users have to change systems software and/or applications software in order to take advantage of more sophisticated microprocessors? What impact do you think faster microprocessors will have on the hardware industry? Will users have to scrap existing hardware?

net Using the Internet

Objective: *In this exercise we describe how to use Netscape Navigator 2.0 or 2.01 to search the Web for general and specific topics.*

Before you continue: *We assume you have access to the Internet through your university, business, or commercial service provider and to the Web browser tool named Netscape Navigator. Additionally, we assume you know how to connect to the Internet and load Netscape Navigator. If necessary, ask your instructor or system administrator for assistance.*

1. Make sure you have started Netscape. The home page for Netscape Navigator should appear on your screen.

2. If you're interested in finding information on a general topic—such as processing hardware or portable computers—use a *directory*. Directories contain lists of Web sites that are organized by topic. For example, Yahoo (*http://www.yahoo.com*) contains a list of over 80,000 Web sites which are divided into 14 categories. After selecting a topic of interest, you're presented with a sub-list of topics. You continue selecting topics until the information you're looking for appears. If you're just becoming familiar with the Web, directories can provide a useful means of seeing what the Web has to offer. Other examples of directories are
   Galaxy (*http://galaxy.einet.net/*),
   InfoSeek (*http://www2.infoseek.com/*),
   Magellan (*http://www.mckinley.com/*),
   and Scott Yanoff's Internet Services Directory
   (*http://www.uwm.edu/Mirror/inet.services.html*).

   To display Yahoo's home page:
   CLICK: Open button (▣)

TYPE: http://www.yahoo.com
PRESS: Enter or CLICK: Open

After a few moments, Yahoo's home page should appear. (*Note:* The progress indicator in the bottom-right corner of the screen indicates how much of a page has been loaded at a given time.) Your screen should appear similar to the following:

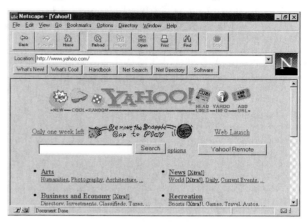

3. Point to some of the links (underlined words and phrases). Notice that the associated URL appears in the status area at the bottom of the screen.

4. Drag the vertical scroll bar downward and notice that additional topics appear.

5. In the next few steps you'll search for information on portable computing. The link that looks the most promising at this point is the CS link that appears in the Science category. Drag the scroll bar until you see the Science category and then click the CS link.

6. Drag the vertical scroll bar downward to see a list of topics, as shown below.

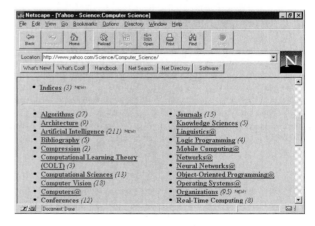

7. To continue with your search on portable computing, click the Mobile Computing@ link on the current page and then the Portable Computing link that is located on the next page.

8. As you can see, with each selection you make, the list of Web pages becomes more and more specific. The idea is that eventually the information you're looking for will appear on the screen. Now that you've had some practice using a directory, return to Netscape's home page and then practice using a search engine in the next few steps.

9. If you're interested in finding information on a specific topic, you may be better off using an *index*, or *search engine* rather than a directory. A search engine is a Web page that contains a form. You type a text string into the form that identifies the topic you want to search for. Whereas directories display lists of topics, search engines display a list of Web sites that match your search criteria. Some examples of general-interest search engines are Lycos (*http://lycos.cs.cmu.edu/*), Excite Netsearch (*http://www.excite.com/*), Open Text (*http://www.opentext.com:8080/*), and Alta Vista (*http://www.altavista.digital.com/*).

   To display Lycos' search engine:
   CLICK: Open button (⊞)
   TYPE: http://lycos.cs.cmu.edu
   PRESS: (**Enter**) or CLICK: Open
   The Lycos search form appears.

10. To display a list of Web sites that relate to the topic of "portable computing:"
    CLICK: in the Find text box
    TYPE: portable computing
    CLICK: Go Get It button

    After a few moments, a list of Web sites that relate to your search criteria will be loaded into the current page. (*Note:* You must drag the vertical scroll bar downward to see the list of Web sites.)

11. Continue dragging the vertical scroll bar downward until you see a list of related Web sites. Although Lycos found many Web sites that match your search criteria, Lycos only displays the first 10 Web sites. To see the next ten sites that match your criteria, click the "Next 10 hits" link on the bottom of the page.

12. Now that you've practiced using a directory and a search engine, display Netscape's home page and then exit Netscape.

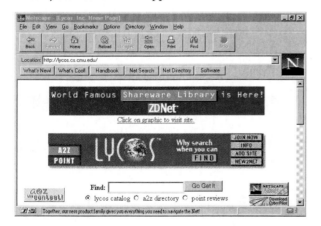

CHAPTER 5

Input
Taking Charge of Computing & Communications

Concepts You Should Know

After reading this chapter, you should be able to:

1. Distinguish between the two types of input hardware.
2. List and explain the types of pointing devices, including operation of a mouse.
3. Identify and explain the different scanning devices.
4. Distinguish between magnetic-stripe, smart, and optical cards.
5. Discuss the operation and limitations of voice-recognition devices.
6. Describe other input devices: audio, video, cameras, sensors, and human-biology devices—biometric systems, line-of-sight systems, and cyber gloves and body suits.
7. Discuss some adverse effects of computers on health—repetitive strain injuries, eyestrain and headaches, and backstrain—and the significance of ergonomics.

For a sizable number of people, it's always "0:00" on the VCR clock.

About one in five owners of a videocassette recorder fails a basic test of the Information Age: setting the VCR's digital clock.[1] Even though 88% of Americans say their family owns a VCR, according to a Washington, D.C., polling firm, apparently 16% of them don't know how to set the time on it. In such households the VCR clock is always blinking "0:00" or "12:00."

Is this a sign that many people are being left behind in the Digital Revolution? If people can't set the time on the VCR, how will they be able to get the machine to automatically record programs from some of those hundreds of TV channels we are supposed to have some day? Or figure out how to download (transfer) crucial information into their "telecomputer" from some far-flung database?

Today, learning to benefit from information technology means becoming comfortable with the input and output devices that constitute its two principal interfaces with people. In this chapter we explain what input devices are.

Input Hardware: Keyboard Entry Versus Source Data Entry

Preview & Review: Input hardware is classified as keyboard entry or source data entry (automation).

Input hardware **consists of devices that take data and programs that people can read or comprehend and convert them to a form the computer can process.** The people-readable form may be words like the ones in these sentences, but the computer-readable form consists of 0s and 1s, or off and on electrical signals.

Input devices are categorized as *keyboard entry* and *source data entry* devices. (■ *See Panel 5.1.*)

Keyboard Entry

In a computer, **a** *keyboard* **is a device that converts letters, numbers, and other characters into electrical signals that are machine-readable by the computer's processor.** The keyboard may look like a typewriter keyboard to which some special keys have been added. Or it may look like the keys on a bank's automated teller machine or the keypad of a pocket computer used by a bread-truck driver.

Source Data Entry

Source data entry devices **refers to the many forms of data-entry devices that are not keyboards.** These devices create machine-readable data on magnetic media or paper or feed it directly into the computer's processor. This is also known as *source-data automation,* the process in which data created while an event is taking place is entered directly into the system in a machine-processable form.

Source data entry devices include the following:

- Pointing devices
- Scanning devices
- Magnetic-stripe, smart, and optical cards

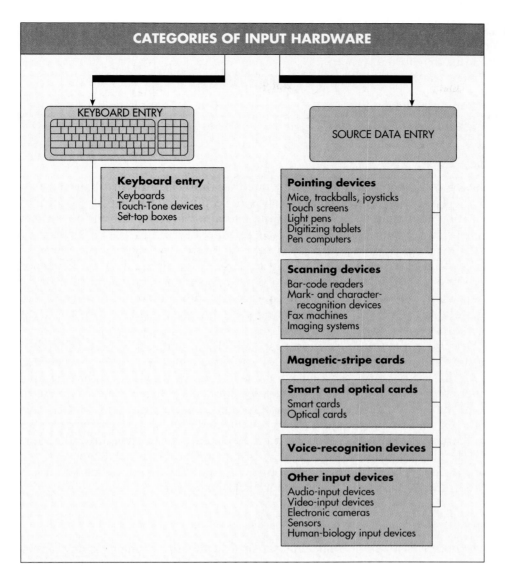

PANEL 5.1

Summary of input devices

- Voice-recognition devices
- Audio-input devices
- Video-input devices
- Electronic cameras
- Sensors
- Human-biology input devices

Often keyboard and source data entry devices are combined in a single computer system. A desktop-publishing (✓ p. 82) system, for example, may use a keyboard, a mouse, and an image scanner.

Keyboard Input

Preview & Review: Keyboard-type devices include computer keyboards, terminals, Touch-Tone devices, and set-top boxes or future "information appliances."

Even if you aren't a ten-finger touch typist, you can use a keyboard. Yale University computer scientist David Gelernter, for instance, lost the use of his right hand and right eye in a mail bombing. However, he expressed not only gratitude at being alive but also recognition that he could continue to use a keyboard even with his limitations. "In the final analysis," he wrote in an online message to colleagues, "one decent typing hand and an intact head is all you really need. . . ."[2]

Here we describe the following keyboard-type devices:

- Computer keyboards
- Terminals
- Touch-Tone devices
- Set-top boxes

Computer Keyboards

Conventional computer keyboards, such as those for microcomputers, have all the keys that typewriter keyboards have plus others unique to computers. People who always thought typing was an overrated skill will find themselves slightly behind in the Digital Age, since learning to use a keyboard is still probably the most important way of interacting with a computer. Of course, compared to a mouse or other source data entry devices, keyboards have one disadvantage: if you're to use them efficiently, you need training. Users who have to use the hunt-and-peck method waste a lot of time. Fortunately, there are software programs available that can help you learn typing skills or improve your existing ones.

You are probably already familiar with a computer keyboard. The illustration opposite provides a review of keyboard functions. (■ See Panel 5.2.)

As the use of computer keyboards has become widespread, so has the incidence of various hand and wrist injuries. Accordingly, keyboard manufacturers have been giving a lot of attention to ergonomics. *Ergonomics* is the study of the physical relationships between people and their work environment. Various attempts are being made to make keyboards more ergonomically sound in order to prevent injuries. One interesting variation is a $99 device that substitutes foot-driven pedals for the Shift, Alt, and Ctrl keys. *(■ See Panel 5.3.)* (We discuss ergonomics in the Experience Box at the end of this chapter.)

Terminals

A *terminal* is a device that consists of a keyboard, a video display screen, and a communications line to a large (usually mainframe) computer system. Terminals are generally used for input; they also display output. Terminals may be dumb or smart.

- **Dumb:** The most common type of terminal is dumb. A *dumb terminal* can be used only to input data to and receive information from a computer system. That is, it cannot do any processing on its own.

  An example of a dumb terminal is the type used by airline clerks at airport ticket and check-in counters.

- **Smart:** A *smart terminal*, also called an *X-terminal*, can do input and output and has some processing capability and RAM. However, a smart terminal is not designed to operate as a stand-alone computer. Thus it cannot be used to do programming—that is, create new instructions.

  Two examples of smart terminals are automated teller machines and point-of-sale terminals.

  An *automated teller machine (ATM)* is used to retrieve information on bank balances, make deposits, transfer sums between accounts, and withdraw cash. Usually the cash is disbursed in $20 bills. Some Nevada gambling casinos have machines that dispense only $100 bills.

  ATMs have become wildly popular throughout the world. You can now use your ATM card, for example, to get cash from machines on cruise ships or local currency in machines in foreign airports. Residents of

Only capital letters will be displayed.

Prints what's currently displayed on the screen.

Prevents the screen from scrolling.

Temporarily suspends the current task.

The Esc key allows you to exit a command or menu and return to the work screen.

These status lights indicate when these functions are on or off.

Function Keys are used to issue commands specific to the software package being used.

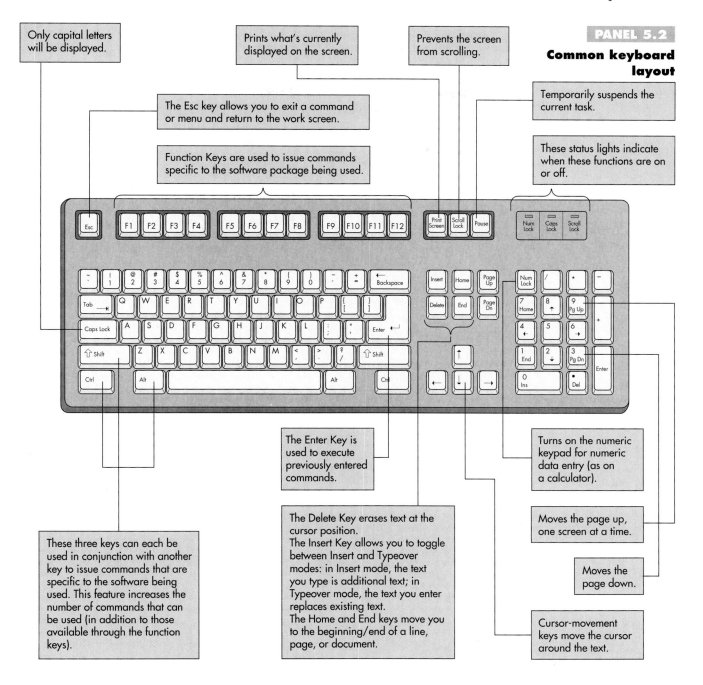

The Enter Key is used to execute previously entered commands.

Turns on the numeric keypad for numeric data entry (as on a calculator).

The Delete Key erases text at the cursor position.
The Insert Key allows you to toggle between Insert and Typeover modes: in Insert mode, the text you type is additional text; in Typeover mode, the text you enter replaces existing text.
The Home and End keys move you to the beginning/end of a line, page, or document.

Moves the page up, one screen at a time.

Moves the page down.

Cursor-movement keys move the cursor around the text.

These three keys can each be used in conjunction with another key to issue commands that are specific to the software being used. This feature increases the number of commands that can be used (in addition to those available through the function keys).

PANEL 5.3

Foot controls

This foot-driven device assigns keystrokes to pedals. It uses three pedals as substitutes for the Shift, Alt, and Ctrl keys. Their purpose is to help people who spend long hours working at computers avoid repetitive strain injuries to arms, wrists, and hands.

Singapore now use ATMs to buy shares of stock.[3] A variant on the ATM is the ETM, or electronic ticket machine, found in airports and railroad stations; you swipe your credit card through a slot, enter the appropriate information, and a ticket pops out.[4,5] The main purpose of the ETM is to help travelers avoid lines when buying tickets.

A *point-of-sale (POS) terminal* is a smart terminal used much like a cash register. It records customer transactions at the point of sale but also stores data for billing and inventory purposes. POS terminals are found in most department stores.[6]

Microcomputers increasingly are being used in business as terminals—and in place of dumb and smart terminals. This trend is occurring not only because their prices have come down but also because they reduce the processing and storage load on the main computer system.

Touch-Tone Devices

The Touch-Tone, or push-button, telephone can be used like a dumb terminal to send data to a computer. FedEx customers, for example, can request pickup service for their packages by pushing buttons on their phones. Another common device is the *card dialer*, or *card reader*, used by merchants to verify credit cards over phone lines with a central computer.

A fairly recent Touch-Tone device is the bank-by-phone telephone, which enables you to do your banking from your home. The device contains not only a telephone keypad but also a small, three-line display screen. A computer voice announces the beginning and end of your banking transactions.[7] (■ *See Panel 5.4.*)

Set-Top Boxes—or "Information Appliances"

If you receive television programs from a cable-TV service instead of free through the air, you may have what is called a *set-top box*. A *set-top box* works with a keypad to allow cable-TV viewers to change channels or, in the case of interactive systems, to exercise other commands. (■ *See Panel 5.5.*) For example, in the 1970s the Warner-Amex interactive Qube system was

---

PANEL 5.4

Bank-by-phone phone

This telephone, designed by Northern Telecom of Canada, enables a bank customer to pay bills, transfer funds, and inquire about balances. The telephone has a liquid-crystal-display (LCD) screen, about the length of four postage stamps, that can display three lines of information. It also has a computer voice that announces the beginning and end of transactions.

PANEL 5.5

Set-top box

Scientific-Atlanta's Model 8600X Home Communications Terminal combines computing and cable-TV functions in a single device.

instituted in Columbus, Ohio. The system's set-top box enabled users to request special programs, participate in quiz shows, order services or products, or respond to surveys.[8] Although Qube was discontinued because it was unprofitable, similar experiments are continuing, such as Time Warner's Full Service Network in Orlando, Florida.

One question about the Information Superhighway is what kind of set-top box—or so-called information appliance or communications appliance or telecomputer—and two-way services will be available. Apple, for instance, is marketing an ezTV cable-box interface. Paul Saffo, of the Institute for the Future in Menlo Park, California, believes that the microcomputer will retain basic traits such as the display screen. However, it will mutate and blend into devices both familiar (such as TVs) and exotic (such as handheld "personal digital assistants" like the Apple Newton, described later). "The true value of a personal computer," Saffo says, "is increasingly determined by what it's connected to, not by the applications it processes."[9]

We turn now from keyboard entry to various types of source data entry, in which data is converted into machine-readable form as it is entered into the computer or other device. Source data entry is important because it cuts out the keyboarding step, thereby reducing mistakes and labor costs. For example, in many places utility companies rely on human meter readers to fight through snowdrifts and brave vicious dogs to read gas and electric meters. They then type the readings into specialized portable computers—a time-consuming and potentially error-producing step. Over the next decade, however, probably 25% of the 160 million gas and electric meters in the United States will be automated. That is, a transmitter plugged into the meter in a customer's home will send data directly to the utility company's computers—a quicker and more accurate process.[10] (Because the ranks of human meter readers will shrink, the transmitters are being installed gradually, and many workers are being retrained or reassigned.)

Practical Matters: Software to Teach or Improve Typing Skills

Fast and accurate typing is a crucial skill for the information age. You need to know more than how to click your mouse buttons.

Like starting an exercise program, developing your typing skills will take at least a few weeks to start paying off. But it'll pay off for the rest of your computing life.

Luckily, you can find plenty of fine software that makes learning to type fun, or at least more fun than in the not-so-good old days of manual and electric typewriters and exercise books. These programs can show you where to put your fingers, encourage you through progressive lessons, time and test you to focus on your weak fingers and keys, and even play games where you have to type to win.

What makes a good typing program?

Look for: low cost; diagrams or animations to show how the fingers touch the keys; progressive lessons taken from both random words and realistic documents; timers, charts, and progress reports to analyze your speed and accuracy; lessons that learn your weaknesses and strengths; and typing games for practice and comic relief. CD-ROM versions typically offer more lessons, animations, and games.

Instructor modes track pupils' progress in classroom situations. For the office, you'll appreciate ergonomic and word-processor tips and hints. . . . Numeric-keypad lessons and mouse practice are nice additions. And some kids-oriented programs offer keyboard familiarity and comfort.

—Phillip Robinson, "Reps to Flex Fingers," *San Jose Mercury News*

Comparing typing software

If your typing skills need updating or are nonexistent, you can get help from a variety of software programs. These are for older children and adults. Prices range from about $14.95 up to $59.95 (for some CD-ROMs).

	Programs					
	Typing Tutor	**Typing Instructor**	**Expert Typing for Windows**	**Mavis Bacon Teaches Typing**	**UltraKey**	**Mario Teaches Typing**
Phone number	800-428-5331	800-822-3522	800-759-2562	800-234-3088	800-465-6428	800-969-4263
Windows	$17.98	N/A	$14.95	$59.95	N/A	N/A
Windows CD-ROM	N/A	$29.95	N/A	$59.95	N/A	N/A
DOS	$17.98	$19.95	N/A	N/A	$49.95	$29.95
DOS CD-ROM	N/A	N/A	N/A	N/A	N/A	$39.95
Mac	$17.98	$19.95	N/A	$49.95	$49.95	$34.95
Suggested age range	N/A	12 up	12 up	9 up	8 up	6 up
Finger position graphics	Yes	No	Yes	Yes	Yes	Yes
Lessons adapt to you	Yes	Yes	No	Yes	No	No
Create your own tests	Yes	Yes	No	Yes	Yes	No
Progress reports	Yes	Yes	Yes	Yes	Yes	Yes
Instructor mode	Yes	No	No	Yes	Yes	No
Mouse practice	Yes	No	No	No	No	No
10-Key practice	Yes	Yes	Yes	Yes	No	No

Pointing Devices

Preview & Review: Pointing devices include mice, trackballs, and joysticks; touch screens; light pens; digitizing tablets; and pen computers.

One of the most natural of all human gestures, the act of pointing, is incorporated in several kinds of input devices. The most prominent ones are the following:

- Mice, trackballs, and joysticks
- Touch screens
- Light pens
- Digitizing tablets
- Pen computers

Mice, Trackballs, & Joysticks

The principal pointing tools used with microcomputers are the mouse, the trackball, and the joystick, all of which have variations. (■ *See Panel 5.6.*)

- **Mouse:** **A *mouse* is a device that is rolled about on a desktop and directs a pointer on the computer's display screen.** The pointer may sometimes be, but is not necessarily the same as, the cursor. **The *mouse pointer* is the symbol that indicates the position of the mouse on the display screen.**

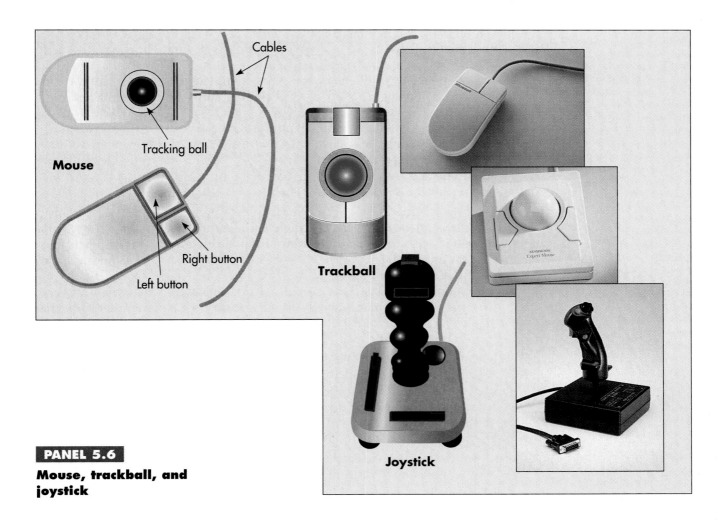

PANEL 5.6

Mouse, trackball, and joystick

Mouse and mouse pointer

Movement of the mouse on the desktop causes a corresponding movement of the mouse pointer on the display screen. A mouse may have one to three buttons.

(■ *See Panel 5.7.*) It may be an arrow, a rectangle, or even a representation of a person's pointing finger. The pointer may change to the shape of an I-beam to indicate that it is a cursor and shows the place where text may be entered.

The mouse has a cable that is connected to the microcomputer's system unit by being plugged into a special port (✓ p. 178). The tail-like cable and the rounded "head" of the instrument are what suggested the name *mouse.*

On the bottom side of the mouse is a ball (trackball) that translates the mouse movement into digital signals. On the top side are one, two, or three buttons. Depending on the software, these buttons are used for such functions as *clicking, dropping,* and *dragging.*

Gently holding the mouse with one hand, you can move it in all directions on the desktop (or on a mouse pad, which may provide additional traction). This will produce a corresponding movement of the mouse pointer on the screen.

- **Trackball:** Another form of pointing device, the trackball, is a variant of the mouse. **A *trackball* is a movable ball, on top of a stationary device, that is rotated with the fingers or palm of the hand.** In fact, the trackball looks like the mouse turned upside down. Instead of moving the mouse around on the desktop, you move the trackball with the tips of your fingers.

   Trackballs are especially suited to portable computers, which are often used in confined places, such as on airline tray tables. Trackballs may appear on the keyboard centered below the space bar, as on the Apple PowerBook, or built into the right side of the screen. On some portables they are a separate device that is clipped to the side of the keyboard.

- **Joystick:** **A *joystick* is a pointing device that consists of a vertical handle like a gearshift lever mounted on a base with one or two buttons.** Named for the control mechanism that directs an airplane's fore-and-aft and side-to-side movement, joysticks are used principally in videogames and in some computer-aided design systems.

- **Other pointing controllers:** The IBM ThinkPad, a portable computer, uses a pointing controller that resembles a small stick. It protrudes through the keyboard amid the G, H, and B keys.

Another pointing controller is an "air mouse," a cordless mouse that works up to 40 feet away from the computer. With this you can roam a conference room during a slide-show presentation while clicking on icons or making other changes.[11]

Despite the mouse's ease of use, there is something that is often even easier to use—the keyboard. Depending on the software, many commands that can be done with a mouse can also be performed through the keyboard. The mouse may make it easy to *learn* the commands for, say, a word processing program. However, you may soon find that you can *more quickly execute* those commands by pressing a combination of keys on the keyboard.

Touch Screens

A *touch screen* is a video display screen that has been sensitized to receive input from the touch of a finger. (■ *See Panel 5.8, next page.*) Because touch screens are easy to use, they can convey information quickly. You'll find touch screens in automated teller machines, tourist directories in airports, and campus information kiosks making available everything from lists of coming events to (with proper ID and personal code) student financial-aid records and grades.[12] Touch screens are also available for personal computers, consisting of an overlay that mounts with adhesive to the front of a monitor.[13]

Touch screen

Making menu choices at a fast-food restaurant.

Light Pens

The *light pen* is a light-sensitive stylus, or pen-like device, connected by a wire to the computer terminal. The user brings the pen to a desired point on the display screen and presses the pen button, which identifies that screen location to the computer. (■ *See Panel 5.9.*) Light pens are used by engineers, graphic designers, and illustrators.

Digitizing Tablets

A *digitizing tablet* consists of a tablet connected by a wire to a stylus or puck. A *stylus* is a pen-like device with which the user "sketches" an image. A *puck* is a copying device with which the user copies an image. (■ *See Panel 5.10.*)

When used with drawing and painting software, a digitizing tablet and stylus allow you to do shading and many other effects similar to those artists achieve with pencil, pen, or charcoal. Alternatively, when you use a puck, you can trace a drawing laid on the tablet, and a digitized copy is stored in the computer.

Digitizing tablets are used primarily in design and engineering.

Pen Computers

In the next few years, students may be able to take notes in class without writing a word, if pen-based computer systems evolve as Depauw University computer science professor David Berque hopes they will. **Pen computers** use a pen-like stylus to allow people to enter handwriting and marks onto a computer screen rather than typing on a keyboard. (■ *See Panel 5.11.*) Berque has developed a prototype for a system that would connect an instructor's electronic "whiteboard" on the classroom wall with students' pen computers, so that students could receive notes directly, without having to copy information word for word. "The idea is that this might free the students up to allow them to think about what's going on," Berque says. "They wouldn't have to blindly copy things that maybe would distract them from what's going on."[14]

Even if this system of electronic note taking never happens, there's a good chance you will use a pen computer system. There are four such types of systems:

Light pen

(Top) This person is using a light pen to create a cartoon. *(Bottom)* Light pen and its adapter card, which goes in an expansion slot inside the system unit.

- **Gesture recognition or electronic checklists:** *Gesture recognition* refers to a computer's ability to recognize various check marks, slashes, or carefully printed block letters and numbers placed in boxes. This type of pen-based system is incorporated in devices that resemble simple forms or checklists on handheld electronic clipboards that have an accompanying electronic pen or stylus. Such pen computers are used by meter readers, package deliverers, and insurance claims representatives.

- **Handwriting stored as scribbling:** A second type of pen-based system recognizes and stores handwriting. The handwriting is stored as a scribble and is not converted to typed text.

  This kind of handwriting recognition is found in Inkwriter from Aha Software, a program that runs on Magic Link or Envoy.[15]

- **Handwriting converted to typed text with training:** Some pen-based devices can recognize your handwriting and transform it into typed text. These systems require that the machine be "trained" to recognize your particular (or even peculiar) handwriting. Moreover, the writing must be neat printing rather than script. The advantage of converting writing to typed text is that after conversion the text can be retrieved and later edited or further manipulated.

  An example of this kind of handwriting recognition is found in IBM's pen-based ThinkPad computer, which consists of a tablet screen and a stylus.

- **Handwriting converted to typed text without training:** The most sophisticated—and still mostly elusive—application of pen computers converts script handwriting to typed text without training. This was the claim originally made by Apple Computer for its Newton MessagePad, a personal digital assistant. As we mentioned in Chapter 4, *personal digital assistants (PDAs)* are small pen-controlled, handheld computers that, in their most developed form, can do two-way wireless messaging. In addition, Newton users were supposed to be able to use handwritten commands to send and receive faxes and electronic mail, to write memos, and to keep an appointment calendar.

  Initial versions of the Newton failed to live up to their advance billing for recognizing handwriting. A later version is better because it recognizes letters independently, whereas the original typically tried to identify words from letter groups. It also allows users to defer translation of their scribbling until later, so that they can take notes at full speed.

Clearly, it is not too difficult to design a pen computer system that will store handwriting as it is scrawled, and several of these exist. What is more difficult is to convert—particularly without training—a person's distinctive script handwriting into typescript. After all, when you're trying to read someone else's notes, you may ask "Is that an *e* or an *a* or an *o*?"

This completes the section on the type of source data entry devices called *pointing devices.* Now we proceed to *scanning devices.*

Scanning Devices

Preview & Review: Scanning devices include bar-code readers, mark- and character-recognition devices, fax machines, and imaging systems.

Mark- and character-recognition devices include magnetic-ink character recognition (MICR), optical-mark recognition (OMR), and optical-character recognition (OCR).

Fax machines may be dedicated machines or fax modems.

Imaging systems convert text and images to digital form.

Scanning devices **translate images of text, drawings, photos, and the like into digital form.** The images can then be processed by a computer, displayed on a monitor, stored on a storage device, or communicated to another computer. Scanning devices include:

* Bar-code readers
* Mark- and character-recognition devices
* Fax machines and fax modems
* Imaging systems

Bar-Code Readers

Bar codes **are the vertical zebra-striped marks you see on most manufactured retail products**—everything from candy to cosmetics to comic books. In North America, supermarkets, food manufacturers, and others have agreed to use a bar-code system called the Universal Product Code. Other kinds of bar-code systems are used on everything from FedEx packages to railroad cars. (■ *See Panel 5.12.*)

Bar codes are read by *bar-code readers,* **photoelectric scanners that translate the bar-code symbols into digital forms.** The price of a particular item is set within the store's computer and appears on the salesclerk's point-of-sale terminal and on your receipt. Records of sales are input to the store's computer and used for accounting, restocking store inventory, and weeding out products that don't sell well.

A recent innovation is the self-scanning bar-code reader, which grocers hope will extend the concept of self-service and help them lower costs. (■ *Refer back to Panel 5.12.*) Here customers bring their groceries to an automated checkout counter, where they scan them and bag them. They then take the bill to a cashier's station to pay. To guard against theft, the bar-code scanner is able to detect attempts to pass off steak as peas.[16]

Self-service might go partway toward helping customers deal with a suspicion that exists even now—namely, are these devices "scanners" or "scammers"? That is, are the prices set by stores in their electronic checkout systems honest—the same ones that appear on store shelves or in newspaper sale ads? The technology is capable of 100% accuracy, but some stores average as low as 85%, according to a weights and measures coordinator at the National Institute of Standards and Technology.[17] Inaccuracies, which result

PANEL 5.12

Bar codes and bar-code readers

(Top left) Bar-coded groceries being scanned at the check-out counter. *(Bottom left)* Customer self-service scanning station. *(Bottom right)* Checking bar code inventory numbers.

probably less from fraud than from human error—as when store managers forget to enter new data when programming scanner computers—occur mainly with products that have frequent price changes, such as sale items. Retailers point out that generally scanners are more accurate than clerks, who punch in wrong prices about 10% of the time. Clearly, however, if you're making a transaction that involves a bar-code scanner, you should keep your eye on the prices as they light up on the display screen.

Mark-Recognition & Character-Recognition Devices

There are three types of scanning devices that sense marks or characters. They are usually referred to by their abbreviations MICR, OMR, and OCR.

- **Magnetic-ink character recognition:** *Magnetic-ink character recognition (MICR)* **reads the strange-looking numbers printed at the bottom of checks.** (■ *See Panel 5.13, next page.*) MICR characters, which are printed with magnetized ink, are read by MICR equipment, producing a digitized signal. This signal is used by a bank's reader/sorter machine to sort checks.

- **Optical-mark recognition:** *Optical-mark recognition (OMR)* **uses a device that reads pencil marks and converts them into computer-usable form.** The most well-known example is the OMR technology used to read the College Board Scholastic Aptitude Test (SAT) and the Graduate Record Examination (GRE).

- **Optical-character recognition:** *Optical-character recognition (OCR)* **uses a device that reads preprinted characters in a particular font (typeface design) and converts them to digital code.** Examples of the use of OCR characters are utility bills and price tags on department-store merchandise. The *wand reader* is a common OCR scanning device. (■ *See Panel 5.14.*)

MICR technology

Checks use magnetized ink that can be read by a bank's magnetic-ink character-recognition equipment.

Optical-character recognition

Various typefaces can be read by a special scanning device called a *wand reader.*

Fax Machines

A *fax machine*—or *facsimile transmission machine*—scans an image and sends it as electronic signals over telephone lines to a receiving fax machine, which re-creates the image on paper. Fax machines are useful to anyone who wants to send images rather than text. Many businesses have found them to also be a fast, reliable way to transmit text documents.

There are two types of fax machines—dedicated fax machines and fax modems:

- **Dedicated fax machines:** The type we generally call "fax machines," *dedicated fax machines* are specialized devices that do nothing except send and receive fax documents. They are found not only in offices and homes but also alongside regular phones in public places such as airports.

For the status-conscious or those trying to work from their cars during a long commute, fax machines can be installed in an automobile. (■ *See Panel 5.15.*) The movie *The Player*, for example, contains a scene in which the stalker of a movie-studio executive faxes a threatening note. It arrives through the fax machine housed beneath the dashboard in the executive's Range Rover.

- **Fax modem:** **A *fax modem* is installed as a circuit board inside the computer's system cabinet. It is a modem with fax capability that enables you to send signals directly from your computer to someone else's fax machine or computer fax modem.** With this device, you don't have to print out the material from your printer and then turn around and run it through the scanner on a fax machine. The fax modem allows you to send information much more quickly than if you had to feed it page by page into a machine.

The fax modem is another feature of mobile computing, although it's more powerful as a receiving device. Fax modems are installed inside portable computers, including pocket PCs and PDAs. (■ *See Panel 5.16.*) You can also link up a cellular phone to the fax modem in your portable computer and thereby receive wireless fax messages. Indeed, faxes may be sent and received all over the world.

The main disadvantage of a fax modem is that you cannot scan in outside documents. Thus, if you have a photo or a drawing that you want to fax to someone, you need an image scanner, as we describe next.

PANEL 5.16
Fax in a PDA
This PDA not only is a notepad and address book but also can send and receive fax messages.

By now, communication by fax has become cheaper than first-class mail for many purposes.

Imaging Systems

An *imaging system*—or *image scanner* or *graphics scanner*—converts text, drawings, and photographs into digital form that can be stored in a computer system and then manipulated using different software programs. The system scans each image with light and breaks the image into light and dark dots, which are then converted to digital code.

An example of an imaging system is the type used in desktop publishing. This device scans in artwork or photos that can then be positioned within a page of text. Other systems are available for turning paper documents into electronic files so that people can reduce their paperwork. (■ *See Panel 5.17.*)

Note: For scanned-in text to be editable, OCR software must be used in conjunction with the scanner.

Imaging-system technology has led to a whole new art or industry called electronic imaging. *Electronic imaging* **is the software-controlled combining of separate images, using scanners, digital cameras, and advanced graphic computers.** This technology has become an important part of multimedia.

It has also led to some serious counterfeiting problems. (■ *See Panel 5.18.*) With scanners, crime rings have been able to fabricate logos and trademarks (such as those of Guess, Ray-Ban, Nike, and Adidas) that can be affixed to clothing or other products as illegal labels.[18] More importantly, electronic imaging has been used to make counterfeit money. As one article points out, "Big changes in technology over the last decade have made it easier to reproduce currency through the use of advanced copiers, printers, electronic digital scanners, color workstations, and computer software."[19] This impelled the U.S. Treasury to redesign the $100 bill—the one with Benjamin Franklin on it, and the most widely used U.S. paper currency throughout the world—and other bills for the first time in nearly 70 years. Besides using special inks and polyester fibers that glow when exposed to ultraviolet light, the new bills use finer printing techniques to thwart accurate copying.[20,21]

The next category of source data entry we'll cover comprises devices the size of credit cards.

Magnetic-Stripe, Smart, & Optical Cards

Preview & Review: Magnetic-stripe cards are encoded with data specific to a particular use. A smart card contains a microprocessor and a memory chip. An optical card is a plastic, laser-recordable card used with an optical card reader.

PANEL 5.18

Electronic imaging and counterfeiting

(Left) $2.8 million in counterfeit U.S. currency seized in New York in 1994. *(Right)* The U.S. $100 bill was overhauled for the first time in nearly 70 years to thwart increasingly sophisticated counterfeiters using imaging systems and other computer technology. On the new bill, Ben Franklin is bigger and slightly left of center.

It has already come to this: just as many people collect stamps or baseball cards, there is now a major worldwide collecting mania for used wallet-size telephone debit cards. These are the cards, called "stored value" cards, by which telephone time is sold and consumed in many countries. Generally the cards are collected for their designs, which bear likenesses of anything from Elvis Presley to Felix the Cat to Martin Luther King, Jr.[22] (■ *See Panel 5.19.*) The cards have been in use in Europe for about 15 years, and many U.S. phone companies are now selling them.

PANEL 5.19

Future collectibles?

Many of these smart-card telephone debit cards are of interest to collectors because of their designs and messages.

Most of these telephone cards are examples of "smart cards." An even more sophisticated technology is the optical card. However, the simplest form of card entry is via the magnetic-stripe card.

Magnetic-Stripe Cards

ATM cards and credit cards are **magnetic-stripe cards, with stripes of magnetically encoded data on their backs.** These cards are used only for specific purposes, such as getting cash or charging a purchase, so the encoded data might include your name, account number, and PIN number (personal identification number).

Smart Cards

The next level of data entry via card is represented by the smart card.

A *smart card* looks like a credit card but contains a microprocessor and memory chip. In France, where the smart card was invented, you can buy telephone debit cards at most cafes and newsstands. You insert the card into a slot in the phone, wait for a tone, and dial the number. The time of your call is automatically calculated on the chip inside the card and deducted from the balance. The French also use smart cards as bank cards and as medical history cards that patients carry.

The United States has been slow to embrace smart cards because of the prevalence of the conventional magnetic-stripe credit card. Moreover, the United States has a large installed base of credit-card readers and phone networks with which merchants can check on cards. However, in some other countries phone lines are scarcer, and merchants cannot as easily check over the phone with a centralized credit database. In these places, stored-value smart cards, sometimes called "electronic purses," make sense because they carry their own spending limits.[23]

Recently, some credit-card companies and banks in North America have begun to promote smart cards, combining ATM, credit, and debit cards in one instrument. When fully implemented, this system will follow the Mondex card system used in England. "Mondex money" looks and works something like a debit card. The difference is that it disburses electronic "cash" that has previously been loaded from the customer's bank account onto the card's microchip by means of an ATM or special smart phone that conveys financial data.[24,25] Customers can use the Mondex cards, which can pay automatically in five currencies, at retail stores, toll booths, pay phones, and even taxis equipped with card readers.[26]

Optical Cards

The conventional magnetic-stripe credit card holds the equivalent of a half page of data. The smart card with a microprocessor and memory chip holds the equivalent of 30 pages. The optical card presently holds about 2000 pages of data. Optical cards use the same type of technology as music compact disks (or CD-ROMs) but look like silvery credit cards. **Optical cards are plastic, laser-recordable, wallet-type cards used with an optical-card reader.** Because they can cram so much data (6.6 megabytes) into so little space, they may become popular in the future. With an optical card, for instance, there's enough room for a person's health card to hold not only his or her medical history and health-insurance information but also digital images. Examples are electrocardiograms, low-resolution chest X-rays, and ultrasound pictures of a fetus. A book containing 1000 pages of text plus 150 detailed drawings could be mailed on an optical card in a 1-ounce first-class letter. One manufacturer of optical library-card systems suggested that people might wish to store personal information on their cards, such as birth certificates and insurance policies.[27]

Digital smart cards and optical cards may be used to solve some present problems, but they may also generate some ethical problems of their own. On the positive side, for instance, by substituting plastic debit cards for paper coupons, Texas and Maryland have successfully reduced the fraudulent exchange of food stamps for cash or drugs. Electronic records make it easier to detect misuse.[28] On the negative side, smart cards may evolve into some sort of universal identity card containing a lot of personal information that you wouldn't want to share but can't control. "We think everything will migrate toward a single card," says Visa chief executive officer Carl Pascarella. "I might want to have my driver's license, frequent-flier miles, medical information, and HMO [health-maintenance organization] data on my card."[29] Privacy experts fear the possible result.

Voice-Recognition Input Devices

Preview & Review: Voice-recognition systems, which convert human speech into digital code, still have several limitations.

London native Caroline Goldie now lives in Washington, D.C., but she still retains her British accent. While in Vermont on a ski trip she tried to place an operator-assisted long-distance call. When she dialed the telephone company's long-distance access number, a tinny recorded voice asked her to speak a word from a number of choices, including the word "operator." Every time Ms. Goldie said "operator," the uncomprehending voice-recognition system said "Sorry, please repeat." In frustration, she finally handed the phone to an American friend, whose flat *"ah-per-aid-er"* finally got her connected to the real thing. Since then Ms. Goldie has worked out an American accent that she uses for the phone company. (Actually, if a caller waits long enough, an operator will come on the line.)[30]

Such are the difficulties with which telephone-company linguists and computer experts must deal in improving voice-recognition technology.[31] A **voice-recognition system converts a person's speech into digital code by comparing the electrical patterns produced by the speaker's voice with a set of prerecorded patterns stored in the computer.**

Voice-recognition systems are finding many uses. Among the more commonplace uses, warehouse workers are able to speed inventory-taking by recording inventory counts verbally and traders on stock exchanges can communicate their trades by speaking to computers. Children can talk to toys, such as the Talkback Picture Phone, which recognizes 25–30 words spoken by children.[32] Among the more exotic uses, blind and other disabled people can give verbal commands to their PCs rather than use a keyboard.[33,34] Nurses can fill out patient charts by talking to a computer.[35] Speakers of Chinese can speak to a machine that will print out Chinese characters.[36] (■ *See Panel 5.20.*)

So far, most voice-recognition technology has been hindered by three basic limitations:

Voice-recognition technology

(Top) A registered nurse tests a voice-recognition system designed to help fill out patient charts. *(Bottom)* Taiwanese scientist Lee Lin-shan displays a computer that can listen to continous speech in Chinese (Mandarin) and then print out the words at the rate of three characters a second.

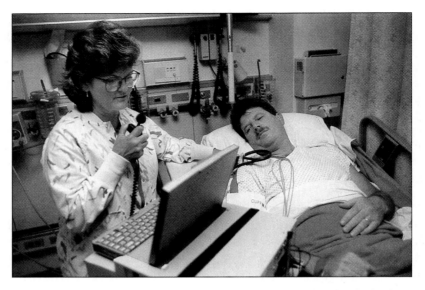

- **Speaker dependence:** Most systems need to be "trained" by the speaker to recognize his or her distinctive speech patterns and even variations in the way a particular word is said. Systems that are "speaker independent" are beginning to appear, but consider the hurdles to be overcome: different voices, pronunciations, and accents.

- **Single words versus continuous speech:** Most systems can handle only single words and have vocabularies of 1000 words or less. However, some technologies, such as DragonDictate, now offer continuous-speech recognition—that is, one need not artificially pause between words when speaking—and 30,000-word dictionaries (vocabularies).[37]

- **Lack of comprehension:** Most systems merely translate sounds into characters. A more useful technology would be one that actually comprehends the *meaning* of spoken words. You could then ask a question, and the system could check a database and formulate a meaningful answer. Some such systems are being developed for the military.

All in all, however, the field of voice recognition is making gigantic strides. For instance, after two decades of research, IBM is now selling a dictation system that adapts to your voice after only 90 minutes of practice. The system can recognize 22,000 words and process up to 100 dictated words a minute.[38]

Other Input Devices: Audio, Video, Cameras, Sensors, & Human-Biology Devices

Preview & Review: Five other types of input devices are audio-input devices, video-input devices, electronic cameras, sensors, and human-biology input devices.

Audio-input devices may digitize sound by means of an audio board or a MIDI board.

Video-input devices may use frame-grabber or full-motion video cards.

Electronic cameras are still-video or digital cameras; the latter use light-sensitive silicon chips to capture photographic images.

Sensors collect specific kinds of data directly from the environment.

Human-biology input devices include biometric systems, line-of-sight systems, cyber gloves and body suits, and brainwave devices.

"The machine never grew impatient, never gave up," reported journalist Jerry Adler about the Miracle Piano Teaching System. This combination of electronic keyboard and software for an IBM-compatible computer promises to teach novices Handel's "Water Music" in only 40 lessons. "It simply devised new exercises, sent me back to practice again, issued wildly inflated praise at the slightest signs of improvement, until at last I made it through all 15 bars of 'Mary Had a Little Lamb' and was rewarded with a heart-felt WHEW! YOU'RE BECOMING A MUSICIAN!"[39]

As this example suggests, there are all kinds of input devices beyond those we have mentioned. Here we describe five types:

- Audio-input devices
- Video-input devices
- Electronic cameras
- Sensors
- Human-biology input devices

Audio-Input Devices

Voice-recognition devices are only one kind of *audio input*, which can include music and other sounds. **An *audio-input device* records or plays analog sound and translates it for digital storage and processing.** You'll recall that an *analog sound signal* represents a continuously variable wave within a certain frequency range (✓ p. 6). Such continuous fluctuations are usually represented with an analog device such as an audiocassette player. For the computer to process them, these variable waves must be converted to digital 0s and 1s.

There are two ways by which audio is digitized:

- **Audio board:** Analog sound from, say, a cassette player goes through a special circuit board called an audio board. **An *audio board* is an add-on circuit board in a computer that converts analog sound to digital sound and stores it for further processing.**

- **MIDI board:** A *MIDI board*—MIDI stands for *Musical Instrument Digital Interface*—is an add-on board that creates digital music. That is, the music is created in digital form as the musician performs, as on a special MIDI keyboard.

The principal use of audio-input devices such as these is to provide digital input for multimedia PCs. A *multimedia system* is a computer that incorporates text, graphics, sound, video, and animation in a single digital presentation. Video input is also often used for this purpose, as we describe next.

Video-Input Devices

As with sound, most film and videotape is in analog form, with the signal a continuously variable wave. Thus, to be used by a computer, the signals that come from a VCR, a videodisk or laserdisk, or a camcorder must be converted to digital form through a special *video card* installed in the computer.

Two types of video cards are frame-grabber video and full-motion video:

- **Frame-grabber video card:** Some video cards, called *frame grabbers,* can capture and digitize only a single frame at a time.
- **Full-motion video card:** Other video cards, called *full-motion video cards,* can convert analog to digital signals at the rate of 30 frames per second, giving the effect of a continuously flowing motion picture.

The main limitation in capturing full video is not input but storage. It takes 15 megabytes to store just 1 second of video.

A number of people have taken what would seem to be the next step by transmitting what their video cameras see to sites on the Internet's World Wide Web. At the Massachusetts Institute of Technology, for instance, graduate student Steve Mann walks around with head-mounted camera, visor with display screen (to see what the camera sees), antenna, and wraparound computer at his waist, looking like someone in a 1950s science-fiction movie. His travels are fed to the Web, where anyone may log on (his home page address is *http://www-white.media.mit.edu/~steve/netcam.html*) and see the world through his eyes.[40,41] (■ *See Panel 5.21.*)

Electronic Cameras

The electronic camera is a particularly interesting piece of hardware because it foreshadows major change for the entire industry of photography. Instead of using traditional (chemical) film, **an *electronic camera* captures images in electronic form for immediate viewing on a television or computer display screen.**

Electronic cameras are of two types: still-video and digital.

- **Still-video cameras:** *Still-video cameras* **are like camcorders. However, they capture only a single video image at a time.** Because the cameras are meant to display images only on a television screen, the pictures must be converted (by a video card) before they can be stored in a computer. In addition, compared to the digital camera, picture resolution and color range are limited.
- **Digital cameras:** More interesting is the digital camera. **A *digital camera* uses a light-sensitive silicon chip to capture photographic images in digital form.** The bits of digital information can then be copied right into a computer's hard disk for manipulation and printing out.

![Netscape browser window showing a man wearing a head-mounted video camera]

Netscape - [GIF image 640x480 pixels]

File Edit View Go Bookmarks Options Directory Window Help

Back | Forward | Home | Reload | Images | Open | Print | Find | Stop

Location: http://www-white.media.mit.edu/~steve/steve_by_claudio.gif

Document: Done

PANEL 5.21

Video camera linked to the Web

MIT graduate student Steve Mann wears a head-mounted video camera, visor, antenna, and wraparound computer at his waist. Thus, whatever Mann sees, the Web sees.

In February 1994, Eastman Kodak and the Associated Press (AP) introduced an electronic camera specifically designed for news photographers, the News-Camera 20. The camera stores 30–50 digitized images in a battery-backed memory. Later a photographer can download (transfer) these images to his or her microcomputer or transmit them by modem or even via satellite uplink to a photo editor at the AP's New York office.

Sensors

A *sensor* **is a type of input device that collects specific kinds of data directly from the environment and transmits it to a computer.** Although you are unlikely to see such input devices connected to a PC in an office, they exist all around us, often in invisible form. Sensors can be used for detecting all kinds of things: speed, movement, weight, pressure, temperature, humidity, wind, current, fog, gas, smoke, light, shapes, images, and so on.

Beneath the pavement, for example, are sensors that detect the speed and volume of traffic. These sensors send data to computers that can adjust traffic lights to keep cars and trucks away from grid-locked areas. In California, sensors have been planted along major earthquake fault lines in an experiment to see whether scientists can predict major earth movements. (■ *See Panel 5.22.*) In aviation, sensors are used to detect ice buildup on airplane wings or to alert pilots to sudden changes in wind direction. Sensors are also used by government regulators to monitor whether companies are complying with air-pollution standards. Recently sensors for monitoring highways have been developed that are wireless and work by remote control.[42]

PANEL 5.22

Earthquake sensor

Sensor instruments of a tele-metered weak motion seismic station (earthquake motion detection)

R E A D M E

Case Study: Future Careers in Journalism with a CamCutter

August 1995: A correspondent and camera crew are dispatched to cover a battle between Bosnians and Serbs. What they send back is limited by what they can hear, see, and smell at the scene. They can't find out about the broader context. Since they have no way to refine what they've shot, once they've shot it, they have to send back an unedited package. They have to trust editors back home to insert any germane information from officialdom or to add other elements that properly belong in the story.

August 1999 (or thereabouts): A one-person crew (correspondent, camera operator, sound recordist, and editor, all combined in one journalist) is dispatched to cover a battle between Grinks and Orinks in, say, Ishmaelia. No longer limited to just what she finds at the scene, she is also able to download text from The Associated Press, video clips from a network news operation in New York City, and audio clips from the BBC in London. She can then make her own decisions about what to include in and what to exclude from her report. She . . . is truly a one-woman band, equipped with a formidable tool: a portable newsroom.

The harbinger of the portable newsroom is here; it's called a CamCutter. It is a portable though still bulky camera attached to a computer. Its magic lies in two words: "digital" and "nonlinear."

As a "digital" recorder, the CamCutter speaks in the parlance of computerdom—ones and zeros. It can record broadcast-quality video on a pop-out, reusable, rewritable disk called a FieldPak. A FieldPak can store almost 20 billion ones and zeros—15 to 20 minutes of video.

The second key word, "nonlinear," means that the CamCutter/FieldPak combination records images to a computer disk and not to film or videotape. The only way to edit film is linearly—by looking down the film for those expendable 10 seconds, those 300 frames, and literally cutting them out, with what amounts to scissors, and splicing back together what's left. Videotape is also linear: one edits tape only by rebuilding the whole segment, top to bottom, linearly.

With digital recording, however, a passage can be inserted or deleted by the equivalent of clicking a mouse—like editing with WordPerfect. . . . And it can be done—only roughly, so far—by the reporter/editor where the news is happening. . . .

The ramifications for the way news will be covered, edited, and processed are fascinating. No longer will the field reporter be "blind." She is suddenly empowered in all sorts of new and interesting ways. One day she'll be able to transmit her reports directly to a satellite and, inevitably, directly to a distribution system, in effect a video Internet. No editors. And anyone with a computer and modem can look at her report, as a supplement—or substitute—for the standard media version. . . .

—Stephen D. Isaacs, "Have Newsroom, Will Travel," *Columbia Journalism Review*

Human-Biology Input Devices

Characteristics and movements of the human body, when interpreted by sensors, optical scanners, voice recognition, and other technologies, can become forms of input. Some examples:

- **Biometric systems:** **Biometrics is the science of measuring individual body characteristics.** *Biometric security devices* identify a person through a fingerprint, voice intonation, or other biological characteristic. For example, retinal-identification devices use a ray of light to identify the distinctive network of blood vessels at the back of one's eyeball.

  Biometric systems are used in lieu of typed passwords to identify people authorized to use a computer system. In Sacramento, California, to make it more difficult for welfare recipients to commit fraud by claiming duplicate benefits, recipients claiming checks must stick their hands in a box in which infrared rays read their "hand geometry."[43]

- **Line-of-sight systems:** *Line-of-sight systems* enable a person to use his or her eyes to "point" at the screen, a technology that allows physically handicapped users to direct a computer. For example, the Eyegaze System from LC Technologies allows you to operate a computer by focusing on particular areas of a display screen. A camera mounted on the computer analyzes the point of focus of the eye to determine where you are looking. You operate the computer by looking at icons on the screen and "press a key" by looking for a specified period.[44]

- **Cyber gloves and body suits:** Special gloves and body suits—often used in conjunction with "virtual reality" games (described in Chapter 6)—use sensors to detect body movements. The data for these movements is sent to a computer system. Similar technology is being used for human-controlled robot hands, which are used in nuclear power plants and hazardous-waste sites.

Perhaps the ultimate input device is that using brainwaves to direct the computer. We describe that in another few pages.

Input Technology & Quality of Life: Health & Ergonomics

Preview & Review: The use of computers and communications technology can have important effects on our health. Some of these are repetitive strain injuries such as carpal tunnel syndrome, eyestrain and headaches, and backstrain.

Negative health effects have increased interest in the field of ergonomics, the study of the relationship of people to a work environment.

Susan Harrigan, a financial reporter for *Newsday,* a daily newspaper based on New York's Long Island, now writes her stories using a voice-activated computer. She does not do so by choice, nor is she as efficient as she used to be. She does it because she is too disabled to type at all. After 20 years of writing articles with deadline-driven flying fingers at the keyboard, she developed a crippling hand disorder. At first the pain was so severe she couldn't even hold a subway token. "Also, I couldn't open doors," she says, "so I'd have to stand in front of doors and ask someone to open them for me."[45]

Health Matters

Harrigan suffers from one of the computer-induced disorders classified as repetitive strain injuries (RSIs). The computer is supposed to make us efficient. Unfortunately, it has made some users—journalists, postal workers, data-entry clerks—anything but. The reasons are repetitive strain injuries, eyestrain and headache, and back and neck pains.[46] In this section we consider these health matters along with the effects of electromagnetic fields and noise.

- **Repetitive strain injuries:** *Repetitive strain injuries (RSIs)* **are several wrist, hand, arm, and neck injuries resulting when muscle groups are forced through fast, repetitive motions.** Most victims of RSI are in meat-packing, automobile manufacturing, poultry slaughtering, and clothing manufacturing. Musicians, too, are often troubled by RSI (because of long hours of practice).

  People who use computer keyboards—some of whom make as many as 21,600 keystrokes *an hour*—account for only 12% of RSI cases that result in lost work time. However, the number of cases is rising.[47] Before computers came along, typists would stop to make corrections or change paper.

These motions had the effect of providing many small rest breaks. Today keyboard users must devise their own mini-breaks to prevent excessive use of hands and wrists.

RSIs cover a number of disorders. Some, such as muscle strain and tendinitis, are painful but usually not crippling. These injuries, often caused by hitting the keys too hard, may be cured by rest, anti-inflammatory medication, and change in typing technique. However, carpal tunnel syndrome is disabling and often requires surgery. *Carpal tunnel syndrome (CTS)* **consists of a debilitating condition caused by pressure on the median nerve in the wrist, producing damage and pain to nerves and tendons in the hands.** (■ *See Panel 5.23.)*

It's important to point out, however, that scientists still don't know what causes RSIs. They don't know why some people operating keyboards develop upper body and wrist pains and others don't. The working list of possible explanations for RSI includes "wrist size, stress level, relationship with supervisors, job pace, posture, length of workday, exercise routine, workplace furniture, [and] job security." Other possible contributors are diabetes, weight, and menopause.[48]

- **Eyestrain and headaches:** Computers compel people to use their eyes at close range for a long time. However, our eyes were made to see most efficiently at a distance. It's not surprising, then, that people develop what's called computer vision syndrome.

  Computer vision syndrome (CVS) **consists of eyestrain, headaches, double vision, and other problems caused by improper use of computer display screens.** By "improper use," we mean not only staring at the screen for too long but also not employing the technology as it should be employed. This includes allowing faulty lighting and screen glare, and using screens with poor resolution.

- **Back and neck pains:** Many people use improper chairs or position keyboards and display screens in improper ways, leading to back and neck

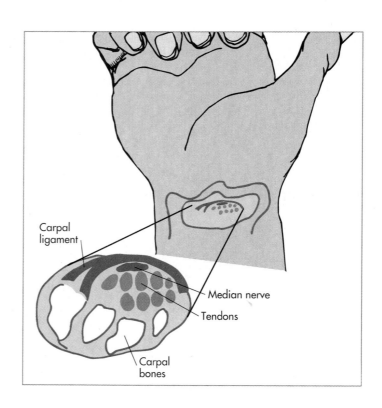

PANEL 5.23

Carpal tunnel syndrome

The carpal ligament creates a tunnel across the bones of the wrist. When the tendons passing through the carpal tunnel become swollen, they press against the median nerve, which runs to the thumb and the first three fingers.

Carpal ligament

Median nerve

Tendons

Carpal bones

pains. All kinds of adjustable, special-purpose furniture and equipment is available to avoid or diminish such maladies.

• **Electromagnetic fields:** Like kitchen appliances, hairdryers, and television sets, many devices related to computers and communications generate low-level electromagnetic field emissions. ***Electromagnetic fields (EMFs) are waves of electrical energy and magnetic energy.***

In recent years, stories have appeared in the mass media reflecting concerns that high-voltage power lines, cellular phones, and CRT-type (✓ p. 155) computer monitors might be harmful. There have been worries that monitors might be linked to miscarriages and birth defects, and that cellular phones and power lines might lead to some types of cancers. There have been suggestions that people with high occupational exposure to EMFs (such as sewing-machine operators) have higher rates of Alzheimer's disease.[49]

Is there anything to this? The answer is: so far no one is sure. The evidence seems scant that weak electromagnetic fields, such as those used for cellular phones and found near high-voltage lines, cause cancer. Still, unlike car phones or older, luggable cellular phones, the handheld cellular phones do put the radio transmitter next to the user's head. This causes some health professionals concern about the effects of radio waves entering the brain as they seek out the nearest cellular transmitter. Customers are usually warned not to let the antenna touch them while they are talking.[50]

As for CRT monitors, "Dangerous monitor emissions are much like the legend of Bigfoot," says one magazine article. "No one has been able to decisively prove that monitor emissions are harmful or that Bigfoot exists. But enough interesting evidence keeps cropping up on both phenomena that people stay interested."[51]

CRT monitors made since the early 1980s produce very low emissions. Even so, users are advised to not work closer than arm's length to a CRT monitor. The strongest fields are emitted from the sides and backs of terminals. Alternatively, you can use laptop computers, because their liquid crystal display (LCD) screens emit negligible radiation.[52]

The current advice from the Environmental Protection Agency is to exercise *prudent avoidance.* That is, we should take precautions when they are easy steps. However, we should not feel compelled to change our whole lives or to spend a fortune minimizing exposure to electromagnetic fields. Thus, we can take steps to put some distance between ourselves and a CRT monitor. However, it is probably not necessary to change residences because we happen to be living near some high-tension power lines.

• **Noise:** The chatter of impact printers or hum of fans in computer power units can be psychologically stressful to many people. Sound-muffling covers are available for impact printers. Some system units may be placed on the floor under the desk to minimize noise from fans.

Ergonomics: Design with People in Mind

Previously workers had to fit themselves to the job environment. However, health and productivity issues have spurred the development of a relatively new field, called ergonomics, that is concerned with fitting the job environment to the worker.

***Ergonomics* is the study of the physical relationships between people and their work environment.** It is concerned with designing hardware and software that is less stressful and more comfortable to use, that blends more smoothly with a person's body or actions. Examples of ergonomic hardware are tilting display screens, detachable keyboards, and keyboards hinged in the middle so that the user's wrists are presumably in a more natural position. (■ *See Panel 5.24.*)

We address some further ergonomic issues in the Experience Box at the end of this chapter.

Ergonomic keyboard

This keyboard is by Kinesis.

The Future of Input

Preview & Review: Increasingly, input will be performed in remote locations and will rely on source data automation.

Future source data automation will include high-capacity bar codes, 3-D scanners, more sophisticated touch devices, smarter smart cards and optical cards, better voice recognition, smaller electronic cameras, and brainwave input devices.

Input technology seems headed in two directions: (1) toward more input devices in remote locations and (2) toward more refinements in source-data automation.

Toward More Input from Remote Locations

When management consultant Steve Kaye of Santa Ana, California, wants to change a brochure or company letterhead, he doesn't have to drop everything and drive over to a printer. He simply enters his requests through the phone line to an electronic bulletin board at the Sir Speedy print shop that he deals with. "What this does is free me up to focus on my business," Kaye says.[53]

The linkage of computers and telecommunications means that input may be done nearly anywhere. Not so long ago, for example, airlines installed telephones in seatbacks for passengers to use. Now they are installing modems and other connections. Recently Delta Air Lines began employing a new technology that includes a Delta "button" in software being built into many microcomputers and modems for home use. The button allows users to communicate voice and data information to book reservations and get other travel information.[54]

Toward More Source Data Automation

The keyboard will always be with us, but increasingly input technology is designed to capture data at its source. This will reduce the costs and mistakes that come with copying or otherwise preparing data in a form suitable for processing.

Some reports from the input-technology front:

• **Higher-capacity bar codes:** Traditional bar codes read only horizontally. A new generation of bar codes has appeared that reads vertically as well, which enables them to store more than 100 times the data of traditional bar codes. With the ability to pack so much more information in such a small space, bar codes can now be used to include digitized photos, along with a person's date of birth, eye color, blood type, and other personal data. Some states are beginning to use the codes on driver's licenses.[55]

- **3-D scanners:** Have difficulty getting the perfect fit in blue jeans? Modern clothes are designed to fit mannequins, which is why it's difficult to get that "sprayed-on" look for people. However, clothing makers (including Levi Strauss, the world's largest jeans maker) have been experimenting with a body scanner that would enable people to buy clothes that fit precisely.

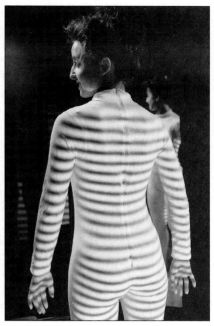

  The device—which doesn't use lasers, to alleviate possible customer health concerns—would allow you to enter a store, put on a body suit, and be measured three-dimensionally all over. You could then select the clothes you're interested in, view them imposed on your body-scanned image on a screen, and then order them custom-manufactured.[56–58] Clearly, this is the first in a generation of scanners being used for more sophisticated purposes.

- **More sophisticated touch devices:** Touch screens are becoming commonplace. Sometime in the near future, futurists suggest, you may be able to use a dashboard touch screen in your car. The screen would be linked to mobile electronic "yellow pages" that would enable you to reserve a motel room for the night or find the nearest Chinese restaurant.[59]

  More interesting, perhaps, is the Phantom from SensAble Devices, which offers a glimpse of what the technology to simulate touch—known as "force feedback"—can do. *Force-feedback devices* have been around since the 1970s, when they were developed for remote-control robots to handle radioactive materials. However, the Phantom offers an incredibly sophisticated tactile response. Surgeons-in-training, for instance, may insert their fingers in a "thimble" at the end of a robotic arm and practice what it feels like to discover a tumor in soft brain tissue. SensAble also wants to adapt the Phantom as a game joystick, which could simulate, for example, the vibrations of a tank on rough terrain.[60]

- **Smarter smart cards and optical cards:** Over the next five years or so, stored-value smart cards with microchips, acting as "electronic purses," will no doubt begin to displace cash in many transactions. Targets for smart cards are not only convenience stores and toll booths but also battery-powered card readers in newspaper racks and similar devices. Colorado, for instance, has tested parking meters containing smart-card readers and photo cells that reset the meter as soon as a car moves so later drivers can't get free parking time.[61]

  Already microchips with identification numbers are being injected into dogs and cats so that, with the help of a scanner, stolen or lost pets can be identified.[62] Although it's doubtful chip implantation for identification purposes would be extended to people (though it could), smart cards and optical cards could evolve into all-purpose cards including biotechnological identifiers. As one writer suggests, these "could contain medical records, insurance information, driver's license data, security codes for the office or membership club, and frequent flier or other loyalty program information."[63]

- **Better voice recognition:** Sensory Circuits of San Jose, California, has built a voice-recognition chip from a simulation of *neural network* technology (discussed in Chapter 12). This technology processes information in a manner similar to the brain's neurons, so that the neural networks are capable of learning. As a result, a chip can be trained to recognize different words (such as "operator") regardless of who speaks them. Indeed, one company, DSP Communications, has announced a chip that can recognize about 128 words with 97% accuracy. DSP hopes the technology can be used to help drivers speak telephone numbers into their car phones, so they won't have to take their eyes off the road when "dialing."[64]

Gil Amelio, chief executive officer of Apple Computer, predicts that by the year 2005 voice recognition will have achieved the dream of world travelers. That is, you'll be able to speak in English and a voice-recognition device will be able to instantly translate it into another language, such as French or Japanese, so that you can carry on a normal conversation.[65]

- **Smaller electronic cameras:** Digital still cameras and video cameras are fast becoming commonplace. The next development may be the *camera-on-a-chip*, which will contain all the components necessary to take a photograph or make a movie. Such a device, called an "active pixel sensor," based on NASA space technology, is now being made by a company called Photobit. Because it can be made on standard semiconductor production lines, the camera-on-a-chip can be made incredibly cheaply, perhaps for $20 apiece.

  Scientists are attempting to build a prototype digital video camera the size of a plastic gambling die. Before long we may see the technology used in toys, portable video phones, baby monitors, and document imaging.[66]

- **Brainwave devices:** Perhaps the ultimate input device analyzes the electrical signals of the brain and translates them into computer commands. Experiments have been successful in getting users to move a cursor on the screen through sheer power of thought. Other experiments have shown users able to type a letter by slowly spelling out the words in their heads. In the future, pilots may be able to direct planes without using their hands, communicating their thoughts through helmet headsets. Totally paralyzed people may gain a way to communicate.[67–71]

  Although there is a very long way to go before brainwave input technology becomes practical, the consequences could be tremendous, not only for handicapped people but for all of us.

◤ Onward

When Stanley Adelman, late of New York City, died at age 72 in November 1995, he took with him some skills that some of his customers will find extremely hard to replace. Adelman was a typewriter repairer considered indispensable by many literary stars—many of whom could not manage the transition to word processors—from novelist Isaac Bashevis Singer to playwright David Mamet. He was able to fix all kinds of typewriters, including even those for languages he could not read, such as Arabic. Now his talent is no more.[72]

Adelman's demise followed by only a few months the near-death of Smith Corona Corporation, which filed for bankruptcy-court protection from creditors after losing the struggle to sell its typewriters and personal word processors in a world of microcomputers.[73] Clearly, the world is changing, and changing fast. The standard interfaces are obsolete. We look forward to interfaces that are more intuitive—computers that respond to voice commands, facial expressions, and thoughts, computers with "digital personalities" that understand what we are trying to do and offer assistance when we need it.[74,75]

Good Habits: Protecting Your Computer System, Your Data, & Your Health

Whether you set up a desktop computer and never move it or tote a portable PC from place to place, you need to be concerned about protection. You don't want your computer to get stolen or zapped by a power surge. You don't want to lose your data. And you certainly don't want to lose your health for computer-related reasons. Here are some tips for taking care of these vital areas.

Protecting Your Computer System

Computers are easily stolen, particularly portables. They also don't take kindly to fire, flood, or being dropped. Finally, a power surge through the power line can wreck the insides.

Guarding Against Hardware Theft & Loss
Portable computers—laptops, notebooks, and subnotebooks—are easy targets for thieves. Obviously, anything conveniently small enough to be slipped into your briefcase or backpack can be slipped into someone else's. *Never* leave a portable computer unattended in a public place.

It's also possible to simply lose a portable, as in forgetting it's in the overhead-luggage bin in an airplane. To help in its return, use a wide piece of clear tape to tape a card with your name and address to the outside of the machine. You should tape a similar card to the inside also. In addition, scatter a few such cards in the pockets of the carrying case.

Desktop computers are also easily stolen. However, for under $25, you can buy a cable and lock, like those used for bicycles, that secure the computer, monitor, and printer to a work area. For instance, you can drill a quarter-inch hole in your equipment and desk, then use a product called LEASH-IT (from Z-Lock, Redondo Beach, California) to connect them together. LEASH-IT consists of two tubular locks and a quarter-inch aircraft-grade stainless steel cable.[76]

If your hardware does get stolen, its recovery may be helped if you have inscribed your driver's license number, Social Security number, or home address on each piece. Some campus and city police departments lend inscribing tools for such purposes. (And the tools can be used to mark some of your other possessions.)

Finally, insurance to cover computer theft or damage is surprisingly cheap. Look for advertisements in computer magazines. (If you have standard tenants' or homeowners' insurance, it may not cover your computer. Ask your insurance agent.)

Guarding Against Heat, Cold, Spills, & Drops
"We dropped 'em, baked 'em, we even froze 'em. Which ones survived?" proclaimed the *PC Computing* cover, ballyhooing a story about its third annual notebook "torture test."[77]

The magazine put eight notebook computers through durability trials. One approximated putting these machines in a car trunk in the desert heat, another with leaving them outdoors in a Buffalo, New York, winter. A third test simulated sloshing coffee on a keyboard, and a fourth dropped them the equivalent of from desktop height to a carpeted floor. All passed the bake test, but one failed the freeze test. Three completely flunked the coffee-spill test, one other revived, and the rest passed. One that was dropped lost the right side of its display; the others were unharmed. Of the eight, half passed all tests unscathed.

This gives you an idea of how durable *some* computers are. Designed for portability, notebooks may be hardier than desktop machines. Pushing your computer off your desk and onto the floor is surely tempting fate, as it might cause your hard-disk drive to fail.

Guarding Against Power Fluctuations Electricity is supposed to flow to an outlet at a steady voltage level. No doubt, however, you've noticed instances when the lights in your house suddenly brighten or, because a household appliance kicks in, dim momentarily. Such power fluctuations can cause havoc with your computer system, although most computers have some built-in protection. An increase in voltage may be a *spike,* lasting only a fraction of a second, or a *surge,* lasting longer. A surge can burn out the power supply circuitry in the system unit. A decrease may be a momentary voltage *sag,* a longer *brownout,* or a complete failure or *blackout.* Sags and brownouts can produce a slowdown of the hard-disk drive or a system shutdown.[78]

Extreme spikes and surges can be handled by plugging your computer, monitor, and other devices into a *surge protector* or *surge suppressor.* This device, which is plugged into the wall outlet and usually has six or more sockets, will diffuse excess voltage before it reaches your hardware.

Voltage brownouts and blackouts can be dealt with by buying a *UPS,* or *uninterruptible power supply* unit. A UPS is essentially a standby battery, which is charged by keeping it constantly plugged into a wall outlet. The UPS contains three or so sockets into which you may plug your computer hardware. The UPS filters incoming power, smoothing out power sags. In the event of a brownout or blackout, the UPS will keep your computer system running a few minutes, giving you time to save whatever work you are currently engaged in and shut your system down.

If you're concerned about a lightning storm sending a surge to your system, simply unplug all your hardware until the storm passes.

Guarding Against Damage to Software Systems software and applications software generally come on diskettes. The unbreakable rule is simply this: *Copy* the original disk, either onto your hard-disk drive or onto another diskette. Then store the original disk in a safe place. If your computer gets stolen or your software destroyed, you can retrieve the original and make another copy.

Protecting Your Data

Computer hardware and commercial software are nearly always replaceable, although perhaps with some expense and difficulty. Data, however, may be major trouble to replace or even be irreplaceable. (A report of an eyewitness account, say, or a complex spreadsheet project might not come out the same way when you try to reconstruct it.) The following are some precautions to take to protect your data.

Backup Backup Backup If your hard-disk drive fails ("crashes"), do you have the same data on a diskette? Or, if you're using just a diskette (no hard disk), do you have a duplicate of the information on it on another diskette?

Almost every microcomputer user sooner or later has the experience of accidentally wiping out or losing material and having no copy. This is what makes people true believers in *backing up* their data—making a duplicate in some form. If you're working on a research paper, for example, it's fairly easy to copy your work onto a second diskette at the end of your work session. You can then store that disk in another location. If your computer is destroyed by fire, at least you'll still have the data (unless you stored your disk right next to the computer).

If you do lose data because your disk has been physically damaged, you may still be able to recover it, using special software. (See the discussion of utility programs in Chapter 3.)

Treating Diskettes with Care Diskettes can be harmed by any number of enemies. These include spills, dirt, heat, moisture, weights, and magnetic fields and magnetized objects. Here are some diskette-maintenance tips:

- Insert the diskette *carefully* into the disk drive.
- Don't manipulate the metal "shutter" on the diskette; it protects the surface of the magnetic material inside.

- Do not place heavy objects on the diskette.
- Do not expose the diskette to excessive heat or light.
- Do not use or place diskettes near a magnetic field, such as a telephone or paper clips stored in magnetic holders. Data can be lost if exposed.
- Do not use alcohol, thinners, or freon to clean the diskette.

Instead of leaving disks scattered on your desk, where they can be harmed by dust or beverage spills, it's best to store them in their boxes.

From time to time it's also best to clean the diskette drive, because dirt can get into the drive and cause data loss. You can buy an inexpensive drive-cleaning kit, which includes a disk that looks like a diskette and which cleans the drive's read/write heads.

Note: No disk lasts forever. Experts suggest that a diskette that is used properly might last 10 years. However, if you're storing data for the long term, you should copy the data onto new disks every 2 years (or use tape for backup).[79]

Guarding Against Viruses Computer *viruses* are programs—"deviant" programs—that can cause destruction to computers that contract them. They are spread from computer to computer in two ways: (1) They may be passed by way of an "infected" diskette, such as one a friend gives you containing a copy of a game you want. (2) They may be passed over a network or an online service. They may then attach themselves to software on your hard disk, adding garbage to or erasing your files or wreaking havoc with the systems software. They may display messages on your screen (such as "Jason Lives") or they may evade detection and spread their influence elsewhere.

Each day, viruses are getting more sophisticated and harder to detect. There are several types of viruses. *Boot sector viruses* attach themselves to the part of a diskette or hard disk called the *boot sector*. *File viruses* attach themselves to your software files. *Polymorphic viruses* change their binary patterns each time they infect a new file in order to keep from being identified. There are several routine activities you can practice to minimize the possibility of viruses, as shown in the accompanying box. (■ *See Panel 5.25.*) Beyond these measures, the best protection is to install antivirus software. Some programs prevent viruses from infecting your system, others detect the viruses that have slipped through, and still others remove viruses or institute damage control. (Some of the major antivirus programs are Central Point Anti-Virus from Central Point Software, Norton AntiVirus from Symantec, Pro-Scan from McAfee Associates, and VirusCare from IMSI.)

PANEL 5.25

Practical tips for avoiding viruses

Be careful about . . .

- *Sharing diskettes.* There are two kinds of disks—data disks and program disks. You're probably safe exchanging data files with friends, such as word processing files, because viruses can't be passed along via data files. However, *do not* borrow a friend's *program* disk, which may be infected.

  Don't share your software with someone else, since it might be returned to you with a virus on it.

  Don't let anyone else use your computer to run their program disks, which might be inadvertently infected.

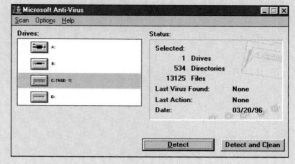

- *Downloading programs from electronic bulletin board systems.* If your computer has a modem that lets you communicate over telephones lines with electronic services such as bulletin board systems (BBSs), you can download (transfer to your computer) all kinds of wonderful programs, often for free. To ensure that a program is virus-free, query the BBS system operator (the "sysop") whether it has been scanned for viruses. If you can, use a commercial virus-protection program to scan the program yourself before using it.

- *Inspecting and write-protecting.* Don't buy new commercial software unless it's still in its shrink-wrapped package. An open package may have been returned by a previous customer or used by a dealer for demonstration purposes—both opportunities for infection.

  Before you use a new program, write-protect the disks. This will prevent them from becoming infected in case your own computer is infected.

- *Backing up your hard disk—often.* If a virus ever erases your hard disk, you may be able to kill the virus with a commercial antivirus program. Then, if you've been conscientious about regularly making backup copies of your programs and data on diskettes, you can restore them to your hard disk.

- *Booting.* When booting from your hard disk, make sure you have removed any diskettes from the diskette drives. Thus you will not inadvertently attempt to boot from a possibly infected diskette. (This helps protect against common boot sector viruses.)

Coping with Airport Security You're standing in front of the metal detector and hand-luggage X-ray machine at airport security. Suddenly it occurs to you, "Could these machines mess up my computer files?"

If you're carrying loose diskettes, it's *possible* they will be damaged. The harm could come not from the metal detector or the X-rays but from the magnetic fields of the powerful AC transformer inside the X-ray machine.[80] However, millions of travelers put their disks through without incident. The best course: Give your diskettes to the airport security personnel to hand-check.

The same precautions apply with a portable PC with a hard disk. It's *possible* the magnetic fields from the X-ray machine's transformer will partly erase the data on your hard disk (in already-weak places on the disk). It's best, then, to ask for a hand inspection of your machine. This means you'll have to take it out, set it up, and turn it on.

The point of the inspection is to show the airport security people that your portable is not a bomb in disguise. "Portable computers look especially fearsome viewed

through an airport X-ray scanner," says frequent traveler and computer expert Jim Seymour. "They're small, flat, have a maze of wires inside, and then, off to one side, there's this suspicious, opaque thing you and I know to be a battery. But which could just as easily be a wad of plastic explosive. All those other wires? The timing and trigger mechanisms, of course."[81] Thus, be prepared to cooperate with airport security in showing that what you've got *really is* just a PC.

Protecting Your Health

More important than any computer system and (probably) any data is your health. What adverse effects might computers cause? As we discussed earlier in the chapter, the most serious are painful hand and wrist injuries, eyestrain and headache, and back and neck pains. Some experts also worry about the long-range effects of exposure to electromagnetic fields and noise. All these matters can be addressed by *ergonomics,* the study of the physical relation-

235

ships between people and their work environment. Let's see what you can do to avoid these problems.

Protecting Your Hands & Wrists   To avoid difficulties, consider employing the following:

- *Hand exercises:* You should warm up for the keyboard just as athletes warm up before doing a sport in order to prevent injury. There are several types of warm-up exercises. You can gently massage the hands, press the palm down to stretch the underside of the forearms, or press the fist down to stretch the top side of the forearm. (■ *See Panel 5.26.*) Experts advise taking frequent breaks, during which time you should rotate and massage your hands.

- *Work-area setup:* Many people set up their computers in the same way as they would a typewriter. However, the two machines are for various reasons ergonomically different. With a computer, it's important to set up your work area so that you sit with both feet on the floor, thighs at right angles to your body. The chair should be adjustable and support your lower back. Your forearms should be parallel to the floor. You should look down slightly at the screen. (■ *See Panel 5.27.*) This setup is particularly important if you are going to be sitting at a computer for hours at a stretch.

- *Wrist position:* To avoid wrist and forearm injuries, you should keep your wrists straight and hands relaxed as you type. Instead of putting the keyboard on top of a desk, therefore, you should put it on a low table or in an underdesk keyboard drawer. Otherwise the nerves in your wrists will rub against the sheaths surrounding them, possibly leading to RSI pains.[82] Some experts also suggest using a padded, adjustable wrist rest, which attaches to the keyboard.

Various kinds of ergonomic keyboards are also available, such as those that are hinged in the middle.

Guarding Against Eyestrain, Headaches, & Back & Neck Pains   Eyestrain and headaches usually arise because of improper lighting, screen glare, and long shifts staring at the screen. Make sure your windows and lights don't throw a glare on the screen, and that your

---

PANEL 5.26

Hand exercises

Positions should be held for 10 seconds or more. Warm-up exercises can prevent injuries.

Gently pull thumb down and back until you feel the stretch.

Grasp fingers and gently bend back wrist.

Massage inside and outside of hand with thumb and fingers.

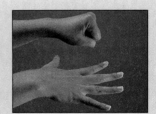

Clench fist tightly, then release, fanning out fingers. Repeat five times.

Gently massage hands; press palm down to stretch underside of forearm; press fist down to stretch top side of forearm. Hold positions for at least 10 seconds.

How to set up your computer work area

HEAD Directly over shoulders, without straining forward or backward, about an arm's length from screen.

NECK Elongated and relaxed.

SHOULDERS Kept down, with the chest open and wide.

BACK Upright or inclined slightly forward from the hips. Maintain the slight natural curve of the lower back.

ELBOWS Relaxed, at about a right angle.

WRISTS Relaxed, and in a neutral position, without flexing up or down.

KNEES Slightly lower than the hips.

CHAIR Sloped slightly forward to facilitate proper knee position.

LIGHT SOURCE Should come from behind the head.

SCREEN At eye level or slightly lower. Use an anti-glare screen.

FINGERS Gently curved.

KEYBOARD Best when kept flat (for proper wrist positioning) and at or just below elbow level. Computer keys that are far away should be reached by moving the entire arm, starting from the shoulders, rather than by twisting the wrists or straining the fingers. Take frequent rest breaks.

FEET Firmly planted on the floor. Shorter people may need a footrest.

computer is not framed by an uncovered window. Headaches may also result from too much noise, such as listening for hours to an impact printer printing out.

Back and neck pains occur because furniture is not adjusted correctly or because of heavy computer use. Adjustable furniture and frequent breaks should provide relief here.

Some people worry about emissions of electromagnetic waves and whether they could cause problems in pregnancy or even cause cancer. The best approach is to simply work at an arm's length from computers with CRT-type monitors.

SUMMARY

What It Is / What It Does	**Why It's Important**
audio board *(p. 223, LO 6)* Type of audio-input device; add-on circuit board in a computer that converts analog sound to digital sound and stores it for further processing.	Audio boards are used to provide digital sound input for multimedia personal computers.
audio-input device *(p. 223, LO 6)* Device that records or plays analog sound and translates it for digital storage and processing.	Audio-input devices, such as audio boards and MIDI boards, are important for multimedia computing.
bar code *(p. 214, LO 3)* Vertical striped marks of varying widths that are imprinted on retail products and other items; when scanned by a bar-code reader, the code is converted into computer-acceptable digital input.	Bar codes may be used to input data from many items, from food products to overnight packages to railroad cars, for tracking and data manipulation.
bar-code reader *(p. 214, LO 3)* Photoelectric scanner, found in many supermarkets, that translates bar code symbols on products into digital code.	With bar-code readers and the appropriate computer system, retail clerks can total purchases and produce invoices with increased speed and accuracy; and stores can monitor inventory with greater efficiency.
biometrics *(p. 226, LO 6)* Science of measuring individual body characteristics.	Biometric systems are used in lieu of typed passwords to identify people authorized to use a computer system.
carpal tunnel syndrome (CTS) *(p. 228, LO 7)* Type of repetitive strain injury; condition caused by pressure on the median nerve in the wrist, producing damage and pain to nerves and tendons in the hands.	CTS is a debilitating, possibly disabling, condition brought about by overuse of computer keyboards; it may require surgery.
computer vision syndrome (CVS) *(p. 228, LO 7)* Computer-related disability; consists of eyestrain, headaches, double vision, and other problems caused by improper use of computer display screens.	Contributors to CVS include faulty lighting, screen glare, and screens with poor resolution.
dedicated fax machine *(p. 216, LO 3)* Specialized machine for scanning images on paper documents and sending them as electronic signals over telephone lines to receiving fax machines or fax-equipped computers; a dedicated fax machine will also received faxed documents.	Unlike fax modems installed inside computers, dedicated fax machines can scan paper documents.
digital camera *(p. 224, LO 6)* Type of electronic camera that uses a light-sensitive silicon chip to capture photographic images in digital form.	Unlike still-video cameras, digital cameras can produce images in digital form that can be transmitted directly to a computer's hard disk for manipulation, storage, and/or printing out.
digitizing tablet *(p. 212, LO 2)* Tablet connected by a wire to a pen-like stylus, with which the user sketches an image, or a puck, with which the user copies an image.	A digitizing tablet can be used to achieve shading and other artistic effects or to "trace" a drawing, which can be stored in digitized form.
electromagnetic fields (EMFs) *(p. 229, LO 7)* Waves of electrical energy and magnetic energy.	Some users of cellular phones and CRT monitors have expressed concerns over possible health (cancer) effects of EMFs, but the evidence is weak.

What It Is / What It Does	**Why It's Important**
electronic camera *(p. 224, LO 6)* Type of camera that captures images in electronic form for immediate viewing on a television or computer display screen. Two types are still-video cameras and digital cameras.	Unlike conventional cameras using chemical film, electronic cameras can convert images to digital form, which can be manipulated by a computer.
electronic imaging *(p. 218, LO 3)* The combining of separate images, using scanners, digital cameras, and advanced graphic computers.	Electronic imaging has become an important part of multimedia.
ergonomics *(p. 229, LO 7)* Study of the physical and psychological relationships between people and their work environment.	Ergonomic principles are used in designing ways to use computers to further productivity while avoiding stress, illness, and injuries.
fax machine *(p. 216, LO 3)* Short for *facsimile transmission machine;* input device for scanning an image and sending it as electronic signals over telephone lines to a receiving fax machine, which re-creates the image on paper. Fax machines may be dedicated fax machines or fax modems.	Fax machines enable the transmission of text and graphic data over telephone lines quickly and inexpensively.
fax modem *(p. 217, LO 3)* Modem with fax capability installed as a circuit board inside a computer; it can send and receive electronic signals via telephone lines directly to/from a computer similarly equipped or to/from a dedicated fax machine.	With a fax modem, users can send information much more quickly than they would if they had to feed it page by page through a dedicated fax machine. However, fax modems cannot scan paper documents for faxing.
imaging system *(p. 217, LO 3)* Also known as *image scanner,* or *graphics scanner;* input device that converts text, drawings, and photographs into digital form that can be stored in a computer system.	Image scanners have enabled users with desktop publishing software to readily input images into computer systems for manipulation, storage, and output.
input hardware *(p. 202, LO 1)* Devices that take data and programs that people can read or comprehend and convert them to a form the computer can process. Devices are of two types: keyboard entry and direct entry.	Input hardware enables data to be put into computer-processable form.
joystick *(p. 210, LO 2)* Pointing device that consists of a vertical handle like a gearshift lever mounted on a base with one or two buttons; it directs a cursor or pointer on the display screen.	Joysticks are used principally in videogames and in some computer-aided design systems.
keyboard *(p. 202, LO 1)* Typewriter-like input device that converts letters, numbers, and other characters into electrical signals that the computer's processor can "read."	Keyboards are the most popular kind of input device.
light pen *(p. 212, LO 2)* Light-sensitive pen-like device connected by a wire to a computer terminal; the user brings the pen to a desired point on the display screen and presses the pen button, which identifies that screen location to the computer.	Light pens are used by engineers, graphic designers, and illustrators for making drawings.
magnetic-ink character recognition (MICR) *(p.215, LO 3)* Type of scanning technology that reads magnetized-ink characters printed at the bottom of checks and converts them to computer-acceptable digital form.	MICR technology is used by banks to sort checks.
magentic-stripe cards *(p. 220, LO 4)* Credit-type cards with encoded magnetic stripes on their backs.	Magnetic-stripe cards are used for specific purposes, such as ATM cards and charge cards.
mouse *(p. 209, LO 2)* Direct-entry input device that is rolled about on a desktop to position a cursor or pointer on the computer's display screen, which indicates the area where data may be entered or a command executed.	For many purposes, a mouse is easier to use than a keyboard for communicating commands to a computer. With microcomputers, a mouse is needed to use most graphical user interface programs and to draw illustrations.

What It Is / What It Does	Why It's Important

mouse pointer *(p. 209, LO 2)* Symbol on the display screen whose movement is directed by movement of a mouse on a flat surface, such as a table top.

The position of the mouse pointer indicates where information may be entered or a command (such as clicking, dragging, or dropping) may be executed. Also, the shape of the pointer may change, indicating a particular function that may be performed at that point.

Musical Instrument Digital Interface (MIDI) board *(p. 224, LO 6)* Type of audio-input device; add-on circuit board in a computer that creates music in digital form as a musician performs, as on a special MIDI keyboard.

MIDI boards are used to provide digital input of music for multimedia personal computers.

optical card *(p. 220, LO 4)* Plastic, wallet-type card using laser technology like music compact disks, which can be used to input data.

Because they hold so much data, optical cards have considerable uses, as for a health card holding a person's medical history, including digital images such as X-rays.

optical-character recognition (OCR) *(p. 215, LO 3)* Type of scanning technology that reads special preprinted characters and converts them to computer-usable form. A common OCR scanning device is the wand reader.

OCR technology is frequently used with utility bills and price tags on department-store merchandise.

optical-mark recognition (OMR) *(p. 215, LO 3)* Type of scanning technology that reads pencil marks and converts them into computer-usable form.

OMR technology is frequently used for grading multiple-choice and true/false tests, such as parts of the College Board Scholastic Aptitude Test.

pen computer *(p. 212, LO 2)* Input system that uses a pen-like stylus to enter handwriting and marks into a computer. The four types of systems are gesture recognition, handwriting stored as scribbling, personal handwriting stored as typed text with training, and standard handwriting "typeface" stored as typed text without training.

Pen-based computer systems benefit people who don't know how to or who don't want to type or need to make routinized kinds of inputs such as checkmarks.

point-of-sale (POS) terminal *(p. 206, LO 1)* Smart terminal used much like a cash register.

POS terminals record customer transactions at the point of sale but also store data for billing and inventory purposes.

puck *(p. 212, LO 2)* Copying device with which the user of a digitizing tablet may copy an image.

With a puck, users may "trace" (copy) a drawing and store it in digitized form.

repetitive strain injuries (RSI) *(p. 227, LO 7)* Several wrist, hand, arm, and neck injuries resulting when muscle groups are forced through fast, repetitive motions.

Computer users may suffer RSIs such as muscle strain and tendinitis, which are not disabling, or carpal tunnel syndrome, which is.

scanning devices *(p. 214, LO 3)* Input devices that translate images such as text, drawings, and photos into digital form.

Scanning devices—bar-code readers, fax machines, imaging systems—simplify the input of complex data.

sensor *(p. 225, LO 6)* Type of input device that collects specific kinds of data directly from the environment and transmits it to a computer.

Sensors can be used for detecting speed, movement, weight, pressure, temperature, humidity, wind, current, fog, gas, smoke, light, shapes, images, and so on.

smart card *(p. 220, LO 4)* Wallet-type card containing a microprocessor and memory chip that can be used to input data.

In some countries, telephone users may buy a smart card that lets them make telephone calls until the total cost limit programmed into the card has been reached.

source data entry device *(p. 202, LO 1)* Also called *source-data automation;* non-keyboard data-entry device. The category includes pointing devices; scanning devices; magnetic-stripe, smart, and optical cards; voice-recognition devices; audio-input devices; video-input devices; electronic cameras; sensors; and human-biology input devices.

Source data entry devices lessen reliance on keyboards for data entry and can make data entry more accurate. Some also enable users to draw graphics on screen and create other effects not possible with a keyboard.

still-video camera *(p. 224, LO 6)* Type of electronic camera that resembles a camcorder; captures images that can be displayed on a television screen.

Unlike digital cameras, still-video cameras must have images converted by a video capture board before they can be stored in a computer; picture resolution and color range are also more limited.

stylus *(p. 212, LO 2)* Pen-like device with which the user of a digitizing tablet "sketches" an image.

With a stylus, users can achieve artistic effects similar to those achieved with pen or pencil.

terminal *(p. 204, LO 1)* Input device that consists of a keyboard, a video display screen, and a communications line to a main computer system.

A terminal is generally used to input data to, and receive visual data from, a mainframe computer system.

touch screen *(p. 211, LO 2)* Video display screen that has been sensitized to receive input from the touch of a finger. It is often used in automatic teller machines and in directories conveying tourist information.

Because touch screens are easy to use, they can convey information quickly and can be used by people with no computer training; however, the amount of information offered is usually limited.

trackball *(p. 210, LO 2)* Movable ball, on top of a stationary device, that is rotated with the fingers or palm of the hand; it directs a cursor or pointer on the computer's display screen, which indicates the area where data may be entered or a command executed.

Unlike a mouse, a trackball is especially suited to portable computers, which are often used in confined places.

voice-recognition system *(p. 221, LO 5)* Input system that converts a person's speech into digital code; the system compares the electrical patterns produced by the speaker's voice with a set of prerecorded patterns stored in the computer.

Voice-recognition technology is useful for inputting data in situations in which people are unable to use their hands or need their hands free for other purposes.

(Selected answers appear at the back of the book.)

Short-Answer Questions

1. What are the two main categories of input hardware?
2. What is *ergonomics*?
3. What is the difference between dumb terminals and smart terminals?
4. What is the main difference between a mouse and a trackball?
5. What is the main disadvantage of using a fax modem instead of a dedicated fax machine?
6. What are some current and future uses of smart cards?
7. What are the three main limitations of voice-recognition technology?
8. What is a point-of-sale terminal used for?
9. What is a digitizer and how is it used?
10. What determines what the function keys on a keyboard do?

Fill-in-the-Blank Questions

1. _____ describes the process in which data created while an event is taking place is entered directly into the computer system in a machine-processable form.
2. A(n) _____ converts a person's speech into digital code that your computer can understand.
3. A(n) _____ can collect information directly from the environment and transmit it to a computer.
4. The _____ combines computing and cable-TV functions in a single device.
5. A(n) _____ scans an image and sends it as electronic signals over telephone lines to a receiving machine, which prints the transmission on paper.
6. A(n) _____ captures images in electronic form for immediate viewing on a television or computer screen.
7. A(n) _____ is rolled about on the desktop to direct a pointer on the computer's display screen.
8. Small pen-controlled, handheld computers that can also do two-way wireless messaging are called _____.

9. A(n) _____ terminal can be used only to input and receive data; it cannot do any independent processing.
10. List three types of computer-induced health disorders.
    a. _____
    b. _____
    c. _____

Multiple-Choice Questions

1. Which of the following doesn't use keyboard entry?
   a. optical cards
   b. terminals
   c. Touch-Tone devices
   d. microcomputers
   e. set-top boxes
2. Which of the following is often used to verify credit cards?
   a. optical card
   b. pen-based system
   c. Touch-Tone device
   d. microcomputer
   e. set-top box
3. Which of the following uses a stylus?
   a. light pen
   b. digitizing tablet
   c. touch screen
   d. joystick
   e. none of the above
4. Which of the following do banks use to sort checks?
   a. bar-code reader
   b. magnetic-ink-recognition device
   c. fax machine
   d. imaging system
   e. all of the above
5. Which of the following is a type of human-biology input device?
   a. biometric system
   b. line-of-sight system
   c. cyber gloves
   d. body suit
   e. all of the above
6. Which of the following would you use to enter handwriting and marks directly onto the display screen?
   a. electronic book
   b. digitizing tablet
   c. pen-based system
   d. joystick
   e. none of the above

7. Which of the following enables a person to use his or her eyes to "point" at the screen?
   a. biometric system
   b. line-of-sight system
   c. cyber gloves
   d. joystick
   e. none of the above

8. Which of the following can convert music into a computer-usable form?
   a. audio-input device
   b. musical sensor
   c. MTV-input device
   d. musical camera
   e. none of the above

True/False Questions

T F 1. Cursor-movement keys are used to execute commands.

T F 2. Trackballs are used to input data when the user has a great deal of space to work in.

T F 3. Many commands that you execute with a mouse can also be performed with a keyboard.

T F 4. A smart card contains a microprocessor and a memory chip.

T F 5. Experiments using brainwave devices to move a cursor on the screen through the power of thought have yet to be successful.

T F 6. Function keys are used the same way with every software application.

T F 7. Electronic cameras can capture images for immediate viewing on a television.

T F 8. The main limitation of capturing full video is storage, not input.

T F 9. Electronic imaging has become an important part of multimedia.

T F 10. A dedicated fax machine does nothing but send and receive faxes.

Projects/Critical-Thinking Questions

1. During the next week, make a list of all the input devices you notice—in stores, on campus, at the bank, in the library, on the job, at the doctor's, and so on. At the end of the week, report on the devices you have listed and name some of the advantages you think are associated with each device.

2. Research image and graphics scanning technology. What differentiates one scanner from another? Is it the clarity of the scanned image? Price? Software? If you were going to buy a scanner, which do you think you would buy? Why? What do you need to run a color scanner?

3. Given the many ways that data can be input to a computer system and that new input technologies will surely become available, do you think that keyboards might become obsolete one day? Why/why not?

4. Research how the legal system is dealing with ergonomic issues in the workplace. Are any ergonomic laws or standards in place in either the private or public sector? Do you think reform is necessary?

5. Identify an input task that you think would greatly benefit from an improved input technology. How is the task currently performed? By whom? In what industry? Is the input technology you describe currently available in some form?

net Using the Internet

Objective: *In this exercise we describe the basics of viewing and finding newsgroups using Netscape Navigator 2.0 or 2.01.*

Before you continue: *We assume you have access to the Internet—through your university, business, or commercial service provider—and to a news server. Additionally, we assume you are using the Web browser tool named Netscape Navigator and know how to connect to the Internet. If necessary, ask your instructor or system administrator for assistance.*

1. Make sure you have started Netscape. The home page for Netscape Navigator should appear on your screen.

2. The name "newsgroup" is somewhat misleading given that newsgroups don't really contain news. Newsgroups provide users the opportunity to share information on just about any subject that interests them such as art, gardening, travel, or zoology. Newsgroup discussions are organized into topics which contain one or more messages relating to the topic. After a user posts a message to a newsgroup, one or more users might respond to the message, evoking additional responses from other users. The entire discussion relating to a topic is called a *thread*.

   To read news via Netscape, you need access to a news server. The default is Netscape's own server, but most users use a local server to improve speed. To configure a news server, you choose Options, Mail and News Preferences from the Menu bar. Then select the Servers tab where you enter the address of a news server. Most users obtain access to a news server through their university, organization, or commercial Internet service provider.

   In the next few steps you'll enter a newsgroup that contains Microsoft Windows-related topics and messages. The newsgroup's URL is *news:comp.os.ms-windows.*

CLICK: Open button (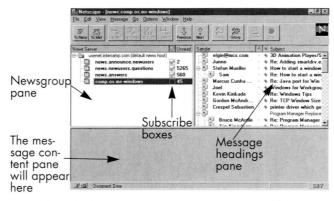)
TYPE: `news:comp.os.ms-windows`
CLICK: Open command button

After maximizing the newsgroup window, your screen should appear similar to the following:

Newsgroup pane

Subscribe boxes

Message headings pane

The message content pane will appear here

4. The newsgroup window is divided into two panes. The left pane is referred to as the *newsgroup pane* and the right pane is referred to as the *message headings pane*. To display a newsgroup's messages, double-click the newsgroup's name in the left pane. (*Note:* To easily find this newsgroup again and display it in the newsgroup window, you can subscribe to it by clicking its Subscribe box.) To display the contents of a message that appears in the right pane, click the message's heading.
   DOUBLE-CLICK: *"comp.os.ms-windows" newsgroup name in the left pane.*
   To display the contents of a message:
   CLICK: *a message heading in the right pane*
   The contents of the message should appear in the lower portion of the screen, called the *message content pane*. (*Note:* To see the message, drag the vertical scroll bar in the message content pane downward.)

5. Notice that the newsgroups toolbar appears at the top of the newsgroups window.

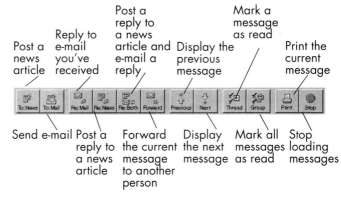

Post a news article — Reply to e-mail you've received — Post a reply to a news article and e-mail a reply — Display the previous message — Mark a message as read — Print the current message

Send e-mail — Post a reply to a news article — Forward the current message to another person — Display the next message — Mark all messages as read — Stop loading messages

6. To move forward and backward through articles in the same topic, use the Previous and Next buttons located on the newsgroups toolbar.

7. Earlier in this exercise we supplied the URL for the Microsoft Windows newsgroup. But what if you are new to newsgroups and don't know what newsgroups are available? You can display a list of newsgroups using the following procedure:
   CHOOSE: Options, Show All Newsgroups from the Menu bar

8. If this is the first time you've connected to the news server you will now need to click the OK button to save the file. Depending on the speed of your modem, the process of saving the file may take a while. A listing of newsgroups will appear, similar to the following:

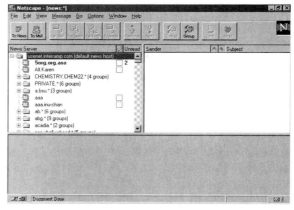

9. Simply double-click a newsgroup category in the newsgroup pane to see what messages are available in the message headings pane.

10. When you've had enough practice navigating through newsgroups, exit Netscape.

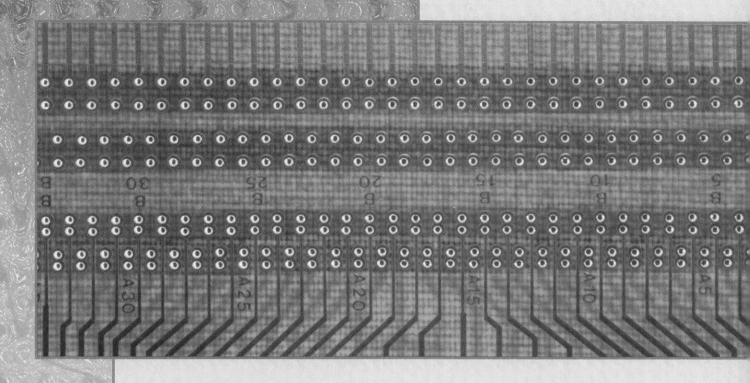

CHAPTER 6

Output
The Many Uses of Computing & Communications

Concepts You Should Know

After reading this chapter, you should be able to:

1. Distinguish between softcopy and hardcopy output.
2. Describe the different types of display screens.
3. List and explain three types of video display adapters.
4. Explain the operations of the different types of printers, plotters, and multifunction devices.
5. Describe other output devices: audio, video, virtual reality, and robots.

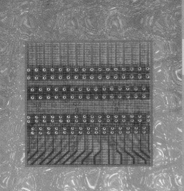

How long until speech robots call you during the dinner hour and use charming voices to get you to part with a charitable donation through your credit card?

Perhaps sooner than you think. Already University of Iowa scientists have created a computer program and audio output that simulates 90% of the acoustic properties made by one man's voice. "What we have now are extremely human-like speech sounds," says Brad H. Story, the lead researcher and also the volunteer whose vocal tract was analyzed and simulated. "We can produce vowels and consonants in isolation." A harder task, still to be accomplished, is to re-create the transitions between key linguistic sounds— what researchers call "running speech."

In the meantime, Mr. Story and a colleague have performed around the country with an electronic device they call "Pavarobotti." The operatic "singing" (such as Puccini arias) is not borrowed from recordings of humans but is created electronically. Singing, however, is easier to simulate than speech. Even singing that would bring an audience to its feet, such as holding a high pitch for a long time, is easy to perform with microchips and speakers.[1]

Output Hardware: Softcopy Versus Hardcopy

Preview & Review: Output devices translate information processed by the computer into a form that humans can understand.

The two principal kinds of output are hardcopy, which is printed, and softcopy, such as material shown on a display screen.

Output devices include display screens; printers, plotters, and multifunction devices; audio-output devices; video-output devices; virtual reality; and robots.

If the foregoing example is any guide, the closing days of the 20th century represent exciting times in the development of computer output. In the following pages we describe present and future types of output devices, ranging from printers to high-definition television.

The principal kinds of output are *hardcopy* and *softcopy*. (■ *See Panel 6.1.*)

- **Hardcopy:** *Hardcopy* **refers to printed output.** The principal examples are printouts, whether text or graphics, from printers. Film, including microfilm and microfiche, is also considered hardcopy output.
- **Softcopy:** *Softcopy* **refers to data that is shown on a display screen or is in audio or voice form.** This kind of output is not tangible; it cannot be touched. Virtual reality and robots might also be considered softcopy devices.

There are several types of output devices. We will discuss the following ones.

- Display screens
- Printers, plotters, multifunction, and microfilm/fiche devices
- Audio-output devices
- Video-output devices
- Virtual-reality devices
- Robots

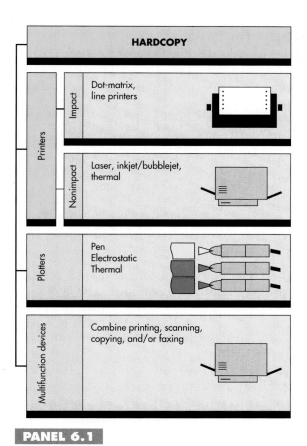

PANEL 6.1

Summary of output devices

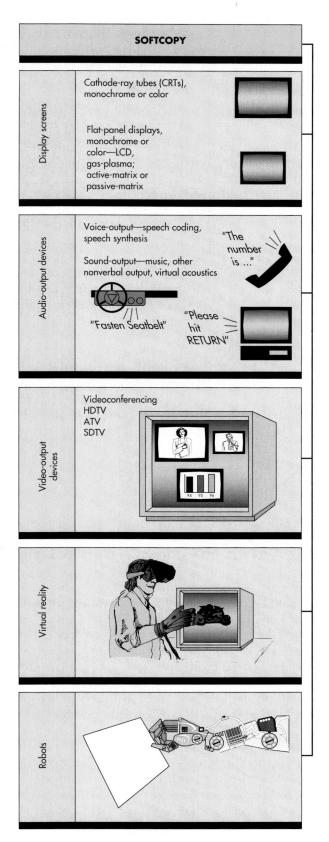

Display Screens: Softcopy Output

Preview & Review: Display screens are either CRT (cathode-ray tube) or flat-panel display.

CRTs use a vacuum tube like that in a TV set.

Flat-panel displays are thinner, weigh less, and consume less power than CRTs but are not as clear. Principal flat-panel displays are liquid-crystal display (LCD) and gas-plasma display.

Users must decide about screen clarity, monochrome versus color, and text versus graphics (character-mapped versus bitmapped).

Various video display adapters (such as VGA, SVGA, and XGA) allow various kinds of resolution and colors.

Lost in San Francisco or Boston? Maybe you could glance at the display screen of your dashboard-mounted electronic map. Introduced in 1985 by Etak of Menlo Park, California, auto navigation systems use global positioning satellites (explained in Chapter 8), so your car knows exactly where it is even if you don't. Tap two points on the screen, and the map will help direct you where you want to go. At $2000, the Etak navigation system is presently found only on the dashboards of cars belonging to the well-off but will no doubt become more affordable in a few years.[2]

Display screens are among the principal windows of information technology. As more and more refinements are made, we can expect to see them adapted to many such innovative uses.

R E A D M E

Case Study: Vehicle Navigation Systems & Computerized Maps

Five minutes after I left the Avis counter at San Jose International Airport, I was convinced that the NavMate guidance system in my rented Oldsmobile 88 Royale LSS was every traveler's dream. It had plotted a route to the nearest bank ATM and was talking to me in reassuring phrases like "freeway entrance ahead on the right."

Five hours later, the dream had turned into every traveler's worst nightmare: Mapless, I was lost deep in the woods, in unfamiliar territory with dusk rapidly approaching.

I had fallen into a black hole in the navigation/information database, according to Zexel USA Corp., which provides the NavMate guidance system used by Avis in its San Jose–San Francisco test market area. Within two years, when the system is offered countrywide as an Oldsmobile option, the holes will have been plugged, according to Zexel. . . .

The system relies on Global Positioning System satellites to transmit data to an antenna mounted on the car's trunk.

The satellites track the car's location, making it possible for the small computer in the trunk, which also receives data from the odometer and a gyroscope, to plot a path to your destination. A four-inch-square backlit display shows an ever-changing map with a scale you can vary between one-eighth of a mile and 16 miles to the inch. At one-eighth of a mile, you can read the name of every street as you pass by.

If you miss a turn, punching one button produces a new route to glance at, or you can wait for the digitized voice to reroute you. Except for the rare black holes, the 12 counties in and around San Francisco are well covered. NavMate normally knows where you are within one-tenth of a mile. Incidentally, my black hole ended half a mile from my house. So you *can* go home again.

—Sandy Reed, "Do You Know the Way to San Jose?" *Popular Science*

Display screens—also variously called *monitors, CRTs,* or simply *screens*—are output devices that show programming instructions and data as they are being input and information after it is processed. Sometimes a display screen is also referred to as a *VDT,* for *video display terminal,* although technically a VDT includes both screen and keyboard. The size of a screen is measured diagonally from corner to corner in inches, just like television screens. For terminals on large computer systems and for desktop microcomputers, 15- to 17-inch screens are a common size. Portable computers of the notebook and subnotebook size may have screens ranging from 7.4 inches to 10.4 inches. Pocket-size computers may have even smaller screens. To give themselves a larger screen size, some portable-computer users buy a larger desktop monitor (or a separate "docking station") to which the portable can be connected. Near the display screen are control knobs that, as on a television set, allow you to adjust brightness and contrast.

Display screens are of two types: *cathode-ray tubes* and *flat-panel displays.*

Cathode-Ray Tubes (CRTs)

The most common form of display screen is the CRT. **A *CRT,* for *cathode-ray tube,* is a vacuum tube used as a display screen in a computer or video display terminal.** (■ *See Panel 6.2.*) This same kind of technology is found

PANEL 6.2

CRT and flat-panel monitors

(Top) CRT monitor. *(Bottom left)* LCD display. *(Bottom right)* Gas-plasma display.

not only in the screens of desktop computers but also in television sets and flight-information monitors in airports.

Images are represented on the screen (whether CRT or flat-panel display) by individual dots or "picture elements" called *pixels*. **A *pixel* is the smallest unit on the screen that can be turned on and off or made different shades.** A stream of bits defining the image is sent from the computer (from the CPU, ✓ p. 172) to the CRT's electron gun, where the bits are converted to electrons. The inside of the front of the CRT screen is coated with phosphor. When a beam of electrons from the electron gun (deflected through a yoke) hits the phosphor, it lights up selected pixels to generate an image on the screen.

Flat-Panel Displays

If CRTs were the only existing technology for computer screens, we would still be carrying around 25-pound "luggables" instead of lightweight notebooks, subnotebooks, and pocket PCs. CRTs provide bright, clear images, but they add weight and consume space and power.

Compared to CRTs, *flat-panel displays* **are much thinner, weigh less, and consume less power. Thus, they are better for portable computers.** Flat-panel displays are made up of two plates of glass with a substance in between them, which is activated in different ways.

Flat-panel displays are distinguished in two ways: (1) by the substance between the plates of glass and (2) by the arrangement of the transistors in the screens.

- **Substances between plates—LCD and gas plasma:** Two common types of technology used in flat-panel display screens are *liquid-crystal display* and *gas-plasma display*. (■ *Refer back to Panel 6.2.*)

  Liquid-crystal display (LCD) **consists of a substance called *liquid crystal*, the molecules of which line up in a way that alters their optical properties. As a result, light—usually backlighting behind the screen—is blocked or allowed through to create an image.**

  Gas-plasma display **is like a neon bulb, in which the display uses a gas that emits light in the presence of an electric current.** That is, the technology uses predominantly neon gas and electrodes above and below the gas. When electric current passes between the electrodes, the gas glows. Although gas-plasma technology has better resolution than LCD technology, it is more expensive and thus is not used as often as LCD. On the other hand, LCDs aren't practical for screens larger than 20 inches and so aren't practical for TV-size screens.[3]

- **Arrangement of transistors—active-matrix or passive-matrix:** Flat-panel screens are either active-matrix or passive-matrix displays.

  In an *active-matrix display*, each pixel on the screen is controlled by its own transistor. Active-matrix screens are much brighter and sharper than passive-matrix screens, but they are more complicated and thus more expensive.

  In a *passive-matrix display*, a transistor controls a whole row or column of pixels. The advantage is that passive-matrix displays are less expensive and use less power than active-matrix screens.

Screen Clarity

Whether CRT or flat-panel display, screen clarity depends on three qualities: *resolution, dot pitch,* and *refresh rate.*

- **Resolution:** **The clarity or sharpness of a display screen is called its *resolution;* the more pixels there are per square inch, the better the resolution.** Resolution is expressed in terms of the formula *horizontal pixels* × *vertical pixels.* Each pixel can be assigned a color or a particular shade of gray. A screen with 640 × 480 pixels multiplied together equals 307,200 pixels. This screen will be less clear and sharp than a screen with 800 × 600 (equals 480,000) or 1024 × 768 (equals 786,432) pixels.
- **Dot pitch:** ***Dot pitch* is the amount of space between the centers of adjacent pixels; the closer the dots, the crisper the image.** For crisp images, dot pitches should be less than .31 millimeter.
- **Refresh rate:** ***Refresh rate* is the number of times per second that the pixels are recharged so that their glow remains bright.** In dual-scan screens, the tops and bottoms of the screens are refreshed independently at twice the rate of single-scan screens, producing more clarity and richer colors. In general, displays are refreshed 45 to 100 times per second.

Monochrome Versus Color Screens

Display screens can be either *monochrome* or *color.*

- **Monochrome:** ***Monochrome display screens* display only two colors**—usually black and white, amber and black, or green and black.
- **Color:** ***Color display screens* can display between 16 and 16.7 million colors,** depending on their type. Most software today is developed for color, and—except for some pocket PCs—most microcomputers today are sold with color display screens.

Text Versus Graphics: Character-Mapped Versus Bitmapped Display

Another distinction in display screens relates to their capacity to display graphics. A screen lacking this capacity is referred to as *character-mapped;* a *bitmapped* screen can display graphics.

- **Character-mapped:** ***Character-mapped display* screens display only text**—letters, numbers, and special characters. They cannot display graphics unless a video display adapter card is installed, as explained next.

  Text is displayed in rows and columns, with rows measuring the height of the screen and columns measuring the width. Most computer screens display 25 rows and 80 columns, which means that a row or line can have up to 80 characters of text.
- **Bitmapped:** ***Bitmapped display* screens permit the computer to manipulate pixels on the screen individually rather than as blocks,** enabling software to create a greater variety of images. Today most screens can display text and graphics—icons, charts, graphs, and drawings.

Video Display Adapters

To display graphics, a display screen must have a video display adapter. **A video display adapter, also called a graphics adapter card, is a circuit board that determines the resolution, number of colors, and how fast images appear on the display screen.** Video display adapters come with their own memory chips, which determine how fast the card processes images and how many colors it can display. A video display adapter with 256 kilobytes of memory will provide 16 colors; one with 1 megabyte will support 16.7 million colors.[4]

The video display adapter is often built into the motherboard (✓ p. 171), although it may also be an expansion card that plugs into an expansion slot (✓ p. 176). Video display adapters embody certain standards. (■ *See Panel 6.3.*) Today's microcomputer monitors commonly use VGA and SVGA standards.

- **VGA:** Perhaps the most common video standard today, *VGA,* **for** *Video Graphics Array,* **will support 16 to 256 colors, depending on resolution.** At 320 × 200 pixels it will support 256 colors; at the sharper resolution of 640 × 480 pixels it will support 16 colors.

- **SVGA:** *SVGA,* **for** *Super Video Graphics Array,* **will support 256 colors at higher resolution than VGA.** SVGA has two graphics modes: 800 × 600 pixels and 1024 × 768.

- **XGA:** Also referred to as *high resolution,* **XGA, for** *Extended Graphics Array,* **supports up to 16.7 million colors at a resolution of 1024 × 768 pixels.** Depending on the video display adapter memory chip, XGA will support 256, 65,536, or 16,777,216 colors.

For any of these displays to work, video display adapters and monitors must be compatible. Your computer's software and the video display adapter must also be compatible. Thus, if you are changing your monitor or your video display adapter, be sure the new one will still work with the old.

PANEL 6.3

Video display adapters

The video display adapter may be built into the motherboard or added as an expansion card.

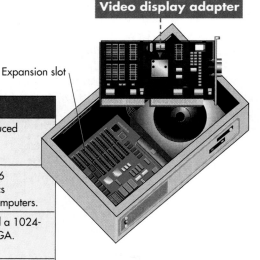

Video display adapter

Expansion slot

Monitor Type	Remarks
EGA (Enhanced Graphics Adapter)	Supports 16 colors in 640-by-350-pixel resolution. Introduced by IBM in 1984. Superseded CGA.
VGA (Video Graphics Array)	Displays 16 colors in 640-by-480-pixel resolution and 256 colors at 320 by 200 pixels. This color bitmapped graphics display standard was introduced in 1987 for IBM PS/2 computers.
SVGA (Super Graphics Array)	Supports 256 colors with 800-by-600-pixel resolution and a 1024-by-768 resolution. This is a higher-resolution version of VGA.
XGA (Extended Graphics Array)	Displays up to 16,777,216 colors at resolutions up to 1024 by 768 pixels.

From the standpoints of protecting the environment and the health of one-self or other users, anyone buying a monitor should be aware of two factors: energy consumption and electromagnetic emissions (✓ p. 229).[5] If a monitor manufacturer says that the display complies with *Energy Star standards*, it means it meets a voluntary federal standard for reduced power consumption in computer equipment. If the monitor also meets so-called *MPR-2 standards*, it means it meets a Swedish government rule, also adopted by many American manufacturers, for reduced electromagnetic emissions, which may be advantageous to health.

Printers, Plotters, Multifunction, & Microfilm/fiche Devices: Hardcopy Output

Preview & Review: Printers, plotters, and multifunction devices produce printed text or images on paper.

Printers may be desktop or portable, impact or nonimpact. The most common impact printers are dot-matrix printers. Nonimpact printers include laser, inkjet, and thermal printers.

Plotters are pen, electrostatic, and thermal.

Multifunction devices combine capabilities such as printing, scanning, copying, and faxing.

Computer output microfilm/fiche can solve storage problems.

It's known as "the hotel facsimile trick," a strategy of desperation for the computer-mobile.

Many travelers carry a portable computer but no printer, to avoid toting along a second device that may weigh as much as their PC. What do they do, then, when on reaching their hotel they find they need to print out a document, but there's no nearby copy shop where they can rent time on a printer? In the hotel facsimile trick, they connect the built-in fax modem in their PC to the phone jack in their hotel room. Then they go down to the hotel's front desk and pick up the paper copy of the document they have just sent to themselves at the hotel's fax telephone number.

Unfortunately, hotel fax machines don't always use high-quality paper of the sort you would want, say, for preparing a presentation to a client. Moreover, the hotel may charge $2 to $5 a page for the fax. Fortunately, however, some new innovations in printers are being made to help travelers.

To get to definitions, **a *printer* is an output device that prints characters, symbols, and perhaps graphics on paper.** Printers are categorized according to whether or not the image produced is formed by physical contact of the print mechanism with the paper. *Impact printers* do have contact; *nonimpact printers* do not.

Desktop Versus Portable Printers

Technologies used for printing range from those that resemble typewriters to those that resemble photocopying machines. An important question you need to ask yourself in this era of mobility is: Will a desktop printer be sufficient, or will you need a printer that is portable?

- **Desktop printers:** Many people, perhaps including most students, find portable computers useful but have no need for a portable printer. You can

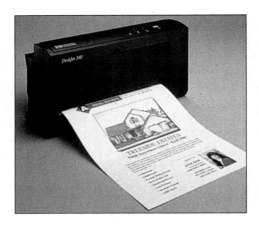

do your writing or computing wherever you can tote your portable PC, then print out documents back at your regular desk. The advantage of desktop printers is the wide range in quality and price available.

* **Portable printers—transportable versus ultraportable:** Portable printers are of two types: "bigger with everything and smaller with less," as one magazine puts it, or transportable versus ultraportable.[6] Transportables are for people who are more concerned about the quality of output. Ultraportables are for frequent travelers who are more concerned about light weight. (■ *See Panel 6.4.*)

Transportable printers usually weigh around 5 pounds, including an adapter that plugs into an AC electrical outlet (but excluding battery weight). They generally offer desktop printer–quality text and graphics printing—even color graphics. Moreover, they come with an envelope-printing feature and an automatic sheet feeder, so that you don't have to sit there and feed in a page at a time.

Ultraportable printers weigh in at around 2 pounds (including AC adapter and battery) and try to deliver reasonably good printing in as small a package as possible. However, you have to give up the ability to do color and print envelopes. Moreover, you'll have to hand-feed sheets one at a time. Finally, you may need special paper, such as thermal paper.

In the final analysis, do you really *need* a portable printer? Even business travelers, says portable-computing expert Jim Seymour, don't usually require an immediate printout for most of the work they're doing while away from the office. Moreover, it's often easy, he says, to simply borrow someone else's printer. "For years, I've carried a little parallel printer cable in the bottom of the case containing the portable PC of the moment," says Seymour. "When I'm traveling and need the occasional quick printout, I just look around for someone in the office or hotel where I'm working who *does* have a PC printer."[7] Having his own printer cable makes it easier to hook up to someone else's printer. (Seymour also carries six basic, widely interchangeable printer *drivers* on disk. This is software that runs most common printers he is likely to encounter.) Another option when traveling is to rent a printer for a couple of days.

What would seem to be the most elegant solution for the traveler is a laptop with a built-in printer—for example, the Canon NoteJet. (■ *See Panel 6.5.*) You feed a sheet into the slot at the front of the computer, and the printed copy comes out the back. When equipped with a fax modem, the PC can also serve as a plain-paper fax receiver and printer.

Impact Printers

An impact printer has mechanisms resembling those of a typewriter. That is, **an *impact printer* forms characters or images by striking a mechanism such as a print hammer or wheel against an inked ribbon, leaving an image on paper.**

For microcomputer users, the most common type of impact printer is a dot-matrix printer. **A *dot-matrix printer* contains a print head of small pins, which strike an inked ribbon against paper, forming characters or images.** Print heads are available with 9, 18, or 24 pins, with the 24-pin head offering the best quality. Dot-matrix printers can print *draft quality*, a coarser-looking 72 dots per inch vertically, or *near-letter-quality (NLQ)*, a crisper-looking 144 dots per inch vertically. (■ *See Panel 6.6.*) Students and others find draft quality acceptable for doing drafts of papers and reports. They then usually switch to near-letter-quality when they are preparing a finished product to be shown to other people.

Dot-matrix printers print 150–300 characters per second and print graphics as well as text. However, they are somewhat noisy.

Another type of impact printer is not used with microcomputers. Large computer installations use high-speed *line printers*, which print a whole line of characters at once rather than a single character at a time. Some of these can print up to 3000 lines a minute. Two types of line printers are *chain printers*, which contain characters on a rotating chain, and *band printers*, which contain characters on a rotating band.

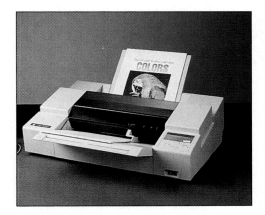

This is a sample of draft quality.
This is a sample of near-letter-quality.

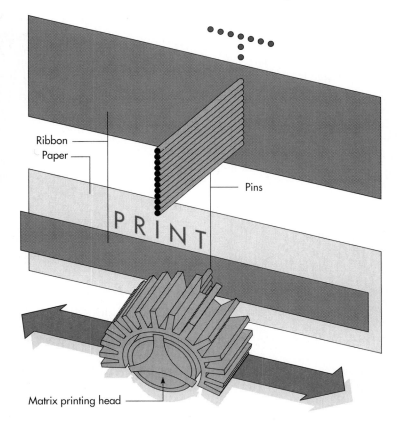

Ribbon
Paper
Pins
PRINT
Matrix printing head

Nonimpact Printers

Nonimpact printers are faster and quieter than impact printers because they have fewer moving parts. **Nonimpact printers form characters and images without making direct physical contact between printing mechanism and paper.**

Two types of nonimpact printers often used with microcomputers are *laser printers* and *inkjet printers.* A third kind, the thermal printer, is seen less frequently, except with ultraportable printers.

- **Laser printer:** Similar to a photocopying machine, **a *laser printer* uses the principle of dot-matrix printers in creating dot-like images. However, these images are created on a drum, treated with a magnetically charged ink-like toner (powder), and then transferred from drum to paper. (■** *See Panel 6.7.)*

There are good reasons that laser printers are the most common type of nonimpact printer. They produce sharp, crisp images of both text and graphics, and they are quieter and faster than dot-matrix printers. Whereas most impact printers are serial printers, printing one character at a time, laser printers print whole pages at a time. They can print 8–20 pages per minute for individual microcomputers (and up to 200 pages per minute for mainframes). They can print in different *fonts—that is, type styles and sizes.* The more expensive models can print in different colors.

One particular group of laser printers has become known as PostScript printers. *PostScript* **is a printer language, or page description language, that has become a standard for printing graphics with laser printers. A *page description language* (software) describes the shape and position of letters and graphics to the printer.** PostScript printers are essential if you are printing a lot of graphics or want to generate fonts in various sizes. Another page description language used with laser printers is **Printer Control Language (PCL), which has resolutions and speeds similar to those of PostScript.**[8]

PANEL 6.7

Laser printer

A small laser beam is bounced off a mirror millions of times per second onto a positively charged drum. The spots where the laser beam hits become neutralized, enabling a special toner (powder) to stick to them and then print out on paper. The drum is then recharged for the next cycle.

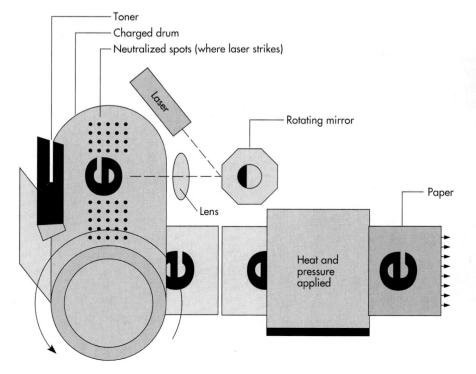

Toner
Charged drum
Neutralized spots (where laser strikes)
Laser
Rotating mirror
Lens
Paper
Heat and pressure applied

- **Inkjet printer:** Like laser and dot-matrix printers, inkjet printers also form images with little dots. *Inkjet printers* **spray small, electrically charged droplets of ink from four nozzles through holes in a matrix at high speed onto paper.** (■ *See Panel 6.8.*)

  Most color printing is done on inkjets because the nozzles can hold four different colors. Moreover, inkjet printers can match the speed of dot-matrix printers. They are even quieter than laser printers and produce an equally high-quality image.

  A variation on inkjet technology is the *bubblejet printer,* **which uses miniature heating elements to force specially formulated inks through print heads with 128 tiny nozzles.** The multiple nozzles print fine images at high speeds. This technology is commonly used in portable printers.

- **Thermal printer:** For people who want the highest-quality color printing available with a desktop printer, thermal printers are the answer. However, they are expensive, and they require expensive paper. Thus, they are not generally used for jobs requiring a high volume of output.

  Thermal printers **use colored waxes and heat to produce images by burning dots onto special paper.** (■ *See Panel 6.9.*) The colored wax sheets are not required for black-and-white output.

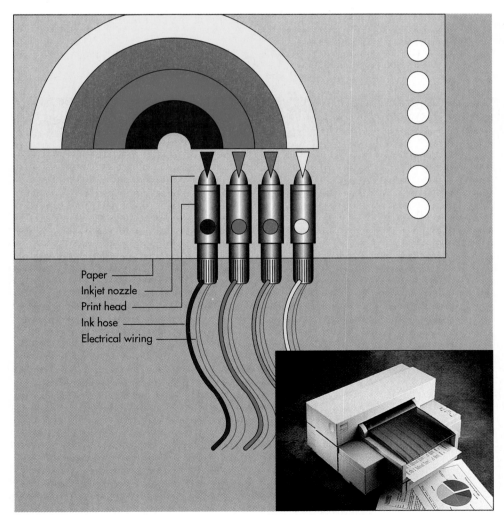

PANEL 6.8

Inkjet printer

Paper
Inkjet nozzle
Print head
Ink hose
Electrical wiring

PANEL 6.9

Thermal printer

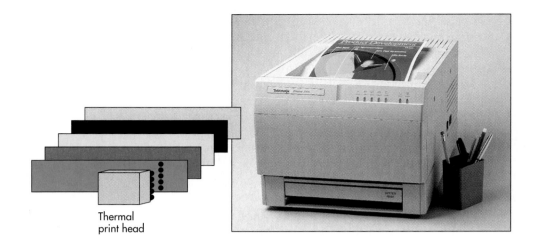

Thermal
print head

It's not uncommon these days to receive a crisply printed letter rendered on a word processor—in an envelope with a handwritten address. Why? It might be because the person writing the letter doesn't have, or doesn't know how to use, a printer envelope feeder for printing addresses. Envelope feeders and label printers are only one of several considerations when buying a printer. (Specialty printers are also available that print nothing but envelopes and labels, useful for businesses that do large mailings.)

The table below compares printer technologies. (■ Panel 6.10.)

PANEL 6.10 **Printer comparisons**

Type	Technology	Advantages	Disadvantages	Typical Speed	Approximate Cost
Dot-matrix	Print head with small pins strikes an inked ribbon against paper	Inexpensive; produces draft quality and near-letter-quality; can output some graphics; can print multi-part forms; low cost per page	Noisy; cannot produce high-quality output of text and graphics; limited fonts	30 to 500+ cps*	$100–$2000
Laser	Laser beam directed onto a drum, "etching" spots that attract toner, which is then transferred to paper	Quiet; excellent quality; output of text and graphics; very high speed	High cost, especially for color	8–200 ppm*	$400–$20,000
Inkjet	Electrostatically charged drops hit paper	Quiet; prints color, text, and graphics; less expensive; fast	Relatively slow; clogged jets; fewer dots per inch	35–400+ cps	$200–$2000
Thermal	Temperature-sensitive paper changes color when treated; characters are formed by selectively heating print head	Quiet; high-quality color output of text and graphics; can also produce transparencies	Special paper required; expensive; slow	11–80 cps	$2000–$22,000

*cps = characters per second; ppm = pages per minute.

Black & White Versus Color Printers

Today prices have plummeted to $400 or less for laser and inkjet printers, so that the cheap, noisy dot-matrix printer may well be going the way of the black-and-white TV set. (Impact printers are still needed, however, for any activity that involves printing multipart forms.) "The choice between a laser printer and an inkjet comes down to how much you print and whether you need color," says one analysis of microcomputer printers.[9]

Lasers, which print a page at a time, can handle thousands of black-and-white pages a month. Moreover, compared to inkjets, laser printers are faster and crisper (though not by much) at printing black-and-white and a cent or two cheaper per page.[10] Finally, a freshly printed page from a laser won't smear, as one from an inkjet might.

However, inkjets, which spray ink onto the page a line at a time, can give you both high-quality black-and-white text and high-quality color graphics. Moreover, color inkjets start at about $400 (black-and-white inkjets at about $200), whereas color lasers cost thousands. Thus, if you think you might have occasion to print multicolor charts for reports or presentations or do colorful children's school projects, a color inkjet would seem to be most desirable.

Some questions to consider when choosing a printer for a microcomputer are given in the accompanying box. (■ *See Panel 6.11.*)

PANEL 6.11 **Questions to consider when choosing a printer for a microcomputer**

Do I want a desktop or portable printer—or both? You'll probably find a desktop printer satisfactory (and less expensive than a portable). If you're on the road enough to warrant using a portable, see whether a *transportable* or an *ultraportable* would best suit you. (See text p. 254.)

Do I need color, or will black-only do? Are you mainly printing text or will you need to produce color charts and illustrations (and, if so, how often)? If you print lots of black text, consider getting a laser printer. If you might occasionally print color, get an inkjet that will accept cartridges for both black and color.

Do I have other special output requirements? Do you need to print envelopes or labels? special fonts (type styles)? multiple copies? transparencies or on heavy stock? Find out if the printer comes with envelope feeders, sheet feeders holding at least 100 sheets, or whatever will meet your requirements.

Is the printer easy to set up? Can you easily put the unit together, plug in the hardware, and adjust the software (the "driver" programs) to make the printer work with your computer?

Is the printer easy to operate? Can you add paper, replace ink/toner cartridges or ribbons, and otherwise operate the printer without much difficulty?

Does the printer provide the speed and quality I want? Will the machine print at least three pages a minute of black text and two pages a minute of color? Are the blacks dark enough and the colors vivid enough?

Will I get a reasonable cost per page? Special paper, ink or toner cartridges (especially color), and ribbons are all ongoing costs. Inkjet color cartridges, for example, may last 100–500 pages and cost $25–$30 new. Laser toner cartridges are cheaper. Ribbons for dot-matrix printers are cheaper still. Ask the seller what the cost per page works out to.

Does the manufacturer offer a good warranty and good telephone technical support? Find out if the warranty lasts at least 2 years. See if the printer's manufacturer offers telephone support in case you have technical problems. The best support systems offer toll-free numbers and operate evenings and weekends as well as weekdays.

Plotters

A *plotter* is a specialized output device designed to produce high-quality graphics in a variety of colors. Plotters are especially useful for creating maps and architectural drawings, although they may also produce less complicated charts and graphs.

The three principal kinds of plotters are *pen, electrostatic,* and *thermal.*[11] (■ *See Panel 6.12.*)

- **Pen plotter:** **A *pen plotter* is designed so that paper lies flat on a table-like surface (flatbed plotter) or is mounted on a drum (drum plotter). Between one and four pens move across the paper, or the paper moves beneath the pens.** This is the most popular type of plotter.

  - **Electrostatic plotter:** Instead of using pens, **an *electrostatic plotter* uses electrostatic charges to create tiny dots on specially treated paper. The paper is then run through a developer to produce the image, which may be four-color.**

  - **Thermal plotter:** **A *thermal plotter* uses electrically heated pins and heat-sensitive paper to create images.** Thermal printers are capable of producing only two colors.

Multifunction Technology: Printers That Do More Than Print

Everything is becoming something else, and even printers are becoming devices that do more than print. For instance, plain-paper fax machines are now available (such as several from Ricoh) that can also function as laser or inkjet printers. Since 1990, Xerox has sold an expensive printer-copier-scanner that can be hooked into corporate computer networks.

A relatively recent type of machine, sometimes called an "office in a box," can do even more. ***Multifunction machines* combine several capabilities, such as printing, scanning, copying, and faxing, all in one device.** An example is Okidata's Doc-It, which combines four pieces of office equipment—photocopier, fax machine, scanner, and laser printer—in one. (■ *See Panel 6.13.*) Similar systems are available from Canon, Brother, Xerox, and Hewlett-Packard. Prices start at $500. Lexmark International sells a multifunction

Plotters

(Left) Pen plotter.
(Right) Electrostatic plotter.

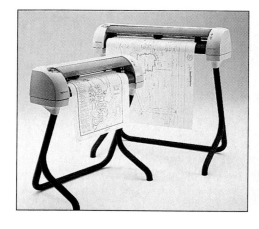

PANEL 6.13

The multifunction device

(Top) Okidata's Doc-It combines four machines in one: printer, copier, fax machine, and scanner. *(Bottom)* Lanier's multifunction machine.

"Four pieces of office equipment—photocopier, fax machine, scanner and laser printer—share much of the same technology. . . .

Now Okidata, a U.S. unit of Oki Electric Industry of Japan, [offers] a product called the Doc-It. It's a well-designed, compact 'desktop document processor,' as the company labels it, that hooks up to a standard IBM-compatible PC and takes advantage of the computer's power to do a wide range of jobs. . . .

Doc-It does the work of four separate office machines acceptably well at a price below the combined cost of buying the devices separately. Thus, it should prove attractive for small offices that now have to buy some or all of these machines, find space to house them and keep each stocked with different parts and supplies."

—Walter S. Mossberg, "'Document Processor' Combines Functions to Turn Out Forms," *The Wall Street Journal*

machine that can print—but not copy—in color.[12–14] By doing the work of four separate office machines at a price below the combined cost of buying these devices separately, the multifunction machine offers budgetary and space advantages.

Computer Output Microfilm/fiche

If you take your time throwing out old newspapers, you know it doesn't take long for them to pile up. No wonder, then, that libraries try to save space by putting newspaper back issues on microfilm or microfiche. One ounce of microfilm can store the equivalent of 10 pounds of paper.

Computer output microfilm/fiche (COM) **is computer output produced as tiny images on rolls/sheets of microfilm.** Indeed, the images are up to 48 times smaller than those produced on a printer. Moreover, they can be recorded far faster and cheaper than the same thing on paper-printed output.

The principal disadvantage, however, is that a microfilm/fiche reader is needed to read COM. It's possible that COM could be made obsolete by developments in secondary-storage techniques, such as the use of removable, high-capacity hard disks. However, at present *computer-assisted retrieval,* which uses microfilm/fiche readers with automatic indexing and data-lookup capabilities, makes COM the preferred technology.

The Paper Glut: Whither the "Paperless Office"?

In producing hardcopy, printers and plotters produce *paper*—great quantities of it. In the past decade the United States' annual rate of paper consumption has nearly doubled. Supposedly, computers were going to make the use of paper obsolete, providing us with "the paperless office," but in fact the opposite has happened. As one writer has observed:

What computers have done, in essence, is bestow independence on thousands of office workers who once needed a secretary to type and produce documents. With their own PCs, most of them can draft memos and reports. They can enlarge charts and make color reproductions on ultra-equipped copier machines. They can use the fax machine to disperse copies or program it to shoot batches of paper all over the country.[15]

Why don't we simply leave information in electronic form, as on a 3½-inch diskette, holding the equivalent of 240 sheets of paper? "People like to hold paper; they like printouts," says Stanford University professor Clifford Nass, who studies social patterns and technology. "If the information exists only in the computer, where is it? You can't tell people it's being stored as a bunch of ones and zeros. They want to touch it."[16]

We make printouts because we are afraid "anything can happen to a computer." Moreover, we get physical comfort from handling paper, from marking it up with colored pens, and from filing it away. As a result, however, there has been a paper explosion such as the world has never seen.

The Theater of Output: Audio, Video, Virtual Reality, & Robots

Preview & Review: Other output hardware includes audio-output devices, video-output devices, virtual-reality devices, and robots.

Audio output includes voice-output technology (speech coding and speech synthesis) and sound-output technology.

Video output includes videoconferencing and digitized television.

Virtual reality is a kind of computer-generated artificial reality.

Robots perform functions ordinarily ascribed to human beings.

What seems to be the leading growth industry today?

The answer, in a word, is *fun.*

According to the U.S. Bureau of Labor Statistics, the entertainment and recreation industries jumped 12% in net new employment in a recent year. This rate outpaced that of the health-care industry, which had been the big job-creation machine of the 1980s.[17] Recreation and entertainment include not only toys, sports, theme parks, magazines, books, and gambling casinos but also movies, videos, games, music, and similar electronic diversions.

Indeed, some observers believe that the entertainment industry is now the driving force for new technology, just as defense used to be.[18] For instance, companies are planning to spend tens of billions of dollars over the next decade on hardware, communications lines, and services just for possible interactive-TV.

Many of the other forms of output technology that remain to be discussed relate—although by no means exclusively—to entertainment and recreation. They include:

- Audio-output devices
- Video-output devices
- Virtual-reality devices
- Robots

Audio-Output Devices

Audio-output devices include those devices that output voice or voice-like sounds and those that output music and other sounds.

- **Voice-output devices:** *Voice-output devices* **convert digital data into speech-like sounds.** By now these devices are no longer very unusual. You hear such forms of voice output on telephones ("Please hang up and dial your call again"), in soft-drink machines, in cars, in toys and games, and recently in vehicle-navigation devices.

  Two types of voice-output technology exist: *speech coding* and *speech synthesis*.

  Speech coding **uses actual human voices speaking words to provide a digital database of words that can be output as voice sounds.** That is, words are codified and stored in digital form. Later they may be retrieved and translated into voices as needed. The drawback of this method is that the output is limited to whatever words were previously entered into the computer system. However, the voice-output message does sound more convincingly like real human speech.

  Speech synthesis **uses a set of 40 basic speech sounds (called *phonemes*, the bases of all speech in English) to electronically create any words.** No human voices are used to make up a database of words; instead, the computer converts stored text into voices. For example, with one Apple Macintosh program, you can type in *Wiyl biy ray5t bae5k*—the numbers elongate the sounds. The computer will then speak the synthesized words, "We'll be right back." Such voice messages are usually understandable, though they don't sound exactly human.

  Some uses of speech output are simply frivolous or amusing. You can replace your computer start-up beep with the sound of James Brown screaming "I feel goooooood!" Or you can attach a voice annotation to a spreadsheet that says "I know this looks high, Bob, but trust me."[19]

  But some uses are quite serious. For the disabled, for example, computers help to level the playing field. A 39-year-old woman with cerebral palsy had little physical dexterity and was unable to talk. By pressing keys on the laptop computer attached to her wheelchair, she was able to construct the following voice-synthesized message: "I can do checkbooks for the first time in my life. I cannot live without my computer."[20]

- **Sound-output devices:** *Sound-output devices* **produce digitized sounds, ranging from beeps and chirps to music.** All these sounds are nonverbal. PC owners can customize their machines to greet each new program with the sound of breaking glass or to moo like a cow every hour. Or they can make their computers issue the distinctive sounds available (from the book/disk combination *Cool Mac Sounds*) under the titles "Arrgh!!!" or "B-Movie Scream."[21] To exercise these possibilities, you need both the necessary software and the sound card, or digital audio circuit board (such as the popular Sound Blaster and Sound Blaster Pro cards). The sound card plugs into an expansion slot in your computer. A sound card is also required in making computerized music.[22]

The various kinds of audio outputs and their devices are important components of multimedia systems (✓ p. 28). Some software developers are working on applications not only for entertainment but also for the business world. For instance, the Microsoft Windows Sound System is a combination of a plug-in sound card and software that works with the sound functions of Windows. It lets executives read out the numbers in a spreadsheet, attach voice messages to electronic mail (✓ p. 75), or use voice commands for computer tasks such as opening a file.

Video-Output Devices

As we explained, CRTs are used both for TV sets and for computer monitors. Innovations in video technology are being seen in a number of areas, including the following:

- **Videoconferencing and video editing:** Want to have a meeting with someone across the country and go over some documents—without having to go there? The answer is videoconferencing. ***Videoconferencing is a method whereby people in different geographical locations can have a meeting—and see and hear one another—using computers and communications.*** We describe this topic in detail elsewhere (Chapter 8), but it's of interest here because of the kind of technology used.

  Videoconferencing was, and still is, a service offered by long-distance telephone companies to large businesses. Now, however, you can do it yourself. Say you're on the West Coast and want to go over a draft of a client proposal with your boss on the East Coast. The first thing you need for such a meeting is a high-capacity telephone line. To this you link your IBM-compatible PC running Windows to which you have added a hardware/software package called ProShare Video System from Intel. ProShare consists of a small video camera that sits atop your display monitor, a circuit board, and software that turns your microcomputer into a personal conferencing system.

  Your boss's image appears in one window on your computer's display screen, and the document you're working on together is in another window. (An optional window shows your own image.) Although the display screen images are choppier than those on a standard TV set, they're clear enough to enable both of you to observe facial expressions and most body language. The software includes drawing tools and text tools for adding comments. Thus, you can go through and edit paragraphs and draw crude sketches on the proposal draft.[23]

- **HDTV, ATV, or SDTV?** *High-definition television (HDTV)* is a television system that features enhanced video and crisp, clear pictures. These pictures are far superior to any seen on television today, appearing within a frame that is twice as wide. They also have twice the present number of scan lines (1100 versus 525 of an average TV screen), thus offering twice the resolution, or sharpness. In addition, HDTV, offers better color and better audio, with sound being of compact-disk quality.

  In the mid-1980s, Japan threatened to make standard television equipment obsolete by being the first out with HDTV. This would have forced American broadcasters to invest in expensive equipment without the prospect of offsetting revenues. To meet the threat, a competition was organized to invent an American HDTV standard. The competition was won by a lab belonging to General Instrument, which abandoned the older *analog* (✔ p. 6) transmission standard (that using radio waves analogous to sound and light waves). Instead, it proposed to create better pictures using *digital* transmission, the 1s and 0s of computer code.[24]

  To exploit the new technology, a consortium of seven companies and labs (including AT&T, the Massachusetts Institute of Technology, and Philips) known as "the Grand Alliance" was formed. But when the alliance unveiled its new technology in December 1994 it was discovered it could do more than show "stunningly realistic images of mountain vistas, springtime flowers, and sun-sparkled lakes," as one article described it.[25] It could also be used to transmit huge amounts of *digital data*—computer files, e-mail, paging messages, and similar services. In the broadcast-spectrum space now used by one analog channel, the Grand Alliance's technology can transmit not just one high-definition image but five or six digital television signals.

Because the new system surpassed all expectations, the Federal Communications Commission also gave it a new name—advanced television. *Advanced television (ATV)* is television that, in place of one analog channel, can transmit not just one high-definition image but five or six digital channels as good as current ones. These channels could be used for non-TV-related services, such as paging signals and e-mail.

Some broadcasters, however, would rather use these additional channels to cram more television programming into the same space. For them the hot buzz words are "standard definition" television. *Standard definition television (SDTV)* allows a broadcaster to pack four or five programs of ordinary quality into a single existing channel. For example, Fox Broadcasting floated the idea of carrying a package of four channels in the area now occupied by one. This would include an all-news channel, an all-children's channel, and a sports channel, in addition to the channel carrying existing Fox programming.

All of these matters have a great many practical and public policy implications (which we consider in Chapters 8 and 12). Among them: Might you have to junk your existing analog TV set? Would you receive TV through your personal computer, and how?[26]

Virtual-Reality & Simulation Devices

Virtual reality (VR) is a kind of computer-generated artificial reality that projects a person into a sensation of three-dimensional space. (■ *See Panel 6.14, next page.*) To achieve this effect, you need the following interactive sensory equipment:

- **Headgear:** The headgear—which is called *head-mounted display (HMD)*—has two small video display screens, one for each eye, that create the sense of three-dimensionality. Headphones pipe in stereophonic sound or even "3-D" sound. Three-dimensional sound makes you think you are hearing sounds not only near each ear but also in various places all around you.

- **Glove:** The glove has sensors that collect data about your hand movements.

- **Software:** Software gives the wearer of this special headgear and glove the interactive sensory experience that feels like an alternative to the realities of the physical world.

You may have seen virtual reality used in arcade-type games, such as Atlantis, a computer simulation of The Lost Continent. You may even have tried to tee off on a virtual golf range. There are also a few virtual-reality home videogames, such as the 7th Sense. However, there are far more important uses, one of them being in simulators for training.

Simulators are devices that represent the behavior of physical or abstract systems. Virtual-reality simulation technologies are applied a great deal in training. For instance, they have been used to create lifelike bus control panels and various scenarios such as icy road conditions to train bus drivers.[27]

Virtual reality

(Top) Man wearing interactive sensory headset and glove. When the man moves his head, the 3-D stereoscopic views change. *(Middle left)* What the man is looking at—a simulation of an office. *(Middle right)* When the man moves his glove, sensors collect data about his hand movements. The view then changes so that the man feels he is "moving" over to the bookshelf and "grasping" a book. *(Bottom)* Playing a virtual reality game.

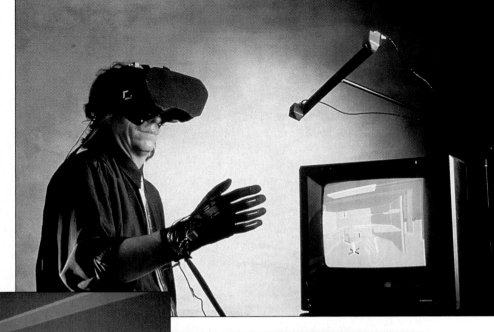

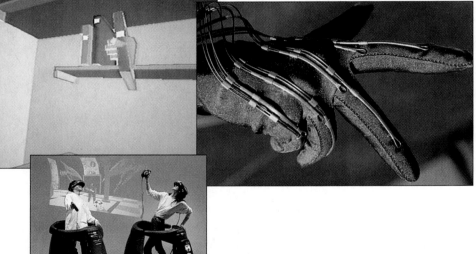

They are used to train pilots on various aircraft and to prepare air-traffic controllers for equipment failures.[28] They also help children who prefer hands-on learning to explore subjects such as chemistry.[29]

Of particular value are the uses of virtual reality in health and medicine. For instance, surgeons-in-training can rehearse their craft through simulation on "digital patients."[30] Virtual-reality therapy has been used for autistic children and in the treatment of phobias, such as extreme fear of public speaking or of being in public places or high places.[31,32] It has also been used to rally the spirits of quadriplegics and paraplegics by engaging them in plays and song-and-dance routines. (■ *See Panel 6.15.*) As one patient said, "When you spend a lot of time in bed, you can go crazy."[33]

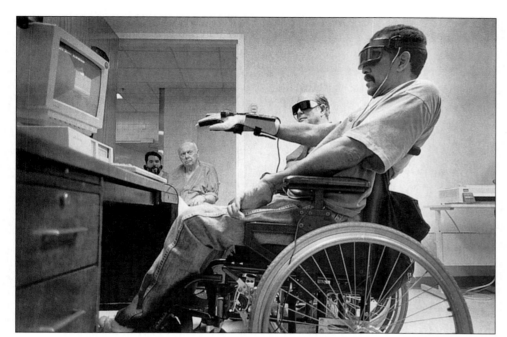

PANEL 6.15

Virtual-reality therapy

Miguel Rivera uses a glove and goggles with virtual-reality computer in the spinal-injury ward of the Veterans Affairs Medical Center, the Bronx, New York. The technology allows the disabled to "move" freely in an electronic world.

Interestingly, an ethical—and potential litigation—problem for makers of virtual-reality equipment is that of certain health side effects.[34] Known as *cybersickness* or *simulator sickness,* symptoms include eyestrain, queasiness, nausea and confusion, and even visual and audio "flashbacks" among some VR users. The disorder sometimes afflicts military pilots training on flight simulators, for example, who are then prohibited from flying. Not all users of virtual-reality equipment will experience all symptoms. Nevertheless, in preparation for the expected wave of VR products, researchers are taking a long look at what kinds of measures can be taken to head off these complaints.

Robots

The first Robot Olympics was held in Toronto in November 1991. "Robots competed for honors in 15 events—jumping, rolling, fighting, climbing, walking, racing against each other, and solving problems," reported writer John Malyon.[35] For instance, in the Micromouse race, robots had to negotiate a standardized maze in the shortest possible time.

We discuss robots—which might be considered complete computer systems with both input and output aspects—in detail in Chapter 12. Here, however, they are of interest to us because as output devices they output *motion* rather than information. They can perform computer-driven electromechanical functions that the other devices so far described cannot.

To get to definitions, **a *robot* is an automatic device that performs functions ordinarily ascribed to human beings or that operates with what appears to be almost human intelligence.** Actually, robots are of several kinds—industrial robots, perception systems, and mobile robots, for example, as we discuss in Chapter 12.

Forty years ago, in *Forbidden Planet*, Robby the Robot could sew, distill bourbon, and speak 187 languages.[36] We haven't caught up with science-fiction movies, but we may get there yet. ScrubMate—a robot equipped with computerized controls, ultrasonic "eyes," sensors, batteries, three different cleaning and scrubbing tools, and a self-squeezing mop—can clean bathrooms.[37] Rosie the HelpMate delivers special-order meals from the kitchen to nursing stations in hospitals.[38] Robodoc—notice how all these robots have names—is used in surgery to bore the thighbone so that a hip implant can be attached.[39] Remote Mobile Investigator 9 is used by Maryland police to flush out barricaded gunmen and negotiate with terrorists.[40] A driverless harvester, guided by satellite signals and artificial vision system, is used to harvest alfalfa and other crops.[41]

Robots are also used for more exotic purposes such as fighting oil-well fires, doing nuclear inspections and cleanups, and checking for mines and booby traps. An eight-legged, satellite-linked robot called Dante II was used to explore the inside of Mount Spurr, an active Alaskan volcano, sometimes without human guidance.[42,43] (■ *See Panel 6.16.*) The robot Lunar Prospector is being built for launching in 1997 to map the chemical composition of the Moon's surface from orbit.[44] TROV—for Telepresence Controlled Remotely Operated Vehicle—has been built and tested for searching for life on Mars, perhaps as early as 2003.[45]

Ⓔ Creative Output: The Manipulation of Truth in Art & Journalism

Preview & Review: Users of information technology must weigh the effects of the digital manipulation of sound, photos, and video in art and journalism.

California artist Joan Heemskerk is called "a Dr. Frankenstein of photography." Into her computer she scans images of technological paraphernalia—wheels, springs, gears, glass beakers, and other gadgets. She then "cuts and pastes," using computer software, and produces huge prints on four-foot sheets of photographic paper. The weird appliances she creates are described as "purposeless machines, part whimsy and part menace" that "loom like soldiers, or big benign robots."[46]

PANEL 6.16

Robots

(Left) Dante II explores Alaska's active Mount Spurr volcano; for much of the time, it was without human guidance. *(Right)* HelpMate robot serves meals at a Connecticut hospital.

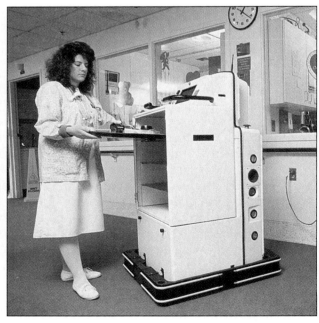

The ability to manipulate digitized output—images and sounds—has brought a wonderful new tool to art. However, it has created some big new problems in the area of credibility, especially for journalism. How can we now know that what we're seeing or hearing is the truth? Consider the following.

Manipulation of Sound

Frank Sinatra's 1994 album *Duets* paired him through technological tricks with singers like Barbra Streisand, Liza Minnelli, and Bono of U2. Sinatra recorded solos in a recording studio. His singing partners, while listening to his taped performance on earphones, dubbed in their own voices. This was done not only at different times but often, through distortion-free phone lines, from different places. The illusion in the final recording is that the two singers are standing shoulder to shoulder.

Newspaper columnist William Safire loves the way "digitally remastered" recordings recapture great singing he enjoyed in the past. However, he called *Duets* "a series of artistic frauds." Said Safire, "The question raised is this: When a performer's voice and image can not only be edited, echoed, refined, spliced, corrected, and enhanced—but can be transported and combined with others not physically present—what is performance? . . . Enough of additives, plasticity, virtual venality; give me organic entertainment."[47] Another critic said that to call the disk *Duets* seemed a misnomer. "Sonic collage would be a more truthful description."[48]

Some listeners feel that the technology changes the character of a performance for the better—that the sour notes and clinkers can be edited out. Others, however, think the practice of assembling bits and pieces in a studio drains the music of its essential flow and unity.

Whatever the problems of misrepresentation in art, however, they pale beside those in journalism. Could not a radio station edit a stream of digitized sound to achieve an entirely different effect from what actually happened?

Manipulation of Photos

When O. J. Simpson was arrested in 1994 on suspicion of murder, the two principal American newsmagazines both ran pictures of him on their covers.[49,50] *Newsweek* ran the mug shot unmodified, as taken by the Los Angeles Police Department. *Time*, however, had the shot redone with special effects as a "photo-illustration" by an artist working with a computer. Simpson's image was darkened so that it still looked like a photo but, some critics said, with a more sinister cast to it.

Should a magazine that reports the news be taking such artistic license? Should *National Geographic* in 1982 have moved two Egyptian pyramids closer together so that they would fit on a vertical cover? Was it even right for *TV Guide* in 1989 to run a cover showing Oprah Winfrey's head placed on Ann-Margret's body?[51] In another case, to show what can be done, a photographer digitally manipulated the famous 1945 photo showing the meeting of the leaders of the wartime Allied powers at Yalta. Joining Stalin, Churchill, and Roosevelt are some startling newcomers: Sylvester Stallone and Groucho Marx. The additions are done so seamlessly it is impossible to tell the photo has been altered. (■ *See Panel 6.17.*)

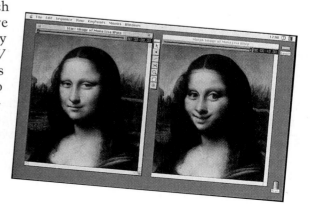

Photo manipulation

In this 1945 photo, World War II Allied leaders Joseph Stalin, Winston Churchill, and Franklin D. Roosevelt are shown from left to right. Digital manipulation has added Sylvester Stallone standing behind the bench and Groucho Marx seated at right.

The potential for abuse is clear. "For 150 years, the photographic image has been viewed as more persuasive than written accounts as a form of 'evidence,'" says one writer. "Now this authenticity is breaking down under the assault of technology."[52] Asks a former photo editor of the *New York Times Magazine*, "What would happen if the photograph appeared to be a straightforward recording of physical reality, but could no longer be relied upon to depict actual people and events?"[53]

Many editors try to distinguish between photos used for commercialism (advertising) versus for journalism, or for feature stories versus for news stories. However, this distinction implies that the integrity of photos applies only to some narrow definition of news. In the end, it can be argued, altered photographs pollute the credibility of all of journalism.

Manipulation of Video

The technique of morphing, used in still photos, takes a quantum jump when used in movies, videos, and television commercials. In *morphing*, a film or video image is displayed on a computer screen and altered pixel by pixel, or dot by dot. The result is that the image metamorphoses into something else—a pair of lips into the front of a Toyota, for example.

Morphing and other techniques of digital image manipulation have had a tremendous impact on filmmaking. Director and digital pioneer Robert Zemeckis (*Death Becomes Her*) compares the new technology to the advent of sound in Hollywood.[54] It can be used to erase jet contrails from the sky in a western and to make digital planes do impossible stunts. It can even be used to add and erase actors. In *Forrest Gump*, many scenes involved old film and TV footage that had been altered so that the Tom Hanks character was interacting with historical figures.

Films and videotapes are widely thought to accurately represent real scenes (as evidenced by the reaction to the amateur videotape of the Rodney King beating by police in Los Angeles). Thus, the possibility of digital alterations raises some real problems. One is the possibility of doctoring videotapes supposed to represent actual events. Another concern is for film archives: Because digital videotapes suffer no loss in resolution when copied, there are no "generations." Thus, it will be impossible for historians and archivists to tell whether the videotape they're viewing is the real thing or not.[55]

Information technology increasingly is blurring humans' ability to distinguish between natural and artificial experience, say Stanford University communications professors Byron Reeves and Clifford Nass.[56] For instance, they have found that showing a political candidate on a large screen (30 or 60 inches) makes a great difference in people's reactions. In fact, you will actually like him or her more than if you watch on a 13-inch screen. "We've found in the laboratory that big pictures automatically take more of a viewer's attention," said Reeves. "You will like someone more on the large screen and pay more attention to what he or she says but remember less." (This is why compelling TV or computer technology may not aid education.) Our visual perception system, they find, is unable to discount information— to say that "this is artificial"—just because it is symbolic rather than real.

If our minds have this inclination anyway, how can we be expected to exercise our critical faculties when the "reality" is not merely artificial but actively doctored?

The Future of Output

Preview & Review: On the horizon are better, cheaper, and larger display screens; higher-fidelity audio using wave-table synthesis and three-dimensional sound; and "real-time" video using digital wavelet theory. 3-D technology is bringing three-dimensionality to computer displays and, through VRML software, 3-D "virtual worlds" to users of the World Wide Web.

As you might guess, the output hardware we've described so far is becoming more and more sophisticated. Let's consider what's coming into view.

Display Screens: Better & Cheaper

Computer screens are becoming crisper, brighter, bigger, and cheaper. For instance, an LCD monitor from NEC that measures 13 inches from corner to corner now has as high a resolution as a conventional 20-inch monitor. The prices of active-matrix screens are becoming more affordable, and those for passive-matrix screens are lower still.[57]

New plasma technology is being employed to build flat-panel screens as large as 50 inches from corner to corner. Using a technique known as *microreplication*, researchers have constructed a thin transparent sheet of plastic prisms that allows builders of portable computer screens to halve the amount of battery power required.[58] Another technique, known as *field-emission display (FED)*, also promises to lower power requirements and to simplify manufacturing of computer screens, thereby lowering the cost.[59,60] All this is good news in an industry in which screens make up about 40% of a portable computer's cost.

Audio: Higher Fidelity

Generally, points out one technology writer, "the sound wafting from the PC has the crackly ring of grandpa's radio—the one he retired to the attic long ago."[61] Yet users' expectations for microcomputer sound are apt to zoom in 1997, when the DVD will be on the market. The DVD, or CD-size digital video disk, can hold an entire movie with terrific audio.

At present, sounds in a PC are merely approximate. The amount of data from real sound is too much for a microcomputer's microprocessor and conventional sound card to handle smoothly and quickly. The next generation of PCs, however, will use a technique known as *wave-table synthesis*, which is much more realistic than the older method. The wave-table is created, using actual notes played on musical instruments, to form a bank, from which digitally synthesized musical sounds can be drawn on and blended. "The quality of sound you can get from a wave-table is astonishing," says a Yamaha executive. "For a few hundred dollars' worth of add-ons, you can do things on your PC now that 10 years ago musicians in the best studios in the world could not have done."[62]

Sound quality will also be boosted by improvements in amplifiers on sound cards and in speakers. Finally, several companies (SRS Labs, QSound Labs, Spatializer Audio Laboratories) are developing so-called *3-D sound*, using just two speakers to give the illusion of three-dimensional sound. Unlike conventional stereo sound, "3-D audio describes an expanded field of sound—a broad arc starting at the left ear and curving around to the right," one writer explains. "In a video game, for example, that means the player could hear an enemy jet fighter approach, take a hit, and explode somewhere off his or her left shoulder."[63] The effect is achieved by boosting certain frequencies that provide clues to a sound's location in the room or by varying the timing of sounds from different speakers.

Video: Movie Quality for PCs

Today the movement of most video images displayed on a microcomputer comes across as fuzzy and jerky, with a person's lip movements out of sync with his or her voice. This is because currently available equipment is capable of running only about eight frames a second.

New technology based on *digital wavelet theory*, a complicated mathematical theory, has led to software that can compress digitized pictures into fewer bytes and do it more quickly than current standards. Indeed, the technology can display 30–38 frames a second—"real-time video," giving images the look and feel of a movie.[64] Although this advance has more to do with software than hardware, it will clearly affect future video output. For example, it will allow two people to easily converse over microcomputer connections in a way that cannot be accomplished easily now. (We discuss compression in more detail in Chapter 7.)

Three-Dimensional Display

In the 1930s, radiologists tried to create three-dimensional images by holding up two slightly offset X-rays of the same object and crossing their eyes.[65] Now the same effects can be achieved by computers. With 3-D technology, flat, cartoon-like images give way to rounded objects with shadows and textures. Artists can even add "radiosity," so that a dog standing next to a red car, for instance, will pick up a red glow.

During the past several years, many people have come to experience 3-D images not only through random-dot stereograms like the Magic Eye books but also through videogames using virtual-reality head-mounted displays.[66,67] Now companies are moving beyond VR goggles to design 3-D displays that produce two different eye views without glasses. In one design, for instance, the display remotely senses the viewer's head movements and moves lenses to change the scene presented to each eye.

In early 1994 the Japanese manufacturer Sanyo demonstrated two types of 3-D systems on television screens. One was an experimental 3-D HDTV system that projected a 120-inch picture on a theater-type screen and required the use of special glasses with polarized lenses. The other, which involved neither HDTV nor special glasses, used what appeared to be a normal 40-inch TV set, with a screen that employed hundreds of tiny prisms/lenses. "The 3-D effect was stunning," wrote one reporter who saw it. "In one scene, water was sprayed from a hose directly at the camera. When watching the replay, I had to control the urge to jump aside."[68]

Virtual Worlds: 3-D in Cyberspace

Virtual reality, as we explained, involves head-mounted displays and data gloves. By contrast, a *virtual world* requires only a microcomputer, mouse, and special software to display and navigate three-dimensional scenes.[69] The 3-D scenes are presented via interlinked, or networked, computers on the World Wide Web part of the Internet. For example, users can meet in three-dimensional fantasy landscapes (using on-screen stand-ins for themselves called *avatars*) and move around while "talking" through the keyboard with others.

Virtual worlds are a feat mostly of software rather than output hardware, although the subject deserves discussion here because of the possible transformations for output this kind of three-dimensionality promises. Internet sites are rendered in 3-D thanks to software known as *VRML*, for *virtual reality markup language* (discussed in Chapter 11). To view a site in 3-D, you need to download a piece of software called a *VRML browser*, such as WebFX or WebSpace. This enables you to get into 3-D environments such as "Habitat," "AlphaWorld," and "WorldsAway."[70-72] In WorldsAway, for example, instead of having a text-based "chat room" or discussion group, you appear "live" as your cartoon character (avatar) in the three-dimensional space. "When you talk, a little balloon appears over your head on the computer screen, containing your words," explains one report. "When you wave, your image waves. . . . Maybe you will go for a walk in the community's virtual woods, with whomever you've befriended online."[73]

Onward

The advances in digitization and refinements in output devices are producing more and more materials in *polymedia* form. That is, someone's intellectual or creative work, whether words, pictures, or animation, may appear in more than one form or medium. For instance, you could be reading this chapter printed in a traditional bound book. Or you might be reading it in a "course pack," printed on paper through some sort of electronic delivery system. Or it might appear on a computer display screen. Or it could be in multimedia form, adding sounds, pictures, and video to text. Thus, information technology changes the nature of how ideas are communicated.

If materials can be input and output in new ways, where will they be stored? That is the subject of the next chapter.

Telling Computer Magazines Apart

Visit a newsstand in an urban or suburban area—even one in a supermarket—and you'll quickly discover there are several popular computer magazines. Visit a computer store or bookseller such as B. Dalton, and you'll be overwhelmed by the number of such periodicals available. Here we try to explain which publications may be most practical for you.[74-78]

What You'll Find in Computer Magazines

Nobody *needs* computer magazines. However, once you find a couple or three you're comfortable with, they can help you keep up on trends, determine what to buy and not buy, and learn things to make your computing tasks easier. Specifically, computer magazines offer the following:

- *First looks and reviews:* Many periodicals review the latest in hardware and software. They thus serve the same purpose as reviewers of books and movies.

  However, Russ Walter, author of the annually revised *The Secret Guide to Computing,* advises "don't take the reviews too seriously: the typical review is written by just one person and reflects just that individual's opinion." Some reviewers are too appreciative, some too critical —about the same product.

- *News and rumors:* Computer magazines report news and gossip about software and hardware companies, the industry buzz on forthcoming products, and what certain computer celebrities are supposed to be up to. They also have articles on what users are doing with their technology. *San Jose Mercury News* computer columnist Phillip Robinson believes that the magazines need lots more of this user coverage, "but it doesn't thrill their advertisers as much as industry news, because it is more likely to include negatives."

- *Features and columns:* Feature articles deal with how-to topics or think pieces, which are not considered "hard news." Examples are how antivirus utilities work, how main memory can be expanded, what the future of multimedia will be. Columns reflect the opinions of the writers and carry disagreements, predictions, generalizations, or thoughts about trends.

- *Ads:* Like most magazines, computer periodicals are principally supported by advertisers. Ads can show you the offerings available, yet the advertisers can have a way of indirectly influencing a magazine's coverage. (A magazine devoted to Macintosh, for example, might not be very aggressive in covering flaws in these machines, for fear that Apple might yank its advertising.)

Categories of Magazines

It's worth noting that many daily newspapers print reviews, columns, and feature articles concerning computing and communications, usually in the business section. The *New York Times,* for instance, regularly prints articles by its computer columnist Peter Lewis, and the *Wall Street Journal* does the same with a column by Walter Mossberg. In addition, the magazines *Fortune* and *Business Week* provide good news coverage on trends in computing and communications. (Perhaps once a year, the last three publications also publish special issues or sections on key aspects of information technology.) Popular magazines on scientific subjects, such as *Technology Review* and *Popular Science,* also publish feature articles on trends in the field.

As for computer magazines, some are published for novices, some are aimed at professionals who buy lots of hardware and software, and some are for users of particular types of microcomputer systems (only Macintoshes or only IBM-compatible computers).

We will describe popular computer publications according to the following categories:

- Magazines for novices
- Somewhat technical magazines and newspapers
- Magazines for users of IBM PCs and compatibles
- Magazines for users of Macintoshes
- Magazines for users with special interests, such as portable computing, online communications, multimedia, and desktop publishing

Also note our list of computer periodicals, with contact phone numbers and subscription rates. (■ *See Panel 6.18.*)

Magazines for Novices A number of magazines designed for nontechnical users have appeared in recent years, and they have expanded greatly as more computers have moved into the home, constituting a growing market for advertisers. The publications include the following:

- *PC Novice:* Beneath the title appears the subtitle "personal computers in plain English." For anyone who knows nothing about microcomputers—particularly IBM and IBM-compatibles—this magazine makes a good beginning because *everything* is explained. Moreover, at 100 or so pages, *PC Novice* is far easier to get into than other computer magazines, which may run to 400 or even 800 pages (most of them ads). The magazine also runs a glossary in its back pages.

PANEL 6.18   Periodicals covering computers and communications

Periodical and Publisher	Single-Issue Price	Yearly Subscription Price (and Any Discounted Price)	Telephone
BBS (Callers Digest)	$3.50	$30	800-822-0437
Byte (McGraw-Hill)	$3.95	$30 ($25)	800-257-9402
CD-ROM Today (GP)	$7.95	$49.95	415-696-1661
Computer Currents (IDG)	$3.00	$20 (free)	508-820-8118
ComputerLife (Ziff)	$2.95	$35.40 ($19.97)	303-665-8930
Computer Shopper (Ziff)	$3.95	$30 ($22)	800-274-6384
Computerworld (IDG)	$6.00	$48 ($40)	800-669-1002
Electronic Entertainment (Infotainment World)	$3.95	$20	800-770-3248
FamilyPC (Ziff and Disney)	$2.95	$9.95	800-413-9749
Home Office Computing (Scholastic)	$2.95	$19.97 ($16.97)	800-288-7812
HomePC (CMP)	$2.95	$21.97	800-829-0119
Information Week (CMP)	$2.95	$63.95 ($47.95)	516-562-5000
Infoworld (IDG)	$3.95	$130 (free*)	415-572-7341
Internet World (Mecklermedia)	$4.95	$29	800-573-3062
MacHome Journal (MacHome Journal)	$2.95	$19.95	800-800-6542
MacUser (Ziff)	$2.95	$27 ($20)	800-627-2247
MacWeek (Ziff)	$6.00	$125 (free*)	415-243-3500
Macworld (IDG)	$3.95	$30 ($24)	800-524-3200
Net Guide	$2.95	$35 ($23)	800-829-0421
New Media (Hypermedia)	$3.95	$38	415-573-5170
Online Access (Chicago Fine Print)	$4.95	$29.95	800-36-MODEM
PC Computing (Ziff)	$2.95	$25 ($17)	800-365-2770
PC Magazine (Ziff)	$3.95	$50 ($35)	800-289-0429
PC Novice (Peed)	$2.95	$24	800-424-7900
PC Today (Peed)	$2.95	$24	800-424-7900
PC Week (Ziff)	$6.00	$195 (free*)	800-451-1032
PC World (IDG)	$3.95	$30 ($20)	800-825-7595
Publish (IDG)	$4.95	$39.90 ($29.95)	800-656-7495
Windows Magazine (CMP)	$2.95	$25 ($17)	516-562-5948
Windows Sources (Ziff)	$2.95	$28 ($20)	800-364-3414
Wired (Wired Ventures)	$4.95	$40	800-SO-WIRED

*Controlled circulation; free to people who qualify.

- *PC Today:* Aimed at "computing for small business," this sister publication of *PC Novice,* and a notch above it in difficulty, is designed for the small businessperson who is getting going in computing.

- *ComputerLife:* Published by Ziff-Davis (one of the two biggest computer magazine conglomerates, the other being IDG), this monthly is "aimed at people who really, really like their computers but regard themselves as non-technical," according to *Wall Street Journal* computer columnist Walter Mossberg. Says Phillip Robinson, *ComputerLife* "is aimed at computing fun for people already tuned in to CPUs, modems, and the like."

- *HomePC:* Aimed at every kind of home (that is, with and without children), this monthly publication has regular sections, ranging from do-it-yourself to education to kids' reviews of children's software.

- *MacHome Journal:* Subtitled "for work, play, and education," this is the only home-oriented magazine for Apple Macintosh owners. Running around 130 pages, it offers reviews of Macintosh products (even citing *other* magazine reviews alongside their own) as well as news of the latest trends.

- *FamilyPC:* Published by Ziff-Davis and Walt Disney, *FamilyPC* is a visually inviting magazine that is designed to help parents figure out how to help their children get the most out of computers. Instead of test labs, the magazine uses panels of kids and parents to evaluate products.

In addition, from time to time what the magazine industry calls "one-shots" appear on newsstands and are useful for beginners. These are irregularly appearing publications such as *Computer Buyer's Guide and Handbook,* with variations such as *Laptop Buyer's Guide* and *Printer Buyer's Guide.*

Somewhat Technical Magazines & Newspapers

The next level of publication requires that readers have some knowledge of computers and communications—at least as much as will have been achieved by those who have read this far in the book. We are not, however, including the kinds of scholarly publications available from professional computing and engineering societies.

- *Monthly magazines:* A recent publishing success story is *Wired,* which may be considered the *Rolling Stone* of computer/communications magazines. Not terribly technical, it features wild graphics and layouts and takes an irreverent view of life. "Since each issue of *Wired* is wild, slick, and expensive," writes Russ Walter, "it's read by the hip rich, so it includes ads for upscale consumer goods such as Jetta cars and Absolut Vodka."

  At the other end of the scale, described as "the *Scientific American* of computing" by Robinson, is *Byte,* the oldest popular computer magazine. This is probably the best source for finding in-depth explanations of the latest technologies, such as uses of the radio-frequency spectrum or the concept of plug and play.

  Computer Shopper, a monthly monster of 800-plus pages, is billed as "the computer magazine for direct buyers," and thus it assumes a certain amount of technical knowledge (mainly of IBMs and compatibles). Because so much of it is advertising from direct sellers of computers, it has the feel of a phonebook-size catalog. It does offer articles of interest to people who follow computers closely, such as "Special Report: CPU Technology." However, Walter faults such articles for being "relentlessly upbeat; they never criticize."

- *Free weekly newspapers:* A weekly newspaper published by IDG is *Computer Currents,* which is distributed free at certain newsstands and in news racks in six regions: Atlanta, Boston, Dallas, Houston, Los Angeles, and San Francisco. Other free newspapers are available in selected cities. If you can't get one free, you can subscribe to them.

  These newspapers, which have a microcomputer orientation, offer much the same coverage as is found in magazines and newspapers you would pay for. However, they have the advantage of offering news and listings for local events, as well as carrying local advertising.

- *Subscription weeklies:* The oldest weekly newspaper covering computers of all types—micros, minis, mainframes, and supers—is IDG's *Computerworld,* which began publishing in 1967. *Computerworld,* which may be found in many college libraries, is intended for computer professionals.

  Perhaps the principal competition to *Computerworld* is CMP's *Information Week,* a slick four-color magazine that is billed as being "for business and technology managers." It can be found on newsstands such as those in airports along with other business magazines.

- *Controlled-circulation weeklies:* Three weekly publications that are more apt to be found in a university library than on a newsstand are *Infoworld, PC Week,* and *MacWeek.* The three publications have good reviews, but they are principally of interest to people who need to follow trends closely and know where information technology is headed.

  All three weeklies are "controlled circulation." This means that you can get them free if you complete an application form successfully enough to assure the publishers that you buy sufficient hardware or software to be of interest to advertisers. Otherwise you have to pay $125 to $195 for a yearly subscription.

Magazines for Users of IBM PCs & Compatibles

Because IBM PCs and compatibles control 85% of the microcomputer market and Windows is the most popular recent operating system, there are, not surprisingly, several magazines exclusively devoted to these areas.

- *PC Magazine:* Appearing every 2 weeks, *PC Magazine* (from Ziff-Davis) is considered the most respected and most important magazine for PC users, although it engages in rather lengthy technical discussions. It has the product reviews that count the most, most of the well-known computer columnists, and is comprehensive and carefully edited. It is well known for its blockbuster issues, such as the annual November issue that compares all new printers. Walter, however, complains that its "editor's choice" for new buys tends to assume that readers are well-heeled and can easily afford to spend $4000 on a new system.

- *PC World:* Formed as the IDG competitor to *PC Magazine, PC World* has its own stable of columnists and its own news and reviews. Robinson finds it less skeptical of new products than its Ziff competitor, but Walter praises it for playing consumer advocate and for publishing complaints about rip-offs and bad service.

- *PC Computing:* Published by Ziff, *PC Computing* is less technical and more irreverent than the above two competitors and has livelier writing, as well as good tutorials. Unless you feel you need encyclopedic knowledge of new products, Robinson thinks users of IBM PCs and compatibles could get by on just this magazine alone.

Windows users may also find two special-interest magazines useful—*Windows Magazine* and *Windows Sources*—both of which offer the usual mix of news, gossip, and product reviews.

Magazines for Users of Macintoshes

Two magazines are available for Apple Macintosh users—one from each of the two publishing conglomerates.

- *MacUser:* This monthly from Ziff-Davis is even more irreverent and humorous than *PC Computing.* In Robinson's view, it is not skeptical enough, but it offers some of the best comparisons in the business, doing a good job of explaining technology and analyzing products in detail.

- *MacWorld:* IDG's monthly is the better skeptic, not being quite so gee-whiz about new products and technology. It also reviews more things every month than its competitor.

Magazines for Users with Special Interests

New magazines seem to appear every month, as new technology gains ascendancy. We have not the space to cover them all, but here are some areas of more general interest:

- *Entertainment, multimedia, and CD-ROM: Electronic Entertainment,* billed as "the entertainment resource for the interactive age," is mostly about multimedia, CD-ROM, and games for Macs, PCs, Nintendo, Sega, 3DO, and their competitors. *New Media* is about multimedia computing and is devoted to digitized art, sounds, games, and news about companies and individuals in this field. *CD-ROM Today,* which calls itself "the leading guide to PC and Mac multimedia," includes a CD-ROM disk, as well as the usual how-to articles and news and reviews of new CD-ROM programs.

- *Communications and portable computing:* Some online services, such as CompuServe, publish their own magazines, which members receive when they join. In addition, there are magazines available on newsstands and elsewhere devoted to the fast-growing subject of telecommunications and its applications.

  Online Access calls itself "the magazine that makes modems work," and contains articles on tips for using online services, browsing the Internet, and finding jobs online. *Internet World,* "the magazine for Internet users," covers Internet subjects for readers ranging from entry-level to expert. *BBS,* "the bulletin board systems magazine," describes various bulletin boards and lists connections. *Home Office Computing* offers news and features on computers and communications for the use of those working at home.

- *Desktop publishing: Publish,* "the magazine for electronic publishing professionals," is an IDG monthly publication that covers the technical nuances (color, scanners, and the like) for both PC and Macintosh users of desktop publishing programs.

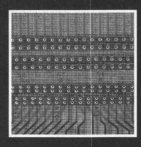

SUMMARY

What It Is / What It Does	**Why It's Important**
active-matrix display *(p. 250, LO 2)* Type of flat-panel display in which each pixel on the screen is controlled by its own transistor.	Active-matrix screens are much brighter and sharper than passive-matrix screens, but they are more complicated and thus more expensive.
audio-output device *(p. 262, LO 5)* Device that outputs voice or voice-like sounds (voice-output technology) or that outputs music and other sounds (sound-output technology).	Audio-output devices are important in speech synthesis and in multimedia computing.
bitmap *(p. 251, LO 2)* In computer graphics, an area in memory that represents an image. Depending on the screen, 1 or several bits represent 1 pixel or several pixels of the image.	Bitmapped display screens permit the computer to manipulate pixels on the screen individually rather than as blocks (character map), enabling software to create a greater variety of images. However, bitmaps take up a lot of storage space.
bubblejet printer *(p. 257, LO 4)* Nonimpact printer that uses miniature heating elements to force specially formulated inks through print heads with 128 tiny nozzles. The multiple nozzles print fine images at high speeds.	Bubblejet technology is commonly used in portable printers.
cathode-ray tube (CRT) *(p. 249, LO 2)* Vacuum tube used as a display screen in a computer or video display terminal. Images are represented on the screen by individual dots or "picture elements" called *pixels*.	This technology is found not only in the screens of desktop computers but also in television sets and flight-information monitors in airports.
character-map *(p. 251, LO 2)* Fixed location on a video display screen where a predetermined character can be placed. Character-mapped display screens display only text—letters, numbers, and special characters (as opposed to bitmapped display screens).	Character-mapped display screens cannot display graphics unless a video adapter card is installed.
color display screen *(p. 251, LO 2)* Display screens that can display between 16 and 16.7 million colors, depending on their type (see EGA, VGA, SVGA, XGA).	Most software today is developed for color, and—except for some pocket PCs—most microcomputers today are sold with color display screens.
computer output microfilm/fiche (COM) *(p. 261, LO 4)* Computer output produced as tiny images on rolls/sheets of microfilm.	COM, which is fast and inexpensive, can store a lot of data in a small amount of space.
display screen *(p. 249, LO 2)* Also variously called *monitor*, *CRT*, or simply *screen*; softcopy output device that shows programming instructions and data as they are being input and information after it is processed. Sometimes a display screen is also referred to as a VDT, for video display terminal, although technically a VDT includes both screen and keyboard. The size of a screen is measured diagonally form corner to corner in inches, just like television screens.	Display screens enable users to immediately view the results of input and processing.
dot-matrix printer *(p. 255, LO 4)* Printer that contains a print head of small pins that strike an inked ribbon, forming characters or images. Print heads are available with 9, 18, or 24 pins, with the 24-pin head offering the best quality.	Dot-matrix printers can print draft quality, a coarser-looking 72 dots per inch vertically, or near-letter-quality (NLQ), a crisper-looking 144 dots per inch vertically. They can also print graphics.
dot pitch *(p. 251, LO 2)* Amount of space between pixels (dots); the closer the dots, the crisper the image.	Dot pitch is one of the measures of display screen capability.

What It Is / What It Does	**Why It's Important**

electrostatic plotter *(p. 260, LO 4)* Instead of using pens, an electrostatic plotter uses electrostatic charges to create tiny dots on specially treated paper. The paper is then run through a developer to produce the image, which may be four-color.

Electrostatic plotters operate much like photocopiers, producing crisp images. Some can handle paper up to 6 feet wide.

Extended Graphics Array (XGA) *(p. 252, LO 3)* Graphics board display standard, also referred to as *high resolution;* supports up to 16.7 million colors at a resolution of 1024 × 768 pixels. Depending on the video display adapter memory chip, XGA will support 256, 65,536, or 16,777,216 colors. For any of these displays to work, video display adapters and monitors must be compatible. The computer's software and the video display adapter must also be compatible.

Extended Graphics Array offers the most sophisticated standard for color and resolution.

flat-panel display *(p. 250, LO 2)* Refers to display screens that are much thinner, weigh less, and consume less power than CRTs. Flat-panel displays are made up of two plates of glass with a substance between them that is activated in different ways. Two common types of technology are used in flat-panel display screens: liquid-crystal display and gas-plasma display. Flat-panel screens are either active-matrix or passive-matrix displays. Images are represented on the screen by individual dots, or picture elements called *pixels.*

Flat-panel displays are used in portable computers.

font *(p. 256, LO 4)* Set of type characters in a particular type style and size.

Desktop publishing programs, along with laser printers, have enabled users to dress up their printed projects with many different fonts.

gas-plasma display *(p. 250, LO 2)* Type of flat-panel display in which the display uses a gas that emits light in the presence of an electric current, like a neon bulb. The technology uses predominantly neon gas and electrodes above and below the gas. When electric current passes between the electrodes, the gas glows.

Gas plasma displays offer better resolution than LCD displays, but they are more expensive.

hardcopy *(p. 246, LO 1)* Refers to printed output (as opposed to softcopy). The principal examples are printouts, whether text or graphics, from printers. Film, including microfilm and microfiche, is also considered hardcopy output.

Hardcopy is convenient for people to use and distribute; it can be easily handled or stored.

impact printer *(p. 255, LO 4)* Type of printer that forms characters or images by striking a mechanism such as a print hammer or wheel against an inked ribbon, leaving an image on paper.

For microcomputer users, the most common impact printers are daisywheel printers and dot-matrix printers.

inkjet printer *(p. 257, LO 4)* Nonimpact printer that forms images with little dots. Inkjet printers spray small, electrically charged droplets of ink from four nozzles through holes in a matrix at high speed onto paper.

Because they produce high-quality images, they are often used by people in graphic design and desktop publishing. However, inkjet printers are slower than laser printers.

laser printer *(p. 256, LO 4)* Nonimpact printer similar to a photocopying machine; images are created on a drum, treated with a magnetically charged ink-like toner (powder), and then transferred from drum to paper.

Laser printers produce much better image quality than dot-matrix printers do and can print in many more colors; they are also quieter. Laser printers, along with page description languages, enabled the development of desktop publishing.

liquid-crystal display (LCD) *(p. 250, LO 2)* Flat-panel display that consists of a substance called *liquid crystal,* the molecules of which line up in a way that alters their optical properties. As a result, light—usually backlighting behind the screen—is blocked or allowed through to create an image.

LCD is useful not only for portable computers but also as a display for various electronic devices, such as watches and radios.

monochrome display screen *(p. 251, LO 2)* Refers to "single color"; a monochrome computer screen displays a single-color image on a contrasting background—usually black on white, amber on black, or green on black.

Monochrome display is suitable for nongraphics applications such as word processing or spreadsheets.

What It Is / What It Does	Why It's Important

multifunction machine *(p. 260, LO 4)* Single hardware device that combines several capabilities, such as printing, scanning, copying, and faxing.

A multifunction machine can do the work of several separate office machines at a price below the combined cost of buying these devices separately.

nonimpact printer *(p. 256, LO 4)* Printer that forms characters and images without making direct physical contact between printing mechanism and paper. Two types of nonimpact printers often used with microcomputers are laser printers and inkjet printers. A third kind, the thermal printer, is seen less frequently.

Nonimpact printers are faster and quieter than impact printers because they have fewer moving parts. They can print text, graphics, and color, but they cannot be used to print on multipage forms.

page description language *(p. 256, LO 4)* Software used in desktop publishing that describes the shape and position of characters and graphics to the printer.

Page description languages, used along with laser printers, gave birth to desktop publishing. They allow users to combine different types of graphics with text in different fonts, all on the same page.

passive-matrix display *(p. 250, LO 2)* Type of flat-panel display in which each transistor controls a whole row or column of pixels.

Although passive-matrix displays are less bright and less sharp than active-matrix displays, they are less expensive and use less power.

pen plotter *(p. 260, LO 4)* Plotter designed so that paper lies flat on a table-like surface (flatbed plotter) or is mounted on a drum (drum plotter). Between one and four pens move across the paper, or the paper moves beneath the pens.

This is the most popular type of plotter and can produce color.

pixel *(p. 250, LO 2)* Short for *picture element;* smallest unit on the screen that can be turned on and off or made different shades. A stream of bits defining the image is sent from the computer (from the CPU) to the CRT's electron gun, where the bits are converted to electrons.

Pixels are the building blocks that allow graphical images to be presented on a display screen.

plotter *(p. 260, LO 4)* Specialized hardcopy output device designed to produce high-quality graphics in a variety of colors. The three principal kinds of plotters are pen, electrostatic, and thermal.

Plotters are especially useful for creating maps and architectural drawings, although they may also produce less complicated charts and graphs.

PostScript *(p. 256, LO 4)* Printer language, or page description language, that has become a standard for printing graphics on laser printers.

PostScript printers are essential for users who need to print a lot of graphics or heavily designed pages or who want to generate different fonts in various sizes.

printer *(p. 253, LO 4)* Output device that prints characters, symbols, and perhaps graphics on paper. Printers are categorized according to whether the image produced is formed by physical contact of the print mechanism with the paper. Impact printers have contact; nonimpact printers do not.

Printers provide one of the principal forms of computer output.

Printer Control Language (PCL) *(p. 256, LO 4)* Page description language that has resolutions and speeds similar to those of PostScript.

See PostScript.

refresh rate *(p. 251, LO 2)* Number of times per second that screen pixels are recharged so that their glow remains bright.

In dual-scan screens, the tops and bottoms of the screens are refreshed independently at twice the rate of single-scan screens, producing more clarity and richer colors.

resolution *(p. 251, LO 2)* Clarity or sharpness of a display screen; the more pixels there are per square inch, the better the resolution. Resolution is expressed in terms of the formula horizontal pixels $\times$ vertical pixels. A screen with 640×480 pixels multiplied together equals 307,200 pixels. This screen will be less clear and sharp than a screen with 800×600 (equals 480,000) or 1024×768 (equals 786,432) pixels.

Users need to know what screen resolution is appropriate for their purposes.

robot *(p. 267, LO 5)* Automatic device that performs functions ordinarily ascribed to human beings or that operate with what appears to be almost human intelligence.

Robots are of several kinds—industrial robots, perception systems, and mobile robots. They are performing more and more functions in business and the professions.

What It Is / What It Does	Why It's Important

softcopy *(p. 246, LO 1)* Refers to data that is shown on a display screen or is in audio or voice form. This kind of output is not tangible; it cannot be touched. Virtual reality and robots might also be considered softcopy devices.

This term is used to distinguish nonprinted output from printed output.

sound-output device *(p. 263, LO 5)* Audio-output device that produces digitized, nonverbal sounds, ranging from beeps and chirps to music. It includes software and a sound card or digital audio circuit board.

PC owners can customize their machines to greet each new program with particular sounds. Sound output is also used in multimedia presentations.

speech coding *(p. 263, LO 5)* Voice-output technology that uses actual human voices speaking words to provide a digital database of words that can be output as voice sounds. That is, words are codified and stored in digital form. Later they may be retrieved and translated into voices as needed.

The drawback of this method is that the output is limited to whatever words were previously entered into the computer system. However, the voice-output message does sound more convincingly like real human speech than other sound-output devices.

speech synthesis *(p. 263, LO 5)* Voice-output technology that uses a set of 40 basic speech sounds (called *phonemes,* the bases of all speech in English) to create words electronically.

No human voices are used to make up a database of words; instead, the computer converts stored text into voices. Such voice messages are usually understandable, though they don't sound exactly human.

Super Video Graphics Array (SVGA) *(p. 252, LO 3)* Graphics board standard that supports 256 colors at higher resolution than VGA. SVGA has two graphics modes: 800 × 600 pixels and 1024 × 768.

Super VGA is a higher-resolution version of Video Graphics Array (VGA), introduced in 1987.

thermal plotter *(p. 260, LO 4)* Plotter that uses electrically heated pins and heat-sensitive paper to create images.

Thermal plotters are capable of producing only two colors.

thermal printer *(p. 257, LO 4)* Nonimpact printer that uses colored waxes and heat to produce images by burning dots onto special paper.

The colored wax sheets are not required for black-and-white output because the thermal print head will register the dots on the paper.

video display adapter *(p. 252, LO 3)* Also called a *graphics adapter card;* circuit board that determines the resolution, number of colors, and how fast images appear on the display screen.

Video display adapters determine how fast the card processes images and how many colors it can display.

video display terminal (VDT) *(p. 249, LO 2)* Computer keyboard and display screen.

Video display terminals are the principal input/output devices for accessing large computer systems such as mainframes.

Video Graphics Array (VGA) *(p. 252, LO 3)* Graphics board standard that supports 16 to 256 colors, depending on resolution. At 320 × 200 pixels it will support 256 colors; at the sharper resolution of 640 × 480 pixels it will support 16 colors.

VGA is the most common video standard used today.

videoconferencing *(p. 264, LO 5)* A method of communicating whereby people in different geographical locations can have a meeting—and see and hear one another—using computers and communications technologies.

Videoconferencing technology enables people to conduct business meetings without having to travel.

virtual reality *(p. 265, LO 5)* Computer-generated artificial reality that projects user into sensation of three-dimensional space. Interactive sensory equipment consists of headgear, headphones, glove, and software. The headgear has small video display screens, one for each eye, to create a three-dimensional sense. The headphones pipe in stereo or 3-D sound. The glove has sensors that collect data about hand movements. The software gives the wearer the interactive sensory experience.

Virtual reality is used most in entertainment, as in arcade-type games, but has applications in architectural design and training simulators.

voice-output device *(p. 263, LO 5)* Audio-output device that converts digital data into speech-like sounds. Two types of voice-output technology exist: speech coding and speech synthesis.

Voice-output devices are a common technology, found in telephone systems, soft-drink machines, and toys and games.

(Selected answers appear at the back of the book.)

Short-Answer Questions

1. What is the difference between hardcopy and soft-copy?

2. What are the two types of display screen?

3. What is the difference between character-mapped displays and bitmapped displays?

4. What do VGA, SVGA, and XGA refer to, and how are they different?

5. What is a multifunction machine?

6. How are active-matrix and passive matrix displays different?

7. What advantages does a laser printer have over other types of printers?

8. How does a display screen's refresh rate relate to screen clarity?

9. What is the difference between a transportable printer and an ultraportable printer?

10. What is a page description language?

Fill-in-the-Blank Questions

1. The more pixels that can be displayed on the screen, the better the _____ of the image.

2. _____ is a computer-generated, 3-D environment that users can enter through the use of special hardware and software.

3. Output is available in two principal forms: _____ and _____.

4. PostScript and PCL are _____; they describe the shape and position of letters and graphics to laser printers.

5. _____ can produce high-quality graphics and are used most often for outputting maps and architectural drawings.

6. Whether for CRT or flat-panel display, screen clarity depends on the following three qualities:
   a. _____
   b. _____
   c. _____

7. _____ is a method whereby people in different geographical locations can have a meeting.

8. A(n) _____ machine incorporates printing, scanning, copying, and faxing capabilities into one device.

9. Voice-output devices convert _____ data into speech-like sounds.

10. A _____ is a computerized device that can operate with almost human intelligence.

Multiple-Choice Questions

1. To display graphics, a display screen must have a(n):
   a. CRT
   b. vacuum tube
   c. video display adapter
   d. plotter
   e. none of the above

2. For microcomputer users, the most common type of impact printer is the:
   a. dot-matrix printer
   b. line printer
   c. chain printer
   d. band printer
   e. all of the above

3. Which of the following printers uses a combination of color waxes and heat to form images?
   a. laser printer
   b. dot-matrix printer
   c. inkjet printer
   d. thermal printer
   e. none of the above

4. Which of the following is a softcopy output device?
   a. display screen
   b. plotter
   c. multifunction device
   d. robot
   e. none of the above

5. Which of the following uses a vacuum tube?
   a. liquid-crystal display
   b. flat-panel display
   c. gas-plasma display
   d. CRT
   e. all of the above

6. Most microcomputers today are sold with a(n):
   a. inkjet printer
   b. color display screen
   c. impact printer
   d. page description language
   e. all of the above

7. Which of the following determines the number of colors that can appear on a display screen?
   a. liquid crystal
   b. bitmapped capability
   c. video display adapter
   d. gas plasma
   e. none of the above

8. Advancements in video technology can be seen in the following area:
   a. videoconferencing
   b. HDTV
   c. ATV
   d. SDTV
   e. all of the above

9. Virtual-reality equipment may induce certain health side effects, called:
   a. cybersickness
   b. VR sickness
   c. 3-D disorientation
   d. sensory sickness
   e. none of the above

10. Which of the following would you use if you need to generate multipart forms?
    a. laser printer
    b. dot-matrix printer
    c. inkjet printer
    d. thermal printer
    e. none of the above

True/False Questions

T F   1. A picture element on the screen is called a *pixel.*

T F   2. *Display screen, CRT,* and *monitor* are different names for the same thing.

T F   3. CRTs are used on portable computers.

T F   4. Screen resolution is measured by vertical and horizontal lines of pixels.

T F   5. Industry observers expect that display screens will get more expensive in the future.

T F   6. Hardcopy sometimes refers to information displaying on a display screen.

T F   7. Laser printers, inkjet printers, and thermal printers are all nonimpact printers that can output color graphics.

T F   8. You need the following special equipment to experience virtual reality: headgear, gloves, shoes, pants, and software.

T F   9. HDTV is a television system that features enhanced video and very clear pictures.

T F   10. Audio-ouput devices can output only music.

Projects/Critical-Thinking Questions

1. Visit a local computer store to compare the output quality of the different printers on display. Then obtain output samples and a brochure on each printer sold. After comparing output quality and price, what printer would you recommend to a friend who needs a printer that can output resumes, research reports, and professional-looking correspondence with a logo? Why?

2. What hardware and software would you need to add to your PC in order for it to support videoconferencing? What companies sell videoconferencing systems? Who sells the best videoconferencing system? How is videoconferencing technology limited? Why do you think that videoconferencing technology isn't built into today's PCs? How could you use videoconferencing in your job, future line of work, or personal area of interest?

3. Computer magazines often sponsor tests to compare laser printers. By reviewing current computer magazines, identify the most highly-rated laser printer. How much does the printer cost? What capabilities does this printer have? Why was this printer rated above other printers in the study? Would you buy this printer for personal use? Why/why not?

4. Explore the state of the art of computer-generated 3-D graphics. What challenges are involved in creating photo-realistic 3-D images? What hardware and software are needed to generate 3-D graphics? Who benefits from this technology?

5. Explore the state of the art of liquid-crystal display (LCD) technology. What are the current limitations of LCD technology? Why do you think that billions of dollars are currently being invested in LCD development? Who will benefit from improved LCD technology?

net Exploring the Internet

Objective: *In this exercise we describe how to use Netscape Navigator to access an online magazine and newspaper.*

Before you continue*: We assume you have access to the Internet through your university, business, or commercial service provider and to the Web browser tool named Netscape Navigator 2.0 or 2.01. Additionally, we assume you know how to connect to the Internet and then load Netscape Navigator. If necessary, ask your instructor or system administrator for assistance.*

1. Make sure you have started Netscape. The home page for Netscape Navigator should appear on your screen.

2. Several magazines and newspapers are available to you online including the Network Observer (*http://communication.ucsd.edu/pagre/tno.html*), which emphasizes issues relating to networks and democracy, and the *Mercury Center* (*http://www.sjmercury.com/*), which carries top stories from the *San Jose Mercury News,* a Silicon Valley-based newspaper.

   To display the *Network Observers'* home page:
   CLICK: Open button ( )
   TYPE: http://communication.ucsd.edu/pagre/tno.html
   PRESS: **[Enter]** or CLICK: Open

   Your screen may appear similar to the following:

   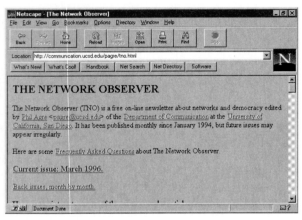

3. Drag the vertical scroll bar downward to see a list of articles. Notice that articles are grouped by section. Identify an article you're interested in reading and then click its link.

4. If you're connected to a printer, you can print the article and then read it at your leisure.

   CHOOSE: File, Print from the Menu bar (or click the Print button located on the Netscape toolbar)
   PRESS: **[Enter]** or CLICK: OK

5. To redisplay the list of topics:
   CLICK: Back button ( )

6. On your own, display additional articles.

7. To display the *Mercury Center's* home page:
   CLICK: Open button ( )
   TYPE: http://www.sjmercury.com/
   PRESS: **[Enter]** or CLICK: Open

   Your screen should appear similar to the following:

   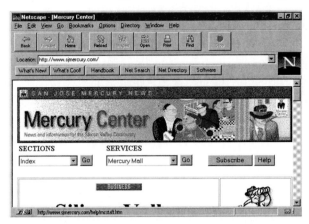

8. Drag the vertical scroll bar downward to see the contents of the home page. On your own, explore the contents of the newspaper by clicking links that are of interest to you. Remember that if you would rather read an article in printed form, you can print it by choosing File, Print from the Menu bar or by clicking the Print button.

9. Display Netscape's home page and then exit Netscape.

CHAPTER 7

Storage

Foundations for Interactivity, Multimedia, & Knowledge

Concepts You Should Know

After reading this chapter, you should be able to:

1. Define and explain data hierarchy and units of storage capacity.
2. Distinguish between primary and secondary storage.
3. Discuss the different kinds of compression and decompression.
4. Identify the basic criteria for rating secondary storage devices.
5. Explain how data is stored on diskette, hard disks, and optical disks and describe the purposes for which these storage media are best used.
6. Describe flash-memory cards.
7. Explain the different types of magnetic-tape storage.
8. Explain some other forms of storage: online storage, CD-ROM jukeboxes, high-density disks, digital VCRs, and molecular electronics.

See it. Hear it. Change it.

These are the promises frequently being heard, whether for today's videogames or tomorrow's smart TVs. Such claims have made "interactivity" and "multimedia" among the most overused marketing words of recent times. *Interactivity* refers to users' ability to have a back-and-forth dialogue with whatever is on their screen. *Multimedia* refers to the use of a variety of media—text, sound, video—to deliver information, whether on a computer disk or (as the phone and cable companies use it) via the anticipated union of computers, telecommunications, and broadcasting technologies.[1]

Interactive programs and multimedia programs require machines that can handle and store enormous amounts of data. For this reason, the capacities of storage hardware, called *secondary storage* or just *storage*, seem to be increasing almost daily. This chapter covers the most common types of secondary storage, as well as their uses. First, however, we need to go over a few concepts that relate to the whole topic of secondary storage: data hierarchy, data capacity, data compression/decompression, and criteria for rating storage devices.

Data Hierarchy & Data Capacity

Preview & Review: Storage is categorized as primary or secondary. Data in storage is organized as a hierarchy: bits, bytes, fields, records, files, and databases.

The capacity of storage devices is measured in bytes or multiples: bytes, kilobytes, megabytes, gigabytes, and terabytes.

You'll recall from Chapters 1 and 4 that **primary storage is also known as memory, main memory, internal memory, or RAM (for random access memory). It is working storage that holds (1) data for processing, (2) instructions (the programs) for processing the data, and (3) processed data (that is, information) that is waiting to be sent to an output or secondary-storage device** such as a printer.

Primary storage is in effect the computer's short-term capacity, determining the total size of the programs and data files it can work on at any given moment. Primary storage, which is contained on RAM chips (✓ p. 174), is also *temporary*. Once the power to the computer is turned off, all the data and programs within memory simply vanish. For this reason, primary storage is said to be *volatile*. **Volatile storage is temporary storage; the contents are lost when the power is turned off.** If you accidentally kick out the power cord underneath your desk, or a storm knocks down a power line to your house, whatever you are currently working on will immediately vanish.

In contrast, **secondary storage consists of devices that store data and programs permanently on disk or tape.** Secondary storage is **nonvolatile—that is, data and programs are permanent, or remain intact when the power is turned off.**

Secondary storage is also needed because computer users require far greater storage capacity than is possible with primary storage. People use the words *storage media* to refer to the material that stores data. Storage media include disk and magnetic tape.

You'll also recall that computers are based on the principle that electricity may be "on" or "off," or "high-voltage" or "low-voltage," or "present" or "absent," or some similar two-state system. Thus, individual items of data are represented by 0 for off and 1 for on. **A 0 or 1 is called a *bit*. A unit of 8 bits is called a *byte*; it may be used to represent a character, digit, or other value,** such as A, ?, or 3.

Bits and bytes are the building blocks for representing data, whether it is being processed, stored, or telecommunicated.

The Data Hierarchy

Data can be grouped into a hierarchy of categories, each increasingly more complex. **The *data storage hierarchy* consists of the levels of data stored in a computer file: bits, bytes, fields, records, and files.** Bits and bytes are what the computer hardware deals with, and you need not be overly concerned with them. You will, however, be dealing with characters, fields, records, files, and databases. (This topic is covered in more detail in Chapter 9.)

- **Character:** **A *character* may be—but is not necessarily—the same as a byte. A character is a single letter, number, or special character** such as ;, $, or %.
- **Field:** **A *field* is a unit of data consisting of one or more characters.** An example of a field is your name, your address, or your Social Security number. Note: One reason the Social Security number is often used in computing—for good or for ill—is that, perhaps unlike your name, it is a *distinctive* field. Thus, it can be used to easily locate information about you.
- **Record:** **A *record* is a collection of related fields.** An example of a record would be your name *and* address *and* Social Security number.
- **File:** **A *file* is a collection of related records.** An example of a file is collected data on everyone employed in the same department of a company, including all names, addresses, and Social Security numbers.
- **Database:** **A *database* is a collection of interrelated files.** A company database might include files on all past and current employees in all departments. There would be various files for each employee: payroll, retirement benefits, sales quotas and achievements (if in sales), and so on.

  A database may be fairly small, contained entirely within your own personal computer. Or it may be massive, available online to you from an information service through computer and telephone connections.

Units of Measurement for Storage

We explained the meanings of kilobytes, megabytes, gigabytes, and terabytes in conjunction with the capabilities of processing hardware. The same terms are also used to measure the data capacity of storage devices. To repeat:

- **Kilobyte:** **A *kilobyte* (abbreviated *K or KB*) is equivalent to 1024 bytes.** Kilobytes are a common unit of measure for storage capacity.
- **Megabyte:** **A *megabyte* (abbreviated *M or MB*) is about 1 million bytes.**
- **Gigabyte:** **A *gigabyte* (*G or GB*) is about 1 billion bytes.**
- **Terabyte:** **A *terabyte* (*T or TB*) is about 1 trillion bytes.**

The amount of data being held in a file or database in your personal computer might be expressed in kilobytes, megabytes, or gigabytes. The amount of data being held by a far-flung database accessible to you over a communications line might be expressed in gigabytes or terabytes.

Because, as we said at the beginning of this chapter, new sophisticated software programs require huge storage capacities, the capacities of storage devices are steadily increasing. There is, however, a way to increase a system's storage capacity without buying new storage devices—that is, through compression programs.

Compression & Decompression

Preview & Review: Compression is a method of removing redundant elements from a computer file so that it requires less storage space. Compression and decompression techniques are often called *codec* techniques. The two principal compression techniques are "lossless" and "lossy." The principal compression schemes are JPEG for still images and MPEG-1, MPEG-2, and MPEG-4 for moving images.

"Like Gargantua, the computer industry's appetite grows as it feeds . . . ," says one writer. "The first symptoms of indigestion are emerging. So the smartest software engineers are now looking for ways to shrink the data-meals computers consume, without reducing their nutritional value."[2]

What this writer is referring to is the "digital obesity" brought on by the requirements of the new multimedia revolution for putting pictures, sound, and video onto disk or sending them over a communications line. For example, a 2-hour movie contains so much sound and visual information that, if stored on a standard CD-ROM, it would require 360 disk changes during a single showing. A broadcast of *Oprah Winfrey* that presently fits into one conventional, or analog, television channel would require 45 channels if sent in digital language.[3]

The solution for putting more data into less space comes from the mathematical process called compression. **Compression, or *digital-data compression,* is a method of removing redundant elements from a file so that the file requires less storage space and less time to transmit.** After the data is stored or transmitted and is to be used again, it is decompressed. **The techniques of compression and decompression are often referred to as *codec* (for compression/decompression) *techniques.***

"Lossless" Versus "Lossy" Compression

There are two principal methods of compressing data:

- **"Lossless" techniques:** *"Lossless" compression techniques* **achieve compression by avoiding repetition but still preserving every bit of data that was input.** That is, the data that comes out is every bit the same as what went in; it has merely been repackaged for purposes of storage or transmission.

  Lossless techniques are used for computer data. Microcomputer users, for example, can double the amount of data they store on their hard disks by using software utility products (✓ p. 129) such as Stacker, Superstor, and DriveSpace. These programs store text by eliminating irrelevant letters and redundant spaces between words. Some techniques can shrink a file of text by 70% or more.

- **"Lossy" techniques:** It is much easier to compress text than to compress sounds, pictures, and videos. "Here, reconstructive surgery is not enough," commented one article. "Some information has to be thrown away for-ever. The trick is to work out what will not be missed."[4] This is the prob-lem for "lossy" techniques.

***"Lossy" compression techniques* permanently discard some data during compression.** Thus, a lossy codec might discard shades of color that a viewer would not notice or soft sounds that are masked by louder ones. In general, most viewers or listeners would not notice the absence of these details.

Standards for Visual Compression

The major difficulty now is that several standards exist for compression, par-ticularly of visual data. If you record and compress in one standard, you can-not play it back in another. The main reason for the lack of agreement is that different industries have different priorities. What will satisfy the users of still photographs, for instance, will not work for the users of moving images.

The principal compression schemes are as follows:

*   **Still images—JPEG:** Techniques for storing and transmitting still pho-tographs require that the data remain of high quality, so a lossless tech-nique is required. The leading standard for still images is *JPEG* (pro-nounced "jay-peg"), for the Joint Photographic Experts Group of the International Standards Organization. The JPEG codec looks for a way to squeeze a single image, mainly by eliminating repetitive pixels, or picture-element dots, within the image. Unfortunately there are more than 30 kinds of JPEG programs. "Unless the decoder in your computer recognizes the version that was used to compress a particular image," noted one reporter, "the result on your computer screen will be multimedia apple-sauce."[5]

*   **Moving images—MPEG:** People who work with videos are less con-cerned with the niceties of preserving details than are those who deal with still images. They are interested mainly in storing or transmitting an enor-mous amount of visual information in economical form. A group called *MPEG* ("em-peg"), for Motion Picture Experts Group, has been formed to set standards for weeding out redundancies between neighboring images in a stream of video.

    Three MPEG standards have been developed for compressing visual information—MPEG-1, MPEG-2, and MPEG-4. (■ *See Panel 7.1, next page.*)

    (1) *MPEG-1* is the standard for providing images for microcomputers and consumer gadgets. The images are of about the same quality as those found on a VHS videocassette.

    (2) *MPEG-2*, a higher standard, is for compressing high-quality video for delivery over cable networks, satellite dishes, and new types of CD-ROMs. The MPEG-2 codecs work by recording just a few (rather than all) frames of video in detail, then describing adjacent frames in terms of how they differ from the detailed ones. For example, when two red dots, or pixels, are adjacent, only one needs to be described.

    (3) *MPEG-4* (MPEG-3 was incorporated into MPEG-2) is a standard for wireless videoconferencing or teleconferencing, in which two or more people confer over a televised communications channel. Clearly these images need to appear smoothly and instantaneously. For this purpose, a codec should be able to compress as it records and be able to play back immediately. The MPEG-4 codec standard is expected to arrive in 1998 or later.

PANEL 7.1 **How video images are compressed, and three MPEG standards**

(Left) MPEG is a method of computerized compression/decompression that can reduce the size of a video signal by 95%. As a result, the signal can be stored or transmitted more efficiently and economically. *(Right)* Three MPEG standards (MPEG-3 has been incorporated into MPEG-2).

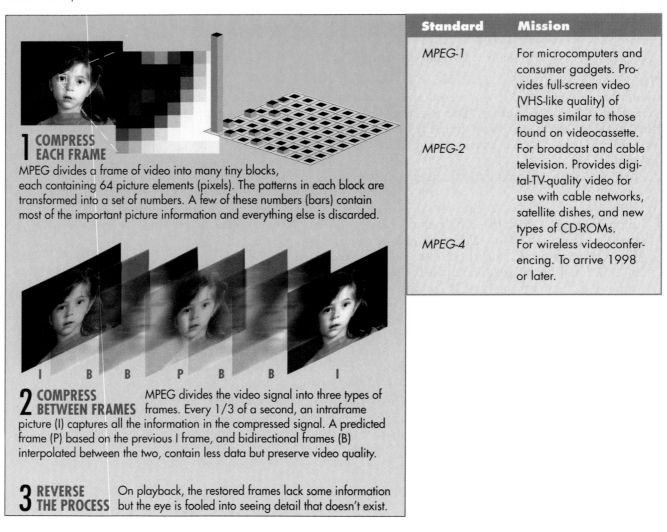

1 COMPRESS EACH FRAME

MPEG divides a frame of video into many tiny blocks, each containing 64 picture elements (pixels). The patterns in each block are transformed into a set of numbers. A few of these numbers (bars) contain most of the important picture information and everything else is discarded.

2 COMPRESS BETWEEN FRAMES MPEG divides the video signal into three types of frames. Every 1/3 of a second, an intraframe picture (I) captures all the information in the compressed signal. A predicted frame (P) based on the previous I frame, and bidirectional frames (B) interpolated between the two, contain less data but preserve video quality.

3 REVERSE THE PROCESS On playback, the restored frames lack some information but the eye is fooled into seeing detail that doesn't exist.

Standard	Mission
MPEG-1	For microcomputers and consumer gadgets. Provides full-screen video (VHS-like quality) of images similar to those found on videocassette.
MPEG-2	For broadcast and cable television. Provides digital-TV-quality video for use with cable networks, satellite dishes, and new types of CD-ROMs.
MPEG-4	For wireless videoconferencing. To arrive 1998 or later.

The vast streams of bits and bytes of text, audio, and visual information threaten to overwhelm us. Compression/decompression has become a vital technology for rescuing us from the swamp of digital data.

◄ Criteria for Rating Secondary-Storage Devices

Preview & Review: Storage capacity, access speed, transfer rate, size, and cost are all factors in rating secondary-storage devices.

Several kinds of secondary-storage devices are available, and they generally differ in *storage capacity, access speed, transfer rate, size,* and *cost.*

• **Storage capacity:** As we mentioned earlier, high-capacity storage devices are desirable or required for many sophisticated programs and large databases. However, as capacity increases, so does price. Some users find codec software to be an economical solution to storage-capacity problems. Hard disks can store more data than diskettes, and optical disks can store more than hard disks.

- **Access speed:** *Access speed* refers to the average time needed to locate data on a secondary storage device. Access speed is measured in milliseconds (thousandths of a second). Hard disks are faster than optical disks, which are faster than diskettes. Disks are faster than magnetic tape. However, the slower media are more economical.

- **Transfer rate:** *Transfer rate* refers to the speed at which data is transferred from secondary storage to main memory (✓ p. 164). It is measured in megabytes per second.

- **Size:** Some situations require compact storage devices, others don't. Users need to know what their options are.

- **Cost:** As we have intimated, the cost of a storage device is directly related to the previous four factors.

Now let's take an in-depth look at the following types of secondary storage devices.

- Diskettes
- Hard disks
- Optical disks
- Flash-memory cards
- Magnetic tape

Diskettes

Preview & Review: Diskettes are round pieces of flat plastic that store data and programs as magnetized spots. A disk drive copies, or reads, data from the disk and writes, or records, data to the disk.

Components of a diskette include tracks and sectors; diskettes come in various densities. All have write-protect features.

Care must be taken to avoid destroying data on disks, and users are advised to back up, or duplicate, the data on their disks.

A *diskette*, or *floppy disk*, is a removable round, flat piece of mylar plastic that stores data and programs as magnetized spots. More specifically, data is stored as electromagnetic charges on a metal oxide film that coats the mylar plastic. Data is represented by the presence or absence of these electromagnetic charges, following standard patterns of data representation (such as ASCII, ✓ p. 168). The disk is contained in a plastic case or square paper envelope to protect it from being touched by human hands. It is called "floppy" because the disk within the case or envelope is flexible, not rigid.

Two sizes of diskettes are commonly used for microcomputers. (■ *See Panel 7.2.*)

- **3½ inches:** The smaller size, now by far the most common, is 3½ inches across. This size comes inside a hard plastic jacket, so that no additional protective envelope is needed.

- **5¼ inches:** The older and larger size is 5¼ inches across. The disk is encased inside a flexible plastic jacket. The 5¼-inch disk is often inserted

PANEL 7.2

Diskettes

(Top) 3½-inch diskettes. *(Bottom)* 5¼-inch diskette.

into a removable paper or cardboard envelope or sleeve for protection when it is not being used.

Larger and smaller sizes of diskettes also exist, although they are not standard on most microcomputers.

Incidentally, you should never try to jam a 3½-inch disk into a 5¼-inch-disk drive slot. Not only will the arrangement not work, but it will damage the disk and possibly the internal drive mechanisms (read/write heads).

The Diskette Drive

To use a diskette, you need a disk drive. **A *disk drive* is a device that holds, spins, and reads data from and writes data to a diskette.**

The words *read* and *write* have exact meanings:

- **Read:** *Read* **means that the data represented in the magnetized spots on the disk (or tape) are converted to electronic signals and transmitted to the primary storage (memory) in the computer.** That is, *read* means the disk drive *copies* data—stored as magnetic spots—from the diskette.
- **Write:** *Write* **means that the electronic information processed by the computer is recorded magnetically onto disk (or tape).** *Write* means that the disk drive *transfers* data—represented as electronic signals within the computer's memory—onto the storage medium.

Whereas *reading* simply makes a copy of the original data, without altering the original, *writing* actually replaces the data underneath it. With writing, it is as though you recorded on an audiotape recorder a new song, obliterating the original song already on the tape.

A diskette drive may be a separate unit attached to the computer, particularly on older models. Usually, however, it is built into the computer's system cabinet (✓ p. 170).

How a Diskette Drive Works

A diskette is inserted into a slot, called the *drive gate* or *drive door*, in the front of the disk drive. (■ *See Panel 7.3.*) Sometimes a door or a latch must be closed after the disk is inserted. This clamps the diskette in place over the spindle of the drive mechanism so the drive can operate. Usually today, however, the diskette is simply pushed into the drive until it clicks into place. An access light goes on when the disk is in use. After using the disk, you can retrieve it either by pressing an eject button beside the drive or by opening the drive gate.

The device by which the data on a disk is transferred to the computer, and from the computer to the disk, is the disk drive's *read/write head*. The diskette spins inside its jacket, and the read/write head moves back and forth over the data access area. **The *data access area* is an opening in the disk's jacket through which data is read or written.**

It is *possible* to operate a microcomputer with just one diskette drive and no hard-disk drive, although it is a bit cumbersome. Copying from one diskette to another, for example, would require constantly switching the two diskettes back and forth in the single drive. (However, virtually all modern software applications require the use of a hard drive.)

Today most microcomputers have a single diskette drive for 3½-inch disks and a hard-disk drive. There may also, however, be an additional diskette drive that accepts the larger size, 5¼-inch disks.

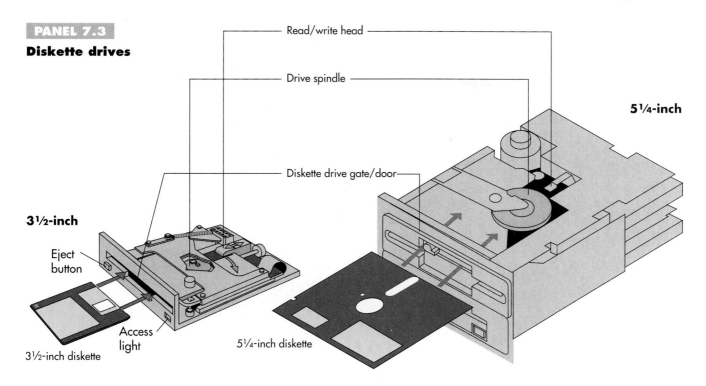

PANEL 7.3

Diskette drives

Characteristics of Diskettes

Both 3½-inch and 5¼-inch disks work in similar ways, although there are some differences. The characteristics of diskettes are as follows:

- **Tracks and sectors:** On a diskette, **data is recorded in rings called** *tracks.* Unlike on a phonograph record, these tracks are neither visible grooves nor a single spiral. Rather, they are closed concentric rings.

  Each track is divided into eight or nine *sectors.* **Sectors are invisible wedge-shaped sections used for storage reference purposes.** When you save data from your computer to a diskette, the data is distributed by tracks and sectors on the disk. That is, the systems software uses the point at which a sector intersects a track to reference the data location in order to spin the disk and position the read/write head.

- **Unformatted versus formatted disks:** When you buy a new box of diskettes to use for storing data, the box may state that it is "unformatted" (or say nothing at all). This means you have a task to perform before you can use the disks with your computer and disk drive. **Unformatted disks are manufactured without tracks and sectors in place. Formatting—** or *initializing,* **as it is called on the Macintosh—means that you must prepare the disk for use so that the operating system can write information on it. This includes defining the tracks and sectors on it.** Formatting is done quickly by using a few simple software commands.

  Alternatively, when you buy a new box of diskettes, the box may state that it is "formatted IBM." This means that you can simply insert a disk into the drive gate of your IBM or IBM-compatible microcomputer and use it without any effort. It's just like plunking an audiotape into a standard tape recorder.

  The following illustration shows the elements of a diskette. (■ *See Panel 7.4.)*

3½-inch disk

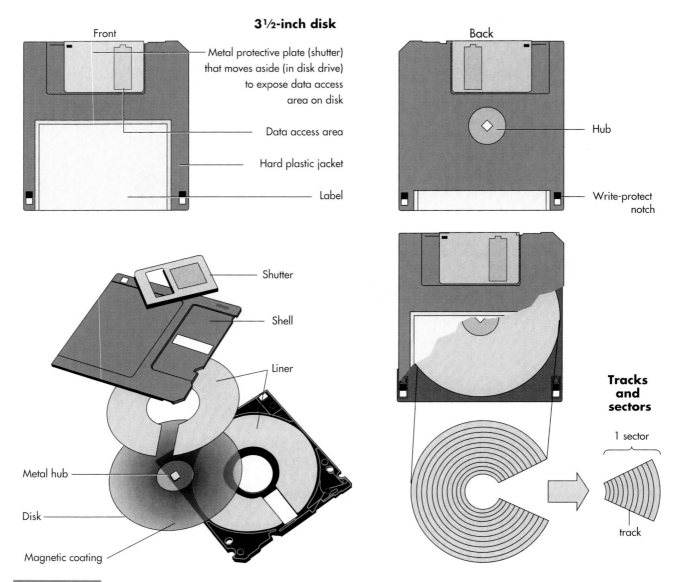

Front

Metal protective plate (shutter) that moves aside (in disk drive) to expose data access area on disk

Data access area

Hard plastic jacket

Label

Back

Hub

Write-protect notch

Shutter

Shell

Liner

Metal hub

Disk

Magnetic coating

Tracks and sectors

1 sector

track

PANEL 7.4

Diskette anatomy

- **Data capacity—sides and densities:** Not all disks hold the same amount of data, because the characteristics of microcomputer disk drives differ.

  The first diskettes were *single-sided,* or diskettes that store data on one side only. Now all diskettes are **double-sided, capable of storing data on both sides.** They therefore hold twice as much data as single-sided disks. For double-sided diskettes to work, the disk drive must have read/write heads that will read both sides simultaneously. This is the case with current disk drives.

  A disk's capacity also depends on its recording density. **Recording density refers to the number of bytes that can be written onto the surface of the disk.** There are three densities: *single-density, double-density,* and *high-density.* A 3½-inch double-sided, double-density disk can store 720 kilobytes. A high-density 5¼-inch disk can store 1.2 megabytes. A high-density 3½-inch disk can store 1.44 megabytes.

- **Write-protect features:** Both 3½-inch and 5¼-inch disks have features to prevent someone from accidentally writing over—and thereby obliterating—data on the disk. (This is especially important if you're working on your only copy of a program or a document that you've transported from somewhere else.) This **write-protect feature allows you to protect a diskette from being written to.**

The write-protect feature works a bit differently for the two sizes of disks. With a 3½-inch disk, you press a slide lever toward the edge of the disk, uncovering a hole (which appears on the lower right side, viewed from the back). With a 5¼-inch disk, there is a write-protect notch, a small, square cutout on the side of the disk. This notch is covered by a piece of tape in order to protect it. (■ *See Panel 7.5.*)

Disks have additional features (such as the index hole, for positioning the disk over a photoelectric sensing mechanism within the disk drive). However, these are of no concern for our present purposes.

Taking Care of Diskettes

Diskettes need at least the same amount of care that you would give to an audiotape or music CD. In fact, they need more care than that if you are dealing with difficult-to-replace data or programs. There are a number of rules for taking care of diskettes:

* **Don't touch disk surfaces:** Don't touch anything visible through the protective jacket, such as the data access area. Don't manipulate the metal shutter on 3½-inch disks.

* **Handle disks gently:** Don't bend them or put heavy weights on them.

Writable

Write-protected

Write-protect
window closed

Write-protect
window open

Writable

Write-protected

Write-protect
notch open

Write-protect
notch covered

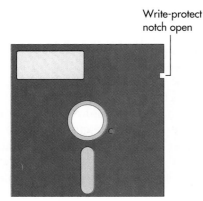

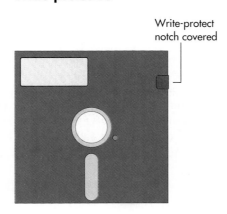

PANEL 7.5

Write-protect features

(Top)—3½-inch diskette: For data to be written to this disk, a small piece of plastic must be closed over the tiny window on one side of the disk. To protect the disk from being written to, you must open the window (using the tip of a pen helps). *(Bottom)* 5¼-inch diskette: For data to be written to this diskette, the write-protect notch must be uncovered, as shown at left. To protect the disk from being mistakenly written over, a small piece of tape must be folded around the notch. (The tape comes with the disks.)

R E A D M E

Practical Matters: Five Steps to a Tidy Desk

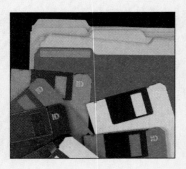

Barbara Hemphill is president of the National Association of Professional Organizers and author of Taming the Paper Tiger (Kiplinger). The following are her ideas for keeping your desk clear—so that you won't lose diskettes, among other items.

Clutter is postponed decisions! Use these tips to help make decisions about the papers that clutter your desk.

1. There are only three actions you can take with regard to any piece of paper: toss it, act on it (including passing it to someone else), or file it.
2. Practice the art of wastebasketry. Research shows that 80% of what goes in most files is never used. Ask yourself, "What's the worst possible thing that would happen if I didn't have this piece of paper?" If you can live without it, toss it.
3. For each piece of paper you want to act on, ask yourself, "What is the *next* action I need to take on this?" Note when you will take that action on your calendar.
4. For each piece of paper you want to file, ask yourself, "If I want or need this information again, what word will I think of first?" Keep a list of the titles in your files. Then you can avoid creating a new file for *Car* when you already have *Auto*.
5. When you are tempted to postpone a decision about a specific piece of paper, ask yourself, "What am I going to know tomorrow that I don't know today?"

—Barbara Hemphill, "Five Steps to a Tidy Desk," *Home Office Computing*

- **Avoid risky physical environments:** Disks don't do well in sun or heat (such as in glove compartments or on top of steam radiators). They should not be placed near magnetic fields (including those created by nearby telephones or electric motors). They also should not be exposed to chemicals (such as cleaning solvents) or spilled coffee or alcohol.

Some specific suggestions for taking care of diskettes are given at the end of Chapter 5. (See the Experience Box, "Good Habits: Protecting Your Computer System, Your Data, & Your Health.")

The Importance of Backup

Having said all this, we hasten to point out that diskettes are surprisingly hardy. Every day people send hundreds of thousands of disks through the mail in cardboard mailing envelopes. They also pass them in luggage through airport-security X-ray machines. They even violate many of the rules we've just cited, and still the disks continue working.

Even with the best of care, however, a disk can suddenly fail for reasons you can't understand. Many computer users have had the experience of being unable to retrieve data from a disk that worked perfectly the day before, because some defect has damaged a track or sector.

Thus, you should always be thinking about backup. **Backup is the name given to a diskette (or tape) that is a duplicate or copy of another form of storage.** The best protection if you're writing, say, a make-or-break research paper is to make *two copies* of your data. One copy is on a diskette, certainly, but a duplicate should be on a second diskette or on your hard disk if you are using one.

Microcomputer Diskette Variations: Removable High-Capacity Diskettes

"I LOVE MY ZIP DRIVE!!!!" exclaims Craig Clarke of Elma, New York, in an online message.[6]

Why the excitement? Because the first law of computing, says one technology journalist, is "You can never have enough disk space."[7] Actually, he means *hard disks,* as we shall discuss. However, the point is this: Today's software takes up so much space (some programs can take up to 100 megabytes of hard-disk space) that it reduces the amount of room on which you can store data files. This is particularly the case for owners of portable computers. Instead of deleting useful files from your hard disk or adding a second hard-disk drive, you can use removable diskettes that hold far more data than conventional diskettes. One of these, the kind enthusiast Clarke likes, is called a "Zip drive."

At present two kinds of removable high-capacity diskette drives seem to be emerging.

- **Zip and EZ diskette drives:** The first of this breed was the *Zip drive* from Iomega, an external drive about the size of a hardcover novel and weighing about 1 pound. It was soon followed by the *EZ135 drive* from SyQuest Technology. The EZ is about the size of two stacked VCR tapes, weighs about 2 pounds, and is twice as fast as the Zip.[8] (■ *See Panel 7.6.*) Unfortunately, the two drives use different technology and incompatible disks.

    Both the Zip and the EZ use hard-shell diskettes about 4 inches square and a quarter-inch thick. The Zip diskette holds 100 megabytes, which is 70 times more than conventional floppy disks. The EZ diskette holds 135 megabytes.

    Both Zip and EZ drives and disks are useful for storing programs and data files that you don't want to lose but that you don't use very often. One significant drawback: Neither drive can be used with older diskettes.

- **Backward-compatible high-capacity diskette drive:** In mid-1995, three computer-industry companies (Compaq, 3M, Matsushita) announced they had developed a 3½-inch diskette drive that could handle 83 times as much data as current diskette drives.[9] More important, the 120-megabyte

PANEL 7.6

Removable high-capacity diskette drive

This Zip drive weighs about 2 pounds.

drive would be backward-compatible and so could read the 720-kilobyte and 1.44-megabyte diskettes that people currently use.

In spite of advances in diskette technology, diskettes still cannot store as much data as hard disks and optical disks, which we cover next.

Hard Disks

Preview & Review: Hard disks are rigid metal platters that hold data as magnetized spots.

Usually a microcomputer hard-disk drive is built into the system unit, but external hard-disk drives are available, as are removable hard-disk cartridges.

Large computers use removable hard-disk packs, fixed-disk drives, or RAID storage systems.

Switching from a microcomputer that uses only diskettes to one containing a hard disk is like discovering the difference between moving your household in several trips in a small sportscar and doing it all at once with a moving van. Whereas a high-density 3½-inch diskette holds 1.44 megabytes, a hard disk in a personal computer may store as many as 9 *gigabytes*. Indeed, at first with a hard disk you may feel you have more storage capacity than you'll ever need. However, after a few months, you may worry that you don't have enough. This feeling may intensify if you're using graphics-oriented programs or multimedia programs, with pictures and other features requiring immense amounts of storage.

Diskettes are made out of flexible material, which makes them "floppy." By contrast, **hard disks are thin but rigid metal platters covered with a substance that allows data to be held in the form of magnetized spots.** Hard disks are also tightly sealed within an enclosed unit to prevent any foreign matter from getting inside. Data may be recorded on both sides of the disk platters.

We'll now describe the following aspects of hard-disk technology:

- Microcomputer hard-disk drives
- Defragmentation to speed up hard disks
- Microcomputer hard-disk variations
- Hard-disk technology for large computer systems

Microcomputer Hard-Disk Drives

In microcomputers, **hard disks are one or more platters sealed inside a *hard-disk drive* that is built into the system unit and cannot be removed.** The drive is installed in a *drive bay,* **a shelf or opening in the computer cabinet.** From the outside of a microcomputer, a hard-disk drive is not visible; it looks simply like part of the front panel on the system cabinet. Inside, however, is a disk or disks on a drive spindle, read/write heads mounted on an actuator (access) arm that moves back and forth, and power connections and circuitry. (■ *See Panel 7.7.)* The disks may be 5¼ inches in diameter, although today they are more often 3½ inches, with some even smaller. The operation is much the same as for a diskette drive, with the read/write heads locating specific pieces of data according to track and sector.

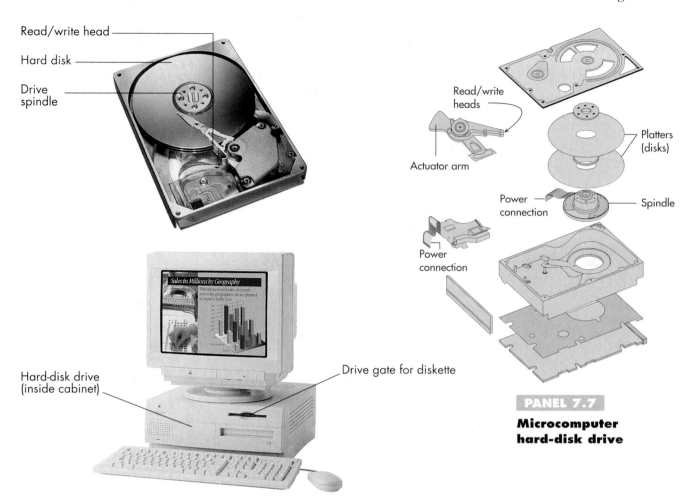

Read/write head
Hard disk
Drive spindle

Read/write heads
Actuator arm
Platters (disks)
Power connection
Spindle
Power connection

Hard-disk drive (inside cabinet)

Drive gate for diskette

Sales in Millions by Geography

PANEL 7.7

Microcomputer hard-disk drive

Hard disks have a couple of real advantages over diskettes—and at least one significant disadvantage.

- **Advantages—capacity and speed:** We mentioned that hard disks have a data storage capacity that is significantly greater than that of diskettes. Microcomputer hard-disk drives typically hold 40–500 megabytes and newer ones hold several gigabytes.

  As for speed, hard disks allow faster access to data than do diskettes because a hard disk spins several times faster than a diskette. (A 2.1-gigabyte hard disk will spin at 7800 revolutions per minute [rpm], compared to 360 rpm for a diskette drive.)

- **Disadvantage—possible "head crash":** In principle a hard disk is quite a sensitive device. The read/write head does not actually touch the disk but rather rides on a cushion of air about 0.000001 inch thick. The disk is sealed from impurities within a container, and the whole apparatus is manufactured under sterile conditions. Otherwise, all it would take is a smoke particle, a human hair, or a fingerprint to cause what is called a head crash.

  A *head crash* happens when the surface of the read/write head or particles on its surface come into contact with the disk surface, causing the loss of some or all of the data on the disk. An incident of this sort could, of course, be a disaster if the data has not been backed up. There are firms that specialize in trying to retrieve (for a hefty price) data from crashed hard disks, though this cannot always be done.

In recent years, computer magazines have evaluated the durability of portable computers containing hard disks by submitting them to drop tests. Most of the newer machines are surprisingly hardy. However, with hard disks—whether in portable or in desktop computers—the possibility of disk failure always exists.

Fragmentation & Defragmentation: Speeding Up Slow-Running Hard Disks

Like diskettes, for addressing purposes hard disks are divided into a number of invisible rings called *tracks* and typically nine invisible pie-shaped wedges called *sectors*. Data is stored within the tracks and sectors in groups of *clusters*. **A *cluster* is the smallest storage unit the computer can access, and it always refers to a number of sectors.** (The number varies among types of computers.)

With a brand-new hard disk, the computer will try to place the data in clusters that are *contiguous*—that is, that are adjacent (next to one another). Thus, data would be stored on track 1 in sectors 1, 2, 3, 4, and so on. However, as data files are updated and the disk fills up, the operating system stores data in whatever free space is available. Thus, files become fragmented. *Fragmentation* **means that a data file becomes spread out across the hard disk in many noncontiguous clusters.**

Fragmented files cause the read/write head to go through extra movements to find data, thus slowing access to the data. This means that the computer runs more slowly than it would if all the data elements in each file were stored in contiguous locations. To speed up the disk access, you must defragment the disk. *Defragmentation* **means that data on the hard disk is reorganized so that data in each file is stored in contiguous clusters.** Programs for defragmenting are available on some operating systems or as separate (external) software utilities (✓ p. 128).

Microcomputer Hard-Disk Variations: Power & Portability

If you have an older microcomputer or one with limited hard-disk capacity, some variations are available that can provide additional power or portability:

- **Miniaturization:** Newer hard-disk drives are less than half the height of older drives (1½ inches versus 3½ inches high) and so are called *half-height drives.* Thus, you could fit two disk drives into the bay in the system cabinet formerly occupied by one.

  In addition, the diameter of the disks has been getting smaller. Instead of 5¼ or 3½ inches, some platters are 2.5, 1.8, or even 1.3 inches in diameter. The half-dollar-size 1.3-inch Kittyhawk microdisk, which is actually designed for use in handheld computers, holds 21 megabytes of data.

- **External hard-disk drives:** An internal disk drive may be the most convenient, but adding an external hard-disk drive is usually easy. Some detached external hard-disk drives, which have their own power supply and are not built into the system cabinet, can store gigabytes of data.

- **Hard-disk cartridges:** The disadvantages of hard disks are that they cannot be easily removed and that they have only a finite amount of storage. *Hard-disk cartridges* **consist of one or two platters enclosed along with read/write heads in a hard plastic case. The case is inserted into a detached external cartridge system connected to a microcomputer.** (■ *See Panel 7.8.*) A cartridge, which is removable and easily transported in a briefcase, may hold several gigabytes of data. An additional advantage of hard-disk cartridges is that they may be used for backing up data.

PANEL 7.8

Removable hard-disk cartridge and portable hard-disk drive

Each cartridge has self-contained disks and read/write heads. The entire cartridge, which may store several gigabytes of data, may be removed for transporting or may be replaced by another cartridge.

Hard-Disk Technology for Large Computer Systems

As a microcomputer user, you may regard secondary-storage technology for large computer systems with only casual interest. However, this technology forms the backbone of the revolution in making information available to you over communications lines. The large databases offered by such organizations as CompuServe, America Online, and Dialog, as well as the predicted movies-on-demand through cable and wireless networks, depend to a great degree on secondary-storage technology.

Secondary-storage devices for large computers consist of the following:

- **Removable packs:** A *removable-pack hard disk system* **contains 6–20 hard disks, of 10½- or 14-inch diameter, aligned one above the other in a sealed unit.** These removable hard-disk packs resemble a stack of phonograph records, except that there is space between disks to allow access arms to move in and out. Each access arm has two read/write heads—one reading the disk surface below, the other the disk surface above. However, only *one* of the read/write heads is activated at any given moment. The disk packs are inserted into receptacles in large, external drive units that can accommodate several disk packs at one time.

  Secondary-storage systems that use several hard disks don't use the sector method to locate data. Rather they use what is known as the cylinder method. Because the access arms holding the read/write heads all move together, the read/write heads are always over the same track on each disk at the same time. **All tracks with the same track number, lined up one above the other, thus form a *cylinder.*** (■ *See Panel 7.9, next page.*)

- **Fixed-disk drives:** *Fixed-disk drives* **are high-speed, high-capacity disk drives that are housed in their own cabinets.** Although not removable or portable, they generally have greater storage capacity and are more reliable than removable packs. A single mainframe computer might have 20 to 100 such fixed-disk drives attached to it.

- **RAID storage system:** A fixed-disk drive sends data to the computer along a single path. **A *RAID storage system*, which consists of over 100 5¼-inch disk drives within a single cabinet, sends data to the computer along several parallel paths simultaneously.** Response time is thereby significantly improved. *RAID* stands for Redundant Array of Inexpensive Disks.

  The advantage of a RAID system is that it not only holds more data than a fixed-disk drive within the same amount of space, but it also is more reliable, because if one drive fails, others can take over.

PANEL 7.9

Disk packs and cylinders

In a stack of disks, such as in a disk pack, access arms slide in and out to specific tracks. They use the cylinder method to locate data—the same track numbers lined up vertically one above the other form a "cylinder."

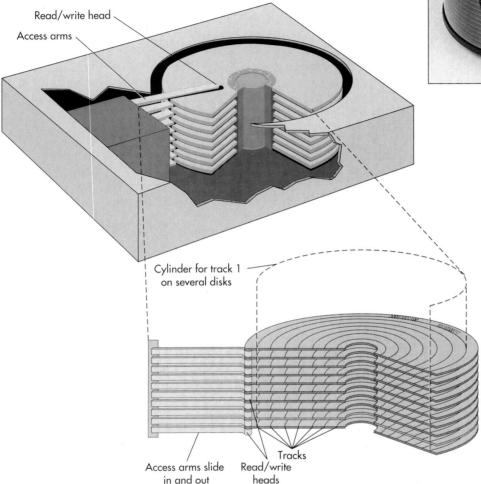

Read/write head

Access arms

Cylinder for track 1 on several disks

Access arms slide in and out

Read/write heads

Tracks

A microcomputer hard disk has about 80 tracks; a mainframe's hard disk about 200 tracks, thus 200 cylinders.

Optical Disks

Preview & Review: Optical disks are removable disks on which data is written and read using laser technology. Six types of optical disks are CD-ROM, CD Plus, CD-R, WORM, erasable, and DVD.

CD-ROM formats may be interactive and multimedia.

By now optical-disk technology is well known to most people. **An *optical disk* is a removable disk on which data is written and read through the use of laser beams.** The most familiar form of optical disk is the one used in the music industry. A *compact disk,* or *CD,* is an audio disk that uses digital code and that looks like a miniature phonograph record. A CD holds up to 74 minutes of high-fidelity stereo sound.

The optical-disk technology that has revolutionized the music business with music CDs is doing the same for secondary storage with computers. A single optical disk of the type called CD-ROM can hold up to about 700 megabytes of data. This works out to about 269,000 pages of text, or more than 7500 photos or graphics, or 20 hours of speech, or 77 minutes of video. Although some disks are used strictly for digital data storage, many combine text, visuals, and sound.

In the principal types of optical-disk technology, a high-power laser beam is used to represent data by burning tiny pits into the surface of a hard plastic disk. To read the data, a low-power laser light scans the disk surface: Pitted areas are not reflected and are interpreted as 0 bits; smooth areas are reflected and are interpreted as 1 bits. (■ *See Panel 7.10.*) Because the pits are so tiny, a great deal more data can be represented than is possible in the same amount of space on a magnetic disk, whether flexible or hard.

The optical-disk technology used with computers consists of six types:

- CD-ROM disks
- CD Plus disks
- CD-R disks
- WORM disks
- Erasable optical disks
- DVD disks

We will also describe some variations on these, such as video CDs, which offer the prospect of putting movies on CD-ROM.

Recording data

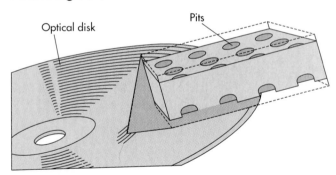

Reading data

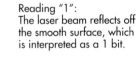

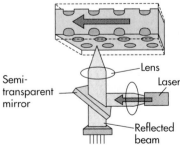

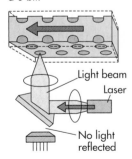

Reading "1":
The laser beam reflects off the smooth surface, which is interpreted as a 1 bit.

Reading "0":
The laser beam enters a pit and is not reflected, which is interpreted as a 0 bit.

PANEL 7.10

Optical disks

(Top) In most cases, the disk producer uses a high-power laser beam in special recording equipment to burn tiny pits, in an encoded pattern, onto the disk's surface. *(Bottom)* The user's optical disk drive uses a low-power laser beam to read the code by reflecting the beam off smooth areas—interpreted as 1 bits. The beam does not reflect off pitted areas—interpreted as 0 bits.

CD-ROM

For microcomputer users, the best-known type of optical disk is the CD-ROM. **CD-ROM, which stands for *compact disk–read-only memory*, is an optical-disk format that is used to hold prerecorded text, graphics, and sound.** Like music CDs, a CD-ROM is a read-only disk. **Read-only means the disk cannot be written on or erased by the user.** You as the user have access only to the data imprinted by the disk's manufacturer.

More and more microcomputers are being made with built-in CD-ROM drives. (■ *See Panel 7.11.*) However, many microcomputer users buy their CD-ROM drives separately and connect them to their computers. This requires installation of an audio circuit board (✓ p. 262) and speakers if you wish to be able to play the kind of CD-ROMs that offer music and sound.

At one time a CD-ROM drive was only a single-speed drive. Now double-speed and quadruple-speed drives—abbreviated 2X and 4X—are standard, and 6X and 8X drives are rapidly becoming affordable. A single-speed drive will access data at 150 kilobytes per second, a double-speed drive at 300 kilobytes per second. This means that a double-speed drive spins the compact disk twice as fast. "Quad-speed" (4X) CD-ROM drives access data at 600 kilobytes per second. The faster the drive spins, the more quickly it can deliver data to the processor.

Some CD-ROMs are designed to run only with Microsoft Windows on IBM-compatible computers or only on Apple Macintoshes. However, some are "hybrid" disks that include versions of the same program for both Macs and Windows.

Originally, computer makers thought that CD-ROMs "would be good for storing databases, documents, directories, and other archival information that would not need to be altered," says one report. "Customers would be libraries and businesses."[10] Although data storage is one use, there are clearly many others:

- **Entertainment and games:** Examples are 25 years of Garry Trudeau's comic strip, *Doonesbury* (on one disk), as well as games such as *Myst, Doom II, Dark Forces,* and *Sherlock Holmes, Consulting Detective.*

PANEL 7.11

CD-ROM drives in microcomputers

(Left) Portable computer. *(Right)* Desktop microcomputer.

- **Music, culture, and films:** Want to experience a little "rock 'n' ROM"? In *Xplora 1: Peter Gabriel's Secret World*, you can not only hear rock star Gabriel play his songs but also create "jam sessions" in which you can match up musicians from around the world and hear the result.[11] Other examples of such CD-ROMs are *Bob Dylan: Highway 61 Interactive*; *Multimedia Beethoven, Mozart, Schubert*; *Art Gallery*; *American Interactive*; *A Passion for Art*; and *Robert Mapplethorpe: An Overview*.

Developers have also released several films on CD-ROM, such as the 1964 Beatles movie, *A Hard Day's Night, This Is Spinal Tap*, and *The Day After Trinity*.

- **Encyclopedias, atlases, and reference works:** The principal CD-ROM encyclopedias are *The Grolier Multimedia Encyclopedia*, *Compton's Interactive Encyclopedia*, and *Microsoft Encarta Encyclopedia*. Each packs the entire text of a traditional multivolume encyclopedia onto a single disk, accompanied by pictures, maps, animation, and snippets of audio and video. All have pull-down menus and buttons to trigger various search functions.

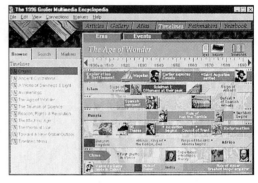

CD-ROMs are also turning atlases into multimedia extravaganzas. Mindscape's *World Atlas & Almanac*, for instance, combines the maps, color photos, geographical information, and demographic statistics of a traditional book-style atlas with video, sounds, and the ability to immediately find what you want. Maps, for example, may include political charts showing countries and cities, three-dimensional maps showing mountain ranges, and even satellite maps. An audio feature lets you hear the pronunciation of place names. There are also street atlases (*Street Atlas U.S.A., StreetFinder*), which give detailed maps that can pinpoint addresses and show every block in a city or town, and trip planners (*TripMaker, Map 'n' Go*), which suggest routes, attractions, and places to eat and sleep.

Examples of other types of CD-ROM reference works are *Eyewitness History of the World, Eyewitness Encyclopedia of Nature, Skier's Encyclopedia*, and *The Way Things Work*. You can get the full text of 1750 great works of literature and other books and documents on *Library of the Future*.

- **Catalogs:** Publishers have also discovered that CD-ROMs can be used as electronic catalogs, or even "megalogs." One, for instance, combines the catalogs of several companies. "A single disk now holds the equivalent of 7000 pages of [text and graphical] information on almost 50,000 different products from salad-bar sneeze-guards to deep-fat fryers," noted one report.[12]

- **Education and training:** Want to learn photography? You could buy a pair of CD-ROMs by Bryan Peterson called *Learning to See Creatively* (about composition) and *Understanding Exposure* (discussing the science of exposure). When you pop these disks in your computer, you can practice on screen with lenses, camera settings, film speeds, and the like.[13] Or you could learn history from such CD-ROMs as *Critical Mass: America's Race to Build the Atomic Bomb* or *The War in Vietnam*. CD-ROMs (*Score Builder for the SAT, Inside the SAT*) are also available to help students raise their scores on the Scholastic Aptitude Test.

- **Edutainment:** *Edutainment software* consists of programs that look like games but actually teach, in a way that feels like fun.[14] An example for children ages 3–6 is *Yearn 2 Learn Peanuts*, which teaches math, geography, and reading. *Multimedia Beethoven: The Ninth Symphony*, an edutainment program for adults, plays the four movements of the symphony while the on-screen text provides a running commentary and allows you to stop and interact with the program.

- **Books and magazines:** Book publishers have hundreds of CD-ROM titles, ranging from *Discovering Shakespeare* and *The Official Super Bowl Commemorative Edition* to business directories such as *ProPhone Select* and *11 Million Businesses Phone Book*.

In 1993, photographer Rick Smolan pioneered the splashy CD-ROM book with *From Alice to Ocean*, which used text, audio, digital photos, and video clips to tell the story of a young woman's 1700-mile lone journey by camel through the Australian Outback. In 1995, Smolan and 70 international photographers produced a CD-ROM and accompanying book called *Passage to Vietnam*. "It's not a guide to the country, or a history of it," said one reviewer. "Instead, it's an ode to the art of photography itself, with Vietnam serving mainly as the backdrop."[15]

Examples of CD-ROM magazines are *Blender*, on the subject of music; *Medio*, a general-interest monthly; *Go Digital*; and *Launch*. Aimed at 18- to 35-year-olds, *Launch* is said to aspire "to be the *Entertainment Weekly* of the magazine-on-a-disk business by offering a look at new music, movies, and computer games."[16] *Launch* has something called the Main Square (a so-called town square), which viewers can visit to select where they want to go. Choices include a listening room to hear music, a movie theater to look at previews of feature films, and an "ominous-looking" statue that is the entrance to the latest videogames.[17]

Clearly CD-ROMs are not just a mildly interesting technological improvement. They have evolved into a full-fledged mass medium of their own, on the way to becoming as important as books or films.

CD Plus/Enhanced CD: The Hybrid CD for Computers & Stereos

In the fall of 1995, disks began to appear with a new technical standard representing the convergence of software and music—*CD Plus*, also called *enhanced CD*. Many people believe the disks "represent a transformation in the way we use music—from a passive activity to an engaging interactive experience," said one writer. "Skeptics contend it's just the latest scheme to separate you and your money."[18]

A *CD Plus*, or *enhanced CD*, **is a digital disk that is a hybrid of audio-only compact disk and multimedia CD-ROM.** It can be played like any music CD in a stereo system, then put in a computer's CD-ROM drive to provide additional audio, video, text, and graphics. Early CD Plus titles, from Sony Music Entertainment, featured Alice in Chains, Toad the Wet Sprocket, Mariah Carey, and Bob Dylan. *Newsweek*'s technology writer Steven Levy says CD Plus is the '90s answer to album covers: "Even those CD Pluses with a rather predictable mix of video clips, lyrics, biographical matter, and interviews," he says, "can offer something we never got with liner notes: insight in an artist's own voice of how he or she went about producing the music."[19]

A problem at present is the number of technical snags.[20,21] With some CD Plus disks, for instance, you may have to skip track 1 manually when playing the disk on an audio CD player because "noise" from the computer data on that track can damage the speakers on your stereo (or your ears). Some CD Plus disks won't perform on users' present CD-ROM drives at all, such as the single-speed models. Indeed, it is variously estimated that CD Plus won't work in one-third to one-half of today's installed CD-ROM drives. The bottom line, then, is that the standard is really designed for *future* computer systems, which will all be CD Plus compatible.

CD-R: Recording Your Own CDs

CD-R, which stands for *compact disk–recordable*, is a CD format that allows you to write data onto a specially manufactured disk that can then be read by a standard CD-ROM drive. Like CD-ROM disks, CD-R disks cannot be erased.

Until quite recently, CD-R technology was far too expensive and complicated for ordinary computer users. Now systems for recording your own CD-ROMs have dropped to under $1000, and by early 1996 blank disks had become available in bulk for about $7 each. Says one article, "With an in-house production system consisting of the CD-R hardware and software, a fast PC, and a dedicated hard drive, an experienced user can record a CD-ROM in an hour or less."[22]

One of the most interesting examples of CD-R technology is the Photo CD system. Developed by Eastman Kodak, **Photo CD is a technology that allows photographs taken with an ordinary 35-millimeter camera to be stored digitally on an optical disk.** (■ *See Panel 7.12, next page.*) You can shoot up to 100 color photographs and take them to a local photo shop for processing. A week later you will receive back not only conventional negatives and snapshots but also the images on a CD-ROM. You can then view the disk on one of Kodak's own Photo CD players, which attaches directly to a television set. Or the disk will play in most CD-ROM-equipped personal computers running Windows or the Macintosh Operating System.

Kodak intended the Photo CD to be a consumer product, but it did not turn out that way. Now it is used mostly by professional photographers, multimedia developers, and designers.[23] Nevertheless, Photo CD is particularly significant for the impact it will have on the manipulation of photographs. With the right software, you can flip, crop, and rotate photos and incorporate them into desktop-publishing materials and print them out on laser printers.[24] Commercial photographers and graphics professionals can manipulate images at the level of pixels (✓ p. 250). For example, photographers can easily do *morphing*—merge images from different sources, such as superimpose the heads of show-business figures in the places of the U.S. presidents on Mount Rushmore. "Because the image is digital," says one writer, "it can be taken apart pixel by pixel and put back together in many ways."[25] This helps photo professionals further their range, although at the same time it presents a danger that photographs will be compromised in their credibility.

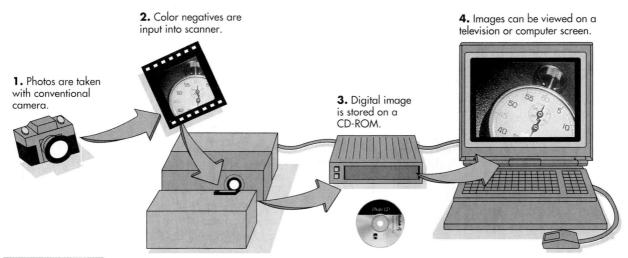

2. Color negatives are input into scanner.

1. Photos are taken with conventional camera.

3. Digital image is stored on a CD-ROM.

4. Images can be viewed on a television or computer screen.

PANEL 7.12

Photo CD

With Kodak's Photo CD system, ordinary snapshots can be stored on CD-ROMs, then viewed on a computer display screen or on a television screen.

R E A D M E

Case Study: The Magic of Morphing

Ever wish you could turn your local congressperson into the rodent of your choice? Give it a whirl with HSC Software's *Digital MORPH*. In fact, you might want to try turning your spouse into a squash or your landlord into an attractive ceiling fixture while you're at it. . . .

Morphing, based on the Greek word *morphe,* meaning "shape," is a type of technology that stretches and distorts one image to another. Digital MORPH is a special-effects program that produces shape-shifting effects such as those seen in the movie *Terminator 2* (where the antagonist shape-shifts into different objects) and Michael Jackson's *Black or White* video in which faces morph from one race to another.

The program was originally created in a higher-end version to help plastic and oral surgeons and orthodontists determine the after-surgery appearances of their patients. Today, it is available from HSC Software, a developer of graphic imaging, special effects, and multimedia software, for those with Microsoft Windows-capable computers and a desire for some creative fun.

Digital MORPH works its magic upon scanned images—images that have been digitally recorded by a scanner (a device that reads images using light beams much like a photocopier). To morph an image, the user need only select source and destination images (a starting and finishing image), select field lines (these determine where on the image changes will take place), and select the number of frames in which the change will take place. The program takes care of the rest.

"It's a lot of fun playing with images on a computer," says HSC CEO John Wilcazk. "Imagine morphing your boss into a gorilla, your sibling into the family dog, or your car into a leopard. The creativity is really endless, not only for individual users but also for multimedia producers, graphic artists, animators, and business-presentation creators."

—Tosca Moon Lee, "Shape-Shifting with Digital MORPH," *PC Novice*

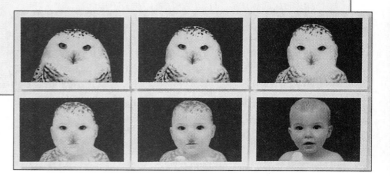

WORM Disks

WORM stands for *write once, read many*. A *WORM disk* can be written, or recorded, onto just once and then cannot be erased; it can be read many times. In principle, WORM technology is similar to CD-R technology; however, it is a bit older and more expensive than CD-R technology. WORM technology is useful for storing data for backup and archival purposes because it can store greater volumes of data than other types of CD-ROMs (and magnetic tape). A mainframe-based WORM disk may hold 200 gigabytes of data; a microcomputer-based WORM disk can hold about 600 megabytes.

Erasable Optical Disks

An *erasable optical disk* allows users to erase data so that the disk can be used over and over again. The most common type of erasable and rewritable optical disk is probably the *magneto-optical (MO) disk,* which uses aspects of both magnetic-disk and optical-disk technologies. Most personal computer users haven't used MO disk drives, but that may change. One recently released MO drive writes on removable disks that hold 4.6 gigabytes of data in all. However, such drives cost about twice as much as a 2-gigabyte hard-disk drive.

DVD & DVD-ROM Disks: The "Digital Convergence" Disk

According to the various industries sponsoring it, *DVD* isn't an abbreviation for anything. The letters used to stand for "digital video disk" and later, when its diverse possibilities became obvious, for "digital versatile disk."[26] But DVD is the designation that Sony/Philips (with its Multi Media CD) and Toshiba/Time Warner (with its Super Density disk) agreed to in late 1995, when they avoided a "format war" by joining forces to meld their two advanced disk designs into one. Suffice it to say that *DVD* is a 5-inch optically readable digital disk that looks like an audio compact disk but can store 4.7 gigabytes of data on a side, allowing great data storage, studio-quality video images, and theater-like surround sound. (■ *See Panel 7.13.)* Actually, the home-entertainment version is called simply the *DVD*. **The computer version of the DVD is called the *DVD-ROM disk*. It represents a new**

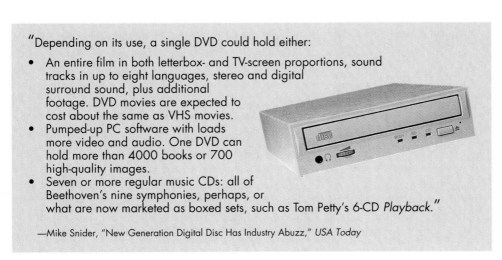

"Depending on its use, a single DVD could hold either:

- An entire film in both letterbox- and TV-screen proportions, sound tracks in up to eight languages, stereo and digital surround sound, plus additional footage. DVD movies are expected to cost about the same as VHS movies.
- Pumped-up PC software with loads more video and audio. One DVD can hold more than 4000 books or 700 high-quality images.
- Seven or more regular music CDs: all of Beethoven's nine symphonies, perhaps, or what are now marketed as boxed sets, such as Tom Petty's 6-CD *Playback*."

—Mike Snider, "New Generation Digital Disc Has Industry Abuzz," *USA Today*

PANEL 7.13

DVD: new era of data storage

Similar to CDs, the disk in this CD-ROM drive can hold 4.7 gigabytes of information on a single side, about seven times the capacity of present CD-ROMs. A whole movie can be offered on one side, with digital video and CD-quality sound. The unit shown here is from Toshiba.

generation of high-density CD-ROM disks, with either write-once or rewritable capabilities.

Though the product of bickering parents, DVD promises to grow into a giant. DVD, says *San Jose Mercury News* computer writer Mike Langberg,

> is clearly destined for greatness as the first information storage technology designed from the ground up to work with all types of electronics, televisions, stereos, computers, and videogame players. That means DVD will finally launch the much-heralded concept of "convergence," where the lines blur between all these devices.[27]

How does a DVD work? Like a CD or CD-ROM, the surface of a DVD contains microscopic pits, which represent the 0s and 1s of digital code that can be read by a laser. The pits on the DVD, however, are much smaller and closer together than those on a CD, allowing far more information to be represented there. Also, the technology uses a new generation of lasers, which allow a laser beam to focus on pits roughly half the size of those on current audio CDs. Another important development is that the DVD format allows for *two layers* of data-defining pits, not just one. Finally, engineers have succeeded in squeezing more data into fewer pits, principally through data compression.[28]

Besides being designed from the outset to work with the full range of electronic, television, and computer hardware, the major characteristics of the DVD are as follows:

- **More storage capacity, faster data transfer:** A single side will hold 4.7 gigabytes of information, or 133 minutes of surround-sound audio playing time. (This compares with 780 megabytes or 74 minutes of playing time on today's single-sided music CD.) A single side with a *dual layer* of data-defining pits can hold 8.5 gigabytes. A double-sided dual-layer DVD can hold a mind-boggling 17 gigabytes, allowing 488 minutes of audio.

  In addition, data can be transferred faster—1.35 megabytes per second, more than twice as fast as today's quadruple-speed CD-ROM players.

- **Better audio:** "A Toshiba demonstration clip of the train wreck scene from *The Fugitive* put me right in the midst of squealing brakes and crunching metal," wrote an observer of a DVD prototype, "thanks to a 5.1-channel digital surround-sound specification."[29] The specification allows for six separate audio tracks.

  One important matter: DVD players will be backward compatible, able to play today's audio compact disks.

- **Better video:** Picture quality can visibly surpass that currently delivered on videocassettes. Viewers will also be able to choose variations in what's known as the "picture aspect-ratio," or screen size: wide-screen format for a wide-screen TV, letterbox format to show the wide-screen version on a conventional TV, or a version cropped to fill a conventional TV screen. Viewers could also employ slow-motion, fast-forward, and freeze-frame features, as they do now on VCRs.

- **DVD-ROM and recordable and rewritable capabilities:** Software publishers now employ so much space-eating video and animation in their programs that they are rapidly running out of room on today's standard 700-megabyte CD-ROM disks. Thus, they and computer makers decided they had to enter the DVD standards fray to push for a single DVD design that could be used for computer data storage as well as for movies and music. They also wanted the new drives to be able to read today's CD-ROMs. The result is the DVD-ROM.

We are now entering one of those periods familiar to computer users in which new technology comes along that simultaneously delights and frustrates consumers. The delight, of course, will be that DVD-ROMs will store a lot more information and enable you (eventually) to see movies and listen to high-quality music on your personal computer. The frustration will be that many present kinds of hardware will soon become obsolete or irrelevant, and there will be lots of changes until everything shakes down.[30]

Interactive & Multimedia CD-ROM Formats

As use of CD-ROMs has burgeoned, so has the vocabulary, creating difficulty for consumers. Much of this confusion arises in conjunction with the words *interactive* and *multimedia*.

As we mentioned earlier, **interactive means that the user controls the direction of a program or presentation on the storage medium.** That is, there is back-and-forth interaction, as between a player and a videogame. You could create an interactive production of your own on a regular 3½-inch diskette. However, because of its limited capacity, you would not be able to present full-motion video and high-quality audio. The best interactive storage media are CD-ROMs and DVD-ROMs.

***Multimedia* (an adjective or noun) refers to technology that presents information in more than one medium, including text, graphics, animation, video, sound, and voice.** As used by telephone, cable, broadcasting, and entertainment companies, *multimedia* also refers to the so-called Information Superhighway. On this avenue various kinds of information and entertainment will be delivered to your home through wired and wireless communication lines.

There are perhaps 20 different CD-ROM formats. The majority of nongame CD-ROM disks are available for Macintosh or for Windows- or DOS-based microcomputers. However, there are a host of other formats—CD-I, CDTV, Video CD, Sega CD, 3DO, CD+G, CD+MIDI, CD-V—that are not mutually compatible. Most will probably disappear as the DVD format takes over.

Now we leave the topic of optical disk storage and move on to a type of storage device touched on earlier in a different context—storage on *PC cards*.

Flash-Memory Cards

Preview & Review: Flash-memory cards consist of circuitry on credit-card-size cards that can be inserted into slots in a microcomputer.

Disk drives, whether for diskettes or for CD-ROMs, all involve moving parts—and moving parts can break. Flash-memory cards, by contrast, are variations on conventional computer-memory chips, which have no moving parts. ***Flash-memory* cards consist of circuitry on PC cards that can be inserted into slots connecting to the motherboard** (✓ p. 171). Each can hold up to 100 megabytes of data.

A videotape produced for Intel, which makes flash-memory cards, demonstrates their advantage, as one report makes clear:

> In it, engineers strap a memory card onto one electric paint shaker and a
> disk drive onto another. Each storage device is linked to a personal computer,

running identical graphics programs. Then the engineers switch on the paint shakers. Immediately, the disk drive fails, its delicate recording heads smashed against its spinning metal platters. The flash-memory card takes the shaking and keeps on going.[31]

Flash-memory cards are not infallible. Their circuits wear out after repeated use, limiting their lifespan. Still, unlike conventional computer memory (RAM or primary storage), flash memory is *nonvolatile*. That is, it retains data even when the power is turned off.

A much older type of storage than flash-memory cards is magnetic tape—indeed, tape used to be the *principal* storage medium.

Magnetic Tape

Preview & Review: Magnetic tape is thin plastic tape on which data can be represented with magnetized spots. On large computers, tapes are used on magnetic-tape units. On microcomputers, tapes are used in cartridge tape units.

Magnetic tape **is thin plastic tape that has been coated with a substance that can be magnetized; data is represented by the magnetized or nonmagnetized spots.** Today "mag tape" is used mainly to provide backup, or duplicate storage. It's a much slower form of secondary storage, but it's also less expensive than disk storage.

Representing Data on Magnetic Tape

Traditional magnetic tape stores each character, or byte, of data in *frames* of magnetic spots running the length of the tape. The 0 or 1 bits making up a byte are represented by a magnetized spot for a 1 bit and a nonmagnetized spot for a 0 bit. (■ *See Panel 7.14.*) (As described earlier, a byte is made up of eight bits, with a ninth bit representing a parity bit to check for errors [✓ p. 169]). The capacity, or storage density, is represented in bytes per inch, or *bpi*.

The two principal forms of tape storage of interest to us are *magnetic-tape units*, traditionally used with mainframes and minicomputers, and *cartridge tape units*, which are often used for backup on microcomputers.

Magnetic-Tape Units for Large Computers

The kind of cassette tapes you use for an audiotape recorder are 200 feet long and record 200 bytes per inch. By contrast, a reel of magnetic tape used in mainframe and minicomputer storage systems is ½ inch wide, 3600 feet long, and can hold 1600–6250 bpi. A traditional 10½-inch tape reel can hold up to 250 megabytes of data.

Tapes are used on *magnetic-tape units* **or** *magnetic-tape drives,* **which consist of a read/write head and spindles for mounting two tape reels: a supply reel and a take-up reel.** (■ *Refer back to Panel 7.14.*) The tape is reeled off the supply reel, fed through pulleys that regulate its speed and hold it still long enough for data to be read from it or written to it by the read/write heads, and then wound up on the take-up reel. During the writing process, any existing data on the tape is automatically written over, or erased.

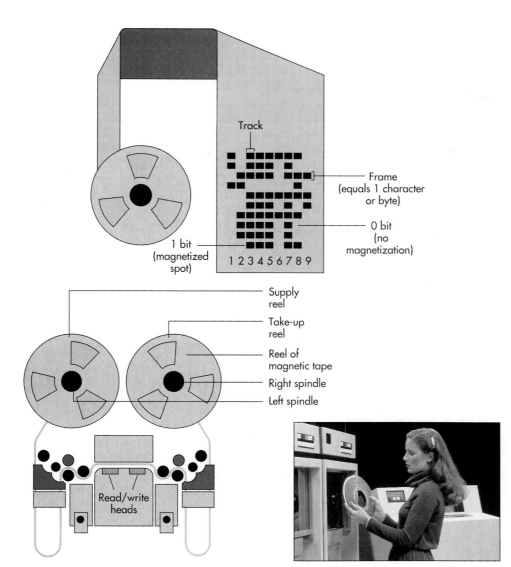

Track

Frame
(equals 1 character
or byte)

0 bit
(no
magnetization)

1 bit
(magnetized
spot)

1 2 3 4 5 6 7 8 9

Supply
reel

Take-up
reel

Reel of
magnetic tape

Right spindle

Left spindle

Read/write
heads

PANEL 7.14

Magnetic-tape storage

(Top) On tape, data is represented by the presence and absence of magnetic spots. *(Bottom)* Tape reels are mounted on spindles, which turn the reels. Tape is unreeled off the supply reel, fed past the read/write heads that read (retrieve) or write (record) data, and wound up on the take-up reel.

Large organizations, such as public utilities, often use reels of mag tape for storing backup records of essential data, such as customer names and account numbers. Usually these reels are housed in *tape libraries,* or special rooms, and there are strict security procedures governing their use. For example, the Omaha computer center for the credit-card processor First Data Corporation is described as being bigger than a football field. "Most of it is filled with data storage silos that contain more than 200,000 computer tapes," says the account. "The silos contain information on 92 million credit cards processed [in 1994]. . . . You can walk blocks before bumping into a human being."[32]

Cartridge Tape Units

"Sometimes I think computing without backup should be against the law," says a computer journalist. "Once you could copy to a few floppy disks. In an era of gigabyte drives and enormous multimedia files, that just isn't practical because that could mean shuffling hundreds of floppies each time you back up."[33]

PANEL 7.15

Cartridge tape unit

This device is used with microcomputers to back up data from a hard disk. The drive is inserted into a drive bay in the system cabinet.

You could use removable disks, such as Iomega's and Syquest's, but even they would require 10 disks each (at $20 apiece) to back up a 1-gigabyte hard-disk drive. Here is the problem for which cartridge tape units are the solution.

Cartridge tape units, **also called** *tape streamers,* **are used to back up data from a microcomputer hard disk onto a tape cartridge.** (■ *See Panel 7.15.*) The dominant standard for tape cartridge units for the last 10 years or so has been *QIC* or quarter-inch cassettes, in which the tape inside the cartridges is one-quarter inch wide. These tapes may hold more than 500 megabytes. The tape drive fits into a computer's drive bay—for example, the recess where you would put a second diskette drive. There are also some external tape units.

A newer tape format is *QIC-Wide.* Another standard, called *Travan,* puts up to 1600 megabytes on a tape. An advanced form of cassette, adapted from technology used in the music industry, is the **digital audiotape (DAT), or R-DAT, which uses 2- or 3-inch cassettes and stores 2 gigabytes or more.** Future R-DATs are expected to hold as much as 8 gigabytes.

Is backup really that important? "Tape drives are similar to life insurance," says a computer columnist, "in that you hope you never need it and resent the premiums—until you need the coverage."[34] (■ *See Panel 7.16.*)

▶ **The Future of Secondary Storage**

Preview & Review: Other developments in secondary storage include online storage, advanced compression schemes, CD-ROM jukeboxes, higher-density disks, digital VCRs and digital videodisks, and molecular electronics.

PANEL 7.16

How much is backup worth?

"'When you lose a disk, you're not only losing the hardware and software,' said John L. Copen, president of Integ, an information protection company in Manhattan. 'The information has to be reproduced, and if you have to reproduce it without a backup. . . .,'

Mr. Copen demonstrates the point by holding up a digital audiotape (DAT) cassette, one of the newer technologies used for backing up data on larger hard disk drives, the kind that act as hubs for networks of personal computers in an office.

'I ask people in the audience what it's worth,' he said. 'It's a little cassette about the size of a credit card. The cassette costs about $16. I ask them to guess how much it can store. Forty megs? Eighty megs? It stores four gigabytes.' A gigabyte is roughly a thousand megabytes, or a billion characters of information.

'How much information can you put in four gigs?' Mr. Copen continued. 'About 20,000 big spreadsheets, which translates to about 100,000 days of work, or 800,000 hours. At $20 an hour, that's $16 million. Never before have people been able to reach down, pick up a cassette and walk out the door with $16 million of data in their pocket.'"

—Peter H. Lewis, "Finding an Electronic Safe-Deposit Data Box," *New York Times*

In the next few years, could a chapter on secondary storage in a book such as this become simply a gigantic waste of time?

Suppose that the network computer, Internet appliance, or "hollow PC" (✓ p. 29) actually becomes as popular as its promoters hope it will. This device consists of a small computer with just a keyboard, display screen, processor, and connecting ports for network or phone cables. The gadget would need not have any secondary storage. Rather, the Internet itself can become your hard disk. This is the notion being pushed by Oracle, Sun, Apple, and other companies producing such machines.

Is such a device just wishful thinking? Computer editor Dan Gillmor, for one, believes the network computer makes plenty of sense. Companies could issue the gadget to on-the-road employees to pull information from corporate databases. Schools and libraries would find NCs ideal information-hunting devices. Families could keep one in the kitchen to input and retrieve recipes and to-do lists. If you think of the network computer as a clever information device to help navigate life's routines, says Gillmor, the idea has a lot of merit.

But there are exceptions. "I don't expect the Net to replace my hard disk," Gillmor says. "For one thing, I care too much about convenient access to my personal information, not to mention privacy, to take such a risk."[35] Writer Robert Rossney points out another problem with storing everything on the Net: "The Internet is a firehose of data," he says. "There's too much stuff out there to keep track of, and it's just going to get worse."[36]

For these and other reasons, then, it seems that most people will continue to want their own secondary-storage devices. What follows are some noteworthy developments to which we should pay close attention—advanced compression schemes, CD-ROM jukeboxes, digital VCRs, and advanced storage technology.

Advanced Compression Schemes

A fully digitized photograph requires about 80 megabytes of storage space. In 1995 Eastman Kodak demonstrated a technique for hypercompressing photographs that reproduced human faces using only about 50 bytes of data. Such images, when stored on the magnetic strip of credit cards or bar codes on bank checks, could be used to counteract fraud. Tellers or retail clerks could pass the card or check through a device that displays the customer's likeness on a display monitor.[37]

The Kodak technology uses a technique called *wavelet compression;* however, other compression methods are also being exploited:

- **Wavelet compression:** In *wavelet compression* an image is compressed as a whole, rather than block by block, as with JPEG and MPEG codecs. This technology is useful for files that must be stored or transmitted with high fidelity, such as medical images or music.

- **Fractal compression:** A *fractal image* is one whose features look similar at different scales of magnification. For example, the features of a coastline are more or less the same no matter how high the altitude from which they are photographed. With *fractal compression,* just one image may be stored and then used over and over to re-create the whole picture. The image is based on information about size and position that is stored somewhere else.

- **Compression by object-oriented programming:** *Object-oriented programming* (discussed in Chapter 11) treats each segment in a software program as an individual unit, called an "object," that can be used repeatedly in different applications and by different programmers. The "objects" might be blocks of data, mathematical procedures, or—for present purposes—video images. With the use of object-oriented programming, an image in a movie might be described once and then reused endlessly. Thus, a plane flying through the air might be defined with as few as two objects—the plane and the sky.

- **Compression using neural networks:** *Neural networks* (discussed in Chapter 12) use physical electronic devices or software to mimic the neurological structure of the human brain. Neural networks, which learn from experience, can be trained to squeeze data down to the irreducible minimum. "The neural net can be fed billions of bits of data—all the pixels in all the frames of a movie or all the text in an online library—and it will generate a cloud-like map of points," says one account. "Because these points exist in a world of higher mathematics with an almost infinite number of dimensions, each point contains a wealth of data, which the net can use to reconstitute every detail of the original information."[38]

CD-ROM Jukeboxes

How does a chemical company with a long history going back to 1861 go about searching for patent information that by 1990 had swelled to 30 million documents? BASF of Germany decided to scan its entire archives and store the contents on a library of 1700 CD-ROM disks. The company then installed what are known as *CD-ROM jukeboxes*—21 of them, each with 100 CD-ROM disks and all interconnected over a network.[39]

Like the coin-operated music machine that entertains patrons in bars, a CD-ROM jukebox uses a robot arm to exchange CD-ROMs between one or more drives. Also called libraries or changers, such jukeboxes can hold as few as six CD-ROM disks and a single drive and as many as 1400 disks and 32 drives. Optical disks may be 3.5, 5.25, and 12 inches.

Now imagine what a jukebox using *DVD-ROM* disks (conceivably up to 17 gigabytes each) will be like!

Higher-Density Diskettes

In 1995, IBM announced it had demonstrated magnetic-data storage density of 3 gigabytes (3 billion bits), which means that the text of 375 average-size novels could be stored in a single square inch of diskette surface.[40,41]

In 10 years, density may well go up by another factor of 10. With such densities, says one writer, "Movie buffs could download entire movies from online services, storing them on diskettes the size of a quarter. Hospitals and doctors could put a patient's records and X-rays on small disks to be carried with the patient."[42]

Digital VCRs & Videodisk Players

In 1994 several companies, including some of the world's leading names, such as IBM, Apple, Sony, and Matsushita, met in Tokyo to agree on standards for a digital videocassette recorder for the home.[43] Machines coming onto the market—called *DVCR* or *DVHS* machines, depending on the maker—are backward compatible, able to play regular VCR tapes.

Digital VCRs could provide a gigantic leap forward in unifying the separate sectors of computers and telecommunications. VCRs in digital format provide better picture quality and smaller and/or longer-playing cassettes. They also have the capability to make perfect copies, which will make it easier for camcorder owners to edit their videotapes. Equally important, digital VCRs will make a better match for future television systems, such as advanced TV, that will process signals in digital form. Indeed, they are specifically designed to record programming signals beamed down directly from the new class of Digital Satellite System. Finally, such VCRs could store not only video and television pictures but also large amounts of computer data. Thus, computer users could use this device to make backup copies of the data stored on their hard disks.[44]

A successor to the digital VCR is the *digital videodisk player*. Video CDs already exist and in fact now outsell VCRs in China. Videodisk players should reach the North American market by 1997.[45]

Video Servers

One of the most talked-about features of the coming Information Superhighway has been the potential ability to deliver interactive video and movies-on-demand through cable or other connections to people's homes. Although the arrival of such offerings seems to recede further into the future, companies such as Oracle are actively working to develop *video servers*, large-scale systems for storing thousands of movies and videos.

This is not an easy challenge. "Broadcasting a digitized film is fairly easy," explains one account. "Serving up 100 different films to 100 houses at 100 different times is a programming nightmare."[46] The solution that Oracle engineers came up with was to use a supercomputer, divide a movie into 1000 different segments, and store each segment on a different disk drive, each with its own microprocessor. One "master" microprocessor controls all the others so that a film flows seamlessly in the correct order.

Molecular Electronics: Storage at the Subatomic Level

An emerging field, molecular electronics, may push secondary storage into another dimension entirely. In what has been called "the world's smallest Etch-a-Sketch," physicists at NEC in Tokyo used a sophisticated probe—a tool called a *scanning tunneling microscope (STM)*—to paint and erase tiny lines roughly 20 atoms thick.[47] This development could someday lead to ultra-high-capacity storage devices for computer data.

Scientists have also reported research involving use of bacteria to store data in three dimensions. Said Robert Birge, who fashioned a 1-centimeter cube made of protein molecules that could store data in three dimensions, "Six of these cubes can store the entire Library of Congress."[48]

Onward: Toward a More Interactive Life

"This is the first generation that has never watched television without a remote control," wrote *Newsweek* writer Michael Rogers. "Neither conventional print nor passive television is really attractive to them anymore."[49]

Rogers was referring to people under age 25. Mass-media experts think this restless, demanding generation in the next century will want information and entertainment that *they* can control—in a word, that is *interactive*.

Also, if people suspect that a piece of information is being slanted, they will want to electronically turn to the sources themselves. In short, electronic multimedia offers a wider base of knowledge.

How to Buy a Multimedia System

Throughout the interactive CD-ROM entitled *Xplora 1: Peter Gabriel's Secret World,* rock star Gabriel's face keeps popping up in a corner of the screen. "Here is a collection of instruments from all over the world," says Gabriel in a soothing voice. Images of musical instruments—percussion, wind, stringed—appear on the screen. "Why not try playing a few of these yourself?" Click your mouse, and you can play one of these instruments. Or click again and get a four-track mixing board so that you can create your own version of Gabriel's "Digging the Dirt."[50] (■ *See Panel 7.17.*)

Has CD-ROM brought forth a new mass medium, like the birth of film or television? It would seem so. The development of multimedia—technology that presents information in more than one medium, including text, graphics, animation, video, music, and voice—is well underway. The number of CD-ROM players installed in North America alone was expected to be 28 million the end of 1995.[51]

Multimedia Standards

What do you need in order to experience an interactive program such as Peter Gabriel's? CD-ROMs can run on Macintosh computers, IBM-DOS and IBM-Windows (and Windows 95) computers, and what are called MPC machines (which are made by various manufacturers).

An *MPC machine* is a multimedia personal computer that adheres to standards set by the Multimedia PC (MPC) Working Group, now part of the Software Publishers Association. The MPC announced its first standard for multimedia personal computers, called *MPC1,* in 1991. As the capabilities of computers increased, the standard has been twice revised—to *MPC2* in 1993 and *MPC3* in 1995. (■ *See Panel 7.18.*) In general, these standards for multimedia computers have gone up even as the price has declined somewhat—to about $2000.

The output equipment required for an MPC3 system is shown in the accompanying illustration. (■ *See Panel 7.19.*)

Output Hardware & Software

The basic hardware and software required to meet MPC3 standards are as follows:

- *Processor, memory, keyboard, and mouse:* Under 1995 standards, the microprocessor should be a 75 MHz Pentium, or equivalent, with 8 megabytes of RAM. Besides the usual 101-key keyboard, you'll need a two-button mouse in order to engage the interactive parts of the multimedia program.

Interactive rock 'n' roll

Peter Gabriel's CD-ROM *Xplora 1: Peter Gabriel's Secret World* allows users to manipulate his music.

XPLORA 1 Peter Gabriel's Secret World

"Click up the image of a four-track mixing board. And here's Gabriel . . . , in the upper left-hand corner to say, 'Play with the balance until you get something you like. . . . Why don't *you* try it?'

You can call up any one of Gabriel's music videos to watch as it was orignally produced. . . . Or click again and make changes. Or listen to Gabriel talk about the inspiration behind some of his songs.

Point the mouse at an image of a suitcase or a photo album to find out about Gabriel's personal life—perhaps more than you want to, including peeking in on his baby photographs. Or use the mouse to make the infant Gabriel move around.

Then shift gears completely and call up material on some of Gabriel's political interests, such as Amnesty International and Witness, a program designed to equip human rights activists with video cameras to record atrocities.

Gabriel's CD-ROM is a fun and challenging program that could keep you chained to your computer for hours, just getting the hang of it."

—Sylvia Rubin, "Interactive Rock and Roll," *San Francisco Chronicle*

Three levels of multimedia specifications

Level 1 (MPC1, 1991) set up a multimedia standard for the first generation of IBM-compatible multimedia microcomputers, listing the hardware and software requirements for playing CD-ROMs with sound and limited video. Level 2 (MPC2, 1993) improved the quality of video and audio playback. Level 3 (MPC3, 1995) is intended to support MPEG-1 CD-ROMs, which can hold 72 minutes of video equal in quality to a regular TV.

	MPC1	MPC2	MPC3
Date introduced	March 1991	May 1993	June 1995
Processor	16 MHz 386SX	25 MHz 486SX	75 MHz Pentium (or similar)
RAM	2 megabytes	4 megabytes	8 megabytes
Hard drive	30 megabytes	160 megabytes	540 megabytes
CD-ROM drive (kilobytes/second)	Single-speed (150)	Double-speed (300)	Quadruple-speed (600)
Sound board	8-bit	16-bit	16-bit with wave-table
System software	Windows 3.0	Windows 3.0	Windows 3.11 and DOS 6.0; Windows 95
Video display	640 X 480, 16 colors	640 X 480, 65,536 colors	MPEG-1 compatible (full screen and full-motion video)

Output components for multimedia

Shown is IBM-compatible hardware required to meet the latest MPC (level 3) standards.

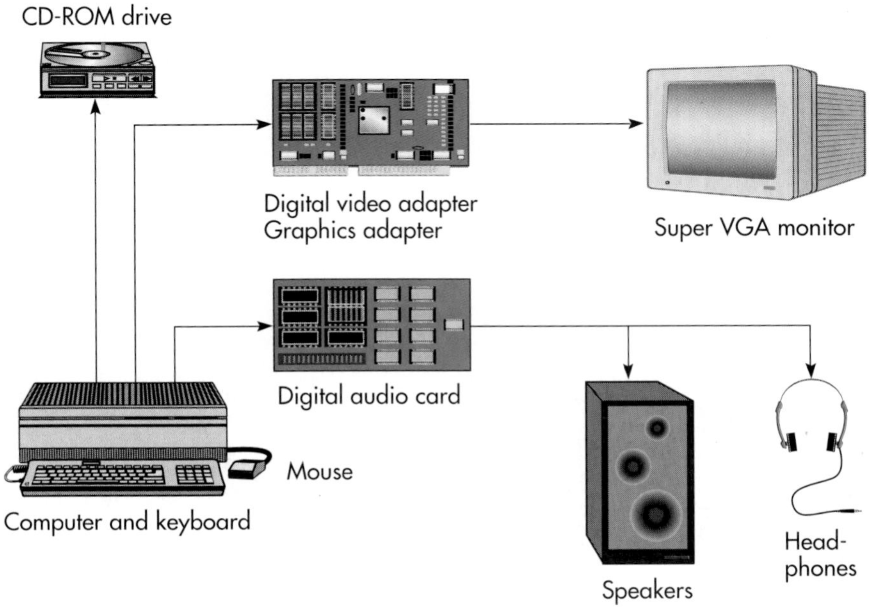

CD-ROM drive

Digital video adapter
Graphics adapter

Super VGA monitor

Digital audio card

Computer and keyboard

Mouse

Speakers

Head-phones

- *Secondary storage:*   Besides a standard diskette drive (3½-inch, high-density), you'll need a hard-disk drive and a CD-ROM drive. The hard disk should be able to store at least 540 megabytes (more is always better). You'll need to gain fast access to many small multimedia files.

  The CD-ROM drive should be quadruple speed, able to transfer data at the rate of 600 kilobytes per second. It should handle 3½-inch optical disks with 600 megabytes.

- *Video output:*   For color output—which will include graphics, animation, and video—you'll need a 65,536-color Super VGA (✓ p. 252) monitor and display adapter card that provide full screen and full-motion video.

- *Audio output:*   Most microcomputers come with a tiny—and tinny—2-inch speaker. Stereo speakers offer the much better sound quality that is appropriate to a multimedia system, assuming you also have a sound board. The sound board is a special circuit board that goes inside the computer. If you don't want to use speakers, you can plug headphones, like those used on a Sony Walkman, into a rear connector on the sound board. The sound board comes with software, which must be used during the installation.

- *Systems software:*   MPC3 standards specify Windows 3.11 and DOS 6.0. Windows 95 and later versions also meet these standards.

If you already have an IBM-style microcomputer system, you may be able to upgrade to the next level of MPC standard—most easily from MPC1 to MPC2. Some CD-ROM drive manufacturers claim their products will run on a microcomputer with '386 processor, but a 25 MHz 486SX processor with 4 megabytes of RAM is actually considered the minimum. In addition, make sure your PC has a 16-bit expansion slot available for the sound card and an empty 5¼-inch drive bay for the CD-ROM drive (otherwise, buy an external CD-ROM drive).

Creating Multimedia

Companies are spending huge amounts of money to create CD-ROMs. (For instance, Software Toolworks spent $500,000 on its five-disk *20th Century Video Almanac*, which covers 100 years of U.S. history.) Thus, the development of multimedia has become as complex as big-time movie production. However, creating multimedia need not be expensive or difficult. Indeed, even children can do it. For instance, pupils in a Hong Kong elementary school, with the assistance of their computer club advisor, produced a multimedia history of their school. They used CD-ROM technology costing the equivalent of $10,000 U.S. The history was produced using a '386 PC with 8 megabytes of RAM, CD-ROM, a SoundBlaster sound card, Kid Pix, Kodak Photo CD, and Microsoft Publisher.[52]

Note: In creating multimedia, one of the most daunting tasks is obtaining rights to reuse copyrighted images and music. For instance, suppose you want to include a 30-second clip from a feature film on a CD-ROM program. You must obtain signed releases from the film studio, the actors and actresses in the scene, the director, and, if there is music, the American Federation of Musicians and the copyright holder of the music. If any one of these refuses, you can't use the clip. Similar problems exist for showing footage from a sports program, or for showing a still photograph or piece of advertising or artwork. Payment for CD-ROM rights may range from nothing to thousands of dollars, if permission is granted at all.[53] Thus, if you have any prospects of having wide distribution of your CD-ROM, especially for commercial purposes, you must obtain permissions. This is true even if you are using photos of your friends or music composed by them.

SUMMARY

What It Is / What It Does	**Why It's Important**
backup *(p. 296, LO 7)* Name given to a diskette (or tape or hard-disk cartridge) that contains duplicates, or copies, of files on another form of storage.	Because secondary storage media can fail or be destroyed, users (and companies) should always make backup copies of their files so that they don't lose them.
bit *(p. 287, LO 1)* Short for binary digit, which is either a 1 or a 0 in the binary system of data representation in computer systems.	The bit is the fundamental element of all data and information stored in a computer system.
byte *(p. 287, LO 1)* Unit of 8 bits; may be used to represent a character, digit, or other value, such as A, ?, or 3.	Bits and bytes—also called *characters*—are the building blocks for representing data, whether it is being processed, stored, or telecommunicated.
cartridge tape units *(p. 314, LO 7)* Also called *tape streamers;* secondary storage used to back up data from a hard disk onto a tape cartridge.	Cartridge tape units are often used with microcomputers.
CD Plus/Enhanced CD *(p. 306, LO 5)* Digital disk that is a hybrid of audio-only compact disk and multimedia CD-ROM.	A CD Plus can be played like any music CD in a stereo system, but it also can be used for multimedia presentations on a computer.
character *(p. 287, LO 1)* May be—but is not necessarily—the same as a byte (8 bits); a single letter, number, or special character such as ;, $, or %.	*See byte.*
cluster *(p. 300, LO 5)* Smallest storage unit the computer can access on a hard disk; refers to a number of sectors.	If a file's data is stored in nonadjacent clusters, the disk becomes fragmented, thus slowing access time.
codec (compression/decompression) techniques *(p. 288, LO 3)* Techniques of eliminating redundant and unnecessary data components to reduce the amount of space needed for data storage or transmission. Two kinds of compression and decompression are "lossless" and "lossy" techniques.	Storage and transmission of digital data—particularly of graphics—requires a huge amount of electronic storage capacity, and transmission is difficult to accomplish over copper wire. Thus compression programs are necessary to reduce the size of these files.
compact disk-read only memory (CD-ROM) *(p. 304, LO 5)* Optical-disk form of secondary storage that holds more data, including photographs, art, sound, and video, than diskettes and many hard disks. Like music CDs, a CD-ROM is a read-only disk. CD-ROM disks will not play in a music CD player.	CD-ROM disks are used in computer systems to create multimedia presentations and do research, among other things.

What It Is / What It Does	Why It's Important

compact disk-recordable (CD-R) *(p. 307, LO 5)* CD format that allows users to write data onto a specially manufactured disk that can then be read by a standard CD-ROM drive.

Home users can do their own recordings in CD format.

compression (digital-data compression) *(p. 288, LO 3)* Mathematical process of removing redundant and unnecessary elements from a file, or collection, of data so that the file requires less storage space or can be easily transmitted. After the data is stored or transmitted and is to be used again, it is decompressed.

See codec.

cylinder *(p. 301, LO 5)* All the tracks in a disk pack with the same track number lined up, one above the other.

Secondary storage systems that use several hard disks don't use the sector method to locate data; they use the cylinder method. Because the access arms holding the read/write heads all move together, the read/write heads are always over the same track on each disk at the same time. All tracks with the same track number, lined up one above the other, thus form a cylinder. Data access is faster because all read/write heads move simultaneously.

data access area *(p. 292, LO 5)* Opening in the disk's jacket through which the read/write heads read and write data.

Without this opening in the diskette jacket, the read/write heads would not be able to access the data on the disk.

data storage hierarchy *(p. 287, LO 1)* Defines the levels of data stored in a computer file: bits, bytes, fields, records, and files.

Bits and bytes are what the computer hardware deals with, so users need not be concerned with them. They will, however, deal with characters, fields, records, files, and databases.

database *(p. 287, LO 1)* Collection of interrelated files in a computer system that is created and managed by database manager software. These files are organized so that those parts with a common element can be retrieved easily.

Businesses and organizations build databases to help them keep track of and manage their affairs. In addition, users with online connections to database services have enormous research resources at their disposal.

defragmentation *(p. 300, LO 5)* Refers to the situation when hard-disk data is reorganized so that data in each file is stored in contiguous clusters.

See fragmentation.

digital audiotape (DAT) (R-DAT) *(p. 314, LO 7)* Advanced form of tape cassette, adapted from technology used in the music industry, expected to hold as much as 8 gigabytes of data.

Holds more data than conventional audio tape. May be used for backup purposes.

disk drive *(p. 292, LO 5)* Computer hardware device that holds, spins, reads from, and writes to magnetic or optical disks.

Users need disk drives in order to use their disks. Disk drives can be internal (built into the computer system cabinet) or external (connected to the computer by a cable).

diskette *(p. 291, LO 5)* Also called *floppy disk;* secondary storage medium; removable round, flexible mylar disk that stores data as electromagnetic charges on a metal oxide film that coats the mylar plastic. Data is represented by the presence or absence of these electromagnetic charges, following standard patterns of data representation (such as ASCII). The disk is contained in a square paper envelope or plastic case to protect it from being touched by human hands. It is called "floppy" because the disk within the envelope or case is flexible, not rigid. The sizes are 5¼ inches, the older size, and 3½ inches, now the most common size.

Diskettes are used on all microcomputers.

What It Is / What It Does	**Why It's Important**

double-sided diskettes *(p. 294, LO 5)* Diskettes that store data on both sides. For double-sided diskettes to work, the disk drive must have read/write heads that will read both sides simultaneously.

Double-sided diskettes hold twice as much data as single-sided diskettes.

drive bay *(p. 298, LO 5)* Slot, or opening, in the computer cabinet for an internal disk drive.

Allows for upgrading (replacing) of disk drives.

DVD-ROM disk *(p. 309, LO 5)* Five-inch optical disk that looks like a regular audio CD but can store 4.7 gigabytes of data on a side.

DVD-ROMs provide great storage capacity, studio-quality images, and theater-like surround sound.

erasable optical disk *(p. 309, LO 5)* Optical disk that allows users to erase data so that the disk can be used over and over again (as opposed to CD-ROMs, which can be read only).

The most common type of erasable and rewritable optical disk is probably the magneto-optical disk, which uses aspects of both magnetic-disk and optical-disk technologies. Such disks are useful to people who need to save successive versions of large documents, handle enormous databases, or work in multimedia production or desktop publishing.

field *(p. 287, LO 1)* Unit of data consisting of one or more characters (bytes). Examples of fields are your name, your address, *or* your Social Security number.

A collection of fields make up a record. *Also see key field.*

file *(p. 287, LO 1)* Collection of related records. An example of a file is collected data on everyone employed in the same department of a company, including all names, addresses, and Social Security numbers.

Interrelated files make up a database.

fixed-disk drive *(p. 301, LO 5)* High-speed, high-capacity disk drive housed in its own cabinet.

Although fixed disks are not removable or portable, these units generally have greater storage capacity and are more reliable than removable disk packs. A single mainframe computer might have 20–100 such fixed disk drives attached to it.

flash-memory *(p. 311, LO 6)* Circuitry on credit-card-size cards (PC cards) that can be inserted into slots in the computer that connect to the motherboard.

Flash-memory cards are variations on conventional computer-memory chips; however, unlike standard RAM chips, flash memory is nonvolatile—it retains data even when the power is turned off. Flash memory can be used not only to simulate main memory but also to supplement or replace hard-disk drives for permanent storage.

formatting (initializing) *(p. 293, LO 5)* Process by which users prepare diskettes so that the operating system can write information on them. This includes defining the tracks and sectors (the storage layout). Formatting is carried out by one or two simple computer commands.

Diskettes cannot be used until they have been formatted.

fragmentation *(p. 300, LO 5)* Refers to the situation when a hard disk's file data is stored in nonadjacent clusters.

Fragmentation slows access time.

gigabyte (G or GB) *(p. 287, LO 1)* Approximately 1 billion bytes (1,073,741,824 bytes); a measure of storage capacity.

Gigabytes are used to express the storage capacity of some microcomputers and many large computers, such as mainframes.

What It Is / What It Does	**Why It's Important**

hard disk *(p. 298, LO 5)* Secondary storage medium; generally nonremovable disk made out of metal and covered with a magnetic recording surface. It holds data in the form of magnetized spots. Hard disks are tightly sealed within an enclosed unit to prevent any foreign matter from getting inside. Data may be recorded on both sides of the disk platters.

Hard disks hold much more data than diskettes do. Nearly all microcomputers now use hard disks as their principal secondary storage medium.

hard-disk cartridge *(p. 300, LO 5)* One or two hard-disk platters enclosed along with read/write heads in a hard plastic case. The case is inserted into an external cartridge system connected to a microcomputer.

A hard-disk cartridge, which is removable and easily transported in a briefcase, may hold gigabytes of data. Hard-disk cartridges are often used for transporting large graphics files and for backing up data.

hard-disk drive *(p. 298, LO 5)* One or more hard-disk platters sealed along with read/write heads inside the computer's system unit; it cannot be removed.

Nearly all microcomputers now use hard disks as their principal secondary storage medium.

head crash *(p. 299, LO 5)* Disk disturbance that occurs when the surface of a read/write head or particles on its surface come into contact with the disk surface, causing the loss of some or all of the data on the disk.

Head crashes can spell disaster if the data on the disk has not been backed up.

interactive *(p. 311, LO 5)* Refers to a situation in which the user is able to make an immediate response to what is going on and modify processes; that is, there is back-and-forth interaction, or "dialogue," between the user and the computer or communications device. The best interactive storage media are those with high capacity, such as CD-ROMs.

Interactive devices allow the user to be an active participant in what is going on instead of just reading to it.

kilobyte (K or KB) *(p. 287, LO 1)* 1024 bytes (often rounded off to 1000 bytes).

Kilobytes are a common unit of measure for storage capacity. The amount of data stored in a file or database might be expressed in kilobytes, megabytes, or gigabytes.

"lossless" compression techniques *(p. 288, LO 3)* Compression techniques that preserve every bit of data that was input; that is, decompressed data is the same as it was before it was compressed.

These programs save storage and transmission space.

"lossy" compression techniques *(p. 289, LO 3)* Compression techniques that permanently discard some data during compression. Decompressed data is not 100% the same as it was before it was compressed. In general, most viewers or listeners would not notice the absence of these details.

Lossy compression techniques can save even more storage and transmission space than lossless techniques.

magnetic tape *(p. 312, LO 7)* Thin plastic tape coated with a substance that can be magnetized; data is represented by the magnetized or nonmagnetized spots. Tape can store files only sequentially.

Tapes are used in reels, cartridges, and cassettes. Today "mag tape" is used mainly to provide backup, or duplicate storage.

magnetic-tape units (magnetic-tape drives) *(p. 312, LO 7)* Consists of a read/write head and spindles for mounting two tape reels: a supply reel and a take-up reel. The tape is reeled off the supply reel, fed through pulleys that regulate its speed and hold it still long enough for data to be read from it or written to it by the read/write heads, and then wound up on the take-up reel. During the writing process, any existing data on the tape is automatically written over, or erased.

Nowadays tape is used mainly to provide duplicate storage (backup), especially for microcomputers. Tape can store data only in sequential order; it is not a direct-access storage medium.

megabyte (M or MB) *(p. 287, LO 1)* Unit for measuring storage capacity; equals approximately 1 million bytes.

The storage capacities of many microcomputer hard disks are measured in megabytes. Users need to know how much data their hard disks can hold and how much space new software programs will take so that they do not run out of disk space.

multimedia *(p. 311, LO 5)* Refers to technology that presents information in more than one medium, including text, graphics, animation, video, sound effects, music, and voice.

Use of multimedia is becoming more common in business, the professions, and education as a means of improving the way information is communicated. Multimedia systems have also added greater depth and variety to presentations, such as those in entertainment and education. As used by telephone, cable, broadcasting, and entertainment companies, *multimedia* also refers to the so-called Information Superhighway. On this avenue various kinds of information and entertainment will be delivered to users' homes through wired and wireless communication lines.

nonvolatile storage *(p. 286, LO 2)* Permanent storage, as in secondary storage.

Data and programs are permanent; they remain intact when the power to the computer is turned off.

optical disk *(p. 302, LO 5)* Removable disk on which data is written and read through the use of laser beams. The most familiar form of optical disk is the one used in the music industry.

Optical disks hold much more data than magnetic disks. Optical disk storage is expected to dramatically affect the storage capacity of microcomputers.

Photo CD *(p. 307, LO 5)* Technology developed by Eastman Kodak that allows photographs taken with an ordinary 35-millimeter camera to be stored digitally on an optical disk.

Users can shoot up to 100 color photographs and take them to a local photo shop for processing. They receive back not only conventional negatives and snapshots but also the images on a CD-ROM disk. Users can then view the disk using any compatible CD-ROM drive.

primary storage *(p. 286, LO 2)* Also known as *memory, main memory, internal memory,* or *RAM (Random Access Memory)*; working storage that holds (1) data for processing, (2) instructions (the programs) for processing the data, and (3) processed data (that is, information) that is waiting to be sent to an output or secondary-storage device such as a printer or diskette drive.

Primary storage is in effect the computer's short-term capacity, determining the total size of the programs and data files it can work on at any given moment. Primary storage, which is contained on RAM chips, is temporary, or *volatile.* Once the power to the computer is turned off, all the data and programs within memory vanish.

RAID (redundant array of inexpensive disks) *(p. 301, LO 5)* Storage system that consists of over 100 5¼-inch disk drives within a single cabinet and that sends data to the computer along several parallel paths simultaneously. Response time is thereby significantly increased.

The advantage of a RAID system is that it not only holds more data than a fixed disk drive within the same amount of space, but it also is more reliable, because if one drive fails, others can take over.

read *(p. 292, LO 5)* Computer activity whereby data represented in the magnetized spots on the disk (or tape) are converted to electronic signals and transmitted to the primary storage (memory) in the computer.

Read means the disk drive copies data—stored as magnetic spots—from the diskette. Whereas reading simply makes a copy of the original data, without altering the original, writing actually replaces the data underneath it.

read-only *(p. 304, LO 5)* Means the storage medium cannot be written on or erased by the user.

With read-only storage media, the user has access only to the data imprinted by the manufacturer.

read/write head *(p. 292, LO 5)* The device that transfers data on a disk to the computer and from the computer to the disk. The diskette spins inside its jacket, and the read/write head moves back and forth over the data access area.

The read/write head locates the specific area on the disk on which the user is seeking to find a file.

record *(p. 287, LO 1)* A collection of related fields. An example of a record would be your name *and* address *and* Social Security number.

Related records make up a file.

recording density *(p. 294, LO 5)* Refers to the number of bytes per inch of data that can be written onto the surface of the disk. There are three diskette densities: single-density, double-density, and high-density.

A double-sided, double-density 5¼-inch diskette contains 360 kilobytes (equal to 260 typewritten pages). A 3½-inch double-sided, double-density disk has 720 kilobytes. A high-density 5¼-inch disk has 1.2 megabytes. A high-density 3½-inch disk has 1.44 megabytes. Users need to know what types of disks their system can use.

removable-pack hard disk system *(p. 301, LO 5)* Secondary storage with 6–20 hard disks, of 10½- or 14-inch diameter, aligned one above the other in a sealed unit. These removable hard-disk packs resemble a stack of phonograph records, except that there is space between disks to allow access arms to move in and out. Each access arm has two read/write heads—one reading the disk surface below, the other the disk surface above. However, only one of the read/write heads is activated at any given moment.

Such secondary storage systems enable a large computer system to store massive amounts of data.

secondary storage *(p. 286, LO 2)* Consists of devices that store data and programs permanently on disk or tape.

Secondary storage is nonvolatile—that is, data and programs are permanent, or remain intact, when the power is turned off. Secondary storage is also needed because computer users require far greater storage capacity than is available through primary storage.

sectors *(p. 293, LO 5)* On a diskette, eight or nine invisible wedge-shaped sections used by the computer for storage reference purposes.

When users save data from computer to diskette, it is distributed by tracks and sectors on the disk. That is, the systems software uses the point at which a sector intersects a track to reference the data location in order to spin the disk and position the read/write head.

terabyte (T or TB) *(p. 287, LO 1)* Unit for measuring storage capacity; equals approximately 1 trillion bytes.

The storage capacities of supercomputers are measured in terabytes, as is also the amount of data being held in remote databases accessible to users over a communications line.

tracks *(p. 293, LO 5)* The rings on a diskette along which data is recorded. Unlike on a phonograph record, these tracks are neither visible grooves nor a single spiral. Rather, they are closed concentric rings. Each track is divided into eight or nine sectors.

See sectors.

unformatted disks *(p. 293, LO 5)* Diskettes manufactured without tracks and sectors in place.

Unformatted diskettes must be formatted (initialized) by users before the disks can be used to store data.

volatile storage *(p. 286, LO 2)* Temporary storage, as in main memory (RAM).

The contents in volatile storage are lost when power to the computer is turned off.

write *(p. 292, LO 5)* Computer activity whereby data processed by the computer is recorded magnetically onto a disk (or tape).

Write means the disk drive transfers data—represented as electronic signals within the computer's memory—onto the disk. Whereas *reading* simply makes a copy of the original data, without altering the original, *writing* replaces the data underneath it.

write once, read many (WORM) *(p. 309, LO 5)* Refers to an optical disk that can be written, or recorded, onto just once and then cannot be erased; it can be read many times. WORM disks hold more data than other types of optical disks.

WORM technology is useful for storing data that needs to remain unchanged, such as that used for archival purposes.

write-protect feature *(p. 295, LO 5)* Feature of 3½-inch and 5¼-inch disks that prevents someone from accidentally writing over—and thereby obliterating—data on a disk.

This feature allows users to protect data on diskettes from accidental change or erasure.

EXERCISES

(Selected answers appear at the back of the book.)

Short-Answer Questions

1. What does "read-only" mean?
2. What can flash-memory cards be used for?
3. Why is the process of formatting (initializing) necessary?
4. What is "codec" short for?
5. What is the difference between primary storage and secondary storage?
6. What is a hard-disk cartridge, and how can it be useful to a microcomputer user?
7. What is a removable-pack hard disk system used for?
8. What is the difference between volatile and non-volatile storage? Give an example of each of these types of storage.
9. What can a user do with a Photo CD?
10. As they apply to disk storage, what is meant by the terms *fragmentation* and *defragmentation*?

Fill-in-the-Blank Questions

1. The two standard diskette sizes are
   _____ and
   _____.
2. The device that transfers data on a disk to the computer, and from the computer to the disk, is called the _____.
3. High-density diskettes can hold
   _____ of data.
4. On a diskette, data is recorded in rings called
   _____.
5. A _____ feature allows you to protect a diskette from being written on.
6. _____ refers to the average time needed for the computer to locate data on a disk.
7. A(n) _____ is a removable disk on which data is written and read through the use of laser beams.
8. _____ consist of storage circuitry on credit-card-size cards that can be inserted into slots connected to the motherboard.
9. _____ is a method of removing redundant elements from a computer file so that the file takes up less space.
10. _____ is the name given to duplicate storage.

Multiple-Choice Questions

1. Which of the following provides you with the most storage?
   a. 3½-inch diskette
   b. hard disk
   c. disk cartridge
   d. CD-ROM
   e. 5¼-inch diskette
2. Which of the following disk devices is used with microcomputers?
   a. removable pack
   b. fixed-disk drive
   c. external hard-disk drive
   d. RAID storage system
   e. none of the above
3. Which of the following can be written on only once?
   a. diskette
   b. hard disk
   c. disk cartridge
   d. CD-ROM
   e. none of the above
4. If you use a notebook computer, it is likely that you are also using a(n) _____.
   a. PC card
   b. cartridge tape unit
   c. RAID storage system
   d. DVD disk
   e. none of the above
5. A(n) _____ holds, spins, and reads data from and writes data to a diskette.
   a. drive gate
   b. sector
   c. disk drive
   d. disk cartridge
   e. none of the above
6. Which of the following rules *doesn't* apply to diskettes?
   a. Don't touch disk surfaces.
   b. Avoid magnetic fields.
   c. Avoid risky physical environments.
   d. Avoid head crashes.
   e. all of the above

7. Which of the following might you use to compress data on a disk?
    a. JPEG standard
    b. MPEG standard
    c. "lossless" compression technique
    d. "lossy" compression technique
    e. all of the above

8. Which of the following is used mainly for backup purposes?
    a. diskette
    b. magnetic tape
    c. CD-ROM
    d. removable-pack hard disk
    e. all of the above

9. Which of the following was designed to work with all types of electronics, televisions, stereos, computers, and videogame players?
    a. hard disk
    b. DVD disk
    c. JPEG standard
    d. flash-memory card
    e. all of the above

10. Which of the following determines the total size of the programs and data files a computer can work with at any given time?
    a. primary storage
    b. secondary storage
    c. auxiliary storage
    d. nonvolatile storage
    e. none of the above

True/False Questions

T F 1. WORM disks are used principally for storing databases, documents, and other archival information.

T F 2. Secondary storage is nonvolatile.

T F 3. Most of today's CD-ROM drives are single-speed.

T F 4. Diskettes are divided into tracks and sectors.

T F 5. Hard-disk cartridges are used mainly with microcomputer systems.

T F 6. Hard disks are read-only.

T F 7. Erasable optical disks are available now that enable the user to erase data so that the disk can be used over and over.

T F 8. Although standards exist for audio compression, none are available for video compression.

T F 9. Cartridge tape units are typically used with mainframe computers.

T F 10. The main disadvantage of using a hard disk over a diskette is the possibility of a head crash.

Projects/Critical-Thinking Questions

1. What types of storage hardware are currently being used in the computer you use at school or at work? What is the storage capacity of this hardware? Would you recommend alternate storage hardware be used? Why? Why not?

2. You want to purchase a hard disk for use with your microcomputer. Because you don't want to have to upgrade your secondary-storage capacity in the near future, you are going to buy one with the highest storage capacity you can find. Use computer magazines or visit computer stores and find a hard disk you would like to buy. What is its capacity? How much does it cost? Who is the manufacturer? What are the system compatibility requirements? Is it an internal or an external drive? Why have you chosen this unit?

3. What would you do if your computer had a "head crash"—a failure of your hard-disk drive—that seemed to wipe out a major project? Assume you had not been foresighted enough to back up your data on diskettes or tape. Go to the library (or use the Internet) and investigate articles on data retrieval or hard-disk salvage methods. Or look in the telephone book Yellow Pages to see if you can find an organization that specializes in retrieving data from damaged hard disks; then call them up and ask what they do and what their rates are.

4. Primary storage is volatile: the data and programs disappear as soon as the power is turned off. Secondary storage is nonvolatile: the data and programs remain intact after the power is turned off. Do you think during your lifetime you might see this distinction vanish in microcomputers as new technology comes along that will allow main memory to preserve what is being worked on after the power is shut off? What kind of technology might this be? (*Hint:* Look at the end of the processing and storage chapters, then follow these leads to expand your research by going to a library or using the Internet.)

5. Do you think books on CD-ROMs will ever replace printed books? Why or why not? Look up some recent articles on this topic and prepare a short report.

net **Using the Internet**

Objective: *In this exercise you use Netscape Navigator to shop on the World Wide Web for a multimedia computer that costs under $2000.*

Before you continue: *We assume you have access to the Internet through your university, business, or commercial service provider and to the Web browser tool Netscape Navigator 2.0 or 2.01 Additionally, we assume you know how to connect to the Internet and then load Netscape Navigator. If necessary, ask your instructor or system administrator for assistance.*

1. Make sure you have started Netscape. The home page for Netscape Navigator should appear on your screen.

2. Although the commercial side of the Web is still in its formative stages, the Web does provide an excellent means for accessing some resources. Today, most Web stores provide the same functionality as a mail-order catalog with an associated toll-free 800 number. Not surprisingly, many mail-order companies use their store on the Web as an additional way to get customers. The following is a small sampling of popular Web shops:

   a. *Noteworthy Music (http://www.netmarket.com/noteworthy/bin/main)*—Noteworthy Music, a discount CD dealer, uses its Web site to supplement its mail-order business. As you order CDs, a running total of your charges appears on the page.

   b. *Internet Shopping Network (http://shop.internet.net/)*—For hardware and software at discounted prices, the Internet Shopping Network (ISN) may be the shop for you. Because you can use key words to search for items, they are easier to find than in a paper catalog.

   c. *Computer Literacy (http://www.clbooks.com/)*— Computer Literacy, a discount computer book dealer, uses its Web site to supplement its mail-order business. Computer Literacy's search feature makes it easy to find the book you want by typing in a topic or an author name.

   d. *The Internet Mall (http://www.mecklerweb.com/imall/)*—The Internet Mall, like the shopping malls we're accustomed to, provides access to more than one company with something to sell. The Internet Mall is organized into floors, each of which is sponsored by a different company. Companies (or individuals) who have a Web site and/or an electronic e-mail address are listed (for free) in the The Internet Mall.

3. In the following steps you'll use the Internet Shopping Network to find a multimedia computer that costs under $2000.

   To display the Internet Shopping Network's home page:
   CLICK: Open button (⊞)
   TYPE: http://shop.internet.net/
   PRESS: **Enter** or CLICK: Open
   Your screen may appear similar to the following:

4. Drag the vertical scroll bar downward until you see the ISN Directory list. The list should appear similar to the following:

5. On your own, shop for a computer by clicking links in the ISN Directory. Remember that the computer should have multimedia capabilities and cost under $2000.

6. Print out a description of the computer. (*Note:* If you were actually purchasing the computer, you would click the Buy link, located after the product's description.)

7. Display Netscape's home page and then exit Netscape.

CHAPTER 8

Communications

Starting Along the Information Highway

Concepts You Should Know

After reading this chapter, you should be able to:

1. Discuss examples of usage of communications technology: telephone-related services, online information services, BBSs, the Internet, videoconferencing, shared resources, portable work, and information appliances.

2. Explain the communications components of a microcomputer, describing analog and digital signals, modems, and communications software.

3. Identify and explain the various communications channels, both wired and wireless.

4. Discuss the three types of networks.

5. List and discuss the factors affecting data transmission.

6. Discuss ethical matters of netiquette, free speech, censorship, and privacy.

"Computers and communications: These are the parents of the Information Age," says one writer. "When they meet, the fireworks begin."[1]

What kind of fireworks are we talking about? Maybe it is that portable information and communications technologies *are changing conventional meanings of time and space.* As one expert pointed out (during a round-table discussion on an online network), "the physical locations we traditionally associate with work, leisure and similar pursuits are rapidly becoming meaningless."[2]

Alaska salmon fisherman Blanton Fortson exemplifies this blurring of traditional boundaries between work and leisure, isolation and availability. Fortson has so much portable technology on his boat and at home that he wears baggy pants to carry it all around. "I find I very rarely need to be tied to any specific location in order to take care of business," he says.[3] (■ *See Panel 8.1.)*

Not everyone is as enthusiastic about being so accessible. Said one observer back in 1991: "The movable office—a godsend for workaholics, a nightmare for those who live with them—has only just begun."[4] Through communications technologies, computers, telephones, and wireless devices are being linked to create invisible networks everywhere.

The Practical Uses of Communications & Connectivity

Preview & Review: Communications refers to (1) the electronic transfer of information from one location to another, as well as (2) the electromagnetic devices and systems for communicating. The ability to connect devices by communications lines to other devices and sources of information is known as *connectivity*.

Users can employ communications technology for a number of purposes, ranging from low-skill to high-skill activities. They can use telephone-related services, such as fax, voice-mail, and e-mail. They can do voice/video communication via teleconferencing and picture phones. They can use online information services for research and news; e-mail, bulletin boards, and chat; games and entertainment; travel services; and teleshopping. They can use electronic bulletin board systems (BBSs), large and small, for special purposes. They can connect with the global Internet for e-mail, Usenet news groups, mailing lists, FTD, Telnet, Gopherspace, WAIS, and the World Wide Web. They can share resources through workgroup computing, electronic data interchange, and intranets. They can make their work portable, with telecommuting, virtual offices, and mobile workplaces. In the future they will employ information or Internet appliances.

Twenty-two-year-old shipping clerk Neal Berry (online moniker: Shylent Cat) had enough money to buy a Toshiba laptop, a cellular phone, and a connection to an electronic bulletin board. But he couldn't afford an apartment in pricey Novato, California, so he lived in a tent near the freeway, hunkered on a mattress rescued from a trash bin. Though homeless, he spent his evenings happily tapping on his laptop, communicating with the online

PANEL 8.1

**How portable technology changes
time and space**

"I live and travel mostly in Alaska. I work hard in the summertime and I'm fairly recreational the rest of the year. This summer in Bristol Bay, I was logging on to The Well [electronic bulletin board] while lying in my bunk aboard the *F/V Glacier Bay* [fishing vessel]. With my PowerBook [microcomputer] on my tummy, I could log on to my favorite services, access my networks and any servers or other devices on my networks, check my e-mail, send and receive faxes, etc.

I'm often in the woods running, hiking, or biking, and almost always reachable by digital pager. My home is wired, and I maintain a dedicated phone-line link between home and office-network zones. I've experimented with logging on from my aircraft, just for fun.

I find a digital pager quite handy for screening calls and preserving battery life on the smallest cellular phone I can find. With the combination of the PowerBook and cellular technology, I find that I very rarely need to be tied to any specific location in order to take care of business."

—Alaska salmon fisherman Blanton Fortson, in "Talking About Portables," *Wall Street Journal*

world. "I made more friends in a month [electronically]," he said, "than I had all year in Novato."[5]

Clearly, communications is extending into every nook and cranny of civilization—the "plumbing of cyberspace," as it has been called. The term *cyberspace* was coined by William Gibson in his novel *Neuromancer* to refer to a futuristic computer network that people use by plugging their brains into it. Today **cyberspace has come to mean the computer online world and the Internet in particular, but it is also used to refer to the whole wired and wireless world of communications in general.**

Communications & Connectivity

Communications, **also called** *telecommunications*, **is the electronic transfer of information from one location to another. It also refers to the electromagnetic devices and systems for communicating.** The data being communicated may consist of voice, sound, text, video, graphics, or all of these together. The instruments sending the data may be telegraph, telephone, cable, microwave, radio, television, or computer. The distance may be as close as the next room or as far away as the outer edge of the solar system.

The television set is an instrument of communications, but it is a low-skill tool. That is, the many people of a mass audience receive one-way communications from a few communicators. This is why television (like AM/FM radio, newspapers, and music CDs) is called one of the *mass media.* Telephone systems are not mass media, since they involve two-way communications of many to many. But they, too, are low-skill communications tools. By contrast, linkages of microcomputers have allowed some people with a fairly high level of skill to achieve two-way communication with other people. **The ability to connect devices by communications technology to other devices and sources of information is known as** *connectivity.*

Tools of Communications & Connectivity

What kinds of options do communications and connectivity give you? Let us consider the possibilities. We will take them in order, more or less, from *low-skill activities* to *high-skill activities.* That is, we will begin with those that demand relatively little training and proceed to those that require more training. (■ *See Panel 8.2.*) They include:

- Telephone-related communications services: fax messages, voice mail, and e-mail
- Video/voice communication: videoconferencing and picture phones
- Online information services
- Bulletin board systems (BBSs)

PANEL 8.2

The world of connectivity

Wired or wireless communications links offer several options for information and communications.

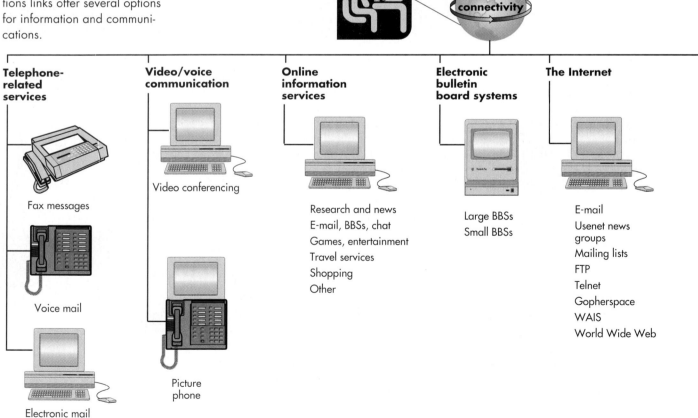

Telephone-related services

Fax messages

Voice mail

Electronic mail

Video/voice communication

Video conferencing

Picture phone

Online information services

Research and news
E-mail, BBSs, chat
Games, entertainment
Travel services
Shopping
Other

Electronic bulletin board systems

Large BBSs
Small BBSs

The Internet

E-mail
Usenet news groups
Mailing lists
FTP
Telnet
Gopherspace
WAIS
World Wide Web

World of connectivity

- The Internet
- Shared resources: workgroup computing, Electronic Data Interchange (EDI), and intranets
- Portable work: telecommuting, mobile workplaces, and virtual offices

We will also discuss possible variations on what are called either *information appliances* or *Internet appliances*—the set-top box, the network personal digital assistant (PDA), and the network PC.

Telephone-Related Communications Services

Preview & Review: Telephone-related communications include fax messages, transmitted by dedicated fax machines and fax modems; voice mail, storing voice messages in digitized form; and e-mail (electronic mail), transmitting written messages.

Phone systems and computer systems have begun to fuse together. Services available through telephone connections, whether the conventional wired kind or the wireless cellular-phone type, include *fax message, voice mail,* and *e-mail.*

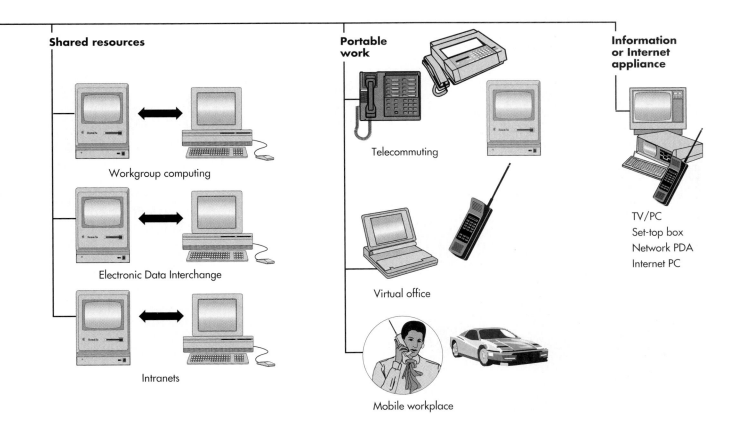

Shared resources

Workgroup computing

Electronic Data Interchange

Intranets

Portable work

Telecommuting

Virtual office

Mobile workplace

Information or Internet appliance

TV/PC
Set-top box
Network PDA
Internet PC

Fax Messages

Asking "What is your fax number?" is about as common a question in the work world today as asking for someone's telephone number. Indeed, the majority of business cards include both a telephone number and a fax number. ***Fax* stands for "facsimile transmission," or reproduction** (✓ p. 216).

A fax may be sent by dedicated fax machine or by fax modem. (■ *See Panel 8.3.)*

*   **Dedicated fax machines:** *Dedicated fax machines* **are specialized devices that do nothing except send and receive copies of documents over transmission lines to and from other fax machines.** These are the stand-alone machines nowadays found everywhere, from offices to airports to instant-printing shops.
*   **Fax modems:** **A** *fax modem,* **which is installed as a circuit board inside a computer's system cabinet, is a modem with fax capability. It enables you to send and receive signals directly between your computer and some-one else's fax machine or fax modem.**

Voice Mail

Like a sophisticated telephone answering machine, ***voice mail* digitizes incoming voice messages and stores them in the recipient's "voice mailbox" in digitized form. It then converts the digitized versions back to voice mes-sages when they are retrieved.**

Voice-mail systems also allow callers to deliver the same message to many people within an organization by pressing a single key. They can for-ward calls to the recipient's home or hotel. They allow the person checking messages to speed through them or to slow them down. He or she can save some messages and erase others and can dictate replies that the system will send out.

PANEL 8.3

Two types of fax hardware

(Left) A fax modem. This circuit board is installed in an expan-sion slot on the motherboard inside the system cabinet of a microcomputer. *(Right)* A dedicated fax machine.

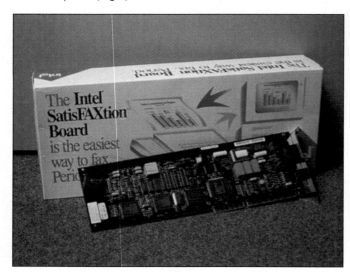

The main benefit of voice mail is that it helps eliminate "telephone tag." More than 80% of phone calls don't reach the intended party. With voice mail, two callers can continue to exchange messages even when they can't reach each other directly. Efficiency is another benefit, because messages eliminate the small talk that often goes with phone conversations.[6]

You don't even need to have a fixed address to use voice mail. Carl Hygrant, a homeless person in New York, found that the technology helped him get work. Earlier, when he put down on his resume the phone number of the Bronx shelter that was his temporary home, prospective employers would lose interest when they called. After a telephone company launched an experimental voice-mail program for homeless people, he landed a job.[7]

E-Mail

"E-mail is so clearly superior to paper mail for so many purposes," writes *New York Times* computer writer Peter Lewis, "that most people who try it cannot imagine going back to working without it."[8] Says another writer, e-mail "occupies a psychological space all its own: It's almost as immediate as a phone call, but if you need to, you can think about what you're going to say for days and reply when it's convenient."[9]

E-mail, or *electronic mail*, **links computers by wired or wireless connections and allows users, through their keyboards, to post messages and to read responses on their display screens.** With e-mail, you dial the e-mail system's telephone number, type in the recipient's "mailbox" address, and then type in the message and click on the "Send" button. The message will be sent to the recipient's mailbox. To gain access to your mailbox, you dial the e-mail system's telephone number and type in the address of your mailbox and your *password*, a secret word or numbers that limit access. You may then read the list of senders and topics, read the messages, print them out, delete them, or download (transfer) to your hard disk.

If you're part of a company, university, or other large organization, you may get e-mail for free. Otherwise you can sign up with a commercial online service (America Online, CompuServe, Prodigy), e-mail service (such as MCI Mail), or Internet access provider (such as Netcom's NetCruiser or PSI's Pipeline USA).

E-mail has both advantages and disadvantages:[10–13]

- **Advantages of e-mail:** Like voice mail, it helps people avoid playing phone tag or coping with paper and stamps. A message can be as simple as a birthday greeting or as complex and lengthy as a report with supporting documents. By reading the list of senders and topics displayed on the screen, you can quickly decide which messages are important. Sending an e-mail message usually costs as little as a local phone call or less, but it can go across several time zones and be read at any time.

- **Disadvantages of e-mail:** Nevertheless there are some problems: You might have to sort through scores or even hundreds of messages a day, a form of junk mail brought about by the ease with which anyone can send duplicate copies of a message to many people. Your messages are far from private and may be read by e-mail system operators and others; thus, experts recommend you think of e-mail as a postcard rather than a private letter. Mail that travels via the Internet often takes a circuitous route, bouncing around various computers in the country, until one of them recognizes the address and delivers the message. Thus, although a lot of messages may go through in a minute's time, others may be hung up because of system overload, taking hours and even days.

R E A D M E

Practical Matters: Managing Your E-Mail

1. Don't use your electronic mailbox as a things-to-do list. Instead, create a second mailbox or folder for e-mail messages that still need to be answered or acted upon. This will prevent important new messages from getting lost.
2. Read all of your e-mail as soon as it arrives and file it away immediately.
3. Don't create too many folders—otherwise, you'll lose messages that you file. Instead, adopt a simple message filing system and stick to it.
4. Don't save long messages with the thought that you will get around to them later. By the time later arrives, you'll have received even more e-mail.
5. Create a new set of folders every year; copy the previous year's correspondence onto a diskette. That way, if you change mail providers (or jobs), you won't lose all of your personal letters.
6. You don't have to reply to every e-mail message that you get. If you do, and your correspondents do as well, then the number of messages you get every day will increase geometrically.
7. Do not send chain-letters. If you get a chain-letter, just delete it. They may seem funny, but they clog mail systems and have shut down networks.
8. If somebody sends a request for help to a mailing list that you are on, send your response directly to that person, rather than to the entire list.
9. If you are on a mailing list that has too much traffic for you, don't make things worse by sending mail to the list asking people to send less mail to the list. Just have yourself taken off.
10. If somebody sends you a flame [an insulting message], don't make things worse by broadening the scope of the disaster. If you feel compelled to send mail back to the flamer, send it just to him or her.

—Simson L. Garfinkel, "Managing Your Mail," *San Jose Mercury News*

Nevertheless, the e-mail boom is only just beginning. In fact, it is perhaps the principal reason for the popularity of the Internet, as we shall discuss. The U.S. Postal Service is planning to offer e-mail with features of first-class mail, including "postmarks" and return receipts.[14,15]

What, however, if you want to meet *face-to-face* with someone who is far away? Then you can use videoconferencing or picture phones.

Video/Voice Communication: Videoconferencing & Picture Phones

Preview & Review: Videoconferencing is the use of television, sound, and computer technology to enable people in different locations to see, hear, and talk with one another.

Videoconferencing could lead to V-mail, or video mail, which allows video messages to be sent, stored, and retrieved like e-mail.

The picture phone is a telephone with a TV-like screen and a built-in camera that allow you to see the person you're calling, and vice versa.

Want to have a meeting with people on the other side of the country or the world but don't want the hassle of travel? You may have heard of or participated in a *conference call*, also known as *audio teleconferencing*, a meeting in which more than two people in different geographical locations talk on the telephone. A variation on this meeting format is e-mail-type *computer conferencing,* sometimes called "chat sessions." Computer conferencing is a keyboard conference among several users at microcomputers or terminals linked through a computer network. Now we have video/voice communication, specifically *videoconferencing* and *picture phones.* (■ *See Panel 8.4.)*

Videoconferencing & V-Mail

Videoconferencing, **also called** *teleconferencing,* **is the use of television video and sound technology as well as computers to enable people in different locations to see, hear, and talk with one another.** At one time, videoconferencing consisted of people meeting in separate conference rooms that were specially equipped with television cameras. Now videoconferencing equipment can be set up on people's desks, with a camera and microphone to capture the people talking, and a monitor and speakers for the listeners.

A relatively new development is an initiative to deliver *V-mail,* or *video mail,* video messages that are sent, stored, and retrieved like e-mail.

Picture Phones

The *picture phone* is a telephone with a TV-like screen and a built-in camera that allow you to see the person you're calling, and vice versa. The idea of the picture phone has been around since 1964, when AT&T showed its Picturephone at the New York World's Fair. However, the main difficulty is that the standard copper wire in what the industry calls POTS—for "plain old telephone service"—has been unable to transmit images rapidly. Thus, present-day picture phones convey a series of jerky, freeze-frame or stop-action still images of the faces of the communicating parties. However, ISDN lines and fiber-optic cables (discussed later in the chapter), rapidly being installed in many places, can better transmit visual information. Moreover, new software can compress images quickly, delivering video as well as audio images in real time even over old-fashioned copper wires.[16]

With the signing into law of the 1996 Telecommunications Act, which permits more competition in telephone service, we will probably see a real free-for-all in the phone business, which could speed up the delivery of picture phones. Already the telcos (telephone companies) are getting competition from unexpected quarters. Many cable companies are jumping into the phone business.[17-19] Technology such as CU-SeeMe software, developed at Cornell University in 1994, is allowing video/voice communication on the Internet.[20]

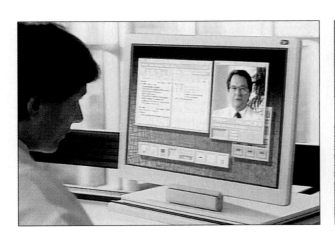

PANEL 8.4

Video/voice communication

(Left) Videoconferencing, or face-to-face communication via desktop computer. *(Right)* A picture phone.

Online Information Services

Preview & Review: Online information services provide computer users access to, among other things, bulletin boards; research and news resources; games, entertainment, and clubs; and travel and shopping services.

For subscribers equipped with telephone-linked microcomputers, *online information services* provide access to all kinds of databases and electronic meeting places. Says one writer:

> Online services are those interactive news and information retrieval sources that can make your computer behave more like a telephone; or a TV set; or a newspaper; or a video arcade, a stock brokerage firm, a bank, a travel agency, a weather bureau, a department store, a grocery store, a florist, a set of encyclopedias, a library, a bulletin board, and more.[21]

Online Services Big & Small

Several online services exist, but those known as the Big Three have the most subscribers and are considered the most mainstream. They are *America Online (AOL)*, with 5 million subscribers; *CompuServe*, with 4.2 million; and *Prodigy*, with an estimated 1.4 million.[22] (These numbers are increasing daily.) Some other online services have been merged into other enterprises (Delphi), are considered special interest (GEnie, oriented toward games), or have redefined their missions (eWorld and Microsoft Network are following new business strategies that involve the Internet). Still others (Dialog, Dow Jones News/Retrieval, Nexis, Lexis) may principally be considered huge collections of databases rather than "department-store"–like online services.

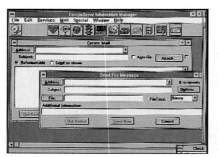

Getting Access

To gain access to online services, you need a *microcomputer* (with hard disk and printer for storing and printing downloaded messages and information). You also need a *modem*, the hardware that, when connected to your phone line, enables data to be transmitted to your computer. Finally, you need *communications software*, so your computer can communicate via modem and interact with distant computers that have modems. (We discuss modems and communications software later in the chapter.) America Online, CompuServe, and Prodigy provide subscribers with their own software for going online, but you can also buy communications programs, such as ProComm Plus, separately.

Opening an account with an online service requires a credit card, and billing policies resemble those used by cable-TV and telephone companies. As with cable-TV, you may be charged a fee for basic service, with additional fees for specialized services. In addition, the online service may charge you for the time spent on the line. Finally, you will also be charged by your telephone company for your time on the line, just as when making a regular phone call. However, most information services offer local access numbers. Thus, unless you live in a rural area, you will not be paying long-distance phone charges. All told, the typical user may pay $10–$20 a month to use an online service.

The Offerings of Online Services

What kinds of things could you use an online service for? Here are a few:

• **People connections—e-mail, bulletin boards, chat rooms:** Online services can provide a community through which you can connect with people with similar interests (without identifying yourself, if you prefer). The primary means for making people connections are via e-mail, bulletin boards, and "chat rooms."

E-mail is basically the same as we described earlier (✓ p. 337).

Bulletin boards, or *message boards*, allow you to post messages on any of thousands of special topics.

Chat rooms are discussion areas in which you may join with others in a real-time "conversation," typed in through your keyboard. The topic may be general or specific, and the collective chat-room conversation scrolls on the screen.

• **Research and news:** The only restriction on the amount of research you can do online is the limit on whatever credit card you are charging your time to. Depending on the online service, you can avail yourself of several encyclopedias.

Many online services provide users with access to huge databases of unabridged text from newspapers, magazines, and journals. The information resources available online are mind-boggling, impossible to describe in this short space.

• **Games, entertainment, and clubs:** Online computer games are extremely popular. In single-player games, you play against the computer. In multiplayer games, you play against others, whether someone in your household or someone overseas. Other entertainments include cartoons, sound clips, pictures of show-business celebrities, and reviews of movies and CDs. You can also join online clubs with others who share your interests, whether science fiction, punk rock, or cooking.

• **Travel services:** Online services offer Eaasy Sabre or Travelshopper, streamlined versions of the reservations systems travel agents use. You can search for flights and book reservations through the computer and have tickets sent to you by FedEx. You can also refer to weather maps, and you can review hotel directories and restaurant guides.

• **Downloading:** Many users obtain download freeware, shareware (✓ p. 55), and commercial demonstration programs from online hosts. (This can also be done via the Internet.)

• **Shopping:** CompuServe, for instance, offers 24-hour shopping with its Electronic Mall. This feature lists products from over 100 retail stores, discount wholesalers, specialty shops, and catalog companies. You can scan through listings of merchandise, order something on a credit card with a few keystrokes, and have the goods delivered by UPS or U.S. mail.

In some cities it's even possible to order groceries through online services. Peapod Inc. is an online grocery service serving 10,000 households in the Chicago and San Francisco areas. Peapod offers more than 18,000 items, from laundry detergent to lettuce, available in Jewel/Osco in Chicago or Safeway in San Francisco. Users can shop by brand name, category, or store aisle, and they can use coupons. Specially trained Peapod shoppers handle each order, even selecting the best produce available. The orders are then delivered in temperature-controlled containers.[23]

Will Online Services Survive the Internet?

In 1995 some new online services—Microsoft Network (MSN), AT&T's Interchange, MCI/News Corporation, and eWorld—sprang up to challenge the Big Three. But four months after it was launched, Microsoft scuttled plans to operate MSN as an online service, restructuring it as an access service (with special Microsoft-only areas) for the World Wide Web, the graphical portion of the Internet. In early 1996, in a major reversal, AT&T scrapped its proprietary system for which it had paid more than $50 million just a year earlier, focusing all its resources on the Internet. About the same time, MCI somewhat backed away from its joint online venture with News Corporation. Also early in that year, Gilbert Amelio, on becoming the new CEO of Apple Computer, wondered aloud whether the world needed another online service, specifically Apple's struggling eWorld.[24-28]

In addition, Prodigy's owners, IBM and retailer Sears, were seeking buyers for their third-place online service even as they were jazzing up content and hiring more aggressive management.[29,30] H&R Block, the tax-preparation company, also announced it was spinning off CompuServe because it lost money owing to heavy spending on the online service.[31,32] Even first-place AOL was looking vulnerable, as investors began quietly anticipating a drop in the stock price.[33]

Why all the bad news? The answer: The Internet and particularly the World Wide Web threaten to swamp the online services.[34] As the Net and the Web have become easier to navigate, online services have begun to lose customers and content providers—even as they have added their own arrangements for accessing the Internet.

Still, the online services have a lot to offer. One survey, for instance, found that *half* of the people on the Net got there through commercial services, which suggests they may be among the easiest ways to get to the Web.[35] In addition, the online services package information so that you can more quickly and easily find what you're looking for. It's also easier to conduct a live "chat" session on an online service than it is on the Web and it is easier for parents to exert control over the kinds of materials their children may view.[36]

Commercial services represent only one way to go online. We will shortly discuss the Internet and the World Wide Web. However, let us now consider yet another way—electronic bulletin board systems.

Bulletin Board Systems (BBSs)

Preview & Review: An electronic bulletin board system (BBS) is a centralized information source and message-switching system for a particular computer-linked interest group. A BBS is usually run by a sysop, or system operator. BBSs are now usually run by commercial online services; however, a few small ones are operated by individuals.

Katie and Gene Hamilton of St. Michaels on Maryland's Eastern Shore had renovated 13 houses and written numerous books and newspaper columns about home repair when in 1991 they decided to start their BBS called *House-Net*. The Hamiltons are **sysops, or system operators, of an electronic bulletin board system. A** *BBS, or bulletin board system,* **is a centralized information**

source and message-switching system for a particular computer-linked interest group. Operated in the midst of perpetual remodeling chaos and consisting of a few modems and personal computers, HouseNet dispenses advice about home repair. It also offers software programs on such matters as how to estimate materials or design a new deck.

The primary difference between an independent BBS and an online service is the single focus. "Typically, BBSs are targeted at a particular single-issue topic," says one expert, "and online services are kind of like department stores that have a thousand different topics that are being talked about."[37]

There are perhaps 100,000 BBSs operating in the United States, covering just about every topic you can imagine, from bird watching to socialism.[38] Most are started in a spare bedroom as a hobby, and the sysops may range from preteens to senior citizens. Some BBSs are mostly file libraries, offering games, software, and data files that may be downloaded (transferred) to users' personal computers. Others emphasize e-mail-type message centers and discussion or "chat" groups (in which users type messages to each other in real time). You can find information on employment, hobbies, and technical information from computer and software companies.

Online-Service BBSs

All the major online services—America Online, CompuServe, Prodigy, and so on—operate bulletin boards. These offer something for everyone. However, BBSs are only *one* of several offerings made available by commercial online services. And, of course, you are charged by the organization for using this particular service.

Independent BBSs

Why would anyone go to an independent BBS when they can go through one of the big online services? Says one writer, "The most obvious answer is cost. The large commercial services charge as much as $4.80 an hour, while many BBSs are free."[39] Indeed, one of the first things many people do when they get a modem installed in their personal computer is to dial up a local BBS, where they can often participate at no charge.

Small BBSs are generally run by individuals, often out of their homes, and, like the Hamiltons' HouseNet, are oriented toward a particular subject. Although such BBSs are generally friendly, the limited phone lines and resources can strain the systems, so busy signals may be common. BBSs are listed at the rate of 500–700 a month in the magazine *Boardwatch*.

As with online services, you need a microcomputer, modem, and communications software to gain access to a BBS.

The Internet

Preview & Review: The Internet, the world's biggest network, uses a protocol called TCP/IP to allow computers to communicate. Users can connect to the Internet via direct connections, online information services, and Internet service providers. There are many tools available to navigate the Internet: e-mail addresses, FAQ files, FTP, Archie, Telnet, gophers, WAIS, and Web browsers.

AT&T MOVE MAY TRIGGER INTERNET WAR, read the *USA Today* headline.[40]

In late February 1996, as an eye-catching lure, the company that used to be known as Ma Bell announced that users who signed up for its phone service would be given 5 hours a month on the Internet free for a year. Eighty percent of the U.S. population would be able to access the service with a local phone call. If ever there was an indicator that the Internet was finally ready to reach a mass audience, perhaps this was it. "When I heard about the AT&T prices, I thought, 'Great!'" said Bud Konheim, chief executive at fashion designer Nicole Miller Ltd. He and his wife had been using an online service to access the Internet, spending $170 a month or more. Added Konheim: "And I can't wait until AT&T gets some competition—then we'll get even better rates."[41] By now you, the reader, will probably be able to take advantage of those rates.

The Internet: What It Is, Where It Came From

Called "the mother of all networks," ***the Internet*, or simply "the Net," is an international network connecting approximately 36,000 smaller networks.** (■ *See Panel 8.5.)* To connect with it, you need pretty much the same things you need to connect with online services and BBSs: a computer, modem, telephone line, and appropriate communications software.

Whereas the number of users of commercial services doubled in 1995, to about 12.5 million, during the same year the number of active users of the World Wide Web jumped from 1 million to 8 million.[42] According to Nielsen Media Research—the company that does the famous Nielsen TV ratings—approximately 37 million people, or 17% of the U.S. and Canadian population 16 and older, have access to the Internet. Some 24 million of them used it during the latter half of 1995, spending an average of 5 hours and 28 minutes on it per week.[43]

Created by the U.S. Department of Defense in 1969 (under the name ARPAnet—ARPA was the department's Advanced Research Project Agency), the Internet was built to serve two purposes. The first was to share research among military, industry, and university sources. The second was to provide a system for sustaining communication among military units in the event of nuclear attack. Thus, the system was designed to allow many routes among many computers, so that a message could arrive at its destination by many possible ways, not just a single path. In 1973 the first international connections were made with England and Norway. By 1977 many more international connections had been made—thus the name *Inter*net.

With the many different kinds of computers being connected, engineers had to find a way for the computers to speak the same language. The solution developed was *TCP/IP*, the standard since 1983 and the standard language of the Internet. ***TCP/IP*, for Transmission Control Protocol/Internet Protocol, is the standardized set of guidelines (protocols) that allow different computers on different networks to communicate with each other efficiently,** no matter how they gained access to the Net, the topic of the next section.

Connecting to the Internet

There are three ways to connect your PC with the Internet:

- **Through school or work:** Many universities, colleges, and large businesses have dedicated, high-speed phone lines that provide a direct connection to the Internet. If you're a student, this may be the best deal because the connection is free or low cost. However, if you live off-campus and want to get this Internet connection from home, you probably

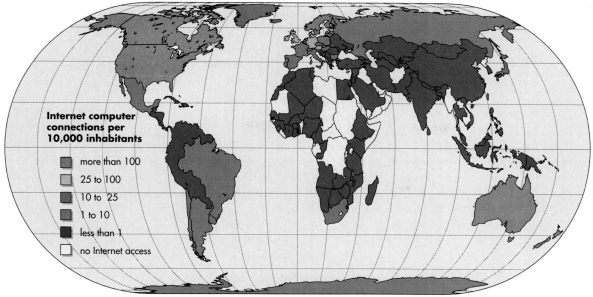

Internet computer connections per 10,000 inhabitants

- more than 100
- 25 to 100
- 10 to 25
- 1 to 10
- less than 1
- no Internet access

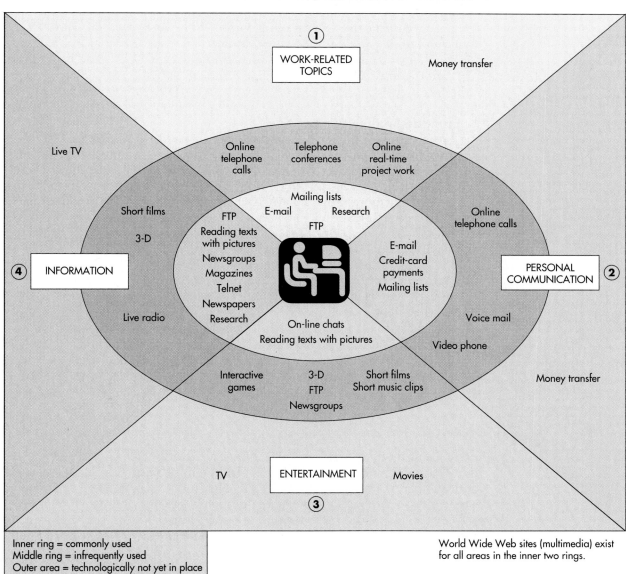

① WORK-RELATED TOPICS

Money transfer

Live TV

Online telephone calls

Telephone conferences

Online real-time project work

Mailing lists

Short films

FTP

E-mail

Research

Online telephone calls

3-D

FTP

Reading texts with pictures

FTP

④ INFORMATION

Newsgroups

Magazines

Telnet

E-mail

Credit-card payments

Mailing lists

PERSONAL COMMUNICATION ②

Live radio

Newspapers

Research

On-line chats

Reading texts with pictures

Voice mail

Video phone

Money transfer

Interactive games

3-D

FTP

Short films

Short music clips

Newsgroups

ENTERTAINMENT

TV

Movies

③

Inner ring = commonly used
Middle ring = infrequently used
Outer area = technologically not yet in place

World Wide Web sites (multimedia) exist for all areas in the inner two rings.

PANEL 8.5

The Internet

(Top) Approximate numbers of users connected to the Internet worldwide computer network (numbers increase daily). *(Bottom)* What's available through the Internet.

won't be able to do so.[44] To use a direct connection, your microcomputer must have TCP/IP software and be connected to the local area network that has the direct-line connection to the Net.

The next two types of connection are called "dial-up" connections.

- **Through online information services:** As mentioned, subscribing to a commercial online information service, which provides you with its own communications software, may not be the cheapest way to connect to the Internet, but it may well be the most trouble-free. AOL, CompuServe, and Prodigy all offer such access—that is, they provide an electronic "gateway" to the Internet. However, these types of connections do not always provide *complete* Internet services.

- **Through Internet service providers (ISPs):** To obtain complete Internet services through a dial-up connection, you use an ISP. ***Internet service providers (ISPs) are local or national companies that provide unlimited public access to the Internet and World Wide Web for a flat rate.*** Using your computer and the ISP's communications software, you dial up your ISP's phone number. The ISP's host computer uses SLIP (serial line Internet protocol) or PPP (point-to-point protocol) software to connect you to the Internet. Forrester Research predicts that ISPs could claim as many as 32 million online subscribers in the U.S. by 2000 versus 12.7 million for commercial online services.[45]

  So far, most ISPs have been small and limited in geographic coverage; the largest national company is Netcom Online Communication Services of San Jose, California. Other established national ISPs are Performance Systems Inc. (PSINet), UUNet Technologies, and BBN Corporation. However, recent competitors are MCI Internet, AT&T WorldNet, and Pacific Bell Internet (all from telephone companies); Microsoft Network (from the software company); and Tele-Communication Inc.'s @Home (from the cable-TV giant and pronounced "At Home"). Clearly, this is an area of fierce competition, but the presence of the phone and cable-TV companies in particular could help expand the mass market for Internet services.

  You can ask someone who is already on the Web to access the world-wide list of ISPs at *http://www.thelist.com*. Besides giving information about each provider in your area, "thelist" provides a rating (on a scale of 1 to 10) by users of different ISPs.

Once you're on the Net, how do you get where you want to go? That topic is next.

Internet Addresses

To send and receive e-mail on the Internet and interact with other networks, you need an Internet address. When Internet e-mail became fashionable, such addresses began to appear on business cards just as fax numbers did a few years earlier. For a while, newsmagazines such as *Time* and *Newsweek* even printed the Net addresses of writers of e-mail letters—until one complained that it exposed them to cranks (like publicizing someone's private postal mail address in a national magazine).

In the *Domain Name System*, the Internet's addressing scheme, an Internet address usually has two sections. For example, consider the address

president@whitehouse.gov.us

The first section, the user ID, tells "who" is at the address—in this case, *president*. The second section, after the @ symbol, tells "where" the address is—subdomain (if required), domain, domain type, and country (if required)—in this case, *whitehouse.gov.us*. Components of the second part of the address are separated by periods (called "dots"). (■ *See Panel 8.6.*)

What an Internet address means

How an e-mail message might find its way to a hypothetical address for Albert Einstein in the Physics Department of Princeton University.

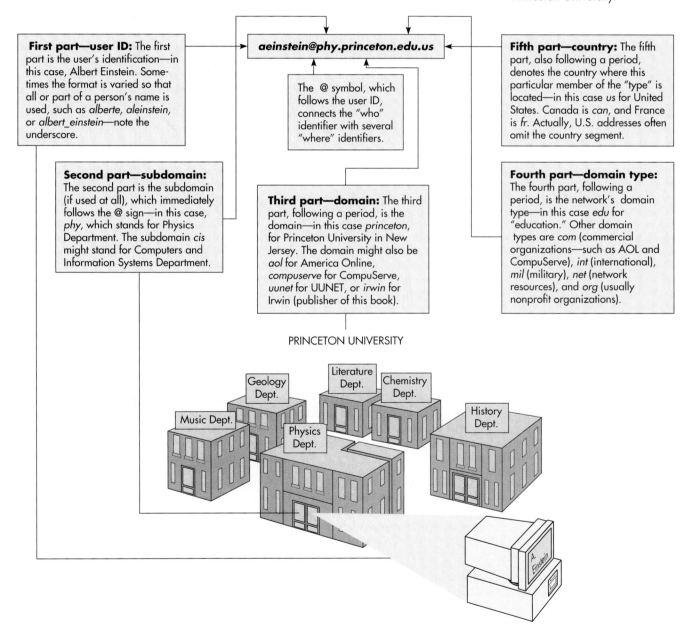

First part—user ID: The first part is the user's identification—in this case, Albert Einstein. Sometimes the format is varied so that all or part of a person's name is used, such as *alberte, aleinstein,* or *albert_einstein*—note the underscore.

aeinstein@phy.princeton.edu.us

The @ symbol, which follows the user ID, connects the "who" identifier with several "where" identifiers.

Fifth part—country: The fifth part, also following a period, denotes the country where this particular member of the "type" is located—in this case *us* for United States. Canada is *can*, and France is *fr*. Actually, U.S. addresses often omit the country segment.

Second part—subdomain: The second part is the subdomain (if used at all), which immediately follows the @ sign—in this case, *phy*, which stands for Physics Department. The subdomain *cis* might stand for Computers and Information Systems Department.

Third part—domain: The third part, following a period, is the domain—in this case *princeton*, for Princeton University in New Jersey. The domain might also be *aol* for America Online, *compuserve* for CompuServe, *uunet* for UUNET, or *irwin* for Irwin (publisher of this book).

Fourth part—domain type: The fourth part, following a period, is the network's domain type—in this case *edu* for "education." Other domain types are *com* (commercial organizations—such as AOL and CompuServe), *int* (international), *mil* (military), *net* (network resources), and *org* (usually nonprofit organizations).

PRINCETON UNIVERSITY

Geology Dept.

Literature Dept.

Chemistry Dept.

Music Dept.

Physics Dept.

History Dept.

A. Einstein

Features of & Tools for Navigating the Internet

The principal features of the Internet are e-mail, discussion groups, file transfer, remote access, and information searches (Gopher, WAIS, World Wide Web):

- **E-mail:** "The World Wide Web is getting all the headlines, but for many people the main attraction of the Internet is electronic mail," says technology writer David Einstein.[46] There are 90 million users of e-mail in the world, and although half of them are on private corporate networks, a great many of the rest are on the Internet.

  Foremost among the Internet e-mail programs is Qualcomm's Eudora software, which has about 10 million users worldwide and is used by 60% of the educational institutions on the Internet.

- **Usenet newsgroups—electronic discussion groups:** One of the Internet's most interesting features goes under the misleading name *newsgroups*, although they don't have much to do with news. Actually, **newsgroups are electronic discussions held by groups of people focusing on a specific topic**, the equivalents of CompuServe's or AOL's forums. They are one of the most lively and heavily trafficked areas of the Net.

  Usenet is a public access network of dispersed newsgroups that exchange e-mail and messages ("news"). "Users post questions, answers, general information, and FAQ files on Usenet," says one online specialist. "The flow of messages, or 'articles,' is phenomenal, and you can get easily hooked."[47] **An *FAQ*, for frequently asked questions, is a file that contains basic information about a newsgroup.** It's always best to read a newsgroup's FAQ before joining the discussion or posting (asking) questions.

  There are perhaps 3000–5000 Usenet newsgroup forums and they cover hundreds of topics. Three examples are *rec.arts.startrek.info*, *soc.culture.african.american*, and *misc.jobs.offered*. The first part is the group—*rec* for recreation, *soc* for social issues, *comp* for computers, *biz* for business, *sci* for science, *misc* for miscellaneous. The next part is the subject—for example, *rec.food.cooking*. The category called *alt* news groups offers more free-form topics, such as *alt.rock-n-roll.metal* or *alt.internet.services*.

- **Mailing lists—e-mail-based discussion groups:** Combining e-mail and newsgroups, mailing lists allow you to subscribe (generally free) to an e-mail mailing list on a particular subject or subjects. The mailing-list sponsor then sends the identical message to everyone on that list. There are probably 3000-plus electronic mailing-list discussion groups.

- **FTP—for copying all the free files you want:** Many Net users enjoy "FTPing"—cruising the system and checking into some of the tens of thousands of so-called FTP sites offering interesting free files to copy (download). **FTP, for file transfer protocol, is a method whereby you can connect to a remote computer and transfer publicly available files to your own PC.** The free files offered cover nearly anything that can be stored on a computer: software, games, photos, maps, art, music, books, statistics. Some 2000-plus FTP sites (so-called *anonymous FTP sites*) are open to anyone; others can be accessed only by knowing a password.

  If you know an FTP file's name or partial name, you can use an Internet tool called *Archie* to help find the file. *Archie* is a program that can find files in over 1000 FTP sites, based on the file name you specify. Once you find a file's location, you can use the FTP feature to go and get it.

- **Telnet—to connect to remote computers:** *Telnet* **is a cooperative system that allows you to connect (log on) to remote computers.** This feature enables you to tap into Internet computers and public-access files as though you were connected directly. It is especially useful for perusing large databases or library card catalogs. There are perhaps 1000 library catalogs accessible through the Internet.

- **Gopherspace, including Veronica and Jughead—the easy menu system:** Several other tools exist to help sift through the staggering amount of information on the Internet, but one of the most important is Gopher. *Gopher* **is a uniform system of menus, or series of lists, that allows users to easily browse through and retrieve files stored on different computers.** Why is it called "Gopher"? Because the first gopher was developed at the home of the Golden Gophers, the University of Minnesota, and it helps you "go fer" the files you seek.

   A classic running joke on the Internet is the use of puns on cartoon character names, such as those in the "Archie" comics. It started with Archie, and then the developers of a search application for Gopherspace concocted *Veronica* (for Very Easy Rodent-Oriented Net-wide Index to Computerized Archives). *Veronica* is a tool to help you search a large collection of Gopher menus for the keyword you specify. *Jughead* does the same thing, only it searches just the Gopher menu at the particular site you are currently visiting. (*Jughead* was developed at the University of Utah by Rhett "Jonzy" Jones and stands for Jonzy's Universal Gopher Hierarchy Excavation And Display.) "Think of Veronica as a good general Gopherspace search tool," advises one writer, "and Jughead as the tool of choice for deep burrowing of a local system."[48]

- **WAIS—ways of searching by content:** Pronounced "wayz," *WAIS*, **for Wide Area Information Server, is a system for searching Internet databases by subject, using specific words or phrases** rather than sorting through a hierarchy of menus. Unfortunately, WAIS is offered only by certain information sites (servers) and so can be applied to only a limited number of files.

- **World Wide Web—for multimedia and hypertext:** The fastest-growing part of the Internet—and many times larger than any online service— the World Wide Web is the most graphically inviting and easily navigable section of the Internet. **The** *World Wide Web,* **or simply "the Web," consists of an interconnected system of sites, or places, all over the world that can store information in multimedia form—sounds, photos, video, as well as text. The sites share a form consisting of a hypertext series of links that connect similar words and phrases.** Web software was developed in 1990 by Tim Berners in Cern, Switzerland.

   Note two distinctive features:

   (1) Whereas Gopher and WAIS deal with text, the Web provides information in *multimedia* form—it contains graphics, video, and audio as well as text.
   (2) Whereas Gopher is a menu-based approach to accessing Net resources, the Web uses a hypertext (✓ p. 88) format. *Hypertext* is a system in which documents scattered across many Internet sites are directly linked, so that a word or phrase in one document becomes a connection to an entirely different document. In particular, the format used on the Web is called *hypertext markup language (HTML)* and swaps information using *hypertext transfer protocol (HTTP).* When you use your mouse to point-and-click on a hypertext link (a highlighted word

or phrase), it may become a doorway to another place within the same document or another document on another computer thousands of miles away.

The places you visit on the Web are called *Web sites,* and the estimated number of such sites throughout the world ranges between 90,000 and 265,000.[49] More specifically, **a *Web site* is a file stored on a computer (server or host computer).** For example, the Parents Place Web site (*http://www.parentsplace.com*) is a resource run by mothers and fathers that includes links to other related sites, such as the Computer Museum Guide to the Best Software for Kids and the National Parenting Center.[50]

Information on a Web site is stored on "pages." **The *home page* is the main page or first screen you see when you access a Web site,** but there are often other pages or screens. *Web site* and *home page* tend to be used interchangeably, although a site may have many pages. (There might be a total of 16 million individual pages.[51] Some of them are simply abandoned because their creators have not updated or deleted them, the online equivalent of space-age debris orbiting the earth.[52])

To find a particular Web site (home page), you need its *URL.* **The *URL,* for Uniform Resource Locator, is an address that points to a specific resource on the Web.** Often it looks something like this: *http://www.blah.blah.html.* (Here *http* stands for "hypertext transfer protocol," *www* for "World Wide Web," and *html* for "hypertext markup language.")

To get to this address, you need a **Web browser—software that helps you get information you want by clicking your mouse pointer on words or pictures on the screen.** Four popular Web browsers are Netscape Navigator, NCSA Mosaic, InterCon WebShark, and Microsoft Internet Explorer. With the browser you can browse (search through) the Web. When you connect with a particular Web site, the screenful of information (the home page) is sent to you. You can easily skip from one page to another by using your mouse to click on the hypertext links. ISPs and online services provide their subscribers with Web browsers.

We offered some suggestions for exploring the Web in the Experience Box at the end of Chapter 3.

What Can You Find on the Net?

"Try as you may," says one writer, "you cannot imagine how much data is available on the Internet."[53] Besides e-mail, chat rooms, bulletin boards, games, and free software, there are thousands of databases containing information of all sorts. Here is a sampling:

> The Library of Congress card catalog. The daily White House press releases. Weather maps and forecasts. Schedules of professional sports teams. Weekly Nielsen television ratings. Recipe archives. The Central Intelligence Agency world map. A ZIP Code guide. The National Family Database. Project Gutenberg (offering the complete text of many works of literature). The Alcoholism Research Data Base. Guitar chords. U.S. government addresses and phone (and fax) numbers. *The Simpsons* archive.[54]

And that's just the beginning.

Practical uses of the Internet are discussed further in the Experience Boxes at the end of Chapters 9 and 12. The next section covers some aspects of connectivity other than online services and the Internet.

Shared Resources: Workgroup Computing, Electronic Data Interchange, & Intranets

Preview & Review: Workgroup computing enables teams of co-workers to use networked microcomputers to share information and cooperate on projects.

Electronic data interchange (EDI) is the direct electronic exchange of standard business documents between organizations' computer systems.

Intranets are internal corporate networks that use the infrastructure and standards of the Internet and the Web.

When they were first brought into the workplace, microcomputers were used simply as another personal-productivity tool, like typewriters or calculators. Gradually, however, companies began to link a handful of microcomputers together on a network, usually to share an expensive piece of hardware, such as a laser printer. Then employees found that networks allowed them to share files and databases as well. Networking using common software also allowed users to buy equipment from different manufacturers—a mix of workstations from both Sun Microsystems and Hewlett-Packard, for example. The possibilities for sharing resources have led to workgroup computing.

Workgroup Computing & Groupware

Workgroup computing, also called *collaborative computing*, **enables teams of co-workers to use networks of microcomputers to share information and cooperate on projects.** Workgroup computing is made possible not only by networks and microcomputers but also by *groupware*. **Groupware is software that allows two or more people on a network to work on the same information at the same time.**

In general, groupware permits office workers to collaborate with colleagues and tap into company information through computer networks. It also enables them to link up with crucial contacts outside their organization—a customer in Nashville, a supplier in Hong Kong, for example.

The best-known groupware is Lotus Notes (purchased by IBM in 1995), which had an installed base of 4.5 million copies at the beginning of 1996.[55] Among its advantages, Notes can run on a variety of operating systems and allows users to send e-mail via several online services. It also lets users create and store all kinds of data—text, audio, video, pictures—on common databases. Notes 4.0, released in early 1996, lets users create documents that can be displayed on the Web and use a built-in "browser" to surf (look through) the Web. In addition, Notes has the advantages of offering better security and the ability to synchronize multiple kinds of databases.[56–58] In April 1996, Microsoft launched its long-awaited Notes competitor called *Exchange Server.*

Electronic Data Interchange

Paper handling is the bane of organizations. Paper must be transmitted, filed, and stored. It takes up much of people's time and requires the felling of considerable numbers of trees. Is there a way to accomplish the same business tasks without using paper?

One answer lies in business-to-business transactions conducted via a computer network. *Electronic data interchange (EDI)* **is the direct electronic exchange between organizations' computer systems of standard business documents,** such as purchase orders, invoices, and shipping documents. For example, Wal-Mart has electronic ties to major suppliers like Procter & Gamble, allowing both companies to track the progress of an order or other document through the supplier company's computer system.

To use EDI, organizations wishing to exchange transaction documents must have compatible computer systems, or else go through an intermediary. For example, more than 500 colleges are now testing or using EDI to send transcripts and other educational records to do away with standard paper handling and its costs.

Intranets

It had to happen: First, businesses found that they could use the World Wide Web to get information to customers, suppliers, or investors. FedEx, for example, saved millions by putting up a server in 1994 that enabled customers to click through Web pages to trace their parcels, instead of having FedEx customer-service agents do it. It was a short step from that to companies starting to use the same technology inside—in internal Internet-like networks called *intranets*.[59,60] **Intranets are internal corporate networks that use the infrastructure and standards of the Internet and the World Wide Web.** "The Web, it turns out, is an inexpensive yet powerful alternative to other forms of internal communications, including conventional computer setups," says one writer. "Because Web browsers run on any type of computer, the same electronic information can be viewed by any employee."[61] Thus, intranets connect all the types of computers, be they PCs, Macs, or workstations.

One of the greatest considerations of an intranet is security—making sure that sensitive company data accessible on intranets is protected from the outside world. The means for doing this is security software called *firewalls*. **A firewall is a security program that connects the intranet to external networks, such as the Internet.** It blocks unauthorized traffic from entering the intranet and can also prevent unauthorized employees from accessing the intranet.

We consider security matters at length in Chapter 12. Now we move on to look at some of the changes that connectivity has brought about in people's work lives.

Portable Work: Telecommuting & Virtual Offices

Preview & Review: Working at home with computer and communications connections between office and home is called *telecommuting.*

The virtual office is a nonpermanent and mobile office run with computer and communications technology.

"In a country that has been moaning about low productivity and searching for new ways to increase it," observed futurist Alvin Toffler, "the single most anti-productive thing we do is ship millions of workers back and forth across the landscape every morning and evening."[62]

Toffler was referring, of course, to the great American phenomenon of physically commuting to and from work. More than 108 million Americans commute to work by car and another 6 million by public transportation. Information technology has responded to the cry of "Move the work instead of the workers!" Computers and communications tools have led to telecommuting and telework centers, the mobile workplace, and the virtual office and "hoteling."

Telecommuting & Telework Centers

Working at home with telecommunications between office and home is called *telecommuting.* In 1994, the number of part-time and full-time telecommuters reached 9.1 million, up 20% from the year before.[63] The figure then dropped by a million in September 1995, with many workers giving up telecommuting partly because severe corporate downsizing made them worry about the stability of their jobs ("out of sight, out of mind").[64] Still, many companies, particularly high-technology ones, are encouraging telecommuting because they have found it boosts morale and improves productivity.

The reasons for telecommuting are quite varied. One may be to eliminate the daily drive, reducing traffic congestion, energy consumption, and air pollution. Another may be to take advantage of the skills of homebound workers with physical disabilities (especially since the passage of the Americans with Disabilities Act). Parents with young children, as well as "lone eagles" who prefer to live in resort areas or other desirable locations, are other typical telecommuter profiles.

Another term for telecommuting is *telework.* However, *telework* includes not only those who work at least part time from home but also those who work at remote or satellite offices, removed from organizations' main offices. Such satellite offices are sometimes called *telework centers.* An example of a telework center is the Riverside Telecommuting Center, in Riverside, California, supported by several companies and local governments. The center provides office space that helps employees who live in the area avoid lengthy commutes to downtown Los Angeles. However, these days an office can be virtually anywhere.

The Virtual Office

The term *virtual office* borrows from "virtual reality" (artificial reality that projects the user into a computer-generated three-dimensional space). **The *virtual office* is an often nonpermanent and mobile office run with computer and communications technology.** Employees work not in a central office but from their homes, cars, and other new work sites. They use pocket pagers, portable computers, fax machines, and various phone and network services to conduct business.

Could you stand not having a permanent office at all? Here's how one variant, called *hoteling,* would work: You call ahead to book a room and speak to the concierge. However, your "hotel" isn't a Hilton, and the "concierge" isn't a hotel employee who handles reservations, luggage, and local tours. Rather, the organization is your employer, and the concierge is an administrator who handles scheduling of available office cubicles—of which there is only one for every three workers.

Hoteling works for Ernst & Young, an accounting and management consulting firm. Its auditors and management consultants spend 50–90% of their time in the field, in the offices of clients. When they need to return to E&Y headquarters, they call a few hours in advance. The concierge consults a computerized scheduling program and determines which cubicles are available on the days requested. He or she chooses one and puts the proper nameplate on the office wall. The concierge then punches a few codes into the phone to program its number and voice mail. When employees come in, they pick up personal effects and files from lockers and then take them to the cubicles they will use for a few days or weeks.[65,66]

What makes hoteling possible, of course, is computer and communications technology. Computers handle the cubicle scheduling and reprogramming of phones. They also allow employees to carry their work around with them stored on the hard drives of their laptops. Cellular phones, fax machines, and e-mail permit employees to stay in touch with supervisors and co-workers.

So-called blue-collar workers are also now working out of virtual offices. "These days, some truckers are more inclined to sport white collars than tank tops," says one reporter. "Once (and still) lumped as rednecks and high-school dropouts, they are now fluent with computers, satellites, and fax machines—all of which can be found in the cabs of their 18-wheelers."[67]

Truckers may now be required to carry laptops with which they keep in touch via satellite with headquarters. They may also have to take on tasks previously never dreamed of. These include faxing sales invoices, hounding late-paying customers, and training people to whom they deliver high-tech office equipment.

Other workers—field service representatives, salespeople, and roving executives—also find that to stay competitive they must bring office technology with them. For instance, Bob Spoer of San Francisco, who started a telecommunications firm that needs constant tending, takes a cell phone to baseball games and on ski lifts ("I actually get some good reception up there").[68] Many people, however, find that technology creates an electronic leash. "I get 20 beeps on a weekend," said Peter Hart, then a supervisor at a California chip manufacturer—before he changed jobs because pagers, cell phones, and e-mail were taking over his life.[69]

As we stated at the outset of this chapter, information technology is blurring time and space, eroding the barriers between work and private life. Some people thrive on it, but others hate it.

The Coming Information, or Internet, Appliance

Preview & Review: The coming convergence of computers, communications, consumer electronics, and mass media will likely produce an "Internet appliance," which will combine PC capabilities with TV, telephone, and cable.

At the beginning of this book, we discussed the phenomenon of *technological convergence,* or *digital convergence,* the merger of several industries—computers, communications, consumer electronics, and mass media—through various devices that exchange information in digital form. We suggested that the embodiment of this convergence is the *TV/PC* or *information appliance.* This gadget would presumably receive a digitized stream of sound, video, text, and data from some sort of electronic delivery system, and we would be able to "talk back" or interact with it.

How close are we to realizing this device? The answer is: Maybe we're practically there now.

The TV/PC

Nicholas Negroponte, a media futurist, has suggested that by 2000 "many Americans will be watching TV in the upper-right-hand corner of their PCs."[70] Actually, if you want to do this, you can do it now, although there are limitations. (■ *See Panel 8.7.*) There is, for instance, a Mac TV, which, as one critic wrote, "doesn't let you work and watch *Wonder Woman* at the same time" but can play the TV's sound while the Macintosh display screen is visible.[71]

There is also the IBM-compatible Compaq Presario 920 CDTV, which, besides a TV, includes a phone, CD-ROM drive, fax modem, and software to turn the computer into a telephone dialer and answering machine. You can pop up a remote control on the display screen to do such things as use the built-in phone, play music CDs, and change TV channels. The TV's picture quality is "nowhere nearly as good as on a standard model" TV set, says a reporter who reviewed the machine.[72] However, the software lets you do a number of things that you can't do with a regular TV. You can make the picture fill the screen. You can resize it to a movable window. You can distort the picture so it is taller or wider than usual. You can freeze the picture while leaving the sound on or hide it while you work. You can capture still images (though not continuous video) from CD-ROM.

Although a lot of refinements need to be made, the TV/PC is clearly a forerunner of future gadgetry. The shape of things to come in such a gadget may be visible in some of the Internet's newest services, as we discuss next.

PANEL 8.7

TV/PC

Personal computer featuring television.

The Multifaceted Net: Phone, Radio, TV, & 3-D

Think the Internet is mainly a network for communicating by computers and keyboards? Think again. You can also use it as a phone line, radio network, television network, and 3-D theater—right now.

- **Telephones on the Net:** With your computer modem, a microphone, and the right kind of software (such as Internet Phone by Vocaltech— *http://www.vocaltec.com*; also available with Netcom's NetCruiser software), you can reach out and touch someone voice-to-voice on the Net.[73-75] The software cannot call a conventional telephone, only one person can talk at a time, and calls are not secure from eavesdroppers. Moreover, both parties must be online at the same time, which means calls must be prearranged. Nevertheless, the technology could have a large market, particularly among long-distance callers trying to hold costs down, such as students and people with families overseas. (Most dial-up connections to the Internet are local calls.)

- **Radio on the Net:** Desktop radio broadcasting is here—both music and spoken programming—and has been since 1995, when RealAudio software was unveiled. RealAudio can compress digital sound signals so they can be played in real time, even though sent over low-capacity telephone lines. You can, for instance, listen to net.radio (*http://www.netradio.net*), which features "vintage rock" 24 hours a day, or English-language services of 19 shortwave outlets from World Radio Network in London (*http://town.hall.org/Archives/radio/Mirrors/WRN/audio.html*).[76-78] "The implications of the new technology are enormous," says one computer writer. "It could provide a global soapbox for political parties, religious movements, and other groups that lack access to broadcast services."[79]

- **Television on the Net:** In late 1996, several television broadcasters (NBC, CNN, MTV, VH1, QVC, and WGBH), using a technology called *Intercast*, plan to insert into a portion of the TV signal data streams that will deliver selected Web pages to your PC screen. (You'll need a TV tuner/decoder card, selling at $200–$300, or a new microcomputer with such circuitry built in.) The main appeal would be that you could access Web home pages related to the television program. For example, as NBC anchor Tom Brokaw talks about a part of Africa, you could get background information about the region from the Web.[80,81]

- **3-D on the Net:** Three-dimensionality may be one of the tougher challenges because so far the computer can't update images as fast as the human eye. Thus, says one writer, "walking through 3-D cyberstores usually feels like staggering, which literally makes some people sick."[82] Still, some companies are trying to bring 3-D to the Web, using new software technology called VRML (for virtual reality markup language, discussed in Chapter 11). Silicon Graphics, for instance, offers a 3-D viewer called WebSpace to supplement browsers like Netscape or Mosaic.[83,84]

As we discuss later in the chapter, most consumers don't yet have the very high speed digital lines or other hardware needed to receive many of these services in real time, but that will no doubt change.

The Information/Internet Appliance

At present there seem to be three possible variations on what is called either an *information appliance* or *Internet appliance:* the set-top box, the network PDA, and the network PC.[85]

- **Set-top box:** As we mentioned in Chapter 1, a *set-top box,* or "Net-top box," is a keypad that allows cable-TV viewers to change channels or, in the case of interactive systems, to exercise other commands. A videogame console could double as a set-top box that would let consumers surf the Web on their TV set over phone lines or cable. Companies such as Philips, Sega, Sony, and Thomson are working on new developments in this area.

  A variant is the *cable modem,* which would allow cable-TV subscribers with PCs to make online connections at speeds faster than those possible with traditional computer modems.

- **Network PDAs:** Different kinds of *network personal digital assistants (PDAs)* (✓ p. 157)—variously known as personal communicators, pocket PCs, and smart phones—could provide low-cost, wireless access to the Internet. With these handheld devices, connections to the Net could be as easy and portable as cell-phone telephone calls are today. Companies such as Apple, Sun, and Toshiba are trying to bring such products to market.

  There is even a gadget in development (probably to be operational in 2002) that resembles cartoon detective Dick Tracy's "2-way wrist communicator." Being worked on by Philips and British Telecom, the wrist communicator will combine such features as video phone, TV and radio reception, clock, electronic notebook, and a personal computer with wireless Web access.[86]

- **Internet PCs:** This is the under-$500 "hollow PC" or "network computer" (also called a "Web PC") that we discussed in Chapter 4 (✓ p. 183). Naysayers insist it is an underpowered, impractical throwback to dumb terminals. Others say it is a visionary device that could threaten Microsoft (eliminating the need for its PC operating systems) and the big microcomputer makers and turn the computer business upside down.[87,88]

  The first $500-range Internet PC to be unveiled, in February 1996, was Oracle Corporation's NC (for "network computer"), available in both desktop and laptop versions. It does not include a monitor (you use one you already own or a TV set) or any disk drives (diskette, hard, or CD-ROM).

Instead, you connect the box to your phone or cable line and get all your software from computers located elsewhere.[89–91] Other low-cost Internet PCs have been announced by Sun Microsystems and Apple Computer (whose Pippin, based on the Power Mac but made by Bandai in Japan, can support existing Macintosh applications).

We are still, of course, not at the point where your microcomputer has become the true information appliance just described. Why has it taken so long for computers and communications systems—telephone, radio, television, and so on—to come together? Why can't voice, data, and images be easily transmitted via a telephone/television/computer "information appliance" between your home and that of a friend? To understand this is to understand why you need the equipment you do to communicate using your present PC.

Using a Microcomputer to Communicate: Analog & Digital Signals, Modems & Datacomm Software, ISDN Lines, & Cable Modems

Preview & Review: To communicate online through a microcomputer, users need a modem to send and receive computer-generated messages over telephone lines. Modems translate the computers' digital signals of discrete bursts into analog signals of continuous waves, and vice versa. A modem may be external to the computer or internal and may have various transmission speeds.

Communications, or datacomm, software is also required.

ISDN lines and cable modems are faster than conventional PC modems.

The principal reason the TV/PC is still not quite here lies in the fact that information is transmitted by two types of signals, each requiring different kinds of communications technology. The two types of signals are *analog* and *digital*. (■ *See Panel 8.8.*) In a way they resemble analog and digital watches. As we illustrated in Chapter 1 (✓ p. 9), an analog watch shows time as a continuum. A digital watch shows time as discrete numeric values.

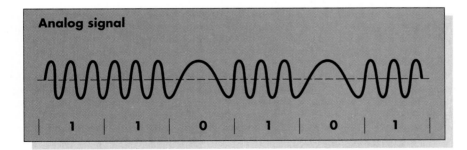

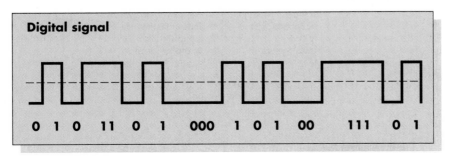

PANEL 8.8

Review of analog and digital signals

An analog signal represents a continuous electrical signal in the form of a wave. A digital signal is discontinuous, expressed as discrete bursts in on/off electrical pulses.

Analog Signals: Continuous Waves

Telephones, radios, and televisions—the older forms of communications technology—were designed to work with an analog signal. **An *analog signal* is a continuous electrical signal in the form of a wave.** The wave is called a *carrier wave.*

Two characteristics of analog carrier waves that can be altered are frequency and amplitude.

- Frequency: *Frequency* **is the number of times a wave repeats during a specific time interval**—that is, how many times it completes a *cycle* in a second.
- Amplitude: *Amplitude* **is the height of a wave within a given period of time.** Amplitude is actually the strength or volume—the loudness—of a signal.

Both frequency and amplitude can be modified by making adjustments to the wave. Indeed, it is by such adjustments that an analog signal can be altered to represent a digital signal, as we shall explain.

Digital Signals: Discrete Bursts

A *digital signal* **uses on/off or present/absent electrical or light pulses in discontinuous, or discrete, bursts, rather than a continuous wave.** This two-state kind of signal is used to represent the two-state binary language of 0s and 1s that computers use. That is, the presence of an electrical/light pulse can represent a 1 bit, its absence a 0 bit.

The Modem: Today's Compromise

Digital signals are better—that is, faster and more accurate—at transmitting computer data. However, many of our present communications connections, such as telephone and microwave, are still analog. To get around this problem, we need a device called a *modem*. **A *modem*—short for *modulater/demodulater*—converts digital signals into analog form (a process known as *modulation*) to send over phone lines. A receiving modem at the other end of the phone line then converts the analog signal back to a digital signal (a process known as *demodulation*).** (■ *See Panel 8.9.*)

Modulation/demodulation does not actually make an analog signal into a digital one. Rather, it changes the shape of the wave to convey digital infor-

How modems work

A sending modem translates digital signals into analog waves for transmission over phone lines. A receiving modem translates the analog signals back into digital signals.

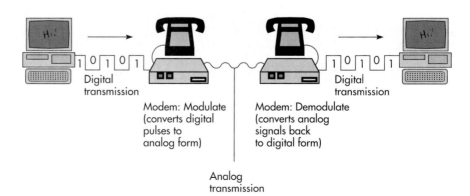

mation. For instance, the frequency might be changed. A normal wave cycle within a given period of time might represent a 1, but more frequent wave cycles within a given period might represent a 0. Or, the amplitude might be changed. A loud sound might represent a 1 bit, a soft sound might represent a 0 bit. That is, a wave with normal height (amplitude) might signify a 1, a wave with smaller height a 0. (■ *See Panel 8.10.*) The wave itself does not assume the boxy on/off shape represented by the true digital signal.

From this we can see that modems are a compromise. They cannot transmit digital signals in a way that delivers their full benefits. As a consequence, communications companies have been developing alternatives, such as the Integrated Services Digital Network (ISDN) and cable modems, discussed shortly.

Choosing a Modem

Two criteria for choosing a modem are whether you want an internal or external one, and what transmission speed you wish:

- **External versus internal:** Modems are either internal or external. (■ *See Panel 8.11, next page.*)

    An *external modem* is a box that is separate from the computer. The box may be large or it may be portable, pocket size. A line connects the modem to a port in the back of the computer. A second line connects the modem to a standard telephone jack. There is also a power cord that plugs into a standard AC wall socket.

    The advantage of the external modem is that it can be used with different computers. Thus, if you buy a new microcomputer, you will probably be able to use your old external modem. Also, external modems help isolate the computer's internal circuitry from phone-line conducted lightning surges.

    An *internal modem* is a circuit board that plugs into a slot inside the system cabinet. Nowadays many new microcomputers come with an internal modem already installed. Advantages of the internal modem are that it doesn't take up extra space on your desk, it is less expensive than an external modem, and it doesn't have a separate power cord.

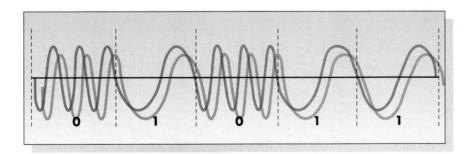

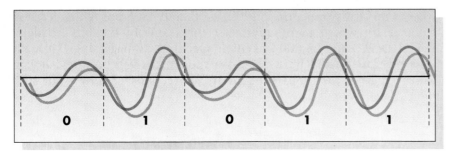

PANEL 8.10

Modifying an analog signal

A modem may modify an analog signal to carry the on/off digital signals of a computer in two ways. *(Top)* The frequency of wave cycles is altered so that a normal wave represents a 1 and a more frequent wave within a given period represents a 0. *(Bottom)* The amplitude (height) of a wave is altered so that a wave of normal height represents a 1 and a wave of lesser height represents a 0.

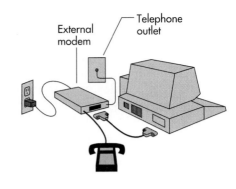

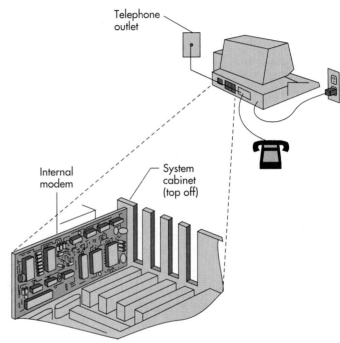

PANEL 8.11

External versus internal modems

An external modem is a box that is outside the computer. An internal modem is a circuit board installed in an expansion slot inside the system cabinet.

- **Transmission speed:** Because most modems use standard telephone lines, users are charged the usual rates by phone companies, whether local or long-distance. Users are also often charged by online services for time spent online. Accordingly, *transmission speed*—the speed at which modems transmit data—becomes an important consideration. The faster the modem, the less time you need to spend on the telephone line.

Today users refer to **bits per second (bps) or, more likely, *kilobits per second (kbps)* to express data transmission speeds.** A 14,400-bps modem, for example, is a 14.4-kbps modem.

Today's modems transmit at 1200, 2400, and 4800 bps (considered slow, and not really worth using anymore); 9600 and 14,400 bps (moderately fast); and 19,200 and 28,800 bps (high speed). A 10-page single-spaced letter can be transmitted by a 2400-bps modem in 2½ minutes. It can be transmitted by a 9600-bps modem in 38 seconds and by a 19,200-bps modem in 19 seconds.

Communications Software

To communicate via a modem, your microcomputer requires communications software. **Communications software, or "datacomm software," manages the transmission of data between computers or video display terminals.** Macintosh users have Smartcom; Windows users have Smartcom, Crosstalk, Wincom, CommWorks, Telix, and HyperAccess; OS/2 Warp users have HyperAccess. Often the software comes on diskettes bundled with (sold along with) the modem.

Besides establishing connections between computers, communications software may perform other functions:

- Error correction: Static on telephone lines can introduce errors, or "noise," into data transmission. **Noise is anything that causes distortion in the signal when it is received.** When acquiring a modem and its accompanying software, you should inquire whether they incorporate error-correction features.

- Data compression: **Data compression reduces the volume of data in a message, thereby reducing the amount of time required to send data from one modem to another** (✓ p. 129). When the compressed message reaches the receiver, the full message is restored. With text and graphics, a message may be compressed to as little as one-tenth of its original size.

- Remote control: **Remote-control software allows you to control a microcomputer from another microcomputer in a different location,** perhaps even thousands of miles away. One part of the program is in the machine in front of you, the other in the remote machine. Such software is useful for travelers who want to use their home machines from afar. It's also helpful for technicians trying to assist users with support problems.

- Terminal emulation: Mainframes and minicomputers are designed to be accessed by terminals, not by microcomputers, which use different operating systems. **Terminal emulation software allows you to use your microcomputer to simulate a mainframe's terminal.** That is, the software "tricks" the large computer into acting as if it were communicating with a terminal. Your PC needs terminal emulation capability to log into computers acting as electronic bulletin board servers or holding databases of research materials.

ISDN Lines & Cable Modems

Users who found themselves banging the table in frustration as their 14.4-kbps modem took 45 minutes to transmit a 1-minute low-quality video from a Web site are about to get some relief. Probably the two most immediate contenders to standard phone modems are *ISDN lines* and *cable modems.* (■ *See Panel 8.12.)*

PANEL 8.12

Comparison of carrying capacity

How long it takes to send a 1-megabyte file:	
Standard phone modem, 28.8 kbps	4.6 minutes
ISDN line, 64–128 kbps	2.1 minutes or less
Cable modem, 3000 kbps or more	2.6 seconds or less

- **ISDN lines:** ISDN stands for Integrated Services Digital Network. *ISDN consists of hardware and software that allow voice, video, and data to be communicated as digital signals over traditional copper-wire telephone lines.* Capable of transmitting up to 128 kbps, ISDN lines are up to five times faster than conventional phone modems.[92]

  ISDN is not cheap, costing perhaps two or three times as much per month as regular phone service. Installation could also cost $200 or more if you need a phone technician to wire your house and install the software in your PC.[93] Nevertheless, with the number of people now working at home and/or surfing the Internet, demand has pushed ISDN orders off the charts. Forecasts are for 7 million U.S. installations by 2000, from the current 450,000 lines today.[94]

  Even so, ISDN's time may have come and gone. The reason: Cable modem and other technologies threaten to render it obsolete.

- **Cable modems:** Cable companies say that a cable modem can carry digital data 1000 times faster than plain old telephone system (POTS) lines, and they've found that usage shoots up when the service is connected. "Some nights I can't get off the thing," said biology professor Grant Balkema, after a cable company installed cable modems at Boston College. "I've started some nights at around 10 and stayed up until 2 A.M. It's— dare I say?—addictive."[95]

  A *cable modem* **is a modem that connects a personal computer to a cable-TV system that offers online services.** The gadgets are still fairly exotic, and it will probably be 1997 before internationally standardized cable modems go on sale.[96] The reason? So far probably 90% of U.S. cable subscribers are served by networks that don't permit much in the way of two-way data communications. "The vast majority of today's . . . cable systems can deliver a river of data downstream," says one writer, "but only a cocktail straw's worth back the other way."[97] Nevertheless, Forrester Research predicts about 6.8 million American homes will have cable modems by 2000.[98]

In addition, competition may also be expected from fiber-optic lines and digital satellites, discussed next.

◣ Communications Channels: The Conduits of Communications

Preview & Review: A channel is the path, either wired or wireless, over which information travels. Various channels occupy various radio-wave bands on the electromagnetic spectrum. Types of wired channels include twisted-pair wire, coaxial cable, and fiber-optic cable. Two principal types of wireless channels are microwave and satellite systems. Some of wireless communications devices are pagers, analog cellular phones, packet radio, and Cellular Digital Packet Data (CDPD).

The next generation of wireless communications will include digital cellular phones, personal communications services (PCS), specialized mobile radio (SMR), and satellite-based systems.

If you are of a certain age, you may recall when two-way individual communications were accomplished mainly in two ways. They were carried by (1) a telephone wire or (2) a wireless method such as shortwave radio. Today there are many kinds of communications channels, although they are still wired or wireless. **A *channel* is the path over which information travels in a telecommunications system from its source to its destination.** (Channels are also called *links, lines,* or *media.*) The basis for all telecommunications channels, both wired and wireless, is the electromagnetic spectrum.

The Electromagnetic Spectrum

Telephone signals, radar waves, and the invisible commands from a garage-door opener all represent different waves on what is called the electromagnetic spectrum. **The *electromagnetic spectrum* consists of fields of electrical energy and magnetic energy, which travel in waves.** (■ *See Panel 8.13.*)

PANEL 8.13

The electromagnetic spectrum

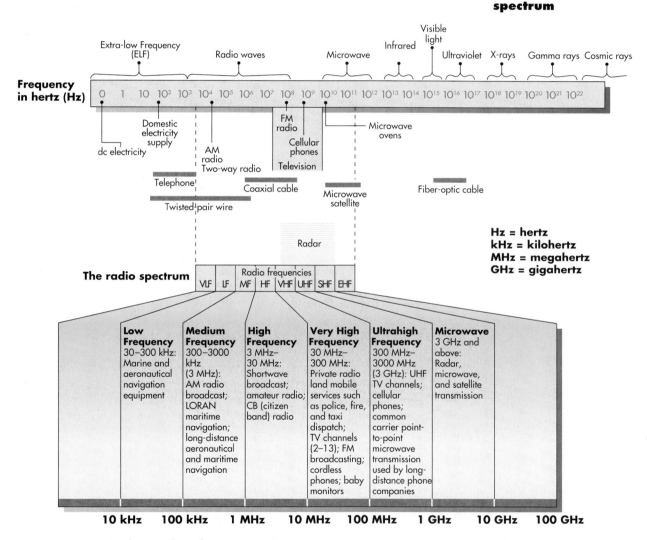

Cellular	Private land mobile	Narrowband PCS	Industrial	Common carrier paging	Point-to-multipoint Point-to-point	PCS	Industrial
824–849 MHz 869–894 MHz	896–901 MHz 930–931 MHz	901–902 MHz 930–931 MHz	902–928 MHz Unlicensed commercial use such as cordless phones and LANs	931–932 MHz Includes national paging services	932–935 MHz 941–944 MHz	1850–1970 MHz 2130–2150 MHz 2180–2200 MHz	2400–2483.5 MHz Unlicensed commercial use such as LANs

Frequencies for wireless data communications

All radio signals, light rays, and X-rays radiate an energy that behaves like rippling waves. The waves can be characterized according to frequency and wavelength:

- **Frequency:** As we've seen, *frequency* is the number of times a wave repeats (makes a cycle) in a second. **Frequency is measured in *hertz (Hz)*, with 1 Hz equal to 1 cycle per second.** One thousand hertz is called a *kilohertz (kHz),* 1 million hertz is called a *megahertz (MHz),* and 1 billion hertz is called a *gigahertz (GHz).*

  Ranges of frequencies are called *bands* or *bandwidths.* The bandwidth is the difference between the lowest and highest frequencies transmitted. Bandwidths are usually referred to by the range they cover. For example, cellular phones are on the 800–900 megahertz band—that is, their bandwidth is 100 megahertz.

- **Wavelength:** Waves also vary according to their length—their *wavelength.* We hear references to wavelength in "shortwave radio" and "microwave oven."

  At the low end of the spectrum, the waves are of low frequency and of long wavelength (such as domestic electricity). At the high end, the waves are of high frequency and short wavelength (such as cosmic rays).

The electromagnetic spectrum can be represented by the appliances and machines that emit or detect particular wavelengths. We could start on the left, at the low-frequency end, with video display terminals and hair dryers. We would then range up through AM and FM radios, shortwave radios, VHF and UHF television, and cellular phones. Next we would proceed through radar, microwave ovens, infrared "nightscope" binoculars, and ultraviolet-light tanning machines. Finally, we would go through X-ray machines and end up with gamma-ray machines for food irradiation at the high-frequency end. The part of the spectrum of interest to us is that area in the middle—between 3 million and 300 billion hertz (3 megahertz to 300 gigahertz). This is the portion that is regulated by the government for communications purposes.

Certain bands are assigned by the Federal Communications Commission (FCC) for certain purposes—that is, to be controlled by different classes or groups of users. Some frequencies traditionally used by railroads, electric utilities, and police and fire departments have in recent times been opened up for new uses. These new applications include personal telephones, mobile data services, and satellite message services. We explain these further in the next few pages.

Let us now look more closely at the various types of channels:

- Twisted-pair wire
- Coaxial cable
- Fiber-optic cable
- Microwave and satellite systems
- Other wireless communications
- The next generation of wireless communications

Twisted-Pair Wire

The telephone line that runs from your house to the pole outside is probably twisted-pair wire. ***Twisted-pair wire* consists of two or more strands of insulated copper wire, twisted around each other in pairs.** They are then covered in another layer of plastic insulation. (■ *See Panel 8.14.*)

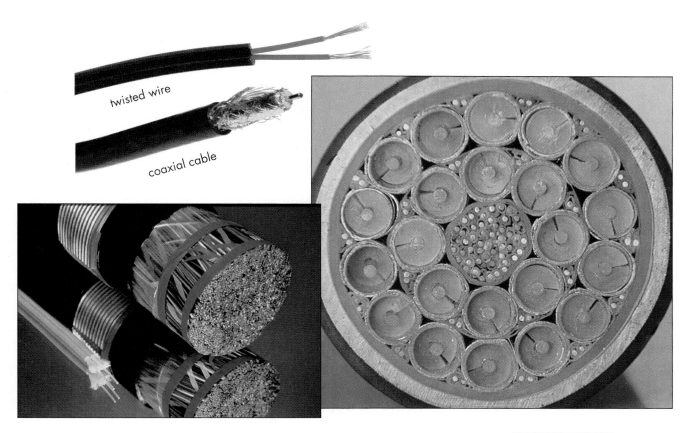

twisted wire

coaxial cable

Because so much of the world is already served by twisted-pair wire, it will no doubt continue to be used for years, both for voice messages and for modem-transmitted computer data. However, it is relatively slow and does not protect well against electrical interference. As a result, it will certainly be superseded by better communications channels, wired or wireless.

Coaxial Cable

Coaxial cable, **commonly called "coax," consists of insulated copper wire wrapped in a solid or braided metal shield, then in an external cover.** Coax is widely used for cable television. Coaxial cable is much better at resisting noise than twisted-pair wiring. Moreover, it can carry voice and data at a faster rate (perhaps 200 megabits per second, compared to 10 megabits per second for twisted-pair wire).

Fiber-Optic Cable

A *fiber-optic cable* **consists of hundreds or thousands of thin strands of glass that transmit not electricity but rather pulsating beams of light.** These strands, each as thin as a human hair, can transmit billions of pulses per second, each "on" pulse representing one bit. When bundled together, fiber-optic strands in a cable 0.12 inch thick can support a quarter- to a half-million voice conversations at the same time. Moreover, unlike electrical signals, light pulses are not affected by random electromagnetic interference in the environment. Thus, they have much lower error rates than normal telephone wire and cable. In addition, fiber-optic cable is lighter and more durable than twisted-pair and coaxial cable.

PANEL 8.14

Three types of wired communications channels

(Top) Twisted-pair wire. This type does not protect well against electrical interference. *(Middle left)* Coaxial cable. This type is shielded against electrical interference. It also can carry more data than twisted-pair wire. *(Right)* When coaxial cable is bundled together, as here, it can carry more than 40,000 conversations at once. *(Bottom left)* Fiber-optic cable. Thin glass strands transmit pulsating light instead of electricity. These strands can carry computer and voice data over long distances.

The main drawbacks until now have been cost and the material's inability to bend around tight corners. In mid-1995, however, new material was announced—called *graded-index plastic optical fiber*—that was cheaper, lighter, and more flexible than glass fibers. The plastic flexible fiber is said to handle loops and curves with ease and thus will be better than glass for curb-to-home wiring.[99]

Microwave & Satellite Systems

Wired forms of communications, which require physical connection between sender and receiver, will not disappear any time soon, if ever. For one thing, fiber-optic cables can transmit data communications 10,000 times faster than microwave and satellite systems can. Moreover, they are resistant to illegal data theft.

Still, some of the most exciting developments are in wireless communications. After all, there are many situations in which it is physically difficult to run wires. Here let us consider microwave and satellite systems.

- **Microwave systems:** *Microwave systems* **transmit voice and data through the atmosphere as super-high-frequency waves.** Microwave systems transmit microwaves, of course. *Microwaves* are the electromagnetic waves that vibrate at 1 gigahertz (1 billion hertz) per second or higher. These frequencies are used not only to operate microwave ovens but also to transmit messages between ground-based earth stations and satellite communications systems.

    Nowadays you see dish- or horn-shaped microwave antennas nearly everywhere—on towers, buildings, and hilltops. (■ *See Panel 8.15.*) Why, you might wonder, do people have to interfere with nature by putting a

PANEL 8.15

Microwave systems

Microwaves cannot bend around corners or around the curvature of the earth. Therefore, microwave antennas must be in "line of sight" of each other—that is, unobstructed. Microwave dishes and relay towers are usually situated atop high places, such as mountains or tall buildings, so that signals can be beamed over uneven terrain.

Line-of-sight signal

Microwave relay station

microwave dish on top of a mountain? The reason is that microwaves cannot bend around corners or around the earth's curvature; they are *line-of-sight*. *Line-of-sight* means that there must be an unobstructed view between transmitter and receiver. Thus, microwave stations need to be placed within 25–30 miles of each other, with no obstructions in between. The size of the dish varies with the distance (perhaps 2–4 feet in diameter for short distances, 10 feet or more for long distances). A string of microwave relay stations will each receive incoming messages, boost the signal strength, and relay the signal to the next station.

More than half of today's telephone system uses dish microwave transmission. However, the airwaves are becoming so saturated with microwave signals that future needs will have to be satisfied by other channels, such as satellite systems.

• **Satellite systems:** To avoid some of the limitations of microwave earth stations, communications companies have added microwave "sky stations"—communications satellites. **Communications satellites are microwave relay stations in orbit around the earth.** (■ *See Panel 8.16.*) Traditionally, the orbit has been 22,300 miles above the earth (although newer systems will be much lower). Because they travel at the same speed as the earth, they appear to an observer on the ground to be stationary in space—that is, they are *geostationary*. Consequently, microwave earth stations are always able to beam signals to a fixed location above. The orbiting satellite has solar-powered receivers and transmitters (transponders) that receive the signals, amplify them, and retransmit them to another earth station. The satellite contains many communications channels and receives both analog and digital signals from earth stations. Note that it can take more than one satellite to get a message delivered, which can slow the delivery process down.

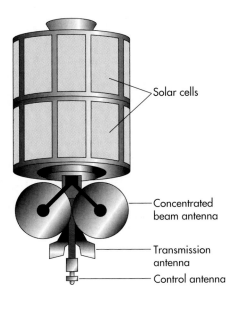

Solar cells

Concentrated
beam antenna

Transmission
antenna

Control antenna

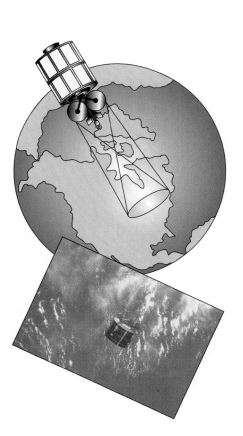

PANEL 8.16

Communications satellite

Other Wireless Communications

Of course, mobile wireless communications have been around for some time. The Detroit Police Department started using two-way car radios in 1921. Mobile telephones were introduced in 1946. Today, however, we are witnessing an explosion in mobile wireless use that is making worldwide changes.

There are essentially four ways to move information through the air long-distance on radio frequencies: (1) via *one-way communications*, as typified by the satellite navigation system known as GPS and by pagers; and via *two-way communications*, which are classified as (2) analog cellular phones, (3) packet radio, and (4) Cellular Digital Packet Data (CDPD).[100] (Other wireless methods operate at short distances.)

- **One-way communications—the Global Positioning System (GPS):** A $10 billion infrastructure developed by the military in the mid-1980s, *GPS*, for Global Positioning System, consists of a series of 24 earth-orbiting satellites that continuously transmit timed radio signals that can be used to identify earth locations. A GPS receiver—handheld or mounted in a vehicle, plane, or boat—can pick up transmissions from any four satellites, interpret the information from each, and calculate to within a few hundred feet or less the receiver's longitude, latitude, and altitude.[101] The accuracy of GPSs is expected to improve to plus or minus a few feet.

    The system is used by the military to tell military units exactly where they are, but it is also being used for civilian purposes such as tracking trucks and taxis, locating stolen cars, orienting hikers, and aiding in surveying. The makers of the film *Forrest Gump* used a GPS device to track the sun and time their sunrise and sunset shots so they didn't get shadows.[102] Some GPS receivers include map software for finding your way around, as with the Guidestar system available with some rental cars in cities such as Miami, Los Angeles, and New York.[103,104]

- **One-way communications—pagers:** In a Clearwater, Florida, child-care center, if one child bites another or otherwise misbehaves, the head of the center can instantly alert the parents. He or she need only dial the number of the pager that parents are given as part of the child-care service.[105]

    Once stereotyped as devices for doctors and drug dealers, pagers are now consumer items. Commonly known as beepers, for the sound they make when activated, *pagers* are simple radio receivers that receive data (but not voice messages) sent from a special radio transmitter. Often the pager has its own telephone number. When the number is dialed from a phone, the call goes by way of the transmitter straight to the designated pager. Paging services include SkyTel (MTel), PageNet, and EMBARC (Motorola). Pagers also do more than beep, transmitting full-blown alphanumeric text (such as four-line, 80-character messages) and other data. Newer ones are mini-answering machines, capable of relaying digitized voice messages.[106]

    Pagers are very efficient for transmitting one-way information—emergency messages, news, prices, stock quotations, mortgage rates, delivery-route assignments, even sports news and scores[107]—at low cost to single or multiple receivers. Recently some companies have introduced what is called *two-way paging* or *enhanced paging*. This technology allows customers to send a preprogrammed acknowledgment ("Will be late—stuck in traffic") that they have received a message. Eventually, paging companies hope, these two-way devices will evolve into full-fledged handheld communicators.[108,109]

- **Two-way communications—analog cellular:** *Analog cellular phones* **are designed primarily for communicating by voice through a system of cells. Each *cell* is hexagonal in shape, usually 8 miles or less in diameter, and is served by a transmitter-receiving tower.** Calls are directed between cells by a mobile telephone switching office (MTSO). Movement between cells requires that calls be "handed off" by the MTSO. (■ *See Panel 8.17.*)

    Handing off voice calls between cells poses only minimal problems. However, handing off data transmission (where every bit counts), with the inevitable gaps and pauses as one moves from one cell to another, is much more difficult. In the long run, data transmissions will probably have to be handled by the technology we discuss next, packet radio.

- **Two-way communications—packet radio:** *Packet-radio-based communications* use a nationwide system of radio towers that send data to handheld computers. Packet radio is the basis for services such as RAM Mobile Data and Ardis. The advantage of packet-radio transmission is that the wireless computer identifies itself to the local base station, which can

PANEL 8.17

Cellular connections

- *Calling from a cellular phone:* When you dial a call on a cellular phone, whether on the street or in a car, the call moves as radio waves to the transmitting-receiving tower that serves that particular cell. The call then moves by wire or microwaves to the mobile telephone switching office (MTSO), which directs the call from there on—generally to a regular local phone exchange, after which it becomes a conventional phone call.

- *Receiving a call on a cellular phone:* The MTSO transmits the number dialed to all the cells it services. Once it finds the phone, it directs the call to it through the nearest transmitting-receiving tower.

- *On the move:* When you make calls to or from phones while on the move, as in a moving car, the MTSO's computers sense when a phone's signal is becoming weaker. The computers then figure out which adjacent cell to "hand off" the call to and find an open frequency in that new cell to switch to.

1. A call originates from a mobile cellular phone.
2. The call wirelessly finds the nearest cellular tower using its FM tuner to make a connection.
3. The tower sends the signal to a Mobile Telephone Switching Office (MTSO) using traditional telephone network land lines.
4. The MTSO routes the call over the telephone network to a land-based phone or initiates a search for the recipient on the cellular network.
5. The MTSO sends the recipient's phone number to all its towers, which broadcast the number via radio frequency.
6. The recipient's phone "hears" the broadcast and establishes a connection with the nearest tower. A voice line is established via the tower by the MTSO.

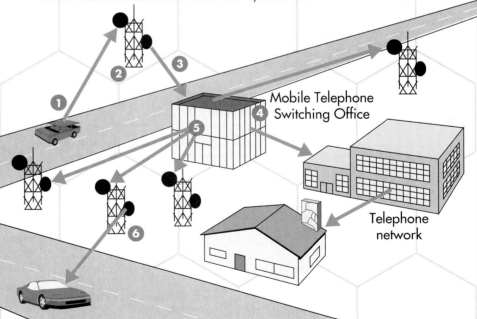

Mobile Telephone Switching Office

Telephone network

transmit over as many as 16 separate radio channels. Packet switching encapsulates the data in "envelopes," which ensures that the information arrives intact.

Packet-radio data networks are useful for mobile workers who need to communicate frequently with a corporate database. For example, National Car Rental System sends workers with handheld terminals to prowl parking lots, recording the location of rental cars and noting the latest scratches and dents. They can thereby easily check a customer's claim that a car was already damaged or find out quickly when one is stolen.[110]

- **Two-way communications—CDPD:** Short for Cellular Digital Packet Data, *CDPD* places messages in packets, or digital electronic "envelopes," and sends them through underused radio channels or between pauses in cellular phone conversations. CDPD is thus an enhancement to today's analog cellular phone systems, allowing packets of data to "hop" between temporarily free voice channels. As a result, a user carrying a CDPD device could have access to both voice and data. One problem with CDPD so far, however, is that it has limited coverage.

The Next Generation of Wireless Communications

Other kinds of wireless data services are on the way, promising to offer us lots of choices. The following are a few such developments:

- **Digital cellular phone:** Cellular telephone companies are trying to rectify the problem of faulty data transmission by switching from analog to digital signals. **Digital cellular phone networks turn your voice message into digital bits, which are sent through the airwaves, then decoded back into your voice by the cellular handset.**

  A digital cell phone costs two or three times more than an analog one, but the monthly bill may be less, especially for heavy users. Digital phone networks promise clearer sound, although some consumers don't agree. They also offer more privacy. So far, however, less than 5% of the nation's estimated 25 million cellular phone users have turned to digital, according to industry estimates.[111]

- **Personal communications services:** Like digital cellular networks, but lower-powered, *personal communications services (PCS)*, or personal communications networks (PCN), are digital wireless services that use a new band of microwave frequencies and transmitter-receivers in thousands of microcells. PCS systems operate at super high frequencies, where the spectrum isn't crowded. The microcells are smaller than the cells of today's cellular phone systems.

- **Specialized mobile radio:** *Specialized mobile radio (SMR)* is a two-way radio voice-dispatching service used by taxis and trucks; it is being converted to a digital system. Nextel Communications is building a nationwide SMR network, putting itself in direct competition with cellular phone services.

- **Satellite-based systems:** More than half the people in the world, mostly in underdeveloped countries, live more than 2 hours from the nearest telephone. (China has only four telephone lines for every 100 people.) These people, as well as business travelers and corporations needing speedy data transmission, will probably demand more phone service than wire-line or cellular service can deliver. As a result, the race is on to build constellations of satellites. "Almost three dozen programs—totaling more than 1500 satellites—are in the works," says one reporter. "That's almost five times the number of commercial communications satellites launched since the first, AT&T's Telstar, in 1962."[112]

Local Networks

Preview & Review: Local networks may be private branch exchanges (PBXs) or local area networks (LANs).

LANs may be client/server or peer-to-peer and include components such as cabling, network interface cards, an operating system, other shared devices, and bridges and gateways.

The topology, or shape, of a network may take five forms: star, ring, bus, hybrid, or FDDI.

Although large networks are useful, many organizations need to have a local network—an in-house network—to tie together their own equipment. Here let's consider the following aspects of local networks:

* Types of local networks—PBXs and LANs
* Types of LANs—client/server and peer-to-peer
* Components of a LAN
* Topology of LANs—star, ring, bus, hybrid, and FDDI
* Impact of LANs

Types of Local Networks: PBXs & LANs

The most common types of local networks are PBXs and LANs.

* **Private branch exchange (PBX):** **A *private branch exchange (PBX)* is a private or leased telephone switching system that connects telephone extensions in-house.** It also connects them to the outside phone system.

    A public telephone system consists of "public branch exchanges"— thousands of switching stations that direct calls to different "branches" of the network. A private branch exchange is essentially the old-fashioned company switchboard. You call in from the outside, the switchboard operator says "How may I direct your call?" and you are connected to the extension of the person you wish to talk to.

    Newer PBXs can handle not only analog telephones but also digital equipment, including computers. However, because older PBXs use existing telephone lines, they may not be able to handle the volume of electronic messages found in some of today's organizations. These companies may be better served by LANs.

* **Local area network (LAN):** PBXs may share existing phone lines with the telephone system. Local area networks usually require installation of their own communication channels, whether wired or wireless. **Local area networks (LANs) are local networks consisting of a communications link, network operating system, microcomputers or workstations, servers, and other shared hardware.** Such shared hardware might include printers, scanners, and storage devices. Unlike larger networks, LANs do not use a host computer.

Types of LANs: Client/Server & Peer-to-Peer

Local area networks are of two principal types: client/server and peer-to-peer. (■ *See Panel 8.18, next page.*)

* **Client/server LANs:** **A *client/server LAN* consists of requesting microcomputers, called *clients*, and supplying devices that provide a service,**

Two types of LANs: client/server and peer-to-peer

(Middle) In a client/server LAN, individual microcomputer users, or "clients," share the services of a centralized computer called a *server*. In this case, the server is a file server, which allows users to share files of data and some programs.
(Bottom) In a peer-to-peer LAN, computers share equally with one another without having to rely on a central server.

Client/server LAN

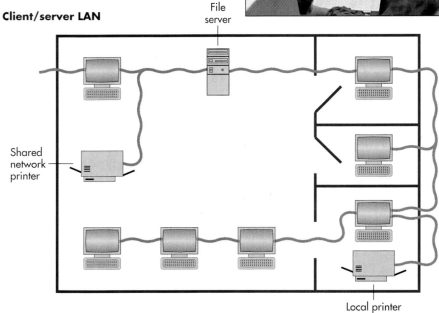

File server

Shared network printer

Local printer

Peer-to-peer LAN

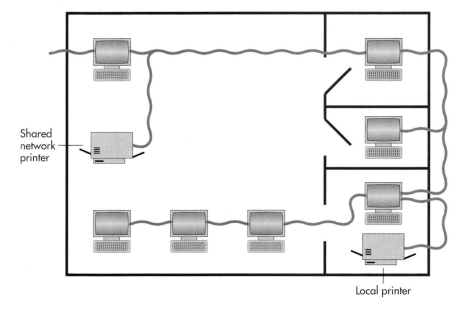

Shared network printer

Local printer

called *servers.* The server is a computer that manages shared devices, such as laser printers. The server microcomputer is usually a powerful one, running on a powerful chip such as a Pentium. Client/server networks, such as those run under Novell's NetWare (✓ p. 126) operating system, are the most common type of LAN.

There may be different servers for managing different tasks—files and programs, databases, printers. The one you may hear about most often is the file server. **A *file server* is a computer that stores the programs and data files shared by users on a LAN.** It acts like a disk drive but is in a remote location.

A *database server* is a computer in a LAN that stores data. Unlike a file server, it does not store programs. A *print server* is a computer in a LAN that controls one or more printers. It stores the print-image output from all the microcomputers on the system. It then feeds the output to the printer or printers one document at a time. *Fax servers* are dedicated to managing fax transmissions, and *mail servers* manage e-mail.

- **Peer-to-peer:** The word *peer* denotes one who is equal in standing with another (as in the phrases "peer pressure" or "jury of one's peers"). **A *peer-to-peer LAN* is one in which all microcomputers on the network communicate directly with one another without relying on a server.** Peer-to-peer networks are less expensive than client/server networks and work effectively for up to 25 computers. Beyond that they slow down under heavy use. They are thus appropriate for networking in small groups, as for workgroup computing.

Many LANs mix elements from client/server and peer-to-peer models.

Components of a LAN

Local area networks are made up of several standard components.

- **Connection or cabling system:** LANs do not use the telephone network. Instead, they use some other cabling or connection system, either wired or wireless. Wired connections may be twisted-pair wiring, coaxial cable, or fiber-optic cable. Wireless connections may be infrared or radio-wave transmission. Wireless networks are especially useful if computers are portable and are moved often. However, they are subject to interference.

- **Microcomputers with interface cards:** Two or more microcomputers are required, along with network interface cards. **A *network interface card*, which is inserted into an expansion slot in a microcomputer, enables the computer to send and receive messages on the LAN.**

- **Network operating system:** The network operating system software manages the activity of the network. Depending on the type of network, the operating system software may be stored on the file server or on each microcomputer on the network. Examples of network operating systems are Novell's NetWare and Apple's LocalTalk.

- **Other shared devices:** Printers, fax machines, scanners, storage devices, and other peripherals may be added to the network as necessary and shared by all users.

- **Bridges and gateways:** A LAN may stand alone, but it may also connect to other networks, either similar or different in technology. Hardware and software devices are used as interfaces to make these connections. **A *bridge* is an interface that enables similar networks to communicate. A *gateway* is an interface that enables dissimilar networks to communicate,** such as a LAN with a WAN. The illustration on page 376 shows the components of a LAN. (■ *See Panel 8.19.*)

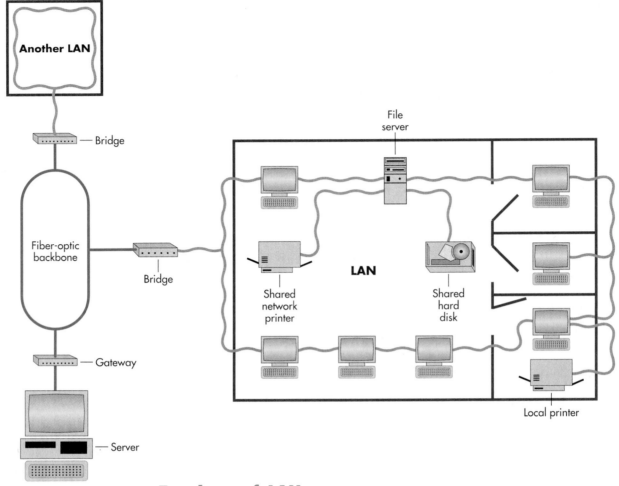

Components of a typical LAN

Topology of LANs

Networks can be laid out in different ways. **The logical layout, or shape, of a network is called a *topology.*** The five basic topologies are *star, ring, bus, hybrid,* and *FDDI.* (■ *See Panel 8.20.*)

- **Star network:** **A *star network* is one in which all microcomputers and other communications devices are connected to a central server.** Electronic messages are routed through the central hub to their destinations. The central hub monitors the flow of traffic. A PBX system is an example of a star network.

    The advantage of a star network is that the hub prevents collisions between messages. Moreover, if a connection is broken between any communications device and the hub, the rest of the devices on the network will continue operating. However, if the hub goes down, the entire network will stop.

- **Ring network:** **A *ring network* is one in which all microcomputers and other communications devices are connected in a continuous loop.** Electronic messages are passed around the ring until they reach the right destination. There is no central server. An example of a ring network is IBM's Token Ring Network, in which a bit pattern (called a "token") determines which user on the network can send information.

    The advantage of a ring network is that messages flow in only one direction. Thus, there is no danger of collisions. The disadvantage is that if a connection is broken, the entire network may stop working.

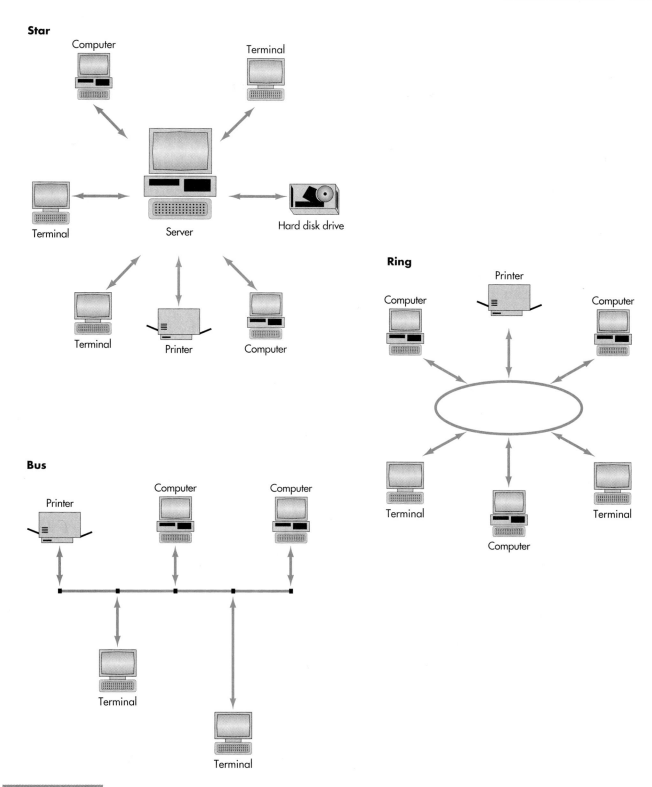

PANEL 8.20

Three LAN topologies: star, ring, bus

(Top) In a star network, all the network's devices are connected to a central server, through which all communications must pass. *(Middle)* In a ring network, the network's devices are connected in a closed loop. If one component fails, the whole system may fail. *(Bottom)* In a bus network, a single channel connects all communications devices.

- **Bus network:** In a *bus network,* **all communications devices are connected to a common channel.** There is no central server. Each communications device transmits electronic messages to other devices. If some of those messages collide, the device waits and tries to retransmit again. An example of a bus network is Xerox's Ethernet.

  One advantage of a bus network is that it may be organized as a client/server or peer-to-peer network. The disadvantage is that extra circuitry and software are needed to avoid collisions between data. Also, if a connection is broken, the entire network may stop working.

- **Hybrid network:** *Hybrid networks* **are combinations of star, ring, and bus networks.** For example, a small college campus might use a bus network to connect buildings and star and ring networks within certain buildings.

- **FDDI network:** A newer and higher-speed network is the FDDI, short for Fiber Distributed Data Interface. Capable of transmitting 100 megabits per second, **an *FDDI network* uses fiber-optic cable with an adaptation of ring topology.** The FDDI network is being used for such high-tech purposes as electronic imaging, high-resolution graphics, and digital video.

The Impact of LANs

Sales of mainframes and minicomputers have been falling for some time. This is largely because companies have discovered that LANs can take their place for many functions, and at considerably less expense. This trend is known as *downsizing.* Still, a LAN, like a mainframe, requires a skilled support staff. Moreover, LANs have neither the great storage capacity nor the security that mainframes have, which makes them inappropriate for some applications.

Factors Affecting Data Transmission

Preview & Review: Factors affecting how data is transmitted include the transmission rate (frequency and bandwidth), the line configuration (point-to-point or multipoint), serial versus parallel transmission, the direction of transmission flow (simplex, half-duplex, or full-duplex), transmission mode (asynchronous or synchronous), packet switching, multiplexing, and protocols.

Several factors affect how data is transmitted. They include the following:

- Transmission rate—frequency and bandwidth
- Line configurations—point-to-point versus multipoint
- Serial versus parallel transmission
- Direction of transmission—simplex, half-duplex, and full-duplex
- Transmission mode—asynchronous versus synchronous
- Packet switching
- Multiplexing
- Protocols

Transmission Rate: Higher Frequency, Wider Bandwidth, More Data

Transmission rate is a function of two variables: frequency and bandwidth.

- **Frequency:** The amount of data that can be transmitted on a channel depends on the wave *frequency*—the cycles of waves per second. Frequency is expressed in hertz: 1 cycle per second equals 1 hertz. The more cycles per second, the more data that can be sent through that channel.
- **Bandwidth:** As mentioned earlier, *bandwidth* is the difference between the highest and lowest frequencies—that is, the range of frequencies. Data may be sent not just on one frequency but on several frequencies within a particular bandwidth, all at the same time. Thus, the greater the bandwidth of a channel, the more frequencies it has available and hence the more data that can be sent through that channel. The rate of speed of data through the channel is expressed in bits per second (bps).

A twisted-pair telephone wire of 4000 hertz might send only 1 kilobyte of data in a second. A coaxial cable of 100 megahertz might send 10 megabytes. And a fiber-optic cable of 200 trillion hertz might send 1 gigabyte.

Line Configurations: Point-to-Point & Multipoint

There are two principal line configurations, or ways of connecting communications lines: point-to-point and multipoint.

- **Point-to-point:** A *point-to-point line* **directly connects the sending and receiving devices,** such as a terminal with a central computer. This arrangement is appropriate for a private line whose sole purpose is to keep data secure while transmitting it from one device to another.
- **Multipoint:** A *multipoint line* **is a single line that interconnects several communications devices to one computer.** Often on a multipoint line only one communications device, such as a terminal, can transmit at any given time.

Serial & Parallel Transmission

Data is transmitted in two ways: serially and in parallel.

- **Serial data transmission:** In *serial data transmission*, **bits are transmitted sequentially, one after the other.** This arrangement resembles cars proceeding down a one-lane road.

  Serial transmission is the way most data flows over a twisted-pair telephone line. Serial transmission is found in communications lines, modems, and mice. The plug-in board for a microcomputer modem usually has a serial port (✓ p. 178).
- **Parallel data transmission:** In *parallel data transmission*, **bits are transmitted through separate lines simultaneously.** The arrangement resembles cars moving in separate lanes at the same speed on a multilane freeway.

  Parallel lines move information faster than serial lines do, but they are only efficient for up to 15 feet. Thus, parallel lines are used, for example, to transmit data from a computer's CPU to a printer.

Direction of Transmission Flow: Simplex, Half-Duplex, & Full-Duplex

When two computers are in communication, data can flow in three ways: simplex, half-duplex, or full-duplex. These are fancy terms for easily understood processes. (■ *See Panel 8.21.*)

- **Simplex transmission:** In *simplex transmission,* **data can travel in only one direction.** An example is a traditional television broadcast, in which the signal is sent from the transmitter to your TV antenna. There is no return signal. Some computerized data collection devices also work this way (such as seismograph sensors that measure earthquakes).

- **Half-duplex transmission:** In *half-duplex transmission,* **data travels in both directions but only in one direction at a time.** This arrangement resembles traffic on a one-lane bridge; the separate streams of cars must take turns. Half-duplex transmission is seen with CB or marine radios, in which both parties must take turns talking. It is also a common transmission method with microcomputers. When you log onto an electronic bulletin board, you may be using half-duplex transmission.

- **Full-duplex transmission:** In *full-duplex transmission,* **data is transmitted back and forth at the same time.** This arrangement resembles automobile traffic on a two-way street. An example is two people on the telephone talking and listening simultaneously. Full-duplex is used frequently between computers in communications systems.

PANEL 8.21

Transmission directions

Simplex, half-duplex, and full-duplex.

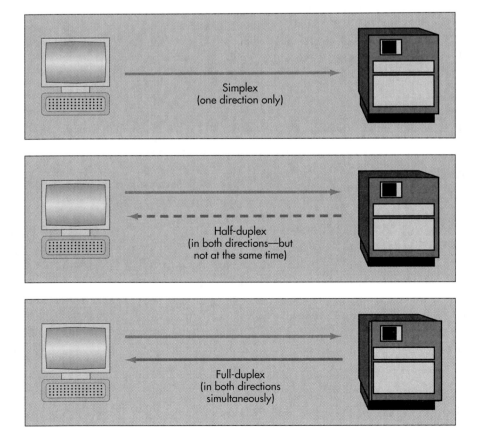

Simplex
(one direction only)

Half-duplex
(in both directions—but
not at the same time)

Full-duplex
(in both directions
simultaneously)

Transmission Mode: Asynchronous Versus Synchronous

Suppose your computer sends the word CONGRATULATIONS! to someone as bits and bytes over a communications line. How does the receiving equipment know where one byte (or character) ends and another begins? This matter is resolved through either *asynchronous transmission* or *synchronous transmission*. (■ *See Panel 8.22.*)

- **Asynchronous transmission:** This method, used with most microcomputers, is also called *start-stop transmission*. **In *asynchronous transmission*, data is sent one byte (or character) at a time. Each string of bits making up the byte is bracketed, or marked off, with special control bits.** That is, a "start" bit represents the beginning of a character, and a "stop" bit represents its end.

    Transmitting only one byte at a time makes this a relatively slow method. As a result, asynchronous transmission is not used when great amounts of data must be sent rapidly. Its advantage is that the data can be transmitted whenever it is convenient for the sender.

- **Synchronous transmission:** Instead of using start and stop bits, ***synchronous transmission* sends data in blocks. Start and stop bit patterns, called sync bytes, are transmitted at the beginning and end of the blocks.** These start and stop bit patterns synchronize internal clocks in the sending and receiving devices so that they are in time with each other.

    This method is rarely used with microcomputers because it is more complicated and more expensive than asynchronous transmission. It also requires careful timing between sending and receiving equipment. It is appropriate for computer systems that need to transmit great quantities of data quickly.

PANEL 8.22

Transmission modes

There are two ways that devices receiving data transmissions can determine the beginnings and ends of strings of bits (bytes, or characters). *(Top)* In asynchronous transmission, each character is preceded by a "start" bit and followed by a "stop" bit. *(Bottom)* In synchronous transmission, messages are sent in blocks, with start and stop patterns of bits, called sync bytes, before and after the blocks. The sync bytes synchronize the timing of the internal clocks between sending and receiving devices.

Asynchronous transmission

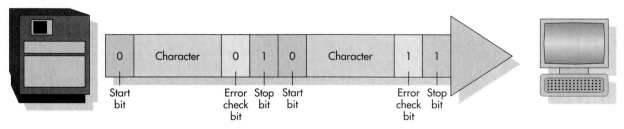

Synchronous transmission

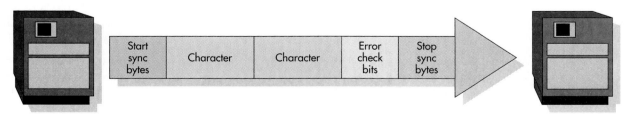

Packet Switching: Getting More Data on a Network

A *packet* is a fixed-length block of data for transmission. The packet also contains instructions about the destination of the packet. ***Packet switching* is a technique for dividing electronic messages into packets for transmission over a network to their destination through the most expedient route.** The benefit of packet switching is that it can handle high-volume traffic in a network. It also allows more users to share a network, thereby offering cost savings. The method is particularly appropriate for sending messages long distances, such as across the country. Accordingly, it is used in large networks such as Telenet, Tymnet, and AT&T's Accunet.

Here's how packet switching works: A sending computer breaks an electronic message apart into packets. The various packets are sent through a communications network—often by different routes, at different speeds, and sandwiched in between packets from other messages. Once the packets arrive at their destination, the receiving computer reassembles them into proper sequence to complete the message.

Multiplexing: Enhancing Communications Efficiency

Communications lines nearly always have far greater capacity than a single microcomputer or terminal can use. Because operating such lines is expensive, it's more efficient if several communications devices can share a line at the same time. This is the rationale for multiplexing. ***Multiplexing* is the transmission of multiple signals over a single communications channel.**

Three types of devices are used to achieve multiplexing—*multiplexers*, *concentrators*, and *front-end processors*:

- **Multiplexers:** **A *multiplexer* is a device that merges several low-speed transmissions into one high-speed transmission.** Depending on the multiplexer, 32 or more devices may share a single communications line. Messages sent by a multiplexer must be received by a multiplexer of the same type. The receiving multiplexer sorts out the individual messages and directs them to the proper recipient. High-speed multiplexers called *T1 multiplexers*, which use high-speed digital lines, can carry as many messages, both voice and data, as 24 analog telephone lines.

- **Concentrators:** Like a multiplexer, a concentrator is a piece of hardware that enables several devices to share a single communications line. However, unlike a multiplexer, **a *concentrator* collects data in a temporary storage area. It then forwards the data when enough has been accumulated to be sent economically.** Often a concentrator is a minicomputer.

- **Front-end processors:** The most sophisticated of these communications-management devices is the front-end processor, a computer that handles communications for mainframes. **A *front-end processor* is a smaller computer that is connected to a larger computer and assists with communications functions.** The front-end processor is itself a minicomputer or even a mainframe. It transmits and receives messages over the communications channels, corrects errors, and relieves the larger computer of routine computational tasks.

Protocols: The Rules of Data Transmission

The word *protocol* is used in the military and in diplomacy to express rules of precedence, rank, manners, and other matters of correctness. (An example would be the protocol for who will precede whom into a formal reception.)

Here, however, **a *protocol*, or *communications protocol*, is a set of conventions governing the exchange of data between hardware and/or software components in a communications network.**

Protocols are built into the hardware or software you are using. The protocol in your communications software, for example, will specify how receiver devices will acknowledge sending devices, a matter called *handshaking*. Protocols will also specify the type of electrical connections used, the timing of message exchanges, error-detection techniques, and so on.

In the past, not all hardware and software developers subscribed to the same protocols. As a result, many kinds of equipment and programs have not been able to work with one another. In recent years, more developers have agreed to subscribe to a standard set of protocols called *OSI*, short for *Open Systems Interconnection*. Backed by the International Standards Organization, **OSI is an international standard that defines seven layers of protocols for worldwide computer communications.**

Cyberethics: Netiquette, Controversial Material & Censorship, & Privacy Issues

E

Preview & Review: Users of communications technology must weigh standards of behavior and conduct in three areas: netiquette, controversial material, and privacy.

Communications technology gives us more choices of nearly every sort. Not only does it provide us with different ways of working, thinking, and playing; it also presents us with some different moral choices—determining right actions in the digital and online universe. Next, let us consider three important aspects of "cyberethics"—netiquette, controversial material and censorship, and matters of privacy.

Netiquette

One morning *New Yorker* magazine writer John Seabrook, who had recently published an article about Microsoft chairman Bill Gates, checked into the e-mail on his computer to find the following reaction to his story:

> Listen, you toadying [*deleted*] scumbag . . . remove your head [*three words deleted*] long enough to look around and notice that real reporters don't fawn over their subjects, pretend that their subjects are making some sort of special contact with them, or, worse, curry favor by TELLING their subjects how great the [*deleted*] profile is going to turn out and then brag in print about doing it. . . .

On finishing the message, Seabrook rocked back in his chair. "Whoa," he said aloud to himself. "I got flamed."[114]

A form of speech unique to online communication, *flaming* is writing an online message that uses derogatory, obscene, or inappropriate language. The attack on Seabrook is probably unusual, since most flaming happens when someone violates online manners or "netiquette."

As mentioned, many online bulletin boards have a set of FAQ's—frequently asked questions—that newcomers, or "newbies," are expected to become familiar with before joining in any chat forums. Most FAQs offer **netiquette, or "net etiquette," guides to appropriate behavior while online.** The commercial online services also have special online sites where the uninitiated can go to learn how to avoid an embarrassing breach of manners. Examples of netiquette blunders are typing with the CAPS LOCK key on—the Net equivalent of yelling—discussing subjects not appropriate to the forum, repetition of points made earlier, and improper use of the software.[115,116] *Spamming,* or sending unsolicited mail, is especially irksome; a *spam* includes chain letters, advertising, or similar junk mail. Something that helps smooth communicating in the online culture is use of *emoticons,* keyboard-produced pictorial representations of expressions. (■ *See Panel 8.23.*)

One of the great things about the Internet is that it has got Americans writing again. However, it shouldn't be sloppy. "Most people who see poor spelling and poor grammar think the writer is a dummy," says Virginia Shea, author of a book called *Netiquette,* who has become the Emily Post or Miss Manners of cyberspace. "You'd think some people never opened a dictionary in their lives from the spelling in their e-mail. They just dash off a note and figure that it doesn't matter how it reads. That is a big mistake."[117]

Controversial Material & Censorship

In Saudi Arabia, officials try to police taboo subjects (sex, religion, politics) on the Internet.[118] In China, Net users must register with the government.[119,120] In Germany, in late 1995, officials persuaded Ohio-based CompuServe to cut off access to 200 sex-related news forums worldwide. (CompuServe reopened them a few months later, offering to provide filtering software to users wanting to block offensive material.)[121,122] In the United States, however, free speech is protected by the First Amendment to the Constitution.

If a U.S. court decides *after* you have spoken that you have defamed or maliciously damaged someone, you may be sued for slander (spoken speech) or libel (written speech) or charged with harassment, but you cannot be stopped beforehand. However, "obscene" material is not constitutionally protected free speech. Obscenity is defined as sexually explicit material that is offensive as measured by "contemporary community standards"—a definition with considerable leeway, depending on localities. In 1996, as part of its broad telecommunications law overhaul, Congress passed legislation called

---

PANEL 8.23

Emoticons

Emoticons enable online users to get in the kind of facial expressions and inflections that are used in normal conversation.

"How do emoticons work? Tilt your head to the left and take a look at this one: **:=)**

Do you see a smiley face with a long nose? A comment or joke followed by a smiley is often a good way to insure that it was taken in good humor. Have sad news? Show it while you tell it. **:-(** Feeling teary-eyed? **:'(** Feeling sarcastic? Show that as well. **:-/** Stick your tongue out at someone. **:=P** Or pucker up for a kiss. **:-***

Emoticons are good for more than just facial expressions, though. You can send hugs. **(())** or **{{}}** And you can send roses. **@- - -^- - -** Placing **<w>** before your words signifies a whisper, and **<g>** a grin."

—Tosca Moon Lee, "Smiling Online," *PC Novice*

:-)	**Happy face**
:-(	**Sorrow or frown**
:-0	**Shock**
:-/	**Sarcasm**
;-)	**Wink**
:--)%	**Boy on a skateboard**

the *Communications Decency Act* that imposed heavy fines and prison sentences on people making available pornography or indecent sexual material over computer networks.[123] This part of the law was temporarily blocked by a federal judge as being constitutionally vague; however, as of this writing, the issue is still unresolved.[124]

Since computers are simply another way of communicating, there should be no surprise that a lot of people use them to communicate about sex. Yahoo!, the Internet directory company (✓ pp. 81), says that the word "sex" is the most popular search word on the Net.[125] All kinds of online X-rated bulletin boards, chat rooms, and Usenet newsgroups exist. A special problem is with children having access to sexual conversations, downloading hard-core pictures, or encountering odious adults tempting them into a meeting. "Parents should never use an online service as an electronic baby-sitter," says computer columnist Lawrence Magid. People online are not always what they seem to be, he points out, and a message seemingly from a 12-year-old girl could really be from a 30-year-old man. "Children should be warned never to give out personal information," says Magid, "and to tell their parents if they encounter mail or messages that make them uncomfortable."[126]

Not only parents are concerned about pornography in electronic form, so are employers. Many companies are concerned about the loss of productivity—and the risk of being sued for sexual harassment—as workers spend time online looking at sexually explicit material. What can be done about all this? Some possibilities:

- **Filtering software:** Some software developers have discovered a golden opportunity in making programs like SurfWatch and Net Nanny. These filtering programs screen out objectionable matter typically by identifying certain nonapproved keywords in a user's request or comparing the user's request for information against a list of prohibited sites.[127] (The screening is sometimes imperfect: The White House Web site was accidentally put off-limits by SurfWatch because it used the supposedly indecent term "couples" in conjunction with the vice president and his wife.)

- **Browsers with ratings:** Another proposal in the works is browser software that contains built-in ratings for Internet, Usenet, and World Wide Web files. Parents could, for example, choose a browser that has been endorsed by the local school board or the online service provider.[128]

- **The V-chip:** The 1996 Telecommunications Law officially launched the era of the V-chip, a device that will be required equipment in most new television sets within 2 years.[129,130] The *V-chip* allows parents to automatically block out programs that have been labeled as high in violence, sex, or other objectionable material. Who will do the ratings (of 600,000 hours of programming currently broadcast per year) and whether the system is really workable remain to be seen. However, as conventional television and the Internet converge, the V-chip could become a concern to Net users as well as TV watchers.

The difficulty with any attempts at restricting the flow of information is the basic Cold War design of the Internet itself, with its strategy of offering different roads to the same place. "If access to information on a computer is blocked by one route," writes the *New York Times*'s Peter Lewis, "a moderately skilled computer user can simply tap into another computer by an alternative route." Lewis points out an Internet axiom attributed to an engineer named John Gilmore: "The Internet interprets censorship as damage and routes around it."[131]

Privacy

Privacy **is the right of people not to reveal information about themselves.** Technology, however, puts constant pressure on this right.

A number of people, for example, have been undone by using cellular phones. The location of football star O.J. Simpson, after he was charged with murdering his ex-wife and a friend, was traced by police through his cellular phone signal.[132] Indeed, anyone using a scanner receiving in the 800–900 megahertz range can listen in to cellular phone conversations.[133] The *Electronic Communications Privacy Act* of 1986 makes eavesdropping on private conversations illegal without a court order. However, authorities have no way to catch people using scanners.

Think you're anonymous online when you don't sign your real name? America Online, in response to a legal petition for discovery, turned over the real name, address, and credit card information of a subscriber named Jenny TRR. Attorneys for a resort were considering suing her for defamation for a critical message she posted on an online bulletin board.[134]

Think the boss can't snoop on your e-mail at work? The Electronic Communications Privacy Act allows employers to "intercept" employee communications if one of the parties involved, such as the employer, agrees to the "interception."[135] Indeed, employer snooping seems to be widespread.

Think your medical records are sacrosanct? Actually, private medical information is bought and sold freely by various companies since there is no federal law prohibiting it. (And they simply ignore the patchwork of varying state laws.)[136]

A great many people are concerned about the loss of their right to privacy. Indeed, a 1995 survey found that 80% of the people contacted worried that they had lost "all control" of the personal information being collected and tracked by computers.[137] Although the government is constrained by several laws on acquiring and disseminating information and listening in on private conversations, there are reasons to be alarmed. We discuss privacy further in Chapter 9 and security in Chapter 12.

Onward

In late 1995, the Federal Communications Commission opened up virgin territory on the electromagnetic spectrum, a section known as *millimeter waves*, those situated above 40 gigahertz.[138] These open up possibilities not only for wireless campuses—so that phones, pagers, and other communications devices could be easily connected—but also for "smart homes," with appliances, heating and cooling, and security controlled by wireless systems. There could also be "smart cars," with radar systems to alert drivers to potential collisions.

Clearly, more roads will be added in cyberspace. We will consider the implications of these developments in the final chapter of this book.

Online Resumes: Career Strategy for the Digital Age

"If you have 8000 resumes and you're looking for a COBOL programmer or a secretary who knows Word-Perfect," commented vice president of recruitment Saundra Banks Loggins of Wells Fargo bank, "this system can save you a lot of time."[139]

Whose time is being saved? That of the employer. What system does this? A high-tech resume-scanning system called Resumix. This technology uses an optical scanner to input 900 pages of resumes a day, storing the data in a computerized database. The system can search for up to 60 key factors, such as job titles, technical expertise, education, geographical location, and employment history. Resumix can also track race, religion, gender, and other factors to help companies diversify their workforce. These descriptors can then be matched with available openings.

Resume scanners can save companies thousands of dollars. They allow organizations to more efficiently search their existing pool of applications before turning to advertising or executive-search ("head-hunter") firms to recruit employees. For applicants, however, resume banks and other electronic systems have turned job hunting into a whole new ball game.

Writing a Computer-Friendly & Recruiter-Friendly Resume

Some of the old rules for presenting yourself in a resume might now not benefit you at all. The latest advice is as follows.

Use the Right Paper & Print In the past, job seekers have used tricks such as colored paper and fancy typefaces in their resumes to try to catch a bored personnel officer's eye. However, optical scanners have trouble reading type on colored or gray paper and are confused by unusual typefaces. They even have difficulty reading underlining and poor-quality dot-matrix printing.[140]

Resumix Inc. suggests observing the following rules of format for resume writing:[141]

- Exotic typefaces, underlining, and decorative graphics don't scan well.

- It's best to send originals, not copies, and not to use a dot-matrix printer.

- Too-small print may confuse the scanner; don't go below 12-point type.

- Use standard 8½ × 11-inch paper and do not fold. Words in a crease can't be read easily.

- Use white or light-beige paper. Blues and grays minimize the contrast between the letters and the background.

- Avoid double columns. The scanner reads from left to right.

Use Keywords for Skills or Attributes Just as important as the format of a resume today are the words used in it. In the past, resume writers tried to clearly present their skills. Now it's necessary to use as many of the buzzwords or keywords of your profession or industry as you can.

Action words ("managed," "created," "developed") should still be used, but they are less important than nouns. Nouns include job titles, capabilites, languages spoken, type of degree, and the like ("vice-president," "systems analyst," "Spanish," "Unix"). The reason, of course, is that a computer will scan for key words applicable to the job that is to be filled.

Because resume-screening programs sort and rank the number of keywords found, those with the most rise to the top of the electronic pile. Thus, careers columnist Joyce Laine Kennedy suggests you pack your resume with every keyword that applies to you. You should especially use keywords of the sort that appear in help-wanted ads.[142]

If you are looking for a job in desktop publishing, for instance, there are a number of specific keywords that will make you stand out. In Kennedy's example, these are *Aldus Pagemaker, Compugraphics, DCF, Harvard Graphics,* and *PagePerfect.* (Newer ones such as *Quark* could also be added.)

Make the Resume Impress People, Too Your resume shouldn't just be pages of keywords. It has to impress a human recruiter, too, who may still have some fairly traditional ideas about resumes.

Some tips for organizing resumes, offered by reporter Kathleen Pender, who interviewed numerous professional resume writers, are as follows.[143]

- *The beginning:* Start with your name, address, and phone number.

    Follow with a clear objective stating what it is you want to do. (Example: "Sales representative in computer furniture industry.")

    Under the heading "Summary" give three compelling reasons why you are the ideal person for the job. (Example on one line: "Experienced sales representative to corporations and small businesses.")

    After the beginning, your resume can follow either a chronological format or a functional format.

- *The chronological resume:* The chronological resume works best for people who have stayed in the same line of work and have moved steadily upward in their careers, with no gaps in work history. Start with your most recent job and work backwards, and say more about your recent jobs than earlier ones.

The format is to list the years you worked at each place down one side of the page. Opposite indicate your job title, employer name, and a few of your accomplishments. Omit accomplishments that have nothing to do with the job you're applying for.

- *The functional resume:* The functional resume works best for people who are changing careers, or re-entering the job market. It also is for people who need to emphasize skills from earlier in their careers or who want to emphasize their volunteer experience. It's recommended, too, for people who have had responsibility but never an important job title.

  The format is to emphasize the skills, then follow with a brief chronological work history emphasizing dates, job titles, and employer names.

- *The conclusion:* Both types of resumes should have a concluding section showing college, degree, and graduation date; professional credentials or licenses; and professional affiliations and awards if they are relevant to the job you're seeking.

- *The biggest mistakes on resumes:* The biggest mistake you can make on a resume is to lie. Sooner or later a lie will probably catch up with you and may get you fired, maybe even sued.

  The second mistake is to have a lot of spelling errors. Spelling mistakes communicate to prospective employers a basic carelessness.

  Other dos and don'ts appear in the accompanying box (■ *See Panel 8.24.*)

Write a Good Cover Letter Write a targeted cover letter to accompany your resume. This advice especially should be followed if you're responding to an ad.

Most people don't bother to write a cover letter focusing on the particular job being advertised. Moreover, if they do, say San Francisco employment experts Howard Bennett and Chuck McFadden, "they tend to talk about what they are looking for in a job. This is a major turn-off for employers."[144] Employers don't care very much about your dreams and aspirations, only about finding the best candidate for the job.

Bennett and McFadden suggest the following strategy for a cover letter:

PANEL 8.24

Resume dos and don'ts

"There are no hard and fast rules to resume writing, but these are a few points on which the majority of experts would agree.

Do
- Start with a clear objective.
- Have different resumes for different types of jobs.
- List as many relevant skills as you legitimately possess.
- Use jargon or buzzwords that are understood in the industry.
- Use superlatives: biggest, best, most, first.
- Start sentences with action verbs (organized, reduced, increased, negotiated, analyzed).
- List relevant credentials and affiliations.
- Limit your resume to one or two pages (unless you're applying for an academic position).
- Use standard-size, white or off-white heavy paper.
- Use a standard typeface and a letter-quality or laser-jet printer.
- Spell check and proofread, several times.

Don't
- Lie
- Sound overly pompous.
- Use pronouns such as I, we.
- Send a photo of yourself.
- List personal information such as height, weight, marital status or age, unless you're applying for a job as an actor or model.
- List hobbies, unless they're directly related to your objective.
- Provide references unless requested. ("References on request" is optional.)
- Include salary information.
- Start a sentence with "responsibilities included:"
- Overuse and mix type styles such as bold, underline, italic, and uppercase."

—Kathleen Pender, "Resume Dos and Don'ts," *San Francisco Chronicle*

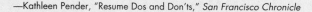

- *Emphasize how you will meet the employer's needs:* Employers advertise because they have needs to be met. "You will get much more attention," say Bennett and McFadden, "if you demonstrate your ability to fill those needs."

  How do you find out what those needs are? You read the ad. By reading the ad closely you can find out how the company talks about itself. You can also discover what attributes it is looking for in employees and what the needs are for the particular position.

- *Use the language of the ad:* In your cover letter, use as much of the ad's language as you can. "Use the same words as much as possible," advise Bennett and McFadden. "Feed the company's language back to them." The effect of this will be to produce "an almost subliminal realization in the company that you are the person they've been looking for."

- *Take care with the format of the letter:* Keep the letter to one page and use bullets or dashes to emphasize the areas where you meet the needs described in the ad. Make sure the sentences read well and—very important—that no word or name is misspelled.

The intent of both cover letter and resume is to get you an interview, which means you are in the top 10–15% of candidates. Once you're into an interview, a different set of skills is needed. You're urged to research these on your own. Richard Bolles, author of the best-selling job-hunting book *What Color Is Your Parachute?* suggests that, aside from looking clean and well-groomed, you need to tell the employer what distinguishes you from the 20 other people he or she is interviewing. "If you say you are a very thorough person, don't just say it," suggests Bolles. "Demonstrate it by telling them what you know about their company, which you learned beforehand by doing your homework."[145]

Resume Database Services

By putting your resume on an online database, you give employers the opportunity to find you. Among the kinds of resources for employers are the following:

- *Databases for college students:* Colleges sometimes have or are members of online database services on which their students or alumni may place electronic resumes. For example, several universities have formed University ProNet, which provides online resumes to interested employers; students and alumni pay a one-time lifetime fee of $35.

  The Career Placement Registry (available through the online service Dialog) allows college students or recent graduates to post resumes for $15.

- *Databases for people with experience:* Resumes for experienced people may be collected by private data-

bases. An example is Connexions, which charges experienced professionals $40 a year to post their resume.

Some databases serve employers in particular geographical areas or those looking for people with particular kinds of experience. HispanData in Santa Barbara, California, specializes in marketing Hispanic professionals to employers seeking to diversify their work forces.

Suggested Resources

Books

Bolles, Richard. *What Color Is Your Parachute?* Berkeley, CA: Ten Speed Press, revised annually. The best-selling job-hunting book of all time, in print for over 20 years.

Kennedy, Joyce Lain, and Thomas J. Morrow. *Electronic Job Search Revolution: Win With the New Technology That's Reshaping Today's Job Market.* New York: Wiley, 1994. Written by a nationally syndicated careers columnist and her coauthor. The book shows you how to reach hundreds of potential employers through resume database services, computerized tracking systems, and online job ads.

Kennedy, Joyce Lain, and Thomas J. Morrow. *Electronic Resume Revolution: Create a Winning Resume for the New World of Job Seeking.* New York: Wiley, 1994. Shows how to write a keyboard resume that will favor you in an electronic job search.

Online Job Banks & Resume Databases

Career Placement Registry. $15 for college students or recent graduates. Telephone: 800-368-3093.

HispanData. $15 one-time listing fee for Hispanic college-educated individuals. Telephone: 805-682-5843.

Job Bank USA. $30 for 12 months. Telephone: 800-296-1872.

kiNexus. Free to many students through their colleges' career centers; everybody else $19.95 for 6 months listing. Telephone: 800-828-0422.

National Resume Bank. $25 for 3 months; $40 for 6 months. Telephone: 813-896-3694.

Peterson's Connexion. Free to many students through their colleges' career centers; everybody else $40 for 12-month listing. Telephone: 800-338-3282.

SkillSearch. $65 one-time fee for college alumni. Telephone: 800-258-6641.

University ProNet. Private company operated by 11 participating universities, including M.I.T., Ohio State, University of Michigan, Stanford, and UCLA. Alumni pay $35 lifetime fee to register and are allowed to update their resumes annually. Telephone: 800-726-0280.

SUMMARY

amplitude *(p. 358, LO 2)* In analog transmission, the height of a wave within a given period of time.

Amplitude refers to the strength or volume—the loudness of a signal.

analog cellular phone *(p. 369, LO 3)* Mobile telephone designed primarily for communicating by voice through a system of cells. Calls are directed to cells by a mobile telephone switching office (MTSO). Moving between cells requires that calls be "handed off" by the MTSO between cells.

Cellular phone systems allow callers mobility.

analog signal *(p. 358, LO 2)* Continuous electrical signal in the form of a wave. The wave is called a *carrier wave*. Two characteristics of analog carrier waves that can be altered are frequency and amplitude. Computers cannot process analog signals.

Analog signals are used to convey voices and sounds over wire telephone lines, as well as in radio and TV broadcasting. Computers, however, use digital signals, which must be converted to analog signals in order to be transmitted over telephone wires.

asynchronous transmission *(p. 381, LO 5)* Also called *start-stop transmission;* data is sent one byte (character) at a time. Each string of bits making up the byte is bracketed with special control bits; a "start" bit represents the beginning of a character, and a "stop" bit represents its end.

This method of communications is used with most microcomputers. Its advantage is that data can be transmitted whenever convenient for the sender. Its drawback is that transmitting only one byte at a time makes it a relatively slow method that cannot be used when great amounts of data must be sent rapidly.

bands (bandwidths) *(p. 364, LO 3, 5)* Ranges of frequencies. The bandwidth is the difference between the lowest and highest frequencies transmitted.

Different telecommunications systems use different bandwidths for different purposes, whether cellular phones or network television.

bits per second (bps) *(p. 360, LO 2)* Measurement of data transmission speeds. Modems transmit at 1200 and 2400 bps (slow), 4800 and 9600 bps (moderately fast), and 14,400, 19,200, and 28,800 bps (high-speed).

A 10-page single-spaced letter can be transmitted by a 2400-bps modem in 2½ minutes. It can be transmitted by a 9600-bps modem in 38 seconds and by a 19,200-bps modem in 19 seconds. The faster the modem, the less time online and therefore less expense.

bridge *(p. 374, LO 4)* Interface that enables similar networks to communicate.

Smaller networks (local area networks) can be joined together to create larger networks.

bulletin board system (BBS) *(p. 343, LO 1)* Centralized information source and message-switching system for a particular computer-linked interest group. A BBS may be operated by an online service or an individual.

The subjects of discussion on BBSs are practically limitless, enabling people with all kinds of special interests to "chat" with each other and to post notices.

bus network *(p. 378, LO 4)* Type of network in which all communications devices are connected to a common channel, with no central server. Each communications device transmits electronic messages to other devices. If some of those messages collide, the device waits and tries to retransmit again.

The advantage of a bus network is that it may be organized as a client-server or peer-to-peer network. The disadvantage is that extra circuitry and software are needed to avoid collisions between data. Also, if a connection is broken, the entire network may stop working.

cable modem *(p. 362, LO 2)* Modem that connects a PC to a cable-TV system that offers online services, as well as TV.

Cable modems transmit data faster than standard modems.

What It Is / What It Does	Why It's Important

cell *(p. 369, LO 3)* Geographical component of a cellular telephone system; a cell is hexagonal in shape, usually 8 miles or less in diameter, and is served by a transmitter-receiving tower. Calls are directed between cells by a mobile telephone switching office (MTSO). Movement between cells requires that calls be "handed off" by the MTSO.

Handing off voice calls between cells poses only minimal problems. However, handing off data transmission (where every bit counts), with the inevitable gaps and pauses as one moves from one cell to another, is much more difficult.

channel *(p. 363, LO 3)* Also called *links, lines,* or *media;* path over which information travels in a telecommunications system from its source to its destination.

There are many different telecommunications channels, both wired and wireless, some more efficient than others for different purposes.

client/server LAN *(p. 373, LO 4)* Type of local area network (LAN); it consists of requesting microcomputers, called *clients,* and supplying devices that provide a service, called *servers.* The server is a computer that manages shared devices, such as laser printers.

Client/server networks are the most common type of LAN. Compare with *peer-to-peer LAN.*

coaxial cable *(p. 365, LO 3)* Type of communications channel; commonly called *coax,* it consists of insulated copper wire wrapped in a solid or braided metal shield, then in an external cover.

Coaxial cable is much better at resisting noise than twisted-pair wiring. Moreover, it can carry voice and data at a faster rate.

communications *(p. 333, LO 1)* Also called *telecommunications;* the electronic transfer of information from one location to another. Also refers to electromagnetic devices and systems for communicating data.

Communications systems have helped to expand human communication beyond face-to-face meetings to electronic connections called the *global village.*

communications protocol *(p. 383, LO 5)* Set of conventions governing the exchange of data between hardware and/or software components in a communications network. Protocols are built into hardware and software. For example, the protocol in communications software will specify how receiver devices will acknowledge sending devices ("handshaking").

In the past, because not all hardware and software developers subscribed to the same protocols, many kinds of equipment and programs did not work with one another. Recently, most developers have agreed to subscribe to a standard of protocols called *OSI.*

communications satellites *(p. 367, LO 3)* Microwave relay stations orbit 22,300 miles above the equator. Because they travel at the same speed as the earth, thus appearing stationary in space, microwave earth stations can beam signals to a fixed location above. The satellite has solar-powered receivers and transmitters (transponders) that receive the signals, amplify them, and retransmit them to another earth station.

An orbiting satellite contains many communications channels and receives both analog and digital signals from ground microwave stations anywhere on earth.

communications software *(p. 361, LO 2)* Software that manages the transmission of data between computers or video display terminals.

Besides establishing connections betwen computers, communications software may perform other functions: error correction, data compression, remote control, terminal emulation.

concentrator *(p. 382, LO 5)* Communications device such as a minicomputer that collects data in a temporary storage area, then forwards the data when enough has been accumulated.

Concentrators enable data to be sent more economically.

connectivity *(p. 334, LO 1)* The state of being able to connect devices by communications technology to other devices and sources of information.

Computers offer greater varieties of connectivity than other communications devices such as telephones or radio systems.

cyberspace *(p. 333, LO 1)* Refers to the computer online world and the Internet in particular, as well as the whole wired and wireless world of communications.

Suggests the vast amount of connections and interactivity now available to users of computer and communications systems.

data compression *(p. 361, LO 2)* Method of reducing the volume of data in a message, thereby reducing the amount of time required to transmit data from sender to receiver over a communications line.

With text and graphics, a message may be compressed to a tenth of its original size, making it much more economical to transmit or to store.

What It Is / What It Does	**Why It's Important**

dedicated fax machine *(p. 336, LO 1)* Specialized device that does nothing except scan in, send, and receive documents over telephone lines to and from other fax machines.

Fax machines have enabled people to instantly transmit graphics and documents for the price of a phone call.

digital cellular phone *(p. 370, LO 3)* Mobile phone system that uses cells like an analog cellular phone system but transmits digital signals.

See analog cellular phone.

digital signal *(p. 358, LO 2)* Type of electrical signal that uses on/off or present/absent electrical pulses in discontinuous, or discrete, bursts, rather than a continuous wave.

This two-state kind of signal works perfectly in representing the two-state binary language of 0s and 1s that computers use.

download *(p. 372, LO 4)* To retrieve files online from another computer and store them in one's own microcomputer. Compare with *upload*.

Downloading enables users of online systems to quickly scan file names and then save the files for later reading; this reduces the time and charges of being online.

electromagnetic spectrum *(p. 363, LO 3)* All the fields of electrical energy and magnetic energy, which travel in waves. This includes all radio signals, light rays, X-rays, and radioactivity.

The part of the electromagnetic spectrum of particular interest is the area in the middle, which is used for communications purposes. Various frequencies are assigned by the federal government for different purposes.

electronic data interchange (EDI) *(p. 351, LO 1)* System of direct electronic exchange between organizations' computer systems of standard business documents, such as purchase orders, invoices, and shipping documents.

EDI allows the companies involved to do away with standard paper handling and its costs.

electronic mail (e-mail) *(p. 337, LO 1)* System in which computer users, linked by wired or wireless communications lines, may use their keyboards to post messages and to read responses on their display screens.

E-mail allows users to send messages to a single recipient's "mailbox"—a file stored on the computer system—or to multiple users. It is a much faster way of transmitting written messages than traditional mail services.

FAQ (frequently asked questions) *(p. 348, LO 1)* Refers to the file that contains basic information about a newsgroup on the Internet.

FAQ files provide users with information they need to decide if a particular newsgroup is right for them.

fax *(p. 336, LO 1, 2)* Stands for *facsimile transmission* or reproduction; a message sent by dedicated fax machine or by fax modem.

A fax message may transmit a copy of text or graphics for the price of a telephone call.

fax modem *(p. 336, LO 1, 2)* Type of modem installed as a circuit board inside a computer; it exchanges fax messages with another fax machine or fax modem.

The benefit of fax modems is that messages can be transmitted directly from a microcomputer; no paper or scanner is required.

FDDI network *(p. 378, LO 4)* Short for Fiber Distributed Data Interface; a type of local area network that uses fiber-optic cable with a dual counter-rotating ring topology.

The FDDI network is being used for such high-tech purposes as electronic imaging, high-resolution graphics, and digital video.

fiber-optic cable *(p. 365, LO 3)* Type of communications channel consisting of hundreds or thousands of thin strands of glass that transmit pulsating beams of light. These strands, each as thin as a human hair, can transmit billions of pulses per second, each "on" pulse representing one bit.

When bundled together, fiber-optic strands in a cable 12 inches thick can support a quarter- to a half-million simultaneous voice conversations. Moreover, unlike electrical signals, light pulses are not affected by random electromagnetic interference in the environment and thus have much lower error rates than telephone wire and cable.

file server *(p. 374, LO 4)* Type of computer used on a local area network (LAN) that acts like a disk drive and stores the programs and data files shared by users of the LAN.

A file server enables users of a LAN to all have access to the same programs and data.

firewall *(p. 352, LO 1)* Software used in internal networks (intranets) to prevent unauthorized people from accessing the network.

Firewalls are necessary to protect an organization's internal network against theft and corruption.

frequency *(p. 358, LO 2)* Number of times a radio wave repeats during a specific time interval—that is, how many times it completes a cycle in a second—1 Hz = 1 cycle per second.

The higher the frequency—that is, the more cycles per second—the more data can be sent through a channel.

What It Is / What It Does	**Why It's Important**

front-end processor *(p. 382, LO 5)* Smaller computer that is connected to a larger computer to assist it with communications functions.

The front-end processor transmits and receives messages over the communications channels, corrects errors, and relieves the larger computer of routine tasks.

FTP (file transfer protocol) *(p. 348, LO 1)* Feature of the Internet whereby users can connect their PCs to remote computers and transfer (download) publicly available files.

FTP enables users to copy free files of software, games, photos, music, and so on.

full-duplex transmission *(p. 380, LO 5)* Type of data transmission in which data is transmitted back and forth at the same time, unlike simplex and half-duplex.

Full-duplex is used frequently between computers in communications systems.

gateway *(p. 374, LO 4)* Interface that enables dissimilar networks to communicate with one another.

With a gateway, a local area network may be connected to a larger network, such as a wide area network.

gopher *(p. 349, LO 1)* Internet program that allows users to use a system of menus to browse through and retrieve files stored on different computers.

Gophers can simplify Internet searches.

groupware *(p. 351, LO 1)* Software that allows two or more people on a network to work on the same information at the same time.

Groupware has become the glue that ties organizations together, permitting office workers to collaborate with colleagues, suppliers, and customers and to tap into company information through computer networks.

half-duplex transmission *(p. 380, LO 5)* Type of data transmission in which data travels in both directions but only in one direction at a time, as with CB or marine radios; both parties must take turns talking.

Half-duplex is a common transmission method with microcomputers, as when logging onto an electronic bulletin board system.

hertz (Hz) *(p. 364, LO 3)* Provides a measure of the frequency of electrical vibrations (cycles) per second.

One million hertz equals 1 megahertz. Bandwidths are defined according to megahertz and gigahertz ranges.

home page *(p. 350, LO 1)* The first page (main page)—that is, the first screen—seen upon accessing a Web site.

The home page provides a menu or explanation of the topics available on that Web site.

host computer *(p. 372, LO 4)* The central computer that controls a network. On a local area network, the host's functions may be performed by a computer called a *server*.

The host is responsible for managing the entire network.

hybrid network *(p. 378, LO 4)* Type of local area network (LAN) that combines star, ring, and bus networks.

A hybrid network can link different types of LANs. For example, a small college campus might use a bus network to connect buildings and star and ring networks within certain buildings.

Integrated Services Digital Network (ISDN) *(p. 362, LO 2)* A set of international communications standards for transmitting voice, video, and data simultaneously as digital signals over twisted-pair telephone lines.

The main benefit of ISDN is speed. It allows people to send digital data ten times faster than most modems can now deliver on the analog voice network.

Internet *(p. 344, LO 1)* International network composed of approximately 36,000 smaller networks. Created as ARPAnet in 1969 by the U.S. Department of Defense, Internet was designed to share research among military, industry, and university sources and to sustain communication in the event of nuclear attack.

Today the Internet is essentially a self-governing and noncommercial community offering both scholars and the public such features as information gathering, electronic mail, and discussion and newsgroups.

Internet service provider (ISP) *(p. 346, LO 1)* Local or national company that provides unlimited public access to the Internet and the Web for a flat fee.

Unless they are connected to the Internet through an online information service or a direct network connection, microcomputer users need an ISP to connect to the Internet.

intranet *(p. 352, LO 1)* Internal corporate network that uses the infrastructure and standards of the Internet and the World Wide Web.

Intranets can connect all types of computers.

What It Is / What It Does	**Why It's Important**

local area network (LAN) *(p. 373, LO 4)* A network consisting of a communications link, network operating system, microcomputers or workstations, servers, and other shared hardware such as printers or storage devices. LANs are of two principal types: client/server and peer-to-peer.

LANs have replaced mainframes and minicomputers for many functions and are considerably less expensive. However, LANs have neither the great storage capacity nor the security of mainframes.

local network *(p. 371, LO 4)* Privately owned communications network that serves users within a confined geographical area. The range is usually within a mile.

Local networks are of two types: private branch exchanges (PBXs) and local area networks (LANs).

metropolitan area network (MAN) *(p. 371, LO 4)* Communications network covering a geographic area the size of a city or suburb. Cellular phone systems are often MANs.

The purpose of a MAN is often to bypass telephone companies when accessing long-distance services.

microwave systems *(p. 366, LO 3)* Communications systems that transmit voice and data through the atmosphere as super-high-frequency radio waves. Microwaves are the electromagnetic waves that vibrate at 1 billion hertz per second or higher.

Microwave frequencies are used to transmit messages between ground-based earth stations and satellite communications systems. More than half of today's telephone system uses microwave transmission.

modem *(p. 358, LO 2)* Short for *modulater/demodulater*. A device that converts digital signals into a representation of analog form (modulation) to send over phone lines; a receiving modem then converts the analog signal back to a digital signal (demodulation).

A modem enables users to transmit data from one computer to another by using standard telephone lines instead of special communications lines such as fiber optic or cable.

multiplexer *(p. 382, LO 5)* Device that merges several low-speed transmissions into one high-speed transmission. Depending on the model, 32 or more devices may share a single communications line.

High-speed multiplexers using high-speed digital lines can carry as many messages, both voice and data, as 24 analog telephone lines.

multipoint line *(p. 379, LO 5)* Single line that interconnects several communications devices to one computer.

Often on a multipoint line only one communications device, such as a terminal, can transmit at any given time.

netiquette *(p. 383, LO 6)* "Net etiquette"; guides to appropriate behavior while online.

Netiquette rules help users to avoid offending other users.

network (communications network) *(p. 371, LO 4)* System of interconnected computers, telephones, or other communications devices that can communicate with one another.

Networks allow users to share applications and data.

network interface card *(p. 374, LO 4)* Circuit board inserted into an expansion slot in a microcomputer that enables it to send and receive messages on a local area network.

Without a network interface card, a microcomputer cannot be used to communicate on a LAN.

newsgroup *(p. 348, LO 1)* Electronic discussions held by groups of people focusing on a specific topic.

Internet newsgroups enable people with similar interests to readily find each other, no matter where they live.

node *(p. 372, LO 4)* Any device that is attached to a network.

A node may be a microcomputer, terminal, storage device, or some peripheral device, any of which enhance the usefulness of the network.

noise *(p. 361, LO 2)* Anything that causes distortion in a communications signal.

Some modems and communications software come with built-in error-correction features to reduce noise.

online information service *(p. 340, LO 1)* Company that provides access to databases and electronic meeting places to subscribers equipped with telephone-linked microcomputers—for example Prodigy, CompuServe, and America Online.

Online information services offer a wealth of services, from electronic mail to home shopping to videogames to enormous research facilities to discussion groups.

Open Systems Interconnection (OSI) *(p. 383, LO 5)* International standard that defines seven layers of protocols for worldwide computer communications.

Creates a set of standards that enables communications hardware manufacturers to build devices that can "talk" to each other.

What It Is / What It Does	**Why It's Important**

packet *(p. 382, LO 5)* Fixed-length block of data for transmission. The packet also contains instructions about the destination of the packet.

By creating data in the form of packets, a transmission system can deliver the data more efficiently and economically, as in packet switching.

packet switching *(p. 382, LO 5)* Technique for dividing electronic messages into packets—fixed-length blocks of data—for transmission over a network to their destination through the most expedient route. A sending computer breaks an electronic message apart into packets, which are sent through a communications network—via different routes and speeds—to a receiving computer, which reassembles them into proper sequence to complete the message.

The benefit of packet switching is that it can handle high-volume traffic in a network. It also allows more users to share a network, thereby offering cost savings.

parallel data transmission *(p. 379, LO 5)* Method of transmitting data in which bits are sent through separate lines simultaneously.

Unlike serial lines, parallel lines move information fast, but they are efficient for only up to 15 feet. Thus, parallel lines are used, for example, to transmit data from a computer's CPU to a printer.

peer-to-peer LAN *(p. 374, LO 4)* Type of local area network (LAN); all microcomputers on the network communicate directly with one another without relying on a server.

Peer-to-peer networks are less expensive than client/server networks and work effectively for up to 25 computers. Thus, they are appropriate for networking in small groups.

point-to-point line *(p. 379, LO 5)* Communications line that directly connects the sending and receiving devices, such as a terminal with a central computer.

This arrangement is appropriate for a private line whose sole purpose is to keep data secure by transmitting it from one device to another.

privacy *(p. 386, LO 6)* Right of people not to reveal information about themselves.

Computer technology and electronic databases have made it more difficult for people to protect their privacy.

private branch exchange (PBX) *(p. 373, LO 4)* Private or leased telephone switching system that connects telephone extensions inhouse as well as to the outside telephone system.

Newer PBXs can handle not only analog telephones but also digital equipment, including computers.

remote-control software *(p. 361, LO 2)* Software that allows a user to control a microcomputer from another microcomputer in a different location.

Such software is useful for travelers who want to use their home computers from afar and for technicians trying to assist computer users with support problems.

ring network *(p. 376, LO 4)* Type of local area network (LAN) in which all communications devices are connected in a continuous loop and messages are passed around the ring until they reach the right destination. There is no central server.

The advantage of a ring network is that messages flow in only one direction and so there is no danger of collisions. The disadvantage is that if a connection is broken, the entire network stops working.

serial data transmission *(p. 379, LO 5)* Method of data transmission in which bits are sent sequentially, one after the other, through one line.

Serial transmission is found in communications lines, modems, and mice.

server *(p. 372, LO 4)* Computer shared by several users in a network.

With servers, users on a LAN can share several devices, as well as data.

simplex transmission *(p. 380, LO 5)* Type of transmission in which data can travel in only one direction; there is no return signal.

Some computerized data collection devices, such as seismograph sensors that measure earthquakes, use simplex transmission.

star network *(p. 376, LO 4)* Type of local area network (LAN) in which all microcomputers and other communications devices are connected to a central hub, such as a file server. Electronic messages are routed through the central hub to their destinations. The central hub monitors the flow of traffic.

The advantage of a star network is that the hub prevents collisions between messages. Moreover, if a connection is broken between any communications device and the hub, the rest of the devices on the network will continue operating.

What It Is / What It Does	Why It's Important

synchronous transmission *(p. 381, LO 5)* Type of transmission in which data is sent in blocks. Start and stop bit patterns, called sync bytes, are transmitted at the beginning and end of the blocks. These start and end bit patterns synchronize internal clocks in the sending and receiving devices so that they are in time with each other.

Synchronous transmission is rarely used with microcomputers because it is more complicated and more expensive than asynchronous transmission. It is appropriate for computer systems that need to transmit great quantities of data quickly.

sysops *(p. 342, LO 1)* The name given to system operators of electronic bulletin board systems (BBSs).

Many BBSs are operated by individual sysops.

TCP/IP (Transmission Control Protocol/Internet Protocol) *(p. 344, LO 1)* Standardized set of guidelines (protocols) that allow computers on different networks to communicate with one another efficiently.

TCP/IP is the standard language of the Internet.

telecommuting *(p. 353, LO 1)* Way of working at home and communicating ("commuting") with the office by phone, fax, and computer.

Telecommuting can help ease traffic and the stress of commuting by car and extend employment opportunities to more people, such as those who need or want to stay at home.

Telnet *(p. 349, LO 1)* Internet feature that allows users to connect (log on) to remote computers.

With Telnet, users can peruse large databases and library card catalogs.

terminal emulation *(p. 361, LO 2)* Communications software that allows users to use their microcomputers to access a mainframe or minicomputer; the software "tricks" the large computer into acting as if it is communicating with a terminal.

Terminal emulation software is necessary for microcomputer users to make full use of the networks and the resources available with mainframes.

topology *(p. 376, LO 4)* The logical layout, or shape, of a local area network. The five basic topologies are star, ring, bus, hybrid, and FDDI.

Different topologies can be used to suit different office and equipment configurations.

twisted-pair wire *(p. 364, LO 3)* Type of communications channel consisting of two strands of insulated copper wire, twisted around each other in pairs.

Twisted-pair wire has been the most common channel or medium used for telephone systems. It is relatively slow and does not protect well against electrical interference.

upload *(p. 372, LO 4)* To send files from a user's microcomputer to another computer. Compare with download.

Uploading allows microcomputer users to easily exchange files with each other over networks.

URL (Uniform Resource Location) *(p. 350, LO 1)* Address that points to a specific resource on the Web.

Addresses are necessary to distinguish among Web sites.

videoconferencing *(p. 339, LO 1)* Also called *teleconferencing;* form of conferencing using video cameras and monitors that allow people at different locations to see, hear, and talk with one another.

Videoconferencing may be done from a special videoconference room or handled with equipment rolled on casters from room to room.

virtual office *(p. 353, LO 1)* A nonpermanent and mobile office run with computer and communications technology.

Employees work not in a central office but from their homes, cars, and customers' offices. They use pocket pagers, portable computers, fax machines, and various phone and network services to conduct business.

voice mail *(p. 336, LO 1)* System in which incoming voice messages are stored in a recipient's "voice mailbox" in digitized form. The system converts the digitized versions back to voice messages when they are retrieved. With voice mail, callers can direct calls within an office using buttons on their touch-tone phone.

Voice mail enables callers to deliver the same message to many people, to forward calls, to save or erase messages, and to dictate replies. The main benefit is that voice mail helps eliminate "telephone tag."

WAIS (Wide Area Information Server) *(p. 349, LO 1)* Feature of the Internet used for searching databases by subject, using specific words or phrases.

WAIS provides one of several tools to simplify searching Internet databases.

What It Is / What It Does	Why It's Important
Web browser *(p. 350, LO 1)* Internet software used to browse through multimedia Websites.	*See World Wide Web.*
Web site *(p. 350, LO 1)* File stored on a computer as part of the World Wide Web.	Each Web site focuses on a particular topic. The information on a site is stored on "pages." The starting page is called the *home page*.
wide area network (WAN) *(p. 371, LO 4)* Type of communications network that covers a wide geographical area, such as a state or a country.	Wide area networks provide worldwide communications systems.
workgroup computing *(p. 351, LO 1)* Also called *collaborative computing;* technology that enables teams of coworkers to use networks of microcomputers to share information and cooperate on projects. Workgroup computing is made possible not only by networks and microcomputers but also by groupware.	Workgroup computing permits office workers to collaborate with colleagues, suppliers, and customers and to tap into company information through computer networks.
World Wide Web *(p. 349, LO 1)* Interconnected system of sites of the Internet that store information in multimedia form.	Web software allows users to view information that includes not just text but graphics, animation, video, and sound.

(Selected answers appear at the back of the book.)

Short-Answer Questions

1. What is the definition of *connectivity*?
2. What are the main disadvantages of using e-mail?
3. List three methods you can use to connect your PC to the Internet.
4. Why is workgroup computing significant?
5. Why would you buy an internal modem rather than an external one?
6. Give the definition of *bandwidth*.
7. What is a virtual office?
8. What is the difference between asynchronous and synchronous transmission?
9. What could you use the Internet for?
10. What advantages are gained by using networks?

Fill-in-the-Blank Questions

1. Two types of fax hardware are
   _____ and
   _____.
2. The Internet is an international
   _____ made up of
   _____.
3. Electronic bulletin board operators are called
   _____.
4. A nonpermanent and mobile office run with computer and communications technology is called a(n)
   _____ office.
5. Workgroup computing uses a type of software called
   _____.
6. In an analog signal, the number of times a wave repeats during a specific time interval is called its
   _____ and the height of a wave within a given period of time is called its
   _____.
7. Software that allows users to use their microcomputers to simulate a mainframe or minicomputer terminal is called
   _____ software.
8. A two-way radio voice-dispatching service used by taxis and trucks that is being converted to digital is called _____ radio.

9. With _____, the user works at home but can communicate, using hardware and software, with the office.
10. When using the World Wide Web,
    _____ software helps you to get the information you want by clicking on underlined words or pictures on the screen.

Multiple-Choice Questions

1. Which of the following allows different computers on different networks to communicate?
   a. e-mail
   b. BBS
   c. LAN
   d. TCP/IP
   e. all of the above
2. Which of the following *isn't* a principal feature of the Internet?
   a. e-mail
   b. file transfer
   c. LANs
   d. information services
   e. all of the above
3. Which of the following would you use if you wanted to copy files for free from a remote computer?
   a. Archie
   b. FTP
   c. Usenet
   d. World Wide Web
   e. none of the above
4. Which of the following provides information in multimedia form?
   a. Archie
   b. FTP
   c. Usenet
   d. World Wide Web
   e. none of the above
5. A communications network that covers a wide geographical area, such as a state or country, is called a(n)_____ network.
   a. wide area network (WAN)
   b. metropolitan area network (MAN)
   c. local area network (LAN)
   d. privatized network (PN)
   e. none of the above
6. A(n) _____ can be either a PBX or a LAN.
   a. MAN
   b. local network
   c. WAN
   d. Internet-based network
   e. none of the above

7. A network in which all communications devices are connected to a common channel is called a _____ network.
   a. star
   b. ring
   c. bus
   d. hybrid
   e. all of the above

8. Which of the following must you have access to before you can use your computer to communicate over the phone lines?
   a. Internet
   b. data compression software
   c. modem
   d. terminal emulation software
   e. all of the above

9. A communications _____ is used in a communications network to govern the exchange of data between hardware and/or software.
   a. protocol
   b. multiplexer
   c. channel
   d. information service
   e. none of the above

True/False Questions

T F 1. Before you can communicate over the phone lines, your computer requires communications software.

T F 2. A metropolitan area network (MAN) typically covers a greater area than a wide area network (WAN).

T F 3. Coaxial cable is better able to resist noise than twisted-pair wiring.

T F 4. Pagers can receive data (but not voice messages) sent from a special radio transmitter.

T F 5. The term *download* refers to retrieving a file from another computer and copying it onto your computer.

T F 6. A private branch exchange (PBX) typically connects telephone extensions in a city or metropolitan area.

T F 7. Local area networks may be client/server or peer-to-peer.

T F 8. A file server is a computer that stores programs and data files for users on a LAN.

T F 9. A gateway is an interface that enables similar networks to communicate.

T F 10. ISDN lines allow voice and video, but not textual data, to be communicated as digital signals over telephone lines.

Projects/Critical-Thinking Questions

1. Are the computers at your school or at work connected to a network? If so, what are the characteristics of the network? What advantages does the network provide in terms of hardware and software support? What types of computers are connected to the network (microcomputers, minicomputers, and/or mainframes)? Specifically, what software/hardware allows the network to function?

2. "Distance learning" or "distance education" uses electronic links to extend college campuses to people who otherwise would not be able to take college courses. In one variant, college instructors using such systems are able to lecture "live" to students in distant locations. Is your school involved in distance learning? If so, research the system's components and uses. What hardware does it use? Software? Communications media?

3. You need to purchase a computer to use at home to perform business-related tasks. You want to be able to communicate with the network at school or work so that you can use its software and access its data. Include the following in a report:
   - A description of the hardware and software used at school or work.
   - A description of the types of tasks you will want to perform at home.
   - The name of the computer you would buy. (Include a detailed description of the computer, such as the RAM capacity and secondary storage capacity.)
   - The communications hardware/software you would need to purchase.
   - A cost estimate.

4. What do you think of the possibility of interactive television? How would you like to see it set up so it would be most useful to you? Explain in a short report.

5. In the world today, what country or group of individuals do you think could benefit the most from using communications technology? Why isn't communications technology used now and what impact do you think this has? How would communications technology be beneficial? What are the barriers to implementing communications technology? Explore the answers to these questions in a two-page report.

Using the Internet

Objective: *In this exercise we describe how to subscribe to a newsgroup and reply to articles using Netscape Navigator.*

Before you continue: *We assume you have access to the Internet through your university, business, or commercial service provider and to the Web browser tool named Netscape Navigator 2.0 or 2.01. Additionally, we assume you know how to connect to the Internet and then load Netscape Navigator. If necessary, ask your instructor or system administrator for assistance.*

1. Make sure you have started Netscape. The home page for Netscape Navigator should appear on your screen.

2. To subscribe to a newsgroup that discusses issues relating to future computer technology (news:comp.society.futures), perform the following steps:
   CLICK: Open button (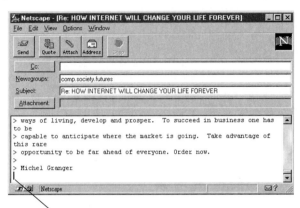).
   TYPE: news:comp.society.futures
   PRESS: (Enter)
   CLICK: *Subscribe box, located next to the "news:comp.society" newsgroup name.* (*Note:* You can easily unsubscribe to a newsgroup by clicking in the newsgroup's associated check box.)

3. To display the messages in the "comp.society. futures" newsgroup:
   DOUBLE-CLICK: *"comp.society.futures" in the newsgroup list*

4. The newsgroup's articles are listed in the right pane. Click an article you're interested in.

5. After reading an article, you may want to use the Re:News button ( ), which sends a copy of your

message to the newsgroup for all subscribers to read, or the Re:Both button ( ), which sends a copy of your message to the newsgroup and an e-mail copy to the sender of the message. A dialog box will appear into which you can type your message. The message dialog box will look different depending on your computer system. On a Windows 95-based system, the following dialog box appeared after clicking the Re:News button ( ) to respond to an article related to how the Internet can change your life.

Begin typing your
message here.

6. Now that you've seen the general process of replying to a newsgroup article, let's exit from the dialog box without posting a message.

7. Display Netscape's home page and then exit Netscape.

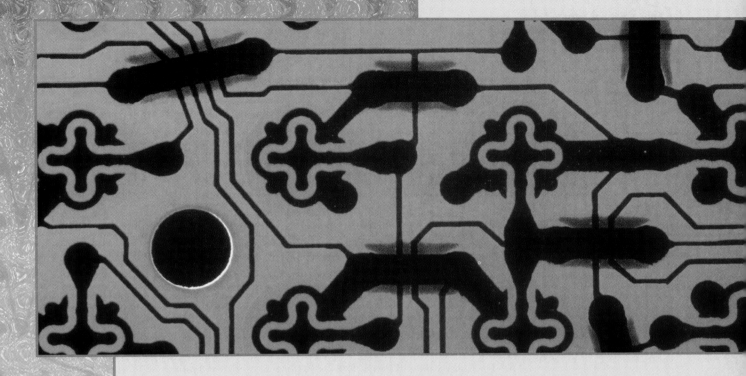

CHAPTER 9

Files & Databases

From Data Organizing to Data Mining

Concepts You Should Know

After reading this chapter, you should be able to:

1. Explain the importance of database administration within an organization.
2. Describe the parts of the data hierarchy and the role of key fields.
3. Distinguish batch from online processing, and online from offline storage.
4. Describe the difference between file management systems and database management systems.
5. Explain the best uses for sequential, direct, and indexed-sequential access storage, and the storage media associated with each.
6. Identify the advantages and disadvantages of the four database models and of database management systems in general.
7. Explain how a data warehouse is set up and what tools are used to sift through it.
8. Describe some ways database users can invade privacy and the methods or laws to fight such invasions.

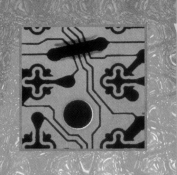

We want to capture the entire human experience throughout history."

So states Corbis Corporation chief executive officer Doug Rowan.[1] Corbis was formed in 1989 by software billionaire Bill Gates to acquire digital rights to fine art and photographic images that can be viewed electronically—in everything from electronic books to computerized wall hangings.[2] In 1995 Corbis acquired the Bettmann Archive of 17 million photographs, for scanning into its digital database.[3] Its founder, Dr. Otto Bettmann, called his famous collection a "visual story of the world," and indeed many of the images are unique. They include tintypes of black Civil War soldiers, the 1937 crash of the *Hindenburg* dirigible, John F. Kennedy, Jr., saluting the casket of his assassinated father. (A few Bettmann images can be viewed electronically at no charge on Corbis's World Wide Web site—*http://www. corbis.com/@@/50Gjud5NsUF9/exhibit/bysubject.chtml*. Some are also visible on the online service CompuServe.)

However, when Rowan says Corbis wants to capture all of human experience, he means not just photos and art works from the likes of the National Gallery in London and the State Hermitage Museum in St. Petersburg, Russia, for which Corbis also owns digital imaging rights. "Film, video, audio," he says. "We are interested in those fields too."

Are there any ethical problems with one company having in its database the exclusive digital rights to our visual and audio history? Like many museums and libraries (such as the Library of Congress), Corbis joins a trend toward democratizing art and scholarship by converting the images and texts of the past into digital form and making them available to people who could never travel to, say, London or St. Petersburg.

However, when Gates acquired the Bettmann images, for example, the move put their future use "into the hands of an aggressive businessman who, unlike Dr. Bettmann, is planning his own publishing ventures," points out one reporter. "While Mr. Gates's initial plans will make Bettmann images more widely accessible, this savvy competitor now ultimately controls who can use them—and who can't."[4] Adds Paul Saffo, of the nonprofit Institute for the Future, "The cultural issue raised by the Bettmann purchase is whether we're seeing history sold to the highest bidder or we'll eventually see history made more accessible to the public as a result."[5] Curators of art museums are afraid that the rights to art works will slip away for less than they are worth or that the images will be pirated or used in silly ways in advertising.[6]

All Databases Great & Small

Preview & Review: Databases are integrated collections of files. Organizations usually appoint a database administrator to manage the database and related activities.

All forms of information technology are affecting our social and business institutions in significant ways. However, as the foregoing example suggests, the arrival of databases promises to stand some institutions on their heads. Databases are not just an interesting new way to computerize filing systems. *Databases* are integrated collections of files, which, as we shall see, makes them usable in more ways than traditional filing systems (computerized or not).

READ ME

Case Study: Using Mobile Computing & Databases in Sports

Using big sports arenas as their factory floors and games as the key gauge of market prowess, pro basketball teams are now employing computers much as other businesses do, especially companies with managers and workers who are constantly on the move.

Armed with notebook computers, fax modems, and specialized software, the teams compile scouting reports, analyze statistics, and create models to predict what players and teams might do in specific game situations. In a matter of minutes and a few keystrokes, the machines permit coaches to do work that used to take hours of laborious sorting through statistics sheets and play diagrams. . . .

The 27 [National Basketball Association] teams are equipped with conventional Thinkpad notebooks. Recently, several teams have been experimenting with a pen-based model, the Thinkpad 700T, loaded with specialized basketball software developed by Information and Display Systems Inc. of Jacksonville, Fla. The pen-based system allows coaches and scouts to diagram plays and store them, up to 99 plays for each of the 27 teams.

Robert Salmi is a 33-year-old assistant coach of the Knicks and one of pro basketball's self-taught computer aces. A 6-foot-7-inch former college basketball player, Mr. Salmi knew little about computers before Pat Riley, the Knicks head coach, came to him early in the 1991–92 season and told him what he wanted. . . .

Today, Mr. Salmi is comfortable with computers, speaking knowledgeably about database programs or pen-based machines. But mostly, he views the computer as a tool that saves time and that can pinpoint problems and opportunities.

For example, Mr. Riley might ask his assistant for additional statistics on their initial playoff opponents, the Indiana Pacers. . . .

The coach, Mr. Salmi says, might want to know the Pacers' won-lost record in games when they were out-rebounded, or when their guard Reggie Miller took more than 15 shots and made more than 50% in a game, or when their forward Rick Smits scored more than 20 points. Or in games when all three happened.

"Two years ago, that would have meant three or four hours of going through files and stat sheets," Mr. Salmi said. "Today, I can snap that out in five minutes."

—Steve Lohr, "Electronics Replacing Coaches' Clipboards," *New York Times*

A database may be small, contained entirely within your own personal computer. Or it may be massive, like those of Corbis, available online through computer and telephone connections. Such online databases are of special interest to us in this book because they offer us phenomenal resources that until recently were unavailable to most ordinary computer users.

Microcomputer users can set up their own databases using popular database management software like that we discussed earlier (in Chapter 2, ✓ p. 71). Examples are Paradox, Access, dBASE 5, and FoxPro. (■ *See Panel 9.1, next page.*) Such programs are used, for example, by graduate students to conduct research, by salespeople to keep track of clients, by purchasing agents to monitor orders, and by coaches to keep watch on other teams and players.

Some databases are so large that they cannot possibly be stored in a microcomputer. Some of these can be accessed by going online through a microcomputer, or other computer. Such databases, sometimes called *information utilities*, represent enormous compilations of data, any part of which is available, for a fee, to the public.

Software for personal databases

These popular microcomputer database packages are available in computer stores.

Examples of well-known information utilities—more commonly known as *online services*—are America Online, CompuServe, and Prodigy. As we described in Chapter 8, these services offer access to news, weather, travel information, home shopping services, reference works, and a great deal more. Some public-access databases are specialized, such as Lexis, which gives lawyers access to local, state, and federal laws.

Other types of large databases are private—collections of records shared or distributed throughout a company or other organization. Generally, the records are available only to employees or selected individuals and not to outsiders.

For example, many university libraries have been transforming drawers of catalog cards into electronic databases for use by their students and faculty. Libraries at Yale, Johns Hopkins, and other universities have contracted with a Virginia company called The Electronic Scriptorium, which employs monks and nuns at six monasteries to convert card catalogs to an electronic system.[7,8] (■ *See Panel 9.2.)*

A database may be *shared* or *distributed*. **A *shared* database is shared by users in one company or organization in one location.** Shared databases can be found in local area networks (✓ p. 373). The company owns the database, which is usually stored on a minicomputer or mainframe. Users are linked to the database through terminals or microcomputer workstations.

A *distributed database* is one that is stored on different computers in different locations connected by a client/server type of network (✓ p. 162). For example, sales figures for a chain of discount stores might be located in computers at the various stores, but they would also be available to executives in regional offices or at corporate headquarters. An employee using the database would not know where the data is coming from. However, all employees still use the same commands to access and use the database.

One thing that large databases have in common is that they must be managed. This is done by the database administrator.

PANEL 9.2

Building a library database

Father Patrick Creeden enters data into a computer at the Monastery of the Holy Cross in Chicago.

"CHICAGO—Father Thomas Baxter stands in his habit, holding a computer print-out. Later today, he'll review the Scriptures in his cell here at the Monastery of the Holy Cross of Jerusalem, but this morning he's carefully proofreading a library's computerized records.

He and two other monks in this community of five are modern-day scribes, using computers to participate in an age-old monastic tradition: preserving knowledge. In this case, they're helping university and public libraries transform drawers of catalogue cards into electronic databases.

Father Baxter is pointing at a line of text highlighted in yellow, indicating a discrepancy between two data fields. One of his jobs as prior, or leader, of the monastery is to look for typographical errors, call them to the attention of the brother responsible for them, and remind him to concentrate more closely on his work.

This monastery is part of an effort to bring religious communities back into the information business. A company calling itself The Electronic Scriptorium—referring to the room where monks would use quills and ink to copy intricate manuscripts long ago—is matching up monastic communities with libraries and others in need of complex data-entry work. The partnerships benefit the monasteries and convents, which need flexible jobs to support themselves, and the libraries, which need their records entered accurately."

—Jeffrey R. Young, "Modern-Day Monastery," *Chronicle of Higher Education*

The Database Administrator

The information in a large database—such as a corporation's patents, formulas, advertising strategies, and sales information—is the organization's lifeblood. Someone, then, needs to manage all activities related to the database. This person is the ***database administrator (DBA), a person who coordinates all related activities and needs for a corporation's database.*** The responsibilities include the following:

• **Database design, implementation, and operation:** At the beginning, the DBA helps determine the design of the database. Later he or she determines how space will be used on secondary-storage devices, how files and records may be added and deleted, and how losses may be detected and remedied.

- **Coordination with users:** The DBA determines user access privileges, assists in establishing priorities for requests, and adjudicates conflicting user needs.

- **System security:** The DBA sets up and monitors a system for preventing unauthorized access to the database.

- **Backup and recovery:** Because loss of data or a crash in the database could vitally affect the organization, the DBA needs to make sure the system is regularly backed up. He or she also needs to develop plans for recovering data or operations should a failure or disaster occur.

- **Performance monitoring:** The DBA monitors the system to make sure it is serving users appropriately. A standard complaint is that the system is too slow, usually because too many users are trying to access it.

As an example of a case of performance monitoring, on Super Bowl Sunday in January 1996, the National Football League's Web site scored 6 million "hits"—a measure of the number of transmissions of text, video, graphics, or audio files. Expecting heavy traffic, managers of the site had used five servers, but network problems still affected visitors, and many reported long waits or just being turned away.[9] Although the Super Bowl is a once-a-year event, it will be the job of the database manager (here called a Webmaster) in charge of the site to make sure things run more smoothly the next time.

The Data Storage Hierarchy & the Concept of the Key Field

Preview & Review: Data in storage is organized as a hierarchy: bits, bytes, fields, records, and files, which are the elements of a database. In data organization, the role of the key field, which uniquely identifies a record, is very important.

How does a database actually work? To understand this, first we need to consider how stored data is structured—the *data storage hierarchy* and the concept of *key field*. We then need to discuss *file management systems*, then *database management systems*.

The Data Storage Hierarchy

Data can be grouped into a hierarchy of categories, each increasingly more complex. **The *data storage hierarchy* consists of the levels of data stored in a computer database: bits, bytes (characters), fields, records, and files.** (■ *See Panel 9.3.*)

Computers, we have said, are based on the principle that electricity may be "on" or "off," or "high-voltage" or "low-voltage," or "present" or "absent," or of some similar two-state system. Thus, individual items of data are represented by 0 for off and 1 for on. As mentioned, **0 or 1 is called a *bit*. A unit of 8 bits is called a *byte*; it may be used to represent a character, digit, or other value,** such as A, ?, or 3. Bits and bytes are the building blocks for representing data, whether it is being processed, stored, or telecommunicated. Bits and bytes are what the computer hardware deals with, but you need not be concerned with them when you are working with databases. You will, however, in this case be dealing with characters, fields, records, files, and databases.

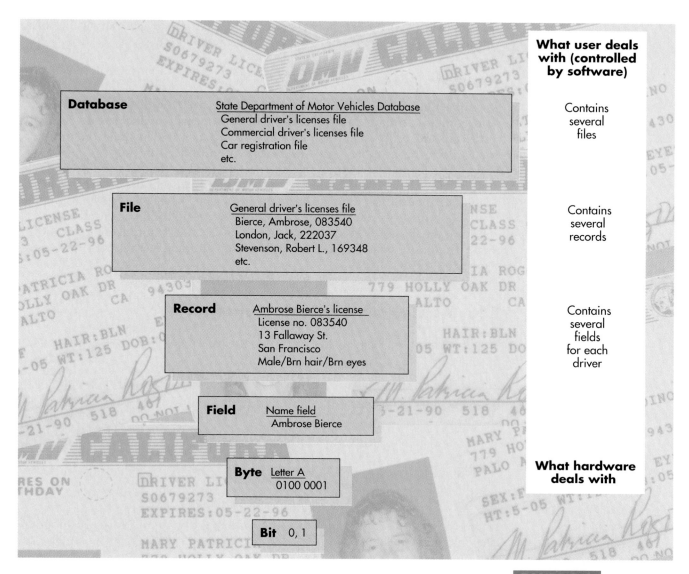

Database
State Department of Motor Vehicles Database
General driver's licenses file
Commercial driver's licenses file
Car registration file
etc.

File
General driver's licenses file
Bierce, Ambrose, 083540
London, Jack, 222037
Stevenson, Robert L., 169348
etc.

Record
Ambrose Bierce's license
License no. 083540
13 Fallaway St.
San Francisco
Male/Brn hair/Brn eyes

Field
Name field
Ambrose Bierce

Byte
Letter A
0100 0001

Bit 0, 1

What user deals with (controlled by software)

Contains several files

Contains several records

Contains several fields for each driver

What hardware deals with

PANEL 9.3

Data storage hierarchy: how data is organized

Bits are organized into bytes, bytes into fields, fields into records, records into files. Related files may be organized into a database.

- **Byte (character):** A *byte* is a group of 8 bits. **A *character* may be—but is not necessarily—the same as a byte. A character is a single letter, number, or special character** such as ;, $, or %.

- **Field:** **A *field* is a unit of data consisting of one or more characters.** An example of a field is your name, your address, or your Social Security number.

     Note: One reason the Social Security number is often used in computing—for good or for ill—is that, perhaps unlike your name, it is a *distinctive* (unique) field. Thus, it can be used to easily locate information about you. Such a field is called a *key field*. More on this below.

- **Record:** **A *record* is a collection of related fields.** An example of a record would be your name *and* address *and* Social Security number.

- **File:** **A *file* is a collection of related records.** An example of a file is collected data on everyone employed in the same department of a company, including all names, addresses, and Social Security numbers.

- **Database:** **A *database* is an integrated collection of files.** A company database might include files on all past and current employees in all departments.

There would be various files for each employee: payroll, retirement benefits, sales quotas and achievements (if in sales), and so on.

The Key Field

An important concept in data organization is that of the *key field*. **A *key field* contains unique data used to identify a record so that it can be easily retrieved and processed.** The key field is often an identification number, Social Security number, customer account number, or the like. As mentioned, the primary characteristic of the key field is that it is *unique*. Thus, numbers are clearly preferable to names as key fields because there are many people with common names like James Johnson, Kim Lee, Susan Williams, Ann Wong, or Roberto Sanchez, whose records might be confused.

Before databases existed there were files. To understand database management, therefore, we first need to understand file management and file management systems.

File Management: Basic Concepts

Preview & Review: Files are either program files or data files; two important data files are master files and transaction files.

Files may be processed by batch processing or by real-time processing. Batch processing tends to favor offline storage; online, or real-time, processing requires online storage.

Three methods of file organization are sequential access, direct access, and indexed-sequential.

Types of Files: Program & Data Files

There are many kinds of files, but the principal division is between program files and data files.

- **Program files:** *Program files* **are files containing software instructions.** In a word processing program, for example, you may see files listed (with names such as INSTALL.EXE, on an IBM PC, or "control panel file" or "applications program," on a Mac) that perform specific functions associated with word processing. These files are part of the software package.

- **Data files:** *Data files* **are files that contain data.** Often you will create and name these files yourself, such as DOCUMENT.1 or PSYCH.RPT, on an IBM PC, or DOCUMENT1 or PSYCHREPORT, on a Mac.

Two Types of Data Files: Master File & Transaction File

Among the several types of data files two are traditionally used to update data: a master file and a transaction file.

- **Master file:** **The *master file* is a data file containing relatively permanent records that are generally updated periodically.** An example of a master file would be the address-label file for all students currently enrolled at your college.

- **Transaction file:** The *transaction file* is a temporary holding file that holds all changes to be made to the master file: additions, deletions, and revisions.

    For example, in the case of the address labels for your college, a transaction file would hold new names and addresses to be added (because over time new students enroll) and names and addresses to be deleted (because students leave). It would also hold revised names and addresses (because students change their names or move). Each month or so, the master file would be *updated* with the changes called for in the transaction file.

Batch Versus Online Processing

Data may be taken from secondary storage (✓ p. 286) and processed in either of two ways: (1) "later," via *batch processing,* or (2) "right now," via *online (real-time) processing.*

- **Batch processing:** In *batch processing,* **data is collected over several days or weeks in a transaction file and then processed all at one time, as a "batch," against a master file.** Thus, if users need to make some request of the system, they must wait until the batch has been processed. Batch processing is less expensive than online processing and is suitable for work in which immediate answers to queries are not needed.

    An example of batch processing is that done by banks for balancing checking accounts. When you deposit a check in the morning, the bank will make a record of it. However, it will not compute your account balance until the end of the day, after all checks have been processed in a batch.

- **Online processing:** *Online processing,* also called *real-time processing,* **means entering transactions into a computer system as they take place and updating the master files as the transactions occur.** For example, when you use your ATM card to withdraw cash from an automated teller machine, the system automatically computes your account balance then and there. Airline reservation systems also use online processing.

Offline Versus Online Storage

Whether it's on magnetic tape or on some form of disk, data may be stored either offline or online.

- **Offline:** *Offline storage* **means that data is not directly accessible for processing until the tape or disk it's on has been loaded onto an input device.** That is, the storage is not under the direct, immediate control of the central processing unit.

- **Online:** *Online storage* **means that stored data is directly accessible for processing.** That is, storage is under the direct, immediate control of the central processing unit. You need not wait for a tape or disk to be loaded onto an input device.

For processing to be online, the storage must be online and *fast.* Generally, this means storage on disk rather than magnetic tape. With magnetic tape, it is not possible to go directly to the required record; instead, the read/write head (✓ p. 163) has to search through all the records that precede it, which takes time. With disk, however, the system can go directly and quickly to the record—just as a CD player can go directly to a particular spot on a music CD.

File Organization: Three Methods

In general, tape storage falls in the category of sequential access storage. *Sequential access storage* **means that information is stored in sequence,** such as alphabetically. Thus, you would have to search a tape past all the information from A to J, say, before you got to K. This process may require running several inches or feet off a reel of tape, which, as we said, takes time.

Disk storage, by contrast, generally falls into the category of direct access storage (although data *can* be stored sequentially). *Direct access storage* **means that the system can go directly to the required information.** Because you can directly access information, retrieving data is much faster with magnetic or optical disk than it is with magnetic tape.

From these two fundamental forms, computer scientists devised three methods of organizing files for secondary storage: *sequential, direct,* and *indexed-sequential.*

- **Sequential file organization:** *Sequential file organization* **stores records in sequence, one after the other.** This is the only method that can be used with magnetic tape. Records can be retrieved only in the sequence in which they were stored. The method can also be used with disk.

  For example, if you are looking for employee record 8888, the computer will have to start with record 0001, then go past 0002, 0003, and so on, until it finally comes to record 8888.

  Sequential file organization is useful, for example, when a large portion of the records needs to be accessed, as when a mail-order house is sending out catalogs to all names on a mailing list. The method also is less expensive than other methods because it uses magnetic tape, which is cheaper than magnetic or optical disk.

  The disadvantage of sequential file organization is that records must be ordered in a particular way and so searching for data is slow.

- **Direct file organization:** Instead of storing records in sequence, *direct file organization,* or *random file organization,* **stores records in no particular sequence. A record is retrieved according to its key field,** or unique element of data. This method of file organization is used with hard disks. It is ideal for applications such as airline reservations systems or computerized directory-assistance operations. In these cases, records need to be retrieved only one at a time, and there is no fixed pattern to the requests for records.

  A mathematical formula, called a *hashing algorithm,* is used to produce a unique number that will identify the record's physical location on the disk. (For example, one hashing algorithm divides the record's key field number by the prime number closest to that of the total number of records stored.)

  Direct file organization is much faster than sequential file organization for finding a specific record. However, because the method requires hard-disk or optical-disk storage, it is more expensive than magnetic tape. Moreover, it is not as efficient as sequential file organization for listing large numbers of records.

- **Indexed-sequential file organization:** A compromise has been developed between the preceding two methods. *Indexed-sequential file organization,* or simply *indexed file organization,* **stores records in sequential order. However, the file in which the records are stored contains an index that lists each record by its key field and identifies its physical location on the disk.** The method requires magnetic or optical disk.

For example, a company could index certain ranges of employee identification numbers—0000 to 1000, 1001 to 2000, and so on. For the computer to find the record with the key field 8888, it would go first to the index. The index would give the location of the range in which the key field appears (for example, 8001 to 9000). The computer would then search sequentially (from 8001) to find the key field 8888.

This method is slower than direct file organization because it requires an index search. The indexed-sequential method is best when large batches of transactions occasionally must be updated, yet users want frequent, rapid access to records. For example, bank customers and tellers want to have up-to-the-minute information about checking accounts, but every month the bank must update bank statements to send to customers.

An illustration of the three file organization methods is shown on the next two pages. (■ *See Panel 9.4.*)

File Management Systems

Preview & Review: Files may be retrieved through a file management system, one file at a time. Disadvantages of a file management system are data redundancy, lack of data integrity, and lack of program independence.

In the 1950s, when commercial use of computers was just beginning, magnetic tape was the storage medium and records and files were stored sequentially. To work with these files, a user needed a file management system.

A ***file management system,*** or ***file manager,*** is software for creating, retrieving, and manipulating files, one file at a time. Traditionally, a large organization such as a university would have different files for different purposes. For you as a student, for example, there might be one file on you for course grades, another for student records, and a third for tuition billing. Each file would be used independently to produce its own separate reports. If you changed your address, someone had to make the change separately in each file.

Disadvantages of File Management Systems

File management systems worked well enough for their time, but they had several disadvantages:

- **Data redundancy:** *Data redundancy* means that the same data fields appear in many different files and often in different formats. Thus, separate files tend to repeat some of the same data over and over. A student's course grades file and tuition billing file would contain some of the same data (name, address, telephone number). When data fields are repeated in different files, they waste storage space.

- **Lack of data integrity:** *Data integrity* means that data is accurate, consistent, and up to date. However, when the same data fields (a student's address and phone number, for example) must be changed in different files, some files may be missed and mistakes will be made. The result is that some reports will be produced with erroneous information.

- **Lack of program independence:** With file management systems, different files were often written by different programmers using different file formats. Thus, the files were not *program-independent*. The arrangement meant more time was required to maintain files. It also prevented a programmer from writing a single program that would access all the data in multiple files.

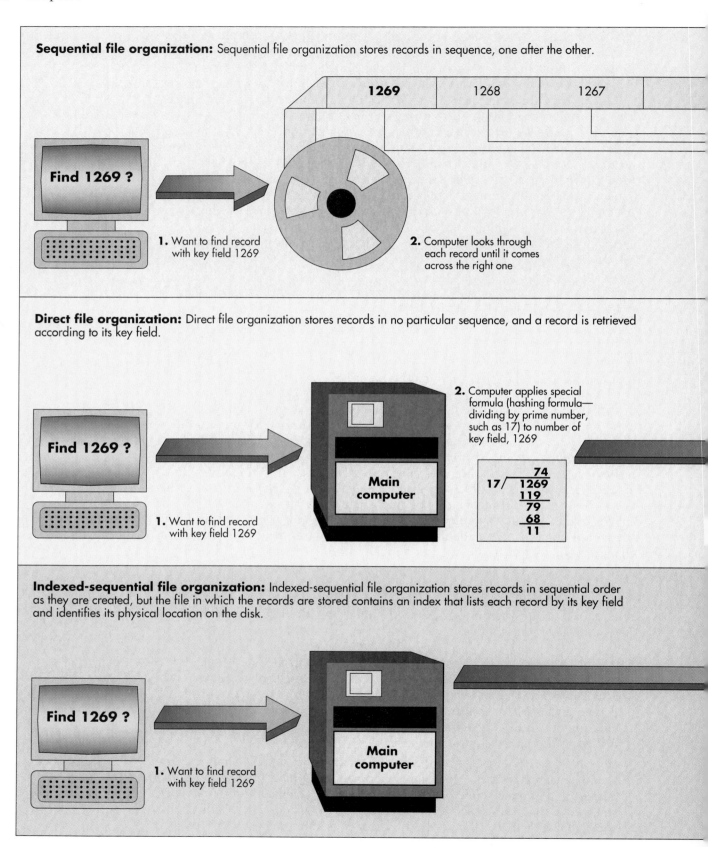

Sequential file organization: Sequential file organization stores records in sequence, one after the other.

1269 | 1268 | 1267

Find 1269 ?

1. Want to find record with key field 1269

2. Computer looks through each record until it comes across the right one

Direct file organization: Direct file organization stores records in no particular sequence, and a record is retrieved according to its key field.

Find 1269 ?

1. Want to find record with key field 1269

Main computer

2. Computer applies special formula (hashing formula—dividing by prime number, such as 17) to number of key field, 1269

$$17\overline{)1269} \; \begin{array}{r} 74 \\ \underline{119} \\ 79 \\ \underline{68} \\ 11 \end{array}$$

Indexed-sequential file organization: Indexed-sequential file organization stores records in sequential order as they are created, but the file in which the records are stored contains an index that lists each record by its key field and identifies its physical location on the disk.

Find 1269 ?

1. Want to find record with key field 1269

Main computer

PANEL 9.4

Three methods of file organization: sequential, direct, and indexed-sequential

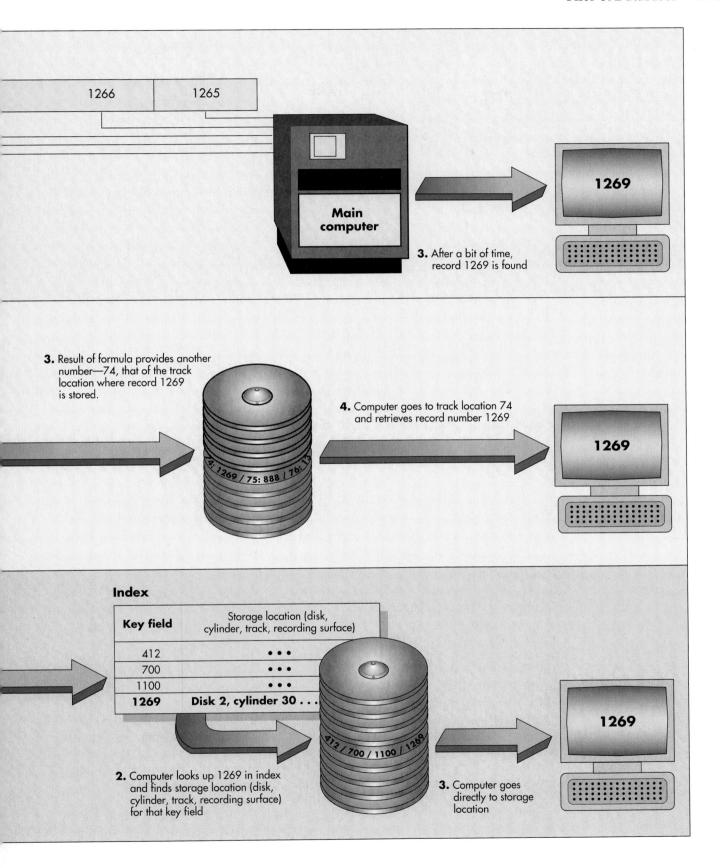

1266   1265

Main computer

3. After a bit of time, record 1269 is found

1269

3. Result of formula provides another number—74, that of the track location where record 1269 is stored.

4. Computer goes to track location 74 and retrieves record number 1269

74: 1269 / 75: 888 / 76: 1...

1269

Index

Key field	Storage location (disk, cylinder, track, recording surface)
412	• • •
700	• • •
1100	• • •
1269	**Disk 2, cylinder 30 . . .**

412 / 700 / 1100 / 1269

2. Computer looks up 1269 in index and finds storage location (disk, cylinder, track, recording surface) for that key field

3. Computer goes directly to storage location

1269

PANEL 9.5

File management system

In the traditional file management system, some of the same data elements, such as addresses, were repeated in different files. Information was not shared among files.

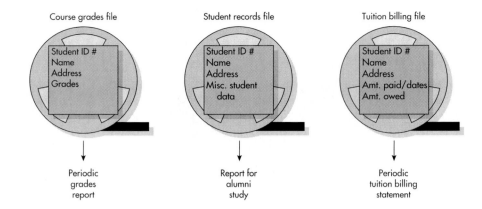

A file management system is illustrated above.(■ *See Panel 9.5.*)

As computers became more and more important in daily life, the frustrations of working with separate, redundant files lacking data integrity and program independence began to be overwhelming. Fortunately, magnetic disk began to supplant magnetic tape as the most popular medium of secondary storage, leading to new possibilities for managing data, which we discuss next.

Database Management Systems

Preview & Review: Database management systems are an improvement over file management systems. They use database management system (DBMS) software, which controls the structure of the database and access to the data. The advantages of databases are reduced data redundancy, improved data integrity, more program independence, increased user productivity, and increased security. However, installing and maintaining a database management system can be expensive.

When magnetic tape began to be replaced by magnetic disk, sequential access storage began to be replaced by direct access storage. The result was a new technology and new software: the database management system.

As mentioned, a *database* is a collection of integrated files, meaning that the file records are logically related, or cross-referenced, to one another. Thus, even though all the pieces of data on a topic are kept in records in different files, they can easily be organized and retrieved with simple requests.

The software for manipulating databases is ***database management system (DBMS) software,*** or a ***database manager, a program that controls the structure of a database and access to the data.*** With a DBMS, then, a large organization such as a university might still have different files for different purposes. As a student, you might have the same files as you would have had in a file management system (one for course grades, another for student records, and a third for tuition billing). However, in the database management system, data elements are integrated (cross-referenced) and shared among different files. (■ *See Panel 9.6.*) Thus, your address data would need to be in only one file, because it can be automatically accessed by the other files.

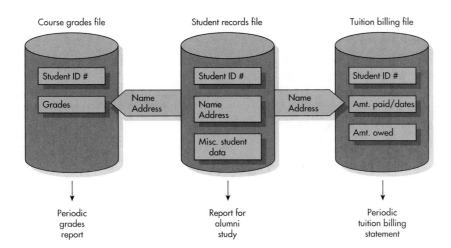

PANEL 9.6

Database management system

In the database management system, data elements are integrated and shared among different files.

Advantages & Disadvantages of a DBMS

The advantages of databases and DBMS software are as follows:

- **Reduced data redundancy:** Instead of the same data fields being repeated in different files, in a database the information appears just once. The single biggest advantage of a database is that the *same* information is available to *different* users. Moreover, reduced redundancy lowers the expense of storage media and hardware, because more data can be stored on the media.

- **Improved data integrity:** Reduced redundancy increases the chances of data integrity—that the data is accurate, consistent, and up to date—because each updating change is made in only one place.

- **More program independence:** With a database management system, the program and the file formats are the same, so that one programmer or even several programmers can spend less time maintaining files.

- **Increased user productivity:** Database management systems are fairly easy to use, so that users can get their requests for information answered without having to resort to technical manipulations. In addition, users don't have to wait for a computer professional to provide what they need.

- **Increased security:** Although various departments may share data in common, access to specific information can be limited to selected users. Thus, through the use of passwords, a student's financial, medical, and grade information in a university database is made available only to those who have a legitimate need to know.

Although there are clear advantages to having databases, there are still some disadvantages:

- **Cost issues:** Installing and maintaining a database is expensive, particularly in a large organization. In addition, there are costs associated with training people to use it correctly.

- **Security issues:** Although databases can be structured to restrict access, it's always possible unauthorized users will get past the safeguards. And when they do, they may have access to *all* the files, not just a few. In addition, if a database is destroyed by fire, earthquake, theft, or hardware or software problems, it could be fatal to an organization's business activities—unless steps have been taken to regularly make backup copies of the files and store them elsewhere.

- **Privacy issues:** Databases may hold information they should not and be used for unintended purposes, perhaps intruding on people's privacy. Medical data, for instance, may be used inappropriately in evaluating an employee for a job promotion. Privacy and other ethical issues are discussed later in this chapter.

Types of Database Organization

Preview & Review: Types of database organization are hierarchical, network, relational, and object-oriented.

Just as files can be organized in different ways (sequentially or directly, for example), so can databases. The four most common arrangements for database management systems are *hierarchical, network, relational,* and *object-oriented.* For installation and maintenance, each of the four types of database requires a database administrator trained in its structure.

Hierarchical Database

In a *hierarchical database,* **fields or records are arranged in related groups resembling a family tree, with lower-level records subordinate to higher-level records.** (■ *See Panel 9.7.*) A lower-level record is called a *child,* and a higher-level record is called a *parent.* The parent record at the top of the database is called the *root record.*

Unlike families in real life, a parent in a hierarchical database may have more than one child, but a child always has only one parent. This is called a one-to-many relationship. To find a particular record, you have to start at the top with a parent and trace down the chart to the child.

PANEL 9.7

Hierarchical database: example of a cruise ship reservation system

Records are arranged in related groups resembling a family tree, with "child" records subordinate to "parent" records. Cabin numbers (A-1, A-2, A-3) are children of the parent July 15. Sailing dates (April 15, May 30, July 15) are children of the parent The Love Boat. The parent at the top, Miami, is called the "root parent."

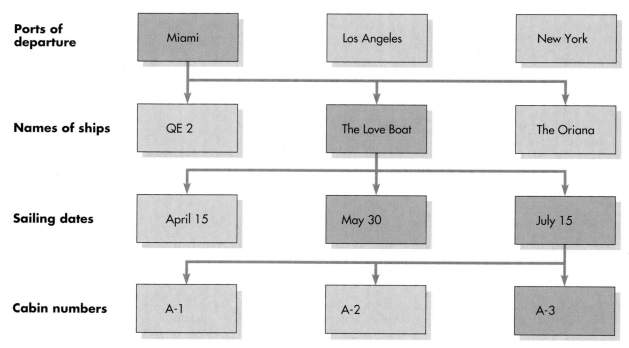

Ports of departure	Miami	Los Angeles	New York
Names of ships	QE 2	The Love Boat	The Oriana
Sailing dates	April 15	May 30	July 15
Cabin numbers	A-1	A-2	A-3

Hierarchical DBMSs are the oldest of the four forms of database organization, and are still used in some reservation systems. Also, accessing or updating data is very fast, because the relationships have been predefined. However, because the structure must be defined in advance, it is quite rigid. There may be only one parent per child and no relationships among the child records. Moreover, adding new fields to database records requires that the entire database be redefined.

Network Database

A *network database* is similar to a hierarchical DBMS, but each child record can have more than one parent record. (■ *See Panel 9.8.*) Thus, a child record, which in network database terminology is called a *member*, may be reached through more than one parent, which is called an *owner*.

This arrangement is more flexible than the hierarchical one, because different relationships may be established between different branches of data. However, it still requires that the structure be defined in advance. Moreover, there are limits to the number of links that can be made among all the records.

Relational Database

More flexible than hierarchical and network database models, **the *relational database* relates, or connects, data in different files through the use of a key field, or common data element.** (■ *See Panel 9.9, next page.*) In this arrangement there are no access paths down through a hierarchy. Instead, data elements are stored in different tables made up of rows and columns. In database terminology, the tables are called *relations*, the rows are called *tuples*, and the columns are called *attributes*.

Within a table, a row resembles a record—for example, a car license-plate number, which is one field, and the car owner's name and address, which is another field. All related tables must have a key field that uniquely identifies

PANEL 9.8

Network database: example of a college class scheduling system

This is similar to a hierarchical database, but each child, or "member," record can have more than one parent, or "owner." For example, Student B's owners are courses Broadcasting 210 and American History 101. The owner History Department has two members—American History 101 and European History 201.

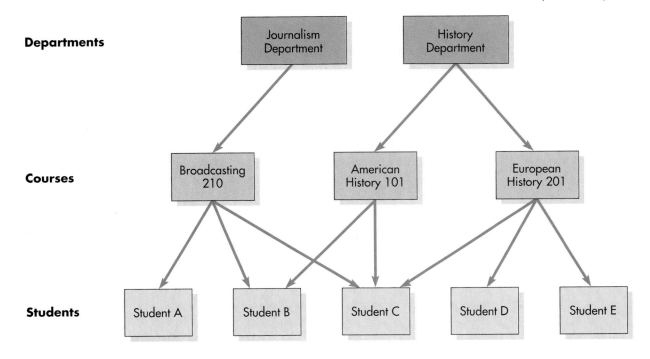

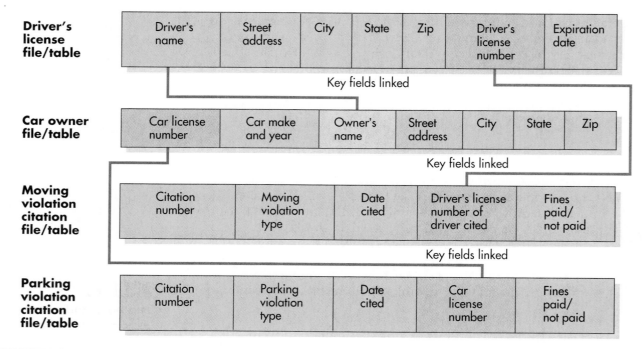

Driver's license file/table	Driver's name	Street address	City	State	Zip	Driver's license number	Expiration date

Key fields linked

Car owner file/table	Car license number	Car make and year	Owner's name	Street address	City	State	Zip

Key fields linked

Moving violation citation file/table	Citation number	Moving violation type	Date cited	Driver's license number of driver cited	Fines paid/ not paid

Key fields linked

Parking violation citation file/table	Citation number	Parking violation type	Date cited	Car license number	Fines paid/ not paid

PANEL 9.9

Relational database: example of a state department of motor vehicles database

This kind of database relates, or connects, data in different files through the use of a key field, or common data element. The relational database does not require predefined relationships.

each row. Thus, another table might have a row consisting of a driver's license number, the key field, and any traffic violations (such as speeding) attributed to the license holder. Another table would have the driver's license number and the bearer's name and address.

The advantage of relational databases is that the user does not have to be aware of any "structure." Thus, they can be used with little training. Moreover, entries can easily be added, deleted, or modified. A disadvantage is that some searches can be time consuming. Nevertheless, the relational model has become popular for microcomputer DBMSs, such as Paradox or Access.

Object-Oriented Database

Relational databases are good for storing and manipulating business data. Object-oriented databases can handle not only numerical and text data but any type of data, including graphics, audio, and video. Object-oriented databases, then, are important in the new world of video servers (✓ p. 317) and technological convergence.

An *object-oriented database* uses "objects" as elements within database files. An object consists of (1) text, sound, video, and pictures and (2) instructions on the action to be taken on the data. A hierarchical, network, or relational database would contain only numeric and text data about a student —identification number, name, address, and so on. By contrast, an object-oriented database might also contain the student's photograph, a "sound bite" of his or her voice, and even a short piece of video. Moreover, the object would store operations, called *methods*, that perform actions on the data— for example, how to calculate the student's grade-point average and how to display or print the student's record.

We describe object-oriented programming, on which this form of database is based, in Chapter 11.

Features of a DBMS

Preview & Review: Features of a database management system include (1) data dictionary, (2) utilities, (3) query language, (4) report generator, (5) access security, and (6) system recovery.

A database management system may have a number of components, including the following. (■ *See Panel 9.10.*)

Data Dictionary

Some databases have a *data dictionary,* **a file that stores the data definitions or a description of the structure of data used in the database.** The data dictionary may monitor the data being entered to make sure it conforms to the rules defined during data definition, such as field name, field size, type of data (text, numeric, date, and so on). The data dictionary may also help protect the security of the database by indicating who has the right to gain access to it.

Utilities

The *DBMS utilities* **are programs that allow the maintenance of the database** by creating, editing, and deleting data, records, and files. The utilities allow people to establish what is acceptable input data, to monitor the types of data being input, and to adjust display screens for data input.

Query Language

Also known as a *data manipulation language,* **a *query language* is an easy-to-use computer language for making queries to a database and for retrieving selected records,** based on the particular criteria and format indicated. Typically, the query is in the form of a sentence or near-English typed command, using such basic words as SELECT, DELETE, or MODIFY. There are several different query languages, each with its own vocabulary and procedures. One

PANEL 9.10 **Some important features of a database management system**

Component	Description
Data dictionary	Describes files and fields of data
Utilities	Help maintain the database by creating, editing, and monitoring data input
Query language	Enables users to make queries to a database and retrieve selected records
Report generator	Enables nonexperts to create readable, attractive on-screen or hardcopy reports of records
Access security	Specifies user access privileges
Data recovery	Enables contents of database to be recovered after system failure

of the most popular is *Structured Query Language*, or *SQL*. An example of an SQL query is as follows:

```
SELECT    PRODUCT-NUMBER,  PRODUCT-NAME
FROM      PRODUCT
WHERE     PRICE  <  100.00
```

This query selects all records in the product file for products that cost less than $100.00 and displays the selected records according to product number and name—for example, like this:

```
A-34     Mirror
C-50     Chair
D-168    Table
```

One feature of most query languages is *query by example*. Often a user will seek information in a database by describing a procedure for finding it. However, in **query by example (QBE), the user asks for information in a database by using a sample record to define the qualifications he or she wants for selected records.**

For example, a university's database of student-loan records of its students all over the United States and the amounts they owe might have the column headings (field names) NAME, ADDRESS, CITY, STATE, ZIP, AMOUNT OWED. When you use the QBE method, the database would display an empty record with these column headings. You would then type in the search conditions that you want in the appropriate columns.

Thus, if you wanted to find all Beverly Hills, California, students with a loan balance due of $3000 or more, you would type *BEVERLY HILLS* under the CITY column, *CA* under the STATE column, and *>=3000* ("greater than or equal to $3000") in the AMOUNT OWED column.

Report Generator

A *report generator* **is a program users may employ to produce an on-screen or printed-out document from all or part of a database.** You can specify the format of the report in advance—row headings, column headings, page headers, and so on. With a report generator, even nonexperts can create attractive, readable reports on short notice.

Access Security

At one point in the Michael Douglas/Demi Moore movie *Disclosure*, the Douglas character, the beleagured division head suddenly at odds with his company, types SHOW PRIVILEGES into his desktop computer, which is tied to the corporate network. To his consternation, the system responds by showing him downgraded from PRIOR USER LEVEL: 5 to CURRENT USER LEVEL: 0, shutting him out of files to which he formerly had access.

This is an example of the use of *access security*, a feature allowing database administrators to specify different access privileges for different users of a DBMS. For instance, one kind of user might be allowed only to retrieve (view) data, whereas another might have the right to update data and delete records. The purpose of this security feature, of course, is to protect the database from unauthorized access and sabotage.

System Recovery

Some advanced database management systems have a *system recovery* feature, which enables the DBA to recover contents of the database in the event of a hardware or software failure. For instance, the feature may recover transactions that appear to have been lost since the last time the system was backed up.

What's New in Database Management: Data Mining, Data Warehouses, & Data "Siftware"

Preview & Review: Data mining is the computer-assisted process of sifting through and analyzing vast amounts of data in order to extract meaning and discover new knowledge. Data taken from many sources is "scrubbed" or cleaned of errors and checked for consistency of formats. The cleaned-up data and a variation called *meta-data* are then sent to a special database called a *data warehouse*. Three kinds of software, or "siftware," tools are used to perform data mining: query-and-reporting tools, multidimensional-analysis tools, and intelligent agents.

A personal database is usually small-scale. But some efforts are going on with databases that are almost unimaginably large-scale, involving records of millions of households and thousands of terabytes of data. Some of these activities require the use of *massively parallel database computers* costing $1 million or more. "These machines gang together scores or even hundreds of the fastest microprocessors around," says one description, "giving them the oomph to respond in minutes to complex database queries."[10]

The efforts of which we speak go under the name *data mining*. Let us see what this is.

Data Mining: What It Is, What It's Used For

Data mining (DM), also called "knowledge discovery," is the computer-assisted process of sifting through and analyzing vast amounts of data in order to extract meaning and discover new knowledge. The purpose of DM is to describe past trends and predict future trends.[11] Although the definition seems simple enough, data mining has overwhelmed traditional query-and-report methods of organizing and analyzing data, such as those previously described in this chapter. The result has been the need for "data warehouses" and for new software tools, as we shall discuss.

Data mining has come about because companies find that, in today's fiercely competitive business environment, they need to turn the gazillions of bytes of raw data at their disposal to new uses for further profitability. However, nonprofit institutions have also found DM methods useful, as in the pursuit of medical and scientific discoveries. For example:[12]

- **Health:** A coach in the U.S. Gymnastics Federation is using a DM system (called *IDIS*) to discover what long-term factors contribute to an athlete's performance, so as to know what problems to treat early on. A Los Angeles hospital is using the same tool to see what subtle factors affect success and failure in back surgery. Another system helps health-care organizations pinpoint groups whose costs are likely to increase in the near future, so that medical interventions can be taken.

* **Science:** DM techniques are being employed to find new patterns in genetic data, molecular structures, global climate changes, and more. For instance, one DM tool (called *SKICAT*) is being used to catalog more than 50 million galaxies, which will be reduced to a 3-terabyte galaxy catalog.

Clearly, short-term payoffs can be dramatic. One telephone company, for instance, mined its existing billing data to identify 10,000 supposedly "residential" customers who spent more than $1000 a month on their phone bills. When it looked more closely, the company found these customers were really small businesses trying to avoid paying the more expensive business rates for their telephone service.[13]

However, the payoffs in the long term could be truly astonishing. Sifting medical-research data or subatomic-particle information may reveal new treatments for diseases or new insights into the nature of the universe.[14]

Preparing Data for the Data Warehouse

Data mining begins with acquiring data and preparing it for what is known as the "data warehouse." (■ *See Panel 9.11.*) This takes the following steps.[15]

* **Data sources:** Data may come from a number of sources: (1) point-of-sale transactions in files (flat files) managed by file management systems on mainframes, (2) databases of all kinds, and (3) other—for example, news articles transmitted over newswires or online sources such as the Internet. To the mix may also be added (4) data from other data warehouses.

* **Data fusion and cleansing:** Data from the diverse sources, whether from inside the company (internal data) or purchased from outside the company (external data), must be fused together, then put through a process known as *data cleansing,* or *scrubbing.*

  Even if the data comes from just one source, such as one company's mainframe, the data may be of poor quality, full of errors and inconsistencies. Therefore, for data mining to produce accurate results, the source data has to be "scrubbed"—that is, cleaned of errors and checked for consistency of formats.

* **Data and meta-data:** Out of the cleansing process comes both the cleaned-up data and a variation of it called *meta-data.* Meta-data shows the origins of the data, the transformations it has undergone, and summary information about it, which makes it more useful than the cleansed but unintegrated, unsummarized data. The meta-data also describes the contents of the data warehouse.

* **The data warehouse:** Both the data and the meta-data are sent to the data warehouse. **A *data warehouse* is a special database of cleaned-up data and meta-data.** It is a replica, or close reproduction, of a mainframe's data. The data warehouse is stored on disk using storage technology such as RAID (redundant arrays of independent disks, ✓ p. 301). Small data warehouses may hold 100 gigabytes of data or less. Once 500 gigabytes are reached, massively parallel processing (MPP) (✓ p. 161) computers are needed. Projections call for large data warehouses holding hundreds of terabytes within 5 years.

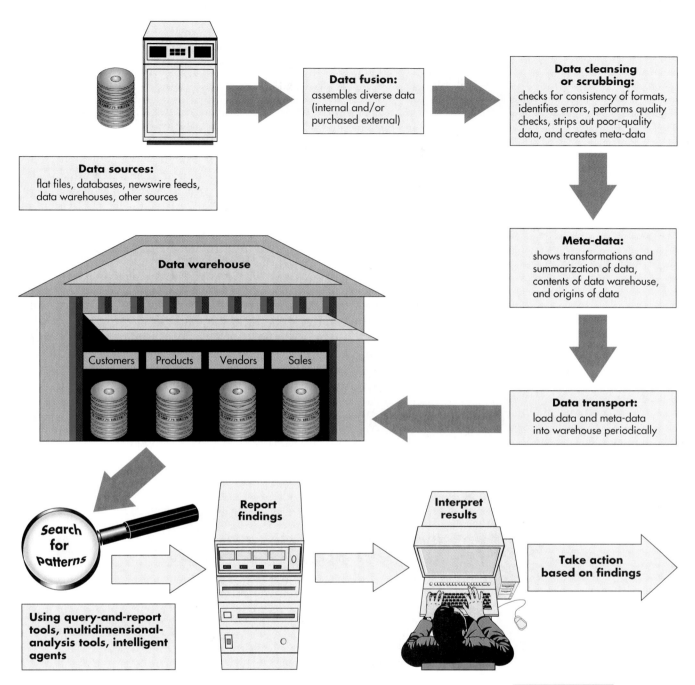

PANEL 9.11

The data-mining process

"Siftware" for Finding & Analyzing

Three kinds of software, or "siftware," tools are used to perform data mining—that is, to do finding and analyzing tasks. They are *query-and-reporting tools, multidimensional-analysis (MDA) tools,* and *intelligent agents.*[16]

- **Query-and-reporting tools:** Query-and-reporting tools (examples are Focus Reporter and Esperant) require a database structure and work well with relational databases. They may have graphical interfaces (✓ p. 115). Their best use is for specific questions to verify hypotheses.

For example, if a company decides to mine its database to find customers most likely to respond to a mail-order promotion, it might use a query-and-reporting tool and construct a query (using SQL): "How many credit-card customers who made purchases of over $100 on sporting goods in August have at least $2000 of available credit?"[17]

- **Multidimensional-analysis tools:** Multidimensional-analysis (MDA) tools (examples are Essbase and Lightship) can do "data surfing" to explore all dimensions of a particular subset of data.

  In one writer's example, "The idea [with MDA] is to load a multidimensional server with data that is likely to be combined. Imagine all the possible ways of analyzing clothing sales: by brand name, size, color, location, advertising, and so on."[18] Using MDA tools, you can analyze this multidimensional database from all points of view.

- **Intelligent agents:** An intelligent agent is a computer program that roams through networks performing complex work tasks for people. There are several kinds of intelligent agents (as we explore in Chapter 12), for instance as those used to prioritize e-mail messages for individuals. However, the kind we are concerned with here (such as DataEngine and Data/Logic) are those used as data-mining tools.

  Intelligent agents are best used for turning up unsuspected relationships and patterns. "These patterns may be so nonobvious as to appear almost nonsensical," says one writer, "such as that people who have bought scuba gear are good candidates for taking Australian vacations."[19]

Is data about you finding its way into data warehouses? No doubt it is. Gathering data isn't difficult. You participate in probably hundreds of transactions a year, recorded in point-of-sale terminals (✓ p. 206), teller machines, credit-card files, and 1-800 telemarketing responses. Sooner or later, some of the records of your past activities will be used, most likely by marketing companies, to try to influence you.

Ⓔ The Ethics of Using Databases: Concerns About Accuracy & Privacy

Preview & Review: Databases may contain inaccuracies or be incomplete. They also may endanger privacy—in the areas of finances, health, employment, and commerce.

"The corrections move by bicycle while the stories move at the speed of light," says Richard Lamm, a former governor of Colorado.

Lamm was lamenting that he was quoted out of context by a Denver newspaper in a speech he made in 1984. Yet even 10 years afterward—long after the paper had run a correction—he still saw the error repeated in later newspaper articles.[20]

How do such mistakes get perpetuated? The answer, suggests journalist Christopher Feola, is the Misinformation Explosion. "Fueled by the growing popularity of both commercial and in-house computerized news databases," he says, "journalists have found it that much easier to repeat errors or rely on the same tired anecdotes and experts."[21]

If news reporters—who are supposed to be trained in careful handling of the facts—can continue to repeat inaccuracies found in databases, what about those without training who have access to computerized facts? How can you be sure that databases with essential information about you—

medical, credit, school, employment, and so on—are accurate and, equally important, are secure in guarding your privacy? We examine the topics of *information accuracy and completeness* and of *privacy* in this section.

Matters of Accuracy & Completeness

Databases—including public databases such as Nexis, Lexis, Dialog, and Dow Jones News/Retrieval—can provide you with *more* facts and *faster* facts but not always *better* facts. Penny Williams, professor of broadcast journalism at Buffalo State College in New York and formerly a television anchor and reporter, suggests there are five limitations to bear in mind when using databases for research:[22]

- **You can't get the whole story:** For some purposes, databases are only a foot in the door. There may be many facts or aspects to the topic you are looking into that are not in a database. Reporters, for instance, find a database is a starting point, but it may take old-fashioned shoe leather to get the rest of the story.

- **It's not the gospel:** Just because you see something on a computer screen doesn't mean it's accurate. Numbers, names, and facts may need to be verified in other ways.

- **Know the boundaries:** One database service doesn't have it all. For example, you can find full text articles from the *New York Times* on Lexis/Nexis, from the *Wall Street Journal* on Dow Jones News/Retrieval, and from the *San Jose Mercury News* on America Online, but no service carries all three.

- **Find the right words:** You have to know which key words (search words) to use when searching a database for a topic. As Lynn Davis, a professional researcher with ABC News, points out, in searching for stories on guns, the key word "can be guns, it can be firearms, it can be handguns, it can be pistols, it can be assault weapons. If you don't cover your bases, you might miss something."[23]

- **History is limited:** Most public databases, Davis says, have information going back to 1980, and a few into the 1970s, but this poses problems if you're trying to research something that happened or was written about earlier.

Matters of Privacy

***Privacy* is the right of people to not reveal information about themselves.** Who you vote for in a voting booth and what you say in a letter sent through the U.S. mail are private matters. However, the ease with which databases and communications lines may pull together and disseminate information has put privacy under extreme pressure.

As you've no doubt discovered, it's no trick at all to get your name on all kinds of mailing lists. Theo Theoklitas, for instance, has received applications for credit cards, invitations to join video clubs, and notification of his finalist status in Ed McMahon's $10 million sweepstakes. Theo is a 6-year-old black cat who's been getting mail ever since his owner sent in an application for a rebate on cat food.[24] A whole industry has grown up of professional information gatherers and sellers, who collect personal data and sell it to fund-raisers, direct marketers, and others.

How easy is it to find out about you or anyone else? Using his home computer, journalist Jeffrey Rothfeder obtained Dan Quayle's credit report. (He buys his clothes mainly at Sears.) All Rothfeder had to do was pay an information seller $50 and type in the former vice president's name. He also found out from another data seller where anchorman Dan Rather shops. "This seller warmed to me quickly," Rothfeder said. "As a bonus, I was sent Vanna White's home phone number for free."[25] In an even worse case of invasion of privacy, a California man, obsessed with a woman he had once known, was able to hatch intricate schemes of harassment—from within a maximum-security prison. He filed post office change-of-address forms so her mail was forwarded to him in prison and obtained a credit report on her. He even sent the IRS forged power-of-attorney forms so he could get her tax returns.[26]

In the 1970s, the Department of Health, Education, and Welfare developed a set of five Fair Information Practices. These rules have since been adopted by a number of public and private organizations. The practices also led to the enactment of a number of laws to protect individuals from invasion of privacy. (■ *See Panel 9.12.*) Perhaps the most important U.S. law is the Federal Privacy Act, or Privacy Act of 1974. **The *Privacy Act of 1974* prohibits secret personnel files from being kept on individuals by government agencies or their contractors. It gives individuals the right to see their records, to see how the data is used, and to correct errors.** Another significant piece of legislation was the Freedom of Information Act, passed in 1970. **The *Freedom of Information Act* allows ordinary citizens to have access to data gathered about them by federal agencies.** Most privacy laws regulate only the behavior of government agencies or government contractors. For example, the *Computer Matching and Privacy Protection Act of 1988* prevents the government from comparing certain records to try to find a match. This law does not affect most private companies.

Of particular concern for privacy are the areas of finances, health, employment, commerce, and communications:

- **Finances:** Banking and credit are two private industries for which there are federal privacy laws on the books. **The *Fair Credit Reporting Act of 1970* allows you to have access to and gives you the right to challenge your credit records.** If you have been denied credit, this access must be given to you free of charge. **The *Right to Financial Privacy Act of 1978* sets restrictions on federal agencies that want to search customer records in banks.**

  In the past, credit bureaus have been severely criticized for disseminating errors and for having reports that were difficult for ordinary readers to understand. Although it may still not be easy to clear up a mistake, the industry has a dispute-resolution process that should make dealing with them less complicated. The major credit bureaus are TRW, Equifax, and Trans Union.

- **Health:** No federal laws protect medical records in the United States (except those related to treatment for drug and alcohol abuse and psychiatric care, or records in the custody of the federal government). Of course, insurance companies can get a look at your medical data, but so can others that you might not suspect. Getting a divorce or suing an employer for wrongful dismissal? A lawyer might subpoena your medical records in hopes of using, say, a drinking problem or medical care for depression against you. When employers have information about personal health, they often use it in making employment-related decisions, according to one study.[27]

  Your best strategy is to not routinely fill out medical questionnaires or histories. You should also not tell any business more than it needs to know about your health. You can ask your doctor to release only the minimum

Fair Information Practices

1. There must be no personal data record-keeping systems whose existence is a secret from the general public.

2. People have the right to access, inspect, review, and amend data about them that is kept in an information system.

3. There must be no use of personal information for purposes other than those for which it was gathered without prior consent.

4. Managers of systems are responsible and should be held accountable and liable for the reliability and security of the systems under their control, as well as for any damage done by those systems.

5. Governments have the right to intervene in the information relationships among private parties to protect the privacy of individuals.

Important Federal Privacy Laws

Freedom of Information Act (1970): Gives you the right to look at data concerning you that is stored by the federal government. A drawback is that sometimes a lawsuit is necessary to pry it loose.

Fair Credit Reporting Act (1970): Bars credit agencies from sharing credit information with anyone but authorized customers. Gives you the right to review and correct your records and to be notified of credit investigations for insurance or employment. A drawback is that credit agencies may share information with anyone they reasonably believe has a "legitimate business need." *Legitimate* is not defined.

Privacy Act (1974): Prohibits federal information collected about you for one purpose from being used for a different purpose. Allows you the right to inspect and correct records. A drawback is that exceptions written into the law allow federal agencies to share information anyway.

Family Educational Rights and Privacy Act (1974): Gives students and their parents the right to review, and to challenge and correct, students' school and college records; limits sharing of information in these records.

Right to Financial Privacy Act (1978): Sets strict procedures that federal agencies must follow when seeking to examine customer records in banks; regulates financial industry's use of personal financial records. A drawback is that the law does not cover state and local governments.

Privacy Protection Act (1980): Prohibits agents of federal government from making unannounced searches of press offices if no one there is suspected of a crime.

Cable Communications Policy Act (1984): Restricts cable companies in the collection and sharing of information about their customers.

Computer Fraud and Abuse Act (1986): Allows prosecution for unauthorized access to computers and databases. A drawback is that people with legitimate access can still get into computer systems and create mischief without penalty.

Electronic Communications Privacy Act (1986): Makes eavesdropping on private conversations illegal without a court order.

Computer Security Act (1987): Makes actions that affect the security of computer files and telecommunications illegal.

Computer Matching and Privacy Protection Act (1988): Regulates computer matching of federal data; allows individuals a chance to respond before government takes adverse actions against them. A drawback is that many possible computer matches are not affected, such as those done for law-enforcement or tax reasons.

Video Privacy Protection Act (1988): Prevents retailers from disclosing video-rental records without the customer's consent or a court order.

PANEL 9.12

The five Fair Information Practices and important federal privacy laws

The Fair Information Practices were developed by the U.S. Department of Health, Education, and Welfare in the early 1970s. They have been adopted by many public and private organizations since.

R E A D M E

Practical Matters: Tactics for Staying Off Mailing Lists

Erik Larson is author of The Naked Consumer, *which describes how consumers unknowingly "shed" information about themselves. That information then is collected and sold to commercial marketers, which is how we end up on mailing lists. Larson says his experience in researching and writing the book made him much more careful about divulging personal information. Here are his six tips for keeping what's left of your private life private.*

1. Don't give out your telephone number. "I'm even reluctant to put my phone number on a check, even if I'm asked to do so. It's none of their business."
2. Don't give your Social Security number to anyone unless required to do so by federal law. "Those have become the key to so many things, including telephone banking. And never, ever give it out to anyone over the phone."
3. Learn to say no to telemarketers. "I've become a lot ruder with them. They traffic in our polite upbringing, because we're all taught to be nice to people on the phone. But as soon as I hear that cheery 'Hi!' I say no and hang up."
4. On mail response cards, "always check the box that says don't send me anything else."
5. When filling out warranty cards (for recently purchased items), don't fill out the lifestyle survey. "They rely on the fact that a lot of people assume their warranty won't be valid unless they do."
6. Never give out your credit card number to any telemarketer who calls you on the telephone. "Always ask them to send you a bill."

—Erik Larson, cited in Martin J. Smith, "Tactics for Evading Nosey Marketers," *San Francisco Examiner*

amount of information that can be released. Finally, ask for a copy of your medical records if you have doubts about the information your doctor or hospital has on you.[28]

- **Employment:** Private employers are the least regulated by privacy legislation. If you apply for a job, for instance, a background-checking service may verify your educational background and employment history. It may also take a look at your credit, driving violations, workers' compensation claims, and criminal record if any.[29]

  Cellular phones, global positioning systems, and "active badges" (clip-on ID cards readable by infrared sensors throughout a building) can tell your employer where you are.[30] Software that counts keystrokes or tracks sales can monitor your productivity. E-mail memos may be read not only by you but perhaps by your boss, if company policy allows.[31]

- **Commerce:** As we've seen, marketers of all kinds would like to get to know you. For example, Virginia Sullivan, a retired school teacher, every month weighed the junk mail she received. She found after 11 months that she had received about 98 pounds worth. Sullivan also noticed that the junk mail companies seemed to know personal details of her private life, such as her age, buying habits, and favorite charities.

  "We constantly betray secrets about ourselves," says Erik Larson, author of *The Naked Consumer*, "and these secrets are systematically collected by the marketers' intelligence network."[32] Larson has a number of suggestions for avoiding putting yourself on mailing lists in the first place.

  With few exceptions, the law does not prohibit companies from gathering information about you for one purpose and using it without your permission for another.[33] This information is culled from both public information sources, such as driver's license records, and commercial transactions, such as warranty cards. One exception is the Video Privacy

Protection Act of 1988. **The *Video Privacy Protection Act* prevents retailers from disclosing a person's video rental records without their consent or a court order.**

"Somewhere along the way," Larson points out, "the data keepers made the arbitrary decision that everyone is automatically on their lists unless they ask to be taken off."[34] Why not have a law that keeps consumers off all lists unless they ask to be included? Congress has considered this approach, but lobbyists for direct marketing companies object that it would put them out of business.[35]

Privacy concerns don't stop with the use or misuse of information in databases. As we have seen (Chapter 8), they also extend to privacy in communications. Although the government is constrained by several laws on acquiring and disseminating information, and listening in on private conversations, privacy advocates still worry. In recent times, the government has tried to impose new technologies that would enable law-enforcement agents to gather a wealth of personal information. Proponents have urged that Americans must be willing to give up some personal privacy in exchange for safety and security. We discuss this matter in Chapter 12.

Onward

When Cynthia Schoenbrun was laid off as a research administrator from a computer software company, she found something even better. She used her own personal computer to link up with people she knew in Russia and become part of a new field known as *information brokering*.

What is an information broker? "Part librarian, part private eye, and part computer nerd," one writer explains, "an information broker searches for everything written and published on a given subject, be it an obscure corner of the biomedical market or the whereabouts of a German engineering expert."[36] Schoenbrun, for instance, searches computer databases and her network of contacts to find business and investment information about Russia and other countries formerly in the Soviet Union. She then sells this information to clients.

The majority of information brokers, who are mainly in one- or two-person firms, are people who have seen a chance to own a business without making a heavy investment. Among other advantages, the profession gives people a lot of flexibility in setting their own hours.

One need not become an information broker, however, to benefit from being able to search a database. Doctors, lawyers, and other professionals are turning to databases in order to keep up with the information explosion within their fields. In the new world of computers and communications, everyone should at least know the rudiments of this skill.

Finding Useful Online Databases: Directories & Search Engines[37–42]

The Web is swarming with *spiders, crawlers,* and *robots.* Can they help you find your way through the confusion of homepages and "what's cool" sites?

In the Experience Box at the end of Chapter 3, we briefly introduced directories and indexes that are helpful for getting useful information out of the 20 million pages contained on 20,000 servers or sites on the Internet's World Wide Web. Here we will show how to apply Web browsers such as Netscape Navigator, Mosaic, or Internet Explorer to use these search tools.

The key to exploring the Net and the Web, says *Fortune* personal computing writer Michael Martin, is to apply two simple concepts, both of which derive from methods we are accustomed to using for finding information in other areas of life: *browsing* and *hunting.*[43]

- *Directories—for browsing:* Browsing, says Martin, "involves looking in a general area of interest, then zooming in on whatever happens to catch your attention." For example, Martin says, a basketball fan would head for the sports section of the newspaper, check the basketball news and scores on the front page, then skim other pages for related sports information.

  Directories—Yahoo! is the best known—arrange resources by subject and thus are best for people who browse.

- *Search engines—for hunting:* Hunting "is what we do when we want specific information," Martin says. In his example, if you were hopelessly nearsighted and wanted to hunt up specifics on the latest advances in laser treatment, you might check with an ophthalmologist, a university library, the National Eye Institute, and so on.

  Search engines, or indexes, such as Lycos and Infoseek, are for those who want specifics.

If you use Netscape Navigator, you can quickly access directories by clicking the NET DIRECTORY button and access search engines by clicking the NET SEARCH button. We explain some of the principal directories and search engines below.

Directories: For Browsing

Directories provide lists of Web sites covering several categories. These are terrific tools if you want to find Web sites pertinent to a general topic you're interested in, such as bowling, heart disease, or the Vietnam War. For instance, in Yahoo! you might click your mouse on one of the general headings listed on the menu, such as Recreation or Health, then proceed to click on menus of subtopics until you find what you need.

Some general directories are the following:

- *Yahoo!* (http://www.yahoo.com/) is one of the most popular Web directories and lists perhaps 85,000 Net sites (not only Web pages but also Usenet newsgroups, Gophers, and FTP sites). Among other things, it features a weekly list of "cool sites" and headline news from the wire service Reuters.

- *The Whole Internet Catalog* (http://nearnet.gnn.com/gnn/wic) is easier to use than Yahoo! but less comprehensive.

- *McKinley Group's Magellan* (http://www.mckinley. com/) offers detailed overviews of more than 40,000 Web sites and brief descriptions of another 1 million sites. Overviews include short review, description of intended audience, and rating on a scale of one to four stars.

- *Argus/University of Michigan Subject-Oriented Clearinghouse* (http://www.lib.umich.edu/chhome.html) is a directory of directories. It provides a list of subject-specific directories on topics ranging from arts and entertainment to social science and social issues. For example, if you're interested in art, you would see it lists *World Wide Arts Resources* (http://www.concourse.com/wwar/default.html), which has links to many museums and galleries and an index of more than 2000 artists. (Besides material on the Web, Argus also lists information in FTP servers and Gopher sites.)

The more comprehensive a directory is, the more unwieldy it is to navigate.

Search Engines: For Specifics

Search engines are best when you're trying to find very specific information—the needle in the haystack. Search engines are Web pages containing forms into which you type keywords to suggest the subject you're searching for. The search engine then scans its database and presents you with a list of Web sites matching your search criteria.

The search engine's database is created by spiders (also known as *crawlers* or *robots*), software programs that scout the Web looking for new sites. When the spider finds a new page, it adds its Internet address (URL, or Universal Resource Locator, title, and usually the headers starting each section to an index in the search engine's database. The principal search engines—such as Lycos and Excite—add index information about new pages every day.

Writer Richard Scoville points out that the bigger the database, the greater your chances for success in your search. For example, he says, he queried several engines with the keywords *recipe wheat beer.* "The massive Lycos database gave us 437 *hits* (matched pages) in return. InfoSeek and Open Text Index gave us around 200 each; others, less than 100."[44]

Some principal search engines are as follows:

- *Lycos (http://www.lycos.com/)* offers a list of interesting Web sites called A2Z, which indicates the most popular pages on the Web, as measured by the number of hypertext links, or "hits," from other Web sites pointing to them. The Lycos database holds about 1.5 million fully indexed Web pages.

- *Excite NetSearch (http://www.excite.com/)* ties with Lycos in also having 1.5 million Web pages. Excite differs from other Web index services in that it returns not only a list of sites and articles in which the keywords you specified appear but also a list of relevant pages based on "concept" by analyzing words in a document. In addition, it ranks the documents as to how well they fit your original search criteria.[45]

- *Open Text Index (http://www.opentext.com:8080)* has 1.3 million Web pages. It returns especially useful results from single word searches, Martin thinks.

- *InfoSeek (http:www.infoseek.com/)* has 400,000 Web pages and ranks results according to relevance to your search criteria. InfoSeek also searches more than the Web, indexing Usenet newsgroups and several non-Internet databases.

The best way to make a search engine useful is to be extremely specific when formulating your keywords. More on this below.

Metasearch Engines

Metasearch engines are search tools that let you use several search engines to track down information, although you are somewhat restricted compared to when you use single search engines. Most metasearch engines also include directories, such as Yahoo! Examples of these "one-stop shopping" sites are Savvy Search (*http://www.cs.colostate.edu/~dreiling/smartform.html*) and Meta-Crawler (*http://www.cs.washington.edu/research/ projects/ ai/metacrawler/www/home.html*).

Tips for Searching

Here are some rules that will help improve your chances for success in operating a search engine:[46,47]

- *Read the instructions!* Every search site has an online search manual. Read it.

- *Make your keywords specific:* The more narrow or distinctive you can make your keywords, the more targeted will be your search. Say *drag racing* or *stock-car racing* rather than *auto racing*, for example. Also try to do more than one pass and try spelling variations: *drag racing, dragracing, drag-racing*. In addition, think of synonyms, and write down related key terms as they come to mind.

- *Use AND, OR, and NOT:* Use connectors as a way of making your keyword requests even more specific. In Martin's example, if you were looking for a 1996 Mustang convertible, you could search on the three terms "1996," "Mustang," and "convertible." However, since you want all three together, try linking them with a connector: "Mustang AND convertible AND 1996." You can also sharpen the keyword request by using the word Not for exclusion—for example, "Mustang NOT horse."

- *Don't bother with "natural language" queries:* Some search engines will let you do *natural language queries*, which means you can ask questions as you might in conversation. For example, you could ask, "Who was the Indianapolis 500 winner in 1996?" You'll probably get better results by entering "Indianapolis 500 AND race AND winner AND 1996."

- *Use more than one search engine:* "We found surprisingly little overlap in the results from a single query performed on several different search engines," writes Scoville. "So to make sure that you've got the best results, be sure to try your search with numerous sites."

All these search tools are constantly adding new features, such as easier interfaces. But whichever you end up using, you'll find that they can turn the Web from a playground or novelty into a source of real value.

SUMMARY

What It Is / What It Does

database management system (DBMS) (database manager) *(p. 414, LO 4)* Software that controls the structure of a database and access to the data; allows users to manipulate more than one file at a time (as opposed to file managers).

DBMS utilities *(p. 419, LO 6)* Programs that allow the maintenance of databases by creating, editing, and deleting data, records, and files.

direct access storage *(p. 410, LO 5)* Storage media that allows the computer direct access to a storage location without having to go through what's in front of it.

direct file organization (random file organization) *(p. 410, LO 5)* Also called *random file organization*; one of the three methods of file organization (along with sequential and indexed-sequential); records are stored in no particular sequence, and each record is retrieved according to its key field, or unique element of data. A mathematical formula, called a *hashing algorithm*, is used to produce a unique number that will identify the record's physical location on the disk.

distributed database *(p.404, LO 1)* Geographically dispersed database (located in more than one physical location). Users are connected to it through a client/server network.

Fair Credit Reporting Act of 1970 *(p. 426, LO 8)* U.S. law that allows people to have access to and gives them the right to challenge their credit records.

field *(p. 407, LO 2)* Unit of data consisting of one or more characters (bytes). An example of a field is your name, your address, *or* your Social Security number.

file *(p. 407, LO 2)* Collection of related records. An example of a file is collected data on everyone employed in the same department of a company, including all names, addresses, and Social Security numbers.

file-management system (file manager) *(p. 411, LO 4)* Software for creating, retrieving, and manipulating files, one file at a time.

Why It's Important

This software enables: sharing of data (same information is available to different users); economy of files (several departments can use one file instead of each individually maintaining its own files, thus reducing data redundancy, which in turn reduces the expense of storage media and hardware); data integrity (changes made in the files in one department are automatically made in the files in other departments); security (access to specific information can be limited to selected users).

DBMS utilities allow people to establish what is acceptable input data, to monitor the types of data being input, and to adjust display screens for data input.

Direct access storage (disk) is much faster than sequential storage (tape).

Direct file organization is much faster than sequential file organization for finding a specific record. However, this method requires direct-access (disk) storage, which is more expensive than magnetic tape. Moreover, it is not as efficient as sequential file organization for listing large numbers of records.

Data need not be centralized in one location.

If a person has been denied credit, the law allows such access free of charge.

A collection of fields make up a record. *Also see key field.*

Integrated files make up a database.

In the 1950s, magnetic tape was the storage medium and records and files were stored sequentially. File managers were created to work with these files. Today, however, database managers are more common.

What It Is / What It Does	**Why It's Important**

Freedom of Information (FOI) Act   *(p. 426, LO 8)*   U.S. law that allows ordinary citizens to have access to data gathered about them by federal agencies.

The FOI Act helps diminish any tendency for government agencies to exercise the kinds of powers over citizens that are found in dictatorships.

hierarchical database   *(p. 416, LO 6)*   One of the three arrangements for database management systems; fields or records are arranged in related groups resembling a family tree, with "child" records subordinate to "parent" records. A parent may have more than one child, but a child always has only one parent. To find a particular record, one starts at the top with a parent and traces down the chart to the child.

Hierarchical DBMSs work well when the data elements have an intrinsic one-to-many relationship, as might happen with a reservations system. The difficulty, however, is that the structure must be defined in advance and is quite rigid. There may be only one parent per child and no relationships among the child records.

indexed-sequential file organization   *(p. 410, LO 5)*   One of the three methods of file organization (along with direct and sequential); records are stored in sequential order. However, the file in which the records are stored contains an index that lists each record by its key field and identifies its physical location on the disk. This type of file organization can be done on disk but not tape. This method is slower than direct file organization because of the time taken for index searching.

The indexed-sequential method is used when large batches of transactions occasionally must be updated, yet users want frequent, rapid access to records. For example, bank customers and tellers want to have up-to-the-minute information about checking accounts, but every month the bank must update bank statements to send to customers.

key field   *(p. 408, LO 2)*   Field that contains unique data used to identify a record so that it can be easily retrieved and processed. The key field is often an identification number, Social Security number, customer account number, or the like. The primary characteristic of the key field is that it is *unique*.

Key fields are needed to identify and retrieve specific records in a database.

master file   *(p. 408, LO 3)*   Data file containing relatively permanent records that are generally updated periodically.

Master files contain relatively permanent information used for reference purposes. Master files are updated through the use of transaction files.

network database   *(p. 417, LO 6)*   One of the three common arrangements for database management systems; it is similar to a hierarchical DBMS, but each child record can have more than one parent record. Thus, a child record may be reached through more than one parent.

This arrangement is more flexible than the hierarchical one. However, it still requires that the structure be defined in advance. Moreover, there are limits to the number of links that can be made among records.

object-oriented database   *(p. 418, LO 6)*   Database structure that uses objects as elements within database files. An object consists of (1) text, sound, video, and pictures and (2) instructions on the action to be taken on the data.

In addition to textual data, an object-oriented database can store, for example, a person's photo, "sound bites" of her voice, and a video clip.

offline storage   *(p. 409, LO 3, 5)*   Refers to data that is not directly accessible for processing until a tape or disk has been loaded onto an input device.

The storage medium and data are not under the immediate, direct control of the central processing unit.

online processing   *(p. 409, LO 3)*   Also called *real-time processing*; means entering transactions into a computer system as they take place and updating the master files as the transactions occur; requires direct access storage.

Online processing gives users accurate information from an ATM machine or an airline reservations system, for example.

What It Is / What It Does	**Why It's Important**
online storage *(p. 409, LO 3, 5)* Refers to stored data that is directly accessible for processing.	Storage is under the immediate, direct control of the central processing unit; users need not wait for a tape or disk to be loaded onto an input device before they can access stored data.
privacy *(p. 425, LO 8)* Right of people to not reveal information about themselves.	The ease with which databases and communications lines may pull together and disseminate information has put privacy under extreme pressure.
Privacy Act of 1974 *(p. 426, LO 8)* U.S. law prohibiting government agencies and their contractors from keeping secret files on personnel.	The law gives individuals the right to see their records, see how the data is used, and correct errors.
program file *(p. 408, LO 3)* File containing software instructions.	This term is used to differentiate program files from data files.
query by example (QBE) *(p. 420, LO 6)* Feature of query-language programs whereby the user asks for information in a database by using a sample record to define the qualifications he or she wants for selected records.	QBE further simplifies database use.
query language *(p. 419, LO 6)* Easy-to-use computer language for making queries to a database and retrieving selected recods.	Query languages make it easier for users to deal with databases. To retrieve information from a database, users make queries—that is, they use a query language. These languages have commands such as SELECT, DELETE, and MODIFY.
record *(p. 407, LO 2)* Collection of related fields. An example of a record would be your name *and* address *and* Social Security number.	Related records make up a file.
relational database *(p. 417, LO 6)* One of the three common arrangements for database management systems; relates, or connects, data in different files through the use of a key field, or common data element. In this arrangement there are no access paths down through a hierarchy. Instead, data elements are stored in different tables made up of rows and columns. The tables are called *relations*, the rows are called *tuples*, and the columns are called *attributes*. Within a table, a row resembles a record. All related tables must have a key field that uniquely identifies each row.	The relational database is the most flexible arrangement. The advantage of relational databases is that the user does not have to be aware of any "structure." Thus, they can be used with little training. Moreover, entries can easily be added, deleted, or modified. A disadvantage is that some searches can be time consuming. Nevertheless, the relational model has become popular for microcomputer DBMSs.
report generator *(p. 420, LO 6)* Database management program users can employ to produce on-screen or printed-out documents from all or part of a database.	Report generators allow users to produce finished-looking reports without much fuss.
Right to Financial Privacy Act of 1978 *(p. 426, LO 8)* U.S. law that sets restrictions on federal agencies that want to search customer records in banks.	The law protects citizens from unauthorized snooping.

sequential access storage *(p. 410, LO 5)* Storage method, like magnetic tape, whereby data is stored in sequence, such as alphabetically.

With sequential access, the system must search through all the preceding data on a tape before reaching the desired item. This process may require running several inches or feet off a reel of tape, which takes time. Disk storage, by contrast, generally falls into the category of direct access storage.

sequential file organization *(p. 410, LO 5)* One of three methods of file organization; stores records in sequence, one after the other.

This is the only form of file organization that can be used with magnetic tape. Records can be retrieved only in the sequence in which they were stored. The method can also be used with disk. Sequential file organization is useful when a large portion of the records needs to be accessed, as when a mail-order house is sending out catalogs to all names on a mailing list. The method also is less expensive than other methods because it uses magnetic tape, which is cheaper than magnetic disk. The disadvantage of sequential file organization is that records must be ordered in a particular way and searched one at a time.

shared database *(p. 404, LO 1)* Database shared by users in one company or organization in one location.

Shared databases give all users in one organization access to the same information.

transaction file *(p. 409, LO 3)* Temporary data file that holds all changes to be made to the master file: additions, deletions, revisions.

The transaction file is used to periodically update the master file.

Video Privacy Protection Act *(p. 428, LO 8)* U.S. law that prevents retailers from disclosing a person's video rental records without their consent or a court order.

Although the prohibition applies to only one kind of industry, it still strengthens the privacy law in the United States.

(Selected answers appear at the back of the book.)

Short-Answer Questions

1. What is a collection of related fields called?
2. What is offline storage? online storage?
3. What is offline storage used for?
4. What is the main advantage of direct access storage (as compared to sequential storage)?
5. What is the main difference between a file-management system and a database management system?
6. What is a distributed database?
7. What is meant by the term *data mining*?
8. What is the difference between batch processing and online processing?
9. What is an object-oriented database?
10. What is a data dictionary? What does it do?

Fill-in-the-Blank Questions

1. The _____ is the person who coordinates all activities and needs relating to the company's database.
2. A 0 or 1 is called a(n) _____; 8 of these make up a _____, which represents a(n) _____.
3. In ascending order, the levels of data stored in a database are _____, _____, _____, _____, and _____.
4. A(n) _____ is a unit of data consisting of one or more characters.
5. A(n) _____ is a collection of integrated files.
6. A(n) _____ is a particular field chosen to uniquely identify a record so that it can be easily retrieved and processed.
7. _____ files contain software instructions; _____ files contain data.
8. Sequential access storage is mainly used on magnetic _____.

9. A(n) _____ is a program that users employ to produce printed or onscreen reports.
10. _____ is the right of individuals to not reveal information about themselves.

Multiple-Choice Questions

1. Which of the following can help speed up data retrieval?
   a. program file
   b. key field
   c. offline storage
   d. data file
   e. all of the above

2. Which of the following storage organizations stores records in sequential order and uses an index?
   a. sequential
   b. indexed
   c. indexed-sequential
   d. directly-indexed
   e. none of the above

3. Which of the following *isn't* a disadvantage of a file-management system?
   a. expensive to develop
   b. data redundancy
   c. lack of data integrity
   d. lack of program independence
   e. all of the above

4. Which of the following falls under the responsibility of the database manager?
   a. database design, implementation, and operation
   b. performance monitoring
   c. system security
   d. backup and recovery
   e. all of the above

5. Which of the following are parts of the data hierarchy?
   a. bit
   b. byte
   c. field
   d. file
   e. all of the above

6. Which of the following is an example of an information utility?
   a. personal database
   b. CompuServe
   c. offline storage
   d. direct-access storage
   e. none of the above

7. To retrieve information from a database, users can use a(n) _____, an easy-to-use computer language.
   a. data dictionary
   b. report generator
   c. data mining structure
   d. query language
   e. none of the above

8. To extract data from a database and display it or print it out in a preformatted form, users can use a(n) _____.
   a. DBMS utility
   b. QBE
   c. report generator
   d. shared database
   e. none of the above

9. Which of the following *doesn't* describe a method whereby files are organized?
   a. sequential
   b. direct
   c. indexed-sequential
   d. batch
   e. all of the above

10. Which of the following would you use to hold all changes to be made to a master data file?
    a. program file
    b. transaction file
    c. batch file
    d. online storage
    e. all of the above

True/False Questions

T F 1. Fields are larger than files.

T F 2. Bits and bytes don't relate to the data storage hierarchy.

T F 3. With online storage, data is directly accessible for processing.

T F 4. One of the disadvantages of database management systems is the cost of installing and maintaining the database.

T F 5. A database is a collection of integrated records.

T F 6. In addition to numerical and text data, relational databases can store sound, video, and pictures.

T F 7. DBMS utilities are programs used for maintaining a database.

T F 8. Information utilities are available online.

T F 9. In a direct file organization, records are stored in no particular sequence.

T F 10. A database administrator must be concerned with backup and recovery issues.

Projects/Critical-Thinking Questions

1. What types of databases are used at your school or business? Who has access to them? Who is responsible for keeping them current? How are the databases organized? Who is responsible for administering the database?

2. Describe a personal database that you would use to organize some aspect of your personal or business life. What fields would you include in the file structure? What types of queries would you like to perform on the database? Would you need to use relational operators? How often would data need to be updated?

3. Conferences dealing with database subjects are held frequently around the world. For example, a Microsoft Access developer's conference was held in February of 1996 in Toronto, Canada. What other conferences are held around the world? Are they geared towards computer professionals or are end-users welcome? Do you find any of the conferences interesting? Why/why not? Perform your research using current computer periodicals and/or the Internet.

4. Pretend you're in the process of obtaining insurance to cover the contents of your home. Using a micro-computer-based database application, create a database to store the list of items that you want to insure. At a minimum, you would want to include fields for an item's name, description, and approximate value. After creating the structure for the database, add at least ten records and print a copy of the database. Then create and print a report that includes a total of the value column.

5. Identify and describe a database used by a corporation, the government, or an educational institution that you think is mismanaged. (For example, the database may contain outdated data.) Describe the purpose of the database. Who are its intended users? Why do you think it is mismanaged? Is there an individual to blame? Will the problem(s) be fixed? Why/why not?

net Using the Internet

Objective: *In this exercise we lead you through using Netscape Navigator to download files from an FTP file server.*

Before you continue: *We assume you have access to the Internet through your university, business, or commercial service provider and to the Web browser tool named Netscape Navigator (depending on your software version, your screens may appear slightly different). We also assume you know how to connect to the Internet and then load Netscape Navigator. If necessary, ask your instructor or system administrator for assistance.*

1. Make sure you have started Netscape. The home page for Netscape Navigator should appear on your screen.

2. FTP (File Transfer Protocol) is an Internet program that enables you to transfer files from a host, or remote, computer to your computer. The process of copying a file from a host computer to your computer is referred to as *downloading*. Hundreds of companies, such as Microsoft Corporation, and other entities set up sites from which you can transfer files for free to your system; these sites are called *anonymous ftp* sites. (*Note:* When you access an anonymous ftp site using ftp software rather than Netscape, you actually have to log in as "anonymous" and then [usually] give your e-mail address as a password.) Many sites charge a fee or, more frequently, require you to have an account on their system in order to gain access to their files.

   In the following steps you use the FTP command to download a text file from Microsoft's FTP site to your computer.

   CLICK: Open button (![Open])

   TYPE: `ftp://ftp.microsoft.com`

   PRESS: [Enter] or CLICK: Open

After a few moments, the following screen should appear:

Directory   Compressed file

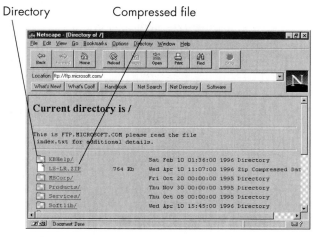

3. In this step you're going to copy to your computer the file named README.TXT that is stored in the Softlib directory.

   CLICK: Softlib/ link to open the Softlib directory

   CLICK: README.TXT file

   The contents of the README.TXT file should appear on the screen (see below).

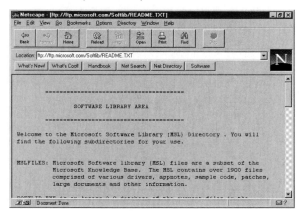

4. To save the file to your computer permanently:

   CHOOSE: File, Save As from the Menu bar

   A dialog box should appear with save options. Notice that the filename (README) already appears in the dialog box.

5. Choose a drive on your computer (for example, drive
   A: or drive C:) and then press (**Enter**). It's that easy!
   The file was copied to your computer.

6. In the previous steps you copied a plain text file to
   your computer. All text files have the extension of
   TXT. You can also copy non-text files. The following
   is a partial list of some of the other extensions you
   might encounter on an FTP file server.

File Extension	Description
.DOC	Document
.EXE	Executable program
.GIF	Graphic
.JPG	Graphic
.PCX	Graphic
.TIF	Graphic
.WAV	Sound Recorder
.AVI	Video
.HTM	Hypertext Markup Language
.ZIP	Compressed file

7. Display Netscape's home page and then exit
   Netscape.

CHAPTER 10

Information Systems

Information Management & Systems Development

Concepts You Should Know

After reading this chapter, you should be able to:

1. Describe six important trends in the workplace.

2. Identify organizational departments, tasks, and levels of managers and the types of information needed by different managers and workers.

3. Define and distinguish among several types of management information systems: TPS, MIS, DSS, EIS, and expert system.

4. Discuss the six phases of the systems development life cycle and how systems analysis and design fit in.

5. Explain several systems analysis tools: data flow diagrams, systems flowcharts, connectivity diagrams, Gantt charts, PERT charts, grid charts, and decision tables.

6. Define the following design tools: prototyping, CASE (computer-aided systems engineering), and project management software.

7. Compare four strategies for converting to a new system.

We're on the verge of what is perhaps the most radical redefinition of the workplace since the Industrial Revolution, with some tremendous benefits involved," says a longtime proponent of flexible work arrangements. "Yet the early signs are that corporations are as likely as not to mess this up."

The speaker, a management consultant, was referring to the changes, a trickle now turning into a tidal wave, brought about by the "mobile office"—also called the *virtual office* (✓ p. 353).[1] Part of the redefinition of the workplace comes with handing employees laptop computers with modems, portable phones, and beepers and telling them to work from their homes, cars, or customers' offices—virtually anywhere. Part of it involves the use of a grab bag of electronic information organizers, personal communicators, personal digital assistants, and similar gadgets that help untether people from a fixed office.

"Flex-time" shift hours and voluntary part-time telecommuting programs have been around for a few years. Unlike them, however, the new high-tech tools are forcing some profound changes in the way people work. Many people, of course, like the flexibility of a mobile office. However, others resent having to work at home or being unable to limit their work hours. One computer-company vice president worries about getting her staff to stop sending faxes to each other in the middle of the night. Some employees may work 90 hours a week and still feel as if they are falling short. In great part, this is because people's skills have not kept pace with technological trends.[2] At some point, a constant work lifestyle becomes counterproductive.

Trends Forcing Reengineering of the Workplace

Preview & Review: The trends of automation, downsizing and outsourcing, total quality management, and employee empowerment, among others, have forced organizations to give considerable thought to reengineering. Reengineering is the search for and implementation of radical change in business processes to achieve breakthrough results.

The virtual office is only one of several trends in recent years that are affecting the way we work. Others, most of which have been under way for some time, have also had a profound effect. They include, but are not limited to, the following:

- Automation
- Downsizing and outsourcing
- Total quality management
- Employee empowerment
- Reengineering

Automation

When John Diebold wrote his prophetic book *Automation* in the 1950s, the computer was nearly new. Yet Diebold predicted that computers would make many changes. First, he suggested, they would change *how* we do our jobs. Second, he thought, they would change the *kind* of work we do.[3] He was right, of course, on both counts. In the 1950s and 1960s, computers changed how factory work, for instance, was done. In the 1970s and 1980s, factory work itself began to decline as Western nations went from manufacturing economies to information economies.

Diebold's third prediction was that the technologies would change the *world* in which we work. "This is the beginning of the next great development in computers and automation," he says, "which has already begun in the 1990s."

Downsizing & Outsourcing

The word *downsizing* has two meanings. **First, *downsizing* means reducing the size of an organization by eliminating workers. Second, it means the movement from mainframe-based computer systems to systems linking smaller computers in networks.** There is some connection between the two.

As a result of automation and a dismal economy, in recent years many companies have had to downsize their staffs—lay off employees. In the process, they have, in business jargon, "flattened the hierarchy," reducing the levels and numbers of middle managers. Since, of course, much of the company's work still remains, this has forced the rest of the staff to take up the slack.

This situation has produced much of the other type of downsizing—the shift from larger computers to smaller ones. For instance, the secretary may be gone, but the secretarial work still remains. The lower-level or middle-level managers found that with personal computers they could accomplish much of this work. They also found that schedules no longer permitted them to ask the people with the mainframes in the "glass house"—the Information Systems Department—to do some of their work. They simply had to do it themselves, again using microcomputers and networks.

Downsizing has also led to another development: outsourcing. ***Outsourcing* is the contracting with outside businesses or services to perform the work once done by in-house departments.** The outside specialized contractors, whether janitors or computer-system managers, often can do the work more cheaply and efficiently.[4]

Total Quality Management

"Total quality management" became the buzzword of the 1980s. ***Total quality management (TQM) is managing with an organization-wide commitment to continuous work improvement and satisfaction of customer needs.*** The group that probably benefited most from TQM principles was the American automobile makers, who had been devastated by better-made foreign imports. However, much of the rest of U.S. industry would probably also have been shut out of competition in the global economy without the quality strides made in the last few years.

In many cases, unfortunately, the push for quality became principally a matter of pursuing the narrow statistical benchmarks favored by TQM experts. This put considerable stress on employees, with no appreciable payoff in customer satisfaction or profitability. For example, originally FedEx pursued speed over accuracy in its sorting operation. However, it found the number of misdirected packages soared as workers scrambled to meet deadlines.[5] Now companies are looking for a better return on quality-management efforts.

Employee Empowerment

Empowerment **means giving employees the authority to act and make decisions on their own.** The old style of management was to give lower-level managers and employees only the information they "needed" to know, which minimized their power to make decisions. As a result, truly good work could not be achieved because of the "If it's not part of my job, I don't do it" attitude. Today's philosophy is that information should be spread widely, not closely held by top managers, to enable employees lower down in the organization to do their jobs better. Indeed, the availability of networks and groupware (✓ p. 78) has enabled the development of task-oriented teams of workers who no longer depend on individual managers for all decisions in order to achieve company goals.

The Virtual Office

As we mentioned earlier, the *virtual office* **is essentially a mobile office.** Using integrated computer and communications technologies, corporations will increasingly be defined not by concrete walls or physical space, but by collaborative networks linking hundreds, thousands, even tens of thousands of people together.[6] Widely scattered workers can operate as individuals or as if they were all at company headquarters. Such "road warriors" break the time and space barriers of the organization, operating anytime, anywhere.

Reengineering

Trends such as the foregoing force—or should force—organizations to face basic realities. Sometimes the organization has to actually *reengineer*—rethink and redesign itself or key parts of it. **Reengineering is the search for and implementation of radical change in business processes to achieve breakthrough results.** Reengineering, also known as *process innovation* and *core process redesign*, is not just fixing up what already exists. Says one description:

> Reengineers start from the future and work backward, as if unconstrained by existing methods, people, or departments. In effect they ask, "If we were a new company, how would we run this place?" Then, with a meat ax and sandpaper, they conform the company to their vision.[7]

Reengineering works best, then, with big processes that really matter, such as new-product development or customer service. Thus, candidates for this procedure include companies experiencing big shifts in their definition, markets, or competition. Examples are information technology companies—computer makers, cable-TV providers, and local and long-distance phone companies—which are wrestling with technological and regulatory change. An expensive software system commonly called SAP, for the company in Germany that produces it, is now available to help companies reengineer and standardize their information systems to give employees the data they need when they need it.

To understand how to bring about change in an organization, we need to understand how organizations work—the first topic of this chapter. We then describe tools for change: systems analysis and design.

Organizations: Departments, Tasks, Management Levels, & Types of Information

Preview & Review: Common departments in an organization are research and development, production, marketing, and accounting and finance.

The tasks of managers are planning, organizing, staffing, supervising (leading), and controlling. Managers occupy three levels of responsibility: top, middle, and lower.

Top managers make strategic decisions, using unstructured information. Middle managers make tactical decisions, using semistructured information. Lower managers make operational decisions, using structured information.

Consider any sizable organization you are familiar with. Its purpose is to perform a service or deliver a product. If it's nonprofit, for example, it may deliver the service of educating students or the product of food for famine victims. If it's profit-oriented, it may, for example, sell the service of fixing computers or the product of computers themselves.

Information—whether computer-based or not—has to flow within an organization in a way that will help managers achieve their goals. To this end, business organizations are often structured according to four departments.

Departments: R&D, Production, Marketing, Accounting

Depending on the services or products they provide, most organizations have departments that perform four functions. They are *research and development (R&D)*, *production*, *marketing*, and *accounting and finance*.

Some authorities add a fifth department—*human resources*, or *personnel*. This department finds and hires people and administers sick leave and retirement matters. It is also concerned with compensation levels and professional development. For simplicity, we will stick with just the first four departments.

- **Research and development:** The research and development (R&D) department does two things: (1) It conducts basic research, relating discoveries to the organization's current or new products. (2) It does product development, developing, testing, and modifying new products or services created by researchers.

- **Production:** The production department makes the product or provides the service. In a manufacturing company, it takes the raw materials and has people turn them into finished goods. In an operations company, it manages the purchasing, handles the inventories, and controls the flow of goods and services.

- **Marketing:** The marketing department oversees advertising, promotion, and sales. The people in this department plan, price, advertise, promote, sell, and distribute the services or goods to customers or clients.

- **Accounting and finance:** The accounting and finance department handles all financial matters. It pays bills, issues paychecks, records payments, and compiles financial statements from time to time. It produces financial budgets and forecasts financial performance.

Whatever the organization—grocery store, computer maker, law firm, hospital, or university—it is likely to have departments corresponding to these.

Each department has managers and employees. Although office automation brought about by computers, networks, and groupware has given employees more decision-making power than they used to have, managers still perform five basic functions.

Management Tasks: Five Functions

Certain specific duties are associated with being a manager. **Management is overseeing the tasks of planning, organizing, staffing, supervising, and controlling business activities.** These five functions, considered the classic tasks of management, are defined as follows:

- **Planning**—setting objectives, both long-term and short-term, and developing strategies for achieving them. Whatever you do in planning lays the groundwork for the other four tasks.
- **Organizing**—making orderly arrangements of resources, such as people and materials.
- **Staffing**—selecting, training, and developing people. In some cases, it may be done by specialists, such as those in the personnel department.
- **Supervising (leading)**—directing or guiding employees to work toward achieving the organization's goals.
- **Controlling**—monitoring the organization's progress toward achieving its goals.

All managers perform all these tasks as part of their jobs. However, the level of responsibility regarding these tasks varies with the level of the manager, as we discuss next.

Management Levels: Three Levels, Three Kinds of Decisions

How do managers carry out the tasks just described? They do it by *making decisions on the basis of the information available to them.* A manager's daily job is to decide on the best course of action, based on the facts known at the time.

For each of the four departments there are three traditional levels of management—top, middle, and lower. These levels are reflected in the organization chart. **An *organization chart* is a schematic drawing showing the hierarchy of formal relationships among an organization's employees.** (■ *See Panel 10. 1.)*

Managers on each of the three levels have different kinds of responsibility and are therefore required to make different kinds of decisions.

- **Top managers—strategic decisions:** The chief executive officer (CEO) or president is the very top manager. However, for our purposes, "top management" refers to the vice presidents, one of which heads each of the four departments.

  Top managers are concerned with long-range planning. Their job is to make strategic decisions. *Strategic decisions* are complex decisions rarely based on predetermined routine procedures, involving the subjective judgment of the decision maker. *Strategic* means that, of the five management tasks (planning, organizing, staffing, supervising, controlling), top managers are principally concerned with *planning.*

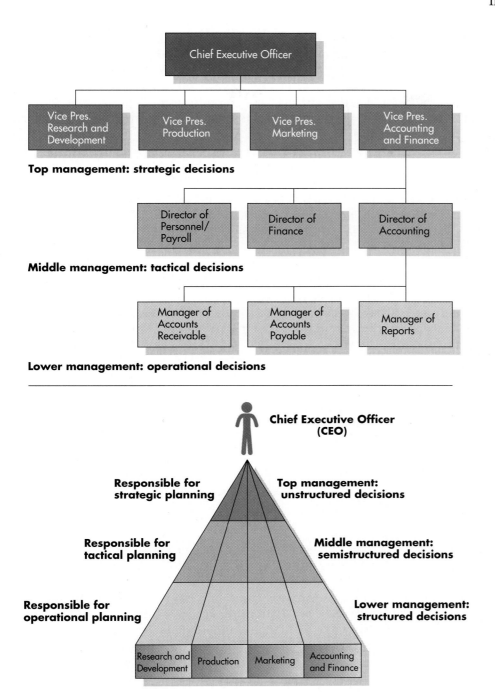

PANEL 10.1

Management levels and responsibilities

(Top) An organization generally has four departments: research and development, production, marketing, and accounting and finance. This organization chart shows the management hierarchy for just one department, accounting and finance. Three levels of management are shown—top, middle, and lower. *(Bottom)* The entire organization can also be represented as a pyramid, with the four departments and three levels of management as shown. Top managers are responsible for strategic decisions, middle management for tactical decisions, and lower management for operational decisions.

Office automation is changing the flow of information in many organizations, thus "flattening" the pyramid, because not all information continues to flow through traditional hierarchical channels.

Besides CEO, president, and vice president, typical titles found at the top management level are treasurer, controller, and senior partner. Examples of strategic decisions are how growth should be financed and what new markets should be tackled first. Other strategic decisions are deciding the company's 5-year goals, evaluating future financial resources, and deciding how to react to competitors' actions.

An AT&T vice president of marketing might have to make strategic decisions about promotional campaigns to sell a new paging service. The top manager who runs an electronics store might have to make strategic decisions about stocking a new line of paging devices.

- **Middle managers—tactical decisions:** *Middle-level managers* implement the goals of the organization. Their job is to oversee the supervisors and to make tactical decisions. A *tactical decision* is a decision that must be made without a base of clearly defined informational procedures, perhaps requiring detailed analysis and computations. *Tactical* means that, of the five management tasks, middle managers deal principally with *organizing* and *staffing.*

    Examples of middle managers are plant manager, division manager, sales manager, branch manager, and director of personnel. An example of a tactical decision is deciding how many units of a specific product should be kept in inventory. Another is whether or not to purchase a larger computer system.

    The director of sales, who reports to the vice president of marketing for AT&T, sets sales goals for district sales managers throughout the country. They in turn feed him or her weekly and monthly sales reports.

- **Lower or supervisory managers—operational decisions:** *Lower-level managers*, or *supervisory managers*, manage or monitor nonmanagement employees. Their job is to make operational decisions. An *operational decision* is a predictable decision that can be made by following a well-defined set of routine procedures. *Operational* means these managers focus principally on *supervising (leading)* and *controlling.* They monitor day-to-day events and, if necessary, take corrective action.

    An example of a supervisory manager is a warehouse manager in charge of inventory restocking. An example of an operational decision is one in which the manager must choose whether or not to restock inventory. (The guideline on when to restock may be determined at the level above.)

    A district sales manager for AT&T would monitor the promised sales and orders for pagers coming in from the sales representatives. When sales begin to drop off, the supervisor would need to take immediate action.

Types of Information: Unstructured, Semistructured, & Structured

To make the appropriate decisions—strategic, tactical, operational—the different levels of managers need the right kind of information: structured, semistructured, and unstructured. *(■ See Panel 10.2.)*

In general, *all* information to support intelligent decision making at all three levels must be correct—that is, accurate. It must also be complete, including *all* relevant data, yet concise, including *only* relevant data. It must be cost effective, meaning efficiently obtained, yet understandable. It must be current, meaning timely, yet also time sensitive, based on historical, current, or future information needs. This shows that information has three distinct properties:

1. Level of summarization
2. Degree of accuracy
3. Timeliness

These properties may vary in the degree to which they are structured or unstructured, depending on the level of management and type of decision making required. **Structured information is detailed, current, concerned with past events; requires highly accurate, nonsubjective data; records a narrow range of facts; and covers an organization's internal activities.** Unstructured information is the opposite. **Unstructured information is summarized, less**

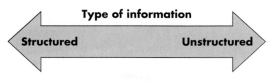

Top management

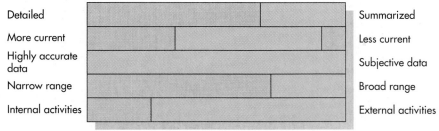

Detailed	Summarized
More current	Less current
Highly accurate data	Subjective data
Narrow range	Broad range
Internal activities	External activities

Time period covered: The future

Middle management

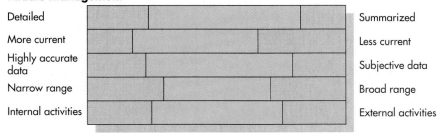

Detailed	Summarized
More current	Less current
Highly accurate data	Subjective data
Narrow range	Broad range
Internal activities	External activities

Time period covered: Comparative

Lower management

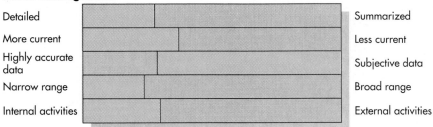

Detailed	Summarized
More current	Less current
Highly accurate data	Subjective data
Narrow range	Broad range
Internal activities	External activities

Time period covered: The past

PANEL 10.2

Types of information: the structured-unstructured continuum

Top managers need information that is unstructured. Unstructured information is summarized, less current, highly subjective; covers a broad range of facts; and is concerned with events outside as well as inside the organization.

Lower-level managers need information that is structured. Structured information is detailed, more current, not subjective; covers a narrow range of facts; and is concerned principally with events inside the organization.

Middle managers require information that is semistructured.

current, concerned with future events; requires subjective data; records a broad range of facts; and covers activities outside as well as inside an organization. *Semistructured information* includes some structured information and some unstructured information.

The illustration on the next page shows the information that the three levels of management might deal with in two businesses. (■ *See Panel 10.3.*)

Now that we've covered some basic concepts about how organizations are structured and what kinds of information are needed at different levels of management, we need to examine what types of management information systems provide the information.

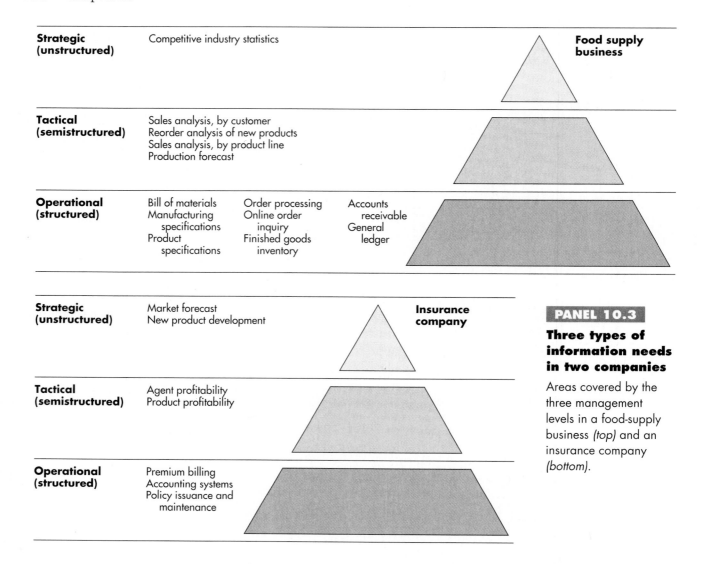

| Strategic (unstructured) | Competitive industry statistics | | | **Food supply business** |

| Tactical (semistructured) | Sales analysis, by customer
Reorder analysis of new products
Sales analysis, by product line
Production forecast | | | |

| Operational (structured) | Bill of materials
Manufacturing specifications
Product specifications | Order processing
Online order inquiry
Finished goods inventory | Accounts receivable
General ledger | |

| Strategic (unstructured) | Market forecast
New product development | | **Insurance company** |

| Tactical (semistructured) | Agent profitability
Product profitability | |

| Operational (structured) | Premium billing
Accounting systems
Policy issuance and maintenance | |

PANEL 10.3

Three types of information needs in two companies

Areas covered by the three management levels in a food-supply business *(top)* and an insurance company *(bottom)*.

Management Information Systems

Preview & Review: Five types of computer-based information systems provide information for decision making.

Transaction processing systems assist lower managers and workers in making operational decisions.

Management information systems help middle managers to make tactical decisions.

Decision support systems and executive information systems support top managers in making strategic decisions.

Expert systems are used at all levels for specific problems.

Top managers make strategic decisions using unstructured information, as we have seen. Middle managers make tactical decisions using semistructured information. Lower-level managers make operational decisions using structured information. The purpose of a computer-based information system is to provide managers (and various categories of employees) with the appropriate kind of information to help them make decisions.

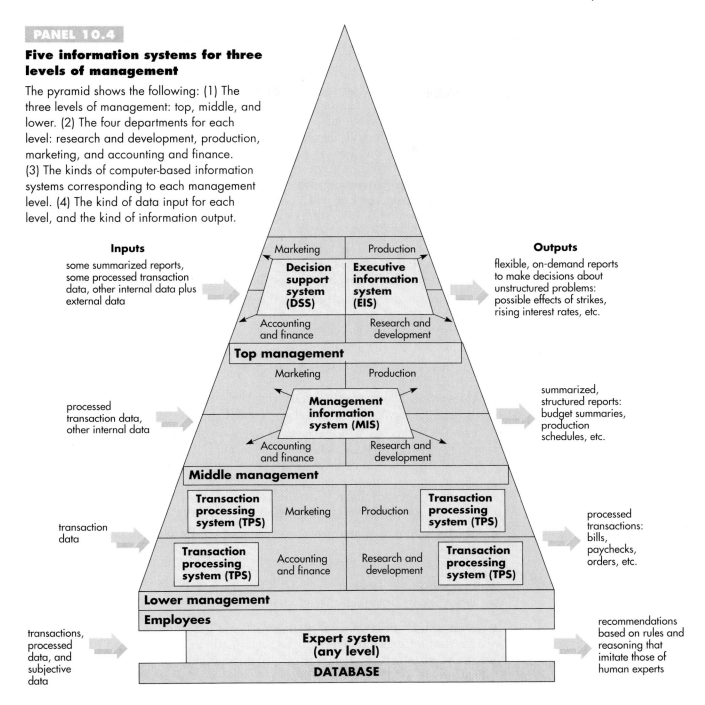

Five information systems for three levels of management

The pyramid shows the following: (1) The three levels of management: top, middle, and lower. (2) The four departments for each level: research and development, production, marketing, and accounting and finance. (3) The kinds of computer-based information systems corresponding to each management level. (4) The kind of data input for each level, and the kind of information output.

Here we describe the following types of computer-based information systems, corresponding to the three management layers and their requirements. Note that we are taking this material *from the bottom up,* because the higher levels build on the lower levels. (■ *See Panel 10.4.)*

* *For lower managers:* Transaction processing systems (TPSs)
* *For middle managers:* Management information systems (MISs)
* *For top managers:* Decision support systems (DSSs) and executive information systems (EISs)
* *For all levels including nonmanagement:* Expert systems

Transaction Processing System: To Support Operational Decisions

In most organizations, particularly business organizations, most of what goes on takes the form of transactions. **A *transaction* is a recorded event having to do with routine business activities.** This includes everything concerning the product or service in which the organization is engaged: production, distribution, sales, orders. It also includes materials purchased, employees hired, taxes paid, and so on. Today in most organizations, the bulk of such transactions are recorded in a computer-based information system.

A *transaction processing system (TPS)* **is a computer-based information system that keeps track of the transactions needed to conduct business.** Some features of a TPS are as follows:

- **Input and output:** The inputs to the system are transaction data: bills, orders, inventory levels, and the like. The output consists of processed transactions: bills, paychecks, and so on.

- **For lower managers:** Because the TPS deals with day-to-day matters, it is principally of use to supervisory managers. That is, the TPS helps in making *operational* decisions. Such systems are not usually helpful to middle or top managers.

- **Produces detail reports:** A lower-level manager typically will receive information in the form of detail reports. **A *detail report* contains specific information about routine activities.** An example might be the information needed to decide whether to restock inventory.

- **One TPS for each department:** Each department or functional area of an organization—Research & Development, Production, Marketing, and Accounting & Finance—usually has its own TPS. For example, the Accounting & Finance TPS handles order processing, accounts receivable, inventory and purchasing, accounts payable, and payroll.

- **Basis for MIS and DSS:** The database of transactions stored in a TPS is used to support a management information system and a decision support system.

Management Information System: To Support Tactical Decisions

A *management information system (MIS)* **is a computer-based information system that derives data from all an organization's departments and produces routine reports of the organization's performance.** Features of a MIS are as follows:

- **Input and output:** Inputs consist of processed transaction data, such as bills, orders, and paychecks, plus other internal data. Outputs consist of summarized, structured reports: budget summaries, production schedules, and the like.

- **For middle managers:** A MIS is intended principally to assist middle managers. That is, it helps them with *tactical* decisions. It enables them to spot trends and get an overview of current business activities.

- **Draws from all departments:** The MIS draws from all four departments or functional areas, not just one.

- **Produces several kinds of reports:** Managers at this level usually receive information in the form of several kinds of reports: *summary, exception, periodic, on-demand.*

Summary reports **show totals and trends.** An example would be a report showing total sales by office, by product, by salesperson, or total overall sales.

Exception reports **show out-of-the-ordinary data.** An example would be an inventory report that lists only those items that number fewer than 10 in stock.

Periodic reports **are produced on a regular schedule.** These may be daily, weekly, monthly, quarterly, or annually. They may contain sales figures, income statements, or balance sheets. Such reports are usually produced on paper, such as computer printouts.

On-demand reports **produce information in response to an unscheduled demand.** A director of finance might order an on-demand credit-background report on a new customer who wants to place a large order. On-demand reports are often produced on a terminal or microcomputer screen rather than on paper.

Decision Support System: To Support Strategic Decisions

A *decision support system (DSS)* **is a computer-based information system that provides a flexible tool for analysis and helps managers with nonroutine decision-making tasks.** To reach the DSS level of sophistication in information technology, an organization must have established a transaction processing system and a management information system.

Some features of a DSS are as follows:

- **Input and output:** Inputs consist of some summarized reports, some processed transaction data, and other internal data. They also include data that is external to that produced by the organization. This external data may be produced by trade associations, marketing research firms, the U.S. Bureau of the Census, and other government agencies.

  The outputs are flexible, on-demand reports with which a top manager can make decisions about unstructured problems.

- **Mainly for top managers:** A DSS is intended principally to assist top managers, although it is now being used by other managers, too. Its purpose is to help them make *strategic* decisions—decisions about unstructured problems, often unexpected and nonrecurring. These problems may involvm the effect of events and trends outside the organization. Examples are rising interest rates or a possible strike in an important materials-supplying industry.

- **Produces analytic models:** The key attribute of a DSS is that it uses *models.* **A model is a mathematical representation of a real system.** The models use a DSS database, which draws on the TPS and MIS files, as well as outside data. The system is accessed through DSS software.

  The model allows the manager to do a simulation—play a "what if" game—to reach decisions. Thus, the manager can simulate an aspect of the organization's environment in order to decide how to react to a change in conditions affecting it. By changing the hypothetical inputs to the model—number of workers available, distance to markets, or whatever—the manager can see how the model's outputs are affected.

Case Study: Silver Screens

United Artists Theatrical Circuit, Inc., uses state-of-the-art network technology to optimize the performance of its information system.

[Recently], it was *Forrest Gump* and *The Lion King*. But [now it was] a heat wave stretching from the Pacific Coast to New England that [was] driving folks to the movies in almost unprecedented torrents. Leave it to the biggest movie-house chain in the nation to harness . . . computing systems to deal with the waves—and the heat.

United Artists Theatrical Circuit, operator of 450 theaters with a total of 2300 viewing screens, has become a subscriber to Accuweather, of College Station, PA, and is taking hourly weather reports for the entire country. The weather data is downloaded into computers at UA headquarters, where it is sliced and diced a half-dozen experimental ways. In one test, movie-house air conditioning is automatically and remotely adjusted based on such variables as the predicted temperature, actual temperature, time-dependent utility rates, tickets sold, and time of film showing. It's just the latest in a series of information strategies UA employs to stay ahead of the pack.

"In the movie-theater business, you have so many unpredictable variables; you try to quantify anything you can," says Madeline Calabrese, vice president of information systems (IS) at the theater chain's Denver headquarters. With a remarkably small information technology staff (25, including Calabrese), her company is building a tightly integrated nationwide computer system that automatically updates popcorn orders through EDI [electronic data interchange, ✓ p. 351], provides theater managers with daily time and attendance data, and turns around new applications within weeks after a business plan is hatched. . . .

Calabrese, who started her career as a computer specialist in the New York University law library, joined UA in 1991 and moved to Denver when the theater chain moved its headquarters in 1992. The 37-year-old leader, an economist by education, hopes to continue growing as a manager and to get broader responsibilities in the coming years.

"UA has helped me cultivate a healthy respect for competitive planning," she says. "That's more glamorous even than a movie."

—Dennis Eskrow, "Silver Screen," *PC Week*

As communications becomes a more important component of an information system, so does a kind of DSS called group decision support systems. A *group decision support system (GDSS)* enables teams of co-workers to use networks of microcomputers to share information and cooperate on projects. A group decision support system is also called *workgroup computing.* By sharing ideas, workers can build consensus and arrive at decisions collaboratively. GDSSs are being found in fields ranging from banking and insurance to architectural design and newspaper publishing.

Executive Information Systems

An *executive information system (EIS)* is an easy-to-use DSS made especially for top managers and specifically supports strategic decision making. An EIS is also called an *executive support system (ESS).* It draws on data not only from systems internal to the organization but also from those outside, such as news services or market-research databases. An EIS might allow senior executives to call up predefined reports from their personal computers, whether desktops or laptops. They might, for instance, call up sales figures in many forms—by region, by week, by fiscal year, by projected increases. The EIS includes capabilities for analyzing data and doing "what if" scenarios.

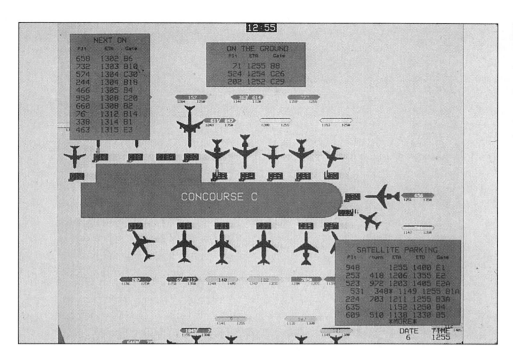

PANEL 10.5

Expert system screen

This screen is from a United Airlines gate assignment expert system, which analyzes airplane traffic to help workers assign gates to incoming planes.

Expert Systems

An *expert system,* **or knowledge system, is a set of computer programs that performs a task at the level of a human expert.** *(■ See Panel 10.5.)* Expert systems are created on the basis of knowledge collected on specific topics from human experts, and they imitate the reasoning process of a human being. Expert systems have emerged from the field of artificial intelligence, the branch of computer science that is attempting to create computer systems that simulate human reasoning and sensation. We describe artificial intelligence in more detail in Chapter 12.

Expert systems are used by both management and nonmanagement personnel to solve specific problems, such as how to reduce production costs, improve workers' productivity, or reduce environmental impact.

We have described how an organization and its managers work, as well as their information needs. Now let us see how organizational changes can be made to keep up with new demands through the use of systems analysis and design.

◣ The Six Phases of Systems Analysis & Design

Preview & Review: Knowledge of systems analysis and design helps you explain your present job, improve personal productivity, and lessen risk of a project's failure. The initiative for suggesting analysis and possibly change in an information system may come from users, managers, or technical staff.

The six phases of systems analysis and design are known as the *systems development life cycle (SDLC).* The six phases are (1) preliminary investigation, (2) systems analysis, (3) systems design, (4) systems development, (5) systems implementation, and (6) systems maintenance.

Organizations can make mistakes, and big organizations can make really big mistakes.

California's state Department of Motor Vehicles' databases needed to be modernized, and in 1988 Tandem Computers said it could do the job. "The fact that the DMV's database system, designed around an old IBM-based platform, and Tandem's new system were as different as night and day seemed insignificant at the time to the experts involved," said one writer investigating the project later.[8] The massive drivers' license database, containing the driving records of more than 30 million people, first had to be "scrubbed" of all information that couldn't be translated into the language used by Tandem Computers. One such scrub yielded 600,000 errors. Then the DMV had to translate all its IBM programs into the Tandem language. "Worse, DMV really didn't know how its current IBM applications worked anymore," said the writer, "because they'd been custom-made decades before by long-departed programmers and rewritten many times since." Eventually the project became a staggering $44 million loss to California's taxpayers.

In Denver, airport officials weren't trying to upgrade an old system but to do something completely new. At the heart of the Denver International Airport was supposed to be a high-tech baggage system. This was intended to whisk bags between terminals and at speeds that would mean that passengers would practically never have to wait for their luggage. As the system failed test after test, airport officials eventually decided they had to *build a manual baggage system*—at an additional cost of $50 million. Spending the money on old technology, it developed, was cheaper than continuing to spend millions paying interest on construction bonds for a nonoperating airport.[9]

Both these examples show how important planning is, especially when an organization is trying to launch a new kind of system. How can we avoid such mistakes? By employing the principles of systems analysis and design.

Why Know About Systems Analysis & Design?

But, you may say, you're not going to have to wrestle with problems on the scale of motor-vehicle departments and airports. That's a job for computer professionals. You're mainly interested in using computers and communications to increase your own productivity. Why, then, do you need to know anything about systems analysis and design?

In many types of jobs, you may find your department or your job the focus of a study by a systems analyst. Knowing how analysis and design works will help you better explain how your job works and what goals your department is supposed to achieve. In progressive companies, management is always interested in employees' suggestions for improving productivity. In some cases, employee input is required.

The Purpose of a System

Suppose you are managing a fleet of delivery trucks for a small family-owned business. When the drivers need to refuel their trucks, they come into the head office and borrow one of a number of gasoline credit cards. These cards are simply kept in an office desk drawer. You suspect that the reason fuel bills are so high is that drivers are also filling up their personal cars and charging the gas to the company. (A better idea would be to open an account with one local gas station. You could then direct the gasoline seller to bill you only for filling the company's trucks.)

Is this a system? It certainly is. **A *system* is defined as a collection of related components that interact to perform a task in order to accomplish a**

goal. A system may not work very well, but it is nevertheless a system. The point of systems analysis and design is to ascertain how a system works and then take steps to make it better.

An organization's computer-based information system consists of hardware, software, people, procedures, and data, as well as communications setups. These work together to provide management with information for running the organization.

From time to time, organizations need to change their information systems. The reasons may be new marketing opportunities, changes in government regulations, introduction of new technology, merger with another company, or other changes. The company may be as big as a cable-TV company trying to set up a billing system for movies on-demand. Or it may be as small as a two-person graphic design business trying to change its invoice and payment system. When this happens, the time is ripe for applying the principles of systems analysis and design.

Getting the Project Going: How It Starts, Who's Involved

All it takes is a single individual who believes that something badly needs changing to get a system development project rolling. An employee may influence a supervisor. A customer or supplier may get the attention of someone in higher management. Top management on its own may decide to take a look at a system that seems to be inefficient. A steering committee may be formed to decide which of many possible projects should be worked on.

Participants in the project are of three types:

- **Users:** The system under discussion should *always* be developed in consultation with users, whether floor sweepers, research scientists, or customers. Indeed, inadequate user involvement in analysis and design can be a major cause of system failure.
- **Management:** Managers within the organization should also be consulted about the system.
- **Technical staff:** Members of the company's information systems department, consisting of systems analysts and software programmers, need to be involved. For one thing, they may well have to carry out and execute the project. Even if they don't, they may have to work with outsiders contracted to do the job.

Complex projects will require a systems analyst. **A *systems analyst* is an information specialist who performs systems analysis, design, and implementation.** His or her job is to study the information and communications needs of an organization and determine what changes are required to deliver better information to people who need it, when they need it. "Better" information means information that is accurate, timely, and useful. The systems analyst achieves this goal through the problem-solving method of systems analysis and design.

The Six Phases of Systems Analysis & Design

Systems analysis and design **is a six-phase problem-solving procedure for examining an information system and improving it.** The six phases make up what is called the *systems development life cycle.* **The systems development** *life cycle (SDLC)* is defined as the step-by-step process that many organizations follow during systems analysis and design.

Whether applied to a Fortune 500 company or a three-person engineering business, the phases in systems analysis and design are the same. *(■ See Panel 10.6.)*

1. **Preliminary investigation:** Conduct preliminary analysis, propose alternative solutions, and describe the costs and benefits of each solution. Submit a preliminary plan with recommendations.
2. **Systems analysis:** Gather data, analyze the data, and make a written report.
3. **Systems design:** Do a preliminary design and then a detailed design, and write a report.
4. **Systems development:** Acquire the hardware and software and test the system.
5. **Systems implementation:** Convert the hardware, software, and files to the new system and train the users.
6. **Systems maintenance:** Audit the system, and evaluate it periodically.

Phases often overlap, and a new one may start before the old one is finished. After the first four phases, management must decide whether to proceed to the next phase. *User input and review is a critical part of each phase.*

PANEL 10.6

The systems development life cycle

An SDLC typically includes six phases.

1. *Preliminary investigation:* Conduct preliminary analysis, propose alternative solutions, describe costs and benefits of each solution, and submit a preliminary plan with recommendations.
2. *Systems analysis:* Gather data, analyze the data, and make a written report.
3. *Systems design:* Make a preliminary design and then a detailed design, and write a report.
4. *Systems development:* Acquire the hardware and software and test the system.
5. *Systems implementation:* Convert the hardware, software, and files to the new system and train the users.
6. *Systems maintenance:* Audit the system, and evaluate it periodically.

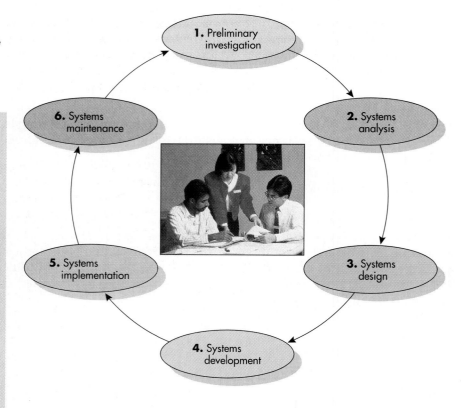

The First Phase: Conduct a Preliminary Investigation

Preview & Review: In the first phase, preliminary investigation, a systems analyst conducts a preliminary analysis, determining the organization's objectives and the nature and scope of the problems. The analyst then proposes some possible solutions, comparing costs and benefits. Finally, he or she submits a preliminary plan to top management, with recommendations.

The objective of Phase 1, *preliminary investigation*, is to conduct a preliminary analysis, propose alternative solutions, describe costs and benefits, and submit a preliminary plan with recommendations. *(■ See Panel 10.7.)*

If you are doing a systems analysis and design, it is safe, even preferable, to assume that you know nothing about the problem at hand. In the first phase, preliminary analysis, it is your job mainly to ask questions, do research, and try to come up with a preliminary plan.

1. Conduct the Preliminary Analysis

In this step, you need to find out what the organization's objectives are and the nature and scope of the problems under study.

- **Determine the organization's objectives:** Even if a problem pertains only to a small segment of the organization, you cannot study it in isolation. You need to find out what the objectives of the organization itself are. Then you need to see how the problem being studied fits in with them.

  To define the objectives of the organization, you can do the following:

  (1) *Read internal documents* about the organization. These can include original corporate charters, prospectuses, annual reports, and procedures manuals.

  (2) *Read external documents* about the organization. These can include news articles, accounts in the business press, reports by securities analysts, audits by independent accounting firms, and similar documents. You should also read reports on the competition (as in trade magazines, investors services' newsletters, and annual reports).

  (3) *Interview important executives* within the company. Within the particular area you are concerned with, you can also interview key users.

---

1. Conduct preliminary analysis. This includes stating the objectives, defining nature and scope of the problem.
2. Propose alternative solutions: leave system alone, make it more efficient, or build a new system.
3. Describe costs and benefits of each solution.
4. Submit a preliminary plan with recommendations.

PANEL 10.7

First phase: preliminary investigation

Some of this may be done face to face. However, if you're dealing with people over a wide geographical area you may spend a lot of time on the phone or using e-mail.

From these sources, you can find out what the organization is supposed to be doing and, to some extent, how well it is doing it.

- **Determine the nature and scope of the problems:** You may already have a sense of the nature and scope of a problem. This may derive from the very fact that you have been asked to do a systems analysis and design project. However, with a fuller understanding of the goals of the organization, you can now take a closer look at the specifics.

Is too much time being wasted on paperwork? on waiting for materials? on nonessential tasks? How pervasive is the problem within the organization? outside of it? What people are most affected? And so on. Your reading and your interviews should give you a sense of the character of the problem.

2. Propose Alternative Solutions

In delving into the organization's objectives and the specific problems, you may have already discovered some solutions. Other possible solutions can come from interviewing people inside the organization, clients or customers affected by it, suppliers, and consultants. You can also study what competitors are doing. With this data, you then have three choices. You can leave the system as is, improve it, or develop a new system.

- **Leave the system as is:** Perhaps the problem really isn't bad enough to take the measures and spend the money required to get rid of it. This is often the case.
- **Improve the system:** Maybe changing a few key elements in the system—upgrading to a new computer or new software, or doing a bit of employee retraining, for example—will do the trick. Efficiencies might be introduced over several months, if the problem is not serious.
- **Develop a new system:** If the existing system is truly harmful to the organization, radical changes may be warranted. A new system would not mean just tinkering around the edges, introducing a new piece of hardware or software. It could mean changes in every part and at every level.

3. Describe Costs & Benefits

Whichever of the three alternatives is chosen, it will have costs and benefits. In this step, you need to indicate what these are.

The changes or absence of changes will have a price tag, of course, and you need to indicate what it is. Costs may depend on benefits, which may offer savings. There are all kinds of benefits that may be derived.[10] A process may be speeded up, streamlined through elimination of unnecessary steps, or combined with other processes. Input errors or redundant output may be reduced. Systems and subsystems may be better integrated. Users may be happier with the system. Customers or suppliers may interact better with the system. Security may be improved. Costs may be cut.

4. Submit a Preliminary Plan

Now you need to wrap up all your findings in a written report. The readers of this report will be the executives (probably top managers) who are in a position to decide in which direction to proceed—make no changes, change a little, or change a lot. You should describe the potential solutions, costs, and benefits and indicate your recommendations.

The Second Phase: Do an Analysis of the System

Preview & Review: In the second phase, systems analysis, a systems analyst gathers data, using the tools of written documents, interviews, questionnaires, observation, and sampling. Next he or she analyzes the data, using data flow diagrams, system flow diagrams, and decision tables. Finally, the analyst writes a report.

The objective of Phase 2, *systems analysis,* is to gather data, analyze the data, and write a report. (■ *See Panel 10.8.*)

In this second phase of the SDLC, you will follow the course that management has indicated after having read your Phase 1 report. We are assuming that they have directed you to perform Phase 2—to do a careful analysis or study of the existing system in order to understand how the new system you proposed would differ. This analysis will also consider how people's positions and tasks will have to change if the new system is put into effect.

1. Gather Data

In gathering data, there are a handful of tools that systems analysts use, most of them not terribly technical. They include written documents, interviews, questionnaires, observation, and sampling.

• **Written documents:** A great deal of what you need is probably available in the form of written documents: reports, forms, manuals, memos, business plans, policy statements, and so on. Documents are a good place to

1. Gather data, using tools of written documents, interviews, questionnaires, observations, and sampling.
2. Analyze the data, using data flow diagrams, systems flowcharts, connectivity diagrams, grid charts, and decision tables.
3. Write a report.

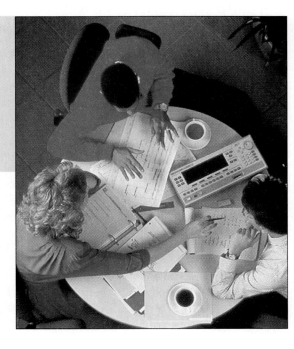

PANEL 10.8

Second phase: systems analysis

start because they at least tell you how things are or are supposed to be. These tools will also provide leads on people and areas to pursue further.

One document of particular value is the organization chart. An *organization chart* shows levels of management and formal lines of authority. We showed an example of an organization chart at the beginning of the chapter. *(Refer back to Panel 10.1, p. 447.)*

- **Interviews:** Interviews with managers, workers, clients, suppliers, and competitors will also give you insights. Interviews may be structured or unstructured.

  Structured interviews include only questions you have planned and written out in advance. By sticking with this script and not asking other questions, you can then ask people identical questions and compare their answers. *Unstructured interviews* also include questions prepared in advance, but you can vary from the line of questions and pursue other subjects if it seems productive.

- **Questionnaires:** Questionnaires are useful for getting information from large groups of people when you can't get around to interviewing everyone. Questionnaires may also yield more information because respondents can be anonymous. In addition, this tool is convenient, is inexpensive, and yields a lot of data. However, people may not return their forms, results can be ambiguous, and with anonymous questionnaires you'll have no opportunity to follow up.

- **Observation:** No doubt you've sat in a coffee shop or on a park bench and just done "people watching." This can be a tool for analysis, too. Through observation you can see how people interact with one another and how paper moves through an organization.

  Observation can be nonparticipant or participant. If you are a *nonparticipant observer*, and people know they are being watched, they may falsify their behavior in some way. If you are a *participant observer*, you may gain more insights by experiencing the conflicts and responsibilities of the people you are working with.

- **Sampling:** If your data-gathering phase involves a large number of people or a large number of events, it may simplify things to study just a sample. That is, you can do a sampling of the work of 5 people instead of 100, or 20 instances of a particular transaction instead of 500.

2. Analyze the Data

Once the data is gathered, you need to come to grips with it and analyze it. A variety of analytical tools, or modeling tools, are available. **Modeling tools enable a systems analyst to present graphic, or pictorial, representations of a system.** Five types of modeling tools are *data flow diagrams, systems flowcharts, connectivity diagrams, grid charts,* and *decision tables,* which are illustrated on the next five pages.

- **Data flow diagrams:** **A *data flow diagram (DFD)* graphically shows the flow of data through a system.** A DFD diagrams the processes that change data into information. (■ *See Panel 10.9.*)

  Data flow diagrams have only four symbols, which makes them easy to use. The first is the *entity* (or external entity) square, which shows the source or destination of data outside the system. The second is the *flow of data* or *vector* arrow, which indicates the path of the data. The third is

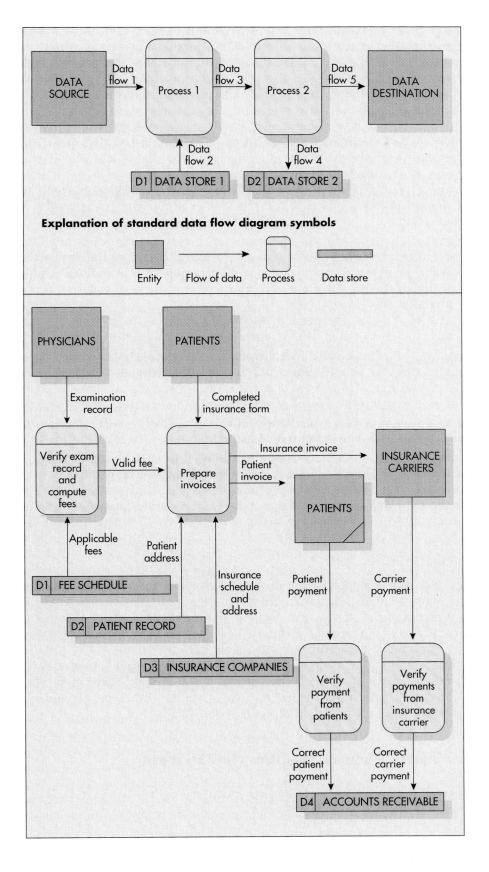

PANEL 10.9

Data flow diagram

(Top) Example of data flow diagram and explanation of symbols. *(Bottom)* Sample of data flow diagram of a physician's billing system.

the *process* rectangle with rounded corners, which indicates how input data is transformed into output. The fourth is the *data store* or *file* rectangle, which shows where data is held, whether filing cabinet or hard disk.

In analyzing the current system and preparing data flow diagrams, the systems analyst must also prepare a data dictionary, which is then used and expanded during all remaining phases of the systems development life cycle. **A *data dictionary* defines all the elements of data that make up the data flow.** Among other things, it records what each data element is by name, how long it is (how many characters), where it is used (files in which it will be found), as well as any numerical values assigned to it. This information is usually entered into a data dictionary (✓ p. 419) software program.

- **Systems flowcharts:** Another tool is the systems flowchart, also called the *system flow diagram.* **A *systems flowchart* diagrams the flow or input of data, processing, and output, or distribution of information.** Unlike a data flow diagram, a systems flowchart graphically depicts all aspects of a system. (■ *See Panel 10.10.*)

  (Note: A *systems* flowchart is not the same as a *program* flowchart, which is very detailed. We describe program flowcharts in Chapter 11.)

- **Connectivity diagrams:** **A *connectivity diagram* is used to map network connections of people, data, and activities at various locations.** (■ *See Panel 10.11, p. 466.*) Because connectivity diagrams are concerned with *communications networks,* we may expect to see these in increasing use.

- **Grid charts:** **A *grid chart* shows the relationship between data on input documents and data on output documents.** (■ *See Panel 10.12, p. 466.*)

- **Decision tables:** **A *decision table* shows the decision rules that apply when certain conditions occur and what actions to take.** That is, a decision table provides a model of a simple, structured decision-making case. It shows which *conditions* must take place in order for which *actions* to occur. (■ *See Panel 10.13, p. 467.*)

3. Write a Report

Once you have completed the analysis, you need to document this phase. This report to management should have three parts. First, it should explain how the existing system works. Second, it should explain the problems with the existing system. Finally, it should describe the requirements for the new system and make recommendations on what to do next.

At this point, not a lot of money will have been spent on the systems analysis and design project. If the costs of going forward seem to be prohibitive, this is a good time for the managers reading the report to call a halt. Otherwise, you will be called on to move to Phase 3.

The Third Phase: Design the System

Preview & Review: In the third phase, systems design, the analyst first does a preliminary design, next a detail design, then writes a report. The preliminary design may use prototyping and CASE tools and project management software. The detail design defines requirements for output, input, storage, and processing, as well as system controls and backup.

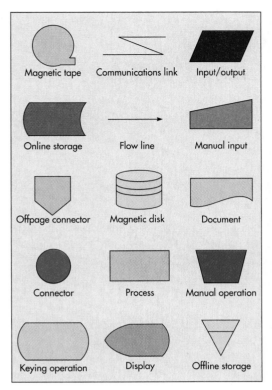

PANEL 10.10

Systems flowchart

(Left) Symbols. *(Right)* Example of a flowchart showing how inventory transactions are reflected in an updated master inventory file.

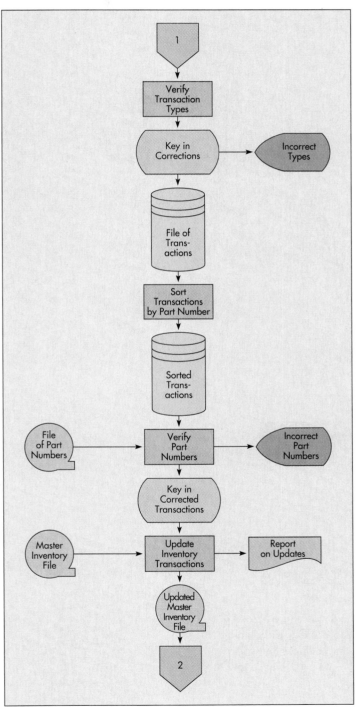

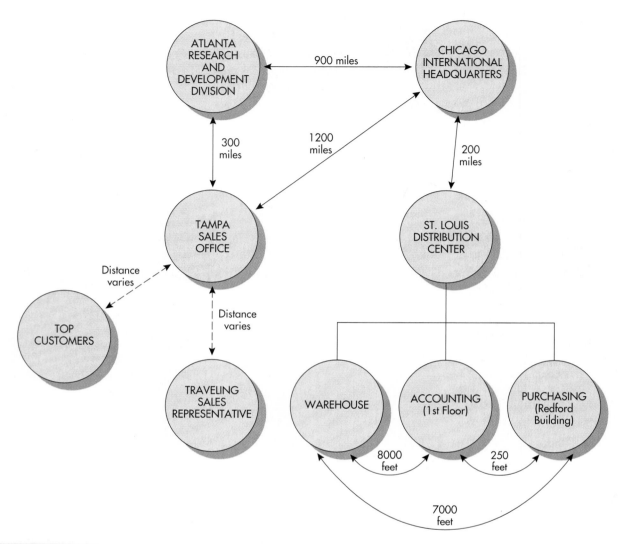

PANEL 10.11

Connectivity diagram

PANEL 10.12

Grid chart

This example shows the relationship between data inputs and data outputs that would appear on particular reports.

Forms (input)	Reports (output)		
	Report A	**Report B**	**Report C**
Form 1	✓	✓	
Form 2			✓
Form 3	✓	✓	

	Decision rules				
	1	**2**	**3**	**4**	**5**
Conditions If . . .	N	Y	Y	Y	N
And if . . .	Y	Y	N	Y	N
And if . . .	Y	Y	N	N	Y
Actions Then do . . .	✓				
Then do . . .			✓		✓
Then do . . .		✓		✓	

PANEL 10.13

Decision table

This type of table shows the decision rules that apply when certain conditions occur and what actions to take. In a real decision table the information in the left column would be specifically spelled out.

The objective of Phase 3, *systems design*, is to do a preliminary design and then a detail design, and write a report. *(■ See Panel 10.14.)*

In this third phase of the SDLC, you will essentially create a "rough draft" and then a "detail draft" of the proposed information system.

1. Do a Preliminary Design

A *preliminary design* describes the general functional capabilities of a proposed information system. It reviews the system requirements and then considers major components of the system. Usually several alternative systems (called *candidates*) are considered, and the costs and the benefits of each are evaluated.

Three tools that may be used in the preliminary design:

- **Prototyping tools:** *Prototyping* refers to building a working model or experimental version of all or part of a system so that it can be quickly tested and evaluated.

1. Do a preliminary design, using prototyping tools, CASE tools, and project management software.
2. Do a detail design, defining requirements for output, input, storage, and processing and system controls and backup.
3. Write a report.

PANEL 10.14

Third phase: systems design

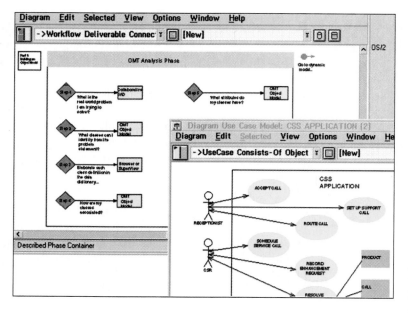

A *prototype* is a limited working system developed to test out design concepts. A prototype allows users to find out immediately how a change in the system might benefit them.

Prototypes are built with *prototyping tools.* These are special software packages that can be used to design screen displays. For example, a systems analyst might develop a menu (✓ p. 111) as a possible screen display, which users could try out. The menu can then be redesigned or fine-tuned, if necessary.

• **CASE tools:** CASE tools are another type of software tool. **CASE (for computer-aided software engineering) tools are software that provides computer-automated means of designing and changing systems.** There are many packages of such specialized software. (Sample names: Application Development Workbench, BACHMAN/Analyst, Excelerator, HyperAnalyst, Information Engineering Facility, PacBASE, System Architect.)

CASE tools may be used at almost any stage of the systems development life cycle, not just design. So-called *front-end CASE tools* are used during the first three phases—preliminary analysis, systems analysis, systems design—to help with the early analysis and design. So-called *back-end CASE tools* are used during two later stages—systems development and implementation—to help in coding and testing, for instance.

• **Project management software:** As we described in Chapter 2, **project management software consists of programs used to plan, schedule, and control the people, costs, and resources required to complete a project on time** (✓ p. 86). Project management software often uses Gantt charts and PERT charts.

A *Gantt chart* uses lines and bars to indicate the duration of a series of tasks. The time scale may range from minutes to years. The Gantt chart allows you to see whether tasks are being completed on schedule.

A *PERT (Program Evaluation Review Technique) chart* shows not only timing but also relationships among the tasks of a project. The relationships are represented by lines that connect boxes describing the tasks.

2. Do a Detail Design

A *detail design* describes how a proposed information system will deliver the general capabilities described in the preliminary design. The detail design usually considers the following parts of the system, in this order: *output requirements, input requirements, storage requirements, processing requirements,* and *system controls and backup.*

• **Output requirements:** What do you want the system to produce? That is the first requirement to determine. In this first step, the systems analyst determines what media the output will be—whether hardcopy and/or softcopy (✓ p. 246). He or she will also design the appearance or format of the output, such as headings, columns, menu, and the like.

• **Input requirements:** Once you know the output, you can determine the inputs. Here, too, you must define the type of input, such as keyboard or

R E A D M E

Case Study: Systematic Lunch & Dinner

They may evoke images of the Jetsons rather than personal service and culinary artistry, but computers and electronic systems are transforming the workings and ambiance of an increasing number of restaurants:

- The Prohost paging network, developed by Dallas-based Rock Systems, equips customers, waiters, cooks, and managers with wireless message devices, worn on the wrist, that Dick Tracy might envy. Among the possibilities: diners buzz their waiters for service, a computer tells the cooks that they're taking too long to prepare food, or the manager sends a "happy birthday" message to a customer.

- At Zoë in Manhattan, a computer is used to post the specials of the day and to keep track of what sold well the previous day. When waiters want to tell the kitchen to pay special attention to a VIP's order, they use the computer to mark the order "Elvis."

- The Dive, Steven Spielberg's new submarine-theme restaurant in Los Angeles, gives diners coasters with little red lights that blink to signal when their tables are ready. . . .

And so it is that a communications revolution is changing the way diners, waiters, kitchen staff, and management interact.

At the heart of these changes is the point-of-sale computer. . . . Typically, after taking a table's order, a waiter goes to a computer terminal and uses a keyboard or touch screen to enter the number of diners and their table location, the dishes selected, and any special instructions (medium rare, spicy, no sauce, etc.). The order is printed out in the kitchen, and the food is routed to the appropriate locations. An expediter, often the executive chef or the sous-chef, coordinates the preparation and assembly of the order, relying on the waiter to signal when it's time to set up each course. At the end of the meal, the waiter tells the computer to print out a check.

A smoothly running point-of-sale system improves efficiency in many ways. Waiters spend more time in the dining room attending to customers. Printed orders eliminate mixups caused by sloppy handwriting or shouted instructions, and when it's rush hour in the kitchen, diminished traffic is a blessing. "I prefer to work with the computer because chefs are often temperamental," said a waiter at La Colombe d'Or who preferred not to be identified.

From the restaurant owner's point of view, point-of-sale systems cut down on give-aways and forgotten charges, since all food must be ordered by computer. The diner gets a legible, accurate check, which can easily be split. Sales, tax, and tips are automatically tabulated and can be linked to back-office systems for accounting, payroll, and inventory. . . .

Designers of point-of-sale systems are increasingly moving to open systems, which allow restaurants to integrate additional features to manage their information flow. Matthew's, a restaurant at 61st Street and Third Avenue in Manhattan, recently started testing a guest-management system that tracks the status of each table and provides a computerized estimate of wait times for walk-ins. The system organizes reservations, keeps track of no-shows, and remembers the preferences of regular diners. . . .

Occasionally, unfamiliar technology causes confusion. At a TGI Friday's in Canton, Ohio, a button underneath the table caught the attention of a party of gangsters. The toughs suspected it was a listening device, yanked it out, stomped on it, and might have stomped on the manager, too, if he had not explained that the button was intended for the busboy to signal the computer when the table was cleared.

—David Karp, "Programming Lunch, From 'Table's Ready' To 'Here's Your Check,'" *New York Times*

---

source data entry (✓ p. 202). You must determine in what form data will be input and how it will be checked for accuracy. You also need to figure what volume of data the system can be allowed to take in.

- **Storage requirements:** Using the data dictionary as a guide, you need to define the files and databases in the information system. How will the files be organized? What kind of storage devices will be used? How will they interface with other storage devices inside and outside of the organization? What will be the volume of database activity?

- **Processing requirements:** What kind of computer or computers will be used to handle the processing? What kind of operating system will be used? Will the computer or computers be tied to others in a network? Exactly what operations will be performed on the input data to achieve the desired output information?

- **System controls and backup:** Finally, you need to think about matters of security, privacy, and data accuracy. You need to prevent unauthorized users from breaking into the system, for example, and snooping in people's private files. You need to have auditing procedures and set up specifications for testing the new system (Phase 4). You need to institute automatic ways of backing up information and storing it elsewhere in case the system fails or is destroyed.

3. Write a Report

All the work of the preliminary and detail designs will end up in a large, detailed report. When you hand over this report to senior management, you will probably also make some sort of presentation or speech.

The Fourth Phase: Develop the System

Preview & Review: The fourth phase, systems development, consists of acquiring software and hardware and then testing the system.

In Phase 4, *systems development,* the systems analyst or others in the organization acquire the software, acquire the hardware, and then test the system. (■ *See Panel 10.15.*)

The fourth phase begins once management has accepted your report containing the design and has "greenlighted" the way to development. This is the phase that will involve the organization in probably spending substantial sums of money. It could also involve spending a lot of time. However, at the end you should have a workable system.

1. Acquire Software

During the design stage, the systems analyst may have had to address what is called the "make-or-buy" decision, but that decision certainly cannot be avoided now. In the *make-or-buy decision,* you decide whether you have to

PANEL 10.15

Fourth phase: systems development

1. Acquire software.
2. Acquire hardware.
3. Test the system.

create a program—have it custom-written—or buy it, meaning simply purchase an existing software package. Sometimes programmers decide they can buy an existing program and modify it rather than write it from scratch.

If you decide to create a new program, then the question is whether to use the organization's own staff programmers or hire outside contract programmers. Whichever way you go, the task could take many months.

(Programming is an entire subject unto itself, and we address it in Chapter 11).

2. Acquire Hardware

Once the software has been chosen, the hardware to run it must be acquired or upgraded. It's possible your new system will not require obtaining any new hardware. It's also possible that the new hardware will cost millions of dollars and involve many items: microcomputers, minicomputers, mainframes, monitors, modems, and many other devices. The organization may find it's better to lease rather than to buy some equipment, especially since chip capability doubles about every 18 months. (The doubling of raw computing power every 18 months is known as *Moore's law*, a formula postulated years ago by Intel cofounder Gordon Moore.)

3. Test the System

With the software and hardware acquired, you can now start testing the system. Testing is usually done in stages called *unit testing*; then *system testing* is done.

- **Unit testing: In *unit testing*, individual parts of the program are tested, using test (made-up) data.** If the program is written as a collaborative effort by multiple programmers, each part of the program is tested separately.

- **System testing: In *system testing*, the parts are linked together, and test data is used to see if the parts work together.** At this point, actual organization data may also be used to test the system. The system is also tested with erroneous and massive amounts of data to see if it can be made to fail ("crash").

At the end of this long process, the organization will have a workable information system, one ready for the implementation phase.

The Fifth Phase: Implement the System

Preview & Review: The fifth phase, systems implementation, consists of converting the hardware, software, and files to the new system and of training the users. Conversion may proceed in four ways: direct, parallel, phased, or pilot.

Phase 5, *systems implementation*, consists of converting the hardware, software, and files to the new system and training the users. (■ *See Panel 10.16.*)

Whether the new information system involves a few handheld computers, an elaborate telecommunications network, or expensive mainframes, the fifth phase will involve some close coordination to make the system not just workable but successful.

1. Convert hardware, software, and files through one of four types of conversions: direct, parallel, phased, or pilot.
2. Train the users.

1. Convert to the New System

Conversion, **the process of converting from an old information system to a new one, involves converting hardware, software, and files.**

Hardware conversion may be as simple as taking away an old PC and plunking a new one down in its place. Or it may involve acquiring new buildings and putting in elaborate wiring, climate-control, and security systems.

Software conversion means making sure the applications that worked on the old equipment also work on the new.

File conversion means converting the old files to new ones without loss of accuracy. For example, can the paper contents from the manila folders in the personnel department be input to the system with a scanner? Or do they have to be keyed in manually, with the consequent risk of errors being introduced?

There are four strategies for handling conversion: *direct, parallel, phased,* and *pilot.* (■ *See Panel 10.17.*)

- **Direct approach:** *Direct implementation* **means the user simply stops using the old system and starts using the new one.** The risk of this method should be evident: What if the new system doesn't work? If the old system has truly been discontinued, there is nothing to fall back on.

- **Parallel approach:** *Parallel implementation* **means that the old and new systems are operated side by side until the new system has shown it is reliable, at which time the old system is discontinued.** Obviously there are benefits in taking this cautious approach. If the new system fails, the organization can switch back to the old one. The difficulty of this method is the expense of paying for the equipment and people to keep two systems going at the same time.

- **Phased approach:** *Phased implementation* **means that parts of the new system are phased in separately**—either at different times (parallel) or all at once in groups (direct).

- **Pilot approach:** *Pilot implementation* **means that the entire system is tried out but only by some users.** Once the reliability has been proved, the system is implemented with the rest of the intended users. The pilot approach still has its risks, since *all* of the users in a particular group are taken off the old system. However, the risks are confined to only a small part of the organization.

In general, the phased and pilot approaches are the most favored methods. Phased is best for large organizations in which people are performing different jobs. Pilot is best for organizations in which all people are performing the same task (such as order takers at a direct-mail house).

2. Train the Users

Back in the beginning of this book (Chapter 1), we pointed out that people are one of the important elements in a computer system. You wouldn't know this, however, to see the way some organizations have neglected their role

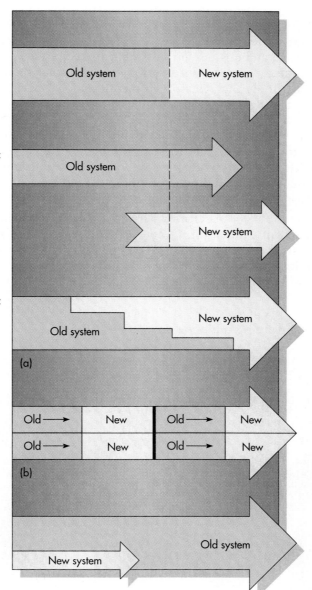

1. Direct implementation: all-at-once change

    Old system | New system

2. Parallel implementation: run at the same time

    Old system

    New system

3. Phased implementation: parts of the system are converted separately—
    (a) gradually or
    (b) in groups

    Old system | New system

    (a)

    Old → New | Old → New
    Old → New | Old → New

    (b)

4. Pilot implementation: tried first in only one part of the company

    New system | Old system

PANEL 10.17

Converting to a new system: four ways

Four strategies for converting to a new system are direct, parallel, phased, and pilot.

when implementing a new computer system. An information system, however, is no better than its users. Hence, training is essential.

Training is done with a variety of tools. They run from documentation (instruction manuals) to videotapes to live classes to one-on-one, side-by-side teacher-student training. Sometimes training is conducted by the organization's own staff; at other times it is contracted out.

The Sixth Phase: Maintain the System

Preview & Review: The last phase, systems maintenance, adjusts and improves the system through system audits and periodic evaluations.

Phase 6, *systems maintenance,* adjusts and improves the system by having system audits and periodic evaluations and by making changes based on new conditions. (■ *See Panel 10.18.*)

Even with the conversion accomplished and the users trained, the system won't just run itself. There is a sixth—and never-ending—phase in which the information system must be monitored to ensure that it is successful. Maintenance includes not only keeping the machinery running but also updating and upgrading the system to keep pace with new products, services, customers, government regulations, and other requirements.

Two tools that are sometimes considered part of the maintenance phase are *auditing* and *evaluation.*

*   **Auditing: *Auditing* means an independent review of an organization's information system to see if all records and systems are as they should be.** Often a systems analyst will design an *audit trail.* An audit trail helps independent auditors trace the record of a transaction from its output back through all processing and storage to its source.
*   **Evaluation:** Auditing, which is usually done by an accountant, is one form of evaluation. Other evaluations may be done by the systems analyst or other systems analysts. Evaluations may also be done by a user or client who is able to compare the workings of the system against some preset criteria.

The six steps in the systems development life cycle are summarized in the accompanying table. (■ *See Panel 10.19.*)

Sixth phase: systems maintenance

The sixth phase is to keep the system running through system audits and periodic evaluations.

Phase	Tasks
Phase 1: Preliminary investigation	1. Conduct preliminary analysis. This includes stating the objectives, defining nature and scope of the problem. 2. Propose alternative solutions: leave system alone, make it more efficient, or build a new system. 3. Describe costs and benefits of each solution.
Phase 2: Systems analysis	1. Gather data, using such tools as written documents, interviews, questionnaires, observation, and sampling. 2. Analyze the data, using data flow diagrams, systems flowcharts, connectivity diagrams, grid charts, and decision tables. 3. Write a report.
Phase 3: Systems design	1. Do a preliminary design, using prototyping tools, CASE tools, and project management software. 2. Do a detail design, defining requirements for output, input, storage, and processing and system controls backup. 3. Write a report.
Phase 4: Systems development	1. Acquire software. 2. Acquire hardware. 3. Test the system.
Phase 5: Systems implementation	1. Convert hardware, software, and files through one of four types of conversion: direct, parallel, phased, or pilot. 2. Train the users.
Phase 6: Systems maintenance	1. Audit the system. 2. Evaluate the system periodically.

PANEL 10.19

Summary of the systems development life cycle

Onward

At the Massachusetts Institute of Technology, a robot called the Collection Machine wanders aimlessly through a university lab at night picking up empty aluminum soda cans and disposing of them in a recycling bin. It does its work without a central computer processor or any human intervention.

Abstracting from this experience, MIT devised some guidelines for developing a system of distributed control: (1) Do simple things first. (2) Learn to do them flawlessly. (3) Add new layers of activity over the results of the simple tasks. (4) Don't change the simple things. (5) Make the new layer work as flawlessly as the simple. (6) Repeat, *ad infinitum*.[11]

Simply and flawlessly. Might these also be used as words to live by for programming? We examine this subject in the next chapter.

Critical Thinking Tools

Critical thinking is part of effective systems analysis and design. But what exactly is critical thinking? *Critical thinking* is the process of sorting out conflicting claims, weighing the evidence for them, letting go of personal biases, and arriving at reasonable views. Critical thinking means actively seeking to understand, analyze, and evaluate information in order to solve specific problems. The major barrier to this kind of thinking is having the habits of thought called *mindsets*—what people mean by a "closed mind."

The Reasoning Tool: Deductive & Inductive Arguments

The tool for breaking through mindsets is reasoning. *Reasoning*—giving reasons in favor of this assertion or that—is essential to critical thinking and solving problems. Reasoning is put in the form of *arguments*, which consist of one or more *premises*, or reasons, logically supporting a result or outcome called a *conclusion*.

An example of an argument is as follows:

Premise 1: All students must pass certain courses in order to graduate.
Premise 2: I am a student who wants to graduate.
Conclusion: Therefore, I must pass certain courses.

Note the tip-off word *therefore*, which signals that a conclusion is coming. In real life, such as arguments on radio and TV shows, in books and magazines and newspapers, and the like, the premises and conclusions are not so neatly labeled. Still, there are clues. The words *because, since,* and *for* usually signal premises. The words *therefore, hence,* and *so* signal conclusions. Not all groups of sentences form arguments. Often they form anecdotes or other types of exposition or explanation.[12]

The two main kinds of correct or valid arguments are inductive and deductive:

- *Deductive argument:* A deductive argument is defined as follows: *If its premises are true, then its conclusions are also true.* In other words, if the premises are true, the conclusions cannot be false.

- *Inductive argument:* An inductive argument is defined as follows: *If the premises are true, the conclusions are PROBABLY true, but the truth is not guaranteed.* An inductive argument is sometimes known as a "probability argument."

An example of a *deductive argument* is as follows:[13]

Premise 1: All students experience stress in their lives.
Premise 2: Reuben is a student.
Conclusion: Therefore, Reuben experiences stress in his life.

This argument is deductive—the conclusion is *definitely* true if the premises are *definitely* true.

An example of an *inductive argument* is as follows:[14]

Premise 1: Stress can cause illness.
Premise 2: Reuben experiences stress in his life.
Premise 3: Reuben is ill.
Conclusion: Therefore, stress may be the cause of Reuben's illness.

Note the word *may* in the conclusion. This argument is inductive—the conclusion is not stated with absolute certainty; rather, it suggests only that stress *may* be the cause. The link between premises and conclusion is not definite because there may be other reasons for Reuben's illness.

Some Types of Incorrect Reasoning

Patterns of incorrect reasoning are known as *fallacies*. Learning to identify fallacious arguments will help you avoid patterns of faulty thinking in your own writing and thinking and identify it in the writing of others.

Jumping to Conclusions Also known as *hasty generalization*, the fallacy called *jumping to conclusions* means that a conclusion has been reached when not all the facts are available.

Example: As a new manager coming into a company that instituted the strategy of total quality management (TQM) 12 months earlier, you see that TQM has not improved profitability this year. Thus, you order TQM junked in favor of more traditional business strategies. However, what you don't know is that the traditional business strategies employed prior to TQM had an even *worse* effect on profitability.

Irrelevant Reason or False Cause The faulty reasoning known as *non sequitur* (Latin for "It does not follow") might be better called *false cause* or *irrelevant reason*. Specifically, it means that the conclusion does not follow logically from the supposed reasons stated earlier. There is no *causal* relationship.

Example: You receive an A on a test. However, because you felt you hadn't been well prepared, you attribute your success to your friendliness with the professor. Or to your horoscope. Or to wearing your "lucky shirt." None of these "reasons" have anything to do with the result.

Irrelevant Attack on a Person or Opponent

Known as an *ad hominem* argument (Latin for "to the person"), the *irrelevant attack on an opponent* disparages a person's reputation or beliefs rather than his or her argument.

Example: Your boss says you may not hire a certain person as a programmer because he has been married and divorced several times. The fallacy here is that the applicant's marital history has no bearing on his skills as a programmer.

Slippery Slope

The *slippery slope* is a failure to see that the first step in a possible series of steps does not lead inevitably to the rest.

Example: The domino theory, based on which the United States waged wars against Communism, was a slippery-slope argument. It assumed that if Communism triumphed in one country (for example, Nicaragua), then it would inevitably triumph in other regions (the rest of Central America), finally threatening the borders of the United States itself.

Appeal to Authority

The *appeal to authority* argument (known in Latin as *argumentum ad verecundiam*) uses authorities in one area to pretend to validate claims in another area in which the person is not an expert.

Example: You see the appeal to authority used all the time in advertising. But how medically qualified is a professional golfer to speak about headache remedies?

Circular Reasoning

The *circular reasoning* argument rephrases a statement to be proven true and then uses the new, similar statement as supposed proof that the original statement is in fact true.

Example: You declare that you can drive safely at high speeds with only inches separating you from the car ahead because you have driven this way for years without an accident.

Straw Man

The *straw man* argument involves misrepresenting your opponent's position to make it easier to attack, or attacking a weaker position while ignoring a stronger one. In other words, you sidetrack the argument from the main discussion.

Example: Politicians use this argument all the time. Attacking a legislator for being "fiscally irresponsible" in supporting funds for a gun-control bill when what you really object to is the *fact* of gun control is a straw man argument.

Appeal to Pity

The *appeal to pity* argument appeals to mercy rather than making an argument on the merits of the case itself.

Example: Begging the dean not to expel you for cheating because your parents are poor and made sacrifices to put you through college represents this kind of argument.

Questionable Statistics

Statistics can be misused in many ways as supporting evidence. The statistics may be unknowable, drawn from an unrepresentative sample, or otherwise suspect.

Example: Stating how much money is lost to taxes because of illegal drug transactions is speculation because such transactions are hidden or underground.

The Importance of Having *No* Opinion

It is not necessary to have an opinion pro or con about everything. Indeed, the basis of the scientific method is that a great deal of what is considered to be "the truth" is established only *tentatively.* This is why scientists talk in terms of probabilities: nothing is definite or 100% certain, only probable and only for the time being. This means always having an awareness that other evidence may come along at some point to change an existing hypothesis or mode of thinking. If you continually take this attitude, you are indeed a critical thinker.

Suggested Resources

Kahane, Howard. *Logic and Contemporary Rhetoric: The Use of Reason in Everyday Life* (6th ed.). Belmont, CA: Wadsworth, 1992. A comprehensive and entertaining look at the most common fallacies used in politics, the mass media, textbooks, and everyday conversation.

Ruchlis, Hy, and Sandra Oddo. *Clear Thinking: A Practical Introduction.* Buffalo, NY: Prometheus, 1990. Simply and entertainingly describes barriers to clear thought: superstitions, stereotypes, prejudices, and conflicting opinions.

SUMMARY

auditing *(p. 474, LO 4)* Refers to an independent review of an organization's information system to see if all records and systems are as they should be; often done during Phase 6 of the SDLC. Often a systems analyst will design an audit trail.

An audit trail helps independent auditors trace the record of a transaction from its output back through all processing and storage to its source.

computer-aided software engineering (CASE) tools *(p. 467, LO 6)* Software that provides computer-automated means of designing and changing systems.

CASE tools may be used in almost any phase of the SDLC, not just design. So-called *front-end CASE tools* are used during the first three phases—preliminary analysis, systems analysis, systems design—to help with the early analysis and design. So-called *back-end CASE tools* are used during two later phases—systems development and implementation—to help in coding and testing, for instance.

connectivity diagram *(p. 464, LO 5)* Modeling tool used to map network connections of people, data, and activities at various locations.

Because connectivity diagrams are concerned with communications networks, we may expect to see these in increasing use.

conversion *(p. 472, LO 7)* Process of converting from an old information system to a new one; involves converting hardware, software, and files. There are four strategies for handling conversion: *direct, phased, parallel,* and *pilot.*

In order to smoothly switch from an old system to a new one, an orderly plan of conversion must be determined ahead of time.

data dictionary *(p. 464, LO 4)* Record of all the elements of data that make up the data flow in a system.

In analyzing a current system and preparing data flow diagrams, systems analysts must prepare a data dictionary, which is used and expanded during subsequent phases of the SDLC.

data flow diagram (DFD) *(p. 462, LO 5)* Modeling tool that graphically shows the flow of data through a system.

A DFD diagrams the processes that change data into information. DFDs have only four symbols, which makes them easy to use: *entity* (or external entity) rectangle, which shows the source or destination of data outside the system; *flow of data* or *vector* arrow, which indicates the path of the data; *process* rectangle with rounded corners, which indicates how input data is transformed into output; and *data store* or *file* open rectangle, which shows where data is held, whether filing cabinet or hard disk.

decision support system (DSS) *(p. 453, LO 3)* Computer-based information system that helps managers with nonroutine decision-making tasks. Inputs consist of some summarized reports, some processed transaction data, and other internal data. They also include data from sources outside the organization—for example, data may be produced by trade associations, marketing research firms, and government agencies. The outputs are flexible, on-demand reports from which a top manager can make decisions about unstructured problems.

A DSS is installed to help top managers and middle managers make *strategic* decisions—decisions about unstructured problems, those involving events and trends outside the organization (for example, rising interest rates). The key attribute of a DSS is that it uses *models*. The DSS database, which draws on the TPS and MIS files, as well as outside data, is accessed through DSS software.

decision table *(p. 464, LO 5)* Modeling tool that shows the decision rules that apply when certain conditions occur and what actions to take.

A decision table provides a model of a simple, structured decision-making case. It shows which conditions must take place in order for which actions to occur.

detail design *(p. 468, LO 4)* Second stage of Phase 3 of the SDLC; describes how a proposed information system will deliver the general capabilities described in the preliminary design phase. The detail design usually considers the following system requirements: *output, input, storage, processing,* and *system controls and backup.*

A new system must be designed in detail before any hardware and software can be developed/purchased.

detail report *(p. 452, LO 3)* Report that contains specific information about routine activities.

Detail reports are produced by transaction processing systems and are commonly used by lower-level managers.

direct implementation *(p. 472, LO 7)* Method of system conversion; the users simply stop using the old system and start using the new one.

The risk of this method is that there is nothing to fall back on if the old system has been discontinued.

downsizing *(p. 443, LO 1)* (1) Reducing the size of an organization by eliminating workers. (2) Moving from mainframe-based computer systems to systems linking smaller computers in networks.

Downsizing is important because as layoffs have shrunk middle and lower management, the remaining people have had to handle more information themselves, leading to greater use of microcomputers and networked systems.

empowerment *(p. 444, LO 1)* Refers to giving others the authority to act and make decisions on their own.

Old-style management gave lower-level managers and employees only the information they "needed" to know, reducing their power to make decisions. Today information is apt to be spread widely, empowering employees lower down in the organization and enabling them to do their jobs better.

exception reports *(p.453, LO 3)* Middle-management reports that show out-of-the-ordinary data—for example, an inventory report listing only those items that number fewer than 10 in stock.

Exception reports highlight matters requiring prompt decisions by management.

executive information system (EIS) *(p. 454, LO 3)* Also called an *executive support system (ESS);* DSS made especially for top managers that specifically supports strategic decision making. It draws on data both from inside and outside the organization (for example, news services, market-research databases).

The EIS includes capabilities for analyzing data and doing "what if" scenarios.

expert system *(p. 455, LO 3)* Set of computer programs that perform a task at the level of a human expert.

Expert systems are used by management and nonmanagement personnel to solve sophisticated problems.

grid chart *(p. 464, LO 5)* Modeling tool that shows the relationship between data on input documents and data on output documents.

Grid charts are used in the systems design phase of the SDLC.

lower-level managers *(p. 448, LO 2)* Also called *supervisory managers;* the lowest level in the hierarchy of the three types of managers. Their job is to make operational decisions, monitoring day-to-day events and, if necessary, taking corrective action.

Lower managers need information that is structured—that is, detailed, current, and past-oriented, covering a narrow range of facts and events inside the organization.

make-or-buy decision *(p. 470, LO 4)* Decision made in Phase 4 (programming) of the SDLC concerning whether the organization has to make a program—have it custom-written—or buy it, meaning simply purchase an existing software package.

The decision taken affects the costs and time required to develop the system.

management *(p. 446, LO 2)* Level of personnel that oversees the tasks of planning, organizing, staffing, supervising, and controlling business activities.

Different levels of managers need different kinds of information on which to make decisions.

What It Is / What It Does	Why It's Important

management information system (MIS) *(p. 452, LO 3)* Computer-based information system that derives data from all an organization's departments and produces *summary, exception, periodic,* and *on-demand* reports of the organization's performance.

A MIS principally assists middle managers, helping them make *tactical* decisions—spotting trends and getting an overview of current business activities.

middle-level managers *(p. 448, LO 2)* One of the three types of managers; they implement the goals of the organization. Their job is to oversee the supervisors and to make tactical decisions.

Middle managers require information that is both structured and unstructured.

model *(p. 453, LO 3)* Mathematical representation of a real system; models are often used in a DSS.

A model allows the manager to do a simulation—play a "what if" game—to reach decisions. By changing the hypothetical inputs to the model, one can see how its outputs are affected.

modeling tools *(p. 462, LO 5)* Charts, tables, and diagrams used by systems analysts. Examples are data flow diagrams, connectivity diagrams, grid charts, Gantt charts, PERT charts, decision tables, and systems flowcharts.

Modeling tools enable a systems analyst to present graphic, or pictorial, representations of a system.

on-demand reports *(p. 453, LO 3)* Middle-management reports that produce information in response to an unscheduled demand, often produced on screen rather than on paper.

On-demand reports help managers make decisions about nonroutine matters (for example, whether to grant credit to a new customer placing a big order).

operational decision *(p. 448, LO 2)* Type of decision made by lower-level managers; predictable decision that can be made by following a well-defined set of routine procedures.

Operational means focusing principally on supervising and controlling instead of on the other management tasks of planning, organizing, and staffing.

organization chart *(p. 446, LO 2, 5)* Schematic drawing showing the hierarchy of relationships among an organization's employees.

Organization charts show levels of management and formal lines of authority.

outsourcing *(p. 443, LO 1)* Contracting with outside businesses or services to perform the work previously done by in-house departments, whether janitorial tasks or systems analysis.

Outside specialized contractors often can do work more cheaply and efficiently.

parallel implementation *(p. 472, LO 7)* Method of system conversion whereby the old and new systems are operated side by side until the new system has shown it is reliable.

If the new system fails, the organization can switch back to the old one. The difficulty is the expense of paying for equipment and people to operate two systems simultaneously.

periodic reports *(p. 453, LO 3)* Middle-management reports that are produced on a regular schedule, such as daily, weekly, monthly, quarterly, or annually. They are usually produced on paper, such as computer printouts.

Periodic reports, such as sales figures, income statements, or balance sheets, help managers make routine decisions.

phased implementation *(p. 472, LO 7)* Method of system conversion whereby parts of the new system are phased in gradually, perhaps over several months, or all at once, in groups.

This conversion strategy is prudent, though it can be expensive.

pilot implementation *(p. 472, LO 7)* Method of system conversion whereby the entire system is tried out by only some users. Once the reliability has been proved, the system is implemented with the rest of the intended users.

The pilot approach has risks, since all the users of a particular group are taken off the old system. However, the risks are confined to only a small part of the organization.

preliminary design *(p. 467, LO 4)* First stage of Phase 3 of the SDLC; describes the general functional capabilities of a proposed information system. Three tools that may be used are *prototyping tools, CASE tools,* and *project management software.*

During the preliminary design phase, staff reviews the system requirements and then considers major components of the system. Usually several alternative systems (called *candidates*) are considered, and the costs and the benefits of each are evaluated.

What It Is / What It Does	**Why It's Important**

preliminary investigation *(p. 459, LO 4)* Phase 1 of the SDLC; the purpose is to conduct a preliminary analysis (determine the organization's objectives, determine the nature and scope of the problem), propose alternative solutions (leave the system as is, improve the efficiency of the system, or develop a new system), describe costs and benefits, and submit a preliminary plan with recommendations.

The preliminary investigation lays the groundwork for the other phases of the SDLC.

project management software *(p. 468, LO 6)* Programs used to plan, schedule, and control the people, costs, and resources required to complete a project on time. Project management software often uses Gantt charts and PERT charts.

Project management software makes it easier for people to plan and develop almost all aspects of a new system.

prototyping *(p. 467, LO 6)* Involves building a model or experimental version of all or part of a system so that it can be quickly tested and evaluated.

Prototyping is part of the preliminary design stage of Phase 3 of the SDLC. (A *prototype* is a limited working system, or part of one. It is developed to test out design concepts. A prototype, which may be constructed in just a few days, allows users to find out immediately how a change in the system might benefit them.) Prototypes are built with *prototyping tools*. These are special software packages that can be used to design screen displays.

reengineering *(p. 444, LO 1)* Also known as *process innovation* and *core process redesign;* refers to the search for and implementation of radical change in business processes to achieve breakthrough results.

Reengineering is not just fixing up what already exists; it works best with big processes that really matter (for example, new-product development or customer service). Thus, candidates for this procedure include companies experiencing big shifts in their definition, markets, or competition. Examples are information technology companies—computer makers, cable-TV providers, and local and long-distance phone companies—which are wrestling with technological and regulatory change.

semistructured information *(p. 449, LO 2)* Information that does not necessarily result from clearly defined, routine procedures. Middle managers need semistructured information that is detailed and more summarized than information for operating managers.

Semistructured information involves review, summarization, and analysis of data to help plan and control operations and implement policy formulated by upper managers.

strategic decision *(p. 446, LO 2)* Type of decision made by top managers; rarely based on predetermined routine procedures but involving the subjective judgment of the decision maker.

Strategic means that, of the five management tasks (planning, organizing, staffing, supervising, controlling), top managers are principally concerned with planning.

structured information *(p. 448, LO 2)* Detailed, current information concerned with past events; it records a narrow range of facts and covers an organization's internal activities.

Lower-level managers need easily defined information that relates to the current status and activities within the basic business functions.

summary reports *(p. 453, LO 3)* Reports that show totals and trends (for example, a report showing total sales by office, by product, and by salesperson).

Summary reports are used by middle managers to make decisions.

system *(p. 456, LO 4)* Collection of related components that interact to perform a task in order to accomplish a goal.

The point of systems analysis and design is to ascertain how a system works and then take steps to make it better.

systems analysis *(p. 461, LO 4, 5)* Phase 2 of the SDLC; the purpose is to gather data (using written documents, interviews, questionnaires, observation, and sampling), analyze the data, and write a report.

The results of systems analysis will determine whether the system should be redesigned.

What It Is / What It Does	Why It's Important

systems analysis and design *(p. 490, LO 4)* See *Systems development life cycle (SDLC)*.

systems analyst *(p. 457, LO 4)* Information specialist who performs systems analysis, design, and implementation.

The systems analyst studies the information and communications needs of an organization to determine how to deliver information that is more accurate, timely, and useful. The systems analyst achieves this goal through the problem-solving method of systems analysis and design.

systems design *(p. 467, LO 4, 5)* Phase 3 of the SDLC; the purpose is to do a preliminary design and then a detail design, and write a report.

Systems design is one of the most crucial phases of the SDLC.

systems development *(p. 470, 4)* Phase 4 of the SDLC; hardware and software for the new system are acquired and tested. The fourth phase begins once management has accepted the report containing the design and has approved the way to development.

This phase may involve the organization in investing substantial time and money.

systems development life cycle (SDLC) *(p. 458, LO 4)* Six-phase process that many organizations follow during systems analysis and design: (1) *preliminary investigation;* (2) *systems analysis;* (3) *systems design;* (4) *systems development;* (5) *systems implementation;* (6) *systems maintenance.* Phases often overlap, and a new one may start before the old one is finished. After the first four phases, management must decide whether to proceed to the next phase. User input and review is a critical part of each phase.

The SDLC is a comprehensive tool for solving organizational problems, particularly those relating to the flow of computer-based information.

systems flowchart *(p. 464, LO 5)* Modeling tool that uses many symbols to diagram the input, processing, and output of data in a system as well the interaction of all the parts in a system.

A systems flowchart graphically depicts all aspects of a system.

systems implementation *(p. 471, LO 4)* Phase 5 of the SDLC; consists of converting the hardware, software, and files to the new system and training the users.

This phase is important because it involves putting design ideas into operation.

systems maintenance *(p. 474, LO 4)* Phase 6 of the SDLC; consists of keeping the system working by having system audits and periodic evaluations.

This phase is important for keeping a new system operational and useful.

system testing *(p. 471, LO 4)* Part of Phase 4 of the SDLC; the parts of a new program are linked together, and test data is used to see if the parts work together.

Test data may consist of actual data used within the organization. Also, erroneous and massive amounts of data may be used to see if the system can be made to fail.

tactical decision *(p. 448, LO 2)* Type of decision made by middle managers that is without a base of clearly defined informational procedures, perhaps requiring detailed analysis and computations.

Tactical means that, of the five management tasks (planning, organizing, staffing, supervising, controlling), middle managers deal principally with organization and staffing.

top managers *(p. 446, LO 2)* One of the three types of managers; also called *strategic managers,* they are concerned with long-range planning and strategic decisions.

Top managers need information that is unstructured—that is, summarized, less current, future-oriented, covering a broad range of facts, and concerned with events outside as well as inside the organization.

total quality management (TQM) *(p. 443, LO 1)* Philosophy of management based on an organization-wide commitment to continuous work improvement and meeting customer needs.

Many industries have benefited from TQM principles, such as American automobile makers.

transaction *(p. 452, LO 3)* Recorded event having to do with routine business activities (for example, materials purchased, employees hired, or taxes paid).

Today in most organizations the bulk of transactions are recorded in a computer-based information system.

transaction processing system (TPS) *(p. 452, LO 3)* Also called an *electronic data processing (EDP) system;* computer-based information system that keeps track of the transactions needed to conduct business. Inputs are transaction data (for example, bills, orders, inventory levels, production output). Outputs are processed transactions (for example, bills, paychecks). Each functional area of an organization—Research and Development, Production, Marketing, and Accounting and Finance—usually has its own TPS.

The TPS helps supervisory managers in making *operational decisions.* The database of transactions stored in a TPS are used to support a management information system and a decision support system.

unit testing *(p. 471, LO 4)* Part of Phase 4 of the SDLC; individual parts of a new program are tested, using test data.

If the program is written as a collaborative effort by multiple programmers, each part of the program is tested separately.

unstructured information *(p. 448, LO 2)* Summarized, less current information concerned with future events; it records a broad range of facts and covers activities outside as well as inside an organization.

Top managers need information in the form of highly unstructured reports. The information should cover large time periods and survey activities outside as well as inside the organization.

virtual office *(p. 444, LO 1)* Mobile office, created by the use of integrated computer and communications technologies.

Widely scattered workers can operate as individuals or as a company, regardless of where they are.

(Selected answers appear at the back of the book.)

Short-Answer Questions

1. What are the three levels of management?

2. What are the five types of management information systems?

3. What is structured information? unstructured information? semistructured information?

4. What is the purpose of the systems development life cycle?

5. During which phase of the SDLC is software and hardware obtained?

6. What are modeling tools used for? In what phase of the SDLC are modeling tools used?

7. What does *outsourcing* mean? *downsizing*?

8. What type of manager uses summary reports, exception reports, periodic reports, and on-demand reports?

9. Why is it important for users to understand the principles of the SDLC?

10. Describe some differences between a management information system and a decision support system.

Fill-in-the-Blank Questions

1. A(n) _____ produces detail reports for lower managers.

2. The trends of automation, downsizing and outsourcing, total quality management, and employee empowerment are forcing the _____ of the workplace.

3. A(n) _____ is a schematic drawing that shows the hierarchy of formal relationships among an organization's employees.

4. A(n) _____ enables teams of co-workers to use networks of microcomputers to share information and cooperate on projects.

5. A(n) _____ is an easy-to-use DSS that is designed to be used by top managers.

6. Project management software often uses _____ charts and _____ charts to help manage projects so they are completed on time.

7. Middle management makes _____ decisions.

8. _____ involves building a model or experimental version of all or part of a system so that it can be quickly tested and evaluated.

9. Whereas a(n) _____ uses four symbols to diagram the flow of data through a system, a(n) _____ uses many symbols to diagram the input, processing, and output of data in all aspects of a system and also shows how parts of the system interact.

Multiple-Choice Questions

1. Which of the following would you use to see out-of-the-ordinary data?
   a. summary report
   b. exception report
   c. periodic report
   d. on-demand report
   e. all of the above

2. Which of the following would you need in order to make decisions about unstructured problems?
   a. transaction processing system
   b. management information system
   c. decision support system
   d. none of the above

3. Which of the following isn't a type of modeling tool used in the SDLC?
   a. data flow diagram
   b. systems flowchart
   c. spreadsheet program
   d. connectivity diagram
   e. decision table

4. In which phase of the SDLC are software and hardware obtained?
   a. preliminary investigation
   b. system analysis
   c. systems design
   d. systems development
   e. systems implementation

5. In which phase of the SDLC is data gathered and analyzed?
   a. preliminary investigation
   b. system analysis
   c. systems design
   d. systems development
   e. systems implementation

6. Which of the following is used to map the connections between people, data, and activities at various locations?
   a. data flow diagram
   b. systems flowchart
   c. spreadsheet program
   d. connectivity diagram
   e. decision table

7. In the systems implementation phase of the SDLC, which of the following conversion approaches is the riskiest?
   a. direct approach
   b. parallel approach
   c. phased approach
   d. pilot approach
   e. none of the above

8. Which of the following best describes the types of tasks performed by a systems analyst?
   a. systems analysis
   b. systems design
   c. systems implementation
   d. all of the above

9. _____ tools are software packages that provide computer-automated means of designing and changing systems.
   a. prototyping
   b. project-management
   c. modeling
   d. CASE
   e. none of the above

10. Modeling tools are used in the _____ phase of systems development.
    a. preliminary investigation
    b. system analysis
    c. systems design
    d. systems development
    e. systems implementation

True/False Questions

T F 1. The lowest level of management makes strategic decisions.

T F 2. Decision support systems are used mainly by upper management.

T F 3. A transaction processing system supports day-to-day business activities.

T F 4. Users are never involved in systems development.

T F 5. The sixth phase of systems development, systems maintenance, is usually completed a few months after the fifth phase.

T F 6. System testing is performed in the fourth phase of systems development.

T F 7. An EIS enables teams of co-workers to use networks of microcomputers to share information.

T F 8. Systems implementation involves converting the hardware, software, and files to the new system.

T F 9. Auditing and evaluation are performed in the sixth phase of systems development, systems maintenance.

Projects/Critical-Thinking Questions

1. Decision support systems often take years to develop. Given this long development period, some experts argue that the system will be obsolete by the time it is complete and that information needs will have changed. Other experts argue that no alternatives exist. By reviewing current computer publications that describe management information systems, formulate an opinion on this issue.

2. Using recent computer publications, research the state of the art of computer-assisted software engineering (CASE) tools. What capabilities do these tools have? What do you think the future holds for CASE tools?

3. Make an appointment to interview a manager in your Campus Computer Services Department. Determine what types of formal management information systems are used at your school, and find out how they were developed and they are maintained. Was outsourcing a factor? Were the systems made or bought? How does the department obtain user feedback?

4. Interview a student majoring in computer science who plans to become a systems analyst. Why is this person interested in this field? What does he or she hope to accomplish in it? What courses must be taken to satisfy the requirements for becoming an analyst?

5. Using the systems flowchart symbols and illustrations shown in this chapter, diagram a small, simple system for managing your budget: what are the inputs? where does the data go? what is done to the data (processing), and at what points? what are the outputs, and where do they go? Try to incorporate all possible activities that could occur within a significant budgetary unit of time, such as a semester.

6. Office automation is changing the flow of information within organizations from the traditional hierarchical up/down manner—from managers down to

employees and vice versa—to a more democratic, all-directional manner. This is happening because networks and groupware are making information available to more people in a company. As a result the traditional definitions of "manager" are changing. Research this topic in publications such as *Business Week*, *Fortune*, *Forbes*, and the *Wall Street Journal*. What are some basic changes in management/employee arrangements that are predicted for the future?

(net) Using the Internet

Objective: *In this exercise you practice researching a topic on the Internet.*

Before you continue: *We assume you have access to the Internet through your university, business, or commercial service provider and to the Web browser tool named Netscape Navigator 2.0 or 2.01. Additionally, we assume you know how to connect to the Internet and then load Netscape Navigator. If necessary, ask your instructor or system administrator for assistance.*

1. Pick one of the following topics. You will then use the Internet to uncover issues and/or state-of-the-art information related to the topic.
   - decision support system (DSS)
   - expert system
   - reengineering
   - project management software
2. Make sure you have started Netscape. The home page for Netscape Navigator should appear on your screen.
3. Use a search engine such as
   Lycos (*http://lycos.cs.cmu.edu/*),
   Excite Netsearch (*http://www.excite.com/*),
   Open Text (*http://www.opentext.com:8080/*),
   and/or Alta Vista (*http://www.altavista.digital.com/*)

to find information about the topic you picked in step 1. Also, consider using Savvy Search (*http://www.cs.colostate.edu/~dreiling/smart form.html*), which allows you to search multiple search engines simultaneously and also allows you to specify the type of information you are looking for (for example, Technical Reports, Software, Commercial, and Academic).

4. Print (File, Print) or save to disk (File, Save as) at least three articles that will support your objective (as stated in step 1). (*Note:* You can also mail the articles to yourself by choosing File, Mail Document from the Menu bar.)
5. To further your research, consider accessing the Library of Congress home page (*http://lcweb.loc.gov/homepage/lchp.html*), which provides access to full-text legislative information and the Library of Congress catalog. (*Note:* You need Telnet software to connect to the Library of Congress catalog. If this is a problem, consider accessing U.C. Berkeley's web site —*http://sunsite.Berkeley.EDU/Libweb/* —which has links to Web libraries around the world.) Other research resources include the Internet Public Library (*http://ipl.sils.umich.edu:80/*) and the WWW Virtual Library (*http://www.w3.org/hypertext/DataSources/bySubject/overview.html*).
6. If you have questions relating to your topic, find a newsgroup that relates to your topic and then post your questions to the newsgroup. Print a copy of your questions and any replies you receive.
7. Now write a one-page paper summarizing your research. Assemble any Internet printouts and attach them to your report.
8. Display Netscape's home page and then exit Netscape.

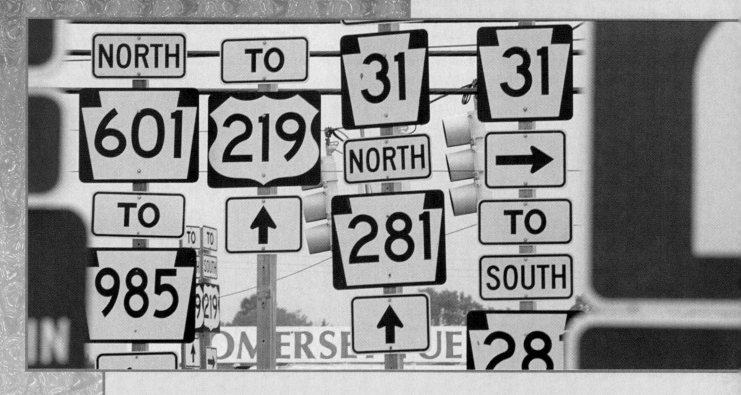

CHAPTER 11

Software Development

Programming & Languages

Concepts You Should Know

After reading this chapter, you should be able to:

1. Explain what a program is and distinguish between packaged and custom software.
2. Explain the five steps in programming.
3. Discuss the concepts of structured program design.
4. Identify and give examples of the five generations of programming languages.
5. Distinguish among assembler, compiler, and interpreter.
6. Identify the major programming languages used today.
7. Explain object-oriented and visual programming.
8. Briefly describe some languages used in Internet programming.

We live in a society that is enlarging the boundaries of knowledge at an unprecedented rate," says James Randi, "and we cannot keep up with more than a small portion of what is made available to us."

A debunker of the paranormal, Randi deplores the uncritical thinking he sees on every hand—people basing their lives on horoscopes, numerology, and similar nonsense. To mix the data available to our senses "with childish notions of magic and fantasy is to cripple our perception of the world around us," he says. "We must reach for the truth, not for the ghosts of dead absurdities."[1]

Mindsets & Critical Thinking

Preview & Review: Mindsets make us comfortable, but they can prevent us from accepting changes. To break past mindsets, we need to learn to think critically. Critical thinking is sorting out conflicting claims, weighing the evidence for them, letting go of personal biases, and arriving at reasonable views.

Reaching for the truth may not come easily; it is a stance toward the world, developed with practice. To achieve this, we have to wrestle with obstacles that are mostly of our own making: mindsets.

Mindsets

By the time we are grown, our minds have become "set" in various patterns of thinking that affect the way we respond to new situations and new ideas. These mindsets are the result of our personal experiences and the various social environments in which we grew up. Such mindsets determine what ideas we think are important and, conversely, what ideas we ignore.

"Because we can't pay attention to all the events that occur around us," points out one book on clear thinking, "our minds filter out some observations and facts and let others through to our conscious awareness."[2] Herein lies the danger: "As a result we see and hear what we subconsciously want to, and pay little attention to facts or observations that have already been rejected as unimportant."

Having mindsets makes life comfortable. However, as the foregoing writers point out, "Familiar relationships and events become so commonplace that we expect them to continue forever. Then we find ourselves completely unprepared to accept changes that are necessary, even when they stare us in the face."[3]

Critical Thinking

To break past mindsets, we need to learn to think critically. *Critical thinking* is sorting out conflicting claims, weighing the evidence for them, letting go of personal biases, and arriving at reasonable views. Critical thinking means actively seeking to understand, analyze, and evaluate information in order to solve specific problems. It is very much a feature of the problem-solving process of the systems development life cycle (SDLC) (✓ p. 457) in general and of programming in particular.

Critical thinking is simply clear thinking, an attribute that can be developed. "Before making important choices," says one writer, clear thinkers "try to clear emotion, bias, trivia and preconceived notions out of the way so they can concentrate on the information essential to making the right decision."[4] All it takes is practice.

We discussed some of the tools for critical thinking in the Experience Box at the end of Chapter 10. We turn now to the type of logical thinking known as *programming.*

Programming: A Five-Step Procedure

Preview & Review: Programming is a five-step procedure for producing a program—a list of instructions—for the computer.

Many people assume that the numbers that appear in a printout of, say, a spreadsheet are probably correct. There is something about the look of a finished product that inspires faith in the result.

However, the numbers are correct only if the data and processing procedures are correct. People often think of programming as simply typing words and numbers into a computer. This is part of it, but only a small part. Basically, programming is a *method of solving problems.*

What a Program Is

To see how programming works, consider what a program is. **A *program* is a list of instructions that the computer must follow in order to process data into information.** The instructions consist of *statements* written in a programming language, such as BASIC.

As we said in Chapter 1, **applications software** is defined as software that **can perform useful work on general-purpose tasks.** Examples are programs that do word processing, desktop publishing, or payroll processing.

Applications software may be *packaged* or *customized.*

- **Packaged software:** This is the kind that you buy from a computer store or mail-order house, which is the kind used by most PC users. ***Packaged software* is an "off-the-shelf," prewritten program developed for sale to the general public.** (We might also include here shareware and freeware, ✓ p. 55, since they come ready to stick in your diskette drive and use right away.)

- **Customized software:** ***Customized software* is software designed for a particular customer.** This is an applications program that is created or custom-made. It is usually written by a professional programmer, although you can do this, too, for some kinds of applications. Customized software is written to perform a task that cannot be done with off-the-shelf packaged software.

The decision whether to buy or create a program is *one* of the phases in the systems development life cycle, as discussed in Chapter 10. Once the decision is made to develop a new system, then some further steps must be taken.

What Programming Is

A program, we said, is a list of instructions that the computer must follow in order to process data into information. ***Programming,* also called *software engineering,* is a five-step process for creating that list of instructions.** Only one of those steps (the step called *coding*) consists of sitting at the keyboard typing words into a computer.

The five steps are as follows. (■ *See Panel 11.1.*)

1. Define the problem.
2. Design a solution.
3. Code the program.
4. Test the program.
5. Document the program.

The First Step: Define the Problem

Preview & Review: Programmers break the definition of the problem down into six mini-steps. They (1) specify program objectives and users; specify (2) output, (3) input, and (4) processing requirements; (5) study the feasibility of implementing the program; and (6) document the analysis. An important part of the process is the make-or-buy decision—whether to buy off-the-shelf software or custom-make a program.

The *problem definition* step requires performing six mini-steps. They include specifying objectives, output, input, and processing tasks, then studying their feasibility and documenting them. (■ *See Panel 11.2.*) Let us consider these six mini-steps.

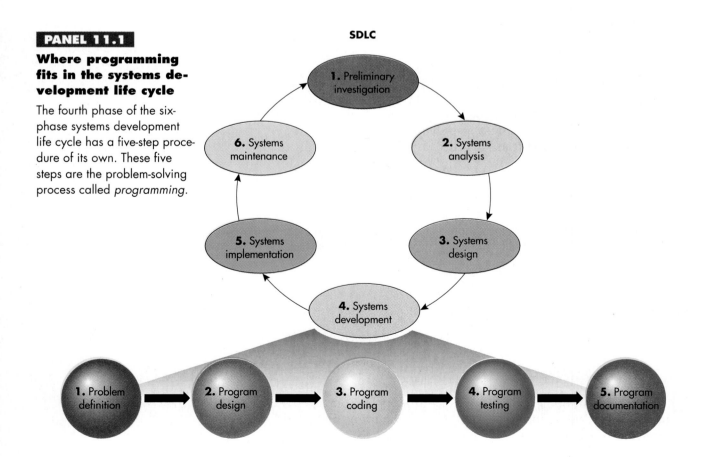

PANEL 11.1

Where programming fits in the systems development life cycle

The fourth phase of the six-phase systems development life cycle has a five-step procedure of its own. These five steps are the problem-solving process called *programming*.

SDLC

1. Preliminary investigation
2. Systems analysis
3. Systems design
4. Systems development
5. Systems implementation
6. Systems maintenance

1. Problem definition
2. Program design
3. Program coding
4. Program testing
5. Program documentation

1. Specify program objectives and users.
2. Specify output requirements.
3. Specify input requirements.
4. Specify processing requirements.
5. Study feasibility of implementing program.
6. Document the analysis.

PANEL 11.2

First step: problem definition

The first step in programming is to have a clear understanding of the problem.

1. Specify Objectives & Users

You solve problems all the time. A problem might be deciding whether to take a required science course this term or next. Or you might try to solve the problem of grouping classes so you can fit in a job. In such cases you are specifying your *objectives.* Programming works the same way. You need to write a statement of the objectives you are trying to accomplish—the problem you are trying to solve. If the problem is that your company needs a new computer-based payroll processing program, you would specify that in your problem definition.

You also need to specify the users of the program. Will it be used by people inside the company, outside, or both? What kind of skills will they bring, or will they have to be trained in some way? An ongoing problem with information technology is determining how "technophobic" or "technoliterate" the people are whom the technology is supposed to serve.

2. Specify the Desired Output

You should specify the outputs—what you want to get out of the system— before you specify the inputs, what will go into it. For example, will you want hardcopy, softcopy (✓ p. 246), or both? What information should the outputs include? This step may require several meetings with users to make sure you're designing what they want.

Sometimes specifying output requires only drawing a picture. For example, you might do a free-hand drawing showing headings, columns, and whatever else you would like the system to produce.

3. Specify the Desired Input

Once you know the kind of output you want, you can then think about input. What kind of input data will be needed? What form should it appear in? What is its source?

4. Specify the Desired Processing

Here you determine the processing tasks that must occur in order for input data to be processed into output data. Should processing be online or batch (✓ p. 409)? Will great processing capabilities be required?

5. Study the Feasibility of Implementing the Program

Is the kind of problem you're trying to solve feasible within the present budget? Will it require hiring a lot more staff? Will it take too long to accomplish?

As we mentioned, a particularly important decision is the "make-or-buy" decision. **In the *make-or-buy decision*, you decide whether you have to *make* a program—have it custom-written—or buy it, meaning simply purchase an existing software package.** Sometimes programmers decide they can buy an existing program and modify it rather than write it from scratch.

6. Document the Analysis

Throughout this first step on problem definition, programmers must document everything they do. This includes writing objective specifications of the entire process being described.

The Second Step: Design the Program

Preview & Review: In the second step, programmers design a solution. This consists of three mini-steps. (1) The program logic is determined through a top-down approach and modularization, using a hierarchy chart. (2) Next the program is designed with certain tools: pseudocode and/or flowcharts with control structures. (3) Finally, the design is tested with a structured walkthrough.

Assuming the decision is to make, or custom-write, the program, you then move on to the solution design. **In the *program design* step, the custom software is designed in three mini-steps. First, the program logic is determined through a top-down approach and modularization, using a hierarchy chart. Then it is designed in detailed form, either in narrative form, using pseudocode, or graphically, using flowcharts with logical tools called control structures. Finally, the design is tested with a structured walkthrough.** (■ *See Panel 11.3.*)

It used to be that programmers took a kind of seat-of-the-pants approach to programming. Programming was considered an art, not a science. Today, however, most programmers use a design approach called structured programming. ***Structured programming* takes a top-down approach that breaks programs into modular forms. It also uses standard logic tools called control structures (sequential, selection, case, and iteration).** The point of structured

PANEL 11.3

Second step: program design

1. Determine program logic through top-down approach and modularization, using hierarchy charts.
2. Design details using pseudocode and/or flowcharts, preferably involving control structures.
3. Test design with structured walkthroughs.

R E A D M E

Case Study: Recruiting Students to Design Software That Explains Tricky Concepts

T. Mack Brown is trying to answer some of the biggest questions in science, mathematics, and the arts at Berry College. He is not involved in cutting-edge research at the tiny liberal arts institution. He is not even on the faculty. Mr. Brown is an entrepreneur who is working with students to design software to help people understand everything from Kirchhoff's voltage law to the central-limit theorem to the intervals between notes in music theory.

"In every discipline, you have that moment when a professor explains a concept and half the class feels left behind," he says. "You can hear the brain go 'splat' against an intellectual hurdle." Mr. Brown says he remembers experiencing that feeling when he was a student, so he wanted to design software to help students recover from the "splat."

Berry doesn't have the reputation of Drexel University or the University of Illinois in software development, but Mr. Brown's activities could change that. He predicts that the college and his company, the Brownstone Research Group, will produce "hundreds and hundreds of tutorials associated with thousands and thousands of questions that have good responses built into them." . . .

He was interested in locating on a college campus because he knew students understood the "splat" concepts and would make the best programmers. He says that moving his operations closer to professors and students also enabled him to harness their creativity and daily feedback. . . .

The students who work with him generally earn between $300 and $900 a semester or can receive academic credit. Berry and Brownstone split the proceeds from sales of the software. . . .

Mr. Brown says he had few requirements for his student employees: they had to be on the dean's list, motivated, and, more important, they had to be visionaries in their academic disciplines.

Fifteen students have worked with Brownstone in the past two years. Some are computer-science majors, but others study philosophy, biology, mathematics, German, education, and physics. "My best programmer is a theology major," he says. . . .

—Amy Wahl, "Entrepreneur Recruits Students at Berry College to Design Software That Explains Tricky Concepts," *Chronicle of Higher Education*

programming is to make programs better organized and to have better notations so that they have clear and correct descriptions.

The three mini-steps of program design are as follows:

1. Determine the Program Logic, Using Top-Down Approach

Logically laying out the program is like outlining a lengthy term paper before you proceed to write it. **Top-down program design proceeds by identifying the top element, or module, of a program and then breaking it down in hierarchical fashion to the lowest level of detail. The top-down program design is used to identify the program's processing steps, or modules.**

The concept of modularization is important. *The beauty of modularization is that an entire program can be more easily developed because the parts can be developed and tested separately.*

A *module* **is a processing step of a program. Each module is made up of logically related program statements.** (Sometimes a module is called a *subprogram* or *subroutine*.) An example of a module might be a programming instruction that simply says "Open a file, find a record, and show it on the display screen." It is best if each module has only a single function, just as an English paragraph should have a single, complete thought. This rule limits the module's size and complexity.

Top-down program design can be represented graphically in a hierarchy chart. **A *hierarchy chart*, or *structure chart*, illustrates the overall purpose of the program, identifying all the modules needed to achieve that purpose and the relationships among them.** (■ *See Panel 11.4.*) The program must move in sequence from one module to the next until all have been processed. There must be three principal modules corresponding to the three principal computing operations—input, processing, and output. (In Panel 11.4 they are "Read input," "Calculate pay," and "Generate output.")

PANEL 11.4

A hierarchy chart

This represents a top-down design for a payroll program. Here the modules, or processing steps, are represented from the highest level of the program down to details. The three principal computing operations—input, processing, and output—are represented by the modules in the second layer: "Read input," "Calculate pay," and "Generate output." Before tasks at the top of the chart can be performed, all the ones below must be performed. Each module represents a logical processing step.

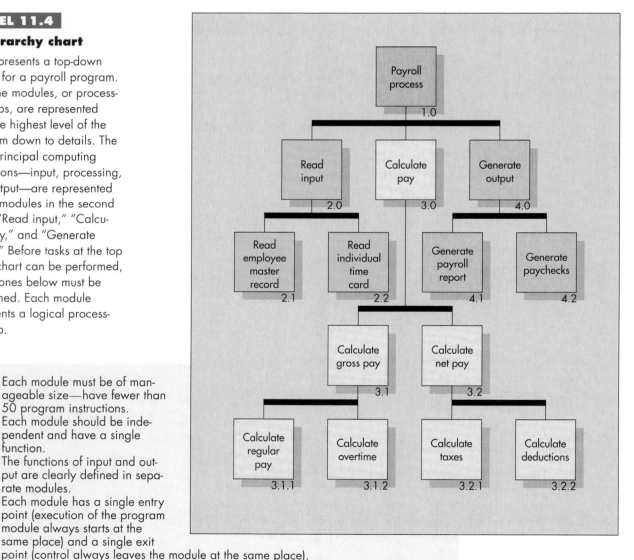

1. Each module must be of manageable size—have fewer than 50 program instructions.
2. Each module should be independent and have a single function.
3. The functions of input and output are clearly defined in separate modules.
4. Each module has a single entry point (execution of the program module always starts at the same place) and a single exit point (control always leaves the module at the same place).
5. If one module refers to or transfers control to another module, the latter module returns control to the point from which it was "called" by the first module.

2. Design Details, Using Pseudocode and/or Flowcharts

Once the essential logic of the program has been determined, through the use of top-down programming and hierarchy charts, you can go to work on the details.

There are two ways to show details—write them or draw them; that is, use *pseudocode* or use *flowcharts*. Normally you would use only one approach, but on large programs sometimes both are used. Flowcharts are commonly used to diagram programs. However, since pseudocode is in narrative form, we discuss it first so you can get a sense of what it's like to write, or *code*, a program.

- **Pseudocode:** ***Pseudocode* is a method of designing a program using English-like statements to describe the logic and processing flow.** (■ *See Panel 11.5.*) Pseudocode is like an outline or summary form of the program you will write.

  There are no real rules for writing pseudocode; however, most organizations apply their own standards. Sometimes pseudocode is used simply to express the purpose of a particular programming module in somewhat general terms. With the use of such terms as IF, THEN, or ELSE, however, the pseudocode follows the rules of *control structures*, an important aspect of structured programming, as we shall explain.

- **Program flowcharts:** We described systems flowcharts in the previous chapter. Here we consider program flowcharts. **A *program flowchart* is a chart that graphically presents the detailed series of steps needed to solve a programming problem.** The flowchart uses standard symbols—called *ANSI symbols*, after the American National Standards Institute, which developed them. (■ *See Panel 11.6.*)

PANEL 11.5

Pseudocode

```
START
DO WHILE (so long as) there are records
        Read a customer billing account record
        IF today's date is greater than 30 days from
        date of last customer payment
                Calculate total amount due
                Calculate 5% interest on amount due
                Add interest to total amount due to calculate
                grand total
                Print on invoice overdue amount
        ELSE
                Calculate total amount due
        ENDIF
        Print out invoice
END DO
END
```

Example of a program flowchart and explanation of flowchart symbols

This example represents a flowchart for a payroll program. Normally the flowchart is read from top to bottom. Arrows are used to show a change from those directions.

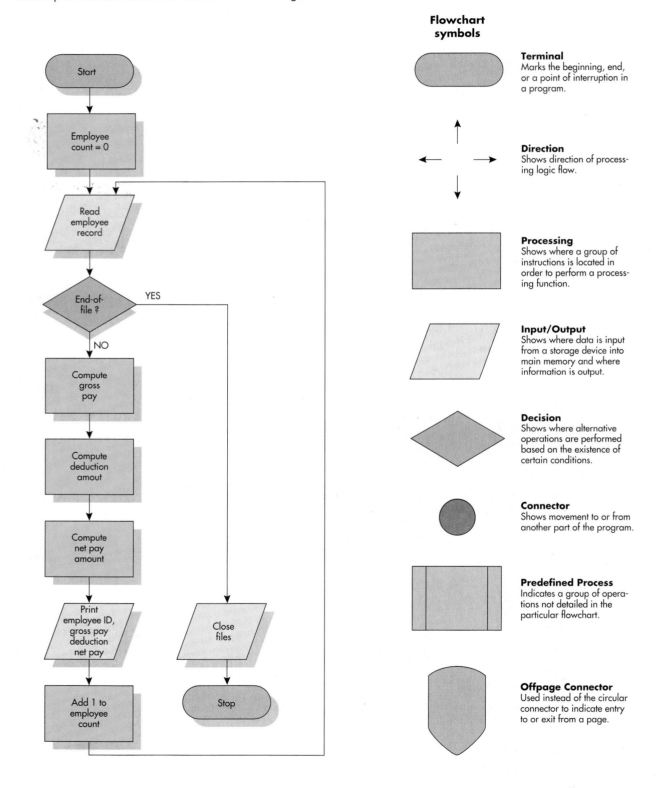

Flowchart symbols

Terminal
Marks the beginning, end, or a point of interruption in a program.

Direction
Shows direction of processing logic flow.

Processing
Shows where a group of instructions is located in order to perform a processing function.

Input/Output
Shows where data is input from a storage device into main memory and where information is output.

Decision
Shows where alternative operations are performed based on the existence of certain conditions.

Connector
Shows movement to or from another part of the program.

Predefined Process
Indicates a group of operations not detailed in the particular flowchart.

Offpage Connector
Used instead of the circular connector to indicate entry to or exit from a page.

The symbols at the left of the drawing might seem clear enough. But how do you know how to express the *logic* of a program? How do you know how to reason it out so it will really work? The answer is to use control structures.

- **Control structures:** When you're trying to determine the logic behind something, you use words like "if" and "then" and "else." (For example, without actually using these exact words, you might reason something like this: "*If* she comes over, *then* we'll go out to a movie, *else* I'll just stay in and watch TV.") Control structures make use of the same words. **A *control structure*, or *logic structure*, is a structure that controls the logical sequence in which computer program instructions are executed. In structured program design, three control structures are used to form the logic of a program: *sequence*, *selection*, and *iteration* (or *loop*). (■ *See Panel 11.7 on the next page.)* These are the tools with which you can write structured programs and take a lot of the guesswork out of programming.

One thing that all three control structures have in common is *one entry* and *one exit*. (Modules, as noted in Panel 11.4, rule 4, also have only one entry and one exit.) The control structure is entered at a single point and exited at another single point. This helps simplify the logic so that it is easier for others following in a programmer's footsteps to make sense of the program. (In the days before this requirement was instituted, programmers could have all kinds of variations, leading to the kind of incomprehensible program known as *spaghetti code.*)

Let us consider the three control structures:

(1) **In the *sequence control structure*, one program statement follows another in logical order.** In the example shown in Panel 11.7, there are two boxes (*statement* and *statement*). One could say "Open file," the other "Read a record." There are no decisions to make, no choices between "yes" or "no." The boxes logically follow one another in sequential order.

(2) **The *selection control structure*—also known as an *IF-THEN-ELSE* structure—is a structure that represents choice. It offers two paths to follow when a decision must be made by a program.** An example of a selection structure is as follows:

   IF a worker's hours in a week exceed 40
   THEN overtime hours equal the number of hours exceeding 40
   ELSE the worker has no overtime hours.

   A variation on the usual selection control structure is the *case control structure.* This offers more than a single yes-or-no decision. The case structure allows several alternatives, or "cases," to be presented. ("IF Case 1 occurs, THEN do thus-and-so. IF Case 2 occurs, THEN follow an alternative course . . . " And so on.) The case control structure saves the programmer the trouble of having to indicate a lot of separate IF-THEN-ELSE conditions.

(3) **The *iteration*, or *loop*, *control structure* is a structure in which a process may be repeated as long as a certain condition remains true.** There are two types of iteration structures—*DO UNTIL* (the most common form) and *DO WHILE.*

   An example of a DO UNTIL structure is as follows:
   DO read in employee records UNTIL there are no more employee records.

   An example of a DO WHILE structure is as follows:
   DO read in employee records WHILE—that is, as long as—there continue to be employee records.

The three control structures

The three structures used in structured program design to form the logic of a program are *sequence, selection,* and *iteration.*

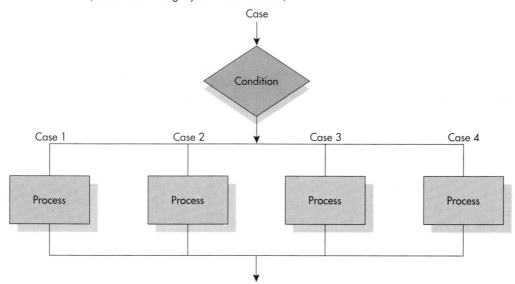

Sequence control structure
(one program statement follows another in logical order)

statement

statement

Selection control structure
(IF-THEN-ELSE)

yes — IF (test condition) — no

THEN (statement)　　ELSE (statement)

**Iteration control structures:
DO UNTIL and DO WHILE**

DO UNTIL

loop statement

DO UNTIL (test condition) — no

yes

DO WHILE

loop statement

DO WHILE (test condition) — yes

no

Variation on selection: the case control structure
(more than a single yes-or-no decision)

Case

Condition

Case 1　　Case 2　　Case 3　　Case 4

Process　　Process　　Process　　Process

What seems to be the difference between the two iteration structures? It is simply this: if there are several statements that need to be repeated, you need to decide when to *stop* repeating them. You can decide to stop them at the *beginning* of the loop, using the DO WHILE structure. Or you can decide to stop them at the *end* of the loop, using the DO UNTIL structure. The DO UNTIL iteration means that the loop statements will be executed at least once. This is because the iteration statements are executed *before* you are asked whether to stop.

3. Do a Structured Walkthrough

No doubt you've had the experience of, after having read over your research paper or project several times, being surprised when a friend (or instructor) pointed out some things you missed. The same thing happens to programmers.

In the ***structured walkthrough*, a programmer leads other people in the development team through a segment of code.** The structured walkthrough is actually an established part of the design phase. It consists of a formal review process in which others—fellow programmers, systems analysts, and perhaps users—scrutinize ("walk through") the programmer's work. They review the parts of the program for errors, omissions, and duplications in processing tasks. Because the whole program is still on paper at this point, these matters are easier to correct now than they will be later. Some programmers get very nervous before a structured walkthrough, treating it as some sort of test of their competence. Others see it merely as a cooperative endeavor.

The Third Step: Code the Program

Preview & Review: Coding the program is actually writing the program, translating the logic of the design into a programming language. It consists of choosing the appropriate programming language and following its rules, or syntax, exactly.

Once the design has been developed and reviewed in a walkthrough, the actual writing of the program begins. ***Writing the program* is called coding.** (■ *See Panel 11.8.*) Coding is what many people think of when they think of programming, although it is only one of the five steps. Coding consists of translating the logic requirements from pseudocode or flowcharts into a

1. Select the appropriate high-level programming language.
2. Code the program in that language, following the syntax carefully.

PANEL 11.8

Third step: program coding

The third step in programming is to translate the logic of the program worked out from pseudocode or flowcharts into a high-level programming language, following its grammatical rules.

programming language—the letters, numbers, and symbols that make up the program.

1. Select the Appropriate Programming Language

A *programming language* is a set of rules that tells the computer what operations to do. Examples of well-known programming languages are BASIC, COBOL, Pascal, FORTRAN, and C. These languages are called "high-level languages," as we explain in a few pages.

Not all languages are appropriate for all uses. Thus, the language needs to be chosen based on such considerations as what purpose the program is designed to serve and what languages are already being used in the organization or field you are in. We consider these matters in the second half of this chapter.

2. Follow the Syntax

For a program to work, you have to follow the *syntax,* **the rules of a programming language.** Programming languages have their own grammar just as human languages do. But computers are probably a lot less forgiving if you use these rules incorrectly.

The Fourth Step: Test the Program

Preview & Review: Testing the program consists of desk-checking, debugging the program of errors, and running real-world data to make sure the program works.

***Program testing* involves running various tests, such as desk-checking and debugging, and then running real-world data to make sure the program works.** (■ *See Panel 11.9.*)

Two principal activities are *desk-checking* and *debugging.*

Fourth step: program testing

The fourth step is to test the program and "debug" it of errors so it will work properly. The word "bug" dates from 1945, when a moth was discovered lodged in the wiring of the Mark I computer. The moth disrupted the execution of the program.

1. Desk-check the program to discover errors.
2. Run the program and debug it.
3. Run real-world data.

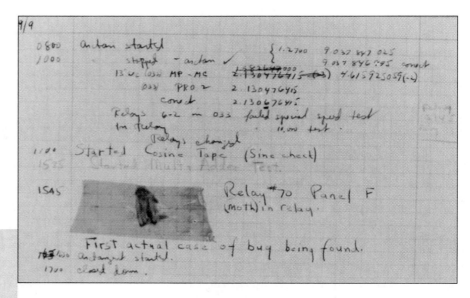

README

Practical Matters: Chasing Killer Computer Bugs

If you're reading this in the sealed comfort of an airplane, you may want to stop right now—or if you've ever felt safe while undergoing any medical treatment, or expect the bank to maintain your correct balance.

The potentially disastrous common ground in all these technologies, and hundreds of others, is not thick-fingered or ill-trained humans, but something much more dangerous: faulty software, says Ivars Peterson, in *Chasing Killer Computer Bugs.*

Computers are an inescapable part of modern life, and almost every hour of our day is touched by high technology. From coffee makers to cancer therapy, highly sophisticated computer technology keeps the operations flowing smoothly and safely. Or so we hope. Peterson argues that we have reached a point in the computer's evolution where we face serious doubts about reliability.

"Most people would be shocked and surprised if their new dining room table collapsed, but they accept as normal that, when they first install a computer system, it will fail frequently and become reliable only after a long sequence of revisions," he writes.

Peterson . . . takes us behind the scenes with those who hunt the computer problems that plague our lives and shows us frightening examples of how these failures can lead to disaster. He intricately explores more than a dozen major computer foul-ups, ranging from deadly crashes involving the computer-reliant A320 airliner to the Therac25, a radiation therapy machine that had an ugly tendency to give cancer patients enormous electrical shocks.

What Peterson discovers is the frequent lack of communication between the computer programmers who design the faulty systems and the systems' users. Take the example of a heart and lung machine used for life support during open-heart surgery. Controlled by a computer, the machine may stop suddenly for a variety of reasons specified by its designers.

But its reason for shutting down may not always be obvious to those charged with keeping the patient alive. An intentional delay may look like a crisis. Some physicians may press every button in sight trying to get the machine to respond. A minor incident like the microwave overcooking your meat loaf isn't tragic, but if a critical piece of machinery malfunctions (or even appears to malfunction) during open-heart surgery, someone's life is on the line. . . .

The problem, he notes, is that computer programming is a highly creative, complex effort requiring many of the same skills that go into writing a good novel. Every computer program is unique. There's no single, correct approach; no single, ideal computer language. Imagine writing a novel with the length and complexity of an encyclopedia, and you have an idea of what's involved.

Computer scientist Frederick P. Brooks, Jr., of the University of North Carolina once observed, "Of all the monsters who fill the nightmares of our folklore, none terrify more than werewolves, because they transform unexpectedly from the familiar into horrors. . . . The familiar software project has something of this character . . . usually innocent and straightforward, but capable of becoming a monster of missed schedules, blown budgets, and flawed products. So we hear desperate cries for a silver bullet."

But there is no silver bullet, writes Peterson, "We can gradually rein in the digital beast that has so rapidly come to dominate our lives or learn how to minimize the risks that flawed computer and misguided information pose." While the former is unthinkable, the latter often seems complex and overwhelming.

The safety and trustworthiness of critical computer systems rest on testing, mathematical review, and verification of personnel and process. Unfortunately, there is no authority in software engineering comparable to those that certify engineers in other areas. No other essential technology, from the building trades to automobile design and manufacture, remains as unregulated as software design.

Although certification of programmers may help, it may only create a false sense of security. The final creepy realization of this gripping book is that the computer only reflects our mental prowess; it is as capable of magnifying our flaws as amplifying our genius. . . .

—James Daly, "Scary Look at Unreliability of Computer Technology," *San Francisco Chronicle*

1. Perform Desk-Checking

***Desk-checking* is simply reading through, or checking, the program to make sure that it's free of errors and that the logic works.** In other words, desk-checking is sort of like proofreading. This step should be taken before the program is actually run on a computer.

2. Debug the Program

Once the program has been desk-checked, further errors, or "bugs," will doubtless surface. (The term "bug" derives from a 1945 incident in which a computer suddenly stopped because a moth got inside the machine.) **To debug means to detect, locate, and remove all errors in a computer program.**

Mistakes may be syntax errors or logic errors. *Syntax errors* **are caused by typographical errors and incorrect use of the programming language.** *Logic errors* **are caused by incorrect use of control structures.** Programs called *diagnostics* exist to check program syntax and display syntax-error messages. Diagnostic programs thus help identify and solve problems.

3. Run Real-World Data

After desk-checking and debugging, the program may run fine—in the laboratory. However, it then needs to be tested with data from the real world. Indeed, it is even advisable to test it with *bad data*—data that is faulty, incomplete, or in overwhelming quantities—to see if you can make the system crash. Many users, after all, may be far more heavy handed, ignorant, and careless than programmers have anticipated.

The testing process may take several trials using different data before the programming team is satisfied the program can be released. Even then, some bugs may remain, but there comes a point at which the pursuit of errors becomes uneconomical. This is one reason many users are nervous about using the first version (version 1.0) of a commercial software package.

The Fifth Step: Document the Program

Preview & Review: Documenting the program consists of writing a description of its purpose and process. Documentation should be prepared for users, operators, and programmers.

Writing the program documentation **is the fifth step in programming.** The resulting *documentation* **consists of written descriptions of what a program is and how to use it.** Documentation is not an end-stage process of programming. It has been (or should have been) going on throughout all previous programming steps. Documentation is needed for people who will be using or involved with the program in the future. (■ *See Panel 11.10.*)

Documentation should be prepared for different kinds of readers—users, operators, and programmers.

1. Write User Documentation

When you buy a commercial software package, such as a spreadsheet, you usually get a manual with it. This is *user documentation.* Programmers need to write documentation to help nonprogrammers use the software.

2. Write Operator Documentation

The people who keep mainframes and minicomputers running are called computer operators. Because they are not always programmers, they need to be told what to do when the program flashes an error message. The *operator documentation* gives them this information.

1. Write user documentation.
2. Write operator documentation.
3. Write programmer documentation.

PANEL 11.10

Fifth step: program documentation

The fifth step is really the culmination of activity that has been going on through all the programming steps—documentation. Developing written descriptions and procedures about a program and how to use it needs to be done for different people—users, operators, and programmers.

3. Write Programmer Documentation

Long after the original programming team has disbanded, the program may still be in use. If, as often happens, one-fifth of the programming staff leaves every year, after 5 years there could be a whole new group of programmers who know nothing about the software. *Program documentation* helps train these newcomers and enables them to maintain the existing system.

A word about maintenance: **Maintenance is any activity designed to keep programs in working condition, error-free, and up to date.** Maintenance includes adjustments, replacements, repairs, measurements, tests, and so on. Modern organizations are changing so rapidly—in products, marketing strategies, accounting systems, and so on—that these changes are bound to be reflected in their computer systems. Thus, maintenance is an important matter, and documentation must be available to help programmers make adjustments in existing systems.

The five steps of the programming process and their substeps are summarized in the accompanying table. (■ *See Panel 11.11, next page.*)

Five Generations of Programming Languages

Preview & Review: Languages are said to have evolved in "generations," from machine language to natural languages. The five generations are machine language, assembly language, high-level languages, very-high-level languages, and natural languages.

As we've said, a *programming language* is a set of rules that tells the computer what operations to perform. Programmers, in fact, use these languages to create *other* kinds of software. Many programming languages have been written, some with colorful names (SNOBOL, HEARSAY, DOCTOR, ACTORS, JOVIAL). Each is suited to solving particular kinds of problems. What is it that all these languages have in common? Simply this: ultimately they must be expressed to the computer in digital form—a 1 or 0, electricity *on* or *off*—because that is all the computer knows.

Step	Activities
Step 1: Problem definition	1. Specify program objectives and program users. 2. Specify output requirements. 3. Specify input requirements. 4. Specify processing requirements. 5. Study feasibility of implementing program. 6. Document the analysis.
Step 2: Program design	1. Determine program logic through top-down approach and modularization, using hierarchy chart. 2. Design details using pseudocode and/or using flowcharts, preferably using control structures. 3. Test design with structured walkthrough.
Step 3: Program coding	1. Select the appropriate high-level programming language. 2. Code the program in that language, following the syntax carefully.
Step 4: Program testing	1. Desk-check the program to discover errors. 2. Run the program and debug it. 3. Run real-world data.
Step 5: Program documentation	1. Write user documentation. 2. Write operator documentation. 3. Write programmer documentation.

To begin to see how this works, it's important to understand that there are five *levels* or *generations* of programming languages, ranging from low-level to high-level. **The five *generations of programming languages* start at the lowest level with (1) machine language. They range up through (2) assembly language, (3) high-level languages, and (4) very-high-level languages. At the highest level are (5) natural languages.** Programming languages are said to be *lower level* when they are closer to the language that the computer itself uses—the 1s and 0s. They are called *higher level* when they are closer to the language people use—more like English, for example.

Beginning in 1945, the five levels have evolved over the years, with later generations gradually coming into greater use with programmers. The births of the generations are as follows. (■ *See Panel 11.12.*)

- First generation, 1945—*Machine language*
- Second generation, mid-1950s—*Assembly language*
- Third generation, early 1960s—*High-level languages:* FORTRAN, COBOL, BASIC, Pascal, C, RPG
- Fourth generation, early 1970s—*Very-high-level languages:* RPG III, SQL, Intellect, NOMAD, FOCUS
- Fifth generation, early 1980s—*Natural languages*

Let us consider these five generations.

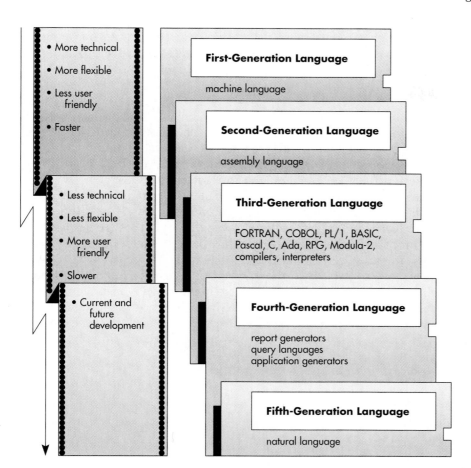

The five generations of programming languages

First Generation: Machine Language

The lowest level of language, ***machine language* is the basic language of the computer, representing data as 1s and 0s** (✓ p. 166). (■ *See Panel 11.13.)* Machine language programs varied from computer to computer; that is, they were *machine dependent.*

These binary digits, which correspond to the on and off electrical states of the computer, clearly are not convenient for people to read and use. Believe

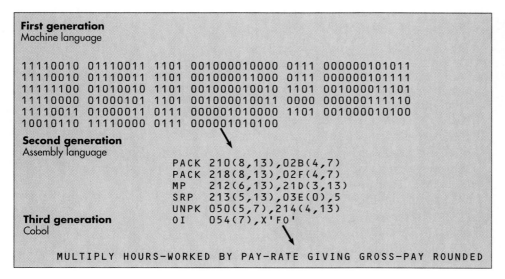

PANEL 11.13

Three generations of programming languages

(Top) Machine language is all binary 0s and 1s—very difficult for people to work with. *(Middle)* Assembly language uses abbreviations for major instructions (such as MP for MULTIPLY). This is easier for people to use, but still quite difficult. *(Bottom)* COBOL, a third-generation language, uses English words that can be understood by people.

it or not, though, programmers *did* work with these mind-numbing digits. When the next generation of programming languages came along, assembly language, there must have been great sighs of relief.

Second Generation: Assembly Language

Assembly language **is a low-level language that allows a programmer to use abbreviations or easily remembered words instead of numbers.** *(Refer to Panel 11.13 again.)* For example, the letters MP could be used to represent the instruction MULTIPLY, and STO to represent STORE.

As you might expect, a programmer can write instructions in assembly language faster than in machine language. Nevertheless, it is still not an easy language to learn, and it is so tedious to use that mistakes are frequent. Moreover, assembly language has the same drawback as machine language in that it varies from computer to computer—it is machine dependent.

We now need to introduce the concept of *language translator.* Because a computer can execute programs only in machine language, a translator or converter is needed if the program is written in any other language. **A** *language translator* **is a type of systems software (✓ p. 106) that translates a program written in a second-, third-, or a higher-generation language into machine language.**

Language translators are of three types:

- Assemblers
- Compilers
- Interpreters

An *assembler,* **or** *assembler program,* **is a program that translates the assembly-language program into machine language.** We describe compilers and interpreters in the next section.

Third Generation: High-Level Languages

A *high-level language* **is an English-like language,** such as COBOL, which is used for business applications. *(Refer again to Panel 11.13.)* A high-level language allows users to write in a familiar notation, rather than numbers or abbreviations. Most high-level languages are not machine dependent—they can be used on more than one kind of computer. Examples of familiar languages of this sort are FORTRAN, COBOL, BASIC, Pascal, C, and Ada.

Assembly language requires an assembler as a language translator. The translator for high-level languages is, depending on the language, either a *compiler* or an *interpreter.*

- **Compiler—execute later:** **A** *compiler* **is a language translator that converts the** *entire* **program of a high-level language into machine language before the computer executes the program.** The high-level language is called the *source code.* The compiler translates it into machine language, which in this case is called the *object code.* The significance of this distinction is that the object code *can be saved.* Thus, it can be executed later rather than run right away. (■ *See Panel 11.14.)*

  Examples of high-level languages using compilers are COBOL, FORTRAN, and Pascal.

- **Interpreter—execute immediately:** **An** *interpreter* **is a language translator that converts each high-level language statement into machine lan-**

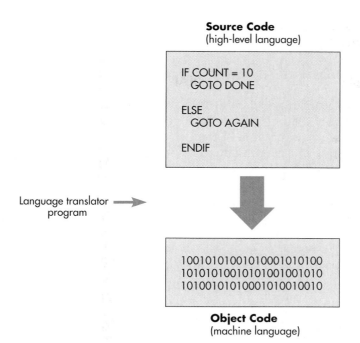

Source Code
(high-level language)

```
IF COUNT = 10
    GOTO DONE

ELSE
    GOTO AGAIN

ENDIF
```

Language translator
program ➙

```
100101010010100010101100
101010100101010010010010
1010010101010001010010010
```

Object Code
(machine language)

PANEL 11.14

Compiler

This language translator converts the high-level language (source code) into machine language (object code) before the computer can execute the program.

guage and executes it immediately, statement by statement. No object code is saved, as with the compiler. However, with an interpreter, processing seems to take less time.

An example of a high-level language using an interpreter is BASIC.

Third-generation, high-level languages are also known as *procedural languages.* That is, programs set forth precise procedures, or series of instructions, so that the programmer had to follow a proper order of actions to solve a problem. This meant the programmer had to have a detailed knowledge of programming and the computer it would run on. For example, say you want to take a taxi to a theater showing a particular movie. If you tell the taxi driver precisely *how* to get to the theater, that's *procedural.* You have to know how to get there yourself, and you will probably get there efficiently. However, if you simply tell the taxi driver to "take me to see movie X," then you're saying only what you *want,* which is nonprocedural. In this case, you may not get to the theater in an efficient manner.

The fourth generation of languages are *nonprocedural languages,* as we shall explain.

Fourth Generation: Very-High-Level Languages

A *very-high-level language* is often called a *4GL,* for 4th-generation language. 4GLs are much more user-oriented and allow programmers to develop programs with fewer commands compared with third-generation languages. 4GLs are called *nonprocedural* because programmers can write programs that need only tell the computer what they want done, not all the procedures for doing it. That is, they do not have to specify all the programming logic or otherwise tell the computer *how* the task should be carried out. This saves programmers a lot of time because they do not need to write as many lines of code as they do with procedural languages.

Fourth-generation languages consist of report generators, query languages, application generators, and interactive database management system (DBMS,

✓ p. 414) programs. Some 4GLs are tools for end-users, some are tools for programmers.

- **Report generators:** A *report generator,* **also called a report writer, is a program for end-users that is used to produce a report.** The report may be a printout or a screen display. It may show all or part of a database file. You can specify the format in advance—columns, headings, and so on— and the report generator will then produce data in that format.

  An example of a report generator is RPG III. Report generators were the precursor to today's query languages.

- **Query languages:** A *query language* **is an easy-to-use language for retrieving data from a database management system.** The query may be expressed in the form of a sentence or near-English command. Or the query may be obtained from choices on a menu.

  Examples of query languages are SQL (for structured query language), QBE (query by example), (✓ p. 420), and Intellect.

- **Application generators:** An *application generator* **is a programmer's tool that allows a person to give a detailed explanation of what data needs to be processed. The software then generates the code needed to create a program to perform the task.** The benefit is that the programmer does not need to specify *how* the data should be processed. The application generator is able to do this because it consists of modules that have been pre-programmed to accomplish various tasks.

  Programmers use application generators to help them create parts of other programs. For example, the software is used to construct on-screen menus or types of input and output screen formats.

  NOMAD and FOCUS, two relational database management systems, include application generators.

4GLs will probably not replace third-generation languages because they are usually focused on specific tasks and hence offer fewer options. Still, they improve productivity because programs are easy to write.

Fifth Generation: Natural Languages

Natural languages are of two types. The first are ordinary human languages: English, Spanish, and so on. The second are programming languages that use human language to give people a more natural connection with computers.

PANEL 11.15

Timeline for development of programming languages

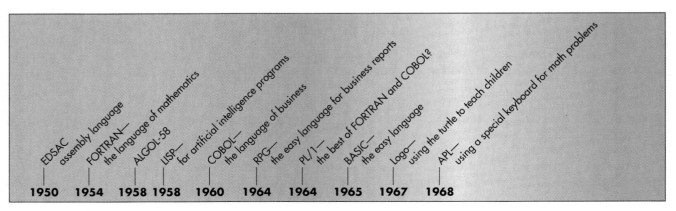

Some of the query languages mentioned above under 4GLs might seem pretty close to human communication, but natural languages—which are still in their infancy—try to be even closer.

With 4GLs, you can type in some rather routinized inquiries. An example of a request in FOCUS might be:

```
SUM SHIPMENTS BY STATE BY DATE.
```

Natural languages allow questions or commands to be framed in a more conversational way or in alternative forms. For example, with a natural language, you might be able to state:

```
I WANT THE SHIPMENTS OF PERSONAL DIGITAL ASSISTANTS FOR
ALABAMA AND MISSISSIPPI BROKEN DOWN BY CITY FOR JANUARY AND
FEBRUARY. ALSO, I NEED JANUARY AND FEBRUARY SHIPMENTS LISTED
BY CITIES FOR PERSONAL COMMUNICATORS SHIPPED TO WISCONSIN AND
MINNESOTA.
```

Natural languages are part of the field of study known as artificial intelligence (discussed in detail in Chapter 12). ***Artificial intelligence (AI) is a group of related technologies that attempt to develop machines to emulate human-like qualities, such as learning, reasoning, communicating, seeing, and hearing.***

The dates of the principal programming languages are shown in the accompanying two-page timeline. (■ *See Panel 11.15.*)

Principal Programming Languages Used Today

Preview & Review: The major third-generation programming languages used today are FORTRAN, COBOL, BASIC, Pascal, C, and Ada.

Let us now turn back and consider some of the major third-generation, or high-level, languages in use today. The most significant are FORTRAN, COBOL, BASIC, Pascal, C, and Ada.

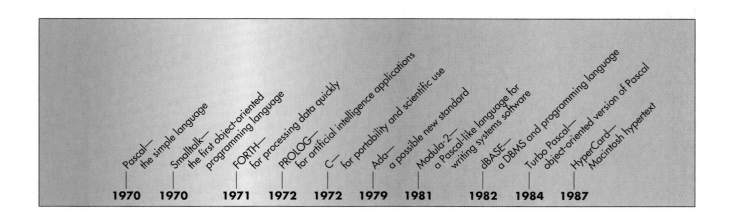

FORTRAN

```
IF (XINVO .GT. 500.00) THEN

        DISCNT = 0.07 * XINVO

ELSE

        DISCNT = 0.0

ENDIF

XINVO = XINVO – DISCNT
```

COBOL

```
OPEN-INVOICE-FILE.
    OPEN I-O INVOICE FILE.

READ-INVOICE-PROCESS.
    PERFORM READ-NEXT-REC THROUGH READ-NEXT-REC-EXIT UNTIL END-OF-FILE.
    STOP RUN.

READ-NEXT-REC.
    READ INVOICE-REC
        INVALID KEY
            DISPLAY 'ERROR READING INVOICE FILE'
            MOVE 'Y' TO EOF-FLAG
            GOTO READ-NEXT-REC-EXIT.
    IF INVOICE-AMT > 500
        COMPUTE INVOICE-AMT = INVOICE-AMT – (INVOICE-AMT * .07)
        REWRITE INVOICE-REC.

READ-NEXT-REC-EXIT.
    EXIT.
```

BASIC

```
10  REM     This Program Calculates a Discount Based on the Invoice Amount
20  REM         If Invoice Amount is Greater Than 500, Discount is 7%
30  REM         Otherwise Discount is 0
40  REM
50  INPUT "What is the Invoice Amount"; INV.AMT
60  IF INV.AMT A> 500 THEN LET DISCOUNT = .07 ELSE LET DISCOUNT = 0
70  REM         Display results
80  PRINT "Original Amt", "Discount", "Amt after Discount"
90  PRINT INV.AMT, INV.AMT * DISCOUNT, INV.AMT – INV.AMT * DISCOUNT
100 END
```

PANEL 11.16

Principal third-generation languages compared: six examples

This shows how six languages handle the same statement. The statement specifies that a customer gets a discount of 7% of the invoice amount if the invoice is greater than $500; if the invoice is lower, there is no discount.

FORTRAN: The Language of Mathematics & the First High-Level Language

Developed in 1954 by IBM, **FORTRAN (for FORmula TRANslator) was the first high-level language.** (■ *See Panel 11.16.*) Originally designed to express mathematical formulas, it is still the most widely used language for mathematical, scientific, and engineering problems. It is also useful for complex business applications, such as forecasting and modeling. However, because it cannot handle a large volume of input/output operations or file processing, it is not used for more typical business problems.

FORTRAN has both advantages and disadvantages:

- **Advantages:** (1) FORTRAN can handle complex mathematical and logical expressions. (2) Its statements are relatively short and simple.

Pascal

```
if INVOICEAMOUNT > 500.00 then

    DISCOUNT := 0.07 * INVOICEAMOUNT

else

    DISCOUNT := 0.0;

INVOICEAMOUNT := INVOICEAMOUNT – DISCOUNT
```

PANEL 11.16

Continued

C

```
if (invoice_amount > 500.00)

    DISCOUNT = 0.07 * invoice_amount;

else

    discount = 0.00;

invoice_amount = invoice_amount – discount;
```

Ada

```
if INVOICE_AMOUNT > 500.00 then

    DISCOUNT := 0.07 * INVOICE_AMOUNT

else

    DISCOUNT := 0.00

endif;

INVOICE_AMOUNT := INVOICE_AMOUNT – DISCOUNT
```

(3) FORTRAN programs developed on one type of computer can often be easily modified to work on other types.

- **Disadvantages:** (1) FORTRAN does not handle input and output operations to storage devices as efficiently as some other higher-level languages. (2) It has only a limited ability to express and process nonnumeric data. (3) It is not as easy to read and understand as some other high-level languages.

COBOL: The Language of Business

Formally adopted in 1960, **COBOL (for COmmon Business Oriented Language) is the most frequently used programming language in business.** *(Refer again to Panel 11.16.)* Its most significant attribute is that it is extremely readable. For example, a COBOL line might read:

```
MULTIPLY HOURLY-RATE BY HOURS-WORKED GIVING GROSS-PAY
```

COBOL is the language used by the majority of mainframe users. There are now compilers that will allow COBOL programs written for mainframes to be run on microcomputers. First standardized in 1968 by the American National Standards Institute (ANSI), the language has been revised several times.

Writing a COBOL program resembles writing an outline for a research paper. The program contains four divisions—Identification, Environment, Data, and Procedure. The divisions in turn are broken into sections, which are divided into paragraphs, which are further divided into sections. The *Identification Division* identifies the name of the program and the author (programmer) and perhaps some other helpful comments. The *Environment Division* describes the computer on which the program will be compiled and executed. The *Data Division* describes what data will be processed. The *Procedure Division* describes the actual processing procedures.

COBOL, too, has both advantages and disadvantages.

- **Advantages:** (1) It is machine independent. (2) Its English-like statements are easy to understand, even for a nonprogrammer. (3) It can handle many files, records, and fields. (4) It easily handles input/output operations.

- **Disadvantages:** (1) Because it is so readable, it is wordy. Thus, even simple programs are lengthy, and programmer productivity is slowed. (2) It cannot handle mathematical processing as well as FORTRAN.

BASIC: The Easy Language

BASIC was developed by John Kemeny and Thomas Kurtz in 1965 for use in training their students at Dartmouth College. By the late 1960s, it was widely used in academic settings on all kinds of computers, from mainframes to PCs. Now its use has extended to business.

BASIC (for *Beginner's All-purpose Symbolic Instruction Code*) is the most popular microcomputer language and is considered the easiest programming language to learn. *(Refer again to Panel 11.16.)* Although it is available in compiler form, the interpreter form is more popular with first-time and casual users. This is because it is interactive, meaning that user and computer can communicate with each other during the writing and running of the program.

Today there is no one version of BASIC. The three most popular versions are *True BASIC* (written by Kemeny and Kurtz), *BASIC* from Microsoft, and *Visual BASIC for Windows.*

The advantages and disadvantages of BASIC are as follows:

- **Advantage:** BASIC is very easy to use.
- **Disadvantages:** (1) Its processing speed is relatively slow, although compiler versions are faster than interpreter versions. (2) There is no one version of BASIC, although in 1987 ANSI adopted a new standard that eliminated portability problems.

Pascal: The Simple Language

Named after the 17th-century French mathematician Blaise Pascal, **Pascal is an alternative to BASIC as a language for teaching purposes and is relatively easy to learn.** *(Refer back to Panel 11.16.)* A difference from BASIC is that Pascal uses structured programming.

Pascal, a compiled language, offers these advantages and disadvantages to users:

- **Advantages:** (1) Pascal is easy to learn. (2) It has extensive capabilities for graphics programming. (3) It is excellent for scientific use.

- Disadvantage: Pascal has limited input/output programming capabilities, which limits its business applications.

C: For Portability & Scientific Use

"C" is the language's entire name, and it does not "stand for" anything. Developed at Bell Laboratories, **C is a general-purpose, compiled language that works well for microcomputers and is portable among many computers.** *(Refer back to Panel 11.16.)* It is useful for writing operating systems, database management software, and some scientific applications. Indeed, C is now the programming language used most commonly in commercial software development, including games, robotics, and graphics. It is now considered a necessary language for programmers to know.

Here are the advantages and disadvantages of C:

- Advantages: (1) C works well with microcomputers. (2) It has a high degree of portability—it can be run without change on a variety of computers. (3) It is fast and efficient. (4) It enables the programmer to manipulate individual bits in main memory.
- Disadvantages: (1) C is considered difficult to learn. (2) Because of its conciseness, the code can be difficult to follow. (3) It is not suited to applications that require a lot of report formatting.

Ada: A Possible New Standard

Ada **is an extremely powerful structured programming language designed by the U.S. Department of Defense to ensure portability of programs from one application to another.** *(Refer back to Panel 11.16.)* Ada was named for Countess Ada Lovelace, considered the world's "first programmer." Based on Pascal, Ada was originally intended to be a standard language for weapons systems. However, it has also been used in commercial applications.

The pluses and minuses of Ada are these:

- Advantages: (1) Ada is a structured language, with a modular design. This means pieces of a large program can be written and tested separately. (2) It has more input and output capability than Pascal, which might make it more favorable to industry. (3) It has features that permit the compiler to check it for errors before the program is run, so programmers using it can write error-free programs more easily.
- Disadvantages: (1) The amount of memory required hinders its use on microcomputers. (2) It has a high level of complexity and difficulty. (3) Business users already have so much invested in COBOL, FORTRAN, and C that they have little motivation to switch over to Ada.

Other Programming Languages

Preview & Review: Other important high-level languages are LISP, PL/1, RPG, Logo, APL, FORTH, PROLOG, Modula-2, and dBASE.

Several other high-level languages exist that, though not as popular or as famous as the foregoing, are well known enough that you may encounter them. Some of them are special-purpose languages.

LISP: For Artificial Intelligence Programs

LISP (for *LISt Processor*) **is a third-generation language used principally to construct artificial intelligence programs.** Developed at the Massachusetts Institute of Technology in 1958 by mathematician John McCarthy, LISP is used to write expert systems and natural language programs. *Expert systems* (✓ p. 455) are programs that are imbued with knowledge by a human expert; the programs can walk you through a problem and help solve it.

PL/1: The Best of FORTRAN & COBOL?

PL/1 was introduced in 1964 by IBM. *PL/1* (for *Programming Language 1*) **is a third-generation language designed to process both business and scientific applications.** PL/1 contains many of the best features of FORTRAN and COBOL and is quite flexible and easy to learn. However, it is also considered to have so many options as to diminish its usefulness. As a result, it has not given FORTRAN and COBOL much competition.

RPG: The Easy Language for Business Reports

RPG was also introduced in 1964 by IBM and has evolved through several important versions since. *RPG* (for *Report Program Generator*) **was a highly structured and relatively easy-to-learn third-generation language designed to help generate business reports.** The user filled out a special form specifying what information the report should include and in what kind of format.

In 1970, improvements were introduced in RPG II. Recently, a successor, RPG III, an interactive fourth-generation language, has appeared that uses menus to give programmers choices.

Logo: Using the Turtle to Teach Children

Logo was developed at MIT in 1967 by Seymour Papert, using a dialect of LISP. *Logo* **is a third-generation language designed primarily to teach children problem-solving and programming skills.** At the basis of Logo is a triangular pointer, called a "turtle," which responds to a few simple commands such as forward, left, and right. The pointer produces similar movements on the screen, enabling users to draw geometric patterns and pictures on screen. Because of its highly interactive nature, Logo is used not only by children but also to produce graphics reports in business.

APL: Using a Special Keyboard for Math Problems

APL was designed in 1968 by Kenneth Iverson for use on IBM mainframes. *APL* (for *A Programming Language*) **is a third-generation language that uses a special keyboard with special symbols to enable users to solve complex mathematical problems in a single step.** The special keyboard is required because the APL symbols are not part of the familiar ASCII (✓ p. 168) character set. Though hard to read, this mathematically oriented and scientific language is still found on a variety of computers.

FORTH: For Processing Data Quickly

FORTH was created in 1971 by Charles Moore. *FORTH* (for *FOuRTH-generation language*) **is actually a third-generation language designed for real-time control tasks, as well as business and graphics applications.** The program is used on all kinds of computers, from PCs to mainframes, and runs very fast because it requires less memory than other programs. Because it runs so fast, it is used in applications that must process data quickly. Thus, it is used to process data acquired from sensors and instruments, as well as in arcade game programs and robotics.

PROLOG: For Artificial Intelligence Applications

Invented in 1972 by Alan Colmerauer of France, PROLOG did not receive much attention until 1979, when a newer version appeared. ***PROLOG* (for *PROgramming LOGic*) is used for developing artificial intelligence applications,** such as natural language programs and expert systems.

Modula-2: A Pascal-Like Language for Writing Systems Software

Invented in 1981 by Swiss professor Niklaus Wirth, the creator of Pascal, Modula-2 is actually an expanded and improved version of Wirth's earlier language. ***Modula-2* (for *MODUlar LAnguage-2*) is a third-generation language similar to Pascal and is designed for writing systems software.** Like Pascal, it is highly structured. Modula-2 is very popular for teaching purposes at colleges and universities.

dBASE: A DBMS & a Programming Language

Created by Wayne Ratliff to manage a company football pool, dBASE was originally named Vulcan. It became dBASE II when the program was acquired by Ashton-Tate in 1982. (Ashton-Tate in turn was acquired by Borland International in 1991, and the most current version is dBASE V.)

Most users are familiar with ***dBASE* as a database management system (DBMS) for controlling the structure of a database and access to the data. However, dBASE is also a Pascal-like, fourth-generation programming language.**

Object-Oriented & Visual Programming

Preview & Review: Object-oriented programming (OOP) is a programming method that combines data and instructions for processing that data into a self-sufficient "object," or block of preassembled programming code, that can be used in other programs.

Three concepts of OOP are encapsulation, inheritance, and polymorphism.

Some examples of OOP languages are Smalltalk, C++, Turbo Pascal, and Hypertalk.

Consider how it was for the computer pioneers, programming in machine language or assembly language. Novices putting together programs in BASIC or Pascal can breathe a collective sigh of relief that they weren't around at the dawn of the Computer Age. Even some of the simpler third-generation languages represent a challenge. Fortunately, two new developments have made things easier—*object-oriented programming* and *visual programming.*

Object-Oriented Programming: Block by Block

Imagine you're programming in a traditional third-generation language, such as BASIC, creating your coded instructions one line at a time. As you work on some segment of the program (such as how to compute overtime pay), you may think, "I'll bet some other programmer has already written something like this. Wish I had it. It would save a lot of time."

Fortunately, a kind of recycling technique now exists. This is object-oriented programming. Let us explain this in four steps:

1. **What OOP is:** *Object-oriented programming (OOP)* **is a programming method that combines data and instructions for processing that data into a self-sufficient "object" that can be used in other programs.** The important thing here is the object.
2. **What an "object" is:** An *object* **is a block of preassembled programming code that is a self-contained module. The module contains, or encapsulates, both (1) a chunk of data and (2) the processing instructions that may be called on to be performed on that data.**
3. **When an object's data is to be processed—sending the "message":** Once the object becomes part of a program, the processing instructions may or may not be activated. That happens only when a "message" is sent. **A** *message* **is an alert sent to the object when an operation involving that object needs to be performed.**
4. **How the object's data is processed—the "methods":** The message need only identify the operation. How it is actually to be performed is embedded within the processing instructions that are part of the object. **These instructions about the operations to be performed on data within the object are called the** *methods.*

Once you've written a block of program code (that computes overtime pay, for example), it can be reused in any number of programs. Thus, unlike with traditional programming, with OOP you don't have to start from scratch—that is, reinvent the wheel—each time. (■ *See Panel 11.17.*)

Object-oriented programming takes longer to learn than traditional programming because it means training yourself to a new way of thinking. Once learned, however, the beauty of OOP is that an object can be used repeatedly in different applications and by different programmers, thereby speeding up development time and lowering costs.

Three Important Concepts of OOP

Object-oriented programming involves three important concepts, which go under the jaw-breaking names of *encapsulation, inheritance,* and *polymorphism.*[5] Actually, these terms are not as fearsome as they look:

* **Encapsulation:** *Encapsulation* **means an object contains (encapsulates) both (1) data and (2) the instructions for processing it,** as we have seen. Once an object has been created, it can be reused in other programs. An object's uses can also be extended through concepts of *class* and *inheritance.*
* **Inheritance:** Once you have created an object, you can use it as the foundation for similar objects that have the same behavior and characteristics. **All objects that are derived from or related to one another are said to form a** *class.* Each class contains specific instructions (methods) that are unique to that group.

Conventional Programs

Object-Oriented Programs

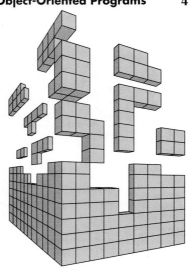

PANEL 11.17

Conventional versus object-oriented programs

(Top) When building conventional programs, programmers write every line of code from scratch. *(Bottom)* With object-oriented programs, programmers can use blocks, or "objects," of preassembled modules containing data and the associated processing instructions.

Classes can be arranged in hierarchies—classes and subclasses. *Inheritance* **is the method of passing down traits of an object from classes to subclasses in the hierarchy.** Thus, new objects can be created by *inheriting* traits from existing classes.

Writer Alan Freedman gives this example: "the object MACINTOSH could be one instance of the class PERSONAL COMPUTER, which could inherit properties from the class COMPUTER SYSTEMS."[6] If you were to add a new computer, such as COMPAQ, you would need to enter only what makes it *different* from other computers. The *general* characteristics of personal computers could be inherited.

- **Polymorphism:** Polymorphism means "many shapes." In object-oriented programming, *polymorphism* **means that a message (generalized request) produces different results based on the object that it is sent to.**

  Polymorphism has important uses. It allows a programmer to create procedures about objects whose exact type is not known in advance but will be at the time the program is actually run on the computer. Freedman gives this example: "a screen cursor may change its shape from an arrow to a line depending on the program mode." The processing instructions "to move the cursor on screen in response to mouse movement would be written for 'cursor,' and polymorphism would allow that cursor to be whatever shape is required at runtime." It would also allow a new cursor shape to be easily integrated into the program.[7]

Four Examples of OOP Languages

Some examples of object-oriented programming languages are Smalltalk, C++, Turbo Pascal, and Hypertalk.

- **Smalltalk—the first OOP language:** Smalltalk was invented by computer scientist Alan Kay in 1970 at Xerox Corporation's Palo Alto Research Center in California. *Smalltalk,* **the first OOP language, uses a keyboard for entering text, but all other tasks are performed with a mouse.**

- **C++—more than C:** **The plus signs in** *C++* **stand for "more than C" because it combines the traditional C programming language with object-oriented capability.** C++ was created by Bjarne Stroustrup. With C++, programmers can write standard code in C without the object-oriented features, use object-oriented features, or do a mixture of both.

- **Turbo Pascal—object-oriented Pascal:** Designed in 1984 by Philippe Kahn of Borland International, *Turbo Pascal* **is an object-oriented version of Pascal.** The language allows a program to use removable modules, or objects, that can be replaced later when the program is changed.

- **Hypertalk—the language for HyperCard:** HyperCard (✓ p. 88), the software introduced for the Apple Macintosh in 1987, is based on the concept of *cards* and *stacks* of cards—just like notecards, only they are electronic. A card is a screenful of data that makes up a single record; cards are organized into related files called stacks. Using a mouse, you can make your way through the cards and stacks to find information or discover connections between ideas.

  HyperCard is not precisely an object-oriented programming language, but a language called Hypertalk is. *Hypertalk,* **which uses OOP principles, is the language used in the HyperCard program to manipulate HyperCard stacks.**

R E A D M E

Case Study: Consumer-Testing Software—The Alpha & the Beta

Debugging programs—whether standard or object-oriented—involves not only various internal checks but also testing them with prospective consumers. Chris Peters, general manager of the Word business unit of Microsoft Corporation, describes how this is done.

Peters says the program is first put through its paces by real users in the Alpha testing stage.

"Alpha is when you're code complete," he says. "In other words, all the features are in, but the program is still very unstable. You really haven't done any quality checking."

Alpha testing is conducted within the software company and uses sophisticated testing labs. Users are placed in rooms set up like typical offices except for the one-way windows that cover a large portion of one of the walls. Programmers sit behind the window and watch as users, probably feeling like rats in a maze, try to figure out how to use a product. Software designers use many variations on the tests such as assigning specific tasks or taking away the users' software manuals.

The designers scrutinize almost every move a user makes during the testing sessions. Cameras record the scene and users are sometimes asked to narrate their actions into a microphone. (For example, "I clicked on Help to find out how to italicize a block of text. But I couldn't find the answer I needed, so I returned to my document.") Some companies compile videotapes of the users' moves and have production technicians that edit the films into records of specific problems.

By watching these "regular guy" testers, programmers quickly learn what features are successful and which ones might die on the cutting room floor. Programmers often get a lesson in humility during the Alpha testing.

"We always screw up, basically," Peters says. "It turns out we're not really quite as smart as we thought we were."

After the problems that showed up in Alpha testing are corrected, the software is sent out to Beta testers. At this point, the software is a fairly solid product; some problems may still exist but, for the most part, the program works. Beta testing uses feedback from users outside the software company. The testers are selected for how they represent various segments of the market, and they are asked for feedback on what they did and didn't like about the software.

—Trevor Meers, "From Statistics to Shrinkwrap: The Development of Software," *PC Novice*

Visual Programming

***Visual programming* is a method of creating programs in which the programmer makes connections between objects by drawing, pointing, and clicking on diagrams and icons.**

The goal of visual programming is to make programming easier for programmers and more accessible to nonprogrammers. It does so by borrowing the object orientation of OOP languages but exercising it in a graphical or visual way. Visual programming enables users to think more about solving the problem than about handling the programming language.

With one example, ObjectVision (from Borland), the user doesn't employ a programming language but simply connects icons and diagrams on screen. Visual BASIC offers another visual environment for program construction, allowing you to build various components using buttons, scroll bars, and menus.

Internet Programming

Preview & Review: Programming languages used to build linked, multimedia sites on the World Wide Web include HTML, VRML, and Java.

As we mentioned in Chapter 8, the Internet connects thousands of data and information sites around the world. Many of these sites are text-based only;

that is, the user sees no graphics, animation, or video, and hears no sound. To build multimedia sites on the World Wide Web, people use some recently developed programming languages: HTML, VRML, and Java.

- **HTML:** *HTML (hypertext markup language) is a type of programming language that embeds simple commands within standard ASCII text documents to provide an integrated, two-dimensional display of text and graphics.* In other words, a document created in any word processor and stored in ASCII format can become a Web page with the addition of a few HTML commands. One of the main features of HTML is its ability to insert hypertext (✓ p. 88) links into a document. Hypertext links enable you to display another Web document simply by clicking on a link area on the current screen. One document may contain links to many other related documents. The related documents may be on the same server (✓ p. 374) as the first document, or they may be on a computer halfway around the world. A link may be a word, a group of words, or a picture.

  Various HTML editors and filters—commercial HTML packages—exist to help people who do not want to learn everything about HTML to create their own Web pages. For example, WebSite Maker from THISoftware Co. is a Windows-based program that lets users build their own Web pages by filling out a template. Netscape Gold, a Web browser (✓ p. 79) also allows users to easily build their own Web sites. (Note that users generally need an ISP, ✓ p. 346, to act as a server.)

- **VRML:** *VRML (virtual reality markup language) is a type of programming language used to create three-dimensional Web pages.* VRML (rhymes with "thermal") is not an extension of HTML; thus, HTML Web browsers cannot interpret it. That is, users need a VRML Web browser to receive VRML Web pages. If they are not on a large computer system, they also need a high-end microcomputer such as a Power Macintosh or a Pentium-based PC. Like HTML, VRML is a document-centered ASCII language. Unlike HTML, it tells the computer how to create 3-D worlds. VRML pages can also be linked to other VRML pages.

  Even though VRML's designers wanted to let nonprogrammers create their own virtual spaces quickly and painlessly, it's not as simple to describe a 3-D scene as it is to describe a page in HTML. However, many existing modeling and CAD (✓ p. 86) tools now offer VRML support, and new VRML-centered software tools are arriving (such as Virtual Home Space Builder).

- **Java:** Java is a new programming language from Sun Microsystems. Basically, *Java is a way to create any conceivable type of software applications that will work on the Internet.* If the use of Java becomes widespread, the Web will be transformed from the information-delivering medium it is today into a completely interactive computing environment. You will be able to treat the Web as a giant hard disk loaded with a never-ending supply of software applications. However, Java is not compatible with most existing microprocessors, such as those from Intel and Motorola. For this reason, users need to use a small "interpreter" program to be able to use Java applications (called "applets"). They also need a recent-version operating system and a Java-capable browser in order to view Java special effects. In the meantime, Sun is working on developing special Java chips designed to run Java software directly.

  Java is a major departure from the HTML coding that makes up most Web pages. It is an interpreted programming environment, and the language is object-oriented and comparable to C++.

README

Case Study: The Net Changes Testing & Selling of Software

On the Internet's World Wide Web, nothing is ever finished—or, at least, beyond improvement. Some commercial software developers are marrying that principle and the Net itself with a distribution technique that might best be termed "Ship Early and Often."

They're in the vanguard of a welcome trend whereby developers publicly post their software. Then, for the cost of your connection, you download it and try it out. If it's prerelease, or beta, software, you tell them what's wrong with it. Eventually, you're supposed to buy it.

This practice is also blurring some supposedly clear lines, with interesting consequences. For example, it's getting harder to tell what is a beta product vs. the "shipping" version. Growing equally tenuous is the distinction between shareware—you try it before you register it, usually for a fee—and regular commercial software.

A case in point: Netscape Communications Corp. I [was] running a beta, or prerelease, copy of the company's Navigator 2.0 browser [before it was released as final]. I downloaded it from the company's Internet site, and [it worked well.] . . .

Netscape is not alone in using this strategy. Oracle Corp., known primarily for its corporate database products, is trying much the same thing with several new packages. So is Starfish, the latest venture of former Borland Chairman Philippe Kahn, with a new Windows software.

Part of this is a page from the shareware book. Shareware authors have been distributing their programs online for years; in that sense, the Netscape browser is just an incredibly popular piece of shareware.

Posting beta software, labeled as such, is another positive outcome of this trend. It has become all too normal for companies to put out products before they're really ready for prime time—and charge for them, anyway. Perhaps this practice is smart marketing because consumers haven't rebelled yet, but it isn't very ethical.

Letting people try out beta software is more than honest labeling. And it's more than a service for curious users who want to stay informed about what's coming along. Companies also get feedback that helps improve the products. And they gain "mind share" among leading-edge users; mind share can translate nicely into market share if the product works well. . . .

—Dan Gillmor, "Net Changes Testing, Selling of Software," *San Jose Mercury News*

Onward

What do object-oriented, visual, and Internet programming imply for the future? Will tomorrow's programmer look less like a writer typing out words and more like an electrician wiring together circuit components, as one magazine suggests?[8] What does this mean for the five-step programming model we have described?

Some institutions are now teaching only object-oriented design techniques, which allow the design and ongoing improvement of working program models. Here programming stages overlap, and users repeatedly flow through analysis, design, coding, and testing stages. Thus, users can test out new parts of programs and even entire programs as they go along. They need not wait until the end of the process to find out if what they said they wanted is what they really wanted.

This new approach to programming is not yet in place in business, but in a few years it may be. If you're interested in being able to communicate with programmers in the future, or in becoming one yourself, the new approaches are worth your attention.

How to Buy Software

You can buy music audiotapes at the 7-Eleven. Sometimes you can buy or rent videocassette tapes at the gas station minimart. Computer software is not quite there yet, but it may be getting close. You *can* buy software at Sears, Circuit City, and Price Club. But should you? Let's consider the various sales outlets.[9-11]

Getting Ready to Buy

Whether shopping for systems software or applications software, you need to be clear on a few things before you buy.

Do You Know Your Needs? Before talking to anyone about software you should have a clear idea of what you want your computer to do for you.

Are you mainly writing research papers? keeping track of performance of employees reporting to you? projecting sales figures? building a mailing list and launching a fund-raising campaign? publishing a newsletter? teaching children about computers? You want the machine to serve you, not vice versa.

Do You Know What Software You Want? The safest course is to pick software used successfully by people you know. Or look for ratings in the leading computer magazines. Brands that consistently get high ratings in magazine reviews are generally likely to be reliable.

Do You Know the Latest Version & Release? If you know the name of the software you want to use, or at least inquire about, so much the better. If you have a particular brand and type in mind, make sure it's the most recent version and release.

A new *version* of a software package resembles a model change on a car. It adds all kinds of new features, generally making the software more powerful and versatile. Versions are usually numbered in ascending order. MS-DOS 6.0 is a later version of Microsoft's Disk Operating System software than DOS 4.1 or DOS 3.3.

A *release number* identifies a specific version of a program. A program labeled 5.2, for example, is the third release of the fifth version. (The first and second releases were 5.0 and 5.1.) New releases generally incorporate routine enhancements and correct the annoying errors called *software bugs*.

Experienced users have a horror of using the very first version, Version 1.0, of anything because the software generally still contains numerous bugs.

Most software manufacturers are continually upgrading their products. Often the upgrades are made available at considerably less expense to purchasers of earlier versions. (Generally you have to have sent in the registration card contained in the software package to qualify for the reduced rate on the upgrade.)

Do You Know If an Upgrade Is Coming Out? Find out if a new version of the software you're interested in is just around the corner. You may want to hold off buying until it's available.

Upgrades are to software as new models are to the auto industry. Every year or so, software manufacturers bring out a new version or release featuring incremental improvements, just like the car makers do. (Microsoft's DOS operating system is in its sixth major incarnation, and Lotus 1-2-3 in its 12th.) If the original cost you $250 or $500, the upgrade will probably cost you only $99. However, you have to have sent in a warranty registration card to the manufacturer in order to qualify for the low price.

Will the Salespeople Speak Your Language? Selling ice cream does not require a lot of product knowledge. Selling software does (or should). Some salespeople know their wares but talk down to newcomers to try to impress them with their knowledge. Others have only the scantest familiarity with their products, although they may be patient with novices' questions. You want someone who is both knowledgeable and helpful. That may require a little investigation on your part.

Software Sellers: The Range of Outlets

The types of software sellers are as follows:

- Small hardware and software retail stores
- Small software-only retail stores
- Computer superstores
- Electronics, office, and department stores and warehouse clubs
- Mail-order companies and direct mail

Small Retail Stores for Both Hardware & Software A small retail store selling both computer hardware and software may be a good place to go if you need to buy a PC as well as software. Prices won't beat those of discounters and superstores, but you may be able to find a well-informed, knowledgeable staff. Examples of small retail stores are those found in chains of dealers such as CompuAdd, MicroAge, and Software City. Many office-products superstores also have software.

Small Retail Stores for Software Only If you already have a microcomputer, you may find the salespeople at software-only retail stores even better able to talk to you about the nuances of programs than those in the previous group. Stores that belong to national chains, such as Egghead Discount Software, Software Etc., and Babbage's, may carry a couple of thousand software titles. Generally such stores have a number of computers on the premises so you can try out software before buying. Prices are often discounted and may match those of computer superstores.

Computer Superstores Computer superstores range up to 30,000 square feet in size, versus, say, 2000 square feet for small retail stores. Not only do they carry all kinds of hardware, including computer furniture, but also upwards of 2000–3000 software titles. Like smaller stores, they offer computers for trying out software. Unlike some smaller stores, they also offer classes to train you in the use of particular software packages. Finally, they have extensive technical departments for installing software and readying and repairing hardware. Examples of chains with superstores are CompuAdd, CompUSA, and Computer City SuperCenters.

Electronics, Office, & Department Stores & Warehouse Clubs You can tell that computer programs have reached the mainstream when they are no longer sold as specialty items in specialty stores. Walk into electronics stores like Radio Shack, Circuit City, and Best Buy and you'll find software there, too. Large office-supply stores often have software departments. Some department stores such as Sears, Montgomery Ward, and Dayton Hudson carry both hardware and software. (Indeed, some Sears stores have special office centers, which carry a large line of computer products.) It's doubtful, however, that the selection and prices are as competitive as those in the other types of stores we've described.

Discount warehouse clubs include stores like Price CostCo, Office Depot, and SAM's Club. These are membership operations that provide consumers with merchandise—often including computer hardware and software—at severely discounted prices. Some warehouse clubs are oriented toward business owners and managers, and so most of their software offerings are most appropriate for business. Some drawbacks are that these stores may not have repair services, customer support, or salespeople with deep product knowledge.

Mail-Order Companies & Direct Mail The quickest and cheapest way to buy computer products, whether software or hardware, is by mail order. You dial a toll-free 800 number, tell the order taker what you want, and charge the purchase to your credit card. The product is delivered to you by delivery service, often overnight. As long as you know what you want and pick a reputable company, mail order is an effective—and cost-effective—way of getting software. Indeed, mail order is probably the cheapest way to get software.

There are two ways to order software by mail order.

- *Mail-order companies:* These companies sell all kinds of software (and perhaps some hardware "peripherals"— equipment like printers and monitors). Examples are CompuAdd Express, Software Unlimited, MicroWarehouse and PC Connection for IBM-compatibles, and MacWarehouse and Mac Connection for Macintoshes.

- *Direct mail:* Some software manufacturers make their products available through direct mail. Examples are Microsoft and IBM.

Sensible Software Shopping: Some Tips

In software shopping, you're concerned not only with getting a good price but also protecting yourself if things go wrong. Here are a few tips.

Ask What Follow-Up Help Is Available It's worth asking what kind of follow-up help the software seller offers. Actually, most retailers don't provide any such assistance, although, for an extra charge, they may provide classes. Many software packages come with some sort of tutorial to help you get started. The tutorial—an instructional book, videotape, or diskette—will lead you through a prescribed sequence of steps to learn the product.

Technical support is generally offered by the software manufacturer through a telephone number. Some manufacturers provide toll-free numbers; some do not.

Confirm the Price Catalogs and ads are frequently revised. Sometimes there are hidden extra fees, such as for shipping charges, "restocking" charges if you return merchandise, or extra fees for credit-card purchases.

Ask About Money-Back Guarantees If you open a software package, try it out, and find you don't like it, can you return it? Be sure to ask. Some sellers will make refunds; others will accept only unopened software. Some will make money-back guarantees only for 14–30 days after purchase; some have no time limits on software returns. In any event, save your receipts and be sure to try out the purchase within the allotted time.

Pay by Credit Card Whether buying in a store or over the phone by mail order, use a credit card. That gives you some leverage to cancel the sale. Some credit-card companies offer added protection in the form of warranties of their own.

SUMMARY

What It Is / What It Does	**Why It's Important**

Ada *(p. 513, LO 6)* Powerful third-generation, structured programming language designed by U.S. Defense Department to ensure portability of programs from one application to another.

Based on Pascal and originally intended to be a standard language for weapons systems, Ada has been used successfully in commercial applications.

APL *(p. 514, LO 6)* *A Programming Language*; third-generation mathematically oriented and scientific language designed in 1968 for use on IBM mainframes.

APL uses a special keyboard with special symbols to enable users to solve complex mathematical problems in a single step.

application generator *(p. 508, LO 4)* Programmer's tool that consists of modules that have been preprogrammed to accomplish various tasks, allowing the user to explain what data needs to be processed; then the software generates the code to create the program needed. Examples are NOMAD and FOCUS.

With an application generator a programmer need not specify how data should be processed. Programmers use application generators to create parts of other programs.

applications software *(p. 489, LO 1)* Software that can perform useful work on general-purpose tasks—for example, word processing, desktop publishing, or payroll processing.

Applications software packages have become commonly used tools for increasing productivity, particularly of microcomputer users.

artificial intelligence (AI) *(p. 509, LO 4)* Group of related technologies that attempt to develop machines to emulate human-like qualities—for instance, learning, reasoning, communicating, seeing, hearing.

AI technologies are important for enabling machines to perform tasks formerly possible only with human effort.

assembler *(p. 506, LO 5)* Also called *assembler program*; language translator program that translates assembly-language programs into machine language.

Computers cannot run assembly language; it must first be translated.

assembly language *(p. 505, LO 4)* Second-generation programming language; it allows a programmer to write a program using abbreviations instead of the 0s and 1s of machine language.

A programmer can write instructions in assembly language faster than in machine language.

BASIC *(p. 512, LO 6)* Beginner's All-purpose Symbolic Instruction Code; developed in 1965, the most popular microcomputer language and the easiest programming language to learn. Most popular versions: True BASIC, QuickBASIC, Visual BASIC for Windows. BASIC was widely used in the late 1960s in academic settings on all kinds of computers and is now used in business.

The interpreter form of BASIC is popular with first-time and casual users because it is interactive—user and computer can communicate during writing and running of a program.

What It Is / What It Does	Why It's Important
C *(p. 513, LO 6)* High-level general-purpose programming language that works well for microcomputers and is portable among many computers.	C is useful for writing operating systems, database management software, and some scientific applications. Increasingly, C has been used in commercial software development, including games, robotics, and graphics.
C++ *(p. 517, LO 7)* High-level programming language that combines the traditional C programming language with object-oriented capability (the plus signs mean "more than C").	With C++, programmers can write standard code in C without the object-oriented features, use object-oriented features, or do a mixture of both.
class *(p. 516, LO 7)* In object-oriented programming, all objects that are derived from or related to one another. Each class contains specific instructions (methods) unique to that group.	Classes can be arranged in hierarchies—classes and subclasses. Once the programmer has created an object, it can be used as the foundation for similar objects that have the same behavior and characteristics.
COBOL *(p. 511, LO 6)* *COmmon Business Oriented Language;* high-level programming language of business. First standardized in 1968, the language has been revised three times, most recently as COBOL-85.	The most significant attribute of COBOL is that it is extremely readable. COBOL is the language used by the majority of mainframe users, although it will also run on microcomputers.
compiler *(p. 506, LO 5)* Language translator that converts the entire program of a high-level language (called the *source code*) into machine language (called the *object code*) for execution later. Examples of compiler languages: COBOL, FORTRAN, Pascal.	Unlike other language translators (assemblers and interpreters), with a compiler program the object code can be saved and executed later rather than run right away. The advantage of a compiler is that, once the object code has been obtained, the program executes faster.
control structure *(p. 497, LO 3)* Also called *logic structure;* in structured program design, the programming structure that controls the logical sequence in which computer program instructions are executed. Three control structures are used to form the logic of a program: sequence, selection, and iteration (or loop).	One thing that all three control structures have in common is one entry and one exit. The control structure is entered at a single point and exited at another single point. This helps simplify the logic so that it is easier for others following in a programmer's footsteps to make sense of the program.
customized software *(p. 489, LO 1)* Applications software created for a particular customer, usually by a professional programmer.	Customized software is written to perform a task that cannot be done with packaged software.
dBASE *(p. 515, LO 6)* Database management system (DBMS) created for controlling the structure of a database and access to the data; became dBASE II in 1982. The most recent version is dBASE IV.	Besides being a DBMS, dBASE is also a Pascal-like, fourth-generation programming language.
debugging *(p. 502, LO 2)* Part of program testing; the detection and removal of syntax and logic errors in a program.	Debugging may take several trials using different data before the programming team is satisfied the program can be released. Even then, some errors may remain, because trying to remove all of them may be uneconomical.
desk-checking *(p. 501, LO 2)* Form of program testing; programmers read through a program to ensure it's error-free and logical.	Desk-checking should be done before the program is actually run on a computer.
documentation *(p. 502, LO 2)* Written descriptions of a program and how to use it; supposed to be done during all programming steps.	Documentation is needed for all people who will be using or involved with the program in the future—users, operators, and programmers.

What It Is / What It Does	Why It's Important

encapsulation *(p. 516, LO 7)* In object-oriented programming, means an object contains (encapsulates) both (1) data and (2) the processing instructions about it.

Once an object has been created, it can be reused in other programs.

FORTH *(p. 514, LO 6)* FOuRTH-generation language; created in 1971, a third-generation language designed for real-time control tasks as well as business and graphics applications. Used on all kinds of computers, it runs very fast because it requires less memory than other languages.

Because FORTH runs so fast, it is used in applications that must process data quickly—for example, sensors, arcade games, robotics.

FORTRAN *(p. 510, LO 6)* FORmula TRANslator; developed in 1954, it was the first high-level language and was designed to express mathematical formulas. The newest version is FORTRAN 90.

The most widely used language for mathematical, scientific, and engineering problems, FORTRAN is also useful for complex business applications, such as forecasting and modeling. Because it cannot handle a large volume of input/output operations or file processing, it is not used for more typical business problems.

generations *(p. 504, LO 4)* Five increasingly sophisticated levels (generations) of programming languages: (1) machine language, (2) assembly language, (3) high-level languages, (4) very high level languages, (5) natural languages.

Programming languages are said to be *lower level* when they are closer to the language used by the computer (0s and 1s) and *higher level* when closer to the language used by people (for example, English).

hierarchy chart *(p. 494, LO 2)* Also called *structure chart;* a diagram used in programming to illustrate the overall purpose of a program, identifying all the modules needed to achieve that purpose and the relationships among them.

In a hierarchy chart, the program must move in sequence from one module to the next until all have been processed. There must be three principal modules corresponding to the three principal computing operations—input, processing, and output.

high-level languages *(p. 506, LO 4)* Also known as *procedural languages* and *third-generation languages;* they somewhat resemble human languages. Examples: FORTRAN, COBOL, BASIC, Pascal.

High-level languages allow programmers to write in a familiar notation rather than numbers or abbreviations. Most can also be used on more than one kind of computer.

HTML (hypertext markup language) *(p. 519, LO 8)* Type of programming language that embeds commands within standard ASCII text documents to provide an integrated, two-dimensional display of text and graphics. Hypertext is used to link the displays.

HTML is used to create Web pages.

Hypertalk *(p. 518, LO 7)* Language that uses principles of object-oriented programming in the Apple Macintosh HyperCard program to manipulate related files called *stacks.* Using a mouse, users can go through the stacks to find information or discover connections between ideas.

Though HyperCard is not precisely an object-oriented language, Hypertalk is.

inheritance *(p. 516, LO 7)* In object-oriented programming, the method of passing down traits of an object from classes to subclasses in the hierarchy.

New objects can be created by inheriting traits from existing classes.

interpreter *(p. 506, LO 5)* Language translator that converts each high-level language statement into machine language and executes it immediately, statement by statement. An example of a high-level language using an interpreter is BASIC.

Unlike with the language translator called the compiler, no object code is saved. The advantage of an interpreter is that programs are easier to develop.

What It Is / What It Does	**Why It's Important**

iteration control structure *(p. 497, LO 3)* Also known as *loop structure;* one of the control structures used in structured programming. A process is repeated as long as a certain condition remains true; the programmer can stop repeating the repetition at the *beginning* of the loop, using the DO WHILE iteration structure, or at the *end* of the loop, using the DO UNTIL iteration structure (which means the loop statements will be executed at least once).

Iteration control structures help programmers write better-organized programs.

Java *(p. 520, LO 8)* New type of programming language used to create any conceivable type of software applications that will work on the Internet.

Java may be able to transform the Internet from just an information-delivering medium into a completely interactive computing environment.

language translator *(p. 506, LO 5)* Type of systems software that translates a program written in a second-, third-, or higher-generation language into machine language. Language translators are of three types: (1) assemblers, (2) compilers, and (3) interpreters.

Because a computer can execute programs only in machine language, a translator is needed if the program is written in any other language.

LISP *(p. 514, LO 6)* LIST Processor; developed in 1958, a third-generation language used principally to construct artificial intelligence programs.

LISP is used to write expert systems and natural language programs. Expert systems are programs imbued with knowledge by a human expert; such programs can walk users through a problem and help them solve it.

logic errors *(p. 502, LO 2)* Programming errors caused by not using control structures correctly.

If a program has logic errors, it will not run correctly or perhaps not run at all.

Logo *(p. 514, LO 6)* Developed in 1967, a third-generation language designed primarily to teach children problem-solving and programming skills, though it is also used to produce graphics reports in business.

Logo uses a triangular pointer, called a *turtle,* which responds to a few simple commands (such as forward, left, right) and produces similar movements on screen, enabling users to draw geometric patterns and pictures.

machine language *(p. 505, LO 4)* Lowest level of programming language; the language of the computer, representing data as 1s and 0s. Most machine language programs vary from computer to computer—they are machine-dependent.

Machine language, which corresponds to the on and off electrical states of the computer, is not convenient for people to use. Assembly language and higher-level languages were developed to make programming easier.

maintenance *(p. 503, LO 2)* Involves any activity designed to keep programs in working condition, error-free, and up to date. Maintenance includes adjustments, replacements, repairs, measurements, and tests.

The rapid changes of modern organizations are reflected in their computer systems. Thus, maintenance is important, and documentation must help programmers keep up with changes.

make-or-buy decision *(p. 492, LO 2)* Decision whether to custom-write a program or buy one already written ("off the shelf").

Whichever route is chosen will affect the cost, schedule, and suitability of the program.

message *(p. 516, LO 7)* In object-oriented programming, an "alert" sent to the object when an operation involving that object needs to be performed.

The message need only identify the operation. How it is actually to be performed is embedded within the processing instructions that are part of the object.

What It Is / What It Does	Why It's Important

methods *(p. 516, LO 7)* In object-oriented programming, instructions about the operations to be performed on data within an object.

The instructions are implemented only when the object is needed by the program.

Modula-2 *(p. 515, LO 6)* Invented in 1981, a highly structured third-generation language that is similar to Pascal and is designed for writing systems software.

Modula-2 is popular for teaching purposes at colleges and universities.

module *(p. 494, LO 2)* A processing step of a program; sometimes called a *subprogram* or *subroutine*. Each module is made up of logically related program statements.

Each module has only a single function, which limits the module's size and complexity.

natural languages *(p. 508, LO 4)* (1) ordinary human languages (for instance, English, Spanish); (2) fifth-generation programming languages that use human language to give people a more natural connection with computers.

Though still in their infancy, natural languages are getting close to human communication.

object *(p. 516, LO 7)* In object-oriented programming, block of preassembled programming code that is a self-contained module. The module contains (encapsulates) both (1) a chunk of data and (2) the processing instructions that may be called on to be performed on that data. Once the object becomes part of a program, the processing instructions may be activated only when a "message" is sent.

The object can be reused and interchanged among programs, thus making the programming process easier, more flexible and efficient, and faster.

object-oriented programming (OOP) *(p. 516, LO 7)* Programming method in which data and the instructions for processing that data are combined into a self-sufficient object—piece of software. Examples of OOP languages: Smalltalk, C++, Turbo Pascal, Hypertalk.

Objects can be reused and interchanged among programs, producing greater flexibility and efficiency than is possible with traditional programming methods.

packaged software *(p. 489, LO 1)* "Off-the-shelf," prewritten applications software developed for sale to the general public.

Packaged software is the kind microcomputer users buy from a computer store or mail-order house.

Pascal *(p. 512, LO 6)* High-level programming language; an alternative to BASIC as a language for teaching purposes that is relatively easy to learn.

Pascal has extensive capabilities for graphics programming and is excellent for scientific use.

PL/1 *(p. 514, LO 6)* Programming Language 1; introduced in 1964, a third-generation language designed to process both business and scientific applications. It is quite flexible and easy to learn.

PL/1 contains many of the best features of FORTRAN and COBOL; however, it also has so many options as to diminish its usefulness and so has not given them much competition.

polymorphism *(p. 517, LO 7)* In object-oriented programming, means that a message (generalized request) produces different results based on the object it is sent to.

Polymorphism allows programmers to create procedures about objects whose exact type is not known in advance but will be when the program is actually run.

problem definition *(p. 490, LO 2)* Step 1 in the programming process. The problem-definition step requires performing six ministeps: specifying objectives, output, input, and processing tasks, then studying their feasibility and documenting them.

Problem definition is the forerunner to step 2, program design, in the programming process.

What It Is / What It Does	Why It's Important

program *(p. 489, LO 1)* List of instructions the computer follows to process data into information. The instructions consist of statements written in a programming language (for example, BASIC).

Without programs, data could not be processed into information by a computer.

program design *(p. 492, LO 2)* Step 2 in the programming process; programs are designed in three mini-steps: (1) the program logic is determined through a top-down approach and modularization, using a hierarchy chart; (2) the program is designed in detail, using pseudocode or flowcharts with logical tools called control structures; (3) the design is tested with a structured walkthrough.

Program design is the forerunner to step 3, writing (coding), in the programming process.

program flowchart *(p. 495, LO 2)* Tool for designing a program in graphical (chart) form; it uses standard symbols called ANSI symbols.

The program flowchart presents the detailed series of steps needed to solve a programming problem.

programming *(p. 489, LO 2)* Five-step process for creating software instructions: (1) define the problem; (2) design a solution; (3) write (code) the program; (4) test the program; (5) document the program.

Programming is one step in the systems development life cycle (*covered in Chapter 10*).

programming language *(p. 500, LO 2)* Set of words and symbols that allow programmers to tell the computer what operations to follow. The five levels (generations) of programming languages are (1) machine language, (2) assembly language, (3) high-level (procedural) languages (FORTRAN, COBOL, BASIC, Pascal, C, RPG, etc.), (4) very high level (nonprocedural) languages (RPG III, SQL, Intellect, NOMAD, FOCUS, etc.), and (5) natural languages.

Not all programming languages are appropriate for all uses. Thus, a language must be chosen to suit the purpose of the program and to be compatible with other languages being used by users.

PROLOG *(p. 515, LO 6)* PROgramming LOGic; invented in 1972, a third-generation language used for developing artificial intelligence applications.

PROLOG is used to develop natural language programs and expert systems.

pseudocode *(p. 495, LO 2)* Tool for designing a program in narrative form using English-like statements to describe the logic and processing flow. Pseudocode is like doing an outline or summary form of the program to be written.

By using such terms as IF, THEN, or ELSE, pseudocode follows the rules of control structures, an important aspect of structured programming.

query language *(p. 508, LO 4)* Fourth-generation languages for retrieving data from a database management system. Examples: SQL, QBE, Intellect.

Query languages are easy to use. The query may be expressed in the form of a sentence or near-English command or obtained from choices on a menu.

report generator *(p. 508, LO 4)* Also called a *report writer;* fourth-generation program for producing reports as printout or screen display, showing all or part of a database file. Example: RPG III.

With a report generator, users can specify the report format in advance (for example, columns, headings) and the program will then produce data in that format. Report generators were the precursor to query languages.

RPG *(p. 514, LO 6)* Report Program Generator; introduced in 1964, a third-generation language that has evolved through several versions (RPG, RPG II, RPG III) and is designed to help generate business reports. Users fill out a special form specifying the information to be included and its format.

RPG is highly structured and relatively easy to use.

selection control structure *(p. 497, LO 3)* Also known as an *IF-THEN-ELSE structure;* one of control structures used in structured programming. It offers two paths to follow when a decision must be made by a program.

Selection control structures help programmers write better organized programs.

sequence control structure *(p. 497, LO 3)* One of control structures used in structured programming; each program statement follows another in logical order. There are no decisions to make.

Sequence control structures help programmers write better organized programs.

Smalltalk *(p. 517, LO 7)* Object-oriented programming (OOP) language, invented in 1970; a keyboard is used to enter text, but all other tasks are performed with a mouse.

Smalltalk was the first OOP language.

structured programming *(p. 492, LO 3)* Method of programming that takes a top-down approach, breaking programs into modular forms and using standard logic tools called control structures (sequence, selection, iteration).

Structured programming techniques help programmers write better-organized programs, using standard notations with clear, correct descriptions.

structured walkthrough *(p. 499, LO 2)* Program review process that is part of the design phase of the programming process; a programmer leads other development team members in reviewing a segment of code to scrutinize the programmer's work.

The structured walkthrough helps programmers find errors, omissions, and duplications, which are easy to correct because the program is still on paper.

syntax *(p. 500, LO 2)* "Grammar" rules of a programming language that specify how words and symbols are put together.

Each programming language has its own syntax, just as human languages do.

syntax errors *(p. 502, LO 2)* Programming errors caused by typographical errors and incorrect use of the programming language.

If a program has syntax errors, it will not run correctly or perhaps not run at all.

testing the program *(p. 500, LO 2)* Step 4 in the programming process; involves doing various tests, such as desk-checking and debugging, and then running real-world data to make sure the program works.

Program testing is the forerunner to step 5, writing the documentation, in the programming process.

top-down program design *(p. 493, LO 2)* Method of program design; a programmer identifies the top or principal processing step, or module, of a program and then breaks it down in hierarchical fashion into smaller processing steps. The design can be represented in a top-down hierarchy chart.

Top-down program design enables an entire program to be more easily developed because the parts can be developed and tested separately.

Turbo Pascal *(p. 518, LO 7)* Object-oriented version of Pascal designed in 1984.

Turbo Pascal allows a program to use removable modules, or objects, that can be replaced later when the program is changed.

What It Is / What It Does	**Why It's Important**

very-high-level languages *(p. 507, LO 1)* Also known as *nonprocedural languages* and *fourth-generation languages (4GLs)*; more user-oriented than third-generation languages, 4GLs require fewer commands. 4GLs consist of report generators, query languages, application generators, and interactive database management system programs. Some 4GLs are tools for end-users, some are tools for programmers.

Programmers can write programs that need only tell the computer what they want done, not all the procedures for doing it, which saves them the time and labor of having to write many lines of code.

visual programming *(p. 518, LO 7)* Method of creating programs; the programmer makes connections between objects by drawing, pointing, and clicking on diagrams and icons. Programming is made easier because the object orientation of object-oriented programming is used in a graphical or visual way.

Visual programming enables users to think more about the problem solving than about handling the programming language.

VRML (virtual reality markup language) *(p. 519, LO 8)* Type of programming language used to create three-dimensional (3-D) Web pages.

VRML expands the information-delivering capabilities of the Web.

writing the program *(p. 528, LO 2)* Also called *coding*; step 3 in the programming process. The programmer translates the logic requirements from pseudocode or flowchart into the letters, numbers, and symbols of the programming language making up the program.

Writing the program is the forerunner to step 4, program testing, in the programming process.

writing the program documentation *(p. 499, LO 2)* Step 5 in the programming process; programmers write procedures explaining how the program was constructed and how it is to be used.

Program documentation is the final stage in the five-step programming process, although documentation should also be an ongoing task accompanying all steps.

(Selected answers appear at the back of the book.)

Short-Answer Questions

1. How would you define *programming language*?
2. Why is it important to document a program?
3. What is the basic difference between procedural and nonprocedural languages?
4. What were the reasons behind the development of high-level programming languages?
5. What is visual programming?
6. What is the main difference between standard programming and object-oriented programming?
7. What are the five basic steps of program development?
8. What is meant by the term *top-down program design*?
9. What is natural language?
10. What is the basic difference between a compiler and an interpreter?

Fill-in-the-Blank Questions

1. _____ is a five-step process for creating a list of instructions that the computer uses to process data into information.
2. The _____ is the formal review process whereby a programmer leads other people from the development team through a segment of code.
3. A(n) _____ is a block of pre-assembled programming code that forms a self-contained module that can be used in creating many different programs.
4. Writing a computer program is called _____.
5. To _____ a program means to detect, locate, and remove all errors in a computer program.
6. Program _____ consists of written instructions on how to use a program.
7. Assembly language is a _____-generation language.
8. In _____ programming, programmers make connections between objects by drawing, pointing, and clicking on diagrams and icons.
9. A(n) _____ uses ANSI symbols to graphically depict a program's design.
10. Smalltalk, C++, and Turbo Pascal are _____ programming languages.

Multiple-Choice Questions

1. Which of the following uses English-like statements to describe the logic and processing flow of a program?
   a. pseudocode
   b. control structure
   c. logic structure
   d. program flowchart
   e. none of the above

2. Which of the following depicts program logic graphically using ANSI symbols?
   a. pseudocode
   b. control structure
   c. logic structure
   d. program flowchart
   e. none of the above

3. Machine language is a(n) _____-generation language.
   a. first
   b. second
   c. third
   d. fourth
   e. fifth

4. Fifth-generation languages are also called _____.
   a. natural languages
   b. report generators
   c. query languages
   d. application generators
   e. none of the above

5. Which of the following doesn't relate to object-oriented programming?
   a. encapsulation
   b. inheritance
   c. polymorphism
   d. application generator
   e. all of the above

6. A(n) _____ is a processing step of a program.
   a. module
   b. structure
   c. flowchart
   d. iteration
   e. all of the above

7. The rules of a programming language are called _____.
   a. documentation
   b. iterations
   c. syntax
   d. pseudocode
   e. all of the above

8. Which of the following languages was developed to help students learn programming?
   a. FORTRAN
   b. COBOL
   c. BASIC
   d. Pascal
   e. none of the above

9. Which of the following was the first high-level language?
   a. FORTRAN
   b. COBOL
   c. BASIC
   d. Pascal
   e. none of the above

10. To convert a high-level language statement into machine language and execute it immediately, you need a(n):
    a. converter
    b. report generator
    c. interpreter
    d. application generator
    e. none of the above

True/False Questions

T F 1. The rules for using a programming language are called *syntax*.

T F 2. A query language allows the user to easily retrieve information from a database using regular human-language statements.

T F 3. Desk-checking involves reading through program code.

T F 4. LISP is used to write systems software.

T F 5. Pascal is a high-level programming language.

T F 6. The iteration control structure is also known as the loop structure.

T F 7. Customized software can be purchased off the shelf of a computer store.

T F 8. Most, but not all, programs have to be converted into machine language before your computer can execute them.

T F 9. The top-down approach to designing a program uses modules and hierarchy charts.

T F 10. Programming is a seven-step process for creating program instructions.

Projects/Critical-Thinking Questions

1. Scan the employment ads in a few major newspapers and professional journals. What programming languages are most in demand? For what types of jobs?

2. If you were a computer programmer, would you rather work on writing applications software or systems software? Think of as many reasons as you can to support your choice, and write a brief report.

3. Interview several students who are majoring in computer science and are studying to become computer programmers. What languages do they plan to master? Why? What kinds of jobs do these people expect to get? What kinds of future developments do they anticipate in the field of software programming?

4. *Java* is a programming language that promises to make sending programs over the Internet as easy as sending e-mail. Why is this programming language receiving so much attention today? How did it become popular? Who is writing Java applications and what are they used for? In the years ahead, what impact might Java have on the computer industry? Expand on the discussion of Java presented near the end of this chapter.

5. Using computer periodicals, magazines, and the Internet (optional), explore the state-of-the-art of visual programming. What applications is visual programming used for? What are the names of some programming languages that support visual programming? If visual programming is intended to make programming more accessible to end-users, where can end-users go to learn more about the method of visual programming?

net Using the Internet

Objective: *In this exercise you practice using Gopher, a text-based menu system that helps you access information on the Internet.*

Before you continue: *We assume you have access to the Internet through your university, business, or commercial service provider and to the Web browser tool named Netscape Navigator 2.0 or 2.01. We also assume you know how to connect to the Internet and then load Netscape Navigator. If necessary, ask your instructor or system administrator for assistance.*

1. Make sure you have started Netscape. The home page for Netscape Navigator should appear on your screen.

2. Gopher is a text-based, hierarchical menu system that lets you search gopher servers on the Internet. In contrast, when you access a Web server, you click on hypertext links to access resources. Gopher servers are public so anyone can access them. They are located all around the world and are referred to collectively as "gopherspace."

   In this step you contact The World gopher server, which you will then use to locate the Boston Public Library for the purpose of displaying information about cities in the United States.

   CLICK: Open button (⬚)
   TYPE: gopher://gopher.std.com
   CLICK: Open command button

   Your screen should appear similar to the following one. Notice that the information is categorized by topic. To find out more about the topic, you click its associated directory or file icon.

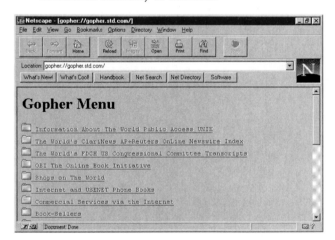

3. In the following steps our objective is to reach the part of the Boston Public Library that enables you to display demographic data about cities in the United States. If you were doing this without our guidance you would have to put your detective hat on and click on directories and files that looked like they

would lead you to the right place. Since we've already done the detective work for you, simply follow along with our instructions.

Use the vertical scroll bar to scroll through the list of directories until you locate the directory named Libraries.

CLICK: the directory's icon (or click the underlined link itself)

4. Locate the Boston Public Library directory.

   CLICK: the directory's icon

   (*Note:* Telnet links are represented by the terminal icon. As long as you have telnet software on your computer, these links make a telnet connection to a remote computer; otherwise, you'll get an error message.)

5. Locate the Ready-reference Shelf directory.

   CLICK: the directory's icon

6. Locate the Geographic Name Server: Search by City or ZIP Code search title. (*Note:* The binoculars icon indicates that the file is searchable.) To find general information about cities in the United States:

   CLICK: once on the Search icon

   Your screen should now be similar to this one:

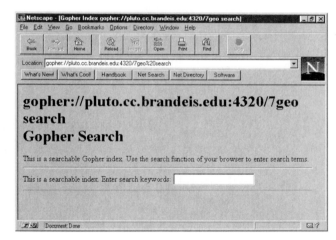

7. To display information about a city in the United States, you can search on either a city name or a zip code. On your own, click in the search text box and then type in a city or zip code. Then press **Enter**. One or more city names should appear depending on what you typed into the search text box. Click on a city name to display demographic data. On your own, practice searching for information on other cities in the United States. The following screen

appeared after typing "94010" into the search text box, pressing **⟨Enter⟩**, and then clicking "Burlingame, CA:"

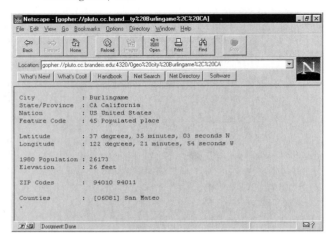

8. To display gopher's main menu, choose Go from the Menu bar and then choose "gopher://gopher.std.com" from the bottom of the drop-down list.

9. On your own, practice using gopher to search for information about one of the following topics:
   - Programming languages
   - Compilers
   - Object-oriented programming
   - Visual programming

10. Print (File, Print) or save (File, Save as) any documents that you find as a result of your efforts in step 9.

11. Display Netscape's home page and then exit Netscape.

CHAPTER 12

Society & the Digital Age
Challenges & Promises

Concepts You Should Know

After reading this chapter, you should be able to:

1. Explain the models for the Information Superhighway.
2. Discuss the implications of legislation that affects an information-based society.
3. Describe security issues—accidents, hazards, crime, viruses—and security safeguards.
4. Discuss quality-of-life issues—environment, mental health, the workplace.
5. Explain the implications of information technology for employment and the future.
6. Describe intelligent agents and avatars and their uses.
7. Define and explain the different types of artificial intelligence.
8. Discuss the promises of the Information Superhighway.

T here's a fundamental shift going on in society right now," says Jane Metcalfe. "... [T]he digital revolution is whipping through our lives like a Bengali typhoon."

Metcalfe, co-founder of the popular information-technology magazine *Wired*, says she is "profoundly optimistic about how we can use technology to change things."[1] Not everyone thinks the future is so rosy. Unemployment, says technology critic Jeremy Rifkin gloomily, "can be expected to climb steadily and inexorably over the next four decades as the global economy makes the transition to the Information Age."[2] Who's right? In this final chapter, we consider the challenges and promises of the Digital Age.

Communications: The Development of the Information Superhighway

Preview & Review: Systems of communications have shrunk society to a "global village."

Technological convergence has led telephone and cable executives to suggest they could establish tree-and-branch networks in which they could provide every household with up to 500 channels of programming. The rising popularity of the Internet and the World Wide Web suggests a more democratic switched-network model may arise instead.

The hallmark of great civilizations has been their great systems of communications. In the beginning, communications was based on transportation: the Roman Empire had its network of roads, the great European powers had their far-flung navies, and the United States and Canada built their transcontinental railroads. Then transportation yielded to the electronic exchange of information. Beginning in 1844, the telegraph ended the short existence of the Pony Express and, beginning in 1876, found itself in competition with the telephone. In 1889, electromagnetic radio waves were transmitted the width of the English Channel. The amplifying vacuum tube, invented in 1906, led to commercial radio, beginning with Pittsburgh's KDKA in 1920. Television came into being in England in 1925; one of the first scheduled television shows, broadcast from New York on July 21, 1931, introduced Kate Smith singing "When the Moon Comes Over the Mountain."

The Global Village & the 500-Channel Dream

During the 1950s and 1960s, as television exploded throughout the world, Canadian communications philosopher Marshall McLuhan posed the notion of a "global village." The *global village* refers to the "shrinking" of the world society because of the ability to communicate. This shrinking has resulted from the universality of mass media such as television.

Recently, technology has become portable, giving individuals even more power over communications and making the global village even more closely knit. A decade ago, cellular phones, pagers, and portable computers with fax and voice-mail links barely existed. Now they are commonplace, offering e-mail, mobile calling, and online information.

In the last few years, computer and communications technology began to go digital, leading to suggestions of technological convergence among the computer, communications, consumer electronics, entertainment, and mass media industries. A major assumption of telephone and cable executives was that the Digital Age would also give every household access to multiple channels of information— 500 seemed to be the number on everyone's mind. Their model, says *Newsweek* technology writer Steven Levy, was that of today's mass-media television but with just two differences:[3] (1) We would have the same programming, only more of it. (2) There would be some interactivity, permitting us to press buttons on a set-top box (✓ p. 206) so that we could choose programs, perhaps play games, and—most important—buy things. To prepare for the 500-channel dream, during 1993–1995 the media and communications giants went into a frenzy of mergers and alliance-building. All along and barely noticed, however, the Internet, and its most interesting part, the World Wide Web, were growing steadily— and the Internet offered not just 500 channels but unlimited channels and complete interactivity.

Tree-and-Branch Versus Switched Network: The Internet as a Participatory System

"If you want an arbitrary date for the burial of the 500-channel dream, Aug. 9, 1995, will do just fine," writes Levy. On that date, investors went into a frenzy at the first opportunity to buy stock in the year-old Netscape Communications Corporation, developer of software for navigating the World Wide Web. What is the significance of this? Says Levy:

> Aug. 9 marked the moment when Wall Street finally realized what had been becoming increasingly apparent to computer users: a set of highly technical but reliably standardized communications protocols known as the Internet had established itself as the real key to the electronic future. That future would be made not by silver-haired telephone- and cable-company executives in Denver, New York and Washington, building an empire around a golden goose called pay-per-view television, but by companies like Netscape and their customers.
> In short, the end of the 500-channel dream. . . .

August 9, then, signaled public recognition of the fact that the roadbed for the Information Superhighway—which the media, entertainment, and telecommunications giants thought they were building—was in fact already here! Moreover, it looked as though the "I-way" would not be under the control of the "silver-haired executives" but could be influenced by nearly anyone with access.

To appreciate the differences, consider that telecommunications can be organized through two kinds of arrangements: the "tree-and-branch" model or the "switched-network" model.[4]

- **Tree-and-branch model:** In the ***tree-and-branch model*, a centralized information provider sends out messages through many channels to thousands of consumers.** This is the model of most mass media, such as radio and TV broadcasting. It is also the model envisioned by cable and entertainment companies looking to provide movies and other services from a centralized library.

- **Switched-network model:** In the ***switched-network model*, people on the system are not only consumers of information but also possible providers.** This is the model of the telephone system and also of most computer networks, including, of course, the Internet.

The significance of the two types is that the tree-and-branch allows only few-to-many communications. The switched-network allows many-to-many communications. Thus, the more the switched-network becomes prominent, the more democratized and participatory our systems of communications become.

The Challenges & Promises of Information Technology

Preview & Review: Information technology offers a number of challenges—the blueprint for the Information Superhighway, security issues, quality-of-life issues, and economic issues. It also promises some benefits, such as those that may be derived from intelligent agents and avatars, artificial intelligence, and the various sectors of the Information Superhighway.

In the rest of this chapter, we consider both the challenges and the promises of computers and communications in relation to society. First let us consider the following challenges of the Digital Age:

- The blueprint for the Information Superhighway
- Security issues—accidents, hazards, crime, viruses—and security safeguards
- Quality-of-life issues—environment, mental health, the workplace
- Economic issues—employment and the haves/have-nots

We will then consider the following promises:

- The roles of intelligent agents and avatars
- Artificial intelligence and its benefits
- The sectors of the Information Superhighway

The Information Superhighway: Is There a Blueprint?

Preview & Review: The Information Superhighway envisions using wired and wireless capabilities of telephones and networked computers with cable-TV. It may evolve following a model backed by the federal government called the *National Information Infrastructure.* Or it may evolve out of competition brought on by the deregulation of long-distance and local telephone companies, cable companies, and television broadcasters created by the 1996 Telecommunications Act.

As we said in Chapter 1, the *Information Superhighway* is a vision or a metaphor. It envisions a fusion of the two-way wired and wireless capabilities of telephones and networked computers with cable-TV's capacity to transmit hundreds of programs. When complete, the I-way would supposedly give us video on-demand, multimedia, fast data exchange, teleconferencing, distance learning, enormous research databases, and teleshopping services. It would also help government by giving us electronic "town hall" democracy.

What shape will the Information Superhighway take? Some government officials hope it will follow a somewhat orderly model, such as that envisioned in the National Information Infrastructure. Others think it will evolve out of competition brought about by the passage of the 1996 Telecommunications Act. Let us look at both of these.

The National Information Infrastructure

As envisioned by various U.S. government officials, the **National Information Infrastructure (NII) would include today's existing networks and technologies as well as technologies yet to be deployed. Its services would be delivered by telecommunications companies, cable-television companies, and the Internet.** Applications would be varied—education, health care, information access, electronic commerce, and entertainment.[5] (■ *See Panel 12.1, next page.*)

Who would put the pieces of the NII together? The present national policy is to let private industry do it, with the government trying to ensure fair competition among phone and cable companies and compatibility among various technological systems.[6] In addition, it strives for open access to people of all income levels.

The 1996 Telecommunications Act

"Let the telecom wars begin," wrote the *Wall Street Journal*.[7] After years of attempts to overhaul the 1934 Communications Act, on February 8 President Clinton signed into law (using a high-tech stylus and electronic tablet) the 1996 Telecommunications Act. The **Telecommunications Act of 1996 undoes 60 years of federal and state communications regulations and lets phone, cable, and TV businesses compete and combine more freely.** A section called the **Communications Decency Act makes it a crime to transmit information that could be interpreted as unsuitable for viewing by children.** Such information includes pornography, of course, but it could also include conversation about breast cancer or photos of Michelangelo's statue *David.*

Supporters compared the law's significance to the fall of Communism, saying it would create jobs, expand consumer choices, and lower phone and cable rates.[8] Opponents—mostly consumer groups—say consolidation of businesses will cost jobs and probably raise rates. Free-speech supporters who opposed the Decency section responded by blackening their Web sites as part of a 48-hour "Thousand Points of Darkness" protest.[9] In any case, the law tears down regulatory barriers and, predicted one article, "is certain to accelerate the convergence of local and long-distance phone businesses with cable operators, cellular companies, broadcast concerns, computer makers, and others."[10]

The act will affect communications companies in a number of ways.[11–14] (■ *See Panel 12.2, page 541.*) It permits greater competition between local and long-distance telephone companies, as well as between the telephone and cable industries. In particular, the legislation allows cable companies (such as TCI) and long-distance carriers (AT&T, Sprint, MCI) to offer local telephone service. It would also allow the seven regional Bell operating companies to offer not just local phone service but also long-distance services. Both local and long-distance phone companies can enter the cable-TV business, offering video via wired or wireless means. Cable services will be allowed to go into cross-ownership with telephone companies in small communities, which is likely to spur new alliances. Cable services will also be permitted to boost their rates. Television stations will get a new broadcast spectrum for advanced TV (✔ p. 265)—possibly without cost, although Congress was supposed to decide this later.

Toward a National Information Infrastructure

Some services that supporters of the NII hope might be offered.

Commerce

With inexpensive access charges, small companies could afford to act like big ones. Boundaries would be erased between company departments, suppliers, and customers. Designs for new products could be tested and exchanged with factories in remote locations. With information flowing faster, goods could be sent to market faster and inventories kept low.

Government

An information highway could extend electronic democracy through electronic voting, allow interactive local-government meetings between electors and elected, and help deliver government services such as administering Social Security forms.

Education

"Virtual" classrooms and distance learning would replace lecture halls and scheduled class times. Students could take video field trips to distant places and get information from remote museums and libraries (such as the Library of Congress).

Home Services

Consumers would be able to receive movies on demand, home shopping, and videogames; do electronic bill paying; and tap into libraries and schools.

Health Care

Through telemedicine, health-care providers and researchers could share medical images, patient records, and research and perform long-distance patient examinations. Interactive, multimedia materials directed to the public would outline health-care options.

Information

Government records, patents, contracts, legal documents, and satellite maps could be made available to the public online. Libraries could also be digitized, with documents available for downloading.

Mobile Communications

Users with handheld personal communicators or personal digital assistants would be able to send and receive voice, fax, text, and video messages anywhere.

- *Long-distance phone service:* Local phone companies can enter long-distance phone markets (after proving they've opened their local phone networks to competitors). Most will probably target their own regions first. Long-distance call rates will probably drop.

  Originally, seven Regional Bell Operating Companies (RBOCs) were created in 1984 when AT&T, in response to a government antitrust suit, spun off its local phone service. *(See map.)* Another local phone giant is GTE, which offers phone service in 28 states.
- *Local phone service:* Long-distance carriers (such as AT&T, MCI, and Sprint, which control 90% of the long-distance market) and cable companies may provide local service.

  Local rates could rise because the act does not specify how much the big carriers will have to pay local companies to connect.
- *Universal phone service:* Phone service is guaranteed everywhere, including remote rural areas. But the Federal Communications Commission and the states must decide how it is to be paid for.
- *Cable service:* Rate regulations are lifted in three years for big cable systems. Rate regulations are eliminated immediately on cable systems with less than 1% of the nation's subscribers. Thus, rates will probably rise.
- *Video delivered by phone:* Phone companies may sell television or video services via phone line or satellite. Competition could reduce cable-TV rates.
- *Lifts on ownership bans:* Cross-ownership between cable and telephone companies is no longer prohibited in small communities. There are no limits on how many TV stations companies may own, if stations don't reach more than 35% (raised from 25%) of the U.S. population.
- *TV broadcast spectrum:* TV stations will get (possibly free) a new broadcast spectrum for advanced TV, to broadcast high-definition (HDTV) images or 4–6 different programs in the same spectrum space. Further legislation to be taken up later.
- *Content regulation on TV sets and Internet:* TV sets required to be equipped with a device, the V-chip, to block violent or sexual programs at parents' discretion. Internet users who transmit indecent material without restricting its access to minors are liable to criminal penalties.

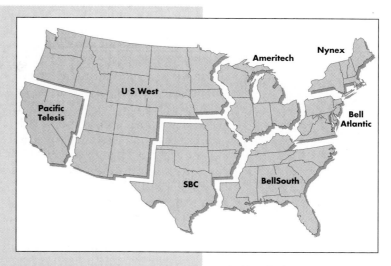

PANEL 12.2

Effects of the 1996 Telecommunications Act

Among other things, the original seven regional Bell local phone companies were allowed to enter the long-distance business. Pacific Telesis and SBC have consolidated as one company. Bell Atlantic and Nynex have merged into another.

The new law will no doubt produce some interesting enterprises. For instance, NBC and Microsoft have already joined together to produce a CNN-like 24-hour all-news cable channel that will allow viewers to watch a television news story from NBC and then call up more material from a Microsoft Network database.[15,16]

If We Build It, Will They Come?

Telecommunications companies and their partners have commissioned inter-active-TV trials—in Denver; New York City; Arlington, Virginia; Castro Valley, California; and Orlando, Florida—to see what people will go for. Yet it's not clear that consumers will be as interested as the companies that stand to profit.

One poll found that only a small percentage of 1000 people interviewed thought that new technology would greatly affect their lives in the next 10 years.[17] Just 15% replied that their "personal and business communications" would be changed greatly by advanced electronics and computer developments. A mere 7% said their "entertainment and leisure time" activities would change much.

Yet the climate may be changing. Today technical, service, and even attitude problems (as in the rudeness shown in certain newsgroups) may make going online still too difficult for a mass audience.[18] However, consumer surveys show that users of the Internet's World Wide Web are broadening to include a more affluent crowd, which holds out hope for commercial enterprises wanting to set up shop on the Web.[19]

Recent evidence indicates that the Internet is also nearing a new stage of global expansion, as companies in Europe and Asia overhaul their telecommunications systems to provide what already exists in the United States: multiple backbones of fiber optics that can move mountains of data. "With continuing international growth," says one report, "the network is becoming less like a U.S. product available globally, and more like a polyglot linkup of local broadcasters."[20]

The final piece of the puzzle could well be the "Internet appliance" (✓ p. 28). The under-$500 network computer envisioned by Oracle, Apple, IBM, Sony, Sun, Toshiba, and others could be the device that turns the Net into a system of truly mass appeal. "For millions of fearful and/or lower-income folks," says computer columnist Gina Smith, "these pared-down PCs will make a lot of sense."[21]

However, no matter what developments occur in computer and communications technology, these systems will remain vulnerable to certain threats, as we describe next.

Security Issues: Threats to Computers & Communications Systems

Preview & Review: Information technology can be disabled by a number of occurrences. It may be harmed by people; by procedural, and software errors; by electromechanical problems; and by "dirty data." It may be threatened by natural hazards and by civil strife and terrorism.

Criminal acts perpetrated against computer systems include theft of hardware, software, time and services, and information; and crimes of malice and destruction.

Computers may be harmed by viruses.

Computers can also be used as instruments of crime. Criminals may be employees, outside users, hackers, crackers, and professional criminals.

Security issues go right to the heart of the workability of computer and communications systems. Here we discuss the following threats to computers and communications systems:

- Errors and accidents
- Natural and other hazards
- Crimes against computers and communications
- Crimes using computers and communications
- Worms and viruses
- Computer criminals

Errors & Accidents

ROBOT SENT TO DISARM BOMB GOES WILD IN SAN FRANCISCO, read the headline.[22] Evidently, a hazardous-duty police robot started spinning out of control when officers tried to get it to grasp a pipe bomb. Fortunately, it was shut off before any damage could be done. Most computer glitches are not so spectacular, although they can be almost as important.

In general, errors and accidents in computer systems may be classified as people errors, procedural errors, software errors, electromechanical problems, and "dirty data" problems.

- **People errors:** Recall that one part of a computer system is the people who manage it or run it (✓ p. 9). For instance, Brian McConnell of Roanoke, Virginia, found that he couldn't get past a bank's automated telephone system to talk to a real person. This was not the fault of the system so much as of the people at the bank. McConnell, president of a software firm, thereupon wrote a program that automatically phoned eight different numbers at the bank. People picking up the phone heard the recording, "This is an automated customer complaint. To hear a live complaint, press. . . ."[23] Quite often, what may seem to be "the computer's fault" is human indifference or bad management.

- **Procedural errors:** Some spectacular computer failures have occurred because someone didn't follow procedures. Consider the 2½-hour shutdown of Nasdaq, the nation's second largest stock market. Nasdaq is so automated that it likes to call itself "the stock market for the next 100 years." In July 1994, Nasdaq was closed down by an effort, ironically, to make the computer system more user-friendly. Technicians were phasing in new software, adding technical improvements a day at a time. A few days into this process, the technicians tried to add more features to the software, flooding the data-storage capability of the computer system. The result was a delay in opening the stock market that shortened the trading day.[24]

- **Software errors:** We are forever hearing about "software glitches" or "software bugs." A *software bug* **is an error in a program that causes it not to work properly** (✓ p. 502). (■ *See Panel 12.3.*)

"In one of the biggest computer errors in banking history, Chemical Bank mistakenly deducted about $15 million from more than 100,000 customers' accounts [one] night, causing consternation among its customers around the New York area.

The problem stemmed from a single line in an updated computer program installed by Chemical on Tuesday in its Somerset, N.J. computer center that caused the bank to process every withdrawal and transfer at its automated teller machines twice. Thus a person who took $100 from a cash machine had $200 deducted, although the receipt only indicated a withdrawal of $100. . . .

[T]he obvious suspect was a small section of new software that had been installed as part of a year-long effort by Chemical to improve the software it uses to operate its ATM's.

The problem line of the computer program was meant to be 'dormant,' until further changes in the system were made. . . . What it did, however, was to send an electronic carbon copy of every ATM withdrawal and transfer that was made to a second computer system used for processing paper checks. That meant money was deducted from customers' accounts once by the ATM system and then a second time by the check system."

—Saul Hansell, "Cash Machines Getting Greedy at a Big Bank," *New York Times*

PANEL 12.3

Software error

A software error caused automated teller machines to deduct more than they should from customers' accounts.

An example of a somewhat small error was when a school employee in Newark, New Jersey, made a mistake in coding the school system's master scheduling program. When 1000 students and 90 teachers showed up for the start of school at Central High School, half the students had incomplete or no schedules for classes. Some classrooms had no teachers while others had four instead of one.[25]

Especially with complex software, there are always bugs, even after the system has been thoroughly tested and debugged. However, there comes a point in the software development process where debugging must stop. That is, the probability of the bugs disrupting the system is considered to be low enough that it is not considered to be cost effective to find them and fix them.

- **Electromechanical problems:** Mechanical systems, such as printers, and electrical systems, such as circuit boards, don't always work. They may be incorrectly constructed, get dirty or overheated, wear out, or become damaged in some other way. Power failures (brownouts and blackouts) can shut a system down. Power surges can burn out equipment.

  Modern systems, argues Yale University sociologist Charles Perrow, are made up of thousands of parts, all of which interrelate in ways that are impossible to anticipate. Because of that complexity, he says, what he calls "normal accidents" are inevitable. That is, it is almost certain that some combinations of minor failures will eventually amount to something catastrophic. Indeed, it was just such a collection of small failures that led to the blowing up of the Challenger space shuttle in 1986 and the near-meltdown of the Three Mile Island nuclear-power plant in 1979.[26] In the Digital Age, "normal accidents" will not be anomalies but are to be expected.

- **"Dirty data" problems:** When keyboarding a research paper, you undoubtedly make a few typing errors. So do all the data-entry people around the world who feed a continual stream of raw data into computer systems. A lot of problems are caused by this kind of "dirty data." **Dirty data is data that is incomplete, outdated, or otherwise inaccurate.**

  A good reason for having a look at your records—credit, medical, school—is so that you can make any corrections to them before they cause you complications. As the president of a firm specializing in business intelligence writes, "Electronic databases, while a time-saving resource for the information seeker, can also act as catalysts, speeding up and magnifying bad data."[27]

  An interesting source of "bad data," this same person notes, is the memos that every executive receives. He points out that "internal memos are rarely footnoted, or their 'facts' supported with proof. Very often the recipient of corporate information has to weigh its validity based on who said it," or on how badly he or she wants to believe it.

Natural & Other Hazards

Some disasters can do more than lead to temporary system downtime, they can wreck the entire system. Examples are natural hazards, and civil strife and terrorism.

- **Natural hazards:** Whatever is harmful to property (and people) is harmful to computers and communications systems. This certainly includes natural disasters: fires, floods, earthquakes, tornadoes, hurricanes, blizzards, and the like. If they inflict damage over a wide area, as have some Florida and Hawaii hurricanes, natural hazards can disable all the electronic systems we take for granted. Without power and communications connections, automated teller machines, credit-card verifiers, and bank computers are useless.

- **Civil strife and terrorism:** We may take comfort in the fact that wars and insurrections seem to take place in other parts of the world. Yet we are not immune to civil unrest, such as the so-called Rodney King riots that wracked Los Angeles in 1992. Nor are we immune, apparently, to acts of terrorism, such as the February 1993 bombing of New York's World Trade Center. In the latter case, companies found themselves frantically having to move equipment to new offices and reestablishing their computer networks.

  In August 1995, *Time* magazine ran a cover story on the Department of Defense plans for "cyberwar," a nonbloody kind of information warfare in which the computer systems of adversaries could be disabled through the use of viruses, phony radio messages, and electronic jamming.[28,29] The Pentagon itself (which has 650,000 terminals and workstations, 100 WANs, and 10,000 LANs) is taking steps to reduce its own systems' vulnerability to intruders.[30]

Crimes Against Computers & Communications

An *information-technology crime* **can be one of two types. It can be an illegal act perpetrated against computers or telecommunications. Or it can be the use of computers or telecommunications to accomplish an illegal act.** Here we discuss the first type.

Crimes against technology include theft—of hardware, of software, of computer time, of cable or telephone services, of information. Other illegal acts are crimes of malice and destruction. Some examples are as follows:

- **Theft of hardware:** Stealing of hardware can range from shoplifting an accessory in a computer store to removing a laptop or cellular phone from someone's car. Professional criminals may steal shipments of microprocessor chips off a loading dock or even pry cash machines out of shopping-center walls.

  Eric Avila, 26, a history student at the University of California at Berkeley, had his doctoral dissertation—involving six years of painstaking research—stored on the hard drive of his Macintosh PowerBook, when a thief stole it out of his apartment. Although he had copied an earlier version of his dissertation (70 pages entitled "Paradise Lost: Politics and Culture in Post-War Los Angeles") onto a diskette, the thief stole that, too. "I'm devastated," Avila said. "Now it's gone, and there is no way I can recover it other than what I have in my head." To make matters worse, he had no choice but to pay off the $2000 loan for a computer he did not have anymore.[32]

  Portable computer thefts, incidentally, rose 39% from 1994 to 1995.[32] Popular sites for laptop larceny are airports and hotels.[33,34] Theft of computers has become a major problem on many campuses, as well. Often the thieves, who may be professionals, don't take the peripheral devices, only the system unit.[35]

- **Theft of software:** Stealing software can take the form of physically making off with someone's diskettes, but it is more likely to be copying of programs. Software makers secretly prowl electronic bulletin boards in search of purloined products, then try to get a court order to shut down the bulletin boards.[36] They also look for companies that "softlift"—buying one copy of a program and making copies for as many computers as they have.

Many pirates are reported by co-workers or fellow students to the "software police," the Software Publishers Association. The SPA has a toll-free number (800-388-7478) on which anyone can report illegal copying, to initiate antipiracy actions. In mid-1994, two New England college students were indicted for allegedly using the Internet to encourage the exchange of copyrighted software.[37]

Another type of software theft involves selling copies or counterfeits of well-known software programs. These pirates often operate in China, Taiwan, Mexico, Russia, and various parts of Asia and Latin America. In some countries, more than 90% of U.S. microcomputer software in use is thought to be illegally copied.[38]

- **Theft of time and services:** The theft of computer time is more common than you might think. Probably the biggest use of it is people using their employer's computer time to play games. Some people also may run sideline businesses.

  Theft of cable and telephone services has increased over the years. Cable-TV Montgomery reported it lost $12 million a year to pirates using illegal set-top converter boxes. Recently, thieves have been able to crack the codes of the fast-growing digital satellite industry, using illegal decoders.[39] Under federal law, a viewer with an illegal decoder box can face up to 6 months in jail and a $1000 fine.[40]

  For years "phone phreaks" have bedeviled the telephone companies. They have found ways to get into company voice-mail systems, then use an extension to make long-distance calls at the company's expense.[41] In addition, they have also found ways to tap into cellular phone networks and dial for free.[42]

- **Theft of information:** "Information thieves" have been caught infiltrating the files of the Social Security Administration, stealing confidential personal records and selling the information.[43] Thieves have also broken into computers of the major credit bureaus and have stolen credit information. They have then used the information to charge purchases or have resold it to other people. On college campuses, thieves have snooped on or stolen private information such as grades.

  The makers of huge sums of illicit profits, such as drug traffickers, are also going high-tech, doing their money laundering by using home-banking software, for example, to zip money across borders. Authorities fear that the rise of cybercash—for instance, use of smart cards (✓ p. 220) containing memory chips that can be filled or emptied with the equivalent of cash—will turn money laundering into a financial crime that will be harder than ever to track. Says a U.S. Treasury Department official, "That's the drug kingpin of the future: The guy walking around with a chip in his pocket worth a few million."[44]

- **Crimes of malice and destruction:** Sometimes criminals are more interested in abusing or vandalizing computers and telecommunications systems than in profiting from them. For example, a student at a Wisconsin campus deliberately and repeatedly shut down a university computer system, destroying final projects for dozens of students. A judge sentenced him to a year's probation, and he left the campus.[45]

There are many devices, principally involving programming tricks, for entering into computer systems and wreaking havoc. Some of these are listed in the accompanying box. (■ *See Panel 12.4.*)

- *Carding:* Obtaining, using, or selling other people's credit card numbers.
- *Data diddling:* Changing the data before or as it enters into the computer system.
- *Data leakage:* Removing copies of confidential data from within a system without any trace.
- *Phreaking:* Any manipulation of phone systems, such as simulating tones to get free calls.
- *Piggybacking:* Using access permission belonging to someone else to gain entry to a computer system.
- *Salami shaving:* Diverting small dollar amounts from larger ones to unauthorized sources without being noticed.
- *Scavenging:* Searching trash cans for printouts, memos, carbons, and so on that contain confidential information not intended for public distribution.
- *Superzapping:* Bypassing all security systems by means of specialized software packages.
- *Trapdoor:* Using systems escapes to gain illegitimate access to a computer system.
- *Trojan horse:* Placing covert instructions that would allow unauthorized access to a computer system from within a legitimate program.
- *Warez trading:* Exchanging or selling pirated software.

PANEL 12.4

Some types of computer crime

Crimes Using Computers & Communications

Just as a car can be used to assist in a crime, so can a computer or communications system. For example, four college students on New York's Long Island who met via the Internet used a specialized computer program to steal credit-card numbers, then, according to police, went on a one-year, $100,000 shopping spree. When arrested, they were charged with grand larceny, forgery, and scheming to defraud.[46]

In addition, investment fraud has come to cyberspace. Many people now use online services to manage their stock portfolios through brokerages hooked into the services. Scam artists have followed, offering nonexistent investment deals and phony solicitations and manipulating stock prices.[47–49]

Information technology has also been used simply to perpetrate mischief. For example, three students at a Wisconsin campus faced disciplinary measures after distributing bogus e-mail messages, one of which pretended to be a message of resignation sent by the university's chancellor.[50]

Worms & Viruses

Worms and viruses are forms of high-tech maliciousness. A *worm* **is a program that copies itself repeatedly into memory or onto a disk drive until no more space is left.** An example is the worm program unleashed by a student at Cornell University that traveled through an e-mail network and shut down thousands of computers around the country.

A *virus* **is a "deviant" program that attaches itself to computer systems and destroys or corrupts data.** Viruses are passed in two ways:

- **By diskette:** The first way is via an infected diskette, such as one you might get from a friend or a repair person. It's also possible to get a virus from a sales demo disk or even (in 3% of cases) from a shrink-wrapped commercial disk.

- **By network:** The second way is via a network, as from e-mail or an electronic bulletin board. This is why, with all the freebie games and other software available online, you should use virus-scanning software to check downloaded files before you open them.

The virus usually attaches itself to your hard disk. It might then display annoying messages ("Your PC is stoned—legalize marijuana") or cause Ping-Pong balls to bounce around your screen and knock away text. More seriously, it might add garbage to or erase your files or destroy your system software. It may evade your detection and create havoc elsewhere.

Viruses take several forms, the two traditional ones being boot-sector viruses and file viruses.[51,52] One recent type is the macro virus. (■ *See Panel 12.5.*) There have been many strains of viruses in recent years, some of them quite well known (Stoned, Jerusalem B, Lehigh, Pakistani Brain, Michelangelo). Some 6000 viruses have been identified, but only a few hundred of them have been found "in the wild," or in general circulation. Although most are benign, some are intended to be destructive. Some virus writers do it for the intellectual challenge or to relieve boredom, but others do it for revenge, typically against an employer.[53] One virus writer calling himself Hellraiser, who in his pre-computer youth used to roam New York City streets with a can of spray paint, says that "Viruses are the electronic form of graffiti."[54]

The fastest-growing virus in history, many experts say, is the Word Concept virus (or simply Concept virus), which worries people because it sneaks past security devices by hitching rides on e-mail and other common Internet files.[55–57] Concept attaches itself to documents created by Microsoft's popular word processing program, Word 6.0 or higher. A virus called *Boza*, though not easily spread, specifically infects programs on the Windows 95 operating system, corrupting them so they can no longer function.[58]

A variety of virus-fighting programs are available at stores, although you should be sure to specify the viruses you want to protect against. ***Antivirus software* scans a computer's hard disk, diskettes, and main memory to detect viruses and, sometimes, to destroy them.** We described some antivirus programs in Chapter 3 (✓ p. 129) and some other antivirus measures in the Experience Box at the end of Chapter 5. (A detailed list of antivirus software can be found on the World Wide Web at *http://www.ncsa.com*. Up-to-date material may also be found at *FTP://mcafee.com/pub/antivirus*.)

Computer Criminals

What kind of people are perpetrating most of the information-technology crime? Over 80% may be employees, and the rest are outside users, hackers and crackers, and professional criminals.

• **Employees:** "Employees are the ones with the skill, the knowledge, and the access to do bad things," says Donald Parker, an expert on computer security at SRI International in Menlo Park, California. "They're the ones, for example, who can most easily plant a 'logic bomb.' . . ." Dishonest or disgruntled employees, he says, pose "a far greater problem than most people realize."[59] Says Michigan State University criminal justice professor David Carter, who surveyed companies about computer crime, "Seventy-five to 80% of everything happens from inside."[60]

Most common frauds, Carter found, involved credit cards, telecommunications, employees' personal use of computers, unauthorized access to confidential files, and unlawful copying of copyrighted or licensed software. In addition, the increasing use of laptops off the premises, away from the eyes of supervisors, concerns some security experts. They worry that dishonest employees or outsiders can more easily intercept communications or steal company trade secrets.

- *Boot-sector virus:* The boot sector is that part of the system software containing most of the instructions for booting, or powering up, the system. The boot-sector virus replaces these boot instructions with some of its own. Once the system is turned on, the virus is loaded into main memory before the operating system. From there it is in a position to infect other files.

  Any diskette that is used in the drive of the computer then becomes infected. When that diskette is moved to another computer, the contagion continues.

- *File virus:* File viruses attach themselves to executable files—those that actually begin a program. (In DOS these files have the extensions .com and .exe.) When the program is run, the virus starts working, trying to get into main memory and infecting other files.

- *Logic bomb:* Logic bombs, or simply *bombs,* differ from other viruses in that they are set to go off at a certain date and time. A disgruntled programmer for a defense contractor created a bomb in a program that was supposed to go off two months after he left. Designed to erase an inventory tracking system, the bomb was discovered only by chance.

- *Trojan horse:* The Trojan horse covertly places illegal, destructive instructions in the middle of a legitimate program, such as a computer game. Once you run the program, the Trojan horse goes to work, doing its damage while you are blissfully unaware.

- *Polymorphic virus:* A polymorphic virus, of which there are several kinds, can mutate and change form just as human viruses can. These are especially troublesome because they can change their profile, making existing antiviral technology ineffective.

- *Macro virus:* These viruses take advantage of a procedure in which miniature programs, known as macros, are embedded inside common data files, such as those created by e-mail or spreadsheets, which are sent over computer networks. Such documents are typically ignored by antivirus software. The Word Concept virus is an example of this type.

Workers may use information technology for personal profit or steal hardware or information to sell. They may also use it to seek revenge for real or imagined wrongs, such as being passed over for promotion. Sometimes they may use the technology simply to demonstrate to themselves that they have power over people. This may have been the case with a Georgia printing-company employee convicted of sabotaging the firm's computer system. As files mysteriously disappeared and the system randomly crashed, other workers became so frustrated and enraged that they quit.

- **Outside users:** Suppliers and clients may also gain access to a company's information technology and use it to commit crimes. With both, this becomes more possibile as electronic connections such as Electronic Data Interchange (✓ p. 351) systems become more commonplace.

- **Hackers and crackers:** *Hackers* **are people who gain unauthorized access to computer or telecommunications systems for the challenge or even the principle of it.** For example, Eric Corley, publisher of a magazine called *2600: The Hackers' Quarterly,* believes that hackers are merely engaging in "healthy exploration." In fact, by breaking into corporate computer systems and revealing their flaws, he says, they are performing a favor and a public service. Such unauthorized entries show the corpora-

README

Practical Matters: Tips for Avoiding Viruses

You can avoid most virus trouble by simply practicing "safer computing."

- Make backups of your disks (you should do this anyway) so you can recover in case a virus erases anything.

- Avoid shared programs from clubs and obscure online services.

- Stick to commercial software in plastic-wrapped boxes or respectable online services—although even those sources can give you infections.

- Write-protect your floppies—slip the little tab to open the hole in the top left corner of 3½-inch floppies. Use CD-ROMs when you can because they are immune—neither viruses nor anything else can record to them.

—Phillip Robinson, "What You Need to Know and Do About Viruses," *San Jose Mercury News*

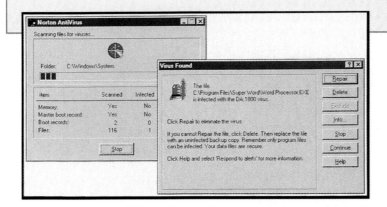

tions involved the leaks in their security systems.[61] Indeed, in late 1995, Netscape launched its so-called Bugs Bounty program, offering a cash reward to the first hacker to identify a "significant" security flaw in its latest Web browser software (✓ p. 79).[62]

Crackers **also gain unauthorized access to information technology but do so for malicious purposes.** (Some observers think the term hacker covers malicious intent, also.) Crackers attempt to break into computers and deliberately obtain information for financial gain, shut down hardware, pirate software, or destroy data.

The tolerance for "benign explorers"—hackers—has waned. Most communications systems administrators view any kind of unauthorized access as a threat, and they pursue the offenders vigorously. Educators try to point out to students that universities can't provide an education for everybody if hacking continues.[63] The most flagrant cases of hacking are met with federal prosecution. A famous instance involved five young New York–area men—with pseudonyms such as Acid Phreak, Phiber Optik, and Scorpion—calling themselves the Masters of Deception who had used cheap computers to break into some of the nation's most powerful information systems.[64] They were indicted for computer trespass in 1992, with Phiber Optik receiving the longest sentence—a year in federal prison.[65]

- **Professional criminals:** Members of organized crime rings don't just steal information technology. They also use it the way that legal businesses do—as a business tool, but for illegal purposes. For instance, databases can be used to keep track of illegal gambling debts and stolen goods. Not surprisingly, the old-fashioned illegal bookmaking operation has gone high-tech, with bookies using computers and fax machines in place of betting slips and paper tally sheets.[66] Drug dealers have used pagers as a link to customers. Microcomputers, scanners, and printers can be used to forge checks, immigration papers, passports, and driver's licenses. Telecommunications can be used to transfer funds illegally. For instance, in 1995, Russian computer crackers broke into a Citibank electronic money transfer system and stole more than $10 million before they were caught.[67]

As information-technology crime has become more sophisticated, so have the people charged with preventing it and disciplining its outlaws. Campus administrators are no longer being quite as easy on offenders and are turning them over to police. Industry organizations such as the Software Publishers Association are going after software pirates large and small. (Commercial software piracy is now a felony, punishable by up to five years in prison and fines of up to $250,000 for anyone convicted of stealing at least 10 copies of a program, or more than $2500 worth of software.) Police departments as far apart as Medford, Massachusetts, and San Jose, California, now have police patrolling a "cyber beat." That is, they cruise online bulletin boards looking for pirated software, stolen trade secrets, child molesters, and child pornography.[68]

In 1988, after one widespread Internet break-in, the U.S. Defense Department created the Computer Emergency Response Team (CERT). Although it has no power to arrest or prosecute, *CERT* provides round-the-clock international information and security-related support services to users of the Internet. Whenever it gets a report of an electronic snooper, whether on the Internet or on a corporate e-mail system, CERT stands ready to lend assistance. It counsels the party under attack, helps them thwart the intruder, and evaluates the system afterward to protect against future break-ins.[69,70]

Security: Safeguarding Computers & Communications

Preview & Review: Information technology requires vigilance in security. Four areas of concern are identification and access, encryption, protection of software and data, and disaster-recovery planning.

Does your Social Security number say anything about you? Indeed it does. The first three numbers, called area numbers, generally tell in which state you were born or, if you recently emigrated, where you lived when you received your working papers. The next six digits can tell government officials whether a Social Security number is fraudulent.[71] Some area numbers were issued for special purposes. For example, railroad workers once received area numbers 700–728. Southeast Asian refugees arriving between April 1975 and November 1979 got numbers beginning with 574, 580, or 586.

It's usually nobody's business but your own where you came from, but you'd like to keep it that way. However, the ongoing dilemma of the Digital Age is balancing convenience against security. *Security* **is a system of safeguards for protecting information technology against disasters, systems failure, and unauthorized access that can result in damage or loss.** We consider four components of security:

- Identification and access
- Encryption
- Protection of software and data
- Disaster-recovery planning

Identification & Access

Are you who you say you are? The computer wants to know.

There are three ways a computer system can verify that you have legitimate right of access. Some security systems use a mix of these techniques. The systems try to authenticate your identity by determining (1) what you have, (2) what you know, or (3) who you are.

- **What you have—cards, keys, signatures, badges:** Credit cards, debit cards, and cash-machine cards all have magnetic strips or built-in computer chips that identify you to the machine. Many require you to display your signature, which someone may compare as you sign their paperwork. Computer rooms are always kept locked, requiring a key. Many people also keep a lock on their personal computers. A computer room may also be guarded by security officers, who may need to see an authorized signature or a badge with your photograph before letting you in.

    Of course, credit cards, keys, and badges can be lost or stolen. Signatures can be forged. Badges can be counterfeited.

- **What you know—PINs, passwords, and digital signatures:** To gain access to your bank account through an automated teller machine (ATM), you key in your PIN. A **PIN, or personal identification number, is the security number known only to you that is required to access the system.** Telephone credit cards also use a PIN. If you carry either an ATM or phone card, never carry the PIN written down elsewhere in your wallet (even disguised).

    A **password is a special word, code, or symbol that is required to access a computer system.** Passwords are one of the weakest security links, says AT&T security expert Steven Bellovin. Passwords can be guessed, forgotten, or stolen. To reduce a stranger's guessing, Bellovin recommends never choosing a real word or variations of your name or birthdate or those of your friends or family. Instead you should mix letters, numbers, and punctuation marks in an oddball sequence of no fewer than eight characters.[72]

    The advice is sound, but the problem today is that many people have to remember several passwords. "Now password overload has become a plague, with every computer online service, voice-mail box, burglar-alarm disarmer, and office computer network demanding a unique string of code," says technology writer William Bulkeley.[73] Still, in line with Bellovin's suggestions, he offers these possibilities as good passwords: 2b/orNOT2b%. Alfred!E!Newman7. He also reports the strategy of Glenn Maxwell, a computer instructor from Farmington, Michigan, who uses an obvious and memorable password, but shifts the position of his hands on the keyboard, creating a meaningless string of characters.

    Skilled hackers may break into national computer networks and detect passwords as they are being used. Or they pose, on the telephone, as computer technicians to cajole passwords out of employees. They may even find access codes in discarded technical manuals in trash bins.[74]

    A relatively new technology is the digital signature, which security experts hope will lead to a world of paperless commerce. A **digital signature is a string of characters and numbers that a user signs to an electronic document being sent by his or her computer.** The receiving computer performs mathematical operations on the alphanumeric string to verify its validity. The system works by using a public-private key system. That is, the system involves a pair of numbers called a private key and a public key. One person creates the signature with a secret private key, and the recipient reads it with a second, public key. "This process in effect notarizes the document and ensures its integrity," says one writer.[75]

For example, when you write your boss an electronic note, you sign it with your secret private key. (This could be some bizarre string beginning with 479XY283 and continuing on for 25 characters.) When your boss receives the note, he or she looks up your public key. Your public key is available from a source such as an electronic bulletin board, the Postal Service, or a corporate computer department. If the document is altered in any way, it will no longer produce the same signature sequence.[76,77]

- **Who you are—physical traits:** Some forms of identification can't be easily faked—such as your physical traits. Biometrics tries to use these in security devices. ***Biometrics* is the science of measuring individual body characteristics.**

  For example, before a number of University of Georgia students can use the all-you-can-eat plan at the campus cafeteria, they must have their hands read. As one writer describes the system, "a camera automatically compares the shape of a student's hand with an image of the same hand pulled from the magnetic strip of an ID card. If the patterns match, the cafeteria turnstile automatically clicks open. If not, the would-be moocher eats elsewhere."[78]

  Besides hand prints, other biological characteristics read by biometric devices are fingerprints (computerized "finger imaging"), voices, the blood vessels in the back of the eyeball, the lips, and even one's entire face.[79–81]

Some computer security systems have a "call-back" provision. In a ***call-back system*, the user calls the computer system, punches in the password and hangs up.** The computer then calls back a certain preauthorized number. This measure will block anyone who has somehow got hold of a password but is calling from an unauthorized telephone.

Encryption

PGP (for Pretty Good Privacy) is a computer program written for encrypting computer messages—putting them into secret code. ***Encryption*, or enciphering, is the altering of data so that it is not usable unless the changes are undone.** PGP is so good that it is practically unbreakable; even government experts can't crack it.[82] (This is because it uses a two-key method similar to that described above under digital signatures.)

Encryption is clearly useful for some organizations, especially those concerned with trade secrets, military matters, and other sensitive data. However, from the standpoint of our society, encryption is a two-edged sword. For instance, police in Sacramento, California, found that PGP blocked them from reading the computer diary of a convicted child molester and finding links to a suspected child pornography ring. Should the government be allowed to read the coded e-mail of its citizens? What about its being blocked from surveillance of overseas terrorists, drug dealers, and other enemies?

Protection of Software & Data

Organizations go to tremendous lengths to protect their programs and data. As might be expected, this includes educating employees about making backup disks, protecting against viruses, and so on. (We discussed these matters in detail elsewhere, especially in the Experience Box at the end of Chapter 5.)

Other security procedures include the following:

- **Control of access:** Access to online files is restricted only to those who have a legitimate right to access—because they need them to do their jobs. Many organizations have a transaction log that notes all accesses or attempted accesses to data.

- **Audit controls:** Many networks have audit controls, which track the programs and servers used, the files opened, and so on. This creates an audit trail, a record of how a transaction was handled from input through processing and output.

- **People controls:** Because people are the greatest threat to a computer system, security precautions begin with the screening of job applicants. That is, resumes are checked to see if people did what they said they did. Another control is to separate employee functions, so that people are not allowed to wander freely into areas not essential to their jobs. Manual and automated controls—input controls, processing controls, and output controls—are used to check that data is handled accurately and completely during the processing cycle. Printouts, printer ribbons, and other waste that may yield passwords and trade secrets to outsiders are disposed of through shredders or locked trash barrels.

Disaster-Recovery Plans

A *disaster-recovery plan* **is a method of restoring information processing operations that have been halted by destruction or accident.** "Among the countless lessons that computer users have absorbed in the hours, days, and weeks after the [New York] World Trade Center bombing," wrote one reporter, "the most enduring may be the need to have a disaster-recovery plan. The second most enduring lesson may be this: Even a well-practiced plan will quickly reveal its flaws."[83]

Mainframe computer systems are operated in separate departments by professionals, who tend to have disaster plans. Mainframes are usually backed up. However, many personal computers, and even entire local area networks, are not backed up. The consequences of this lapse can be great. It has been reported that, on average, a company loses as much as 3% of its gross sales within eight days of a sustained computer outage. In addition, the average company struck by a computer outage lasting more than 10 days never fully recovers.[84]

A disaster-recovery plan is more than a big fire-drill. It includes a list of all business functions and the hardware, software, data, and people to support those functions. It includes arrangements for alternate locations, either hot sites or cold sites. A *hot site* is a fully equipped computer center, with everything needed to resume functions. A *cold site* is a building or other suitable environment in which a company can install its own computer system. The disaster-recovery plan includes ways for backing up and storing programs and data in another location, ways of alerting necessary personnel, and training for those personnel.

How involved should the alternate arrangements be? The best way to judge is to figure out how long it would take to duplicate the present arrangements. "At the World Trade Center," said an executive with a company offering recovery services, "traders had two or three terminals on each desk. That is a lot of communications to duplicate."[85]

Quality-of-Life Issues: The Environment, Mental Health, & the Workplace

Preview & Review: Information technology can create problems for the environment, people's mental health (isolation, gambling, Net addiction, and stress), and the workplace (misuse of technology and information overload).

Earlier in this book, we pointed out some of the worrisome effects of technology on intellectual property rights and truth in art and journalism (✓ p. 52, 268), censorship (✓ p. 384), on health matters and ergonomics (✓ p. 227), on environmental matters (✓ p. 180), and on privacy (✓ pp. 386, 425). Here are some other quality-of-life issues.

Environmental Problems

"This county will do peachy fine without computers," says Micki Haverland, who has lived in rural Hancock County, Tennessee, for 20 years.[86] Telecommunications could bring jobs to an area that badly needs them, but several people moved there precisely because they like things the way they are—pristine rivers, unspoiled forests, and mountain views.

But it isn't just people in rural areas who are concerned. Suburbanites in Idaho and Utah, for example, worry that lofty metal poles topped by cellular-transmitting equipment will be eyesores that will destroy views and property values.[87] City dwellers everywhere are concerned that the federal government's 1996 decision to deregulate the telecommunications industry will lead to a rat's nest of roof antennas, satellite dishes, and above-ground transmission stations.[88] As a result, telecommunications companies are now experimenting with hiding transmitters in the "foliage" of fake trees made of metal.[89]

Political scientist James Snider of Northwestern University points out that the problems of the cities could expand well beyond the cities, if telecommuting triggers a massive movement of people to rural areas. "If all Americans succeed in getting their dream homes with several acres of land," he writes, "the forests and open lands across the entire continental United States will be destroyed" as they become carved up with subdivisions and roads.[90]

Mental-Health Problems: Isolation, Gambling, Net-Addiction, Stress

"People need things to count on, to believe in, to go home to, figuratively speaking," says journalist Bruce Weber.[91] So how do they feel when their telephone area codes are suddenly changed, as is happening at a tremendous rate as new technologies—fax machines, pagers, cell phones, modems—require millions of additional telephone numbers? (U.S. area codes went from 126 in 1984 to 144 a decade later.) "It's like an eviction from one's psychic address," Weber says, "no small disruption in a world that seems ever more bent on isolating its inhabitants."

From a mental-health standpoint, will being wired together really set us free? Consider:

- **Isolation:** Automation allows us to go days without actually speaking with or touching another person, from buying gas to playing games. Even the friendships we make online in cyberspace, some believe, "are likely to be trivial, short lived, and disposable—junk friends." Says one writer, "We may be overwhelmed by a continuous static of information and casual acquaintance, so that finding true soul mates will be even harder than it is today."[92]

- **Gambling:** Gambling is already widespread in North America, but information technology could make it almost unavoidable. Although gambling by wire is illegal in the U.S., all kinds of moves are afoot to get around that. For example, host computers for Internet casinos and sports books are being set up in Caribbean tax havens, and satellites, decoders, and remote-control devices are being used so TV viewers can do racetrack wagering from home.[93–95]

  Some mental-health professionals are concerned with the long-range effects. "About 5% [of the users or viewers] will be compulsive gamblers and another 10% to 15% will be problem gamblers," says Kevin O'Neill of New Jersey's Council on Compulsive Gambling. "Compulsive gamblers want action, which is what interactive television [or computers] can give you."[96] In Congress, legislation has been introduced making it a crime to transmit money wagers or gambling information over the Internet.[97]

- **Net-addiction:** Don't let this happen to you: "A student e-mails friends, browses the World Wide Web, blows off homework, botches exams, flunks out of school."[98] This is the downward spiral of the "Net addict," often a college student—because schools give students no-cost/low-cost linkage to the Internet—though it can be anyone. Some become addicted (although some mental-health professionals feel "addiction" is too strong a word) to chat groups, some to online pornography, some simply to escape from real life.[99,100] Indeed, sometimes the computer replaces one's spouse or boyfriend/girlfriend in the user's affections. In one instance, a man sued his wife for divorce for having an "online affair" with a partner who called himself The Weasel.[101–104]

- **Stress:** In a 1995 survey of 2802 American PC users, three-quarters of the respondents (whose ages ranged from children to retirees) said personal computers had increased their job satisfaction and were a key to success and learning. However, many found PCs stressful: 59% admitted getting angry at them within the previous year. And 41% said they thought computers have reduced job opportunities rather than increased them.[105]

Psychologist and sociologist Sherry Turkle of MIT believes that, when it comes to mental health, information technology is neither a blessing nor a curse. In fact, she holds, "The Internet is not a drug." Rather, people who seem addicted may be "working through important personal issues in the safety of life on the screen."[106]

In her book *Life on the Screen: Identity in the Age of the Internet*, she suggests that cyberspace can make people's lives more fulfilling by getting them to face issues of identity and relationships that they have never had to confront before.[107–109] In particular, people exploring the computer-based fantasy worlds known as MUDs (for Multi-User Domains) on the Net can communicate anonymously with others and try on different roles, playing at being different genders, animals, or even beings from another planet. "The way we used to think about identity is that people had a core self, a one," she says. Now, in the behavior of people playing Internet role-playing games, she sees evidence that identity itself consists of different constructions of personae.

Workplace Problems

First the mainframe computer, then the desktop stand-alone PC, and lately the networked computer were all brought into the workplace for one reason only: to improve productivity. How is it working out? Let's consider two aspects: the misuse of technology and information overload.

- **Misuse of technology:** "For all their power," says an economics writer, "computers may be costing U.S. companies tens of billions of dollars a year in downtime, maintenance and training costs, useless game playing, and information overload."[110]

  Consider games. Employees may look busy, staring into their computer screens with brows crinkled. But often they're just hard at work playing Doom or surfing the Net. Workers with Internet access average 10 hours a week online.[111] However, fully 23% of computer game players use their office PCs for their fun, according to one survey.[112] A study of employee online use at one major company concluded that the average worker wastes 1½ hours each day.[113]

  Another reason for so much wasted time is all the fussing that employees do with hardware and software. Says one editor, "Back in the old days, when I toiled on a typewriter, I never spent a whole morning installing a new ribbon.... I did not scan the stores for the proper cables to affix to my typewriter or purchase books that instructed me on how to get more use from my liquid white-out."[114] A 1992 study estimated that microcomputer users waste 5 billion hours a year waiting for programs to run, checking computer output for accuracy, helping co-workers use their applications, organizing cluttered disk storage, and calling for technical support.[115]

  Many companies don't even know what kind of microcomputers they have, who's running them, or where they are. The corporate customer of one computer consultant, for instance, swore it had 700 PCs and 15 users per printer. An audit showed it had 1200 PCs with one printer each.[116]

  A particularly interesting misuse is the continual upgrade. Ask yourself, Do I really need that slick new product? "I use an old version of Word-Perfect on my PC," says Ron Erickson, former chairman of the Egghead Software stores and now a technology consultant. "The one thing I don't need is a new version of WordPerfect—or of any of the other software on my machine for that matter."[117] Erickson's advice to consumers: "don't get the new version if the old one is working O.K." As for many businesses, he says, the rule should be: "stop buying new software, and train employees on what you have."

- **Information overload:** "My boss basically said, 'Carry this pager seven days a week, 24 hours a day, or find another job,'" says the chief architect for a New Jersey school system. (He complied, but pointedly notes that the pager's "batteries run out all the time.")[118] "It used to be considered a status symbol to carry a laptop computer on a plane," says futurist Paul Saffo. "Now anyone who has one is clearly a working dweeb who can't get the time to relax. Carrying one means you're on someone's electronic leash."[119]

  The new technology is definitely a two-edged sword. Cellular phones, pagers, fax machines, and modems may unleash employees from the office, but they tend to work longer hours under more severe deadline pressure than do their tethered counterparts who stay at the office, according to one study.[120] Moreover, the gadgets that once promised to do away with irksome business travel by ushering in a new era of communications have done the opposite—created the office-in-a-bag that allows business travelers to continue to work from airplane seats and hotel-room desks.[121]

What does being overwhelmed with information do to you, besides inducing stress and burnout? One result is that because we have so many choices to entice and confuse us, we become more averse to making decisions. Home buyers now take twice as long as a decade earlier to sign a contract on a new house, organizations take months longer to hire top executives, and managers tend to consider worst-case scenarios rather than benefits when considering investing in a new venture.[122]

"The volume of information available is so great that I think people generally are suffering from a lack of meaning in their lives," says Neil Postman, chair of the Department of Culture and Communication at New York University. "People are just adrift in the sea of information, and they don't know what the information is about or why they need it."[123]

People and businesses are beginning to realize the importance of coming to grips with these problems. Some companies are employing GameCop, a software program that catches unsuspecting employees playing computer games on company time.[124] Some are installing asset-management software that tells them how many PCs are on their networks and what they run. Some are imposing strict hardware and software standards to reduce the number of different products they support.[125] To avoid information overload, some people—those who have a choice—no longer carry cell phones or even look at their e-mail. Others are installing so-called Bozo filters, software that screens out trivial e-mail messages and cellular calls and assigns priorities to the remaining files.[126] Still others are beginning to employ programs called intelligent agents to help them make decisions, as we discuss shortly.

But the real change may come as people realize that they need not be tied to the technological world in order to be themselves, that solitude is a scarce resource, and that seeking serenity means streamlining the clutter and reaching for simpler things.[127]

Economic Issues: Employment & the Haves/Have-Nots

Preview & Review: Many people worry that jobs are being reduced by the effects of information technology. They also worry that it is widening the gap between the haves and have-nots.

"If you'd had any brains," Yale University professor David Gelernter read, in the letter sent by the Unabomber, whose explosive device had savagely disfigured the computer scientist, "you would realize that there are a lot of people out there who resent bitterly the way techno-nerds like you are changing the world."[128]

People who don't like technology in general, and today's information technology in particular, have been called neo-Luddites. The original Luddites were a group of weavers in northern England who, while proclaiming their allegiance to a legendary King Ludd, also called Ned Ludd, in 1812–1814 went about smashing modern looms that moved cloth production out of the hands of in-home peasant weavers and into inhumane factories. Although the term now seems to connote a knee-jerk antagonism to technology, in actuality the 19th-century Luddites were desperate, brave people, with no other means of employment and no future after being stripped of their livelihoods.[129]

Nevertheless, in recent times a number of books (such as Clifford Stoll's *Silicon Snake Oil*, Stephen Talbott's *The Future Does Not Compute*, and Mark Slouka's *War of the Worlds*), have appeared that try to provide a counterpoint to the hype and overselling of information technology to which we have long been exposed. Some of these strike a sensible balance, but some make the alarming case that technological progress is actually no progress at all—indeed, it is a curse. The two biggest charges (which are related) are, first, that information technology is killing jobs, and second, that it is widening the gap between the rich and the poor.

- **Technology, the job killer?**   There's probably no question that technological advances play an ambiguous role in social progress. But is it true, as Jeremy Rifkin says in *The End of Work*, that intelligent machines are replacing humans in countless tasks, "forcing millions of blue-collar and white-collar workers into temporary, contingent, and part-time employment and, worse, unemployment"?[130]

    This is too large a question to be fully considered in this book. Many factors are responsible for the decline in economic growth, the downsizing of companies, the rise of unemployment among many sectors of society. The U.S. economy is undergoing powerful structural changes, brought on not only by the widespread diffusion of technology but also by the growth of international trade, the shift from manufacturing to service employment, the weakening of labor unions, more rapid immigration, and other factors.[131,132]

    Many economists seem to agree that the boom times of economic growth that the United States enjoyed in the 1950s and 1960s won't return until there is more public investment and more personal saving instead of spending—savings that could be used for machinery and other tools of a thriving economy. Investment in recent years has been concentrated in computers. Surprisingly, however, as one economics writer points out, "so far computers have not yielded the rapid growth in production that came from investments in railroads, autos, highways, electric power, and aircraft—all huge outlays, involving government as well as the private sector, that changed the way Americans lived and worked."[133]

- **Gap between rich and poor:**   "In the long run," says Stanford University economist Paul Krugman, "improvements in technology are good for almost everyone. . . . Unfortunately, what is true in the long run need not be true over shorter periods."[134] We are now, he believes, living through one of those difficult periods in which technology doesn't produce widely shared economic gains but instead widens the gap between those who have the right skills and those who don't.

    A U.S. Department of Commerce survey of "information have-nots" reveals that about 20% of the poorest households in the U.S. do not have telephones. Moreover, only a fraction of those poor homes that do have phones will be able to afford the information technology that most economists agree is the key to a comfortable future.[135] The richer the family, the more likely it is to have and use a computer.

    Schooling—especially college—makes a great difference. Every year of formal schooling after high school adds 5–15% to annual earnings later in life.[136] Being well educated is only part of it, however; one should also be technologically literate. Employees who are skilled at technology "earn roughly 10–15% higher pay," according to the chief economist for the U.S. Labor Department.[137]

Advocates of information access for all find hope in the promises of NII proponents for "universal service" and the wiring of every school to the Net. But this won't happen automatically. Ultimately we must become concerned with the effects of growing economic disparities on our social and political health. "Computer technology is the most powerful and the most flexible technology ever developed," says Terry Bynum, chair of the American Philosophical Association's Committee on Philosophy and Computing. "Even though it's called a technical revolution, at heart it's a social and ethical revolution because it changes everything we value."[138]

Now that we've considered the challenges, let us discuss some of the promises of information technology not described so far. Some of these are truly awesome. They include agents and avatars, artificial intelligence, and the promises of the Information Superhighway.

Intelligent Agents & Avatars

Preview & Review: Intelligent agents are computer programs that act as electronic secretaries, e-mail filters, and electronic news clipping services. Types of Internet agents called *spiders* roam the Web assembling page information to put into databases for later searching by search engines. *Avatars* are (1) a graphical image of you or someone else on a computer screen or (2) a graphical personification of a computer or process that's running on a computer.

What the online world really needs is a terrific librarian. "What bothers me most," says Christine Borgman, chair of the UCLA Department of Library and Information Science, "is that computer people seem to think that if you have access to the Web, you don't need libraries. But what's on the Web now is just a fraction of what's in the average-sized library."[139] And what's in a library is standardized and well organized, and what's on the Web is unstandardized and chaotic.

As a solution, scientists have been developing so-called *intelligent agents* to send out on computer networks to find and filter information. And to make them more friendly they are inventing graphical on-screen personifications called *avatars*. Let's consider both of these.

Intelligent Agents

An ***intelligent agent* is a computer program that performs work tasks on your behalf, including roaming networks and compiling data.** It acts as an electronic assistant that will perform, in the user's stead, such tasks as filtering messages, scanning news services, and similar secretarial chores. It will also travel over communications lines to nearly any kind of computer database, collecting files to add to a database.

Examples of agents are the following:[140–145]

- **Electronic secretaries:** Wildfire is a voice-recognition system (✓ p. 221), an electronic secretary that will answer the phone, take messages, track you down on your cell phone and announce the caller, place calls for you, and remind you of appointments. In development are plans to have Wildfire handle e-mail and faxes in much the same way as phone calls.

- **E-mail filters:** BeyondMail will filter your e-mail, alerting you to urgent messages, telling you which require follow-up, and sorting everything according to priorities. For people whose e-mail threatens to overwhelm them with "cyberglut," such an agent is a great help.

- **Electronic clipping services:** Several companies offer customized electronic news services (Heads Up, I-News, Journalist, Personal Journal, News Hound, the Personal Internet Newspaper) that will scan online news sources and publications looking for information that you have previously specified using keywords. Some will rank a selected article according to how closely it fits your request. Others will pull together articles in the form of a condensed electronic newspaper.

- **Internet agents—spiders, crawlers, and robots:** ***Search engines* are Web pages containing forms into which you type text on the topic you want to search for.** The search engine then looks through its database and presents a list of Web sites matching your search criteria. Examples of search engines are Lycos, InfoSeek, Excite, and WebCrawler, as described in the Experience Box at the end of Chapter 9.

  What interests us here, however, are the intelligent agents used to assemble the database that a search engine searches. Most such databases are created by spiders. ***Spiders*—also known as *crawlers* or *robots*—are software programs that roam the Web, looking for new Web sites by following links from page to page.** When a spider finds a new page, it adds information about it—its title, address, and perhaps summary of contents—to the search engine's database.[146]

Avatars

Want to see yourself—or a stand-in for yourself—on your computer screen? Then try using a kind of cyberpersona called an *avatar.* An ***avatar* is either (1) a graphical image of you or someone else on a computer screen or (2) a graphical personification of a computer or a process that's running on a computer.**[147]

- **Avatar as yourself or others:** The on-screen version of yourself could be "anything from a human form to a pair of cowboy boots with lips," writes technology columnist Denise Caruso. "Users move them around while talking (via keyboard) with other avatars on the same screen."[148] In CompuServe's Worlds Chat, subscribers participating in online "chat rooms," which are furnished like cartoon stage sets, can get together with other users, each of whom can construct an avatar from a variety of heads, clothing, shoes, and even animal identities.[149] (■ *See Panel 12.6.*)

- **Avatars representing a process:** "The driving force behind avatars is the ongoing search for an interface that's easier and more comfortable to use," says one writer, "especially for the millions of people who are still computerphobic."[150] This was, of course, the motivation behind Microsoft's Bob (✓ p. 106), the add-on interface for Windows that uses animals

 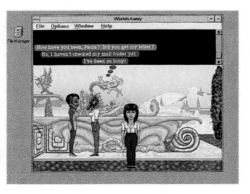

PANEL 12.6

Avatars

Made-up personas representing participants in a CompuServe Worlds Chat chat room. Conversations may be carried on in speech balloons.

and other fanciful creatures to help novice computer users. One difficulty with designing computer-controlled avatars—called agents, characters, and bots—is making sure that they don't make people react negatively to them. Thus, instead of faces or personifications, it may be better to use pictures of notepads, checkbooks, and similar objects.

Agents and avatars are still in their infancy. In time, however, the promise is that they will make information technology much easier to use—by helping to tame the "cyberglut" and by helping us deal more effectively with all the on-screen choices available to us. No doubt their improved versions will draw on the field of artificial intelligence, as we discuss next.

Artificial Intelligence

Preview & Review: Artificial intelligence (AI) is a research and applications discipline that includes the areas of robotics, perception systems, expert systems, natural language processing, fuzzy logic, neural networks, and genetic algorithms. Another area, artificial life, is the study of computer instructions that act like living organisms.

The Turing test has long been used as a standard to determine whether a computer possesses "intelligence."

Behind all aspects of AI are ethical questions.

You're having trouble with your new software program. You call the customer "help desk" at the software maker. Do you get a busy signal or get put on hold to listen to music (or, worse, advertising) for several minutes? Technical support lines are often swamped, and waiting is commonplace. Or, to deal with your software difficulty, do you find yourself dealing with . . . other software?

This event is not unlikely. Programs that can walk you through a problem and help solve it are called *expert systems* (✓ p. 455). As the name suggests, these are systems that are imbued with knowledge by a human expert.[151] Expert systems are one of the most useful applications of an area known as artificial intelligence.

***Artificial intelligence (AI)* is a group of related technologies that attempt to develop machines to emulate human-like qualities, such as learning, reasoning, communicating, seeing, and hearing.** Today the main areas of AI are:

- Robotics
- Perception systems
- Expert systems
- Natural language processing
- Fuzzy logic
- Neural networks
- Genetic algorithms

We will consider these and also an area known as *artificial life.*

Robotics

As we discussed earlier (Chapter 6), *robotics* is a field that attempts to develop machines that can perform work normally done by people. The machines themselves, of course, are called robots. **A *robot* is an automatic device that performs functions ordinarily ascribed to human beings or that operates with what appears to be almost human intelligence.** Dante II, for instance, is an eight-legged, 10-foot-high, satellite-linked robot used by scientists to explore the inside of Mount Spurr, an active volcano in Alaska (✓ p. 268).[152] Robots may be controlled from afar, as in an experiment at the University of Southern California in which Internet users thousands of miles away were invited to manipulate a robotic arm to uncover objects in a sandbox.[153]

Robots that resemble R2D2 in the movie *Star Wars* are some way from being realized. Such "personal robots" as have been developed are like B.O.B. (for Brains On Board), a device sold by Visual Machines. B.O.B. speaks prerecorded phrases and maneuvers around objects by using ultrasonic sound.

Perception Systems

***Perception systems* are sensing devices that emulate the human capabilities of sight, hearing, touch, and smell.** Clearly, perception systems are related to robotics, since robots need to have at least some sensing capabilities. Examples of perception systems are vision systems, used for pattern recognition. Vision systems are used, for example, to inspect products for quality control in factory assembly lines. To discriminate among parts or shapes, a robot measures varying intensities of light off each part. Each intensity has a numbered value that is compared to a similar palette of intensities stored in the system's memory. If the intensity is not recognized, the part is rejected. An example of this kind of perception system is the Bin Vision Systems used by General Electric to pick up specific parts. (■ *See Panel 12.7.*)

PANEL 12.7

Perception systems

(Right) A vision system. *(Top left and bottom left)* A touch system that inserts objects into various parts of machinery.

Expert Systems

We mentioned one example of an expert system in the "help desk" software. An *expert system* **is an interactive computer program that helps users solve problems that would otherwise require the assistance of a human expert.**

The programs simulate the reasoning process of experts in certain well-defined areas. That is, professionals called *knowledge engineers* interview the expert or experts and determine the rules and knowledge that must go into the system. Programs incorporate not only surface knowledge ("textbook knowledge") but also deep knowledge ("tricks of the trade"). What, exactly, is this latter kind of knowledge? "An expert in some activity has by definition reduced the world's complexity by his or her specialization," say some authorities. One result is that "much of the knowledge lies outside direct conscious awareness. . . ."[154] Expert systems exist in many areas. MYCIN helps diagnose infectious diseases. PROSPECTOR assesses geological data to locate mineral deposits. DENDRAL identifies chemical compounds. Home-Safe-Home evaluates the residential environment of an elderly person. Business Insight helps businesses find the best strategies for marketing a product. REBES (Residential Burglary Expert System) helps detectives investigate crime scenes. CARES (Computer Assisted Risk Evaluation System) helps social workers assess families for risks of child abuse. CLUES (Countrywide Loan Underwriting Expert System) evaluates home-mortgage-loan applications. Muckraker assists journalists with investigative reporting. Crush takes a body of expert advice and combines it with worksheets reflecting a user's business situation to come up with a customized strategy to "crush competitors."

An expert system consists of three components. (■ *See Panel 12.8.*)

- **Knowledge base:** A *knowledge base* **is an expert system's database of knowledge about a particular subject.** This includes relevant facts, information, beliefs, assumptions, and procedures for solving problems. The basic unit of knowledge is expressed as an IF-THEN-ELSE rule ("IF this happens, THEN do this, ELSE do that"). Programs can have as many as 10,000 rules. A system called ExperTAX, for example, which helps accountants figure out a client's tax options, consists of over 2000 rules.

- **Inference engine:** The *inference engine* **is the software that controls the search of the expert system's knowledge base and produces conclusions.** It takes the problem posed by the user of the system and fits it into the rules in the knowledge base. It then derives a conclusion from the facts and rules contained in the knowledge base.

  Reasoning may be by a forward chain or backward chain. In the forward chain of reasoning, the inference engine begins with a statement of the

PANEL 12.8

Components of an expert system

The three components are the knowledge base, inference engine, and user interface.

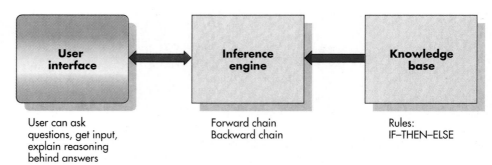

User interface	Inference engine	Knowledge base
User can ask questions, get input, explain reasoning behind answers	Forward chain Backward chain	Rules: IF–THEN–ELSE

R E A D M E

Case Study: Developing "Muckraker: An Expert System for Journalists"

Journalist Steve Weinberg, former executive director of Investigative Reporters and Editors, agreed to provide his knowledge of investigative journalism as the basis for Muckraker, an expert system for journalists. The following excerpt explains how he was interviewed about his knowledge by an authority on expert systems.

It was Louanna Furbee . . . who worked hardest in the early stages to puzzle out the underlying logic (if any, I worried) of how I worked on an investigation. She explained that, after interviewing me, she would try to reduce what I had said into concepts. She would write each concept separately on an index card, then ask me to sort the cards into groupings. From those groupings, she hoped to sketch a tree of knowledge, which the computer programmers could then translate into electronic impulses.

When I viewed the tree a week later, I was amazed at how Furbee had managed to translate my words into a graphic that would be the basis of a computer program. She had sketched fifty-seven connected branches. The two main trunks were "paper trails" and "people trails," a distinction I had made when she interviewed me. (When conducting an investigation, I almost always consult paper first, then find the people to help explain the paper.)

On the "paper trail" trunk, Furbee sketched my distinctions between primary-source documents and secondary-source accounts. She also captured my thinking about how the type of subject (Is the story primarily about an individ-

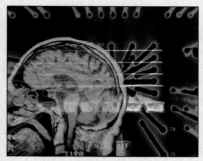

ual, an institution, or an issue?) determines which documents I will seek first. . . .

On the "people" trunk, Furbee focused on the two main problems I had found of most concern to journalists: getting in the door and, once inside, conducting the interview successfully. She worked in branches reflecting my thinking on when to request an interview by letter or telegram rather than by telephone, on dealing with secretaries and other potential bars to access, on how to bring an off-the-record source back on the record.

Using her tree, the rest of the Expert Systems team began to imprint my thinking onto a computer disk. . . .

—Steve Weinberg, "Steve's Brain," *Columbia Journalism Review*

When the system was unveiled, the first screen after the title read: "Muckraker's purpose is to provide advice on following the paper trail and interviewing sources. After a series of questions, Muckraker will make a recommendation. Use Muckraker to help plan your investigation."

problem from the user. It then proceeds to apply any rule that fits the problem. In the backward chain of reasoning, the system works backward from a question to produce an answer.

- **User interface:** The user interface is the display screen that the user deals with. It gives the user the ability to ask questions and get answers. It also explains the reasoning behind the answer.

Natural Language Processing

Natural languages are ordinary human languages, such as English. (A second definition is that they are programming languages, called *fifth-generation languages* (✓ p. 508), that give people a more natural connection with computers.) **Natural language processing is the study of ways for computers to recognize and understand human language, whether in spoken or written form.**

Think how challenging it is to make a computer translate English into another language. In one instance, the English sentence "The spirit is willing, but the flesh is weak" came out in Russian as "The wine is agreeable, but the meat is spoiled." The problem with human language is that it is often ambiguous and often interpreted differently by different listeners.

Today you can buy a handheld computer that will translate a number of English sentences—principally travelers' phrases ("May I see a menu, please?")—into another language. This trick is similar to teaching an English-speaking child to sing "Frère Jacques." More complex is the work being done by AI scientists trying to discover ways to endow the computer with an "understanding" of how human language works. This means working with ideas about the instinctual instructions or genetic code that babies are born with for understanding language.

Still, some natural-language systems are already in use. Intellect is a product that uses a limited English vocabulary to help users orally query databases. LUNAR, developed to help analyze moon rocks, answers questions about geology from an extensive database. Verbex, used by the U.S. Postal Service, lets mail sorters read aloud an incomplete address and then replies with the correct zip code.

In the future, natural-language comprehension may be applied to incoming e-mail messages, so that such messages can be filed automatically. However, this would require that the program understand the text rather than just look for certain words.[155]

Fuzzy Logic

A relatively new concept being used in the development of natural languages is fuzzy logic. The traditional logic behind computers is based on either/or, yes/no, true/false reasoning. Such computers make "crisp" distinctions, leading to precise decision making. *Fuzzy logic is a method of dealing with imprecise data and uncertainty, with problems that have many answers rather than one.* Unlike classical logic, fuzzy logic is more like human reasoning: it deals with probability and credibility. That is, instead of being simply true or false, a proposition is mostly true or mostly false, or more true or more false.

A frequently given example of an application of fuzzy logic is in running elevators. How long will most people wait in an elevator before getting antsy? About a minute and a half, say researchers at the Otis Elevator Company. The Otis artificial intelligence division has thus done considerable research into how elevators may be programmed to reduce waiting time.[156] Ordinarily when someone on a floor in the middle of the building pushes the call button, the system will send whichever elevator is closest. However, that car might be filled with passengers, who will be delayed by the new stop (perhaps making them antsy). Another car, however, might be empty. In a fuzzy-logic system, the computer assesses not only which car is nearest but also how full the cars are before deciding which one to send.

Neural Networks

Fuzzy logic principles are being applied in another area of AI, neural networks. *Neural networks use physical electronic devices or software to mimic the neurological structure of the human brain.* Because they are structured to mimic the rudimentary circuitry of the cells in the human brain, they learn from example and don't require detailed instructions.

To understand how neural networks operate, let us compare them to the operation of the human brain. (■ *See Panel 12.9.*)

- **The human neural network:** The word *neural* comes from *neurons,* or nerve cells. The neurons are connected by a three-dimensional lattice called *axons.* Electrical connections between neurons are activated by synapses.

PANEL 12.9

Neural network

(Top) Neurons, or human brain cells, and their connections, where synapse occurs. *(Bottom)* A neural network is a web of densely interconnected processing elements, called *neurons,* or nodes, because of their functional resemblance to the basic nerve cells of the human brain. Like their biological counterparts, the neurons in a neural network can send information to, and receive information from, thousands of fellow processors at once.

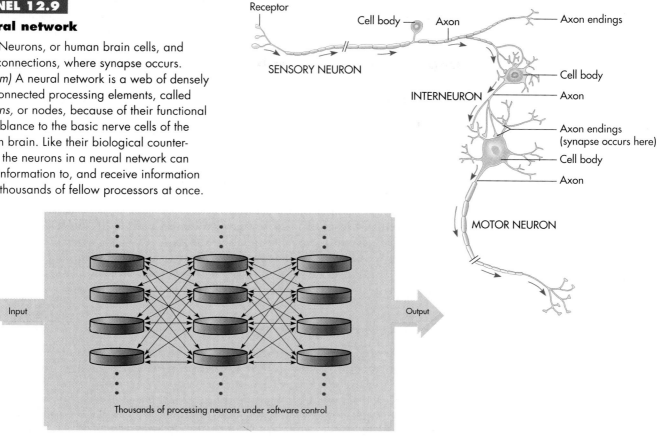

Thousands of processing neurons under software control

The human brain is made up of about 100 billion neurons. However, these cells do not act as "computer memory" sites. No cell holds a picture of your dog or the idea of happiness. You could eliminate any cell—or even a few million—in your brain and not alter your "mind." Where do memory and learning lie? In the electrical connections between cells, the synapses. Using electrical pulses, the neurons send "on/off" messages along the synapses.

- **The computer neural network:** In a hardware neural network, the nerve cell is replaced by a transistor, which acts as a switch. Wires connect the cells (transistors) with each other. The synapse is replaced by an electronic component called a resistor, which determines whether a cell should activate the electricity to other cells. A software neural network emulates a hardware neural network, although it doesn't work as fast.

  Computer-based neural networks use special AI software and complicated fuzzy-logic processor chips to take inputs and convert them to outputs with a kind of logic similar to human logic.

Ordinary computers mechanically obey instructions according to set rules. However, neural-network computers, like children, learn by example, problem solving, and memory by association. The network "learns" by fine-tuning its connections in response to each situation it encounters. (In the brain, learning takes place through changes in the synapses.) If you're teaching a neural network to speak, for example, you train it by giving it sample words and sentences as well as the pronunciations. The connections between the electronic "neurons" gradually change, allowing more or less current to pass.

Using software from a neural-network producer, Intel has developed a neural-network chip that contains many more transistors than the Pentium. Other chip makers are also working on neural-network chips. Over the next few years, these chips will begin to bring the power of these silicon "brains" not only to your PC but also to such tasks as automatically balancing shifting laundry loads in washing machines.

Neural networks are already being used in a variety of situations. One such program learned to pronounce a 20,000-word vocabulary overnight.[157] Another helped a mutual-fund manager to outperform the stock market by 2.3–5.6 percentage points over three years.[158] At a San Diego hospital emergency room in which patients complained of chest pains, a neural-network program was given the same information given doctors. It correctly diagnosed patients with heart attacks 97% of the time, compared to 78% for the human physicians.[159] In Chicago, a neural-net system has also been used to evaluate patient X-rays to look for signs of breast cancer. It outperformed most doctors in distinguishing malignant tumors from benign ones.[160] Banks use neural-network software to spot irregularities in purchasing patterns associated with individual accounts, thus often noticing when a credit card is stolen before its owner does.[161]

Genetic Algorithms

A *genetic algorithm* **is a program that uses Darwinian principles of random mutation to improve itself.** The algorithms are lines of computer code that act like living organisms. Different sections of code haphazardly come together, producing programs. Like Darwin's rules of evolution, many chunks of code compete with each other to see which can best perform the desired solution—the aim of the program. Some chunks will even become extinct. Those that survive will combine with other survivors and will produce offspring programs.[162-164]

Expert systems can capture and preserve the knowledge of expert specialists, but they may be slow to adapt to change. Neural networks can sift through mountains of data and discover obscure causal relationships, but if there is too much or too little data they may be ineffective—garbage in, garbage out. Genetic algorithms, by contrast, use endless trial and error to learn from experience—to discard unworkable approaches and grind away at promising approaches with the kind of tireless energy of which humans are incapable.[165]

The awesome power of genetic algorithms has already found applications. Organizers of the 1992 Paralympic Games used it to schedule events. LBS Capital Management Fund of Clearwater, Florida, uses it to help pick stocks for a pension fund it manages. In something called the FacePrints project, witnesses use a genetic algorithm to describe and identify criminal suspects. Texas Instruments is drawing on the skills that salmon use to find spawning grounds to produce a genetic algorithm that shipping companies can use to let packages "seek" their own best routes to their destinations. A hybrid expert system–genetic algorithm called Engeneous was used to boost performance in the Boeing 777 jet engine, a feat that involved billions of mind-boggling calculations.

Computer scientists still don't know what kinds of problems genetic algorithms work best on. Still, as one article pointed out, "genetic algorithms have going for them something that no other computer technique does: they have been field-tested, by nature, for 3.5 billion years."[166]

Artificial Life, the Turing Test, & AI Ethics

Genetic algorithms would seem to lead us away from mechanistic ideas of artificial intelligence and into more fundamental questions: "What is life, and how can we replicate it out of silicon chips, networks, and software?" We are dealing now not with artificial intelligence but with artificial life. *Artificial life,* or *A-life,* is a field of study concerned with "creatures"—computer instructions, or pure information—that are created, replicate, evolve, and die as if they were living organisms.[167]

Of course, "silicon life" does not have two principal attributes associated with true living things—it is not water- and carbon-based. Yet in other respects such "creatures" mimic life: if they cannot learn or adapt, they perish.

How can we know when we have reached the point where computers have achieved human intelligence? How will you always know, say, when you're on the phone, whether you're talking to a human being or to a computer? Clearly, with the strides made in the fields of artificial intelligence and artificial life, this question is no longer just academic.

Interestingly, this matter was addressed back in 1950 by Alan Turing, an English mathematician and computer pioneer. Turing predicted that by the end of the century computers would be able to mimic human thinking and converse so naturally that their communications would be indistinguishable from a person's. Out of these observations came the Turing test. The *Turing test* is a test or game for determining whether a computer is considered to possess "intelligence" or "self-awareness."

In the Turing test, a human judge converses by means of a computer terminal with two entities hidden in another location. One entity is a person typing on a keyboard. The other is a software program. As the judge types in and receives messages on the terminal, he or she must decide whether the entity is human. In this test, intelligence and the ability to think is demonstrated by the computer's success in fooling the judge. (■ *See Panel 12.10.*)

PANEL 12.10

The Turing test

In this test, a human judge converses via computer terminal with two entities in another location. One is a human being, the other a computer program. The task of the judge is to figure out whether the "intelligence" he or she is conversing with is human or not. Often the computer just gives itself away by being illogical, as in this "conversation."

"**Computer:** As a psychoanalyst, I am familiar with discussing such marital troubles with you as the patient on the couch. If you don't have a bad marriage just pretend that you do. Please start with 'My wife . . .' or 'My husband . . .'.
Judge: Ok, I'll pretend I have a bad marriage. Let's say the reason is because he never picks up his dirty underwear. Is this marriage irreparable?
Computer: Well, if you think I'm hopeless, you can always just swear at me."

—Judith Anne Gunther, "An Encounter with A.I.," *Popular Science*

Judith Anne Gunther participated as one of eight judges in the third annual Loebner Prize Competition, which is based on Turing's ideas.[168] (There have been two other competitions since.) The "conversations" are restricted to predetermined topics, such as baseball. This is because today's best programs have neither the databases nor the syntactical ability to handle an unlimited number of subjects. Conversations with each entity are limited to 15 minutes. At the end of the contest, the program that fools the judges most is the one that wins.

Gunther found that she wasn't fooled by any of the computer programs. The winning program, for example, relied as much on deflection and wit as it did on responding logically and conversationally. (For example, to a judge trying to discuss a federally funded program, the computer said: "You want logic? I'll give you logic: shut up, shut up, shut up, shut up, shut up, now go away! How's that for logic?") However, Gunther *was* fooled by one of the five humans, a real person discussing abortion. "He was so uncommunicative," wrote Gunther, "that I pegged him for a computer."

Behind everything to do with artificial intelligence and artificial life—just as it underlies everything we do—is the whole matter of ethics. In his book *Ethics in Modeling*, William A. Wallace, professor of decision sciences at Rensselaer Polytechnic Institute, points out that many users are not aware that computer software, such as expert systems, is often subtly shaped by the ethical judgments and assumptions of the people who create it.[169] In one instance, he points out, a bank had to modify its loan-evaluation software after it discovered that it tended to reject some applications because it unduly emphasized old age as a negative factor. Another expert system, used by health maintenance organizations (HMOs), instructs doctors on when they should opt for expensive medical procedures, such as magnetic resonance imaging tests. HMOs like the systems because they help control expenses, but critics are concerned that doctors will have to base decisions not on the best medicine but simply on "satisfactory" medicine combined with cost cutting.[170] Clearly, there is no such thing as completely "value-free" technology. Human beings build it, use it, and have to live with the results.

The Promises of the Information Superhighway

Preview & Review: The Information Superhighway promises great benefits in the areas of education and information, health, commerce and electronic money, entertainment, and government and electronic democracy.

"Where can I find the on-ramp to the information highway?" When a video crew asked this single question of several passers-by on the streets of New York City, the answers showed how vague people are about the concept. ("Take a left on Houston Street, and keep going straight," one man said. "Ask Reynaldo, the doorman," said another.)[171] Companies, too, are searching for the on-ramp to the Information Highway or Superhighway, the catch phrase for the convergence of computer, telephone, and television technologies that is supposed to deliver text, video, and sound to the home screen. Ultimately, the notion is that you would be able to hook up your "information appliance"—whether desktop model or mobile personal communicator—and access numerous services. Here business, government, and educators all see different priorities and needs. Nevertheless, potential services include the following.

Education & Information

The government is interested in reforming education, and technology can assist that effort. Presently the United States has more computers in its classrooms than other countries, but the machines are older and teachers aren't as computer-literate. A recent study shows that 61.2% of urban schools have phone lines they could use for Internet access, while 42% own modems. The poorer the school district, the less likely it is to have modems.[172] President Clinton proposed a "high-tech barn-raising," a government-industry collaboration to put every school in the nation on the Internet by the year 2000.[173]

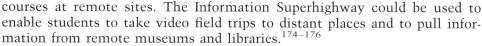
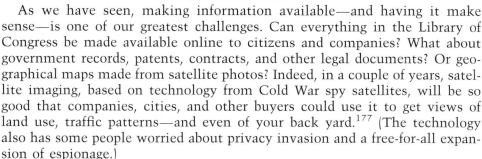

Computers can be used to create "virtual" classrooms not limited by scheduled class time. Some institutions (Stanford, MIT) are replacing the lecture hall with forms of learning featuring multimedia programs, workstations, and television courses at remote sites. The Information Superhighway could be used to enable students to take video field trips to distant places and to pull information from remote museums and libraries.[174-176]

As we have seen, making information available—and having it make sense—is one of our greatest challenges. Can everything in the Library of Congress be made available online to citizens and companies? What about government records, patents, contracts, and other legal documents? Or geographical maps made from satellite photos? Indeed, in a couple of years, satellite imaging, based on technology from Cold War spy satellites, will be so good that companies, cities, and other buyers could use it to get views of land use, traffic patterns—and even of your back yard.[177] (The technology also has some people worried about privacy invasion and a free-for-all expansion of espionage.)

Of particular interest is distance learning, or the "virtual university." *Distance learning* is the use of computer and/or video networks to teach courses to students outside the conventional classroom. At present, distance learning is largely outside the mainstream of campus life. That is, it concentrates principally on part-time students, those who cannot easily travel to campus, those interested in noncredit classes, or those seeking special courses in business or engineering. However, part-timers presently make up about 45% of all college enrollments. This, says one writer, is "a group for whom 'anytime, anywhere' education holds special appeal."[178]

Health

Another goal of the Information Superhighway is to improve health care. The government is calling for an expansion of "telemedicine." *Telemedicine* is the use of telecommunications to link health-care providers and researchers, enabling them to share medical images, patient records, and research. Of particular interest would be the use of networks for "teleradiology" (the exchange of X-rays, CAT scans, and the like), so that specialists could easily confer. Telemedicine would also allow long-distance patient examinations, using video cameras and, perhaps, virtual-reality kinds of gloves that would transmit and receive tactile sensations.[179,180]

Commerce & Electronic Money

Businesses clearly see the Information Superhighway as a way to enhance productivity and competitiveness. However, the changes will probably go well beyond this.

The thrust of the original Industrial Revolution was separation—to break work up into its component parts to permit mass production. The effect of computer networks in the Digital Revolution, however, is unification—to erase boundaries between company departments, suppliers, and customers.[181]

Indeed, the parts of a company can now as easily be global as down the hall from one another. Thus, designs for a new product can be tested and exchanged with factories in remote locations. With information flowing faster, goods can be sent to market faster and inventories kept reduced. Says an officer of the Internet Society, "Increasingly you have people in a wide variety of professions collaborating in diverse ways in other places. The whole notion of 'the organization' becomes a blurry boundary around a set of people and information systems and enterprises."[182]

The electronic mall, in which people make purchases online, is already here. Record companies, for instance, are making sound and videos of new albums available on Web sites, which you can sample and then order as a cassette or CD.[183] Banks in cyberspace are allowing users to adopt avatars or personas of themselves in three-dimensional virtual space on the World Wide Web, where they can query bank tellers and officers and make transactions.[184] Wal-Mart Stores and Microsoft have developed a joint online shopping venture that allows shoppers to browse online and buy merchandise.[185]

Cybercash, or e-cash, will change the future of money. Whether it takes the form of smart cards or of electronic blips online, cybercash will probably begin to displace (though not completely supplant) checks and paper currency. This would change the nature of how money is regulated as well as the way we spend and sell.[186,187]

Entertainment

Among the future entertainment offerings could be movies on-demand, videogames, and gaming ("telegambling"). *Video on-demand* would allow viewers to browse through a menu of hundreds of movies, select one, and start it when they wanted. This definition is for true video on-demand, which is like having a complete video library in your house. (An alternative, simpler form could consist of running the same movie on multiple channels, with staggered starting times.) True video on-demand will require a server (✓ p. 372), a storage system with the power of a supercomputer that would deliver movies and other data to thousands of customers at once.

Government & Electronic Democracy

Will information technology help the democratic process? There seem to be two parts to this. The first is its use as a campaign tool, which may, in fact, skew the democratic process in some ways. The second is its use in governing and in delivering government services.

Santa Monica, California, established a computer system called Public Electronic Network (PEN), which residents may hook into free of charge. PEN gives Santa Monica residents access to city council agendas, staff reports, public safety tips, and the public library's online catalog. Citizens may also enter into electronic conferences on topics both political and non-political; this has been by far the most popular attraction.[188]

PEN could be the basis for wider forms of electronic democracy. For example, electronic voting might raise the percentage of people who vote. Interactive local-government meetings could enable constituents and town council members to discuss proposals.

The Information Superhighway could also deliver federal services and benefits. In 1994, the government unveiled a program in which Social Security pensioners and other recipients of federal aid without bank accounts could use an automated-teller-machine card to walk up to any ATM and withdraw the funds due them.[189]

Onward

The four principal information technologies—computer networks, imaging technology, massive data storage, and artificial intelligence—will affect probably 90% of the workforce by 2010.[190] Clearly, information technology is driving the new world of jobs, leisure, and services, and nothing is going to stop it.

Where will you be in all this? People pursuing careers find the rules are changing very rapidly. Up-to-date skills are becoming ever more crucial. Job descriptions of all kinds are metamorphosing, and even familiar jobs are becoming more demanding. Today, experts advise, you need to prepare to continually upgrade your skills, prepare for specialization, and prepare to market yourself. In a world of breakneck change, you can still thrive. The most critical knowledge, however, may turn out to be self-knowledge.

READ ME

Practical Matters: Does Your Computer Know What Year It Is?

It's not uncommon for people to wake up with hangovers on New Year's Day. But on January 1, 2000, many computers may also find themselves dazed and confused.

While humans know that the morning after 12-31-99 should bring 01-01-00, many computers—especially older mainframes—don't. Programmers of the 1960s lopped off the first two digits of the year to save what was then extremely valuable memory.

Experts foresee two scenarios for computer users who don't address the problem: in one, the computer resets to 01-01-00 but assumes that means January 1, 1900. Jumping back a century would wreak havoc with a wide range of time-sensitive computer applications: hypothetically, a Los Angeleno calling New York at 2 A.M. eastern standard time on New Year's Day 1000 could be billed for 99 years of service. The other scenario is that the system crashes.

Sounds a tad apocalyptic? It's not, according to Bill Goodwin, 62, who calls his consulting firm 2000AD and publishes a newsletter (*Tick, Tick, Tick . . .*) devoted to the problem. "This is going to be the biggest maintenance job ever done," he says. "It's getting to the panic stage." He estimates that fixing the error will eventually cost somewhere between $50 billion and $75 billion. . . .

Solutions do exist. Manufacturers like Apple Computer and IBM now ship computers whose operating systems can handle the plunge into 2000. And mainframe users are starting to turn to software packages like System Vision Year 2000, made by San Francisco-based Adpac. System Vision scans the computer's software, does a cost-benefit analysis of what needs to be fixed, and even rewrites code to correct the problem. Still, many managers are wary, given that the package costs $47,500. . . .

Job Searching on the Internet & World Wide Web

If you haven't done a job search in a while, you will find many changes in a modern-day, high-quality search for a new position," says Mary Anne Buckman, consultant at Career Directions Inc.[191] Indeed, technological change has so affected the whole field of job hunting that futurists refer to it as a *paradigm shift*. This means that , in one definition, the change is of a magnitude in which the "prevailing structure is radically, rapidly, and unalterably transformed by new circumstances."[192]

Within five years online services will become the most prevalent means of nonlocal hiring and recruitment, says Tom Jackson, a career development expert in Woodstock, New York, who created one of the first computerized job banks. Moreover, he predicts, multimedia resumes and online interviews will become commonplace. "Anyone who doesn't know how to use these services in the next 12 to 18 months will lose out," Jackson says.[193] Even if you never hunt for a job by computer, states Martin Yate, a career consultant, online networking—making friends and exchanging news with others in your field—will become "imperative for your professional survival."[194]

In the Experience Box at the end of Chapter 8 we described how to prepare a computer-friendly resume and put it on a database to showcase yourself to prospective employers. Let's take this further and describe how you can use the Internet and the World Wide Web to help you search for jobs. Online areas of interest for the job seeker include:

- Resources for career advice
- Ways for you to find employers
- Ways for employers to find you

Resources for Career Advice

It's 3 A.M. Still, if you're up at this hour (or indeed at any other time) you can find job-search advice, tips on interviewing and resume writing, and postings of employment opportunities around the world. One means for doing so is through Catapult (*http://www. wm.edu/catapult/ catapult.html*), which was developed by Leo J. Charette, director of career services at the College of William and Mary in Williamsburg, Virginia, but has links to other colleges and potential employers.[195] For instance, through Catapult you could access a database located at Hartwick College called Barterbase, which unifies the expertise of 25 colleges in different employment areas. Barterbase offers, with "one-stop shopping," far more information than would probably be available through your own college's career center.

Another route is to use your Web browser to access a directory such as Yahoo! (*http://www.yahoo.com/*) to obtain a list of popular Web sites. In the menu, you can click on Business, then Employment, then Jobs. This will bring up a list of sites that offer career advice, resume postings, job listings, research about specific companies, and other services. [196] (Caution: As might be expected, there is also a fair amount of junk out there: get-rich-quick offers, resume-preparation firms, and other attempts to separate you from your money.)

Advice about careers, occupational trends, employment laws, and job hunting is also available through on-line chat groups and bulletin boards, such as those on the Big Three online services—America Online, CompuServe, and Prodigy.[197,198] For instance, CompuServe offers career-specific discussion groups, such as the PR Marketing Forum. Through these groups you can get tips on job searching, interviewing, and salary negotiations.

Ways for You to Find Employers

As you might expect, companies seeking people with technical backgrounds and technical people seeking employment pioneered the use of cyberspace as a job bazaar. However, as the public's interest in commercial services and the Internet has exploded, the technical orientation of online job exchanges has changed. Now, says one writer, "interspersed among all the ads for programmers on the Internet are openings for English teachers in China, forest rangers in New York, physical therapists in Atlanta, and models in Florida."[199] Most Web sites are free to job seekers, although some may require you to fill out an online registration form.

Some jobs are posted on Usenets by individuals, companies, and universities or colleges, such as computer networking company Cisco Systems of San Jose, California, and the University of Utah in Salt Lake City. Others are posted by professional or other organizations, such as the American Astronomical Society, Jobs Online New Zealand, and Volunteers in Service to America (VISTA).

Some of the principal organizations posting job listings are listed below.[100–105] Among the most established and largest are America's Job Bank, Career Mosaic, Career Path, E-Span, JobTrak, Job Web, and Online Career Center.

- *American Employment Weekly:* This employment tabloid (*http://branch.com/aew/aew/html*) features ads from the Sunday editions of 50 leading American newspapers.

- *America's postedJob Bank:* A joint venture of the New York State Department of Labor and the federal Employment and Training Administration, America's Job Bank (*http://www.ajb.dni.us/index.html*) advertises more than 100,000 jobs of all types. There are links to each state's employment office. More than a quarter of the jobs posted are sales, service, or clerical. Another quarter are managerial, professional, and technical. Other major types are construction, trucking, and manufacturing. The companies listed have their company Web links included.

- *Career Mosaic:* A service run by Bernard Hodes Advertising, Career Mosaic (*http://www.careermosaic.com/*) offers links to nearly 200 major corporations, most of them high-technology companies. One section is aimed at college students and offers tips on resumes and networking. A major strength is the J.O.B.S database, which lets you fill out forms to narrow your search, then presents you with a list of jobs meeting your criteria. A *New York Times* reporter who did this said a search for "writer" turned up 45 job listings, the oldest less than a month old.

- *Career Path:* Career Path (*http://www.careerpath.com/*) is a classified-ad employment listing from six of the countries largest newspapers, which you can search either individually or all at once. The papers are the *Boston Globe,* the *Chicago Tribune,* the *Los Angeles Times,* the *New York Times,* the *San Jose Mercury News,* and the *Washington Post.*

- *Employment Edge:* Containing both job listings and links to other recruiting Web sites, Employment Edge (*http://www.employmentedge.com/employment.edge/*) lists jobs by category (accounting, management, and so on.) It also offers links to sites with interviewing tips and resume help.

- *E-Span:* One of the oldest and biggest services, the E-Span Interactive Employment Network (*http://www.espan.com*) features all-paid ads from employers.

- *FedWorld:* This bulletin board (*http://www.fedworld.gov*) offers job postings from the U.S. Government.

- *Internet Job Locator:* Combining all major job-search engines on one page, the Internet Job Locator (*http://www.joblocator.com/jobs/*) lets you do a search of all of them at once.

- *Job Hunt:* Started by Dr. Dane Spearing, a geologist at Stanford University, the well-organized Job Hunt page (*http://rescomp.stanford.edu/jobs/*) contains a list of more than 200 sites related to online recruiting.

- *JobLinks:* This resource (*http://www.brandeis.edu/hiatt/web_data/Job_Listings.html*) offers job listings for business, government, health, law, science, technology, and other fields.

- *JobTrak:* The nation's leading online job listing service, JobTrak (*http://www.jobtrak.com*) claims to have been used by more than a million students and alumni, with more than 150,000 employers and 300 college career centers posting new jobs daily.

- *JobWeb:* Operated by the National Association of Colleges and Employers, Job Web (*http://www.jobweb.org/*) is a college placement service.

- *The Monster Board:* Not just for computer techies, the Monster Board (*http://www.monster.com*) offers real jobs for real people, although a lot of the companies listed are in the computer industry.

- *NationJob Network:* Despite the name, NationJob Network (*http://www.nationjob.com*) lists job opportunities primarily in the Midwest.

- *Online Career Center:* Based in Indianapolis, Online Career Center (*http://www.occ.com/occ/*) is a non-profit national recruiting service listing jobs at more than 3000 companies. About 30% of the jobs are nontechnical, with many in sales and marketing and in health care.

- *Workplace:* An employment resource offering staff and administrative positions in colleges and universities, government, and the arts (*http://galaxy.einet.net/galaxy/Community/Workplace.html/*).

The difficulty with searching through these resources is that it can mean wading through thousands of entries in numerous databanks, with many of them not being suitable for or interesting to you. An alternative to trying to find an employer is to have employers find you.

Ways for Employers to Find You

Because of its low (or zero) cost and wide reach, do you have anything to lose by posting your resume on line for prospective employers to view? Certainly you might if the employer happens to be the one you're already working for. In addition, you have to be aware that you lose control over anything broadcast into cyberspace—you're putting your credentials out there for the whole world to see, and you need to be somewhat concerned about who might gain access to them.

Posting your resume with an electronic jobs registry is certainly worth doing if you have a technical background, since technology companies in particular find this an efficient way of screening and hiring. However, it may also benefit people with less-technical backgrounds. Online recruitment "is popular with companies because it pre-screens applicants for at least basic computer skills," says one writer. "Anyone who can master the Internet is likely to know something about word processing, spreadsheets, or database searches, knowledge required in most good jobs these days."[106]

Resumes may be prepared as we described in the Experience Box at the end of Chapter 8. The latest variant, however, is to produce a resume with hypertext links and/or clever graphics and multimedia effects, then put it on a Web site to entice employers to chase after you.[107] If you don't know how to do this, there are many companies that—for a fee—can convert your resume to HTML (✓ p. 519) and publish it on their own Web sites. Some of these services can't dress it up with fancy graphics or multimedia, but since complex pages take longer for employers to download anyway, the extra pizzazz is probably not worth the effort. In any case, for you the bottom line is how much you're willing to pay for these services. For instances, OneWayResume (*http://www2.connectnet.com/users/blorincz*) charges $35 to write a resume and nothing to post it on a Web site. Actors can post their resumes and head shots for free on ActorsPavilion *(http://www.ios.com/~unisoft/act.html)*. You can post your own resume for free on Intellimatch and the Internet Employment Network. (Further information on preparing a hypertext resume may be found in Scott Grusky, "Winning Resume," *Internet World,* February 1996.)

Some of the principal places for posting your online resume are as follows:

- *E-Span:* Featuring paid ads from employers, the E-Span Interactive Employment Network (*http://www.espan.com*) also allows job seekers to post their resumes.

- *Intellimatch:* A free resume posting service, Intellimatch (*http://www.intellimatch.com*) allows applicants to fill out a structured resume, as well as to search for posted jobs.

- *Internet Employment Network:* This free resume referral service also allows you to search a database of all occupational categories (*http:// garnet.msen.com:70/1/vendor/napa/jobs*).

- *JobTailor Employment Online Service:* This resume posting service is free, but your resume but follow a certain structure (*http://www.jobtailor.com*).

- *123 Resume Distribution Service:* A free service that submits resume information to employers (*http://www.webplaza.com/pages/Careers/123 Careers/123Careers.html*).

- *Online Career Center:* This nonprofit job registry allows job searchers to post their resumes for free (*http://www.occ.com/occ/*).

- *Skill Search:* An online employment service that creates an applicant profile, Skill Search (*http://www.internet is.com/skillsearch/*) works with 60 alumni groups.

Companies are also beginning to replace their campus visits by recruiters with online interviewing. For example, the firm VIEWnet Inc. of Madison, Wisconsin, offers first-round screenings or interviews for summer internships through its teleconferencing "InterVIEW" technology, which allows video signals to be transmitted (at 17 frames per second) via telephone lines.[108]

Suggested Resources

Dixon, Pam, and Sylvia Tiersten. *Be Your Own Headhunter Online: Get the Job You Want Using the Information Superhighway.* New York: Random House, 1995. Explains how to conduct a national or international job search, connect with employers, and use online resources to prepare for interviews.

Glossbrenner, Alfred, and Emily Glossbrenner. *Finding a Job on the Internet.* New York: McGraw-Hill Computing, 1995.

Gonyea, James C. *The On-Line Job Search Companion.* New York: McGraw-Hill, 1995.

Kennedy, Joyce Lain. *Hook Up, Get Hired! The Internet Job Search Revolution.* New York: Wiley, 1995.

SUMMARY

antivirus software *(p. 548, LO 3)* Program that scans a computer's hard disk, diskettes, and main memory to detect viruses and, sometimes, to destroy them.

Computer users must find out what kind of antivirus software to install in their systems to protect them against damage or shut-down.

artificial intelligence (AI) *(p. 562, LO 7)* Group of related technologies that attempt to develop machines to emulate human-like qualities, such as learning, reasoning, communicating, seeing, and hearing.

AI is important for enabling machines to do things formerly possible only with human effort.

avatar *(p. 561, LO 6)* (1) Graphical image of you or someone else on a computer screen; (2) graphical personification of a computer or process that's running on a computer.

It is hoped that avatars will make information technologies easier to use.

biometrics *(p. 553, LO 3)* Science of measuring individual body characteristics.

Biometrics is used in some computer security systems—for example, to verify individuals' fingerprints before allowing access.

call-back system *(p. 553, LO 3)* Security measure whereby the user calls the computer system, punches in the password, and hangs up. The computer then calls back on a certain preauthorized number.

Call-back systems prevent access by unauthorized users who have learned the password but are calling from an unauthorized telephone.

Communications Decency Act *(p. 539, LO 2)* Section of the Telecommunications Act of 1996 that makes it a crime to transmit information that could be interpreted as unsuitable for viewing by children.

This act may protect children from exposure to offensive materials; however, many people point out that defining what is "offensive" to everyone is almost impossible and may involve illegal censorship.

crackers *(p. 550, LO 3)* People who gain unauthorized access to information technology for malicious purposes.

Crackers attempt to break into computers and deliberately obtain information for financial gain, shut down hardware, pirate software, or destroy data.

digital signature *(p. 552, LO 3)* String of characters and numbers that a user signs to an electronic document being sent by his or her computer.

Digital signatures are intended to ensure security in the sending of electronic documents.

dirty data *(p. 544, LO 3)* Data that is incomplete, outdated, or otherwise inaccurate.

Dirty data can cause information to be inaccurate.

disaster-recovery plan *(p. 554, LO 3)* Method of restoring information processing operations that have been halted by destruction or accident.

A disaster-recovery plan is important if a company desires to resume its computer and business operations in short order.

What It Is / What It Does	**Why It's Important**
encryption *(p. 553, LO 3)* Also called *enciphering;* the altering of data so that it is not usable unless the changes are undone.	Encryption is useful for users transmitting trade or military secrets or other sensitive data.
expert system *(p. 564, LO 7)* Interactive computer program that helps users solve problems that would otherwise require the assistance of a human expert.	Expert systems allow users to solve problems without assistance of a human expert; they incorporate both surface knowledge ("textbook knowledge") and deep knowledge ("tricks of the trade").
fuzzy logic *(p. 566, LO 7)* Method of dealing with imprecise data and uncertainty, with problems that have many answers rather than one.	Unlike traditional "crisp," yes-no digital logic, fuzzy logic deals with probability and credibility.
genetic algorithm *(p. 568, LO 7)* Program that uses Darwinian principles of random mutation to improve itself.	Genetic algorithms use trial and error to learn from experience, thus constantly improving themselves.
hackers *(p. 549, LO 3)* People who gain unauthorized access to computer or telecommunications systems for the challenge or even the principle of it.	The acts of hackers create problems not only for the institutions that are victims of break-ins but also for ordinary users of the systems.
inference engine *(p. 564, LO 7)* Software that controls the search of the expert system's knowledge base and produces conclusions.	An inference engine fits the user's problem into the knowledge base and derives a conclusion from the rules and facts it contains.
information-technology crime *(p. 545, LO 3)* Crime of two types: an illegal act perpetrated against computers or telecommunications; or the use of computers or telecommunications to accomplish an illegal act.	Information-technology crimes cost billions of dollars every year.
intelligent agent *(p. 560, LO 6)* Computer program that performs work tasks on your behalf, including roaming networks and compiling data.	Agents scan databases and electronic mail; clerical agents answer telephones and send faxes, user-interface agents learn individual work habits.
knowledge base *(p. 564, LO 7)* Expert system's database of knowledge about a particular subject.	A knowledge base includes relevant facts, information, beliefs, assumptions, and procedures for solving problems. Programs can contain up to 10,000 rules.
National Information Infrastructure (NII) *(p. 539, LO 2)* The Information Superhighway; services will be delivered via networks and technologies of several information providers—the telecommunications networks, cable-TV networks, and the Internet—to users' "information appliances."	Services could include education, health, information, commerce, government, home services, and mobile communications.
natural language processing *(p. 565, LO 7)* Study of ways for computers to recognize and understand human language, whether in spoken or written form.	Natural language processing could further reduce the barriers to human/computer communications.
neural networks *(p. 566, LO 7)* Field of artificial intelligence; networks that use physical electronic devices or software to mimic the neurological structure of the human brain, with, for instance, transistors for nerve cells and resistors for synapses.	Neural networks are able to mimic human learning behavior and pattern recognition.

What It Is / What It Does	**Why It's Important**
password *(p. 552, LO 3)* Special word, code, or symbol that is required to access a computer system.	One of the weakest links in computer security, passwords can be guessed, forgotten, or stolen.
perception system *(p. 563, LO 7)* Sensing device that emulates the human capabilities of sight, hearing, touch, and smell.	Perception systems such as vision systems are used in factories for inspecting quality of products.
PIN (personal identification number) *(p. 552, LO 3)* Security number known only to you that is required to access the system.	PINs are required to access many computer systems and automated teller machines and to charge telephone calls.
robot *(p. 562, LO 7)* Automatic device that performs functions ordinarily ascribed to human beings or that operates with what appears to be almost human intelligence.	Robots are performing more and more functions in business and the professions.
search engines *(p. 561, LO 6)* Web pages containing forms into which you type text on the topic you want to search for.	Search engines simplify the task of finding needed information.
security *(p. 551, LO 3)* System of safeguards for protecting information technology against disasters, systems failure, and unauthorized access that can result in damage or loss.	With proper security, organizations and individuals can minimize losses caused to information technology from disasters, system failures, and unauthorized access.
software bug *(p. 543, LO 3)* Error in a program caused by incorrect use of the programming language or faulty logic.	An error in a program will cause it not to work properly.
spiders *(p. 561, LO 6)* Also known as *crawlers* or *robots;* software programs that roam the Web, looking for new Web sites by following linking from page to page.	Like search engines, these software programs simplify information searches.
switched-network model *(p. 537, LO 1)* Telecommunications model whereby people on the system are not only consumers of information but also possible providers; this is many-to-many communications.	This model is used by telephone companies and most computer networks. This system is more democratic and participatory than the tree-and-branch model.
Telecommunications Act of 1996 *(p. 539, LO 2)* Recent U.S. law that undoes 60 years of federal and state communications regulations and lets phone, cable, and TV businesses compete and combine more freely.	This law tore down many regulatory barriers to the convergence of the telephone, cable, TV, and computer industries.
tree-and-branch model *(p. 537, LO 1)* Telecommunications model whereby a centralized information provider sends out messages through many channels to thousands of consumers; this is few-to-many communication.	Most current mass media use this form of telecommunications. *See switched-network model.*
virus *(p. 547, LO 3)* Deviant program that attaches itself to computer systems and destroys or corrupts data.	Viruses can cause users to lose data or files or even shut down entire computer systems.
worm *(p. 547, LO 3)* Program that copies itself repeatedly into memory or onto a disk drive until no more space is left.	Like viruses, worms can shut down a user's computer system.

(Selected answers appear at the back of the book.)

Short-Answer Questions

1. What is the significance of the 1996 Telecommunications Act?

2. How would you define *information-technology crime*?

3. What is the difference between the tree-and-branch and switched-network models of communications?

4. What is the purpose of the Computer Emergency Response Team (CERT)?

5. What is an intelligent agent?

6. What is artificial life?

7. Why is fuzzy logic important?

8. What is the Turing test used for?

9. What is the difference between a hacker and a cracker?

10. What is an avatar?

Fill-in-the-Blank Questions

1. A program that contains a(n) _____ won't function properly.

2. The purpose of_____ is to scan a computer's disk devices and memory to detect viruses and, sometimes, to destroy them.

3. Data that is incomplete, outdated, or otherwise inaccurate is referred to as _____.

4. _____ is a system of safeguards for protecting information technology.

5. List four areas in which the Information Superhighway promises great benefits.

   a. _____

   b. _____

   c. _____

   d. _____

6. The shrinking of the world society because of the ability to communicate is referred to as the _____.

7. A(n) _____ is a special word, code, or symbol that the user must provide before obtaining access to a computer system.

8. So that information processing operations can be restored after destruction or accident, a company should adopt a(n) _____.

9. When telecommunications technology is used to link healthcare providers and researchers, the field is referred to as _____.

10. _____ use physical electronic devices or software to mimic the structure of the human brain.

Multiple-Choice Questions

1. Which of the following is an example of an information-technology crime?
   a. theft of hardware
   b. theft of software
   c. theft of time and services
   d. theft of information
   e. all of the above

2. Which of the following copies itself into memory or a disk drive until no more space is left?
   a. virus
   b. worm
   c. dirty data
   d. antiviral agent
   e. none of the above

3. Which of the following destroys or corrupts data?
   a. virus
   b. worm
   c. dirty data
   d. antiviral agent
   e. none of the above

4. What are the terms *spiders* and *crawlers* most closely related to?
   a. Internet
   b. World Wide Web
   c. electronic news services
   d. artificial intelligence
   e. none of the above

5. Which of the following is an interactive computer program that helps users solve problems?
   a. robot
   b. perception system
   c. expert system
   d. natural language
   e. all of the above

6. Which of the following *isn't* necessarily a component of an expert system?
   a. natural language
   b. knowledge base
   c. inference engine
   d. user interface
   e. all of the above

7. Which of the following involves the study of ways for computers to understand human language?
   a. natural language
   b. knowledge base
   c. inference engine
   d. user interface
   e. all of the above

8. Which of the following can disable information technology?
   a. people
   b. software errors
   c. dirty data
   d. crackers
   e. all of the above

9. Which of the following groups perpetrate over 80% of information technology crime?
   a. hackers
   b. crackers
   c. professional criminals
   d. employees
   e. none of the above

10. Which of the following is the science of measuring individual body characteristics?
    a. encryption
    b. neuralistics
    c. biometrics
    d. none of the above

True/False Questions

T F 1. A digital signature looks the same as your signature on a check.

T F 2. Information technology has been associated with people's mental health.

T F 3. Distance learning involves the use of a computer and/or a video network.

T F 4. A virus is a combination of hardware and software.

T F 5. In the switched-network model of communications, a centralized information provider sends out messages to thousands of consumers.

T F 6. Viruses can be passed to another computer by a diskette or through a network.

T F 7. Encrypted data isn't directly usable.

T F 8. An avatar is a computer program that can roam networks and compile data.

T F 9. The 1996 Telecommunications Act permits greater competition between local and long-distance telephone companies.

T F 10. Artificial intelligence comprises a group of related technologies.

Projects/Critical-Thinking Questions

1. Write a few paragraphs about the challenges and obstacles facing the completion of the Information Superhighway.

2. In addition to *2600: The Hacker's Quarterly*, where do hackers find new information about their field? Are support groups available? In what ways do hackers help companies? Does a hacker underground exist? Research your answers using current computer periodicals and/or the Internet.

3. Aside from helping you do all of your chores, what do you think is a good application for a robot? For example, a snake-like robot was developed in GMD's Institute for System Design Technology that is ideal for inspection tasks that deal with tight tubes. Would your robot be hard to develop? Is it likely that it will be developed someday? If so, when? By whom? Is anything similar to your robot under development today?

4. Assuming you have a microcomputer in your home that includes a modem, what security threats, if any, should you be concerned with? List as many ways as you can think of to ensure that your computer is protected.

5. Explore the National Information Infrastructure (NII) in more detail. Create an executive report describing the objectives for the NII, its guiding principles, and its agenda for action. Does the NII exist today or is it a plan for the future? Or both? Research your answers using current periodicals and/or the Internet.

net Using the Internet

Objective: *In this exercise you practice using Veronica, a frequently-updated index system for searching gopher-based servers.*

Before you continue: *We assume you have access to the Internet through your university, business, or commercial service provider and to the Web browser tool named Netscape Navigator 2.0 or 2.01. We also assume you know how to connect to the Internet and then load Netscape Navigator. If necessary, ask your instructor or system administrator for assistance.*

1. Make sure you have started Netscape. The home page for Netscape Navigator should appear on your screen.

2. Gopher is a text-based menu system that lets you search gopher servers on the Internet. Gopher servers are located all around the world and are often referred to collectively as "gopherspace." Veronica is an index system that can save you time when searching for information in gopherspace.

   In this step you contact The World gopher server. You will then practice searching for the text of the *Magna Carta* (the ancient constitution) with and without the help of Veronica.
   CLICK: Open button ( )
   TYPE: `gopher://gopher.std.com`
   CLICK: Open command button
   The main gopher menu should appear. Notice that the information is categorized by topic. To find out more about the topic, you click its associated directory or file icon.

3. Use the vertical scroll bar to scroll through the list of directories until you locate the directory named "Other Gopher and Information Services."
   CLICK: the directory's icon

4. Locate the search title "Search All the Gopher Servers in the World."
   CLICK: the search icon
   Your screen should now appear similar to the following:

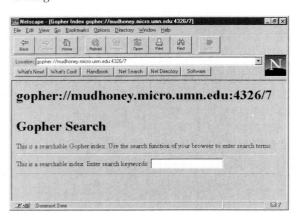

5. CLICK: in the search text box
   TYPE: `Magna Carta`
   PRESS: Enter

Notice that gopher couldn't find a match for this phrase. It's really no wonder that your search came up empty because gopher searched for gopher servers that have "Magna Carta" in the title of their main menu. Now, if you had searched for "North Dakota" instead, gopher would pull up a list of gopher servers in that state. Using the gopher utility named Veronica, your search extends to all of gopherspace rather than to just titles. In the following steps you practice using Veronica.

6. To display gopher's main menu, choose Go from the Menu bar and then choose "gopher://gopher.std.com" from the bottom of the drop-down list.

7. Use the vertical scroll bar to scroll through the list of directories until you locate the directory named "Other Gopher and Information Services."
   CLICK: the directory's icon

8. Locate the search title "Search titles in Gopherspace using Veronica."
   CLICK: the search icon
   Your screen should appear similar to the following:

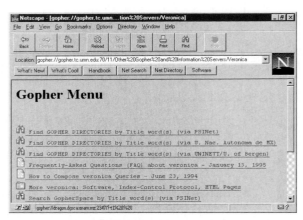

(*Note:* If you keep getting a message indicating that the Veronica server is busy, try a different server such as PSINET.)

9. Locate the search title "Find GOPHER DIRECTORIES by Title word(s) (via PSINet)."
   CLICK: the search icon

10. CLICK: in the search text box
    TYPE: Magna Carta
    PRESS: (Enter)
    With the help of Veronica, some matches were
    found. Your screen should now appear similar to the
    following:

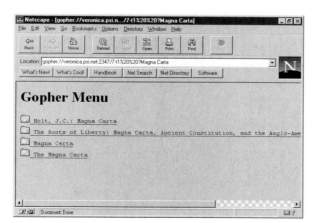

11. Click on the Magna Carta directory, as indicated
    above.

12. CLICK: the Magna Carta document icon to view the
    text of the Magna Carta

    Review this ancient constitution and congratulate
    yourself on a successful search through gopherspace!
    Your screen should now appear similar to the fol-
    lowing:

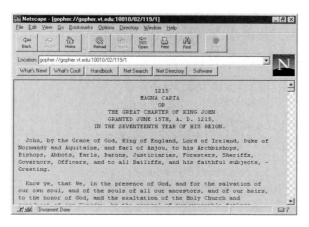

13. To display gopher's main menu, choose Go from the
    Menu bar and then choose
    "gopher://gopher.std.com" from the bottom of the
    drop-down list.

14. On your own, practice using gopher and Veronica to
    search for information about one of the following
    topics:
    • Artificial intelligence
    • Robotics
    • Neural networks

15. Print (File, Print) or save (File, Save as) any docu-
    ments that you find as a result of your efforts in step
    14.

16. Display Netscape's home page and then exit
    Netscape.

Answers

Following are answers to the odd-numbered fill-in-the-blank, multiple-choice, and true/false questions.

Chapter 1
Fill-in-the-Blank Questions
1. computer   3. end-user   5. analog, digital   7. storage hardware   9. telecommuting

Multiple-Choice Questions
1. c   3. b   5. d   7. d   9. c

True/False Questions
1. t   3. f   5. t   7. t   9. f

Chapter 2
Fill-in-the-Blank Questions
1. applications software   3. electronic spreadsheet
5. dialog box   7. electronic mail, or e-mail   9. systems software

Multiple-Choice Questions
1. d   3. b   5. a   7. e   9. d

True/False Questions
1. f   3. t   5. t   7. f   9. t

Chapter 3
Fill-in-the-Blank Questions
1. operating systems, utility programs, language translators   3. format   5. utility programs   7. supervisor
9. plug and play

Multiple-Choice Questions
1. b   3. c   5. c   7. d   9. c

True/False Questions
1. f   3. t   5. f   7. f   9. t

Chapter 4
Fill-in-the-Blank Questions
1. bus line, or bus   3. system clock   5. instruction cycle, execution cycle   7. notebook, subnotebook, pocket PC, electronic organizer, palmtop computer, pen computer
9. game port

Multiple-Choice Questions
1. a   3. e   5. d   7. b   9. d

True/False Questions
1. t   3. f   5. f   7. t   9. t

Chapter 5
Fill-in-the-Blank Questions
1. Source data entry system (source data automation)
3. sensor   5. fax machine   7. mouse   9. dumb

Multiple-Choice Questions
1. a   3. b   5. e   7. b

True/False Questions
1. f   3. t   5. f   7. t   9. t

Chapter 6
Fill-in-the-Blank Questions
1. resolution   3. hardcopy, softcopy   5. plotters
7. videoconferencing   9. digital

Multiple-Choice Questions
1. c   3. d   5. d   7. c   9. a

True/False Questions
1. t   3. f   5. f   7. t   9. t

Chapter 7
Fill-in-the-Blank Questions
3½-inch, 5¼-inch   3. 1.44 MB   5. write/protect
7. optical disk   9. compression

Multiple-Choice Questions
1. d   3. d   5. c   7. e   9. b

True/False Questions
1. t   3. f   5. t   7. t   9. f

Chapter 8
Fill-in-the-Blank Questions
1. dedicated fax machine, fax modem   3. sysops
5. groupware   7. terminal emulation   9. telecommuting

Multiple-Choice Questions
1. d   3. b   5. a   7. c   9. a

True/False Questions
1. t   3. t   5. t   7. t   9. f

Chapter 9
Fill-in-the-Blank Questions
1. database administrator   3. byte/character, field, record, file, database   5. database   7. program, data
9. report generator

Multiple-Choice Questions
1. b   3. a   5. e   7. d   9. d

True/False Questions
1. f   3. t   5. f   7. t   9. t

Chapter 10
Fill-in-the-Blank Questions
1. transaction processing system   3. organization chart 5. executive information system   7. tactical decisions
9. data flow diagram, systems flowchart

Multiple-Choice Questions
1. b   3. c   5. b   7. a   9. d

True/False Questions
1. f   3. t   5. f   7. f   9. t

Chapter 11
Fill-in-the-Blank Questions
1. programming   3. object   5. debug   7. second
9. program flowchart

Multiple-Choice Questions
1. a   3. a   5. d   7. c   9. a

True/False Questions
1. t   3. t   5. t   7. f   9. t

Chapter 12
Fill-in-the-Blank Questions
1. software bug   3. dirty data   5. [*answers will vary*]
7. password   9. telemedicine

Multiple-Choice Questions
1. e   3. a   5. c   7. a   9. d

True/False Questions
1. f   3. t   5. f   7. t   9. t

Notes

Chapter 1

1. Thomas A. Stewart, "The Information Age in Charts," *Fortune,* April 4, 1994, pp. 75–79.
2. Donald Spencer, *Webster's New World Dictionary of Computer Terms,* 4th ed. (New York: Prentice Hall, 1992), p. 206.
3. Robin Nelson, "Swept Away by the Digital Age," *Popular Science,* November 1993, pp. 92–97, 107–109.
4. We are grateful to Prof. John Durham for contributing these ideas.
5. Sandra D. Atchison, "The Care and Feeding of 'Lone Eagles,'" *Business Week,* November 15, 1993, p. 58.
6. Link Resources, cited in Carol Kleiman, "At-Home Employees Multiplying," *San Jose Mercury News,* November 12, 1995, pp. 1PC–2PC.
7. Susan N. Futterman, "Quick-Hit Research of a Potential Employer," *CompuServe Magazine,* September 1993, p. 36.
8. Lee Gomes, "Hollow Dreams," *San Jose Mercury News,* November 12, 1995, pp. 1D, 3D.
9. Bloomberg Business News, "Motorola Links PC to Cable," *San Francisco Chronicle,* November 30, 1995, p. D1.
10. Bloomberg Business News, "New Technology Ties Internet, TV to PCs," *San Francisco Chronicle,* October 24, 1995, p. C3.
11. Catherine Arnst, Paul M. Eng, Richard Brandt, and Peter Burrows, "The Information Appliance," *Business Week,* November 22, 1993, pp. 98–110.
12. Tom Forester and Perry Morrison, *Computer Ethics: Cautionary Tales and Ethical Dilemmas in Computing* (Cambridge, MA: The MIT Press, 1990), pp. 1–2.
13. Barbara Simmons and Gary Chapman, "Information Highway Has Many Potholes," *San Francisco Chronicle,* January 17, 1994, p. B3.
14. Debbie G. Longman and Rhonda H. Atkinson, *College Learning and Study Skills,* 2nd ed. (St. Paul, MN: West, 1992), p. 4.
15. Mervill Douglass and Donna Douglass, *Manage Your Time, Manage Your Work, Manage Yourself* (New York: American Management Association, 1980).
16. W. M. Beneke and M. B. Harris, "Teaching Self-Control of Study Behavior," *Behavior Research and Therapy,* 1972, *10,* 35–41.
17. E. B. Zechmeister and S. E. Nyberg, *Human Memory: An Introduction to Research and Theory* (Pacific Grove, CA: Brooks/Cole, 1982).
18. B. K. Bromage and R. E. Mayer, "Quantitative and Qualitative Effects of Repetition on Learning from Technical Text," *Journal of Educational Psychology,* 1982, *78,* 271–78.
19. Longman and Atkinson, 1992, pp. 148–53.
20. H. C. Lindgren, *The Psychology of College Success: A Dynamic Approach* (New York: Wiley, 1969).
21. R. J. Palkovitz and R. K. Lore, "Note Taking and Note Review: Why Students Fail Questions Based on Lecture Material," *Teaching of Psychology,* 1980, *7,* 159–61.
22. Palkovitz and Lore, 1980, pp. 159–61.
23. F. P. Robinson, *Effective Study,* 4th ed. (New York: Harper & Row, 1970).
24. J. Langan and J. Nadell, *Doing Well in College: A Concise Guide to Reading, Writing, and Study Skills* (New York: McGraw-Hill, 1980), pp. 93–110.
25. Langan and Nadell, 1980, p. 104.

Chapter 2

1. John Markoff, "A Free and Simple Computer Link," *New York Times,* December 8, 1993, p. C1.
2. Alan Deutschman, "Mac vs. Windows: Who Cares?" *Fortune,* October 4, 1993, p. 114.
3. Barbara Kantrowitz, Andrew Cohen, and Melinda Lieu, "My Info Is NOT Your Info," *Newsweek,* July 18, 1994, p. 54.
4. Teresa Riordan, "Writing Copyright Law for an Information Age," *New York Times,* July 7, 1994, pp. C1, C5.
5. David Edelson, "What Price Superhighway Information?" (letter), *New York Times,* January 16, 1994, p. 16.
6. David L. Wheeler, "Computer Networks Are Said to Offer New Opportunities for Plagiarists," *Chronicle of Higher Education,* June 30, 1993, pp. A17, A19.
7. Denise K. Magner, "Verdict in a Plagiarism Case," *Chronicle of Higher Education,* January 5, 1994, pp. A17, A20.
8. Robert Tomsho, "As Sampling Revolutionizes Recording, Debate Grows Over Aesthetics, Copyrights," *Wall Street Journal,* November 5, 1990, p. B1.
9. William Grimes, "A Question of Ownership of Images," *New York Times,* August 20, 1993, p. B7.
10. Jennifer Larson, "Try Before You Buy," *PC Novice,* March 1993, p. 41.
11. Andy Ihnatko, "Right-Protected Software," *MacUser,* March 1993, pp. 29–30.
12. John Pallatto, "Software and the Law," *PC Week,* October 7, 1986, pp. 79–84.
13. Rick Tetzeli, "Videogames: Serious Fun," *Fortune,* December 27, 1993, pp. 110–116.
14. Steve G. Steinberg, "Back in Your Court, Software Designers," *Los Angeles Times,* December 7, 1995, pp. D2, D11.
15. Tetzeli, 1993.
16. Laura Evenson, "Hot CD-ROM Sales Made for Multimedia Christmas," *San Francisco Chronicle,* December 30, 1993, p. D1.
17. Stephen Manes, "What Women Want," *New York Times,* December 12, 1995, p. B8.
18. Herb Brody, "Video Games That Teach?" *Technology Review,* November/December 1993, pp. 50–57.
19. Jay Sivin-Kachala, Interactive Educational Systems Design, quoted in Nicole Carroll, "How Computers Can Help Low-Achieving Students," *USA Today,* November 20, 1995, p. 5D.
20. Edward Baig, "A Hyperlink Feast: New CD-ROM Encyclopedias," *Business Week,* October 31, 1994, p. 134.
21. Chris O'Malley, "Are You Stumped? Ask Your Software," *Popular Science,* November 1995, p. 37.
22. "Typewriters & Word Processors," *Consumer Reports,* November 1991, pp. 763–767.
23. Peter H. Lewis, "In a Battle of the Spreadsheet, Borland Acts 'the Barbarian,'" *New York Times,* August 29, 1993, sec. 3, p. 8.
24. Charles Bermant, "Databases for Everyone," *San Francisco Examiner,* February 28, 1993, p. E-16.
25. Alan Freedman, *The Computer Glossary,* 6th ed. (New York: AMACOM, 1993), p. 406.
26. Walter S. Mossberg, "Organizer Program Takes a Leaf from Date Books," *Wall Street Journal,* January 21, 1993, p. B1.
27. Chris O'Malley, "The New Works," *Popular Science,* November 1995, p. 43.
28. Stephen H. Wildstrom, "Will the 'Works' Do the Job?" *Business Week,* December 5, 1994, p. 15.
29. Steve Lohr, "Microsoft Seeks to Pad Wide Lead in PC Suites," *New York Times,* October 15, 1993, p. C3.
30. Lohr, 1993, p. C3.
31. Karen Ann Hargrove, quoted in Peter H. Lewis, "Everything Intelligent, and All of It Connected," *New York Times,* June 13, 1993, sec. 1, p. 12.
32. David Kirkpatrick, "Groupware Goes Boom," *Fortune,* December 27, 1993, p. 100.
33. Kirkpatrick, 1993, pp. 99–106.
34. Gary McWilliams, "Lotus 'Notes' Get a Lot of Notice," *Business Week,* March 29, 1993, pp. 84–85.
35. Kirkpatrick, 1993, pp. 100–101.
36. Edward Rothstein, "Between the Dream and the Reality Lies the Shadow. Or Is It the Interface?" *New York Times,* December 11, 1995, p. C3.
37. "Brouhaha in Browserland," *Business Week,* March 25, 1996, p. 37.
38. James Kim, "Businesses Bet on the Future," *USA Today,* November 13, 1995, pp. 1E–2E.
39. Margaret Trejo, quoted in Richard Atcheson, "A Woman for *Lear's,*" *Lear's,* November 1993, p. 87.
40. Barbara Kantrowitz, "In Quicken They Trust," *Newsweek,* May 2, 1994, pp. 65–66.
41. Stacey Richardson, quoted in Peter H. Lewis, "Pairing People Management with Project Management," *New York Times,* April 11, 1993, sec. 3, p. 12.
42. Walter S. Mossberg, "PC Program Lets Machines Help Bosses Manage People," *Wall Street Journal,* December 24, 1993, p. B1.
43. Glenn Rifkin, "Designing Tools for the Designers," *New York Times,* June 18, 1992, p. C6.
44. Susan Kuchinskas, "Designing Ways Going Digital," *San Francisco Examiner,* January 16, 1994, pp. F1, F6.
45. Bernie Ward, "Computer Chic," *Sky,* April 1993, pp. 84–90.
46. Alan Freedman, *The Computer Glossary,* 6th ed. (New York: AMACOM, 1993), pp. 196–97.
47. Trevor Meers, "College Computing 101," *PC Novice,* September 1993, pp. 18–22.
48. Daniel Tynan and Christina Wood, "Most Likely to Succeed . . . or Fail," *PC World,* June 1994, pp. 119–38.

Chapter 3

1. Tim Eckles, quoted in Jim Carlton, "Computer Firms Try to Make PCs Less Scary," *Wall Street Journal,* May 6, 1994, pp. B1, B7.
2. Carlton, 1994.
3. Peter H. Lewis, "Champion of MS-DOS, Admirer of Windows," *New York Times,* April 4, 1993, sec. 3, p. 11.
4. Ken Wasch, quoted in Lisa Green, "Windows 95 Drops the Curtain on DOS," *USA Today,* August 22, 1995, p. 2B.
5. David Kirkpatrick, "Mac vs. Windows," *Fortune,* October 4, 1993, pp. 107–114.
6. "The Basic Choice in Computers," *Consumer Reports,* September 1993, pp. 570–573.
7. BIS, cited in Kirkpatrick, 1993.
8. Peter H. Lewis, "A Strong New OS/2, with an Uncertain Future," *New York Times,* June 20, 1993, sec. 3, p. 8.
9. Amy Cortese, "IBM Rides into Microsoft Country," *Business Week,* June 6, 1994, pp. 111–112.
10. "For Microsoft, Nothing Succeeds Like Excess," *Wall Street Journal,* August 25, 1995, pp. B1, B2.
11. Jeffrey Kagan, quoted in James R. Healey and Deborah Sharp, "Worldwide Wait for Windows 95 Ends," *USA Today,* August 24, 1995, p. 1A.
12. "For Microsoft, Nothing Succeeds Like Excess," 1995.
13. Michael J. Himowitz, "Windows 95: Do You Need the Most Overhyped Product of the Decade?" *Fortune,* September 18, 1995, pp. 191–196.
14. Paul A. David, quoted in Steve Lohr, "Business Often Goes to the Swift, Not the Best," *New York Times,* August 6, 1995, sec. 4, p. 3.
15. Stuart Card, quoted in Kevin Maney, "Computer Windows May Be Obsolete," *USA Today,* August 25, 1995, p. 2B.
16. Dan Gillmor, "Best 32-bit Software Is Yet to Hit Market," *San Jose Mercury News,* August 27, 1995, p. 4E.
17. Stephen H. Wildstrom, "Move Over, Netscape," *Business Week,* April 8, 1996, p. 19.
18. Don Clark, "Personal Computer Would Get Big Shift Under Gate's Plan," *Wall Street Journal,* March 29, 1996, p. B3.
19. Steve Lohr, "Standard Set for Unix Interface," *New York Times,* August 31, 1993, p. C4.
20. Lawrence M. Fisher, "New Crusader in Software's Holy War," *New York Times,* October 3, 1993, sec. 3, p. 7.
21. Lawrence M. Fisher, "Windows 95's Big Value May Be as a Lure to Network System," *New York Times,* August 7, 1995, p. C4.
22. Richard Buck, "Novell Is Trying to Connect All of Us," *San Jose Mercury News,* May 21, 1995, pp. 1E, 6E; reprinted from *Seattle Times.*
23. Lawrence M. Fisher, "Novell Readies a Response to Windows," *New York Times,* September 18, 1995, pp. C1, C10.
24. Robert Frankenberg, quoted in Buck, 1995.
25. Fisher, September 18, 1995.
26. Tosca Moon Lee, "Utility Software: Your PC's Life Preserver," *PC Novice,* March 1993, pp. 68–73.
27. Gillian Coolidge, "Investigating the Lost Files of Peter Norton, PC Pioneer," *PC Novice,* May 1992, pp. 14–18.
28. Dan Kuznetsky, cited in Laurie Hays, "IBM Announces New Software Code that Is Universal," *Wall Street Journal,* October 31, 1995, p. B4.
29. Peter Burrows, "New! Improved! Not Here Yet!" *Business Week,* December 18, 1995, pp. 80–84.
30. Dan Gillmor, "It Needs a Hit, but Apple Stays in the Ballgame," *San Jose Mercury News,* September 24, 1995, p. E1.
31. Jim Carlton, "What's Eating Apple? Computer Maker Hits Some Serious Snags," *Wall Street Journal,* September 21, 1995, pp. A1, A5.
32. Peter H. Lewis, "Copland Promises MacOS delights," *San Jose Mercury News,* September 3, 1995, pp. 1E, 7E; reprinted from *New York Times.*
33. Burrows, December, 1995.
34. G. Pascal Zachary and Laurence Hooper, "IBM and Apple Open New Front in PC Wars with Strategic Alliance," *Wall Street Journal,* July 5, 1991, p. A1.
35. Jim Carlton, "Apple-IBM Goal of Universal Computer Faces Hurdles," *Wall Street Journal,* November 9, 1994, p. B4.
36. Jim Carlton, "Oracle to Market Software with IBM, Apple in Bid to Slow Microsoft's Advance," *Wall Street Journal,* July 27, 1995, p. B14.
37. Jim Carlton, "Apple, IBM, Motorola PowerPC Group Issues Blueprint for Common Computer," *Wall Street Journal,* November 15, 1995, p. B6.
38. William Gates, quoted in Laura Evenson, "A New Era of Computing," *San Francisco Chronicle,* November 17, 1993, p. C3.
39. "A New Model for Personal Computing," *San Jose Mercury News,* August 13, 1995, p. 27A.
40. "A New Model for Personal Computing," 1995.
41. Lee Gomes, "Hollow Dreams," *San Jose Mercury News,* November 12, 1995, pp. 1D, 3D.
42. Joseph Jennings, "The End of Wintel?" *San Francisco Examiner,* December 17, 1995, pp. B-5, B-7.
43. Jared Sandberg, "Sun and Netscape Are Forming Alliance Against Microsoft on Internet Standard," *Wall Street Journal,* December 4, 1995, p. B3.
44. Gomes, November, 1995.
45. Jennings, December, 1995.
46. Mark Fleming [letter] and Mike McGowan [letter], "Present at the Creation of the Net," *Business Week,* December 25, 1995, p. 12.
47. Phillip Robinson, "Access to Web Varies in Price, Ease, Speed," *San Jose Mercury News,* September 24, 1995, p. 2E.
48. Tom Abate, "Internet Overload," *San Francisco Examiner,* December 24, 1995, pp. D-1, D-4, D-6.
49. Mike Langberg, "Impatient Web Users Bog Down, Tune Out, Log Off," *San Jose Mercury News,* December 24, 1995, pp. 1A, 14A.
50. Alfred and Emily Glossbrenner, *Finding a Job on the Internet* (New York: McGraw-Hill, 1995).
51. Alan Phelps, "The World Wide Web: Your Mouse's License to Netcruise," *PC Novice,* June 1995, pp. 79–81.
52. Peter H. Lewis, "The Mac's High Road to a Spot on the Internet," *New York Times,* December 5, 1995, p. B10.
53. Robinson, September, 1995.

Chapter 4

1. Hank Roberts, quoted in "Talking About Portables," *Wall Street Journal,* November 16, 1992, p. R19.
2. Hadas Dembo, "The Way Things Were," *Wall Street Journal,* November 16, 1992, pp. R16–R17.
3. C. D. Renmore, *Silicon Chips and You* (New York: Beaufort Books, 1980).
4. Bay Arinze, *Microcomputers for Managers* (Belmont, CA: Wadsworth, 1994), p. 41.
5. Michael S. Malone, "The Tiniest Transformer," *San Jose Mercury News,* September 10, 1995, pp. 1D–2D; excerpted from *The Microprocessor: A Biography* (New York: Telos/Springer Verlag, 1995).
6. Malone, 1995.
7. Laurence Hooper, "No Compromises," *Wall Street Journal,* November 16, 1992, p. R8.

8. Christopher Elliott, "Longest-Lasting Models Run on Nickel Batteries," *USA Today*, April 11, 1995, p. 5E.

9. Amal Kumar Naj, "Latest Version of Zinc-Air Batteries Promises to Show Long-Lasting Results," *Wall Street Journal*, November 10, 1995, p. B6.

10. Peter H. Lewis, "PC Buyers Will Soon Face Choices Among Work Stations," *New York Times*, February 14, 1993, sec. 3, p. 8.

11. Lee Gomes, "Not Necessarily PC," *San Jose Mercury News*, October 15, 1995, pp. 1E–2E.

12. Michael Allen, "Workstations Go from Desk Jobs to Role Once Played Solely by Supercomputers," *Wall Street Journal*, August 11, 1992, p. B1.

13. Laurie Hays, "IBM Tries to Keep Mainframes Afloat against Tide of Cheap, Agile Machines," *Wall Street Journal*, August 12, 1993, pp. B1, B7.

14. Peter Nulty, "When to Murder Your Mainframe," *Fortune*, November 1, 1993, pp. 109–120.

15. John Markoff, "A Remade I.B.M. Reinvents the Mainframe," *New York Times*, January 29, 1993, pp. C1, C5.

16. Study by Gartner Group, cited in Michael Meyer, "Rethinking Your Mainframe," *Newsweek*, June 6, 1994, p. 49.

17. Erick Schonfeld, "Why Mainframes Aren't Dead Yet," *Fortune*, April 17, 1995, p. 19.

18. Richard Preston, "The Mountains of Pi," *New Yorker*, March 2, 1992, pp. 36–37.

19. Lawrence M. Fisher, "Intel Wins Contract to Develop World's Fastest Supercomputer," *New York Times*, September 8, 1995, p. C2.

20. David Einstein, "Intel to Build World's Fastest Supercomputer," *San Francisco Chronicle*, September 8, 1995, pp. A1, A15.

21. Lawrence M. Fisher, "Shifting Lead at Forefront of Computing," *New York Times*, September 21, 1995, pp. C1, C10.

22. John W. Verity and Julie Flynn, "Call It Superbig Blue," *Business Week*, June 29, 1992, pp. 74–75.

23. John Markoff, "A Crucial Linkup in the U.S. Data Highway," *New York Times*, September 30, 1992, p. C6.

24. Bloomberg Business News, "MCI Setting Up Speed Network," *New York Times*, April 25, 1995, p. C6.

25. Peter Coy, "Call It the Supernet," *Business Week*, May 8, 1995, pp. 93–94.

26. John W. Verity, "This Year, Servers Will Be King," *Business Week*, January 8, 1996, p. 94.

27. Verity, January, 1996.

28. Bart Ziegler, "IBM to Expand Its Presence in 'Server' Area," *Wall Street Journal*, October 10, 1995, p. B6.

29. Scott McCartney, "PC Servers Making Inroads as Their Power Accelerates," *Wall Street Journal*, January 17, 1995, p. B3.

30. James L. Pappas, quoted in Laurie Flynn, "Which Peripheral Plug Goes Where? Help Is On the Drawing Board," *New York Times*, March 27, 1995, p. C9.

31. Flynn, 1995.

32. Odyssey Homefront Survey, July 1995, reported in "Home PCs," *Newsweek*, October 9, 1995, p. 10.

33. Dataquest Inc., reported in "Personal Computer Market Seen Doubling in 5 Years," *Wall Street Journal*, June 29, 1995, p. B13.

34. Brian Nadel, "Power to the PC," *PC Magazine*, April 26, 1994, pp. 114–183.

35. Nadel, 1994, p. 116.

36. Dan Gillmor, "Old Computer Will Mean a Lot to Those in Need," *San Jose Mercury News*, December 24, 1995, p. 1F.

37. Merritt Jones, quoted in "FYI," *Popular Science*, November 1995, pp. 88–89.

38. Don Clark, "Meet 'Media Processors,' Computer Chips for the Multimedia Age," *Wall Street Journal*, October 10, 1995, pp. B1, B4.

39. David Einstein, "Here's Pentium Pro, and Then Some," *San Francisco Chronicle*, November 2, 1995, p. D1.

40. Joe McGarvey, "Specialty Chips Could Change PC Landscape," *Inter@ctive Week*, October 23, 1995, p. 16.

41. Clark, 1995.

42. Andrew Pollack, "A Japan Offer of Billion-Bit Chips for '98," *New York Times*, February 14, 1995, p. C4.

43. Molly Williams, Bloomberg Business News, "Developing the 1-Gigabit Chip," *San Francisco Chronicle*, October 26, 1995, pp. D1, D2.

44. Lawrence M. Fisher, "Intel Wins Contract to Develop World's Fastest Supercomputer," *New York Times*, September 8, 1995, p. C2.

45. Lawrence M. Fisher, "Nearing the $500 Computer for Internet Use," *New York Times*, December 4, 1995, p. C5.

46. David Einstein, "Really Cheap Computers to Debut," *San Francisco Chronicle*, January 9, 1996, p. C4.

47. Joan E. Rigdon, "Oracle Nearing Finished Design for Internet PC," *Wall Street Journal*, January 12, 1996, p. B9.

48. Kevin Maney and Julie Schmit, "Computer Firm's Disarray Places CEO in Jeopardy," *USA Today*, January 17, 1996, pp. 1B–2B.

49. Bloomberg Business News, "New Technology Ties Internet, TV to PCs," *San Francisco Chronicle*, October 24, 1995, p. C3.

50. David Stipp, "Scientists Boost Superconductor Current Density," *Wall Street Journal*, April 19, 1995, p. B8.

51. Charles Petit, "State Lab Creates Super-fast Conductor," *San Francisco Chronicle*, April 19, 1995, p. A3.

52. Otis Port, "How to Soup Up Superconductors," *Business Week*, May 1, 1995, p. 128.

53. Jane E. Allen, Associated Press, "New Computers May Use DNA Instead of Chips," *San Francisco Chronicle*, May 3, 1995, p. B2.

54. Steven Levy, "Computers Go Bio," *Newsweek*, May 1, 1995, p. 63.

55. Gina Kolata, "A Vat of DNA May Become Fast Computer of the Future," *New York Times*, April 11, 1995, pp. B5, B8.

56. Sharon Begley and Gregory Beals, "Computing Is in Their Genes," *Newsweek*, April 29, 1995, p. 5.

57. Kyle Pope, "Changing Work Habits Fuel Popularity of Notebooks," *Wall Street Journal*, November 11, 1993, pp. B1, B6.

58. Amy Grenier, quoted in Thomas J. DeLoughry, "Portable Computers, Light and Powerful, Gain Popularity on College Campuses," *Chronicle of Higher Education*, October 6, 1993, pp. A21, A24.

59. Dean Takahashi, "Faster, Better, Cheaper PCs on the Move," *San Jose Mercury News*, November 13, 1995, pp. 1E, 4E.

60. Peter H. Lewis, "Portable Machines Get More Portable," *New York Times*, November 16, 1993, p. B7.

61. "Battery Life by Notebook Category," *PC Magazine*, August 1995, p. 131.

62. Jennifer deJong, "Battery Technology Charges Ahead," *Computer Shopper*, November 1995, p. 278.

63. Phillip Robinson, "Subnotebooks Have Surfaced as the Portable of Choice," *San Jose Mercury News*, February 13, 1994, p. 2E.

64. "Can Notebooks Do Desktop Duty?" *PC/Computing*, January 1994, p. 173.

65. Andrew Pollack, "Organizers Captivate Japanese Girls," *New York Times*, December 22, 1994, pp. C1, C4.

66. Joseph Pereira, "Organizers Help Girls Pass Notes Without Paper," *Wall Street Journal*, December 21, 1994, pp. B1, B5.

67. Phillip Robinson, "Newton and Beyond," *San Jose Mercury News*, May 7, 1995, pp. 1E, 4E.

68. Arthur Leyenberger, "Using a PDA," *PC LapTop Computers Magazine*, February 1995, pp. 44–46.

69. Leyenberger, 1995.

Chapter 5

1. Survey by The Wirthlin Group, Washington, D.C., reported in: Reuters, "It's Always 12:00 for VCR Incompetents," *San Francisco Chronicle*, February 2, 1994, p. A4.

2. David Gelernter, quoted in Associated Press, "Bombing Victim Says He's Lucky to Be Alive," *San Francisco Chronicle*, January 28, 1994, p. A15.

3. Helen Change, "Singaporeans Smitten by ATMs," *San Francisco Chronicle*, October 31, 1995, p. C11.

4. Keith Alexander and Andrea Rothman, "Will Travel Agents Get Bumped by These Gizmos?" *Business Week*, July 28, 1993, p. 72.

5. Keith L. Alexander, "Technology Lets Travelers Avoid Lines, at Times," *USA Today*, August 28, 1995, pp. 1B, 2B.

6. Zachary Coile, "'Free' Computer Revolution Now Has a Price Tag," *San Francisco Examiner*, January 30, 1994, p. B-3.

7. Anthony Ramirez, "Bank to Help Customers Keep Fingers on Dollars," *New York Times*, October 20, 1992, p. C3.

8. Frederick Williams, *Technology and Communication Behavior* (Belmont, CA: Wadsworth, 1987), p. 30.

9. Paul Saffo, quoted in "Old Familiar PC Must Make Room for the Future," *Computer Shopper*, October 1993, p. 170.

10. Jeff Pelline, "CellNet Sees a Big Business in Meter Reading," *San Francisco Chronicle*, September 28, 1995, pp. D1, D2.

11. "Cordless Mouse Works 40 Feet Away," *San Jose Mercury News*, October 8, 1995, p. 4E; reprinted from *Chicago Tribune*.

12. Thomas J. DeLoughry, "Computerized Kiosks," *Chronicle of Higher Education*, February 24, 1995, pp. A25, A26.

13. "Touch and Go," *Newsweek*, March 6, 1995, p. 12.

14. David Berquel, quoted in Mary Geraghty, "Pen-Based Computer Seen as Tool to Ease Burden of Note Taking," *Chronicle of Higher Education*, November 9, 1994, p. A22.

15. Lawrence M. Fisher, "Rethinking Approaches to Digital Handwriting," *New York Times*, March 20, 1995, p. C9.

16. John Pierson, "Do-It-Yourself Grocery Checkout," *Wall Street Journal*, January 31, 1994, p. B1.

17. Ken Butcher, cited in Catherine Young and Willy Stern, "Maybe They Should Call Them 'Scammers,'" *Business Week*, January 16, 1995, pp. 32–33.

18. George James, "Agents Raid Brand-Name Counterfeiters," *New York Times*, September 28, 1995, p. C5.

19. "Will Ben's New Look Stop Counterfeits?" *New York Times*, September 28, 1995, p. C5.

20. Barbara Rudolph, "Some Like Them Hot," *Time*, November 14, 1994, p. 76.

21. "New Bill to Beat Counterfeiters," *San Francisco Chronicle*, September 28, 1995, p. A1.

22. Burr Snyder, "A Calling to Collect," *San Francisco Examiner*, February 27, 1994, pp. E-1, E-10.

23. William M. Bulkeley, "Get Ready for 'Smart Cards' in Health Care," *Wall Street Journal*, May 3, 1993, p. B7.

24. Wilton Woods, "Cashless Society Becomes Reality," *Fortune*, July 12, 1995, p. 24.

25. Nicholas Bray, "Future Shop: 'No Cash Accepted; Microchip-Card Purchases Only,'" *Wall Street Journal*, July 13, 1995, pp. B1, B6.

26. Peter Sinton, "Visa Wants to Kill Cash," *San Francisco Chronicle*, October 11, 1995, pp. B1, B2.

27. "Food Stamps Are Replaced by Plastic Cards in Texas," *New York Times*, November 27, 1995, p. A14.

28. Carl Pascarella, quoted in Marc Levinson and Adam Rogers, "The End of Money?" *Newsweek*, October 30, 1995, pp. 62–65.

29. Bulkeley, 1993.

30. Michael M. Phillips, "Voice Recognition Systems Work, but Only If You Speak American," *Wall Street Journal*, February 7, 1995, p. B1.

31. Michael T. Kaufman, "Sorry Ma'am, No Listing for 'enry 'iggins," *New York Times*, June 26, 1995, pp. C1, C4.

32. Dean Takahashi, "Tell It to the Machines," *San Jose Mercury News*, October 8, 1995, pp. 1D, 2D.

33. Michelle Vranizan, "A New Vision for Blind PC Users," *San Francisco Examiner*, February 26, 1995, pp. B-5, B-6; reprinted from *Orange County Register*.

34. Lawrence J. Magid, "Tech Helps Computers Hear the Disabled," *San Jose Mercury News*, November 5, 1995, p. 9E; reprinted from *Los Angeles Times*.

35. David Einstein, "Computer Age Entering Its Final Frontier," *San Francisco Chronicle*, September 19, 1994, pp. B1, B12.

36. Reuters, "Computer Listens to, Writes Chinese," *San Francisco Chronicle*, March 8, 1995, p. B3.

37. David Haskin, "Speech Recognition: Closer Than You Think," *Windows Sources*, September 1993, p. 46.

38. Takahasi, 1995.

39. Jerry Adler, "The Miracle of the Keys," *Newsweek*, December 23, 1991, p. 67.

40. Sam Vincent Meddis, "Web Cams Capture Lizards, Lunchrooms and More," *USA Today*, December 15, 1995, p. 9B.

41. Associated Press, "MIT Student Has the World Watching," *San Francisco Chronicle*, January 10, 1996, p. A6.

42. Robert Langreth, "Wireless Bridge Watchers," *Popular Science*, August 1995, p. 25.

43. Dana Milbank, "Measuring and Cataloging Body Parts May Help to Weed Out Welfare Cheats," *Wall Street Journal*, December 4, 1995, pp. B1, B6.

44. Magid, 1995.

45. Diana Hembree and Ricardo Sandoval, "The Lady and the Dragon," *Columbia Journalism Review*, August 1991, pp. 44–45.

46. A. Hopkins, "The Social Recognition of Repetition Strain Injuries: An Australian/American Comparison," *Social Science and Medicine*, 1990, 30, pp. 365–372.

47. Edward Felsenthal, "An Epidemic or a Fad? The Debate Heats Up Over Repetitive Stress," *Wall Street Journal*, July 14, 1994, pp. A1, A4.

48. Felsenthal, 1994.

49. "New Suspect in Alzheimer's Risk," *San Jose Mercury News*, July 31, 1994, p. 23A; reprinted from *Los Angeles Times*.

50. Bart Ziegler and Peter Coy, "The Cellular Cancer Risk: How Real Is It?" *Business Week*, February 8, 1993, pp. 94–95.

51. "Myth and Reality in Monitor Emissions," *PC Novice*, July 1993, p. 43.

52. David Kirkpatrick, "Do Cellular Phones Cause Cancer?" *Fortune*, March 8, 1993, pp. 82–89.

53. Steve Kaye, quoted in Elisa Williams, "Making Your Home Work," *San Francisco Examiner*, October 8, 1995, pp. B-1, B-8; reprinted from *Orange County Register*.

54. "PC Users to Be Able to Book Flights, See Data from Delta Air," *Wall Street Journal*, January 11, 1996, p. B5.

55. Erick Schonfeld, "Bar Codes: The Latest Investor Scan," *Fortune*, June 12, 1995, p. 141.

56. Louise Lee, "Garment Scanner Could Be a Perfect Fit," *Wall Street Journal*, September 20, 1995, pp. B1, B6.

57. Glenn Rifkin, "Digital Blue Jeans Pour Data and Legs into Customized Fit," *New York Times*, November 8, 1994, pp. A1, C4.

58. Barbara Ertegun, "Levi's Custom-Fits Jeans by Computer," *San Francisco Chronicle*, November 9, 1994, pp. A1, A12.

59. "Shopping on Your Dashboard," *Futurist*, January–February 1994, p. 8.

60. Geoffrey Smith, "Call It Palpable Progress," *Business Week*, October 9, 1995, pp. 93, 96.

61. Peter Sinton, "Big Banks Launching Smart Card," *San Francisco Chronicle*, August 17, 1995, pp. C1, C3.

62. Dylan Loeb McClain, "Dogs and Cats with Chips on Their Shoulders," *New York Times*, January 22, 1996, p. C5.

63. Sinton, "Visa Wants to Kill Cash," 1995.

64. Takahashi, 1995.

65. Takahashi, 1995.

66. Larry Armstrong and Larry Holyoke, "NASA's Tiny Camera Has a Wide-Angle Future," *Business Week*, March 6, 1995, pp. 54–55.

67. Malcolm W. Browne, "How Brain Waves Can Fly a Plane," *New York Times*, March 7, 1995, pp. B1, B10.

68. Don Clark, "Mind Games: Soon You'll Be Zapping Bad Guys Without Lifting a Finger," *Wall Street Journal*, June 16, 1995, p. B12.

69. Tom Abate, "Marin Investor Bets on an Impulse," *San Francisco Examiner*, July 2, 1995, pp. B-1, B-7.

70. Malcolm W. Browne, "Neuron Talks to Chip and Chip to Nerve Cell," *New York Times*, August 22, 1995, pp. B1, B9.

71. Mary Madison, "Mind Control for Computers," *San Francisco Chronicle*, December 2, 1995, pp. A1, A13.

72. James Barron, "Stanley Adelman, 72, Repairer of Literary World's Typewriters," *New York Times*, December 1, 1995, p. C19.

73. Jonathan Auerbach, "Smith Corona Seeks Protection of Chapter 11," *Wall Street Journal*, July 6, 1995, p. B6.

74. Tom Foremski, "Conquering the Computer Connection," *San Francisco Examiner*, February 12, 1995, pp. D-5, D-7.

75. Federico Faggin, "A Call for Humanized Computing," *San Francisco Examiner*, August 20, 1995, pp. B-5, B-6.

76. "Tying Up Your Computer," *PC Novice*, May 1992, p. 9.

77. Marty Jerome, "Killer Color Notebooks," *PC Computing*, January 1994, pp. 142–177.

78. Lori Beckmann Johnson, "Power Protection: Saving Your Computer from Surges, Spikes, and Sags," *PC Novice*, May 1992, pp. 72–73.

79. Jennifer Larson, "Maintaining Your Diskette Collection," *PC Novice*, February 1994, pp. 36–37.

80. Jim Seymour, *Jim Seymour's On the Road: The Portable Computing Bible* (New York: Brady, 1992), pp. 137–138.

81. Seymour, 1992, p. 138.

82. Robert Schmidt, "Comfortable Keyboards," *PC Novice,* October 1993, pp. 44–46.

Chapter 6

1. David L. Wheeler, "Recreating the Human Voice," *Chronicle of Higher Education,* January 19, 1996, pp. A8–A9.
2. Steve Rubenstein, "Dashboard Gizmo Will Guide Drivers," *San Francisco Chronicle,* December 8, 1995, pp. A25, A28.
3. Larry Light and Oluwabunmi Shabi, "'Plasma' Could Be Some Screen Gem," *Business Week,* July 24, 1995, p. 6.
4. Suzanne Weixel, *Easy PCs,* 2nd ed. (Indianapolis: Que Corp., 1993), p. 52.
5. "Bulletin Board," *Business Week,* December 26, 1994, p. 24.
6. Bruce Brown, "Portable Printers: Fewer Trade-offs," *PC Magazine,* August 1995, pp. 215–225.
7. Jim Seymour, *On the Road: The Portable Computing Bible* (New York: Brady, 1992), pp. 184–185.
8. Weixel, 1993, pp. 54–55.
9. Larry Armstrong, "The Price Is Right for Printers," *Business Week,* November 6, 1995, pp. 132–134.
10. Phillip Robinson, "Color Comes In," *San Jose Mercury News,* October 1, 1995, pp. 1F, 6F.
11. Stephanie LaPolla, "A Plotter Technology for Every Need," *PC Week,* December 2, 1991.
12. Edward Baig, "A Croner Office that Fits in a Corner," *Business Week,* April 17, 1995, pp. 108 E2–E4.
13. Larry Armstrong, "The One-Man Band of Home Offices," *Business Week,* November 6, 1995, p. 134.
14. Phillip Robinson, "It Prints, It Scans, It Faxes, but There Are Compromises," *San Jose Mercury News,* November 12, 1995, pp. 1E, 5E.
15. Liz Spayd, "Taming the Paper Jungle," *San Francisco Chronicle, Sunday Punch,* December 19, 1993, p. 2; reprinted from *The Washington Post.*
16. Clifford Nass, quoted in Spayd, 1993.
17. Michael J. Mandel, Mark Landler, Ronald Grover, et al., "The Entertainment Economy," *Business Week,* March 14, 1994, pp. 58–64.
18. Edward R. McCracken, cited in Mandel et al., 1994, p. 60.
19. Peter H. Lewis, "So the Computer Talks. Does Anyone Want to Listen?" *New York Times,* October 4, 1992, sec. 3, p. 9.
20. Diana Berti, quoted in Timothy L. O'Brien, "Aided by Computers, Many of the Disabled Form Own Businesses," *Wall Street Journal,* October 8, 1993, pp. A1, A5.
21. L. R. Shannon, "Making Your Mac Say Arrgh!" *New York Times,* July 21, 1992, p. B9.
22. Larry Armstrong, "Sweet Music with Ominous Undertones for Yamaha," *Business Week,* November 15, 1993, pp. 119–120.
23. Lawrence J. Magid, "ProShare Lets Pair Video-edit Document," *San Jose Mercury News,* February 13, 1994, p. 5E.
24. Edmund L. Andrews and Joel Brinkley, "The Fight for Digital TV's Future," *New York Times,* January 22, 1995, sec. 3, pp. 1, 6.
25. Edmund L. Andrews, "Quest for Sharper TV Is Likely to Produce More TV Instead," *New York Times,* July 10, 1995, pp. C1, C8.
26. Edmund L. Andrews, "New TV System Is Endorsed, but Its Future Is Questioned," *New York Times,* November 29, 1995, pp. A1, C4.
27. Paul M. Eng, "Virtual Buses for Novice Drivers," *Business Week,* January 29, 1996, p. 68D.
28. Matthew L. Wald, "FAA Will Use a Simulator to Train Controllers on Breakdowns," *New York Times,* September 24, 1995, sec. 1, p. 19.
29. Belinda Thurston, "Virtual Reality Lessons Hit Home," *USA Today,* April 11, 1995, p. 6D.
30. Laurie Flynn, "Virtual Reality and Virtual Spaces Find a Niche in Real Medicine," *New York Times,* June 5, 1995, p. C3.
31. Donna Horowitz, "Virtual Reality Takes Acrophones to Greater Heights," *San Francisco Examiner,* March 26, 1995, pp. C-1, C-6.
32. Daniel Goleman, "In Virtual Reality, Phobias Cease to Exist," *San Francisco Chronicle,* September 2, 1995, p. A7; reprinted from *The New York Times.*
33. Osvaldo Arias, quoted in N. R. Kleinfield, "Stepping into Computer, Disabled Savor Freedom," *New York Times,* March 12, 1995, sec. 1, p. 17.
34. Neil Gross, Dori Jones Yang, and Julia Flynn, "Seasick in Cyberspace," *Business Week,* July 10, 1995, pp. 110–111.
35. John Malyon, *Whole Earth Review,* Spring 1992, pp. 80–84.
36. John Holusha, "Down on the Farm with R2D2," *New York Times,* October 7, 1995, pp. 21, 24.
37. Gene Bylinsky, "High-Tech Help for the Housekeeper," *Fortune,* November 2, 1992, p. 117.
38. Tahree Lane, "Robots Can Fill Jobs Nobody Wants," *San Francisco Examiner,* April 25, 1993, p. E-3.
39. Stephen Baker, "A Surgeon Whose Hands Never Shake," *Business Week,* October 4, 1993, pp. 111–114.
40. Jack Cheevers, "Real-Life 'Robocops' Gain Popularity," *San Francisco Chronicle,* August 13, 1994, p. A9; reprinted from *Los Angeles Times.*
41. Holusha, 1995.
42. Associated Press, "Dante II Bids Farewell to Volcano Purgatory," *San Jose Mercury News,* August 14, 1994, p. 11A.
43. Judith Anne Gunther, "Dante's Inferno," *Popular Science,* November 1994, pp. 66–68, 96.
44. Associated Press, "U.S. to Send Robot Craft to Map Moon," *New York Times,* March 5, 1995, sec. 1, p. 11.
45. Associated Press, "TROV Searches for Treasures," *San Jose Mercury News,* September 10, 1995, p. 3B.
46. Chiori Santiago, "The Art Is the Message," *Arts, San Jose Mercury News,* February 19, 1995, pp. 4–5.
47. William Safire, "Art Vs. Artifice," *New York Times,* January 3, 1994, p. A11.
48. Hans Fantel, "Sinatra's 'Duets': Music Recording or Wizardry?" *New York Times,* January 1, 1994, p. 13.
49. Cover, *Newsweek,* June 27, 1994.
50. Cover, *Time,* June 27, 1994.
51. Cox News Service, "Computers Manipulate Old Photographs," *San Jose Mercury News,* February 19, 1995, p. 5H.
52. Jonathan Alter, "When Photographs Lie," *Newsweek,* July 30, 1990, pp. 44–45.
53. Fred Ritchin, quoted in Alter, 1990.
54. Robert Zemeckis, cited in Laurence Hooper, "Digital Hollywood: How Computers Are Remaking Movie Making," *Rolling Stone,* August 11, 1994, pp. 55–58, 75.
55. Woody Hochswender, "When Seeing Cannot Be Believing," *New York Times,* June 23, 1992, pp. B1, B3.
56. Kathleen O'Toole, "High-Tech TVs, Computers Blur Line Between Artificial, Real," *Stanford Observer,* November–December 1992, p. 8.
57. Stephen H. Wildstrom, "Passing the Screen Test," *Business Week,* January 29, 1996, p. 16.
58. Barnaby J. Feder, "Where Repetition Is Exciting," *New York Times,* November 8, 1995, pp. C1, C10.
59. Sabra Chartrand, "Trying to Establish a U.S. Flat-Screen Market with a More Efficient Technology for Cathodes," *New York Times,* April 10, 1995, p. C1.
60. Gail Edmondson and Neil Gross, "The Grand Alliance in Flat Panels," *Business Week,* August 28, 1995, pp. 73–74.
61. Lawrence B. Johnson, "PC Makers Are Focusing on Fine-Tuning the Sound," *New York Times,* December 11, 1995, p. C3.
62. Gerald W. Tschetter, quoted in Johnson, 1995.
63. Larry Armstrong, "Can You Believe Your Ears?" *Business Week,* May 29, 1995, pp. 88–90.
64. Scott Ritter, "Wavelet Theory Spiffs Up Video in Computers," *Wall Street Journal,* May 25, 1995, pp. B1, B7.
65. Peter Coy and Robert D. Hof, "3-D Computing," *Business Week,* September 4, 1995, pp. 70–77.
66. Stephen Manes, "3-D Capabilities Create a New Dimension for Electronic Games," *San Jose Mercury News,* July 2, 1995, p. 6F; reprinted from *New York Times.*
67. Don Clark, "Comdex Trade Show Presents 3D Effects," *Wall Street Journal,* November 13, 1995, p. B4.
68. Harry Somerfield, "3-D Is Headed for Home Television," *San Francisco Chronicle,* March 2, 1994, p. Z-5.
69. Denise Caruso, "Virtual-World Users Put Themselves in a Sort of Electronic Puppet Show," *New York Times,* July 10, 1995, p. C5.
70. Michael Meyer, "Surfing the Internet in 3-D," *Newsweek,* May 15, 1995, pp. 68–69.
71. Associated Press, "S.F. Firm's 'Virtual World,' *San Francisco Chronicle,* November 30, 1995, p. D3.
72. Joe Kilsheimer, "Next Step on the Net Could Be 3-D," *San Jose Mercury News,* December 31, 1995, p. 5F.
73. Meyer, 1995.
74. Phillip Robinson, "Periodically, Check Out Computer Magazines," *San Jose Mercury News,* October, 17, 1993, p. 4F.
75. Russ Walter, *The Secret Guide to Computing,* 19th ed., p. 17. (Available from Russ Walter, 22 Ashland St., Floor 2, Somerville, MA 02144–3202.)
76. Phillip Robinson, "New Crop of Magazines Offers News, Reviews, Tutorials—and Fun," *San Jose Mercury News,* September 18, 1994, p. 5E.
77. Walter S. Mossberg, "Three Magazines Put Everyday Perspective on Computers," *The Wall Street Journal,* September 22, 1994, p. B1.
78. David Armstrong, "In the Computer Age, Ziff Rules," *San Francisco Examiner,* June 5, 1994, pp. C–1, C–4.

Chapter 7

1. Richard L. Hudson, "Europeans No Longer Scoff at Interactive Multimedia," *Wall Street Journal,* March 2, 1994, p. B6.
2. "Gargantua's 'Lossless' Compression," *The Australian,* March 22, 1994, p. 32; reprinted from *The Economist.*
3. Peter Coy, "Invasion of the Data Shrinkers," *Business Week,* February 14, 1994, pp. 115–116.
4. "Gargantua's 'Lossless' Compression," 1994.
5. Coy, 1994.
6. Craig Clarke, quoted in Leslie Miller, "Zippy New Disk Drive Pumps Up Storage," *USA Today,* May 4, 1995, p. 6D.
7. Stephen H. Wildstrom, "Cheap and Easy Storage Space," *Business Week,* May 22, 1995, p. 26.
8. Phillip Robinson, "Between a Floppy and a Hard Drive," *San Jose Mercury News,* October 8, 1995, pp. 1E, 6E.
9. "Team of 3 Companies in Computer Industry Improves Disk Drives," *Wall Street Journal,* May 9, 1995, p. B7.
10. Evan I. Schwartz, "CD-ROM: A Mass Medium at Last," *Business Week,* July 19, 1993, pp. 82–83.
11. Scott Rosenberg, "Rock 'n' ROM," *San Francisco Examiner,* May 1, 1994, pp. B-13, B-14.
12. William M. Bulkeley, "Publishers Deliver Reams of Data on CDs," *Wall Street Journal,* February 22, 1993, p. B6.
13. Rick Sammon, Associated Press, "Learn Photography via CD-ROMs, Videotapes," *Chicago Tribune,* September 15, 1995, sec. 7, p. 78.
14. Catherine Arnst, Paul M. Eng, and Richard Brandt, "Multimedia: Joyful and Triumphant?" *Business Week,* December 6, 1993, pp. 167–169.
15. Walter S. Mossberg, "Picture Perfect: Book Explores Vietnam Through Camera's Eye," *Wall Street Journal,* June 8, 1995, p. B1.
16. Steve Lohr, "A CD-ROM Magazine with a Difference: More Ads," *New York Times,* May 8, 1995, p. C6.
17. Kevin Goldman, "Firms Click on Ads in CD-ROM Magazines," February 27, 1995, *Wall Street Journal,* p. B8.
18. Donn Menn, "More Than Music Rock 'n' ROM," *Multimedia World,* August 1995, pp. 58–63.
19. Steven Levy, "In Search of Liner Notes," *Newsweek,* November 13, 1995, p. 91.
20. Mike Langberg, "The Minuses of CD Plus," *San Jose Mercury News,* November 12, 1995, pp. 1E, 7E.
21. Edward Baig, "Music to Your Ears—and Eyes," *Business Week,* January 22, 1996, p. 99.
22. Jeffrey Gordon Angus and Carla Thornton, "Do-It-Yourself CD-ROMs," *PC World,* January 1996, pp. 173–175.
23. L. R. Shannon, "Help for Picturing Pictures on Screen," *New York Times,* August 1, 1995, p. B7.
24. L. R. Shannon, "Will Putting All Your Snapshots on a Disk Replace the Photo Album?" *New York Times,* November 16, 1993, p. B7.
25. Peter H. Lewis, "Besides Storing 1,000 Words, Why Not Store a Picture Too?" *New York Times,* October 11, 1992, sec. 3, p. 8.
26. Stephen Manes, "If DVD Is the Next Medium, When Will Future Arrive?" *New York Times,* January 16, 1996, p. B8.
27. Mike Langberg, "The Dawning of the DVD Age," *San Jose Mercury News,* January 14, 1996, pp. 1F, 4F.
28. Dennis Normile, "Get Set for the Super Disc," *Popular Science,* February 1996, pp. 55–58.
29. Normile, 1996.
30. Manes, 1996.
31. Paul M. Eng, Robert D. Hof, and Hiromi Uchida, "It's a Whole New Game: The Hards vs. the Cards," *Business Week,* June 8, 1992, pp. 101–103.
32. Janet L. Fix, "Deal Secures First Data's Credit Lead," *USA Today,* September 18, 1995, pp. 1B, 2B.
33. Phillip Robinson, "With a Tape Drive, Backing Up Isn't Quite So Hard to Do," *San Jose Mercury News,* October 16, 1995, pp. 1F, 6F.
34. Peter H. Lewis, "Revering Redundancy," *San Jose Mercury News,* February 13, 1994, p. 1E; reprinted from *The New York Times.*
35. Dan Gillmor, "A Computer Whose Time Has Come—Internet Appliance," *San Jose Mercury News,* January 28, 1996, p. 1F.
36. "Getting Digital Imaging in Focus," *Fortune,* May 1, 1995, p. 82.
37. Mark Maremont, "A Magnetic Mug Shot on Your Credit Card?" *Business Week,* April 24, 1995, p. 58.
38. Otis Port, "Sifting Through Data with a Neural Net," *Business Week,* October 30, 1995, p. 70.
39. Tom Dellecave Jr., "Jukeboxes: High, Low Notes," *InformationWeek,* March 6, 1995, pp. 40–44.
40. Peter Coy, "Dense and Denser on IBM Disks," *Business Week,* April 17, 1995, p. 93.
41. "IBM Reports Record in Stored-Data Density," *Wall Street Journal,* March 31, 1995, p. B7.
42. Charles Petit, "Tomorrow's Disk to Store Even More," *San Francisco Chronicle,* April 18, 1995, pp. B1, B4.
43. Matthew May, "Digital Videos on the Way," *The Times* (London), April 15, 1994, p. 34.
44. Lawrence B. Johnson, "Videotape's Best Years May Lie in the Future," *New York Times,* August 20, 1995, sec. 2, p. 21.
45. Erick Schonfeld, "Getting in Real Early on Digital Video," *Fortune,* February 5, 1996, p. 136.
46. Richard Brandt, "Can Larry Beat Bill?" *Business Week,* May 15, 1995, pp. 88–96.
47. David P. Hamilton, "NEC Physicists Develop Method to Store and Erase Information Using Atoms," *Wall Street Journal,* June 3, 1993, p. B5.
48. Robert Birge, quoted in Amai Kumar Naj, "Researchers Isolate Bacteria Protein that Can Store Data in 3 Dimensions," *Wall Street Journal,* September 4, 1991, p. B4.
49. Michael Rogers, quoted in Tod Oppenheimer, "*Newsweek*'s Voyage Through Cyberspace," *Columbia Journalism Review,* December 1993, pp. 34–37.
50. Sylvia Rubin, "Interactive Rock and Roll," *San Francisco Chronicle,* December 18, 1993, pp. E1, E6.
51. Don Clark, "Multimedia's Hype Hides Virtual Reality: An Industry Shakeout," *Wall Street Journal,* A1, A9.
52. Neil Taylor, "CD-ROM Maps Out School's Past," *South China Morning Post,* March 29, 1994, Technology Post, p. 1.
53. John Eckhouse, "Rigors of Trying to Obtain Rights," *San Francisco Chronicle,* August 30, 1993, p. D5.

Chapter 8

1. Thomas A. Stewart, "The Information Age in Charts," *Fortune,* April 4, 1994, pp. 75–79.
2. Tom Mandel, in "Talking About Portables," *Wall Street Journal,* November 16, 1992, p. R18.
3. Blanton Fortson, in "Talking About Portables," *Wall Street Journal,* November 16, 1992, pp. R18–R19.
4. Jacob M. Schlesinger, "Get Smart," *Wall Street Journal,* October 21, 1991, p. R18.
5. Elizabeth Fernandez, "Homeless but Wired," *San Francisco Examiner,* January 28, 1996, pp. A-1, A-14.
6. Gautam Naik, "Voice Messaging: It's Efficient and Cheap but Still Too Limited," *Wall Street Journal,* October 26, 1995, pp. B1, B2.
7. N. R. Kleinfield, "For Homeless, Free Voice Mail," *New York Times,* January 30, 1995, p. A12.
8. Peter H. Lewis, "The Good, the Bad and the Truly Ugly Faces of Electronic Mail," *New York Times,* September 6, 1994, p. B7.
9. Robert Rossney, "E-Mail's Best Asset—Time to Think," *San Francisco Chronicle,* October 5, 1995, p. E7.
10. Walter S. Mossberg, "With E-Mail, You'll Get the Message without the Hang-ups," *Wall Street Journal,* April 20, 1995, p. B1.

11. Lewis, 1994.
12. Michelle Quinn, "E-Mail Is Popular—but Far from Perfect," *San Francisco Chronicle,* April 21, 1995, pp. B1, B8.
13. David Cay Johnston, "Not So Fast: E-Mail Sometimes Slows to a Crawl," *New York Times,* January 7, 1996, sec. 4, p. 5.
14. "'Postmarks' on E-Mail," *Business Week,* April 24, 1995, p. 47.
15. Johnston, 1996.
16. Paul M. Eng, "Git Along, Little Video," *Newsweek,* January 29, 1996, p. 68D.
17. "Time Warner Cable to Provide Phone Service," *San Jose Mercury News,* May 18, 1994, p. 1G; reprinted from *New York Times.*
18. Jeff Pelline, "Cable TV Preparing Its Transformation," *San Francisco Chronicle,* July 14, 1995, pp. B1, B2.
19. Jeff Pelline, "Viacom Confirms Phone Trial," *San Francisco Chronicle,* October 12, 1995, pp. B1, B3.
20. Steve Meloan, "Tuning in to V-Mail," *San Francisco Examiner,* August 27, 1995, pp. C-6, C-7.
21. Mike Branigan, "The Cost of Using an Online Service," *PC Novice,* January 1992, pp. 65–71.
22. Tammi Wark, "Online Service Subscribers," *USA Today,* February 9, 1996, p. 4B.
23. Susan Chandler, "The Grocery Cart in Your PC," *Business Week,* September 11, 1995, pp. 63–64.
24. Paul M. Eng, "War of the Web," *Business Week,* March 4, 1996, pp. 71–72.
25. Steven Levy, "Dead Men Walking?" *Newsweek,* January 22, 1996, p. 66.
26. Thomas E. Weber, "AT&T to Move On-line Service onto the Web," *Wall Street Journal,* January 5, 1996, p. B11.
27. David J. Lynch, "On-line Service Shake-out Ahead," *USA Today,* January 5, 1996, p. 1B.
28. Julie Schmit, "On-line Services Experience Static," *USA Today,* February 22, 1996, p. 3B.
29. David J. Lynch, "Prodigy Tries to Upgrade Its Stodgy Image," *USA Today,* February 9, 1996, p. 4B.
30. Peter H. Lewis, "Sears Moves to Shed Stake in Prodigy," *New York Times,* February 22, 1996, pp. C1, C11.
31. Peter Elstrom, "H&R Block May Be Going Off-line," *Business Week,* February 26, 1996, p. 40.
32. Schmit, 1996.
33. Steve Lohr, "As America Online's Share Price Has Soared, So Has the Mushrooming Number of Short-Sellers," *New York Times,* February 12, 1996, p. C4.
34. Jared Sandberg and Bart Ziegler, "Internet's Popularity Threatens to Swamp the On-line Services," *Wall Street Journal,* January 18, 1996, pp. A1, A8.
35. Jesse Kornbluth, "The Truth About the Web," *San Francisco Chronicle,* January 23, 1996, p. C4.
36. Schmit, 1996.
37. Marshall Toplansky, quoted in Jennifer Larson, "Telecommunications and Your Computer," *PC Novice,* March 1993, pp. 14–19.
38. Mark Shapiro, "Bulletin Board Systems Cover Just About Every Topic Imaginable," *San Jose Mercury News,* July 23, 1995, pp. 1F, 4F.
39. Hafner, 1995.
40. "AT&T Move May Trigger Internet War," *USA Today,* February 18, 1996, p. 1B.
41. Bud Konheim, quoted in Jared Sandberg, "The Internet Pulsates with All the Pros and Cons of AT&T's Free-Access Offer," *Wall Street Journal,* February 29, 1996, pp. B1, B3.
42. International Data Corp., cited in Sandberg and Ziegler, 1996.
43. Nielsen Media Research, cited in Julian Dibbell, "Nielsen Rates the Net," *Time,* November 13, 1995, p. 121.
44. David L. Wilson, "Internet@home," *Chronicle of Higher Education,* June 16, 1995, pp. A20, A22.
45. Forrester Research Inc., cited in Eng, 1996.
46. David Einstein, "What They Want Is E-mail," *San Francisco Chronicle,* February 20, 1996, pp. B1, B6.
47. Elizabeth P. Crowe, "The News on Usenet," *Bay Area Computer Currents,* August 8–21, 1995, pp. 94–95.
48. Mary Ann Pike, *Using the Internet,* 2nd ed. (Indianapolis, IN: Que Corp., 1995), p. 638.
49. Trip Gabriel, "The Meteoric Rise of Web Site Designers," *New York Times,* February 12, 1996, pp. C1, C5.
50. David Plotnikoff, "The Mercury News Family Guide to Cyberspace," *San Jose Mercury News,* February 11, 1996, pp. 3H–6H.
51. Gabriel, 1996.
52. Peter H. Lewis, "Home Pages Never Die; You Must Kill Them," *New York Times,* January 2, 1996, p. C15.
53. Rick Tetzeli, "The Internet and Your Business," *Fortune,* March 7, 1994, pp. 86–96.
54. David Landis, "Exploring the Online Universe," *USA Today,* October 7, 1993, p. 4D.
55. International Data Corp., cited in Paul C. Judge, "Lotus Is Learning to Live with the Net," *Business Week,* January 29, 1996, pp. 70–71.
56. Bart Ziegler, "IBM's Gerstner Vows Funds for Lotus in Microsoft Duel and Internet Battle," *Wall Street Journal,* January 23, 1996, p. B7.
57. James Kim, "Lotus Hits High 'Notes' with New PC Program," *USA Today,* January 22, 1996, p. 1B.
58. Judge, 1996.
59. Laurie Flynn, "Companies Use Web Hoping to Save Millions," *New York Times,* July 17, 1995, p. C5.
60. Alison L. Sprout, "The Internet Inside Your Company," *Fortune,* November 27, 1995, pp. 161–168.
61. Amy Cortese, "Here Comes the Intranet," *Business Week,* February 26, 1996, pp. 76–84.
62. Alvin Toffler, quoted in Marianne Roberts, "Computers Replace Commuters," *PC Novice,* September 1992, p. 27.
63. American Information User 1994 survey, cited in Sherri Merl, "Resisting the Call to Telecommute," *New York Times,* October 22, 1995, sec. 3, p. 14.
64. FIND/SVP survey, cited in John Holusha, "Telecommuting Options Are Gaining Sophistication," *New York Times,* January 2, 1996, p. C15.
65. Jonathan Marshall, "Eliminating the Permanent Office," *San Francisco Chronicle,* March 10, 1994, pp. D1, D3.
66. Alison L. Sprout, "Moving Into the Virtual Office," *Fortune,* May 2, 1994, p. 103.
67. Laurie M. Grossman, "Truck Cabs Turn into Mobile Offices as Drivers Take on White-Collar Tasks," *Wall Street Journal,* August 3, 1993, pp. B1, B5.
68. Bob Spoer, quoted in Marilyn Lewis, "Tethered to Work," *San Jose Mercury News,* October 1, 1995, pp. 1A, 18A.
69. Peter Hart, quoted in Lewis, 1995.
70. Nicholas Negroponte, quoted in Stephen Manes, "Combination Computer-TV Set Provides a Low-Quality Picture," *San Jose Mercury News,* March 26, 1995, p. 6F; reprinted from *New York Times.*
71. Jim Heid, "Macintosh TV," *Macworld,* April 1994, pp. 57–59.
72. Manes, 1995.
73. Todd Copileitz, "Dialing Long-Distance on the 'Net," *San Francisco Examiner,* March 5, 1995, pp. C-5, C-7; reprinted from *Dallas Morning News.*
74. Lawrence M. Fisher, "Long-Distance Phone Calls on the Internet," *New York Times,* March 14, 1995, p. C6.
75. William M. Bulkeley, "Top Internet Service to Offer Capability of Phone and Voice," *Wall Street Journal,* June 2, 1995, p. B6.
76. Joshua Quittner, "Radio Free Cyberspace," *Time,* May 1, 1995, p. 91.
77. Leslie Miller, "Tuning in to Brave New Sounds of the Internet," *USA Today,* November 1, 1995, p. 1D.
78. David Colker, "Bringing World News to Your Ears," *Los Angeles Times,* December 8, 1995, pp. E3, E12.
79. David Einstein, "New Era Dawns as Radio Comes to the Internet," *San Francisco Chronicle,* May 18, 1995, pp. B1, B2.
80. Don Clark, "Intel Makes Plan for Integrating TV and Internet," *Wall Street Journal,* October 24, 1995, p. B6.
81. Michael Antonoff, "Broadcasting the Web," *Popular Science,* February 1996, p. 31.
82. Joan E. Rigon, "Coming Soon to the Internet: Tools to Add Glitz to the Web's Offerings," *Wall Street Journal,* August 16, 1995, pp. B1, B2.
83. Michael Meyer, "Surfing the Internet in 3-D," *Newsweek,* May 15, 1996, pp. 68–69.
84. Michelle Quinn, "Silicon Graphics Bringing 3-D to Internet," *San Francisco Chronicle,* March 31, 1995, pp. B1, B2.
85. Neil Gross, Peter Burrows, and Robert D. Hof, "Internet Lite: Who Needs a PC?" *Business Week,* November 13, 1995, pp. 102–103.
86. Mariette DiChristina and David Scott, "Calling Dick Tracy!" *Popular Science,* March 1996, p. 35.
87. Tom R. Halfhill, "Inside the Web PC," *Byte,* March 1996, pp. 44–56.
88. Gina Smith, "Is '$500 Computer' Worth Waiting For?" *San Francisco Examiner,* March 3, 1996, p. D-5.
89. Jeff Pelline, "Oracle's 'Magic Box,'" *San Francisco Chronicle,* January 18, 1996, p. B1.
90. Don Clark, "Oracle Is Demonstrating Network Computer Today," *Wall Street Journal,* February 26, 1996, p. A3.
91. Lee Gomes, "Stripped-down Computing," *San Jose Mercury News,* February 27, 1996, pp. 1C, 2C.
92. Denise Caruso, "On-line Browsing Got You Down? Don't Get Mad, Get Cable," *New York Times,* June 5, 1995, p. C3.
93. Lawrence J. Magid, "ISDN Lines Boost Speed Limit on Net," *San Jose Mercury News,* May 7, 1995, pp. 1E, 6E; reprinted from *Los Angeles Times.*
94. Michael J. Himowitz, "Agony, Ecstasy, and ISDN," *Fortune,* February 19, 1996, p. 107.
95. Leslie Cauley, "Baby Bells Rediscover Fast ISDN Service," *Wall Street Journal,* January 22, 1996, p. B5.
96. Grant Balkema, quoted in Mark Robichaux, "Cable Modems Are Tested and Found to Be Addictive," *Wall Street Journal,* December 27, 1995, pp. 13, 17.
97. Peter Coy, "The Big Daddy of Data Haulers?" *Business Week,* January 29, 1996, pp. 74–76.
98. Lucien Rhodes, "The Race for More Bandwidth," *Wired,* January 1996, pp. 140–145.
99. Forrester Research Inc., cited in Coy, 1996.
100. "Flexible Fiber," *Newsweek,* June 12, 1995, p. 8.
101. David A. Harvey and Richard Santalesa, "Wireless Gets Real," *Byte,* May 1994, pp. 90–96.
102. Carolyn Nielson, "GPS Ready to Take Off," *San Francisco Examiner,* June 4, 1995, pp. B-5, B-6.
103. Nielson, 1995.
104. Betsy Wade, "Navigating New Turf Made Easy," *San Jose Mercury News,* March 5, 1995, p. 2G; reprinted from *New York Times.*
105. Anthony Ramirez, "Cheap Beeps: Across Nation, Electronic Pagers Proliferate," *New York Times,* July 19, 1993, pp. A1, C2.
106. Mark Lewyn, "Beep If Your Pager Sends Voice Mail," *Business Week,* September 25, 1995, p. 147.
107. Kevin Maney, "Pager Offers Sports Fans Scores, News," *USA Today,* November 2, 1995, p. 1B.
108. Quentin Hardy, "SkyTel Is Set to Launch Two-Way Paging," *Wall Street Journal,* September 18, 1995, p. B4.
109. Gautam Naik, "Lowly Beeper May Finally Get Respect as Two-Way Paging Services Emerge," *Wall Street Journal,* September 19, 1995, pp. B1, B11.
110. Lois Therrien and Chuck Hawkins, "Wireless Nets ARen't Just for Big Fish Anymore," *Business Week,* March 9, 1992, pp. 84–85.
111. Jeff Pelline, "Tough Call on Cell Phones," *San Francisco Chronicle,* August 8, 1995, p. B1.
112. David J. Lynch, "Telecom Giants Enter Crowded, High-Cost Race," *USA Today,* November 21, 1995, pp. 1B, 2B.
113. C. Michael Armstrong, quoted in Cole, 1995.
114. John Seabrook, "My First Flame," *The New Yorker,* June 6, 1994, pp. 70–79.
115. Jared Sandberg, "Up in Flames," *Wall Street Journal,* November 15, 1993, p. R12.
116. Ramon G. McLeod, "Netiquette—Cyberspace's Cryptic Social Code," *San Francisco Chronicle,* March 6, 1996, pp. A1, A10.
117. Virginia Shea, quoted in McLeod, 1996.
118. Faiza S. Ambah, "An Intruder in the Kingdom," *Business Week,* August 21, 1995, p. 40.
119. Joseph Kahn, Kathy Chen, and Marcus W. Brauchli, "Beijing Seeks to Build Version of the Internet that Can Be Censored," *Wall Street Journal,* January 31, 1996, pp. A1, A14.
120. Associated Press, "China Tells Internet Users to Register with the Police," *San Francisco Chronicle,* February 16, 1996, p. A15.
121. John Markoff, "On-line Service Blocks Access to Topics Called Pornographic," *New York Times,* December 29, 1995, pp. A1, C4.
122. Peter H. Lewis, "On-line Service Ending Its Ban of Sexual Materials on Internet," *New York Times,* February 14, 1996, pp. A1, C2.
123. Edmund L. Andrews, "Telecommunications Bill Signed, and New Round of Battles Starts," *New York Times,* February 2, 1996, pp. A1, C16.
124. Peter H. Lewis, "Judge Temporarily Blocks Law that Bans Indecency on Internet," *New York Times,* February 16, 1996, pp. A1, C16.
125. Yahoo!, cited in Del Jones, "Cyber-porn Poses Workplace Threat," *USA Today,* November 27, 1995, p. 1B.
126. Lawrence J. Magid, "Be Wary, Stay Safe in the On-line World," *San Jose Mercury News,* May 15, 1994, p. 1F.
127. David Einstein, "SurfWatch Strikes Gold as Internet Guardian," *San Francisco Chronicle,* March 7, 1996, pp. D1, D2.
128. Peter H. Lewis, "Limiting a Medium without Boundaries," *New York Times,* January 15, 1996, pp. C1, C4.
129. Richard Zoglin, "Chips Ahoy," *Time,* February 19, 1996, pp. 58–61.
130. Frank Rich, "The Idiot Chip," *New York Times,* February 10, 1996, p. 15.
131. Lewis, "Limiting a Medium without Boundaries," 1996.
132. Mike Mills, "'Cell' Phones Betraying Their Owners," *San Jose Mercury News,* June 26, 1994, p. 14A.
133. Jonathan Marshall, "Why Crime, Cellular Phones Don't Mix," *San Francisco Chronicle,* June 21, 1994, pp. D1, D3.
134. Constance Johnson, "Anonymity On-Line? It Depends Who's Asking," *Wall Street Journal,* November 24, 1995, pp. B1, B3.
135. Carol Kleiman, "The Boss May Be Listening," *San Jose Mercury News,* February 25, 1996, pp. 1PC, 2PC.
136. Gina Kolata, "When Patients' Records Are Commodities for Sale," *New York Times,* November 15, 1995, pp. A1, B7.
137. Survey by Equifax and Louis Harris & Associates, cited in Bruce Horovitz, "80% Fear Loss of Privacy to Computers," *USA Today,* October 31, 1995, p. 1A.
138. Associated Press, "Companies Looking to Harness Airwaves," *San Francisco Chronicle,* January 30, 1996, p. A4.
139. Saundra Banks Loggins, quoted in Kenneth Howe, "Firm Turns Hiring Into a Science," *San Francisco Chronicle,* September 19, 1992, pp. B1, B2.
140. William M. Bulkeley, "Employers Use Software to Track Resumes," *Wall Street Journal,* June 23, 1992, p. B6.
141. Resumix Inc., cited in Margaret Mannix, "Writing a Computer-Friendly Resume," *U.S. News & World Report,* October 26, 1992, pp. 90–93.
142. Joyce Lain Kennedy and Thomas J. Morrow, *Electronic Resume Revolution: Create a Winning Resume for the New World of Job Seeking* (New York: Wiley, 1994).
143. Kathleen Pender, "Jobseekers Urged to Pack Lots of 'Keywords' Into Resumes," *San Francisco Chronicle,* May 16, 1994, pp. B1, B4.
144. Howard Bennett and Chuck McFadden, "How to Stand Out in a Crowd," *San Francisco Sunday Examiner & Chronicle,* October 17, 1993, help wanted section, p. 29.
145. Richard Bolles, quoted in Sylvia Rubin, "How to Open Your Job 'Parachute' After College," *San Francisco Chronicle,* February 24, 1994, p. E9.

Chapter 9

1. Doug Rowan, quoted in Ronald B. Lieber, "Picture This: Bill Gates Dominating the Wide World of Digital Content," *Fortune,* December 11, 1995, p. 38.
2. Kathy Rebello, "The Ultimate Photo Op?" *Business Week,* October 23, 1995, p. 40.
3. Steve Lohr, "Huge Photo Archive Bought by Software Billionaire Gates," *New York Times,* October 11, 1995, pp. A1, C5.
4. Wendy Bounds, "Bill Gates Owns Otto Bettmann's Life-work," *Wall Street Journal,* January 17, 1996, pp. B1, B2.
5. Paul Saffo, quoted in Lohr, 1995.
6. Phil Patton, "The Pixels and Perils of Getting Art On Line," *New York Times,* August 7, 1994, sec. 2, pp. 1, 31.

7. Amanda Kell, "Modern Monks—Holy, High Tech," *San Francisco Chronicle,* January 9, 1995, p. B3.
8. Jeffrey R. Young, "Modern-Day Monastery," *Chronicle of Higher Education,* January 19, 1996, pp. A21–A22.
9. Mike Snider and Kevin Maney, "Patience Is a Plus as System Keeps Evolving," *USA Today,* February 16, 1996, pp. 1D, 2D.
10. Jonathan Berry, John Verity, Kathleen Kerwin, and Gail DeGeorge, "Database Marketing," *Business Week,* September 5, 1994, pp. 56–62.
11. Sara Reese Hedberg, "The Data Gold Rush," *Byte,* October 1995, pp. 83–88.
12. Hedberg, 1995.
13. Cheryl D. Krivda, "Data-Mining Dynamite," *Byte,* October 1995, pp. 97–103.
14. Edmund X. DeJesus, "Data Mining," *Byte,* October 1995, p. 81.
15. Krivda, 1995.
16. Karen Watterson, "A Data Miner's Tools," *Byte,* October 1995, pp. 91–96.
17. Watterson, 1995.
18. Watterson, 1995.
19. Watterson, 1995.
20. Richard Lamm, quoted in Christopher J. Feola, "The Nexis Nightmare," *American Journalism Review,* July/August 1994, pp. 39–42.
21. Feola, 1994.
22. Penny Williams, "Database Dangers," *Quill,* July/August 1994, pp. 37–38.
23. Lynn Davis, quoted in Williams, 1994.
24. Associated Press, "Many Companies Are Willing to Give a Cat a Little Credit," *San Francisco Chronicle,* January 8, 1994, p. C1.
25. Jeffrey Rothfeder, "What Happened to Privacy?" *New York Times,* April 13, 1993, p. A15.
26. Ken Hoover, "Prisoner's Long-Distance Victims," *San Francisco Chronicle,* June 1, 1993, pp. A1, A6.
27. David Linowes, cited in Joseph Anthony, "Who's Reading Your Medical Records?" *American Health,* November 1993, pp. 54–58.
28. Anthony, 1993.
29. Eugene Carlson, "Business of Background Checking Comes to the Fore," *Wall Street Journal,* August 31, 1993, p. B2.
30. Peter Coy, "Big Brother, Pinned to Your Chest," *Business Week,* August 17, 1992, p. 38.
31. Michelle Vranizan, "Can Your Boss Read Your E-mail?" *San Jose Mercury News,* January 30, 1994, pp. 1F, 5F.
32. Erik Larson, quoted in Martin J. Smith, "Marketers Want to Know Your Secrets," *San Francisco Examiner,* November 21, 1993, pp. E-3, E-8.
33. Deborah L. Jacobs, "They've Got Your Name. You've Got Their Junk," *New York Times,* March 13, 1994, sec. 3, p. 5.
34. Erik Larson, quoted in Martin J. Smith, "Tactics for Evading Nosey Marketers," *San Francisco Examiner,* November 21, 1993, p. E-3.
35. Jacobs, 1994.
36. John R. Emshwiller, "Firms Find Profits Searching Databases," *Wall Street Journal,* January 25, 1993, pp. B1, B2.
37. Michael H. Martin, "Digging Data Out of Cyberspace," *Fortune,* April 1, 1996, p. 147.
38. John W. Verity, "What Hath Yahoo Wrought?" *Business Week,* February 12, 1996, pp. 88–90.
39. Michelle V. Rafter, Reuters, "Web Beginning to Crawl with Net-Surfing 'Spiders,'" *San Francisco Examiner,* March 3, 1996, pp. D-5, D-7.
40. Richard Scoville, "Find It on the Net," *PC World,* January 1996, pp. 125–130.
41. Clay Shirky, "Finding Needles in Haystacks," *NetGuide,* October 1995, pp. 87–90.
42. Tracy LeBlanc, "A Map to the Internet," *PC Novice,* March 1996, pp. 78–79.
43. Martin, 1996.
44. Scoville, 1996.
45. Flynn, 1995.
46. Martin, 1996.
47. Scoville, 1996.

Chapter 10

1. Gil Gordon, cited in Sue Shellenberger, "Overwork, Low Morale Vex the Mobile Office," *Wall Street Journal,* August 17, 1994, pp. B1, B4.
2. Shellenberger, 1994.
3. John Diebold, "The Next Revolution in Computers," *The Futurist,* May–June 1994, pp. 34–37.
4. Jonathan Marshall, "Contracting Out Catching On," *San Francisco Chronicle,* August 22, 1994, pp. D1, D3.
5. David Greising, "Quality: How to Make It Pay," *Business Week,* August 8, 1994, pp. 54–59.
6. Thomas A. Stewart, "Reengineering: The Hot New Managing Tool," *Fortune,* August 23, 1993, pp. 40–48.
7. Samuel E. Bleecker, "The Virtual Organization," *The Futurist,* March–April 1994, p. 9.
8. Gary Webb, "Potholes, Not 'Smooth Transition,' Mark Project," *San Jose Mercury News,* July 3, 1994, p. 18A.
9. Dirk Johnson, "Denver May Open Airport in Spite of Glitches," *New York Times,* July 17, 1994, p. A12.
10. K. Kendall and J. Kendall, *Systems Analysis and Design* (Englewood Cliffs, NJ: Prentice Hall, 1992), p. 39.
11. Kevin Kelly, *Out of Control,* cited in Rick Tetzeli, "Managing in a World Out of Control," *Fortune,* September 5, 1994, p. 111.
12. Howard Kahane, *Logic and Contemporary Rhetoric: The Uses of Reason in Everyday Life,* 6th ed. (Belmont, CA: Wadsworth, 1992).
13. J. Rasool, C. Banks, and M.-J. McCarthy, *Critical Thinking: Reading and Writing in a Diverse World* (Belmont, CA: Wadsworth, 1993), p. 132.

Chapter 11

1. James Randi, "Help Stamp Out Absurd Beliefs," *Time,* April 13, 1992, p. 80.
2. Hy Ruchlis and Sandra Oddo, *Clear Thinking: A Practical Introduction* (Buffalo, NY: Prometheus, 1990), p. 109.
3. Ruchlis and Oddo, 1990.
4. R. Wild, "Maximize Your Brain Power," *Men's Health,* April 1992, pp. 44–49.
5. Alan Freedman, *The Computer Glossary,* 6th ed. (New York: AMACOM, 1993), p. 370.
6. Freedman, 1993.
7. Freedman, 1993.
8. Peter D. Varhol, "Visual Programming's Many Faces," *Byte,* July 1994, pp. 187–188.
9. Gretchen Boehr, "Where to Buy Software," *PC Novice,* June 1992, pp. 20–25.
10. Walter S. Mossberg, "For PC Shoppers Who Know What They Want, Mail Order Delivers," *Wall Street Journal,* March 25, 1993, p. B1.
11. Walter S. Mossberg, "Talk Is Cheap? Not if You're Calling for Software Support," *Wall Street Journal,* October 14, 1993, p. B1.

Chapter 12

1. Jane Metcalfe, quoted in Laurie Flynn, "Tracking High-Tech Culture," *New York Times,* July 10, 1994, sec. 3, p. 10.
2. Jeremy Rifkin, "The End of Work," *San Jose Mercury News,* May 21, 1995, pp. 1C, 4C; adapted from his book *The End of Work: The Decline of the Global Labor Force and the Dawn of the Post Market Era* (Los Angeles: Jeremy P. Tarcher/Putnam, 1995).
3. Steven Levy, "How the Propeller Heads Stole the Electronic Future," *New York Times Magazine,* September 24, 1995, pp. 58–59.
4. "Just a Dataway," *The New Yorker,* May 16, 1994, pp. 6–8.
5. Patricia Schnaidt, "The Electronic Superhighway," *LAN Magazine,* October 1993, pp. 6–8.
6. Al Gore, reported in "Toward a Free Market in Telecommunications," *Wall Street Journal,* April 19, 1994, p. A18.
7. Wall Street Journal News Roundup, "Likely Mergers Herald an Era of Megacarriers," *Wall Street Journal,* February 2, 1996, pp. B1, B3.
8. Steven Levy, "Now for the Free-for-All," *Newsweek,* February 12, 1996, pp. 42–44.
9. Ramon G. McLeod and Reynolds Holding, "President Signs Telecom Overhaul," *San Francisco Chronicle,* February 9, 1996, pp. A1, A19.
10. Bryan Gruley and Albert R. Karr, "Bill's Passage Represents Will of Both Parties," *Wall Street Journal,* February 2, 1996, pp. B1, B3.
11. *Wall Street Journal* News Roundup, 1996.
12. Levy, "Now for the Free-for-All," 1996.
13. Richard Zoglin, "We're All Connected," *Time,* February 12, 1996, p. 52.
14. Gautam Naik, "Bell Companies Ready to Charge into Long-Distance," *Wall Street Journal,* February 5, 1996, p. B4.
15. Richard Turner, "The News Rush," *New York Times,* January 29, 1996, pp. 22–23.
16. Ken Auletta, "The News Rush," *The New Yorker,* March 18, 1996, pp. 42–45.
17. Poll by Institute of Electrical and Electronics Engineers, cited in Steve Lohr, "Americans See Future and Say, 'So What?'" *New York Times,* October 7, 1993, p. C3.
18. Walter S. Mossberg, "Going On-line Is Still Too Difficult to Lure a Mass Audience," *Wall Street Journal,* February 22, 1996, p. B1.
19. Survey by Sunil Gupta, University of Michigan, reported in Steve Lohr, "Out, Damned Geek! The Typical Web User Is No Longer Packing a Pocket Protector," *New York Times,* July 3, 1995, p. 15.
20. Douglas Lavin, "Internet Is Assuming Global Proportions," *Wall Street Journal,* March 15, 1996, p. A8.
21. Gina Smith, "The Next Home Computer," *Popular Science,* April 1996, p. 32.
22. Associated Press, "Robot Sent to Disarm Bomb Goes Wild in San Francisco," *New York Times,* August 28, 1993, p. 7.
23. "Frustrated Bank Customer Lets His Computer Make Complaint," *Los Angeles Times,* October 20, 1993, p. A28.
24. Arthur M. Louis, "Nasdaq's Computer Crashes," *San Francisco Chronicle,* July 16, 1994, pp. D1, D3.
25. Joseph F. Sullivan, "A Computer Glitch Causes Bumpy Start in a Newark School," *New York Times,* September 18, 1991, p. A25.
26. Malcolm Gladwell, "Blowup," *The New Yorker,* January 22, 1996, pp. 32–36.
27. Leonard M. Fuld, "Bad Data You Can't Blame on Intel," *Wall Street Journal,* January 9, 1995, p. A12.
28. Douglas Waller, "Onward Cyber Soldiers," *Time,* August 21, 1995, pp. 38–44.
29. Mark Thompson, "If War Comes Home," *Time,* August 21, 1995, pp. 44–46.
30. John J. Fialka, "Pentagon Studies Art of 'Information Warfare,' to Reduce Its Systems' Vulnerability to Hackers," *Wall Street Journal,* July 3, 1995, p. A10.
31. Henry K. Lee, "UC Student's Dissertation Stolen with Computer," *San Francisco Chronicle,* January 27, 1994, p. A15.
32. Safeware and The Insurance Agency, reported in "Laptop Thefts," *USA Today,* February 6, 1996, p. 9B.
33. Davis Stipp, "Laptop Larceny Is Taking Off at Airports," *Fortune,* February 5, 1996, pp. 30–31.
34. "Laptop Larcency on Rise at Airports and Hotels," *USA Today,* February 6, 1996, p. 9B.
35. David L. Wilson, "Devastating Wave of Computer Theft Pushes Universities to Compare Notes and Search for Ways to Boost Security," *Chronicle of Higher Education,* June 9, 1993, pp. A17–A18.
36. G. Pascal Zachary, "Software Firms Keep Eye on Bulletin Boards," *Wall Street Journal,* November 11, 1991, p. B1.
37. Thomas J. DeLoughry, "2 Students Are Arrested for Software Piracy," *Chronicle of Higher Education,* April 20, 1994, p. A32.
38. Suzanne P. Weisband and Seymour E. Goodman, "Subduing Software Pirates," *Technology Review,* October 1993, pp. 31–33.
39. Jeffrey A. Trachtenberg and Mark Robichaux, "Crooks Crack Digital Codes of Satellite TV," *Wall Street Journal,* January 12, 1996, pp. B1, B9.
40. Mark Robichaux, "Cable-TV Pirates Become More Brazen, Forcing Industry to Seek New Remedies," *Wall Street Journal,* May 7, 1992, pp. B1, B8.
41. John J. Keller, "Hackers Open Voice-Mail Door to Others' Phone Lines," *Wall Street Journal,* March 15, 1991, pp. B1, B3.
42. George James, "3 Accused in Operation that 'Cloned' Cell Phones," *New York Times,* October 19, 1995, p. A13.
43. William Barnhill, "Privacy Invaders," *AARP Bulletin,* May 1992, pp. 1, 10.
44. Stanley E. Morris, quoted in David E. Sanger, "Money Laundering, New and Improved," *New York Times,* December 24, 1995, sec. 4, p. 4.
45. David L. Wilson, "Gate Crashers," *Chronicle of Higher Education,* October 20, 1993, pp. A22–A23.
46. John T. McQuiston, "4 College Students Charged with Theft Via Computer," *New York Times,* March 18, 1995, p. 38.
47. David Einstein, "Crooks Swindle On-Line Investors," *San Francisco Chronicle,* July 1, 1994, pp. B1, B2.
48. Gary Weiss, "The Hustlers Queue Up on the Net," *Business Week,* November 20, 1995, pp. 146–148.
49. Neil Roland, "Scams Abound on the Net," *San Francisco Chronicle,* November 1, 1995, p. B2.
50. Jeremy L. Milk, "3 U. of Wisconsin Students Face Punishment for Bogus E-Mail Messages," *Chronicle of Higher Education,* October 20, 1993, p. A25.
51. Jeffrey Hsu, "Computer Viruses, Technological Poisons," *PC Novice,* October 1993, pp. 41–43.
52. Peter H. Lewis, "The Virus: Threat or Menace," *New York Times,* June 15, 1993, p. B6.
53. Peter H. Lewis, "Cybervirus Whodunit: Who Creates This Stuff?" *New York Times,* September 4, 1995, p. 20.
54. Julian Dibbell, "Viruses Are Good for You," *Wired,* February 1995, pp. 126–135, 175–180.
55. Bloomberg Business News, "Microsoft Says Word Has Virus," *New York Times,* August 31, 1995, p. C6.
56. Peter H. Lewis, "Computers Beware! A Virus Is on the Loose," *New York Times,* September 14, 1995, pp. 1, 20.
57. James Kim, "Understanding Concept Virus Is Key to Prevention," *USA Today,* March 11, 1996, p. 5B.
58. Charles Arthur, "Windows 95 a Victim of Special Virus," *San Francisco Examiner,* February 4, 1996, p. B-8; reprinted from London Independent.
59. Donald Parker, quoted in William M. Carley, "Rigging Computers for Fraud or Malice Is Often an Inside Job," *Wall Street Journal,* August 27, 1992, pp. A1, A5.
60. David Carter, quoted in Associated Press, "Computer Crime Usually Inside Job," *USA Today,* October 25, 1995, p. 1B.
61. Eric Corley, cited in Kenneth R. Clark, "Hacker Says It's Harmless, Bellcore Calls It Data Rape," *San Francisco Examiner,* September 13, 1992, p. B-9; reprinted from Chicago Tribune.
62. Philip Elmer-Dewitt, "Bugs Bounty," *Time,* October 23, 1995, p. 86.
63. Wilson, 1993.
64. Michelle Slatalla and Joshua Quittner, *Masters of Deception: The Gang That Rules Cyberspace* (New York: HarperCollins, 1995).
65. Joshua Quittner, "Hacker Homecoming," *Time,* January 23, 1995, p. 61.
66. Selwyn Raab, "New York Bookies Go Computer Age but Wind up Being Raided Anyhow," *New York Times,* August 25, 1995, p. A8.
67. Associated Press, "Russian Hackers Caught," *San Francisco Chronicle,* August 19, 1995, p. D1.
68. Timothy Ziegler, "Elite Unit Tracks Computer Crime," *San Francisco Chronicle,* May 27, 1994, pp. A1, A17.
69. John Markoff, "Keeping Things Safe and Orderly in the Neighborhoods of Cyberspace," *New York Times,* October 24, 1993, sec. 4, p. 7.
70. Joshua Cooper Ramo, "A SWAT Team in Cyberspace," *Newsweek,* February 21, 1994, p. 73.
71. "What a Social Security Number Reveals About Your Background," *Wall Street Journal,* December 27, 1995, p. 13; based on data from *Privacy Newsletter.*
72. Steven Bellovin, cited in Jane Bird, "More than a Nuisance," *The Times* (London), April 22, 1994, p. 31.
73. William M. Bulkeley, "To Read This, Give Us the Password . . . Ooops! Try It Again," *Wall Street Journal,* April 19, 1995, pp. A1, A8.
74. Anthony Ramirez, "How Hackers Find the Password," *New York Times,* July 23, 1992, p. A12.
75. Robert Lee Hotz, "Sign on the Electronic Dotted Line," *Los Angeles Times,* October 19, 1993, pp. A1, A16.
76. William M. Bulkeley, "Electronic Signatures Boost Security of PCs," *Wall Street Journal,* June 7, 1993.

77. Stephen Wildstrom, "Digital Signatures That Can't Be Forged," *Business Week,* July 4, 1994, p. 13.
78. Eugene Carlson, "Some Forms of Identification Can't Be Handily Faked," *Wall Street Journal,* September 14, 1993, p. B2.
79. Kimberly J. McLarin, "Fingerprinting, Without the Ink, Is Introduced in New York City as a Bar to Welfare Fraud," *New York Times,* July 13, 1995, p. A16.
80. Dana Milbank, "Measuring and Cataloging Body Parts May Help to Weed Out Welfare Cheats," *Wall Street Journal,* December 4, 1995, pp. B1, B6.
81. Evan Perez, "Security System Uses Your Face as the Password," *San Francisco Chronicle,* January 5, 1996, p. C1.
82. William M. Bulkeley, "Popularity Overseas of Encryption Code Has the U.S. Worried," *Wall Street Journal,* April 28, 1994, pp. A1, A7.
83. John Holusha, "The Painful Lessons of Disruption," *New York Times,* March 17, 1993, pp. C1, C5.
84. The Enterprise Technology Center, cited in "Disaster Avoidance and Recovery Is Growing Business Priority," special advertising supplement in *LAN Magazine,* November 1992, p. SS3.
85. John Painter, quoted in Holusha, 1993.
86. Micki Haverland, quoted in Fred R. Bleakley, "Rural County Balks at Joining Global Village," *Wall Street Journal,* January 4, 1996, pp. B1, B2.
87. David Ensunsa, "Proposed Cell-Phone Pole Faces Challenge," *Idaho Statesman,* June 23, 1995, p. 4B.
88. Mary Anne Ostrom, "Coming Distractions," *San Jose Mercury News,* February 18, 1996, pp. 1A, 24A.
89. Andrew Kupfer, "The Trouble with Cellular," *Fortune,* November 13, 1995, pp. 179–188.
90. James H. Snider, "The Information Superhighway as Environmental Menace," *The Futurist,* March–April 1995, pp. 16–21.
91. Bruce Weber, "Lost Amid Change? Take a Number," *New York Times,* August 28, 1995, p. C4.
92. Andrew Kupfer, "Alone Together," *Fortune,* March 20, 1995, pp. 94–104.
93. William M. Bulkeley, "Electronics Is Bringing Gambling into Homes, Restaurants, and Planes," *Wall Street Journal,* August 16, 1995, pp. A1, A5.
94. Linda Kanamine, "Despite Legal Issues, Virtual Dice Are Rolling," *USA Today,* November 17, 1995, p. 1A, 2A.
95. James Sterngold, "Imagine the Internet as Electronic Casino," *New York Times,* October 22, 1995, sec. 4, p. 3.
96. Kevin O'Neill, quoted in David Lieberman, "Racetracks Are Betting on Interactive TV," *USA Today,* October 25, 1995, p. 1B.
97. "Move Afoot to Ban Internet Gambling," *San Francisco Chronicle,* February 14, 1996, p. B2.
98. Marco R. della Cava, "Are Heavy Users Hooked or Just Online Fanatics?" *USA Today,* January 16, 1996, pp. 1A, 2A.
99. Kennth Howe, "Diary of an AOL Addict," *San Francisco Chronicle,* April 5, 1995, pp. D1, D3.
100. Kendall Hamilton and Claudia Kalb, "They Log On, but They Can't Log Off," *Newsweek,* December 18, 1995, pp. 60–61.
101. Nanci Hellmich, "When the Computer Replaces the Spouse Bit by Bit," *USA Today,* February 14, 1995, p. 8D.
102. Laura Evenson, "Losing a Mate to the Internet," *San Francisco Chronicle,* August 9, 1995, pp. A1, A13.
103. Associated Press, "Husband Accuses Wife of Having Online Affair," *San Francisco Chronicle,* February 2, 1996, p. A3.
104. Karen S. Peterson and Leslie Miller, "Cyberflings Are Heating Up the Internet," *USA Today,* February 6, 1996, pp. 1D, 2D.
105. Survey by Microsoft Corporation, reported in Don Clark and Kyle Pope, "Poll Finds Americans Like Using PCs but May Find Them to Be Stressful," *Wall Street Journal,* April 10, 1995, p. B3.
106. Sherry Turkle, quoted in Susan Wloszczyna, "MIT Prof Taps into Culture of Computers," *USA Today,* November 27, 1995, pp. 1D, 2D.
107. Sherry Turkle, *Life on the Screen: Identity in the Age of the Internet* (New York: Simon & Schuster, 1995).
108. Tom Abate, "Exploring the World of Cyber-Psychology," *San Francisco Examiner,* December 3, 1995, pp. B-1, B-3.
109. Christopher Lehmann-Haupt, "The Self in Cyberspace, Decentered and Opaque," *New York Times,* December 7, 1995, p. B4.
110. Jonathan Marshall, "Some Say High-Tech Boom Is Actually a Bust," *San Francisco Chronicle,* July 10, 1995, pp. A1, A4.
111. Yahoo!/Jupiter Communications survey, reported in Del Jones, "On-line Surfing Costs Firms Time and Money," *USA Today,* December 8, 1995, pp. 1A, 2A.
112. Coleman & Associates survey, reported in Julie Tilsner, "Meet the New Office Party Pooper," *Business Week,* January 29, 1996, p. 6.
113. Webster Network Strategies survey, reported in Jones, 1995.
114. Steven Levy, quoted in Marshall, 1995.
115. STB Accounting Systems 1992 survey, reported in Jones, 1995.
116. Ira Sager and Gary McWilliams, "Do You Know Where Your PCs Are?" *Business Week,* March 6, 1995, pp. 73–74.
117. Ron Erickson, "More Software. Gee," *New York Times,* August 4, 1995, p. A19.
118. Alex Markels, "Words of Advice for Vacation-Bound Workers: Get Lost," *Wall Street Journal,* July 3, 1995, pp. B1, B5.
119. Paul Saffo, quoted in Laura Evenson, "Pulling the Plug," *San Francisco Chronicle,* December 18, 1994, "Sunday" section, p. 53.
120. Daniel Yankelovich Group report, cited in Barbara Presley Noble, "Electronic Liberation or Entrapment," *New York Times,* June 15, 1994, p. C4.
121. Leslie Wayne, "If It's Tuesday, This Must Be My Family," *New York Times,* May 14, 1995, sec. 3, pp. 1, 12.
122. Annette Kornblum, "Maybe. Maybe Not." *San Jose Mercury News,* March 10, 1996, pp. 1C, 5C; reprinted from the *Washington Post.*
123. Neil Postman, quoted in Evenson, 1994.
124. Mike Snider, "Keeping PC Play Out of the Office," *USA Today,* January 26, 1995, p. 3D.
125. Sager & McWilliams, 1995.
126. Matthew Rothschild, "When You're Gagging on E-Mail," *Forbes,* June 6, 1994, pp. S25, S26.
127. Evenson, 1994.
128. Unabomber letter to David Gelernter, quoted in Bob Ickes, "Die, Computer, Die!" *New York Times,* July 24, 1995, pp. 22–26.
129. Steven Levy, "The Luddites Are Back," *Newsweek,* June 12, 1995, p. 55.
130. Jeremy Rifkin, "Technology's Curse: Fewer Jobs, Fewer Buyers," *San Francisco Examiner,* December 3, 1995, p. C-19.
131. Michael J. Mandel, "Economic Anxiety," *Business Week,* March 11, 1996, pp. 50–56.
132. Bob Herbert, "A Job Myth Downsized," *New York Times,* March 8, 1996, p. A19.
133. Louis Uchitelle, "It's a Slow-Growth Economy, Stupid," *New York Times,* March 17, 1996, sec. 4, pp. 1, 5.
134. Paul Krugman, "Long-Term Riches, Short-Term Pain," *New York Times,* September 25, 1994, sec. 3, p. 9.
135. Department of Commerce survey, cited in "The Information 'Have Nots'" [editorial], *New York Times,* September 5, 1995, p. A12.
136. Beth Belton, "Degree-based Earnings Gap Grows Quickly," *USA Today,* February 16, 1996, p. 1B.
137. Alan Krueger, quoted in LynNell Hancock, Pat Wingert, Patricia King, Debra Rosenberg, and Allison Samuels, "The Haves and the Have-Nots," *Newsweek,* February 27, 1995, pp. 50–52.
138. Terry Bynum, quoted in Lawrence Hardy, "Tapping into New Ethical Quandaries," *USA Today,* August 1, 1995, p. 6D.
139. Christine Borgman, quoted in Gary Chapman, "What the On-line World Really Needs Is an Old-Fashioned Librarian," *San Jose Mercury News,* August 21, 1995, p. 3D; reprinted from *Los Angeles Times.*
140. John W. Verity and Richard Brandt, "Robo-Software Reports for Duty," *Business Week,* February 14, 1994, pp. 110–113.
141. Katie Hafner, "Have Your Agent Call My Agent," *Newsweek,* February 27, 1995, pp. 76–77.
142. Laurie Flynn, "Electronic Clipping Services Cull Cyberspace and Fetch Data for You. But They Can't Think for You," *New York Times,* May 8, 1995, p. C6.
143. Peter Wayner and Alan Joch, "Agents of Change," *Byte,* March 1995, pp. 94–95.
144. Kurt Indermaur, "Baby Steps," *Byte,* March 1995, pp. 97–104.
145. Peter Wayner, "Free Agents," *Byte,* March 1995, pp. 105–114.
146. Richard Scoville, "Find It on the Net," *PC World,* January 1996, pp. 125–130.
147. Tom R. Halfhill, "Agents and Avatars," *Byte,* February 1996, pp. 69–72.
148. Denise Caruso, "Virtual-World Users Put Themselves in a Sort of Electronic Puppet Show," *New York Times,* July 10, 1995, p. C5.
149. Laura Evenson, "Chat Away in Character Online," *San Francisco Chronicle,* March 6, 1995, pp. E1, E3.
150. Halfhill, 1996.
151. Lawrence M. Fisher, "Trouble with the Software? Ask Other Software," *New York Times,* July 10, 1994, sec. 3, p. 10.
152. Charles Petit, "8-Legged Robot to Crawl into Volcano," *San Francisco Chronicle,* July 28, 1994, p. A3.
153. David L. Wilson, "On-Line Treasure Hunt," *Chronicle of Higher Education,* March 17, 1995, pp. A19–A20.
154. Robert Benfer Jr., Louanna Furbee, and Edward Brent Jr., quoted in Steve Weinberg, "Steve's Brain," *Columbia Journalism Review,* February 1991, pp. 50–52.
155. Tom Foremski, "Read It and Weep: E-Mail Overload Ahead," *San Francisco Examiner,* April 9, 1995, pp. B-5, B-6.
156. Jeanne B. Pinder, "Fuzzy Thinking Has Merits When It Comes to Elevators," *New York Times,* September 22, 1993, pp. C1, C7.
157. Gene Bylinsky, "Computers That Learn by Doing," *Fortune,* September 6, 1993, pp. 96–102.
158. Robert McGrough, "Fidelity's Bradford Lewis Takes Aim at Indexes with His 'Neural Network' Computer Program," *Wall Street Journal,* October 27, 1992, pp. C1, C21.
159. Michael Waldholz, "Computer 'Brain' Outperforms Doctors Diagnosing Heart Attack Patients," *Wall Street Journal,* December 2, 1991, p. B78.
160. Otis Port, "A Neural Net to Snag Breast Cancer," *Business Week,* March 13, 1995, p. 95.
161. Otis Port, "Computers That Think Are Almost Here," *Business Week,* July 17, 1995, pp. 68–72.
162. Port, 1995.
163. Sharon Begley and Gregory Beals, "Software au Naturel," *Newsweek,* May 8, 1995, pp. 70–71.
164. Gautam Naik, "In Sunlight and Cells, Science Seeks Answers to High-Tech Puzzles," *Wall Street Journal,* January 16, 1996, pp. A1, A5.
165. Port, 1995.
166. Begley and Beals, 1995.
167. Peter H. Lewis, "'Creatures' Get a Life, Although It Is an Artificial One," *New York Times,* October 13, 1993, p. B7.
168. Judith Anne Gunther, "An Encounter With A.I.," *Popular Science,* June 1994, pp. 90–93.
169. William A. Wallace, Ethics in Modeling (*New York: Elsevier Science, Inc.,* 1994).
170. Laura Johannes, "Meet the Doctor: A Computer That Knows a Few Things," *Wall Street Journal,* December 18, 1995, p. B1.
171. Steve Lohr, "The Elusive Information Highway," *New York Times,* September 23, 1994, pp. C1, C2.
172. Tamara Henry, "Many Schools Can't Access Net Offer," *USA Today,* November 3, 1995, p. 7D.
173. Susan Yoachum and Edward Epstein, "Clinton Goal—Internet in Every School," San Francisco Chronicle, September 22, 1995, pp. A1, A19.
174. Bill Workman, "Media Revolution in Stanford Future," *San Francisco Chronicle,* February 16, 1994, p. A12.
175. Edward Barrett, "Collaboration in the Electronic Classroom," *Technology Review,* February/March 1993, pp. 51–55.
176. Louis Freedberg, "A Plan to Make Books Obsolete," *San Francisco Chronicle,* July 15, 1993, pp. A1, A15.
177. Jeff Cole, "New Satellite Imaging Could Transform the Face of the Earth," *Wall Street Journal,* November 30, 1995, pp. A1, A5.
178. Robert L. Johnson, "Extending the Reach of 'Virtual' Classrooms," *Chronicle of Higher Education,* July 6, 1994, pp. A19–A23.
179. Ronald Smothers, "New Video Technology Lets Doctors Examine Patients Many Miles Away," *New York Times,* September 16, 1992, p. B6.
180. John Eckhouse and Ken Siegmann, "A Medical Version of the Superhighway," *San Francisco Chronicle,* January 19, 1994, p. B4.
181. Myron Magnet, "Who's Winning the Information Revolution," *Fortune,* November 30, 1992, pp. 110–117.
182. Tony Rutkowski, quoted in Schnaidt, 1993.
183. Bruce Haring, "Record Companies, Retailers Set Up Shop in Cyberspace," *USA Today,* May 10, 1995, p. 5D.
184. Jared Sandberg, "Virtual Bank Branches Are Coming Via Visa and Worlds Inc.," *Wall Street Journal,* August 9, 1995, p. B3.
185. Ellen Neuborne, "Wal-Mart Going On Line Via Microsoft," *USA Today,* January 29, 1996, p. 1B.
186. Steven Levy, "The End of Money?" *Newsweek,* October 30, 1995, pp. 62–65.
187. Kelley Holland and Amy Cortese, "The Future of Money," *Business Week,* June 12, 1995, pp. 66–78.
188. Pamela Varley, "Electronic Democracy," *Technology Review,* November/December 1991, pp. 43–51.
189. "Federal, State Aid to Go On-line," *San Jose Mercury News,* June 1, 1994, pp. 1A, 12A; reprinted from *Los Angeles Times.*
190. Andy Hines, "Jobs and Infotech," *The Futurist,* January–February 1994, pp. 9–13.
191. Mary Anne Buckman, quoted in Carol Kleiman, "Tailor Your Resume for Inclusion in a Company Database," *San Jose Mercury News,* April 14, 1996, pp. PC1–PC2.
192. David Borchard, "Planning for Career and Life," *The Futurist,* January–February 1995, pp. 8–12.
193. Tom Jackson, quoted in Jonathan Marshall, "Surfing the Internet Can Land You a Job," *San Francisco Chronicle,* July 17, 1995, pp. D1, D3.
194. Martin Yate, quoted in Kathleen Murray, "Plug In. Log On. Find a Job," *New York Times,* January 2, 1994, sec. 3, p. 23.
195. Lisa Levenson, "High-Tech Job Searching," *Chronicle of Higher Education,* July 14, 1995, pp. A16–A17.
196. "Online Job Search Resources," *San Francisco Chronicle,* July 17, 1995, p. D3.
197. Hal Lancaster, "How to Pick Up Job Tips Strewn Along the Infobahn," *Wall Street Journal,* June 27, 1995, p. B1.
198. Jane Easter Bahls, "Courting Your Career," *CompuServe Magazine,* November 1995, pp. 24–26.
199. Marshall, 1995.
200. Stephen C. Miller, "High-Tech Job Hunting, Even for Low-Tech Positions," *New York Times,* February 19, 1996, p. C6.
201. Cynthia Chin-Lee and Comet, "Surfing the Web for Your Next Job," *High Technology Careers Magazine,* August/September 1995, p. 94.
202. Scott Grusky, "Winning Resume," *Internet World,* February 1996, pp. 58–64.
203. Marshall Loeb, "Getting Hired by Getting Wired," *Fortune,* November 13, 1995, p. 252.
204. Marshall, 1995.
205. Levenson, 1995.
206. Marshall, 1995.
207. Grusky, 1996.
208. Levenson, 1995.

Index

Sources & Credits

Page numbers for Panels and README boxes are indicated in **boldface**.

Text and Art Sources

27 Data from Link Resources Corp. in Chart from USA Today Snapshots, *USA Today*, November 19, 1993. **34–38, 476–477** Adapted from *Healthy for Life: Wellness & the Art of Living*, by B. K. Williams and S. M. Knight. Copyright © 1994 by Brooks/Cole Publishing Company, a division of International Thomson Publishing Inc., Pacific Grove, CA 93950. Reprinted by permission of the Publisher. **36** Data from H. C. Lindgren, *The Psychology of College Success: A Dynamic Approach* (New York: Wiley, 1969). **58** *Fortune*, December 27, 1993, p. 116. © 1993 Time Inc. All rights reserved. **80** Adapted from *New York Times*, August 14, 1995, p. C5. **88** *Fortune*, June 14, 1993, p. 137. © 1993 Time Inc. All rights reserved. **89** Adapted from illustration on pp. 6–7, *Syllabus*, 1988, published quarterly by Apple Computer, Inc. © 1994 Syllabus Press. Adapted from Syllabus magazine. **93** By Richard Williams as it appeared in *Globe & Mail*, April 10, 1993, p. B13. **107** *Wall Street Journal*, May 6, 1994, pp. B1, B7. Reprinted by permission of The Wall Street Journal, © 1994 Dow Jones & Company, Inc. All Rights Reserved Worldwide. **117** (top) Copyright 1993 by Consumers Union of U.S., Inc., Yonkers, NY 10703-1057. Excerpted by permission from *Consumer Reports*, September 1993, p. 570. **117** (bottom) *Fortune*, October 4, 1993, p. 108. © 1993 Time Inc. All rights reserved. **120** *San Francisco Examiner*, November 8, 1992, p. E-14. © 1992 John Dvorak. **121** *San Francisco Chronicle*, January 25, 1995, pp. B1, B8. © San Francisco Chronicle. **124** (top) Adapted from illustration by Julie Stacey, *USA Today*, November 24, 1995, p. 3B. Copyright 1995 USA TODAY. Reprinted with permission. **124** (bottom) *New York Times Magazine*, November 5, 1995, pp. 50–57, 64. Copyright © 1995 The New York Times Company. Reprinted by permission. **133** Drawing adapted from Sun Microsystems, Inc., in "A New Model for Personal Computing," *San Jose Mercury News*, August 13, 1995, p. 27A. **134** *San Francisco Examiner*, July 30, 1995, pp. C-1, C-4. Reprinted with permission of Lisa Alcalay Klug. **154 top** From Michael S. Malone, *The Microprocessor: A Biography* (New York: Telos/Springer-Verlag, 1995); excerpted in Michael S. Malone, "The Tiniest Transformer," *San Jose Mercury News*, September 10, 1995, pp. 1D–2D. **154** (bottom) *Wall Street Journal*, November 16, 1992, p. R-8. Reprinted by permission of The Wall Street Journal, © 1992 Dow Jones & Company, Inc. All Rights Reserved Worldwide. **160** *Fortune*, January 10, 1994, p. 85. © 1994 Time Inc. All rights reserved. **162** Data from the National Science Foundation Metacenter. Adapted from chart in John Markoff, "A Crucial Linkup in the U.S. Data Highway," *New York Times*, September 30, 1992, p. C6. Copyright © 1992 The New York Times Company. Reprinted by permission. Also adapted from data in Bloomberg Business News, "MCI Setting Up Speed Network," *New York Times*, April 25, 1995, p. C6, and in Peter Coy, "Call It the Supernet," *Business Week*, May 8, 1995, pp. 93–94. **167, 169, 458, 490** Adapted from T. O'Leary, B. Williams, & L. O'Leary, *McGraw-Hill Computing Essentials* (New York: McGraw-Hill, 1990), pp. 58, 60, 155, 172. **172** From the book *Easy PCs, Second Edition*. Copyright © 1994. Published by Que Publishing, a division of Macmillan Computer Publishing. Used by permission of the publisher. **173** (bottom) Adapted from art prepared by Intel Corp. **181** Reprinted from *PC Magazine*, April 26, 1994, pp. 114–183. Copyright © 1994 Ziff-Davis Publishing Company. **182** (top) *San Jose Mercury News*, December 24, 1995, p. 1F. **183** Adapted from table in Don Clark, "Meet 'Media Processors,' Computer Chips for the Multimedia Age," *Wall Street Journal*, October 10, 1995, p. B1. Reprinted by permission of The Wall Street Journal. © 1995 Dow Jones & Company. All Rights Reserved Worldwide. **189** Adapted from special advertising section in *Newsweek*, November 23, 1992, in "Mobile Computing," by the Editors of *PC World* magazine, sponsored by Intel, p. N26. **208** *San Jose Mercury News*, May 14, 1995, pp. 1F, 5F. Table adapted from table, p. 5F. Copyright San Jose Mercury News. Reprinted with permission. All rights reserved. **211** Adapted from Lisa Bow, *How to Use Your Computer* (Emeryville, CA: Ziff-Davis Press, 1993), p. 126; and Suzanne Weixel, *Easy PCs*, 2nd ed. (Indianapolis: Que Corp., 1993), p. 47. **226** Reprinted from the *Columbia Journalism Review*, September/October 1995, p. 51. © 1995 by Columbia Journalism Review. **235** Adapted from Chris Foreman, "Lions and Tigers and . . . Computer Viruses?" Reprinted from *PC Novice*, May 1992, pp. 34–36. PC Novice, 120 W. Harvest Dr., Lincoln, NE 68521. For subscription information, please call 800-472-4100. Please mention code number 4110. **236** Adapted from Megan Jaegerman, "Choosing the Right Angles for Keyboard Safety," *New York Times*, March 4, 1992; and Janice M. Horowitz, "Crippled by Computers," *Time*, October 12, 1992, pp. 70–72. **237** Adapted from Janice M. Horowitz, "Crippled by Computers," *Time*, October 12, 1992, pp. 70–72. **248** *Popular Science*, June 1994, p. 43. Reprinted with permission of *Popular Science* Magazine, copyright 1994 Times Mirror Magazines, Inc. Distributed by the L. A. Times Syndicate. **259** Adapted from Phillip Robinson, "Color Comes In," *San Jose Mercury News*, October 1, 1995, pp. 1F, 6F; Larry Armstrong, "The Price Is Right for Printers," *Business Week*, November 6, 1995, pp. 132–134; and Chris O'Malley, "A More Perfect Printer," *Computer Shopper*, January 1996, pp. 343–352. **261** *Wall Street Journal*, January 7, 1993, p. B1. Reprinted by permission of The Wall Street Journal, © 1993 Dow Jones & Company, Inc. All Rights Reserved Worldwide. **296** *Home Office Computing*, August 1994, p. 53. Reprinted with permission of *Home Office Computing* magazine. Copyright © 1996. For subscription information, call 800-288-7812. **308 top** Adapted from Examiner Graphics, "How Kodak Photo CDs Work," *San Francisco Examiner*, October 27, 1991, p. E-14. **308** (bottom) Reprinted from *PC Novice*, June 1994, pp. 32–33. PC Novice, 120 W. Harvest Dr., Lincoln, NE 68521. For subscription information, please call 800-472-4100. Please mention code number 4110. **309** *USA Today*, January 3, 1996, pp. 1A, 2A. Copyright 1996, USA TODAY. Reprinted with permission. **314** *New York Times*, January 11, 1994, p. B10. Copyright © 1994 by The New York Times Company. Reprinted by permission. **318** *San Francisco Chronicle*, December 18, 1993, pp. E1, E6. © SAN FRANCISCO CHRONICLE. Reprinted by permission. **333** *Wall Street Journal*, November 16, 1992, pp. R18–R19. Reprinted by permission of The Wall Street Journal, © 1992 Dow Jones & Company, Inc. All Rights Reserved Worldwide. **338** *San Jose Mercury News*, July 17, 1994, p. 1E. From Simson L. Garfinkel. © 1994 Simon Garfinkel. Reprinted with permission. **345** (top) Network Wizards; (bottom) Adapted from *Der Spiegel*, November, 1996; **347** Adapted from "How Internet Mail Finds Its Way," *PC Magazine*, October 11, 1994, p. 121. **361** Adapted from table "Bandwidth Battle," in Mark Landler, "Jingling the Keys to Cyberspace, Cable Officials Sing a New Tune," *New York Times*, May 9, 1995, pp. C1, C10. **369** Adapted from Jared Schneidman, "How It Works," *Wall Street Journal*, February 11, 1994, p. R5, and *Popular Science*. **384**

Reprinted from *PC Novice,* December 1993, p. 72. PC Novice, 120 W. Harvest Dr., Lincoln, NE 68521. For subscription information, please call 800-472-4100. Please mention code number 4110. **388** *San Francisco Chronicle,* May 16, 1994, p. B4. © SAN FRANCISCO CHRONICLE. Reprinted by permission. **403** *New York Times,* May 5, 1993, pp. C1, C5. Copyright © 1993 by The New York Times Company. Reprinted by permission. **405** *Chronicle of Higher Education,* January 19, 1996, pp. A21–A22. Copyright 1996, The Chronicle of Higher Education. Reprinted with permission. **423** Adapted from "Data-mining process," illustration by Victor Gad © 1995, p. 84 in Sara Reese Hedberg, "The Data Gold Rush," *Byte,* October 1995, pp. 83–88. **428** *San Francisco Examiner,* November 21, 1993, p. E-3. From Martin J. Smith; reprinted from *Orange County Register.* From *The Naked Consumer* by Erik Larson (New York: Viking/Penguin USA). **454** Reprinted from *PC Week,* August 7, 1995, p. E1. Copyright © 1995 Ziff-Davis Publishing Company. **469** *New York Times,* August 24, 1994, sec. 3, p. 1. Copyright © 1994 The New York Times Company. Reprinted by permission. **493** *Chronicle of Higher Education,* November 7, 1995, p. A20. Copyright 1995, The Chronicle of Higher Education. Reprinted with permission. **505 bottom** Adapted from T. O'Leary & B. Williams, *Computers & Information Systems* (Redwood City, CA: Benjamin/Cummings, 1985), p. 171. **517** From S. Hutchinson & S. Sawyer, *Computers & Information Systems: 1994–1995 Edition* (Burr Ridge, IL: Irwin, 1994), Fig. 7, p. 13.11; adapted from *PC Novice,* November 1992, p. 42. **518** Reprinted from *PC Novice,* April 1993, pp. 33–36. PC Novice, 120 W. Harvest Dr., Lincoln, NE 68521. For subscription information, please call 800-472-4100. Please mention code number 4110. **520** *San Jose Mercury News,* December 17, 1995, p. F1. Copyright San Jose Mercury News. Reprinted with permission. All rights reserved. **540** Adapted from artwork by Kris Strawser/The Chronicle, appearing in article by Don Clark, "New Vision of Communications," *San Francisco Chronicle,* November 23, 1992, p. B1. © SAN FRANCISCO CHRONICLE. **543** *New York Times,* February 18, 1994, pp. A1, C16. Copyright © 1994 by The New York Times Company. Reprinted by permission. **550** *San Jose Mercury News,* September 10, 1995, pp. 1E, 4E. Copyright San Jose Mercury News. Reprinted with permission. All rights reserved. **565** Reprinted from *Columbia Journalism Review,* January/February 1991, pp. 50–52. © 1991 by Columbia Journalism Review. **567** Biological art adapted from Cecie Starr and Ralph Taggart, *Biology: The Unity and Diversity of Life,* 6th ed. (Belmont, CA: Wadsworth, 1992), p. 549. **569** *Popular Science,* June 1994, pp. 90–93. Reprinted with permission from Popular Science Magazine, copyright 1994, Times Mirror Magazines Inc. Distributed L.A. Times Syndicate. **573** Fortune © 1995 Time Inc.

Photo Sources

Page 9 IBM; **16** (middle) Apple Computer, Inc.; (bottom) Hewlett-Packard; **18** (top, middle) Hewlett Packard; (bottom left) © Frank Bevans; (bottom right) Quantum; **23** (clockwise from top left) Intel Corp., IBM, Hewlett-Packard, Hewlett-Packard, Intel Corp., IBM, Adastra Systems; **24** Ted Morrison / Still Life Stock; **25** (top) Ted Morrison / Still Life Stock; (bottom) Sanyo-Fisher USA; **27** Paul Chesley Photographers / Aspen; **28** (top) © Monica Limas; (bottom) Frank Pryor / Apple Computer, Inc.; **49** Luciano Galiardi / The Stock Market; **51** Toshiba; **55** San Francisco Chronicle; **59** IBM; **62–63** Microsoft Corp.; **72** Intersolv Multilink; **74** Microsoft Corp.; **75** ProComm Plus; **76** IBM; **83** Q Software Sous Chef; **85** Hewlett-Packard; **86** (left) Autodesk; (right) IBM; **93** © Frank Bevans; **107** Packard Bell Navigator; **108** © Frank Bevans; **109** © Monica Limas; **111** IBM; **117** (top background) Bradbury Building / G. E. Kidder Smith; (top insert) Tandy / Radio Shack; (bottom insert) Apple Computer, Inc.; **121** © Jerry Telfert / San Francisco Chronicle; **124** Microsoft Corp.; **129** Microsoft Corp.; **139** © Frank Bevans; **151** IBM Archives; **151** (top, middle) IBM; (bottom) Intel Corp.; **154** (top) Intel Corp.; (bottom insert) Tandy / Radio Shack; **156–157** (left to right) Sun, Toshiba, Sharp, Monibook, Apple Computer NEC, Apple Computer, NEC, IBM, Cray; **158** (top) SUN; (middle) © Disney Enterprises, Inc.; (bottom) Digital Equipment Corp.; **159** IBM; **160** IBM; **161** Cray; **172** © Frank Bevans; **175** Centon Electronics; **179** Hayes; **182** © Everett Collection, New York; **183** Power Computing; **186** © Bob Mahoney; **189** (top) John Greenleigh / Apple Computer, Inc.; (middle) Casio; (bottom) Hewlett-Packard; **190** (left) Psion; **204** IBM; **205** Bilbo Innovations, Inc.; **206** (insert) Northern Telecom of Canada; **209** (top) Microsoft Corp.; (middle) Kensington; (bottom) Thrustmaster; **210** (background) Mouse Man; **211** IBM; **212** (top) Microtouch Systems; (middle) FTE Data Systems; (bottom) FTG; **213** (top left) Hewlett-Packard; (top right) Wacon Technology; (middle) Sharp Electronics Corp.; (bottom) Apple Computer, Inc.; **214** (background) Uniform Code Council, Inc.; **215** (top, bottom left) NCR; (bottom right) © Frank Bevans; **216** (bottom) NCR / IBM; **217** (top inset) NEC; (bottom) Frank Pryor / Apple Computer, Inc.; **218** (left) Apple Computer, Inc.; (right) Visioneer; **219** (top) AP / Wide World Photos, Inc.; (bottom) © Brilliant Color Cards; **222** (top) © Liz Hafalia / San Francisco Chronicle; (bottom) Reuters / Archive Photos; **224** Eastman Kodak; **225** (top) Steve Mann Web site screen dump; (bottom) John Burdette / U.S. Dept. of the Interior / Geology Survey. Office of Earthquakes, Volcanoes, and Engineering; **231** Robert A. Flynn, Inc.; **251** Microsoft Corp.; **236** © Frank Bevans; **237** © Frank Bevans; **249** (top) © Frank Bevans; (middle) IBM; (bottom left) Multisync; **251** IBM; **254** (top left) Hewlett-Packard; (top right) Citizen; (bottom) Canon Computer Systems, Inc.; **255** John Greenleigh / Apple Computer, Inc.; **256** Hewlett-Packard; **257** Hewlett-Packard; **258** Tektronix; **260** (top) © Frank Bevans; (bottom left) Calcomp; (bottom right) Hewlett-Packard; **261** (top) Okidata; (bottom) Lanier Worldwide, Inc.; **264** Peter Fox / Apple Computer, Inc.; **266** (top, middle left) Autodesk; (middle right) UPL Research, Inc.; (bottom) Senes 2000; **267** Edward Keating / New York Times; **268** (left) NASA; (right) Transitions Research Corp.; **269** Gryphon Software; **270** Paul Higdon / New York Times; **290** Business Week Magazine; **291** IBM; **299** (top) Maxtor; (bottom) John Greenleigh / Apple Computer, Inc.; **301** (top left) Mountain Gate; (top middle, right) Syquest; (bottom) APS; **302** NCR Corp.; **304** (top) images © the estate of Robert Mapplethorpe; © Digital Collections, Inc. 1995; (bottom left) Toshiba; (bottom right) Peter Fox / Apple Computer, Inc.; **305** GME Timeline screen; **306** © Monica Limas; (bottom) © 95 Passage to Vietnam; **307** © Greenlar / The Image Works; **308** Elastic Reality, Inc.; **309** Toshiba America Information Systems, Inc.; **311** San Disk; **313** IBM; **314** (top) IBM; (bottom) Exabyte; **318** Interplay Productions; **333** (inset) Apple Computer Inc.; (bottom) J. L. Bersuder / © Sipa-Press; **336** (left) Intel Corp.; (right) Sharp Electronics; **339** (left) Hewlett-Packard; (right) AT&T; **340** (top) America Online; (middle) CompuServe; (bottom) Prodigy; **343** © 1994 Bob Mahoney; **348** Network Wizards; **353** Lakewood; **355** Packard Bell Electronics; **360** (top) Hayes; (bottom) Multitech Systems; **365** (top, bottom right) AT&T; (bottom left) U.S. Sprint; **366** Newfoundland Telephone Co.; **375** © Hank Morgan / Rainbow; **388** IBM; **403** NBA Properties; **404** (both) © Frank Bevans; **405** AP / Wide World Photos; **407** Pat Rogondino; **444** Lakewood; **455** United Airlines Gate Assignment Display System; **468** Intersolv; **493** © Frank Bevans; **500** Naval Surface Warfare Center, Dahlgren, VA; **537** © Andrew Bordwin / Stock Photo; **549** Symantec; **550** Symantec; **558** © Frank Bevans; **561** CompuServe; **563** (left) © Russ Kinne / Comstock; **571** (top) IBM; (bottom) © Thomas Dallas.